Digital Integrated Circuit Design
From VLSI Architectures to CMOS Fabrication

Hubert Kaeslin

ETH Zürich

CAMBRIDGE UNIVERSITY PRESS
Cambridge, New York, Melbourne, Madrid, Cape Town, Singapore, São Paulo, Delhi

Cambridge University Press
The Edinburgh Building, Cambridge CB2 8RU, UK

Published in the United States of America by Cambridge University Press, New York

www.cambridge.org

First published 2008

Printed in Malaysia

A catalog record for this publication is available from the British Library

ISBN 978-0-521-88267-5 hardback

Digital Integrated Circuit Design
From VLSI Architectures to CMOS Fabrication

- A top-down guide to the design of digital integrated circuits.

- Reflects industry design methods, moving from VLSI architecture design to CMOS fabrication.

- Practical hints and tips, case studies, and checklists provide a how and when guide to design.

- Key concepts and theory are explained when needed to provide the student with an insightful, practical guide.

- Homework problems test the student's understanding.

- Lecture slides help instructors deliver this course.

This unique guide to designing digital VLSI circuits takes a top-down approach, reflecting the nature of the design process in industry. Starting with architecture design, the book explains the why and how of digital design, using the physics that designers need to know, and no more.

Covering system and component aspects, design verification, VHDL modelling, clocking, signal integrity, layout, electrical overstress, field-programmable logic, economic issues, and more, the scope of the book is singularly comprehensive.

With a focus on CMOS technology, numerous examples — VHDL code, architectural concepts, and failure reports — practical guidelines, and design checklists, this engaging textbook for senior undergraduate and graduate courses on digital ICs will prepare students for the realities of chip design.

Practitioners will also find the book valuable for its insights and its practical approach.

Instructor-only solutions and lecture slides are available at www.cambridge.org/9780521882675.

HUBERT KAESLIN is head of the Microelectronics Design Center at the ETH Zürich, where he is also a lecturer in the Department of Information Technology and Electrical Engineering. He was awarded his Ph.D. in 1985 from the ETH and has 20 years' experience teaching VLSI to students and professionals.

Contents

Chapter 6 | Clocking of Synchronous Circuits   315

Chapter 12 | Design Verification

Chapter 14 | A Primer on CMOS Technology

Chapter 15 | Outlook

Appendix A | Elementary Digital Electronics   732

Preface

Why this book?

Designing integrated electronics has become a multidisciplinary enterprise that involves solving problems from fields as disparate as

- Hardware architecture
- Software engineering
- Marketing and investment
- Solid-state physics
- Systems engineering
- Circuit design
- Discrete mathematics
- Electronic design automation
- Layout design
- Hardware test equipment and measurement techniques

Covering all these subjects is clearly beyond the scope of this text and also beyond the author's proficiency. Yet, I have made an attempt to collect material from the above fields that I have found to be relevant for deciding whether or not to develop digital Very Large Scale Integration (VLSI) circuits, for making major design decisions, and for carrying out the actual engineering work.

The present volume has been written with two audiences in mind. As a textbook, it wants to introduce engineering students to the beauty and the challenges of digital VLSI design while preventing them from repeating mistakes that others have made before. Practising electronics engineers should find it appealing as a reference book because of its comprehensiveness and the many tables, checklists, diagrams, and case studies intended to help them not to overlook important action items and alternative options when planning to develop their own hardware components.

What sets this book apart from others in the field is its top-down approach. Beginning with hardware architectures, rather than with solid-state physics, naturally follows the normal VLSI design flow and makes the material more accessible to readers with a background in systems engineering, information technology, digital signal processing, or management.

Highlights

- Most aspects of digital VLSI design covered
- Top-down approach from algorithmic considerations to wafer processing
- Systematic overview on architecture optimization techniques
- Scalable concepts for simulation testbenches including code examples
- Emphasis on synchronous design and HDL code portability
- Comprehensive discussion of clocking disciplines
- Key concepts behind HDLs without too many syntactical details
- A clear focus on the predominant CMOS technology and static circuit style
- Just as much semiconductor physics as digital VLSI designers really need to know
- Models of industrial cooperation

- What to watch out for when purchasing virtual components
- Cost and marketing issues of ASICs
- Avenues to low-volume fabrication
- Largely self-contained (required previous knowledge summarized in two appendices)
- Emphasis on knowledge likely to remain useful in the years to come
- Many illustrations that facilitate recognizing a problem and the options available
- Checklists, hints, and warnings for various situations
- A concept proven in classroom teaching and actual design projects

A note to instructors

Over the past decade, the capabilities of field-programmable logic devices, such as FPGAs and CPLDs, have grown to a point where they have become invaluable ingredients of many electronic products, especially of those designed and marketed by small and medium-sized enterprises. Beginning with the higher levels of abstraction enables instructors to focus on those topics that are equally relevant irrespective of whether a design eventually gets implemented as a mask-programmed custom chip or from components that are just configured electrically. This material is collected in chapters 1 to 5 of the book and best taught as part of the Bachelor degree for maximum dissemination. No prior introduction to semiconductors is required. For audiences with little exposure to digital logic and finite state machines, the material can always be complemented with appendices A and B.

Learning how to design mask-programmed VLSI chips is then open to Master students who elect to specialize in the field. Designing electronic circuits down to that level of detail involves many decisions related to electrical, physical, and technological issues. An abstraction to purely logical models is no longer valid since side effects may cause an improperly designed circuit to behave differently than anticipated from digital simulations. How to cope with clock skew, metastability, layout parasitics, ground bounce, crosstalk, leakage, heat, electromigration, latch-up, electrostatic discharge, and process variability in fact makes up much of the material from chapter 6 onwards.

Again, the top-down organization of the book leaves much freedom as to where to end a class. A shorter course might skip chapter 8 as well as all material on detailed layout design that begins with section 11.5 on the grounds that only few digital designers continue to address device-level issues today. A similar argument also applies to the CMOS semiconductor technology introduced in chapter 14. Chapter 13, on the other hand, should not be dropped because, by definition, there are no engineering projects without economic issues playing a decisive role.

For those primarily interested in the business aspects of microelectronics, it is even possible to put together a quick introductory tour from chapters 1, 13, and 15 leaving out all the technicalities associated with actual chip design.

The figure below explains how digital VLSI is being taught by the author and his colleagues at the ETH. Probably the best way of preparing for an engineering career in the electronics and microelectronics industry is to complete a design project where circuits are not just being modeled and simulated on a computer but actually fabricated. Provided they come up with a meaningful project proposal, our students are indeed given this opportunity, typically working in teams of two. Following tapeout at the end of the 7th term, chip fabrication via an external multi-project wafer service takes roughly three months. Circuit samples then get systematically tested by their very developers in their 8th and final term. Needless to say that students accepting this offer feel very motivated and that industry highly values the practical experience of graduates formed in this way.

The technical descriptions and procedures in this book have been developed with the greatest of care; however, they are provided as is, without warranty of any kind. The author and editors of the book make no warranties, expressed or implied, that the equations, programs, and procedures in this book are free of error, or are consistent with any particular standard of merchantability, or will meet your requirements for any particular application. They should not be relied upon for solving a problem whose incorrect solution could result in injury to a person or loss of property.

Syllabus of ETH Zurich in
Digital VLSI Design and Test

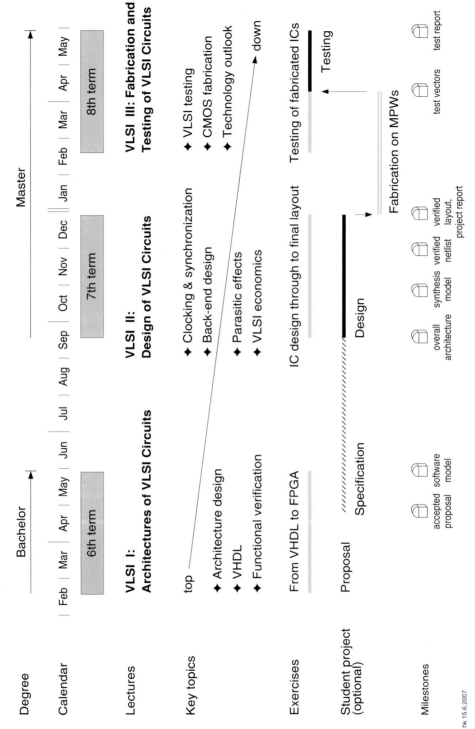

hk 15.6.2007

Acknowledgements

This volume collects the insight and the experience that many persons have accumulated over the last couple of years. While I was fortunate enough to compile the text, I am indebted to all those who have been willing to share their expertise with me.

My thanks thus go not only to my superiors at the ETH Zürich, Switzerland, Professors Wolfgang Fichtner, Qiuting Huang, and Rüdiger Vahldieck, but also to many past and present colleagues of mine including Dr. Dölf Aemmer, Professor Helmut Bölcskei, Dr. Heinz Bonnenberg, Dr. Andreas Burg, Felix Bürgin, Professor Mike Ciletti, Dr. Flavio Carbognani, Dr. Andreas Curiger, Stefan Eberli, Dr. Nobert Felber, Dr. Steffen Gappisch, Dr. Ronald Gull, Dr. Simon Häne, Kurt Henggeler, Dr. Lucas Heusler, Peter Lüthi, Dr. Chiara Martelli, Dieter Müller, Stephan Oetiker, Dr. David Perels, Dr. Robert Rogenmoser, Andreas Romer, Dr. Fritz Rothacher, Dr. Thomas Röwer, Dr. Manfred Stadler, Dr. Andreas Stricker, Christoph Studer, Thomas Thaler, Dr. Markus Thalmann, Jürg Treichler, Dr. Thomas Villiger, Dr. Jürgen Wassner, Dr. Marc Wegmüller, Markus Wenk, Dr. Rumi Zahir, and Dr. Reto Zimmermann. Most of these experts have also reviewed parts of my manuscript and helped to improve its quality. Still, the only person to blame for all errors and other shortcomings that have remained in the text is me.

Next, I would like to express my gratitude towards all of the students who have followed the courses on Digital VLSI Design and Testing jointly given by Dr. Nobert Felber and myself. Not only their comments and questions, but also results and data from many of their projects have found their way into this text. André Meyer and Thomas Peter deserve special credit as they have conducted analyses specifically for the book.

Giving students the opportunity to design microchips, to have them fabricated, and to test physical samples is a rather onerous undertaking that would clearly have been impossible without the continuous funding by the ETH Zürich. We are also indebted to Professor Fichtner for always having encouraged us to turn this vision into a reality and to numerous Europractice partners and industrial companies for access to EDA software, design kits, and low-volume manufacturing services.

The staff of the Microelectronics Design Center at the ETH Zürich and various IT support people at the Integrated Systems Laboratory, including Christoph Balmer, Matthias Brändli, Thomas Kuch, and Christoph Wicki, do or did a superb job in setting up and maintaining the EDA infrastructure and the services indispensable for VLSI design in spite of the frequent landslides caused by rapid technological evolution and by unforeseeable business changes. I am particularly grateful to them for occasionally filling all sorts of gaps in my technical knowledge without making me feel too badly about it.

I am further indebted to Dr. Frank Gürkaynak and Professor Yusuf Leblebici of the EPFL in Lausanne, Switzerland, for inciting me to turn my lecture notes into a textbook and for their advice. In a later phase, Dr. Julie Lancashire, Anna Littlewood, and Dawn Preston of Cambridge University Press had to listen to all my silly requests before they managed to get me acquainted with the realitities of printing and publishing. Umesh Vishwakarma of TeX Support is credited for preparing bespoke style files for the book, Christian Benkeser and Yves Saad for contributing the cover graphics.

Finally, I would like to thank all persons and organizations who have taken the time to answer my reprint requests and who have granted me the right to reproduce illustrations of theirs.

Chapter 1

Introduction to Microelectronics

1.1 | Economic impact

Let us begin by relating the worldwide sales of semiconductor products to the world's gross domestic product (GDP).[1] In 2005, this proportion was 237 GUSD out of 44.4 TUSD (0.53%) and rising.

Assessing the significance of semiconductors on the basis of sales volume grossly underestimates their impact on the world economy, however. This is because microelectronics is acting as a technology driver that enables or expedites a range of other industrial, commercial, and service activities. Just consider

- The computer and software industry,
- The telecommunications and media industry,
- Commerce, logistics, and transportation,
- Natural science and medicine,
- Power generation and distribution, and — last but not least —
- Finance and administration.

Microelectronics thus has an enormous economic leverage as any progress there spurs many, if not most, innovations in "downstream" industries and services.

A popular example . . .

After a rapid growth during the last three decades, the electric and electronic content of passenger cars nowadays makes up more than 15% of the total value in simpler cars and close to 30% in well-equipped vehicles. What's more, microelectronics is responsible for the vast majority of improvements that we have witnessed. Just consider electronic ignition and injection that have subsequently been combined and extended to become electronic engine management. Add to that anti-lock brakes and anti-skid stability programs, trigger circuits for airbags, anti-theft equipment,

---

[1] The GDP indicates the value of all goods and services sold during some specified year.

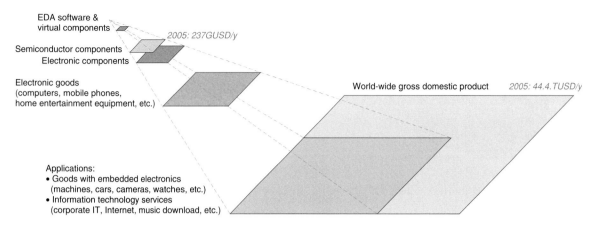

EDA software &
virtual components

2005: 237GUSD/y

Semiconductor components

Electronic components

Electronic goods
(computers, mobile phones,
home entertainment equipment, etc.)

World-wide gross domestic product   *2005: 44.4.TUSD/y*

Applications:
• Goods with embedded electronics
  (machines, cars, cameras, watches, etc.)
• Information technology services
  (corporate IT, Internet, music download, etc.)

Fig. 1.1 Economic leverage of microelectronics on "downstream" industries and services.

automatic air conditioning, instrument panels that include a travel computer, remote control of locks, navigation aids, multiplexed busses, electronically controlled drive train and suspension, audio/video information and entertainment, and upcoming night vision and collision avoidance systems. And any future transition to propulsion by other forms of energy is bound to intensify the importance of semiconductors in the automotive industry even further.

Forthcoming innovations include LED illumination and headlights, active "flywheels", hybrid propulsion, electronically driven valve trains, brake by wire, drive by wire, and, possibly, 42 V power supply to support the extra electrical load.

... and its less evident face

Perhaps less obvious but as important are the many contributions of electronics to the processes of development, manufacturing, and servicing. Innovations behind the scenes of the automotive industry include computer-aided design (CAD) and finite element analysis, virtual crash tests, computational fluid dynamics, computer numeric-controlled (CNC) machine tools, welding and assembly robots, computer-integrated manufacturing (CIM), quality control and process monitoring, order processing, supply chain management, and diagnostic procedures.

This almost total penetration has been made possible by a long-running drop of **cost per function**. Historically, costs have been dropping at a rate of 25% to 29% per year according to [1]. While computing, telecommunication, and entertainment products existed before the advent of microelectronics, today's anywhere, anytime information and telecommunication society would not have been possible without it; just compare the electronic devices in fig.1.2.

Observation 1.1. *Microelectronics is <u>the</u> enabler of information technology.*

On the positive side, microelectronics and information technology improve speed, efficiency, safety, comfort, and pollution control of industrial products and commercial processes, thereby bringing competitive advantages to those companies that take advantage of them.

Fig. 1.2 Four electronic products that take advantage of microelectronics opposed to analogous products that do not. The antiquated devices operate with vacuum tubes, discrete solid-state devices, and other electronic components but include no large-scale integrated circuits. Also observe that, were it not for display size and audio volume, one might replace all four devices with Apple's iPhone that has brought seamless system integration to even higher levels (photos courtesy of Alain Kaeslin).

On the negative side, the rapid progress, most of which is ultimately fueled by advances in semiconductor manufacturing technology, also implies a rapid obsoletion of hardware and software products, services, know-how, and organizations. A highly cyclic economy is another unfortunate trait of the semiconductor industry [2].

1.2 | Concepts and terminology

An **integrated circuit** (IC) is an electronic component that incorporates and interconnects a multitude of miniature electronic devices, mostly transistors, on a single piece of semiconductor material, typically **silicon**.[2] Many such circuits are jointly manufactured on a thin semiconductor wafer with a diameter of 200 or 300 mm before they get cut apart to become (naked) **dies**. The sizes of typical dies range between a pinhead and a large postage stamp. The vast majority of ICs, or **(micro)chips** as they are colloquially referred to, gets individually encapsulated in a hermetic package before being soldered onto **printed circuit boards** (PCB).

The rapid progress of semiconductor technology in conjunction with marketing activities of many competing companies — notably trademark registration and eye catching — has led to a plethora of terms and acronyms, the meaning of which is not consistently understood by all members of the microelectronics community. This section introduces the most important terms, clarifies what they mean, and so prepares the ground for more in-depth discussions.

Depending on perspective, microchips are classified according to different criteria.

1.2.1 The Guinness book of records point of view

In a world obsessed with records, a prominent question is "How large is that circuit?"

Die size is a poor metric for design complexity because the geometric dimensions of a circuit greatly vary as a function of technology generation, fabrication depth, and design style.

Transistor count is a much better indication. Still, comparing across logic families is problematic as the number of devices necessary to implement some given function varies.[3]

Gate equivalents attempt to capture a design's hardware complexity independently from its actual circuit style and fabrication technology. One gate equivalent (GE) stands for a two-input NAND gate and corresponds to four MOSFETs in static CMOS; a flip-flop takes roughly 7 GEs. Memory circuits are rated according to storage capacity in bits. Gate equivalents and memory capacities are at the basis of the naming convention below.

---

[2] This is a note to non-Angloamerican readers made necessary by a tricky translation of the term silicon.

English	German	French	Italian	meaning
silicon	Silizium	silicium	silicio	Si, the chemical element with atomic number 14
silicone	Silikon	silicone	silicone	a broad family of polymers of Si with hydrocarbon groups that comprises viscous liquids, greases, and rubber-like solids

[3] Consistent with our top-down approach, there is no need to know the technicalities of CMOS, TTL, and other logic families at this point. Interested readers will find a minimum of information in appendix 1.6.

circuit complexity	GEs of logic + bits of memory
small-scale integration (SSI)	1–10
medium-scale integration (MSI)	10–100
large-scale integration (LSI)	100–10 000
very-large-scale integration (VLSI)	10 000–1 000 000
ultra-large-scale integration (ULSI)	1 000 000 . . .

Clearly, this type of classification is a very arbitrary one in that it attempts to impose boundaries where there are none. Also, it equates one storage bit to one gate equivalent. While this is approximately correct when talking of static RAM (SRAM) with its four- or six-transistor cells, the single-transistor cells found in dynamic RAMs (DRAMs) and in ROMs cannot be likened to a two-input NAND gate. A better idea is to state storage capacities separately from logic complexity and along with the memory type concerned, e.g. 75 000 GE of logic + 32 kibit SRAM + 512 bit flash $\approx$ 108 000 GE overall complexity.[4]

One should not forget that circuit complexity per se is of no merit. Rather than coming up with inflated designs, engineers are challenged to find the most simple and elegant solutions that satisfy the specifications given in an efficient and dependable way.

1.2.2 The marketing point of view

In this section, let us adopt a market-oriented perspective and ask "How do functionality and target markets relate to each other?"

GENERAL-PURPOSE ICS

The function of a general-purpose IC is either so simple or so generic that the component is being used in a multitude of applications and typically sold in huge quantities. Examples include gates, flip-flops, counters, and other components of the various 7400 families but also RAMs, ROMs, microcomputers, and most digital signal processors (DSPs).

APPLICATION-SPECIFIC INTEGRATED CIRCUITS

Application-specific integrated circuits (ASICs) are being specified and designed with a particular purpose, equipment, or processing algorithm in mind. Initially, the term had been closely associated with **glue logic**, that is with all those bus drivers, decoders, multiplexers, registers, interfaces, etc. that exist in almost any system assembled from highly integrated parts. ASICs have evolved from substituting a single package for many such ancillary functions that originally had to be dispersed over several SSI/MSI circuits.

Today's highly-integrated ASICs are much more complex and include powerful systems or subsystems that implement highly specialized tasks in data and/or signal processing. The term

---

[4] Kibi- (ki), mebi- (Mi), gibi- (Gi), and tebi- (Ti) are binary prefixes recommended by various standard bodies for 2^{10}, 2^{20}, 2^{30}, and 2^{40} respectively because the more common decimal SI prefixes kilo- (k), mega- (M), giga- (G) and tera- (T) give rise to ambiguity as $2^{10} \neq 10^3$. As an example, 1 MiByte = 8 Mibit = $8 \cdot 2^{20}$ bit.

system-on-a-chip (SoC) has been coined to reflect this development. Overall manufacturing costs, performance, miniaturization, and energy efficiency are key reasons for opting for ASICs.

Still from a marketing point of view, ASICs are subdivided further into application-specific standard products and user-specific ICs.

Application-specific standard product (ASSP). While designed and optimized for a highly specific task, an application-specific standard product circuit is being sold to various customers for incorporation into their own products. Examples include graphics accelerators, multimedia chips, data compression circuits, forward error correction devices, ciphering/deciphering circuits, smart card chips, chip sets for cellular radio, serial-ATA and Ethernet interfaces, wireless LAN chips, and driver circuits for power semiconductor devices, to name just a few.[5]

User-specific integrated circuit (USIC). As opposed to ASSPs, user-specific ICs are being designed and produced for a single company that seeks a competitive advantage for their products; they are not intended to be marketed as such. Control of innovation and protection of proprietary know-how are high-ranking motivations for designing circuits of this category. Parts are often fabricated in relatively modest quantities.

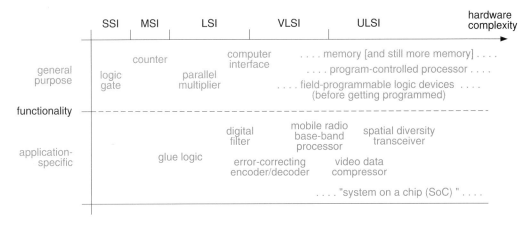

Fig. 1.3 ICs classified as a function of functionality and hardware complexity.

1.2.3 The fabrication point of view

Another natural question is
"To what extent is a circuit manufactured according to user specifications?"

---

[5] Microprocessors that have their instruction sets, input/output capabilities, memory configurations, timers, and other auxiliary features tailored to meet specific needs also belong to the ASSP category.

FULL-CUSTOM ICS

Integrated circuits are manufactured by patterning multiple layers of semiconductor materials, metals, and dielectrics. In a full-custom IC, all such layers are patterned according to user specifications. Fabricating a particular design requires wafers to go through all processing steps under control of a full set of lithographic **photomasks** all of which are made to order for this very design, see fig.1.4. This is relevant from an economic point of view because mask manufacturing is a dominant contribution to non-recurring VLSI fabrication costs. A very basic CMOS process featuring two layers of metal requires some 10 to 12 fabrication masks, any additional metal layer requires two more masks. At the time of writing (late 2007), one of the most advanced CMOS processes comprises 12 layers of metal and involves some 45 lithography cycles.

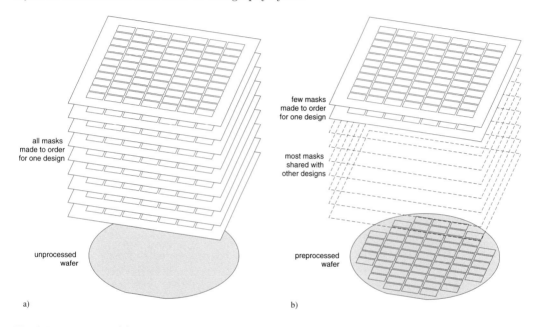

all masks
made to order
for one design

unprocessed
wafer

a)

few masks
made to order
for one design

most masks
shared with
other designs

preprocessed
wafer

b)

Fig. 1.4 Full-custom (a) and semi-custom (b) mask sets compared.

SEMI-CUSTOM ICS

Only a small subset of fabrication layers is unique to each design. Customization starts from preprocessed wafers that include large quantities of prefabricated but largely uncommitted primitive items such as transistors or logic gates. These so-called **master wafers** then undergo a few more processing steps during which those primitives get interconnected in such a way as to complete the electrical and logic circuitry required for a particular design. As an example, fig.1.5 shows how a logic gate is manufactured from a few pre-existing MOSFETs by etching open contact holes followed by deposition and patterning of one metal layer.

In order to accommodate designs of different complexities, vendors make masters available in various sizes ranging from a couple of thousands to millions of usable gate equivalents. Organization and customization of semi-custom ICs have evolved over the years.

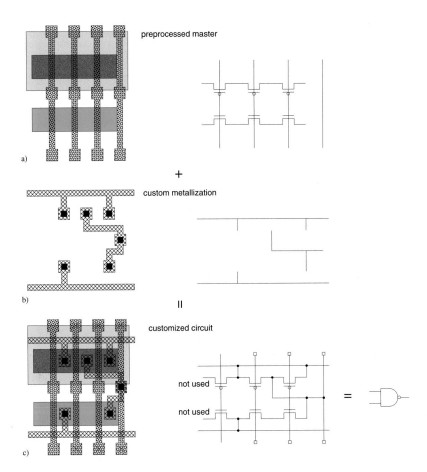

Fig. 1.5 Customization of a gate array site (simplified). A six-pack of prefabricated MOS transistors (a), metal pattern with contact openings (b), and finished 2-input NAND gate (c).

Gate array, aka channeled gate array. Originally, sites of a few uncommitted transistors each were arranged in long rows that extended across most of the die's width. Metal lines were then used to connect the prefabricated transistors into gates and the gates into circuits. The number of custom photomasks was twice that of metal layers made to order. As long as no more than two layers of metal were available, special routing channels had to be set aside in between to accommodate the necessary intercell wiring, see fig.1.6a.

Sea-of-gates. When more metals became available in the early 1990s, those early components got displaced by channelless sea-of-gate circuits because of their superior layout density. The availability of higher-level metals allowed for routing over gates and bistables customized on the layers underneath, so dispensing with the waste of routing channels, see fig.1.6b. More metals further made it possible to insulate adjacent transistors electrically where needed, doing away with periodic gaps in the layout. Sea-of-gates also afforded more flexibility for accommodating highly repetitive structures such as RAMs and ROMs.

Structured ASIC. A decade later, the number of metal layers had grown to a point where it became uneconomical to customize them all. Instead, transistors are prefabricated and pre-connected into small generic subcircuits such as NANDs, MUXes, full-adders, and bistables with the aid of the lower layers of metal. Customization is confined to interconnecting those subcircuits on the top two to four metal layers. What's more, the design process is accelerated as supply and clock distribution networks are largely prefabricated.

Fabric. Exploding mask costs and the limitations of sub-wavelength lithography currently work against many custom-made photomasks. The idea behind fabrics is to standardize the metal layers as much as possible. A subset of them is patterned into fixed segments of predetermined lengths which get pieced together by short metal straps, aka jumpers, on the next metal layer below or above to obtain the desired wiring. Customization is via the vertical contact plugs, called vias, that connect between two adjacent layers.

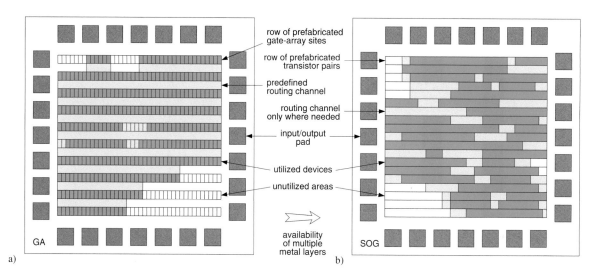

Fig. 1.6 Floorplan of channeled gate-array (a) versus channelless semi-custom circuits (b).

Due to the small number of design-specific photomasks and processing steps, semi-custom manufacturing significantly reduces the non-recurring costs as well as the turnaround time.[6] Conversely, prefabrication necessarily results in non-optimal layouts. Note the unused transistor pair in fig.1.5, for instance, or think of the extra parasitic capacitances and resistances caused by standardized wiring. Prefabrication also implies a self-restraint to fixed transistor geometries, thereby further limiting circuit density, speed, and energy efficiency. Lastly, not all semi-custom masters accommodate on-chip memories equally well.

---

[6] **Turnaround time** denotes the time elapsed from coming up with a finalized set of design data until physical samples become available for testing.

Incidentally, be informed that the concept of metal customization is also applied to analog and mixed-signal circuits. Prefabricated masters then essentially consist of uncommitted transistors (MOSFETs and/or BJTs) and of passive devices.[7]

FIELD-PROGRAMMABLE LOGIC

Rather than manufacturing dedicated layout structures, a generic part is made to assume a user-defined circuit configuration by purely electrical means. Field-programmable logic (FPL) devices are best viewed as "soft hardware". Unlike semi- or full-custom ASICs, FPL devices offer turnaround times that range from a few seconds to a couple of minutes; many product families even allow for in-system configuration (ISC).

The key to obtaining various gate-level networks from the same hardware resources is the inclusion of electrical links that can be done — and in many cases also undone — long after a device has left the factory. Four configuration technologies coexist today; they all have their roots in memory technology (SRAM, PROM, flash/EEPROM, and EPROM). For the moment, you can think of a programmable link as some kind of fuse.

A second dimension in which commercially available parts differ is the organization of on-chip hardware resources. **Field-programmable gate arrays** (FPGAs), for instance, resemble mask-programmed gate arrays (MPGAs) in that they are organized into a multitude of logic sites and interconnect channels. In this text, we will be using the term **field-programmable logic** (FPL) as a collective term for any kind of electrically configurable IC regardless of its capabilities, organization, and configuration technology.[8]

FPL was initially confined to glue logic applications, but has become an extremely attractive proposition for smaller volumes, for prototyping, when a short time to market is paramount, or when frequent modifications ask for agility. Its growing market share affords FPL a more detailed discussion in section 1.4. What also contributed to the success of FPL is the fact that many issues that must be addressed in great detail when designing a custom circuit are implicitly solved when opting for configuring an FPL device instead, just consider testability, I/O subcircuits, clock and power distribution, embedded memories, and the like.

STANDARD PARTS

By standard part, aka commercial off-the-shelf (COTS) component, we mean a catalog part with no customization of the circuit hardware whatsoever.

1.2.4 The design engineer's point of view

Hardware designers will want to know the answer to
"Which levels of detail are being addressed during a part's design process?"

---

[7] Microdul MD300 and Zetex 700 are just two examples.

[8] Referring to all such parts as "field-configurable" would be preferable as this better reflects what actually happens. This would also avoid confusion with program-controlled processors. Yet, the term "programmable" has gained so much acceptance in acronyms such as PLA, PAL, CPLD, FPGA, etc. that we will stay with it.

HAND LAYOUT

In this design style, an IC or some subblock thereof gets entered into the CAD database by delineating individual transistors, wires, and other circuit elements at the layout level. To that end, designers use a **layout editor**, essentially a color graphics editing tool, to draw the desired geometric shapes to scale, much as in the illustration of fig.1.5c. Any design so established must conform with the layout rules imposed by the target process. Porting it to some other process requires the layout to be redesigned unless the new set of rules is obtained from the previous one by simple scaling operations. Editing **geometric layout** is slow, cumbersome, and prone to errors. Productivity is estimated to lie somewhere between 5 and 10 devices drawn per day, including the indispensable verification, correction, and documentation steps, which makes this approach prohibitively expensive.

Conversely, manual editing gives designers full control over their layouts when in search of maximum density, performance, and/or electrical matching. Geometric layout, which in the early days had been the only avenue to IC design, continues to play a dominant role in memory and analog circuit design. In digital design, it is considered archaic, although a fully handcrafted circuit may outperform a synthesis-based equivalent by a factor of three or more.

CELL-BASED DESIGN BY MEANS OF SCHEMATIC ENTRY

Design capture here occurs by drawing circuit diagrams where subfunctions — mostly logic gates — are instantiated and interconnected by wires as illustrated in fig.1.9c. All the details of those elementary subcircuits, aka cells, have been established before and are collected in **cell libraries** that are made available to VLSI designers; see section 1.3.3 for more on this. For the sake of economy, cell libraries are shared among numerous designs. A **schematic editor** differs from a standard drawing tool in several ways.

- Circuit connectivity is maintained when components are being relocated.
- A schematic editor is capable of reading and writing both circuit diagrams and netlists.[9]
- It supports circuit concepts such as connectors, busses, node names, and instance identifiers.

The resulting circuits and netlists are then verified by simulation and other means. Compared with manual layout entry, cell-based design represented a marked step towards abstracting from process-dependent details.

Whether the circuit is eventually going to be fabricated as a full-custom IC or as a semi-custom IC is, in principle, immaterial. In either case, physical design does not go beyond **place and route** (P&R) where each cell is assigned a geometric location and connected to other cells by way of metal lines. As this is done by automatic tools, the resulting layouts are almost always correct by construction and **design productivity** is much better than for manual layout. Another advantage is that any engineer familiar with electronics design can start to develop cell-based ASICs with little extra training.

Library elements are differentiated into standard cells, macrocells, and megacells.

Standard cells are small but universal building blocks such as logic gates, latches, flip-flops, multiplexers, adder slices, and the like with pre-established layouts and defined electrical

---

[9] The difference between a netlist and a circuit diagram, aka schematic (drawing), is explained in section 1.7.

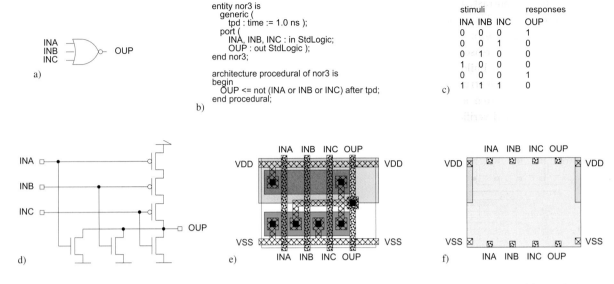

entity nor3 is
  generic (
    tpd : time := 1.0 ns);
  port (
    INA, INB, INC : in StdLogic;
    OUP : out StdLogic);
end nor3;

architecture procedural of nor3 is
begin
  OUP <= not (INA or INB or INC) after tpd;
end procedural;

stimuli			responses
INA	INB	INC	OUP
0	0	0	1
0	0	1	0
0	1	0	0
1	0	0	0
0	0	0	1
1	1	1	0

Fig. 1.7 Views of a library cell or of any other subcircuit shown for a 3-input NOR gate. Icon (a), simulation model (b), test vector set (c), transistor-level schematic (d), detailed layout (e), and cell abstract (f) (simplified).

characteristics.[10] They are the preferred means for implementing random logic as there is virtually no restriction on the functionality that can be assembled from them. Commercial libraries include between 300 and 500 standard cells with logic complexities ranging from 1/2 to some 60 gate equivalents; the collection of datasheets pertaining thereto typically occupies some 400 to 800 pages.

On the semiconductor die, standard cells get arranged in adjoining parallel rows with the interconnecting wires running over the top of them. This so-called **over-the-cell routing** style has been being practiced ever since three and more layers of metal became available.[11]

Megacells also come with a ready-to-use layout. What sets them apart from standard cells is their larger size and complexity. Typical examples include microprocessor cores and peripherals such as direct memory access controllers, various serial and parallel communication interfaces, timers, A/D and D/A converters, and the like. Megacells are ideal for piecing together a microcomputer or an ASIC with comparatively very little effort. Typical application areas are in telecommunications equipment, automotive equipment, instrumentation, and control systems.

---

[10] Standard cells are also termed "books" (within IBM) and **macros** (in the context of semi-custom ICs).

[11] Older processes did not afford that much routing resources and the wires had to be inserted between the rows such as to form well-defined **routing channels**. The resulting separation between adjacent cell rows obviously made a poor usage of silicon. In fact, it was not uncommon that routing channels occupied twice or even three times as much area as the active cells themselves.

Macrocells, in contrast, have their layout assembled on a per case basis according to designer specifications. The software tool that does this is called a **macrocell generator** and is also in charge of providing a simulation model, an icon, a datasheet, and other views of the macrocell. For reasons of area and design efficiency, this approach is essentially limited to a few common building blocks of medium complexity such as RAMs and ROMs. This is because all such structures show fairly regular geometries that lend themselves well to being put together from a limited collection of layout tiles. Those tiles are manually designed, optimized, and verified before being stored as part of the generator package.

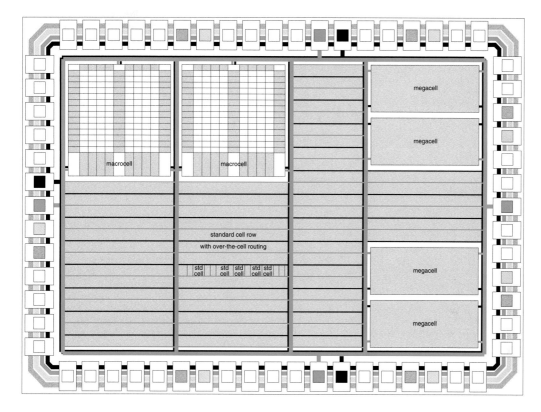

Fig. 1.8 Typical cell mix in a full-custom IC.

As standard cells, macrocells, megacells, and hand layout all have their specific merits and drawbacks, they are often combined in the design of full-custom ICs.[12] The resulting mix of cells is illustrated in fig.1.8. While design productivity in terms of transistors instantiated per day is clearly higher for megacells and macrocells than for standard cells, expect an average of some 15 to 20 GEs per day from cell-based design. Schematic entry at the gate level, and even more so at the transistor

---

[12] In a microcomputer, for instance, the datapath might be implemented in hand layout, data RAM and program ROM generated as macrocells, and the controller as a network of standard cells obtained from automatic synthesis, while a serial interface from an earlier design might get reused as a megacell.

level, should be confined to functions that are neither available as library items nor amenable to automatic synthesis.

The entry level here is a formal description of an entire chip or of a major subblock therein. Most such **synthesis models** are established using a **text editor** and look like software code. Yet, they are typically written in a **hardware description language** (HDL) such as VHDL or Verilog; see fig.1.9b. The output from the automatic synthesis procedure is a gate-level netlist. That netlist then forms the starting point for place and route (P&R) or for preparing a bit stream that will eventually serve to configure an FPL device.

Logic synthesis implies the generation of combinational networks and — as an extension — of fairly simple finite state machines (FSMs). A synthesis tool accepts logic equations built from operators such as NOT, AND, OR, XOR, etc., truth tables, state graphs, and the like. Automatic tools for logic synthesis and optimization have been in routine use for a long time; they have been completely absorbed in more advanced EDA flows.

Register transfer level (RTL) **synthesis** goes one step further in that an entire circuit is viewed as a network made up of storage elements — registers and possibly also RAMs — that are held together by combinational building blocks, see fig.1.9a. Also, behavioral specifications are no longer limited to simple logic operations but are allowed to include arithmetic functions (e.g. comparison, addition, subtraction, multiplication), string operations (e.g. concatenation), arrays, enumerated types, and other more powerful constructs.

The synthesis process essentially begins with the registers that are necessary to store the circuit's state. Next, the combinational networks required to process data words while they are moving back and forth between those registers are generated and optimized. Command on a circuit's structure is otherwise left to the designer who has to decide himself on the number of registers, on the concurrency of operations, on the necessary computational resources, etc.

RTL synthesis became very popular in the early 1990s with the advent of adequate HDLs and computer tools. It dispenses with the need for manually assembling a given functionality from primitive logic gates and, therefore, greatly facilitates design parametrization and maintenance. Synthesis further enables engineers to render their work portable, that is to capture all relevant characteristics of a circuit design in a form that is virtually technology-independent. It so becomes possible to defer the commitment to a specific silicon foundry, to a particular cell library, or to subordinate idiosyncrasies of some FPL family until late in the design process. As fabrication processes are frequently being upgraded, making designs portable and reusable is extremely valuable.

Architecture synthesis, which is also referred to as high-level synthesis in VLSI circles, starts from a data or signal processing algorithm such as a C++ program or a MATLAB model, for instance. As opposed to an RTL model, the source description is purely behavioral and includes no explicit indications for how to marshal data processing operations and the necessary hardware resources. Rather, these elements must be obtained in an automatic process that essentially works in five major phases.

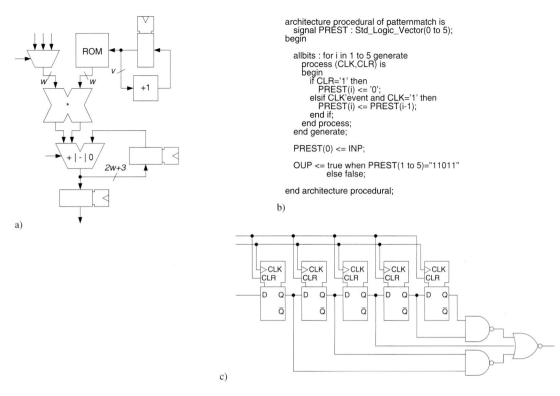

```
architecture procedural of patternmatch is
    signal PREST : Std_Logic_Vector(0 to 5);
begin

    allbits : for i in 1 to 5 generate
        process (CLK,CLR) is
        begin
            if CLR='1' then
                PREST(i) <= '0';
            elsif CLK'event and CLK='1' then
                PREST(i) <= PREST(i-1);
            end if;
        end process;
    end generate;

    PREST(0) <= INP;

    OUP <= true when PREST(1 to 5)="11011"
        else false;

end architecture procedural;
```

b)

a)

c)

Fig. 1.9 Formats for capturing designs at intermediate levels of abstraction. RTL diagram (a), RTL synthesis model (b), and gate-level schematic (c) (simplified, note that (a) and (b) refer to different circuits).

1. Identify the computational and storage requirements of the algorithm.
2. From a virtual library of common hardware building blocks, select a suitable item for each kind of processing and storage operation.
3. Establish a cycle-based schedule for carrying out the algorithm with those resources. Where there is a choice, indicate which building block is to process what data item.
4. Decide on a hardware organization able to execute the resulting work plan. Specify the architecture in terms of combinational logic blocks, data registers, on- and off-chip memories, busses, switches, signals, and finite state machines.
5. Keeping track of data moves and operations for each clock cycle, translate all this into the necessary instructions for synthesis at the RTL level.

Generating a close-to-optimum architecture under performance, power, cost, and further constraints represents a formidable optimization problem, especially if a tool is expected to work well for arbitrary applications. To get an idea, consult the more detailed lists of issues to be addressed in section 1.3.2. Apart from a couple of specialized areas, automatic architecture synthesis does not — up to now — produce results comparable to those of inspired and experienced engineers. Nonetheless architecture synthesis continues to be an active field of research as VLSI design can no longer afford to deal with low-level details.

Even an experienced RTL code writer cannot be expected to complete much more than 40 lines of code per day. Estimates say that design productivity ranges from 20 to 400 GE per working day.[13] Albeit quite impressive, these figures are actually insufficient to keep pace with the rapid advances of fabrication technology.

DESIGN WITH VIRTUAL COMPONENTS

In the late 1990s, synthesis technology together with HDL standardization opened the door for an entirely new approach to designing digital VLSI circuits. A **virtual component** (VC)[14] is essentially a HDL synthesis package that is made available to others on a commercial basis for incorporation into their own ICs. VLSI design teams across the electronics industry are thus put in a position to purchase hardware designs for major subfunctions on the commercial market, dispensing with the need to write too much HDL source code on their own. The licensees just remain in charge of synthesis, place and route (P&R), and overall verification.

Though of highly specific nature, most VCs implement fairly common subfunctions; some degree of parametrization is sought to cover more potential applications. Examples include, but are not limited to, microprocessor and signal processor cores, all sorts of filters, audio and/or video en/decoders, cipher functions, error correction en/decoders, USB, FireWire, and many other interfaces.

While hard modules such as standard cells, macrocells, and megacells had freed most IC designers from addressing transistor-level issues and detailed layout by the mid 1980s, the soft VCs have extended these benefits to higher levels of abstraction in a natural way. New business opportunities have opened up and companies that specialize in marketing synthesis models have emerged. Yet, as is to be explained in section 13.4, the concept has proved more difficult than anticipated for reasons related to quality, adaptation, interfacing, licensing, and liability.

A classification scheme depicted in table 1.1 nicely complements the one of fig.1.3.

ELECTRONIC SYSTEM-LEVEL (ESL) DESIGN AUTOMATION

More recently, the competitive pressure towards shorter and leaner design cycles has incited the industry to look at design productivity from a wider perspective. ESL is a collective term for efforts that take inspirations from numerous ideas.

- Enforce a correct-by-construction methodology by supporting progressive refinement starting with a virtual prototype of the system to be.

---

[13] Be warned that design productivity is extremely dependent on circumstances.
- The effort per transistor is not the same for memories, logic, and mixed-signal designs.
- The more circuit blocks that have been validated before can be reused, the better.
- Skilled engineering teams not only work faster but also manage with fewer design iterations.
- Powerful EDA tools can work out many minor circuit and layout details automatically.
- The existence of an established and proven design flow benefits the design process.
- Tight timing, power, and layout density budgets ask for more human attention.
- Unstable specifications and rapidly changing teams are detrimental to productivity.

[14] Virtual components are better known as **intellectual property modules** or IP modules for short. We prefer the term "virtual component" because IP does not point to electronics in any way and because the acronym might easily be misunderstood as "Internet protocol". Other synonyms include "core" and "core ware".

Table 1.1 | IC families as a function of fabrication depth and design abstraction level

Fabrication depth	Electrical configuration	Semi-custom fabrication	Full-custom fabrication	
Design level	Cell-based as obtained from ∘ synthesis with VCs in HDL form, ∘ synthesis from captive HDL code, ∘ schematic entry, or a mix of these			Hand layout
Product name	Field-programmable logic device (FPGA, CPLD)	Gate-array, sea-of-gates, or structured ASIC	Std. cell IC (with or w/o macrocells and megacells)	Full-custom IC

- Resort to architecture synthesis to explore the solution space more systematically and more rapidly than with conventional, e.g. RTL synthesis, methods.
- Support hardware–software co-design by making it possible to start software development before hardware design is completed.
- Improve the coverage and efficiency of functional verification by dealing with system-level transactions and by taking advantage of formal verification techniques where possible.

1.2.5 The business point of view

Our final question relates to business.
"How are the industrial activities shared between business partners?"

Integrated device manufacturer (IDM) is the name for a company that not only designs and markets microchips but also operates its own wafer processing line, aka fab.
  Examples: Intel, Samsung, Toshiba, ST-Microelectronics, Infineon, NXP Semiconductors.

Fabless vendor. A company that develops and markets proprietary semiconductor components but has their manufacturing subcontracted to an independent silicon foundry rather than operating any wafer processing facilities of its own.
  Examples: Altera (FPL), Actel (FPL), Broadcom (networking components), Cirrus Logic-Crystal (audio and video chips), Lattice Semiconductor (FPL), Nvidia (graphics accelerators), PMC-Sierra (networking components), Qualcomm (CDMA wireless communication), Ramtron (non-volatile memories), Sun Microystems (UltraSPARC processors), and Xilinx (FPL). Please check [3] for a more complete picture.

Silicon foundry, albeit technically incorrect, has become the name for a company that operates a complete wafer processing line and that offers its manufacturing services to others.
  Examples: TSMC, UMC, etc.

Virtual component vendor. A fabless company that makes it a business to develop synthesis packages and to license them to others for incorporation into their ICs.
  Examples: ARM, Sci-worx, Synopsys (formerly InSilicon).

Originally, all IC business had been confined to vertically integrated semiconductor companies that designed and manufactured standard parts for the markets they perceived. Opening VLSI to other companies was essential to instilling new and highly successful fabless business models. Three factors came together in the 1980s to make this possible.

- Generous integration densities at low costs.
- Proliferation of high-performance engineering workstations and EDA software.
- Availability of know-how in VLSI design outside IC manufacturing companies.

This text is intended to contribute to the third item with a focus on synthesis-based design.

1.3 | Design flow in digital VLSI

1.3.1 The Y-chart, a map of digital electronic systems

The Y-chart by Gajski is very convenient for situating the various stages of digital design and the numerous attempts to automate them. Three axes stand for three different ways to look at a digital system and concentric circles represent various levels of abstraction, see fig.1.10.

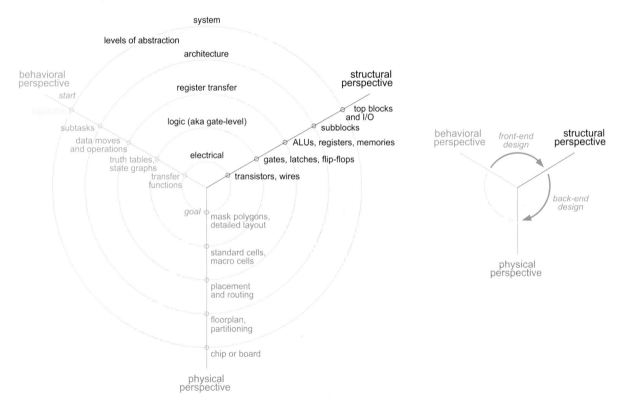

Fig. 1.10 The Y-chart of digital electronic systems.

From a **behavioral** perspective we are interested only in what a circuit or system does, not in how it is actually built. Put differently, the design is viewed as a black box that processes information by producing some output symbols in response to some input symbols. What matters most is the dependency of the output from past and present inputs, but timing relationships between input data, output data, and some clock signal are also of interest.

A **structural** way of looking at electronic circuits is concerned with connectivity, that is with the building blocks from which a circuit is composed and with how they connect to each other. Given some behavioral specification, it is almost always possible to come up with more than one network for implementing it. Structural alternatives typically differ in terms of circuit complexity, performance, energy efficiency, and in other characteristics of practical interest such as parts list, fabrication technology, testability, etc.

What counts from a **physical** point of view is how the various hardware components and wires are arranged in the space available in a cabinet, on a board, or on a semiconductor chip. Again, there is a one-to-many relationship between structural description and physical arrangement.

Examples of circuits viewed at different levels of abstraction and from all three perspectives have been given in figs.1.7 and 1.9. Figure 1.11 adds more illustrations not presented so far. In addition, table 1.2 lists the objects that are of interest for the individual views. It is interesting to note that different time units are used depending on the abstraction level on which behavior is described.

Table 1.2 | Views and levels of abstraction in digital design.

level of abstraction	behavioral	structural	physical	concept of time
system	input/output relationship	system with input/output	chip, board, or cabinet	sequence, throughput
architecture	bus functional model (BFM)	organization into subsystems	partitioning, floorplan	partial ordering relationships
register transfer	data transfers and operations	ALUs, muxes, and registers	placement and routing	clock cycles (cycle true)
logic	truth tables, state graphs	gates, latches, and flip-flops	standard cells or components	events, delays, timing parameters[a]
electrical	transfer functions	transistors, wires, R, L, C	detailed layout, mask polygons	continuous

[a] Such as t_{pd}, t_{su}, t_{ho}. Glitches are also accounted for at this level of abstraction.

1.3.2 Major stages in VLSI design

The development cycle of VLSI circuits comprises a multitude of steps that are going to be explained in more detail next. The interplay of all such steps is illustrated by way of two drawings that partially overlap. Figure 1.12 focusses on system-level issues and reduces all activities that are related to

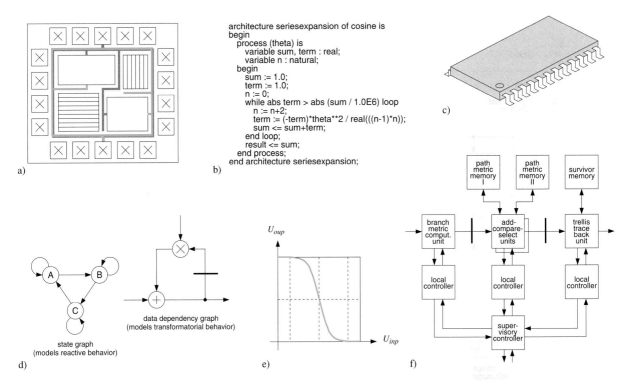

```
architecture seriesexpansion of cosine is
begin
    process (theta) is
        variable sum, term : real;
        variable n : natural;
    begin
        sum := 1.0;
        term := 1.0;
        n := 0;
        while abs term > abs (sum / 1.0E6) loop
            n := n+2;
            term := (-term)*theta**2 / real(((n-1)*n));
            sum <= sum+term;
        end loop;
        result <= sum;
    end process;
end architecture seriesexpansion;
```

Fig. 1.11 More design views. Floorplan of a VLSI chip (a), software model (b), encapsulated chip (c), graphical formalisms (d), transfer characteristic of an inverter (e), and block diagram (f) (simplified).

actual IC design to their simplest expression while fig.1.13 does the opposite. Again, figs.1.7, 1.9, and 1.11 help to clarify what is meant. Also keep in mind that this text focuses on the design of hardware modules in a system and ignores all steps towards implementing its software components.

System-level design. The decisions taken during this stage are most important as they determine the final outcome more than anything else does.

- Specify the functionality, operating conditions, and desired characteristics (in terms of performance, power, form factor, costs, etc.) of the system to be.
- Partition the system's functionality into subtasks.
- Explore alternative hardware and software tradeoffs.
- Decide on make or buy for all major building blocks.
- Decide on interfaces and protocols for data exchange.
- Decide on data formats, operating modes, exception-handling procedures, and the like.
- Define, model, evaluate, and refine the various subtasks from a behavioral perspective.

It is a characteristic trait of this stage that acceptance criteria, design procedures, design expertise, and the software tools that are being put to service vary greatly with the nature of the overall application and of the subsystem currently being considered.

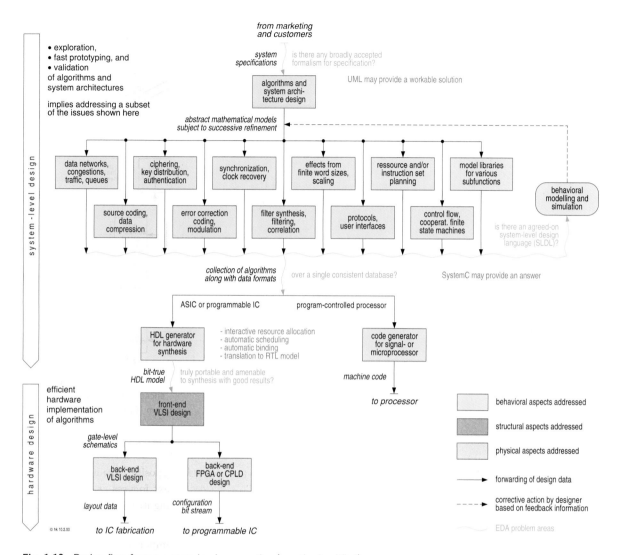

Fig. 1.12 Design flow from a system-level perspective (greatly simplified).

Examples

	packet router	audio compressor	microprocessor
mode of operation	reactive	transformatorial	must fit both
data manipulation	shallow	deep	must fit both
real-time processing	yes	typically yes	maybe, maybe not
numerical precision	known a priori	to be determined	must fit both
design focus	architecture	algorithm	instruction set
background	queuing theory	human perception	code analysis
evaluation tool	traffic simulation	algorithmic simulation	benchmark programs

Figure 1.12 exposes another difficulty of system-level design that has its roots in the highly heterogeneous nature of electronic systems. At various points, some fairly abstract design description must be propagated from one software tool to the next. Yet, there are no mathematical formalisms and agreed-on computer languages of sufficient scope to capture a sufficient portion of a system, let alone a system as a whole. The practical consequences are that some specifications need to be manually restated several times, that simulations do not extend over the entire system, and that certain aspects are being lost in the process.

Algorithm design. The central theme is to meet the data and/or signal processing requirements defined before with a series of computations that are streamlined in view of their implementation in hardware. The subsequent assignments are part of algorithm design.

- Coming up with a collection of suitable algorithms or computational paradigms.[15]
- Cut down computational burden and memory requirements.
- Find acceptable compromises between computational complexity and accuracy.
- Analyze and contain effects of finite word-length computation.
- Decide on number representation schemes.
- Evaluate alternatives and select the one best suited for the situation at hand.
- Quantify the minimum required computational resources (in terms of memory, word widths, arithmetic and logic operations, and their frequencies of occurrence).

Algorithm design culminates in a bit-true software model which is indispensable for checking figures of merit relevant for the application at hand, e.g. signal-to-noise ratio, coding gain, data compression factor, error rate, and the like against specifications.

Architecture design. VLSI architects essentially decide on the necessary hardware resources and organize their interplay in such a way as to implement a known computational algorithm under the performance, cost, power, and other constraints imposed by the target application. The hardware arrangement they have to come up with must capture the essential structural characteristics of the future circuit but, at the same time, abstracts from implementation details. Still, architecture design also implies selecting a target technology and taking into account its possibilities and limitations.[16]

Architecture design starts from fairly abstract notions of a circuit's functionality and gradually proceeds to more detailed representations. The process is understood to happen in two substages, namely high-level architecture design and register transfer-level design. The former involves the following.

---

[15] The term "computational paradigm" has been chosen to include finite state machines, cellular automata, neural networks, fuzzy logic, and other computational schemes that are not necessarily covered by the word "algorithm" as it is normally understood in the context of software engineering.

[16] Take this as an analogy from everyday life. Assume you were given the recipe for a fantastic cake by your grandmother and you were now to make a business out of it by setting up a bakery to mass-produce the cake. The recipe corresponds to the algorithm or software model that specifies how the various ingredients must be processed in order to obtain the final product. Architecture design can then be likened to deciding on the mixers, kneaders, ovens, and other machines for processing the ingredients, and to planning the material flow in an industrial bakery. Observe that you will arrive at different factory layouts depending on the quantity of cakes that you intend to produce and depending on the availability and costs of labor and equipment.

- Partition a computational task in view of a hardware realization.
- Organize the interplay of the various subtasks.
- Decide on the hardware resources to allocate to each subtask (**allocation**).
- Define datapaths and controllers.[17]
- Decide between off-chip RAMs, on-chip RAMs, and registers.
- Decide on communication topologies and protocols (parallel, serial).
- Define how much parallelism to provide in hardware.
- Decide where to opt for pipelining and to what degree.
- Decide on a circuit style, fabrication technology, and manufacturing process.
- Decide what abstraction level to design at and what cell libraries to use, if any.
- Get a first estimate of the circuit's size and cost.
- etc.

The result is captured in a high-level block diagram that includes datapaths, controllers, memories, interfaces, and key signals. A preliminary floorplan is also being established. Verification of an architecture typically occurs by way of simulations, where each major building block is represented by a behavioral model of its own.

The work is then carried down to the more detailed **register transfer level** (RTL) where the circuit gets modelled as a collection of storage elements interconnected by purely combinational subcircuits. Relevant issues at this stage include

- How to implement arithmetic and logic units
  (e.g. ripple-carry, carry-lookahead, carry-select).
- Whether to use hardwired logic or microcode to implement a controller.
- When to use a ROM rather than random logic.
- What operations to perform during which clock cycle (**scheduling**).
- What operations to carry out on which processing unit (**binding**).
- Where to insert pipelining and shimming registers.
- How to balance combinational depth between registers.
- What clocking discipline to adopt.
- What time interval to use as the basic clock period.
- Where to use a bidirectional or a unidirectional bus, and where to prefer three-state bus drivers over multiplexers.
- By what test strategy is testability to be ensured.
- How to initialize the circuit.
- etc.

The outcome is a set of more detailed diagrams that include every single register, memory, and major block of combinational logic. As opposed to gate-level schematics, however, combinational functions are specified in behavioral rather than structural terms. Simulations are instrumental in debugging the RTL code. The floorplan is refined on the basis of the more detailed data that are now available and compared against the die size and cost targets for the final product. This is also the point to decide on the most appropriate design level — synthesis, schematic entry, hand layout — for each circuit block.

---

[17] These and other circuit-related terms are explained in section 1.7.

Logic design. The translation into a gate-level netlist and its Boolean optimization are largely automatic. The design is now definitively being committed to

- A fabrication depth (e.g. full-custom vs. semi-custom vs. FPL),
- One or more cell libraries (e.g. by Artisan vs. LSI Logic vs. Xilinx),
- A circuit style (e.g. static vs. dynamic CMOS logic),
- A fabrication technology (e.g. CMOS vs. BiCMOS), and
- A manufacturing process (e.g. L130 by UMC vs. HCMOS9gp by ST).

The delays and energy-dissipation figures associated with the various computational and storage operations are being calculated. Subcircuits that are found to limit performance during pre-layout analysis are identified and redesigned or reoptimized where possible. The result is a complete set of gate-level schematics and/or netlists validated by **electrical rule check (ERC)**, **logic simulation**, **timing verification**, and power estimation.

Improvement of testability. A malfunctioning IC is the result of design flaws, fabrication defects, or both. Special provisions are necessary to ascertain the correct operation of millions of transistors enclosed in a package with a couple of hundred pins at most. **Design for test (DFT)** implies improving the controllability and observability of inner circuit nodes by adding auxiliary circuitry on top of the payload logic.[18]

In addition, a **test vector set** is generated for distinguishing faulty circuits from correct ones. Such a vector set typically includes thousands or millions of stimuli and expected responses. In a procedure referred to as **fault grading**, testability is rated by relating the number of fabrication defects that can in fact be detected with a test vector set under consideration to the total number of conceivable faults. Both the test circuitry and the test patterns are iteratively refined until a satisfactory fault coverage is obtained.

Physical design. Physical design addresses all issues of arranging the multitude of subcircuits and devices along with their interconnections on a piece of semiconductor material. **Floorplanning** is concerned with organizing the major circuit blocks into a rectangular area as small as possible while, at the same time, limiting the effects of interconnect delays on the chip's performance.[19] Chip-level power and clock distribution are also to be dealt with. A padframe must be generated to hold the bond pads and the top-level layout blocks. During the subsequent **place and route (P&R)** steps, each cell gets assigned a specific location on the die before the courses of myriads of metal wires that are to carry electrical signals between those cells get defined. It is often necessary to reoptimize the circuit logic as a function of the estimated interconnect delays that become available during the process. The final phase

---

[18] Standard techniques include block isolation, scan testing, and BIST. Block isolation makes major circuit blocks accessible from outside a chip with the aid of extra multiplexers so that stimuli can be applied and responses evaluated via package pins while in test mode. Scan testing is to be outlined in section 6.2.2. The idea behind built-in self-test (BIST) is to move stimuli generation and response checking onto the chip itself, and to essentially output a "go/no go" result [4]. BIST and block isolation are popular for testing on-chip memories. As DFT, test vector preparation, and automated test equipment (ATE) are not part of this text, the reader is referred to the specialized literature such as [5], for instance.

[19] Floorplanning makes part of physical design much as layout design does. What is the difference then? As an analogy, floorplanning is concerned with the partitioning of a flat into rooms and hallways whereas layout design deals with tiny geometric patterns on a carpet.

where the global wires running between padframe and core get routed is also known as **chip assembly**.

As the inner layout details of the cells do not really matter for floorplanning, place, and route, cells are typically abstracted to their outlines up to this point. To prepare for IC manufacturing, detailed layout data must be filled in for those abstract views. The outcome is a huge set of polygons that involves all mask layers. Prior to fabrication, the complete layout data need to be checked carefully to protect against fatal mishaps. **Physical design verification** relies on a number of software tools.

- Layout rule check — better known as **design rule check** (DRC) — examines conformity of layout with geometric rules imposed by the target process.
- **Manufacturability analysis** searches for layout patterns likely to be detrimental to the fabrication yield.
- **Layout extraction** (re-)obtains the actual circuit netlist in preparation for
- **layout versus schematic** (LVS) where it gets compared against the desired one.
- Post-layout timing verification.
- Post-layout simulation.

Sign-off. By accepting a design for prototype fabrication, an IC vendor commits himself to delivering circuits that behave like the post-layout simulation model (identical functionality for the test vector set provided by the customer, same or better speed, same or lower power). As no customer is willing to pay for fabricated parts that do not conform with this requirement, the vendor wants to make sure the design is consistent with good engineering practice and with company-specific guidelines before doing so. DRC, manufacturability, ERC, LVS, post-layout simulation, and fault coverage are routinely examined. Inspection often extends to timing verification, clocking discipline, power and clock distribution, circuit design style, test structures, and more.

A couple of comments are due after this rather general overview.

- In reality, the separation into individual subtasks is not as nice and clear as in fig.1.13. Various side effects of deep submicron technologies and the quest for optimum results make it necessary for most software tools to work across several levels of abstraction. As an example, it is no longer possible to place and route a gate-level netlist without adapting the circuit logic as a function of the resulting layout parasitics and interconnect delays. In the drawing, this gets reflected by the joint refinement of layout data and netlists.

- Only ideally does design occur as a linear sequence of steps. Some back and forth between the various subtasks is inevitable to obtain a truly satisfactory result. Also, not all design stages are explicitly covered in every IC development project. Depending on the circuit's nature, fabrication depth, and design level, some of the design stages are skipped or outsourced, i.e. delegated to specialists at third-party companies.[20]

---

[20] The design of a simple glue logic chip, for instance, begins at the logic level as there are no algorithmic or architectural questions to deal with. Models of industrial collaboration are to be discussed in section 13.2.

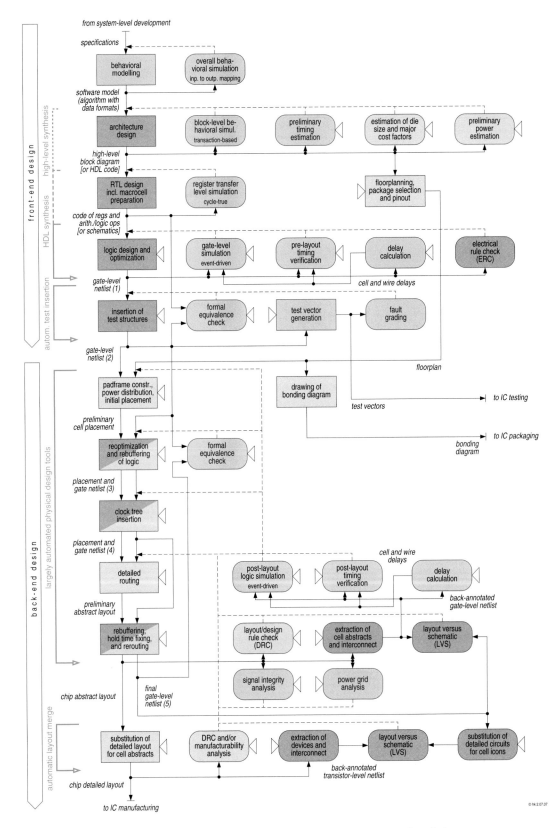

Fig. 1.13 Digital VLSI design flow (simplified). See fig.1.14 for an explanation of symbols.

- Note the presence of angular and rounded boxes in fig.1.13. While angular boxes refer to construction activities, the rounded ones stand for analysis and verification steps. A backward arrow implies that any problem uncovered during such an analysis triggers corrective action by the designer. The results from construction steps are subject to immediate verification, which is typical for VLSI.

  The reason is that correcting a mistake becomes more and more onerous the further the design process has progressed. Correcting a minor functional bug after layout design, for instance, would require redoing several design stages and would waste many hours of labor and computer time. Also, a functional bug can be uncovered more effectively from a behavioral or RTL model than from a post-layout transistor-level netlist because simulation speed is orders of magnitude higher and because automatic response checking is much easier to implement for logic and numeric data types than for analog waveforms.

- A critical point is reached when first silicon is going to be produced. While it is possible to cut and add wires using advanced and expensive equipment such as focused ion-beam (FIB) technology to patch a malfunctioning prototype, there is virtually no way to fix bugs in volume production. Depending on the circuit's size, fabrication depth, process, and manufacturer, expenses somewhere between 12 kUSD and 1 MUSD are involved with preparation of photomasks, tooling, wafer processing, preparation of probe cards and evaluation of pre-production samples. Any design flaw found after prototype fabrication thus implies the waste of important sums of money.

  To make things worse, with turnaround times ranging between two weeks and three months, a product's arrival on the market is delayed so much that the chip is likely to miss its window of opportunity.

Observation 1.2. *Redesigns are so devastating for the business that the entire semiconductor industry has committed itself to "first-time-right" design as a guiding principle. To avoid them, VLSI engineers typically spend much more time verifying a circuit than actually designing it.*

- Figure 1.13 also includes a number of forward arrows that bypass one or two construction steps. They suggest how electronic design automation, cell libraries, and purchased know-how help speed up the design process. Keeping pace with the breathtaking progress of fabrication technology is in fact one of the major challenges for today's VLSI designers.

- While there is not too much of a difference in the front-end flow, back-end design for field-programmable logic (FPL) differs somewhat from that depicted in fig.1.13. The preliminary gate-level netlist obtained from HDL synthesis is mapped onto configurable blocks available in the target FPGA or CPLD device. After the EDA software has decided how to run all necessary interconnects using the wires, switches, and drivers available, the result is converted into a **configuration bit stream** for download into the FPL device. As FPGAs and CPLDs come with many diverse architectures, product-specific back-end tools made available by the FPL vendor are used for this procedure.

Observation 1.3. *Whoever has learned to design full-custom ICs is in an excellent position for designing semi-custom ICs and to design with field-programmable logic, but not necessarily the other way round.*

1.3.3 Cell libraries

Library development occurs quite separately from actual IC design as cell-based circuits largely dominate VLSI.[21] Cell libraries are typically licensed to IC developers by specialized library vendors since silicon vendors have largely withdrawn from this business.

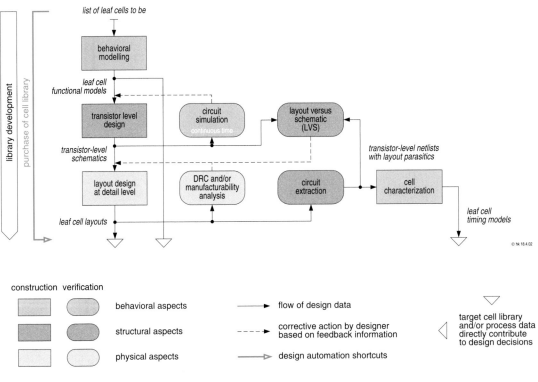

Fig. 1.14 Library design flow (simplified).

Once the set of prospective library cells has been defined functionally, library development proceeds in three major phases, see fig.1.14. **Electrical design** deals with implementing logic functions as transistor-level networks and with sizing the individual devices such as to find an optimum trade-off between performance, circuit complexity, and energy efficiency.

During the subsequent **layout design**, the locations and geometric shapes of individual devices are defined along with the shapes of the wires running in between. The goal is to obtain leaf cells that are compact, fast, energy-efficient, suitable for automatic place and route (P&R), and that can be manufactured with maximum yield.

Verification includes the customary ERC, DRC, manufacturability analysis, extraction, and LVS procedures. Next the electrical and timing parameters that are to be included in data sheets and simulation models of the cells are determined. This **library characterization** step typically relies

---

[21] Semi-custom ICs and FPL rely on prefabricated primitives anyway.

on repeated continuous-time continuous-value simulations under varying load, ramp, and operating conditions.[22]

Designing, characterizing, documenting, and maintaining a cell library is a considerable effort as multiple **design views** must be prepared for each cell, including

- A **datasheet** with functional, electrical, and timing specifications.
- A graphical **icon** or symbol for inclusion into schematic drawings.
- An accurate behavioral model for simulation and timing analysis.
- A set of simulation and test vectors.
- A transistor-level netlist or schematic.
- A detailed layout.
- A simplified layout view showing cell outline and connector locations for the purpose of place and route known as **cell abstract**, floorplanning abstract, or phantom cell.

Please refer back to fig.1.7 for illustrations.

In order to protect their investments, most library vendors consider their library cells to be proprietary and are not willing to disclose how they are constructed internally. They supply datasheets, icons, simulation models, and abstracts, but no transistor-level schematics and no layouts. Under this scheme, detailed layouts are to be substituted for all cell abstracts by the vendor before mask preparation can begin. Note this extra step is reflected in fig.1.13.

1.3.4 Electronic design automation software

The VLSI industry long ago became entirely dependent on electronic design automation (EDA) software. There is not one single step that could possibly be brought to an end without the assistance of sophisticated computer programs. The sheer quantity of data necessary to describe a multi-million transistor chip makes this impossible. The design flow outlined in the previous section gives a rough idea of the variety of CAE/CAD programs that are required to pave the way for VLSI and FPL design. Almost each box in fig.1.13 stands for yet another tool.

While a few vendors can take pride in offering a range of products that covers all stages from system-level decision making down to physical layout, much of their effort tends to focus on relatively small portions of the overall flow for reasons of market penetration and profitability. Frequent mergers and acquisitions are another characteristic trait of the EDA industry. Truly integrated design environments and seamless design flows are hardly available off the shelf.

Also, the idea of integrating numerous EDA tools over a common design database and with a consistent user interface, once promoted as front-to-back environments, aka frameworks, has lost momentum in the marketplace in favor of point tools and the "best in class" approach. Design flows are typically pieced together from software components of various origins.[23] The presence of software tools, design kits, and cell libraries from multiple sources in conjunction with the absence of agreed-on standards adds a lot of complexity to the maintainance of a coherent design environment. Many of the practical difficulties with setting up efficient design flows are left to EDA customers

---

[22] More details are to follow in section 12.7.

[23] A very nice review of the evolution of the EDA industry is given in [6].

and can sometimes become a real nightmare. It is to be hoped that this trend will be reversed one day when customers are willing to pay more attention to design productivity than to layout density and circuit performance.

1.4 | Field-programmable logic

The general idea behind programmable logic has been introduced in section 1.2.3. The goal of this section is to explain the major differences that separate distinct product families from each other. Key properties of any FPL device are fixed by decisions along two dimensions taken at development time. A first choice refers to how the device is being configured and how its configuration is stored electrically while a second choice is concerned with the overall organization of the hardware resources available to customers. Customers, in this case, are design engineers who want to implement their own circuits in an FPL device.

1.4.1 Configuration technologies

Static memory. The key element here is an electronic switch — such as a transmission gate, a pass transistor, or a three-state buffer — that gets turned "on" or "off" under control of a configuration bit. Unlimited reprogrammability is obtained from storing the configuration data in SRAM cells or in similar on-chip subcircuits built from two cross-coupled inverters, see fig.1.15a. As a major drawback, the circuit must (re)obtain its entire configuration from outside whenever it is being powered up. The problem is solved in one of three possible ways, namely

(a) by reading from a dedicated bit-serial or bit-parallel off-chip ROM,
(b) by downloading a bit stream from a host computer, or
(c) by long-term battery backup.

Reconfigurability is very helpful for debugging. It permits one to probe inner nodes, to alternate between normal operation and various diagnostic modes, and to patch a design once a flaw has been located. Many RAM-based FPL devices further allow reconfiguring of their inner logic during operation, a capability known as **in-system configuration** (ISC) that opens a door towards configurable computing.

UV-erasable memory. Electrically programmable read-only memories (EPROM) rely on special MOSFETs where a second gate electrode is sandwiched between the transistor's bulk material underneath and a control gate above, see fig.1.15b. The name **floating gate** captures the fact that this gate is entirely surrounded by insulating silicon dioxide material. An electrical charge trapped there determines whether the MOSFET, and hence the programmable link too, is "on" or "off".[24]

---

[24] More precisely, the presence or absence of an electrical charge modifies the MOSFET's threshold voltage and so determines whether the transistor will conduct or not when a voltage is applied to its control gate during memory readout operations.

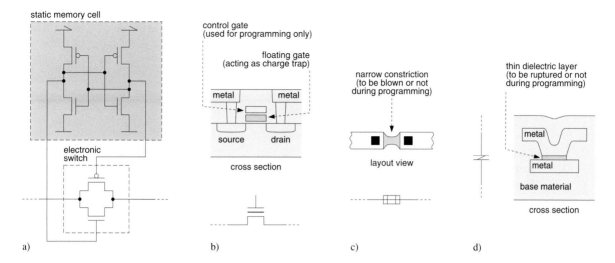

Fig. 1.15 FPL configuration technologies (simplified, programming circuitry not shown). Switch steered by static memory cell (a), MOSFET controlled by a charge trapped on a floating gate (b), fuse (c), and antifuse (d).

Charging occurs by way of hot electron injection from the channel. That is, a strong <u>lateral</u> field applied between source and drain accelerates electrons to the point where they get injected through the thin dielectric layer into the floating gate. The necessary programming voltage in the order of 5 to 20 V is typically generated internally by an on-chip charge pump.

Erasure of the charge is obtained by shining ultraviolet (UV) radiation on the chip, thereby causing the charges to leak away from the floating gate. The necessary quartz window in the plastic or ceramic package gives UV-erasable devices their unmistakable appearance but also renders the package rather expensive.

UV-erasable devices are non-volatile and immediately live at power-up, thereby doing away with the need for any kind of configuration-backup apparatus. Reprogramming necessitates removing the component from the circuit board and placing it into a special UV eraser, however, which is undesirable and often altogether impossible. This explains why EPROM-based FPL devices — much like the memories themselves — have been superseded by parts that are more convenient to reconfigure.

Electrically erasable memory. EEPROM technology borrows from UV-erasable memories. The difference is that the electrons trapped on the floating gate are removed electrically by having them tunnel through the oxide layer underneath the floating gate without exposure to ultra-violet light, thereby making it possible to manufacture FPL devices that are non-volatile but nevertheless reconfigurable through their package pins. The secret is a quantum-mechanical effect known as Fowler–Nordheim tunneling that comes into play when a strong <u>vertical</u> field (8–10 MV/cm or so) is applied across the gate oxide.

Early electrically erasable devices were penalized by the fact that an EEPROM cell occupies about twice as much area as its UV-erasable counterpart because each bit cell includes a select transistor connected in series with the storage transistor. The **flash memory** technology prevalent today manages with a single floating-gate transistor per bit. The fact that erasure must occur in chunks, that is to say many bits at a time, is perfectly adequate in the context of FPL. Data retention times vary between 10 and 40 years. Endurance of flash FPL is typically specified with 100 to 1000 configure–erase cycles, which is much less than for flash memory chips.

Fuse or **antifuse.** Fuses, which were used in earlier bipolar PROMs and SPLDs, are narrow bridges of conducting material that blow in a controlled fashion when a programming current is forced through. Antifuses, such as those employed in today's FPGAs, are thin dielectrics separating two conducting layers that are made to rupture upon applying a programming voltage, thereby establishing a conductive path of low impedance.

In either case, programming is permanent. Whether this is desirable or not depends on the application. Full factory testing prior to programming of one-time programmable links is impossible for obvious reasons. Special circuitry is incorporated to test the logic devices and routing tracks at the manufacturer before the unprogrammed devices are being shipped. On the other hand, antifuses are only about the size of a contact or via and, therefore, allow for higher densities than reprogrammable links, see fig.1.15c and d. Antifuse-based FPL is also less sensitive to radiation effects, offers superior protection against unauthorized cloning, and does not need to be configured following power-up.

Table 1.3 FPL configuration technologies and their key characteristics.

Configuration technology	Non-vola-tile	Live at power-up	Reconfi-gurable	Unlimited endu-rance	Radiation tolerance of config.	Area occupation per link	Extra fabr. steps
SRAM	no	no	in circuit	yes	poor	large	0
EPROM	yes	yes	out of circuit	no	good	small in array	3
Electr. erasable	yes	yes	in circuit		good		>5
EEPROM				no		2·EPROM	
Flash memory				no		≈EPROM	
Antifuse PROM	yes	yes	no	n.a.	best	small	3

1.4.2 Organization of hardware resources

Simple programmable logic devices (SPLDs). Historically, FPL has evolved from purely combinational devices with just one or two programmable levels of logic such as ROMs, PALs, and PLAs. Flip-flops and local feedback paths were added later to allow for the construction of finite state machines, see fig.1.16a and b. Products of this kind continue to be commercially available for glue logic applications. Classic SPLD examples include the 18P8 (combinational) and the 22V10 (sequential).

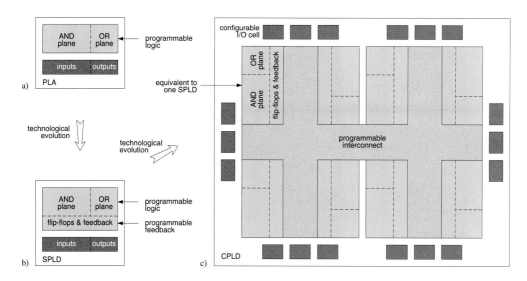

Fig. 1.16 General architecture of CPLDs (c) along with precursors (a,b).

The rigid two-level-logic-plus-register architecture in conjunction with the limited numbers of inputs, outputs, product terms, and flip-flops always restricted SPLDs to small applications. More scalable and flexible architectures had thus to be sought, and the spectacular progress of VLSI technology has made their implementation economically feasible from the late 1980s onwards. Two broad classes of hardware organization prevail today.

Complex programmable logic devices (CPLDs) expand the general idea behind SPLDs by providing many of them on a single chip. Up to hundreds of identical subcircuits, each of which conforms to a classic SPLD, are combined with a large programmable interconnect matrix or network, see fig.1.16c. A difficulty with this type of organization is that a partitioning into a bunch of cooperating SPLDs has to be imposed artificially on any given computational task, which benefits neither hardware nor design efficiency.

Depending on the manufacturer, products are known as complex programmable logic device (CPLD), programmable large-scale integration (PLSI), erasable programmable logic device (EPLD), and the like in the commercial world.

Field-programmable gate arrays (FPGAs) have their overall organization patterned after that of gate arrays. Many **configurable logic cells** are arranged in a two-dimensional array with bundles of parallel wires in between. A switchbox is present wherever two wiring channels intersect, see fig.1.17.[25] Depending on the product, each logic cell can be configured so as to carry out some not-too-complex combinational operation, to store a bit or two, or both.

---

[25] While it is correct to think of alternating cells and wiring channels from a conceptual point of view, you will hardly be able to discern them under a microscope. The reason is that logic and wiring resources are superimposed for the sake of layout density in modern FPGA chips.

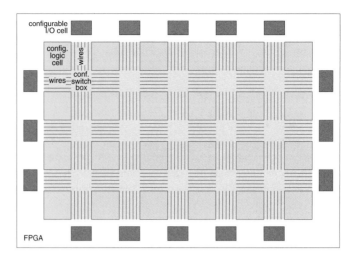

Fig. 1.17   General architecture of FPGAs.

As opposed to traditional gate arrays, it is the state of programmable links rather than fabrication masks that decides on logic functions and signal routing.

Parts with this organization are being promoted under names such as field-programmable gate array (FPGA), logic cell array (LCA), and programmable multilevel device (PMD). The number of configurable logic cells greatly varies between products, with typical figures ranging between a few dozens and hundreds of thousands.

FPGA architectures are differentiated further depending on the granularity and capabilities of the configurable logic cells employed. One speaks of a **fine-grained** architecture when those cells are so simple that they are capable of implementing no more than a few logic gates and/or one bistable. In the example depicted in fig.1.18a, for instance, each logic cell can be configured into a latch, or a flip-flop, or into almost any 3-input gate.

As opposed to this, cells that are designed to implement combinational functions of four to six input variables and that are capable of storing two or more bits at a time are referred to as **coarse-grained**. The logic cell of fig.1.18b has 16 inputs and 11 outputs, and includes two programmable look-up tables (LUTs), two generic bistables that can be configured either into a latch or a flip-flop, a bunch of configurable multiplexers, a fast carry chain, plus other gates. Of course, the superior functional capabilities offered by a coarse-grained cell are accompanied by a larger area occupation.[26]

The gate-level netlists produced by automatic synthesis map more naturally onto fine-grained architectures. The fact that fine-grained FPGAs and semi-custom ICs provide similar primitives further supports extensive reuse of design flows, HDL code, building blocks, and design

---

[26] Incidentally note that FPL vendors refer to configurable logic cells by proprietary names. "Logic tile" is Actel's term for their fine-grained cells whereas Xilinx uses the name "configurable logic block" (CLB) for their coarse-grained counterparts. Depending on the product family, one CLB consists of two or three LUTs plus two flip-flops or of several "slices", each of which includes one LUT and one bistable. "Module" and "eCell" are commercial names used by other vendors.

know-how. It thus becomes practical to move back and forth between field- and mask-programmed circuits with little overhead and to postpone any final commitment until fairly late in the design cycle. Conversely, fine-grained FPGAs tend to be more wasteful in terms of configuration bits and routing resources.

Another reason that contributed to the popularity of coarse-grained FPGAs is that on-chip RAMs come at little extra cost when that architectural concept is combined with configuration from static memory. In fact, a reprogrammable LUT is nothing else than a tiny storage array. It is thus possible to bind together multiple logic cells in such a way as to make them act collectively like a larger RAM. As opposed to many other types of FPGAs, there is no compelling need to set aside special die areas for embedded SRAMs. In the occurrence of fig.1.18b, each of the two larger LUTs in each logic tile contributes another 16 bits of storage capacity.

1.4.3 Commercial products

Table 1.4 classifies major CPLD and FPGA product families along the two dimensions configuration technology and hardware organization. These are not the only features that distinguish the numerous commercial products from each other, however. Most vendors combine field-programmable

Table 1.4 | Commercial field-programmable logic device families.

configuration technology	CPLD	overall organization of resources FPGA coarse-grained	fine-grained
static memory (SRAM)		Xilinx Spartan, Virtex. Lattice SC, EC, ECP. Altera FLEX, APEX, Stratix, Cyclone. eASIC Nextreme SL[a]	Atmel AT6000, AT40K.
UV-erasable (EPROM)	Cypress MAX340[b]		
electrically erasable (flash)	Xilinx XC9500, CoolRunner-II. Altera MAX3000, 7000. Lattice MACH 1,...,5. Cypress Delta39K, Ultra37000.	Lattice XP[c], MACH XO.	Actel ProASIC ProASIC[PLUS], Fusion,[d] Igloo.
antifuse (PROM)		QuickLogic Eclipse II, PolarPro.	Actel MX, Axcelerator AX.

[a] Combines RAM-configurable LUTs with e-beam single via-layer customization for interconnect.
[b] Remaining inventory transferred to Arrow/Zeus Electronics in 2006.
[c] Combines on-chip flash memory with an SRAM-type configuration memory.
[d] Mixed-signal FPGAs with on-chip analog-to-digital converters and optional processor core.

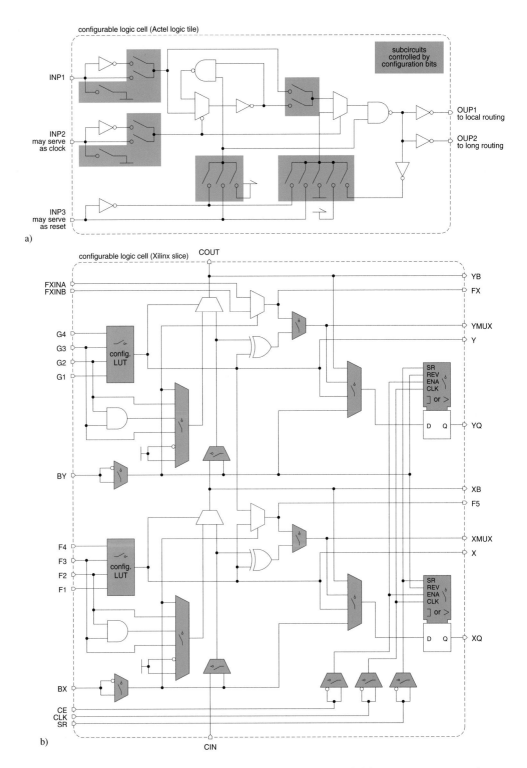

Fig. 1.18 Fine-grained vs. coarse-grained FPGAs. A small (Actel ProASIC) (a) and a large logic cell (Xilinx Virtex-4, simplified) (b).

logic with major hardwired subfunctions on a single die. SRAMs, FIFOs, phase-locked loops (PLLs), processor cores (e.g. PowerPC, ARM), and standard interfaces (PCI, USB, FireWire, Ethernet, WLAN, JTAG, LVDS, clock recovery circuits, etc.) are typical building blocks embedded within FPGA chips. The intention behind all such extensions is to help customers reduce time to market [7].

In addition to FPL, **field-programmable analog array**s (FPAAs) began to appear on the market in the late 1990s. The next logical step was the extension to mixed-signal applications. Advanced products that combine configurable analog building blocks with a micro- or digital signal processor and with analog-to-digital and digital-to-analog converters come quite close to the vision of field-programmable systems on a chip. Vendors of field-programmable analog and mixed-signal arrays include Anadigm, Actel, Cypress, Lattice, and Zetex FAS.

Technical details on commercial FPL devices are distributed over thousands of datasheets, [8] [9] help to keep track of products and manufacturers. More condensed background information is available from references such as [10] [11] [12].

Capacity figures of semi-custom ICs and FPL may be confusing. As opposed to full-custom ICs, manufactured gates, usable gates, and actual gates are not the same. **Manufactured gates** indicate the total number of GEs that are physically present on a silicon die. A substantial fraction thereof is not usable in practice because the combinational functions in a given design do not fit into the available look-up tables exactly, because an FPL device only rarely includes combinational and storage resources with the desired proportions, and because of limited interconnect resources. The percentage of **usable gates** thus depends on the application. The **actual gate** count, finally, tells how many GEs are indeed put to service by a given design. The three figures frequently get muddled up, all too often in a deliberate attempt to make one product look better than its competitors in advertisements, product charts, and datasheets. Some FPL vendors prefer to specify the available resources using their own proprietary capacity units rather than in gate equivalents.

Hint: It often pays to conduct benchmarks with a few representative designs before undertaking serious cost calculations and making a misguided choice. This also helps to obtain realistic timing figures that take into account interconnect delays.

1.5 | Problems

1. Various examples of design views have been given in figs.1.7, 1.9, and 1.11. Locate them in the Y-chart of fig.1.10.

2. Think of some industrial product family of your own liking (record player/MP3 player, mobile phone, (digital) camera, TV set/video recorder; car, locomotive, airplane; computer, photocopier, building control equipment, etc.). Discuss what microelectronics has contributed towards making these products possible in their present form. How has the microelectronic content evolved over the years? Where do you see challenges for improving these products and their microelectronic content?

1.6 | Appendix I: A brief glossary of logic families

A logic family is a collection of digital subfunctions that

- support the construction of arbitrary logic, arithmetic, and storage functions,
- are compatible among themselves electrically, and
- share a common fabrication technology.

A logic family must be available either as physical parts (SSI/MSI/LSI components for board design) or in virtual form as a set of library cells to be instantiated and manufactured together on a die of semiconductor material (for IC design).

Table 1.5 | Major semiconductor technologies and logic families with their acronyms.

Acronym	Meaning
MOS	Metal Oxide Semiconductor.
FET	Field Effect Transistor (either of n- or p-channel type).
BJT	Bipolar Junction Transistor (either of npn or pnp type).
NMOS	n-channel MOS (transistor, circuit style, or fabrication technology).
PMOS	p-channel MOS (transistor, circuit style, or fabrication technology).
CMOS	Complementary MOS (circuit style or fabrication technology) where pairs of n- and p-channel MOSFETs cooperate in each logic gate; features zero quiescent power dissipation, or almost so; supply voltages have evolved from up to 15 V down to 1 V and less.
static CMOS	circuit style that supports suspending all switching activities indefinitely and in any state with no loss of state or data.
dynamic CMOS	circuit style where data and/or state are kept as electrical charges that need to be refreshed or computed anew at regular intervals as data and/or state are otherwise lost.
TTL	Transistor Transistor Logic, made up of BJTs and passive devices; first logic family to gain wide-spread acceptance as SSI/MSI parts, has evolved over many generations, all of which share a 5 V supply.
ECL	Emitter-Coupled Logic, non-saturating current switching circuits built on the basis of BJTs, provides complementary outputs with a mere 0.5 V swing; exhibits prohibitive static power dissipation.
BiCMOS	CMOS subcircuits combined with bipolar devices on a single chip.

Originally a low-power but slow alternative to TTL, CMOS has become the technology that almost totally dominates VLSI today. This is essentially because layout density, operating speed, energy efficiency, and manufacturing costs have benefited and continue to benefit from the geometric downscaling that comes with every process generation. In addition, the simplicity and comparatively low power dissipation of CMOS circuits have allowed for integration densities not possible on the basis of BJTs, also see fig.1.19.

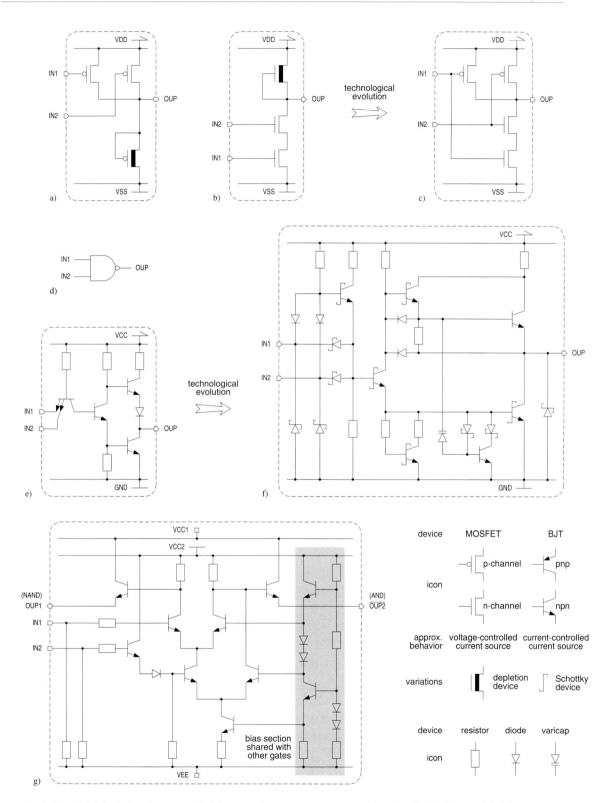

Fig. 1.19 Major logic families exemplified by way of a 2-input NAND gate. Icon (d), PMOS (a), NMOS (b), static CMOS (c), TTL (e,f), and ECL circuits (g). (e) shows the original multi-emitter structure that gave TTL its name whereas (f) refers to a more recent F generation part that includes many auxiliary devices for clamping and speed-up. Observe how disparate one gate equivalent (GE) can be.

The focus of this text is on static circuits in CMOS technology. However, as designing digital VLSI systems and developing with PLDs only loosely depend on technology, the discussion of further details is postponed to forthcoming chapters, notably 8 and 14.

1.7 | Appendix II: An illustrated glossary of circuit-related terms

Table 1.6 lists important terms from digital circuits, microelectronics, and electronic design automation (EDA). Two illustrations follow. Figure 1.20 identifies most of the underlying concepts by way of a circuit diagram while fig.1.21 shows how they reflect in a hardware description language (HDL) model. Although those concepts are applied throughout the EDA community, the terms being used and their meanings vary from one company to the next.

Note the difference between a **schematic** and a **netlist**. Either one unambiguously specifies a circuit as a collection of components with their interconnections. On top of this, schematic data include information that indicate where and how to draw icons, wires, busses, and the like on a computer screen or on a piece of paper. While totally irrelevant from an electrical or functional point of view, the graphical arrangement matters when humans want to grasp a circuit's organization and understand its operation. A netlist is easily derived from a schematic, but the converse is not obvious. Except for trivial examples, circuit diagrams obtained from netlists by automatic means lack the clarity and expressiveness of human-made schematics.

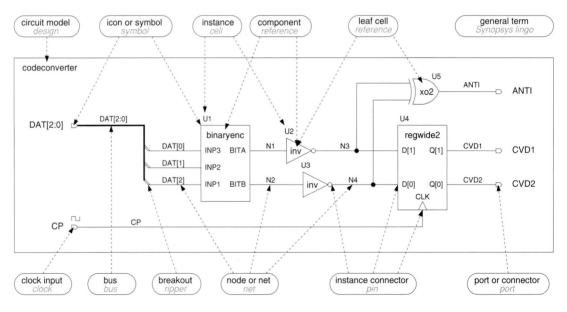

Fig. 1.20 Circuit-related terms illustrated by way of a schematic drawing. Make sure you understand why U2 and U3 relate to the same component but to distinct instances. Also note that "inv" and "xo2" are leaf cells whereas "binaryenc" and "regwide2" are not.

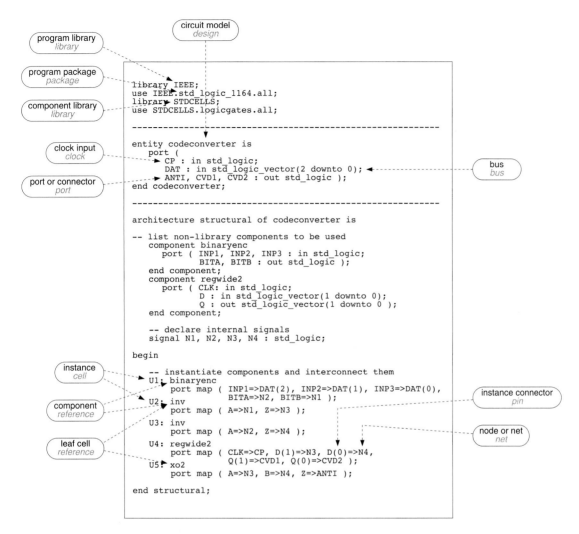

Fig. 1.21 Circuit- and software-related terms in a structural VHDL model. Note that the only way to identify a clock in a port clause is by way of its name. Similarly, the lexical name is the only way to distinguish between leaf cells and other components in an architecture body.

Relating to information-processing hardware, **datapath** is a generic term for all those subcircuits that manipulate payload data, see fig.1.22. That is, a datapath is not confined to arithmetic/logic units (ALUs) that carry out operations on data words, but also includes short-term data storage (accumulators, registers, FIFOs) plus the necessary data routing facilities (busses and switches). Datapaths tend to be highly regular as similar functions are carried out on multiple bits at a time.

Datapath operation is governed by a control section that also coordinates activities with surrounding circuits. The **controller** does so by interpreting various status signals and by piloting datapath operation via control signals in response. A controller is implemented as a hardwired finite state

Table 1.6 A glossary of terms from electronic design.

General term	Synopsys lingo	Meaning
		Circuit elements
circuit model	design	a description of an electronic circuit or subcircuit
component	reference	a self-contained subcircuit of well-defined functionality
component library	library	a named collection of components
(leaf) cell	reference	an atomic component typically available from a library that cannot be decomposed into smaller components
instance	cell	one specific copy of a subcircuit that is being used as part of a larger circuit
		Interconnect
node aka net	net	an electrical node or — which is the same thing — a wire that runs between two or more (instance) connectors
port aka terminal aka connector	port	a node that can be electrically contacted from the next higher level of circuit hierarchy
instance connector	pin	a connector of an instance
clock input	clock	a connector explicitly defined as clock source
bus	bus	a named set of nodes with cardinality > 1
special net		a net not shown but tacitly implied in schematics, examples: ground and power
		Circuit drawings
icon aka symbol	symbol	a graphical symbol for a component or a connector
schematic diagram	schematic	a drawing of a (sub)circuit that is made up of icons and of wires where the latter are graphically shown as lines
netlist	netlist	a data structure that captures what instances make up a (sub)circuit and how they are interconnected
breakout	ripper	a special icon that indicates where a net or a subbus leaves or enters the graphical representation for a bus
		Integrated circuits
die aka chip		a fully processed but unencapsulated IC
package		the encapsulation around a die
(package) pin		a connector on the outside of an IC package
pad	pad	a connector on a die that is intended to be wired or otherwise electrically connected to a package pin; the term is often meant to include interface circuitry
		HDL software
program package	package	a named collection of data types, subprograms, etc.
program library	library	a named repository for compiled program packages
		Functional verification
model under test	design ...	a circuit model subject to simulation
circuit under test		a physical circuit, e.g. a chip, subject to testing

Table 1.6 (*Cont.*)

General term	Synopsys lingo	Meaning
testbench		HDL code written for driving the simulation of a model under test not meant to be turned into a physical circuit
Layout items		
layout		a 2D drawing that captures a component's detailed geometry layer by layer and that guides IC fabrication
(cell) row		many standard cells arranged in a row such as to share common ground lines, power lines, and wells
well		a volume that accommodates MOSFETs of identical polarity; doping is opposite to the source and drain islands embedded
row end cell		a special cell void of functionality to be instantiated at either end of a cell row to properly end the wells
filler cell		a special cell void of functionality to be instantiated between two regular cells to add decoupling capacitance typically where dense wiring asks for extra room anyway
tie-off cell		a special cell void of functionality to be instantiated where a regular net must connect to ground or power
cell outline aka abstract		a simplified view where a cell's layout is reduced to the outline and the locations of all of its connectors
routing channel		space set aside between adjacent cell rows for wiring, no longer needed with today's multi-metal processes
contact		a galvanic connection between a metal and a silicon layer
via		a galvanic connection between two superimposed metal layers
bonding area		a square opening in the protective overglass exposing a die's top-level metal for connecting to a package pin

machine (FSM), as a stored program (program counter plus microcoded instruction sequence), or as a combination of the two. In a computer-type architecture, all facilities dedicated to the sole purpose of address processing must be considered part of the controller, not of the datapath, even if they are ALUs or registers by nature.

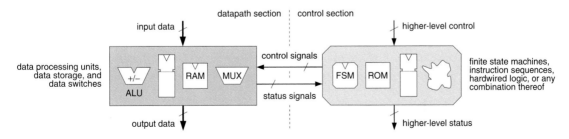

Fig. 1.22 Interplay of datapath and controller in a typical information-processing circuit.

Chapter 2

From Algorithms to Architectures

2.1 | The goals of architecture design

VLSI architecture design is concerned with deciding on the necessary hardware resources for solving problems from data and/or signal processing and with organizing their interplay in such a way as to meet target specifications defined by marketing.

The foremost concern is to get the desired **functionality** right. The second priority is to meet some given performance target, often expressed in terms of data **throughput** or operation rate. A third objective, of economic nature this time, is to minimize **production costs**. Assuming a given fabrication process, this implies minimizing **circuit size** and maximizing fabrication yield so as to obtain as many functioning parts per processed wafer as possible.[1]

Another general concern in VLSI design is **energy efficiency**. Battery-operated equipment, such as hand-held cellular phones, laptop computers, digital hearing aids, etc., obviously imposes stringent limits on the acceptable power consumption. It is perhaps less evident that energy efficiency is also of interest when power gets supplied from the mains. The reason for this is the cost of removing the heat generated by high-performance high-density ICs. While the VLSI designer is challenged to meet a given performance figure at minimum power in the former case, maximizing performance within a limited power budget is what is sought in the latter.

The ability to change from one mode of operation to another in very little time, and the flexibility to accommodate evolving needs and/or to upgrade to future standards are other highly desirable qualities and subsumed here under the term **agility**. Last but not least, two distinct architectures are likely to differ in terms of the overall **engineering effort** required to work them out in full detail and, hence also, in their respective times to market.

---

[1] The problems and methods associated with making sure functionality is implemented correctly are addressed in chapter 3. Yield and cost models are discussed in chapter 13 along with other business issues that relate to VLSI design and manufacturing.

2.1.1 Agenda

Driven by dissimilar applications and priorities, hardware engineers have, over the years, devised a multitude of very diverse architectural concepts which we will try to put into perspective in this chapter. Section 2.2 opposes program-controlled and hardwired hardware concepts before showing how their respective strengths can be combined into one architecture. After the necessary groundwork for architectural analysis has been laid in section 2.3, subsequent sections will then discuss how to select, arrange, and improve the necessary hardware resources in an efficient way with a focus on dedicated architectures. Section 2.4 is concerned with organizing computations of combinational nature. Section 2.6 extends our analysis to nonrecursive sequential computations before timewise recursive computations are addressed in section 2.7. Finally, section 2.8 generalizes our findings to other than word-level computations on real numbers. Inserted in between is section 2.5 that discusses the options available for temporarily storing data and their implications for architectural decisions.

2.2 | The architectural antipodes

Given some computational task, one basically has the choice of writing program code and running it on a program-controlled machine, such as a microprocessor or a digital signal processor (DSP), or of coming up with a hardwired electronic circuit that carries out the necessary computation steps. This fundamental dichotomy, which is described in more detail in table 2.1, implies that a systems engineer has to make a choice:

a) Select a processor-type general-purpose architecture and write program code for it, or
b) Tailor a dedicated hardware architecture for the specific computational needs.

Deciding between a general-purpose processor and an architecture dedicated to the application at hand is a major decision that has to be made before embarking on the design of a complex circuit. A great advantage of commercial microprocessors is that developers can focus on higher-level issues such as functionality and system-level architecture right away. There is no need for them to address all those exacting chores that burden semi- and — even more so — full-custom design.[2] In addition, there is no need for custom fabrication masks.

Observation 2.1. *Opting for commercial instruction-set processors and/or FPL sidesteps many technical issues that absorb much attention when a custom IC is to be designed instead. Conversely, it is precisely the focus on the payload computations, and the absence of programming and configuration overhead together with the full control over every aspect of architecture, circuit, and layout design that make it possible to optimize performance and energy efficiency.*

Circuit examples where dedicated architectures outperform instruction set computers follow.

---

[2] Such as power distribution, clock preparation and distribution, input/output design, physical design and verification, signal integrity, electrical overstress protection, wafer testing, and package selection, all to be discussed in forthcoming chapters. Setting up a working CAE/CAD design flow typically also is a major stumbling block, to say nothing of estimating sales volume, hitting a narrow window of opportunity, finding the right partners, and providing the necessary resources, in-house expertise, and investments. Also note that field-programmable logic (FPL) frees developers from dealing with many of these issues too.

Table 2.1 The architectural antipodes compared.

	Hardware architecture	
	General purpose	Special purpose
Algorithm	any, not known a priori	fixed, must be known
Architecture	instruction set processor, von Neumann or Harvard style	dedicated design, no single established pattern
Execution model	fetch–load–execute–store cycle "instruction-oriented"	process data item and pass on "dataflow-oriented"
Datapath	universal operations, ALU(s) plus memory	specific operations only, customized design
Controller	with program microcode	typically hardwired
Performance indicator	instructions per second, run time of various benchmark programs	data throughput, can be anticipated analytically
Paradigm from manufacturing	craftsman in his machine shop working according to different plans every day	division of labor in a factory set up for smooth production of a few closely related goods
Possible hardware implementations	standard µC|DSP components or ASIC with on-chip µC|DSP	ASIC of dedicated architecture or FPL (FPGA|CPLD)
Engineering effort	mostly software design	mostly hardware design
Strengths	highly flexible, immediately available, routine design flow, low up-front costs	room for max. performance, highly energy-efficient, lean circuitry

Upon closer inspection, one finds that dedicated architectures fare much better in terms of performance and/or dissipated energy than even the best commercially available general-purpose processors in some situations, whereas they prove a dreadful waste of both hardware and engineering resources in others.

Algorithms that are very irregular, highly data-dependent, and memory-hungry are unsuitable for dedicated architectures. Situations of this kind are found in **electronic data processing** such as databank applications, accounting, and **reactive systems**[3] like industrial control,[4] user interfaces,

---

[3] A system is said to be **reactive** if it interacts continuously with an environment, at a speed imposed by that environment. The system deals with <u>events</u> and the mathematical formalisms for describing them aim at capturing the complex ordering and causality relations between events that may occur at the inputs and the corresponding reactions — events themselves — at the outputs. Examples: elevators, protocol handlers, anti-lock brakes, process controllers, graphical user interfaces, operating systems.

As opposed to this, a **transformatorial** system accepts new input values — often at regular intervals — uses them to compute output values, and then rests until the subsequent data items arrive. The system is essentially concerned with arithmetic/logic processing of <u>data values</u>. Formalisms for describing transformatorial systems capture the numerical dependencies between the various data items involved. Examples: filtering, data compression, ciphering, pattern recognition, and other applications colloquially referred to as number crunching but also compilers and payroll programs.

[4] Control in the sense of the German "programmierte Steuerungen" not "Regelungstechnik".

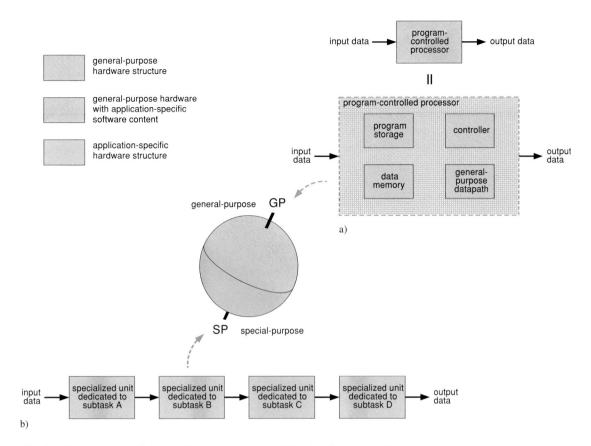

Fig. 2.1 Program-controlled general-purpose processor (a) and dedicated (special-purpose) hardware structure (b) as architectural antipodes.

and others. In search of optimal architectures for such applications, one will invariably arrive at hardware structures patterned after instruction set processors. Writing code for a standard microcomputer — either bought as a physical part or incorporated into an ASIC as a megacell or as a virtual component — is more efficient and more economic in this case.

Situations where data streams are to be processed in fairly regular ways offer far more room for coming up with dedicated architectures. Impressive gains in performance and energy efficiency over solutions based on general-purpose parts can then be obtained, see tables 2.2, 2.3, 2.4, and 2.5 among other examples.

Generally speaking, situations that favor dedicated architectures are often found in real-time applications from **digital signal processing** and **telecommunications** such as

- Source coding (i.e. data, audio, and video (de)compression),
- (De)ciphering (primarily for secret key ciphers),
- Channel coding (i.e. error correction),
- Digital (de)modulation (for modems, wireless communication, and disk drives),

Example

Table 2.2 Comparison of architectural alternatives for a Viterbi decoder (code rate $\frac{1}{2}$, constraint length 7, soft decision decoding, Euclidean distance metric). DSPs are at their best for sustained multiply–accumulate operations and offer word widths of 32 bit or so. However, as the Viterbi algorithm can be arranged to make no use of multiplication and to make do with word widths of 6 bit or less, DSPs cannot take advantage of these resources. A pipeline of tailor-made stages optimized for branch metric computation, path metric update, and survivor path traceback operations, in contrast, makes it possible to exploit the parallelism inherent in the Viterbi algorithm. Diverse throughput requirements can be accommodated by trading the number of computational units in each stage for throughput. Sophisticated DSPs, such as the C6455, include an extra coprocessor to accelerate path metric update and survivor traceback.

Architecture	General purpose		Special purpose	
Key component	DSP		ASIC	
	TI TMS320C6455		sem03w6	sem05w1
	without	with	ETH	ETH
	Viterbi coprocessor VCP2			
Number of chips	1	1	1	1
CMOS process	90 nm	90 nm	250 nm	250 nm
Program code	187 kiByte	242 kiByte	none	none
Circuit size	n.a.	n.a.	73 kGE	46 kGE
Max. throughput	45 kbit/s	9 Mbit/s	310 Mbit/s	54 Mbit/s
@ clock	1 GHz	1 GHz	310 MHz	54 MHz
Power dissipation	2.1 W	2.1 W	1.9 W	50 mW
Year	2005	2005	2004	2006

☐

Example

Table 2.3 Comparison of architectural alternatives for a secret-key block encryption/decryption algorithm (IDEA cipher as shown in fig.2.14, block size 64 bit, key length 128 bit). The clear edge of the VINCI ASIC is due to a high degree of parallelism in its datapath and, more particularly, to the presence of four pipelined computational units for multiplication modulo ($2^{16} + 1$) designed in full-custom layout that operate concurrently and continuously. The more recent IDEA kernel combines a deep submicron fabrication process with four highly optimized arithmetic units. Full-custom layout was no longer needed to achieve superior performance.

Architecture	General purpose		Special purpose	
Key component	DSP	RISC Workst.	ASSP	ASSP
	Motorola 56001	Sun Ultra 10	VINCI [13]	IDEA Kernel
Number of chips	1 + memory	motherboard	1	1
CMOS process	n.a.	n.a.	1.2 μm	250 nm
Max. throughput	1.25 Mbit/s	13.1 Mbit/s	177 Mbit/s	700 Mbit/s
@ clock	40 MHz	333 MHz	25 MHz	100 MHz
Year	1995	1998	1992	1998

☐

Example

Table 2.4 Comparison of architectural alternatives for lossless data compression with the Lempel-Ziv-77 algorithm that heavily relies on string matching operations [14]. The dedicated hardware architecture is implemented on a reconfigurable coprocessor board built around four field-programmable gate-array components. 512 special-purpose processing elements are made to carry out string comparison subfunctions in parallel. The content-addressed symbol memory is essentially organized as a shift register, thereby giving simultaneous access to all entries. Of course, the two software implementations obtained from compiling C source code cannot nearly provide a similar degree of concurrency.

Architecture	General purpose		Special purpose
Key component	RISC Workst. Sun Ultra II	CISC Workst. Intel Xeon	FPGA Xilinx XC4036XLA
Number of chips	motherboard	motherboard	4 + config.
CMOS process	n.a.	n.a.	n.a.
Max. throughput	3.8 Mbit/s	5.2 Mbit/s	128 Mbit/s
@ clock	300 MHz	450 MHz	16 MHz
Year	1997	1999	1999

☐

Example

Table 2.5 Comparison of architectural alternatives for a secret-key block encryption/decryption algorithm (AES cipher, block size 128 bit, key length 128 bit). The Rijndael algorithm makes extensive use of a so-called S-Box function and its inverse; the three hardware implementations include multiple look-up tables (LUTs) for implementing that function. Also, (de)ciphering and subkey preparation are carried out concurrently by separate hardware units. On that background, the throughput of the assembly language program running on a Pentium III is indeed impressive. This largely is because the Rijndael algorithm has been designed with the Pentium architecture in mind (MMX instructions, LUTs that fit into cache memory, etc.). Power dissipation remains daunting, though.

Architecture	General purpose		Special purpose		
Key component	RISC Proc. Embedded Sparc	CISC Proc. Pentium III	FPGA Virtex-II Amphion	ASIC CryptoFun ETH	ASIC core only UCLA [15]
Number of chips	motherboard	motherboard	1 + config.	1	1
Programming	C	Assembler	none	none	none
Circuit size	n.a.	n.a.	n.a.	76 kGE	173 kGE
CMOS process	n.a.	n.a.	150 nm	180 nm	180 nm
Max. throughput	133 kbit/s	648 Mbit/s	1.32 Gbit/s	2.00 Gbit/s	1.6 Gbit/s
@ clock	120 MHz	1.13 GHz	n.a.	172 MHz	125 MHz
Power dissipation	120 mW	41.4 W	490 mW	n.a.	56 mW[a]
@ supply			1.5 V	1.8 V	1.8 V
Year	n.a.	2000	≈2002	2007	2002

[a] Most likely specified for core logic alone, that is without I/O circuitry.

☐

- Adaptive channel equalization (after transmission over copper lines and optical fibers),
- Filtering (for noise cancellation, preprocessing, spectral shaping, etc.),
- Multipath combiners in broadband wireless access networks (RAKE, MIMO),
- Digital beamforming with phased-array antennas (Radar),
- Computer graphics and video rendering,
- Multimedia (e.g. MPEG, HDTV),
- Packet switching (e.g. ATM, IP),
- transcoding (e.g. between various multimedia formats),
- Medical signal processing,
- Pattern recognition, and more.

Observation 2.2. *Processing algorithms and hardware architectures are intimately related. While dedicated architectures outperform program-controlled processors by orders of magnitude in many applications of predominantly transformatorial nature, they cannot rival the agility and economy of processor-type designs in others of more reactive nature.*

More precise criteria for finding out whether a dedicated architecture can be an option or not from a purely technical point of view follow in section 2.2.1 while fig.2.2 puts various applications from signal and data processing into perspective.[5]

2.2.1 What makes an algorithm suitable for a dedicated VLSI architecture?

Costs in hardware are not the same as those in software. As an example, permutations of bits within a data word are time-consuming operations in software as they must be carried out sequentially. In hardware, they reduce to simple wires that cross while running from one subcircuit to the next. Look-up tables (LUTs) of almost arbitrary size, on the other hand, have become an abundant and cheap resource in any microcomputer while large on-chip RAMs and ROMs tend to eat substantial proportions of the timing and area budgets of ASIC designs.

In an attempt to provide some guidance, we have collected ten criteria that an information processing algorithm should <u>ideally</u> meet in order to justify the design of a special-purpose VLSI architecture and to take full advantage of the technology. Of course, very few real-world algorithms satisfy all of the requirements listed. It is nevertheless safe to say that designing a dedicated architecture capable of outperforming a general-purpose processor on the grounds of performance and costs will prove difficult when too many of these criteria are violated. The list below begins with the most desirable characteristics and then follows their relative significance.

1. **Loose coupling between major processing tasks.** The overall data processing lends itself to being decomposed into tasks that interact in a simple and unmutable way. Whether those tasks are to be carried out consecutively or concurrently is of secondary importance at this point; what counts is to come up with a well-defined functional specification for each task and with manageable interaction between them. Architecture design, functional verification, optimization, and reuse otherwise become real nightmares.

2. **Simple control flow.** The computation's control flow is simple. This key property can be tracked down to two more basic considerations:

---

[5] The discussion of management-level decision criteria is deferred to chapter 13.

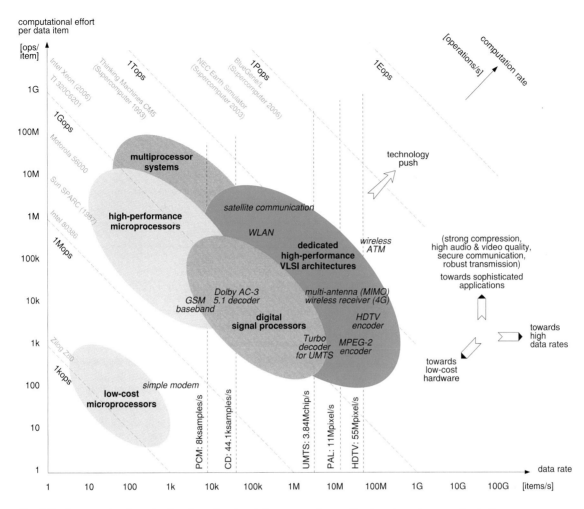

Fig. 2.2 Computational needs of various signal and data processing applications (grossly approximate figures, exact meaning of operation and data item left unspecified; 16 bit-by-16 bit multiply–accumulate (MAC) operations on 16 bit samples are often considered typical in a context of digital signal processing).

a) The course of operation does not depend too much on the data being processed; for each loop the number of iterations is a priori known and constant.[6]

b) The application does not ask for computations to be carried out with overly many varieties, modes of operations, data formats, distinct parameter settings, and the like.

The benefit of a simple control flow is twofold. For one thing, it is possible to anticipate the datapath resources required to meet a given performance goal and to design the chip's architecture accordingly. There is no need for statistical methods in estimating the computational burden or in sizing data memories and the like. For another thing, datapath control can be

---

[6] Put in different terms, the target algorithm is virtually free of branchings and loops such as if...then[...else], while...do, and repeat...until that include data items in their condition clauses.

handled by counters and by simple finite state machines (FSMs) that are small, fast, energy-efficient and — most important — easy to verify.

An overly complicated course of operations, on the other hand, that involves much data-dependent branching, multitasking, and the like, favors a processor-type architecture that operates under control of stored microcode. Most control operations will then translate into a sequence of machine instructions that take several clock cycles to execute.

3. **Regular data flow.** The flow of data is regular and their processing is based on a recurrence of a fairly small number of identical operations; there are no computationally expensive operations that are called only occasionally. Regularity opens a door for sharing hardware resources in an efficient way by applying techniques such as iterative decomposition and time-sharing, see subsections 2.4.2 and 2.4.5 respectively. Conversely, multiple data streams that are to be processed in a uniform way lend themselves to concurrent processing by parallel functional units. A regular data flow further helps to reduce communications overhead in terms of both area and interconnect delay as the various functional units can be made to exchange data over fixed local links. Last but not least, regularity facilitates reuse and reduces design and verification effort.

As opposed to this, operations that are used infrequently either will have to be decomposed into a series of substeps to be executed one after the other on a general-purpose datapath, which is slow, or will necessitate dedicated functional units bound to sit idle for most of the time, which inflates chip size. Irregular data flow requires long and flexible communication busses which are at the expense of layout density, operating speed, and energy efficiency.

4. **Reasonable storage requirements.** Overall storage requirements are modest and have a fixed upper bound.[7] Memories that occupy an inordinate amount of chip area, say more than half or so, cannot be incorporated into ASICs in an economic way and must, therefore, be implemented off-chip from standard parts, see subsection 2.5. Massive storage requirements in conjunction with moderate computational burdens tend to place dedicated architectures at a disadvantage.

5. **Compatible with finite precision arithmetics.** The algorithm is insensitive to effects from finite precision arithmetics. That is, there is no need for floating-point arithmetics; fairly small word widths of, say, 16 bit or less suffice for the individual computation steps. Standard microprocessors and DSPs come with datapaths of fixed and often generous width (24, 32, 64 bit, or even floating-point) at a given price. No extra costs arise unless the programmer has to resort to multiple precision arithmetics.

As opposed to this, ASICs and FPL offer an opportunity to tune the word widths of datapaths and on-chip memories to the local needs of computation. This is important because circuit size, logic delay, interconnect length, parasitic capacitances, and energy dissipation of addition, multiplication, and other operations all tend to grow with word width, combining into a burden that multiplies at an overproportional rate.[8]

---

[7] Which precludes the use of dynamic data structures.

[8] Processor datapaths tend to be fast and area efficient because they are typically hand-optimized at the transistor level (e.g. dynamic logic) and implemented in tiled layout rather than built from standard cells. These are only rarely options for ASIC designers.

6. **Nonrecursive linear time-invariant computation.** The processing algorithm describes a nonrecursive linear time-invariant system over some algebraic field.[9] Each of these properties opens a door to reorganizing the data processing in one way or another, see sections 2.4 through 2.9 for details and table 2.11 for an overview. High throughputs, in particular, are much easier to obtain from nonrecursive computations as will become clear in section 2.7.

7. **No transcendental functions.** The algorithm does not make use of roots, logarithmic, exponential, or trigonometric functions, arbitrary coordinate conversions, translations between incompatible number systems, and other transcendental functions as these must either be stored in large look-up tables (LUT) or get calculated on-line in lengthy and often irregular computation sequences. Such functions can be implemented more economically provided that modest accuracy requirements allow approximation by way of lookups from tables of reasonable size, possibly followed by interpolation.

8. **Extensive usage of data operations unavailable from standard instruction sets.** Of course, there exist many processing algorithms that cannot do without costly arithmetic/-logic operations. It is often possible to outperform traditional program-controlled processors in cases where such operations need to be assembled from multiple instructions. Dedicated datapaths can then be designed to do the same computation in a more efficient way. Examples include complex-valued arithmetics, add–compare–select operations, and many ciphering operations. It also helps when part of the arguments are constants because this makes it possible to apply some form of preprocessing. Multiplication by a variable is more onerous than by a constant, for instance.[10]

9. **Throughput rather than latency is what matters.** This is a crucial prerequisite for pipelined processing, see subsection 2.4.3.

10. **No divisions and multiplications on very wide data words.** Multiplications involving wide arguments are not being used

   The algorithm does not make extensive use of multiplications and even less so of divisions as their VLSI implementation is much more expensive than that of addition/subtraction when the data words involved are wide.

2.2.2 There is plenty of land between the architectural antipodes

Most markets ask for performance, agility, low power, and a modest design effort at the same time. In the face of such contradictory requirements, it is highly desirable to combine the throughput and the

---

[9] Recursiveness is to be defined in section 2.7. **Linear** is meant to imply the principle of superposition $f(x(t) + y(t)) \equiv f(x(t)) + f(y(t))$ and $f(c\,x(t)) \equiv cf(x(t))$. **Time-invariant** means that the sole effect of delaying the input is a delay of the output by the same amount of time: if $z(t) = f(x(t))$ is the response to $x(t)$ then $z(t - T)$ is the response to $x(t - T)$. Fields and other algebraic structures are compared in section 2.11.

[10] Dropping unit factors and/or zero sum terms (both at word and bit levels), substituting integer powers of 2 as arguments in multiplications and divisions, omitting insignificant contributions, special number representation schemes, taking advantage of symmetries, precomputed look-up tables, and distributed arithmetic, see subsection 2.8.3, are just a few popular measures that may help to lower the computational burden in situations where parts of the arguments are known ahead of time.

energy efficiency of a dedicated VLSI architecture for demanding but highly repetitive computations with the convenience and flexibility of an instruction set processor for more control-oriented tasks. The question is

"How can one blend the best of both worlds into a suitable architecture design?"

Five approaches for doing so are going to be presented in sections 2.2.3 through 2.2.7 with diagrammatic illustrations in figs.2.3 to 2.6.

2.2.3 Assemblies of general-purpose and dedicated processing units

The observation below forms the starting point for the conceptually simplest approach.

Observation 2.3. *It is often possible to segregate the needs for computational efficiency from those for flexibility.*

This is because those parts of a system that ask for maximum computation rate are not normally those that are subject to change very often, and vice versa. Examples abound, see table 2.6. The finding immediately suggests a setup where a software-controlled microcomputer cooperates with one or more dedicated hardware units. Separating the quest for computational efficiency from that for agility makes it possible to fully dedicate the various functional units to their respective tasks and to optimize them accordingly. Numerous configurations are possible and the role of the instruction set microcomputer varies accordingly.

Example

Table 2.6 Some digital systems and the computing requirements of major subfunctions thereof.

Application	Subfunctions primarily characterized by	
	irregular control flow and/or need for flexibility	repetitive control flow and need for comput. efficiency
DVD player	user interface, track seeking, tray and spindle control, processing of non-video data (directory, title, author, subtitles, region codes)	16-to-8 bit demodulation, error correction, MPEG-2 decompression (discrete cosine transform), video signal processing
Cellular phone	user interface, SMS, directory management, battery monitoring, communication protocol, channel allocation, roaming, accounting	intermediate frequency filtering, (de)modulation, channel (de)coding, error correction (de)coding, (de)ciphering, speech (de)compression
Pattern recognition (e.g. as part of a defensive missile)	pattern classification, object tracking, target acquisition, triggering of actions	image stabilization, redundancy reduction, image segmentation, feature extraction

□

In fig.2.3a, three dedicated and one program-controlled processing units are arranged in a chain. Each unit does its data processing job and passes the result to the downstream unit. While offering ample room for optimizing performance, this structure cannot accommodate much variation if everything is hardwired and tailor-made. Making the specialized hardware units support a limited degree of parametrization (e.g. wrt data word width, filter order, code rate, data exchange protocol, and the like) renders the overall architecture more versatile while, at the same time, keeping the overhead in terms of circuit complexity and energy dissipation fairly low. The term **weakly programmable satellites** has been coined to reflect the idea. An optional parametrization bus suggests this extension of the original concept in fig.2.3a.

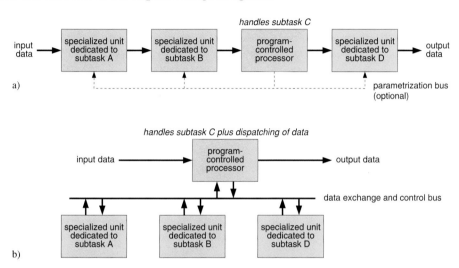

Fig. 2.3 General-purpose processor and dedicated satellite units working in a chain (a), a host computer with specialized coprocessors (b).

2.2.4 Coprocessors

Figure 2.3b is based on segregation too but differs in how the various components interact. All specialized hardware units now operate under control of a software-programmable **host**. A bidirectional bus gives the necessary liberty for transferring data and control words back and forth. Each coprocessor, or helper engine as it is sometimes called, has a rather limited repertoire of instructions that it can accept. It sits idle until it receives a set of input data along with a start command. As an alternative, the data may be kept in the host's own memory all the time but get accessed by the coprocessor via direct memory access (DMA). Once local computation has come to an end, the coprocessor sets a status flag and/or sends an interrupt signal to the host computer. The host then accepts the processed data and takes care of further action.

2.2.5 Application-specific instruction set processors

Patterning the overall architecture after a program-controlled processor affords much more flexibility. Application-specific features are largely confined to the data processing circuitry itself. That is, one or more datapaths are designed and hardwired so as to support specific data manipulations while operating under control of a common microprogram. The number of ALUs, their instruction sets, the data formats supported, the capacity of local storage, etc. are tailored to the computational problems to be solved. What's more, the various datapaths can be made to operate simultaneously on different pieces of data, thereby providing a limited degree of concurrency. The resulting architecture is that of an application-specific instruction set processor (ASIP) [16], see fig.2.4a.

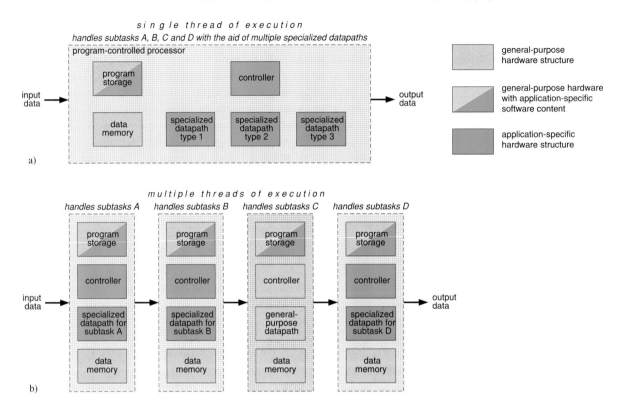

Fig. 2.4   Application-specific instruction set processor (ASIP) (a), multiple cooperating ASIPs (b).

The hardware organization of an ASIP bears much resemblance to architectural concepts from general-purpose computing. As more and more concurrent datapath units are added, what results essentially is a **very-long instruction word** (VLIW) architecture. An open choice is that between a **multiple-instruction multiple-data** (MIMD) machine, where an individual field in the overall instruction word is set apart for each datapath unit, and a **single-instruction multiple-data** (SIMD) model, where a bunch of identical datapaths works under control of a single instruction word. Several data items can thus be made to undergo the same operation at the same time.[11]

---

[11] In an effort to better serve high-throughput video and graphics applications, many vendors enhanced their microprocessor families in the late 1990s by adding special instructions that provide some degree of concurrency. During each such instruction, the processor's datapath gets split up into several smaller subunits. A datapath of 64 bit can be made to process four 16 bit data words at a time, for instance, provided the operation is the

Example

Table 2.7 | An ASIP implementation of the Rijndael algorithm, compare with table 2.5.

Architecture	ASIP
Key component	Cryptoprocessor core UCLA [17]
Number of chips	1
Programming	Assembler
Circuit size	73.2 kGE
CMOS process	180 nm 4Al2Cu
Throughput	3.43 Gbit/s
@ clock	295 MHz
Power dissip.	86 mW[a]
@ supply	1.8 V
Year	2004

[a] Estimate for core logic alone, that is without I/O circuitry, not a measurement.

□

While the mono-ASIP architecture of fig.2.4a affords flexibility, it does not provide the same degree of concurrency and modularity as the multiple processing units of fig.2.3a and b do. A multiprocessor system built from specialized ASIPs, as shown in fig.2.4b, is, therefore, an interesting extension. In addition, this approach facilitates the design, interfacing, reuse, test, and on-going update of the various building blocks involved.

However, always keep in mind that defining a proprietary instruction set makes it impossible to take advantage of existing compilers, debugging aids, assembly language libraries, experienced programmers, and other resources that are routinely available for industry-standard processors. Industry provides us with such a vast selection of micro- and signal processors that only very particular requirements justify the design of a proprietary CPU.[12]

Example

While generally acknowledged to produce more realistic renderings of 3D scenes than industry-standard raster graphics processors, ray tracing algorithms have long been out of reach for real-time applications due to the myriad floating-point computations and the immense memory bandwidth they require. Hardwired custom architectures do not qualify either as they cannot be programmed and as ray tracing necessitates many data-dependent recursions and decisions.

---

same for all of them. The technique is best described as **sub-word parallelism**, but is better known under various trademarks such as multimedia extensions (MMX), streaming SIMD extensions (SSE) (Pentium family), Velocity Engine, AltiVec, and VMX (PowerPC family).

[12] [18] reports on an interesting approach to expedite ASIP development whereby assembler, linker, simulator, and RTL synthesis code are generated automatically by system-level software tools. Product designers can thus essentially focus on defining the most appropriate instruction set for the processor in view of the target application.

Ray tracing may finally find more general adoption in multi-ASIP architectures that combine multiple ray processing units (RPUs) into one powerful rendering engine. Working under control of its own program thread, each RPU operates as a SIMD processor that follows a subset of all rays in a scene. The independence of light rays allows a welcome degree of scalability where frame rate can be traded against circuit complexity. The authors of [19] have further paid attention to defining an instruction set for their RPUs that is largely compatible with pre-existing industrial graphics processors.

□

2.2.6 Configurable computing

Another crossbreed between dedicated and general-purpose architectures did not become viable until the late 1990s but is now being promoted by FPL manufacturers and researchers [20] [21]. The IEEE 1532 standard has also been created in this context. The idea is to reuse the same hardware for implementing subfunctions that are mutually exclusive in time by reconfiguring FPL devices on the fly.

As shown in fig.2.5, the general hardware arrangement bears some resemblance to the coprocessor approach of fig.2.3b, yet **in-system configurable** (ISC) devices are being used instead of hardwired logic. As a consequence, the course of operations is more sophisticated and requires special action from the hardware architects. For each major subtask, the architects must ask themselves whether the computations involved

- Qualify for being delegated to in-system configurable logic,
- Never occur at the same time — or can wait until the FPL device becomes free —, and
- Whether the time for having the FPL reconfigured in between is acceptable or not.

Typically this would be the case for repetitive computations that make use of sustained, highly parallel, and deeply pipelined bit-level operations. When designers have identified some suitable subfunction, they devise a hardware architecture that solves the particular computational problem with the resources available in the target FPGA or CPLD, prepare a configuration file, and have that stored in a configuration memory. In some sense, they create a large **hardware procedure** instead of programming a software routine in the customary way.

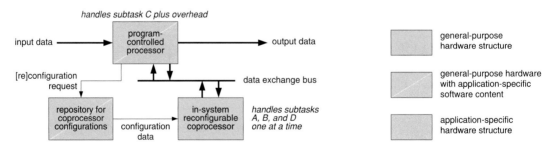

Fig. 2.5 General-purpose processor with juxtaposed reconfigurable coprocessor.

Whenever the host computer encounters a call to such a hardware procedure, it configures the FPL accordingly by downloading the pertaining configuration file. From now on, all the host has to do is to feed the "new" coprocessor with input data and to wait until the computation is completed. The host then fetches the results before proceeding with the next subtask.[13]

It thus becomes possible to support an assortment of data processing algorithms each with its optimum architecture — or almost so — from a single hardware platform. What often penalizes this approach in practice are the dead times incurred whenever a new configuration is being loaded. Another price to pay is the extra memory capacity for storing the configuration bits for all operation modes. Probably the most valuable benefit, however, is the possibility of being able to upgrade information processing hardware to new standards and/or modes of operation even after the system has been fielded.

Examples

Transcoding video streams in real time is a good candidate for reconfigurable computing because of the many formats in existence such as DV, AVI, MPEG-2, DivX, and H.264. For each conversion scheme, a configuration file is prepared and stored in local memory, from where it is transferred into the reconfigurable coprocessor on demand. And should a video format or variation emerge that was unknown or unpopular at the time when the system was being developed, extra configuration files can be made available in a remote repository from where they can be fetched much like software plug-ins get downloaded via the Internet.

The results from a comparison between Lempel–Ziv data compression with a reconfigurable coprocessor and with software execution on a processor [14] have been summarized in table 2.4. A related application was to circumvent the comparatively slow PCI bus in a PC [23].
☐

2.2.7   Extendable instruction set processors

This latest and most exotic approach pioneered by Stretch borrows from ASIPs and from configurable computing. Both a program-controlled processor and electrically reconfigurable logic are present on a common hardware platform, see fig.2.6.

The key innovation is a suite of proprietary EDA tools that allows system developers to focus on writing their application program in C or C++ as if for a regular general purpose processor. Those tools begin by profiling the software code in order to identify sequences of instructions that are executed many times over. For each such sequence, reconfigurable logic is then synthesized into a dedictated and massively parallel computation network that completes within one clock cycle — ideally at least. Finally, each occurrence of the original computation sequence in the machine code gets replaced by a simple function call that activates the custom-made datapath logic.

In essence, the base processor gets unburdened from lengthy code sequences by augmenting his instruction set with a few essential additions that fit the application and that get tailor-made

---

[13] As an extension to the general procedure described here, an extra optimization step can be inserted before the coprocessor is configured [22]. During this stage, the host would adapt a predefined generic configuration to take advantage of particular conditions of the specific situation at hand. Consider pattern recognition, for instance, where the template remains unchanged for a prolonged lapse of time, or secret-key (de)ciphering, where the same holds true for the key. As stated in subsection 2.2.1 item 1, it is often possible to simplify arithmetic and logic hardware a lot provided that part of the operands have fixed values.

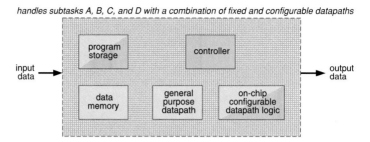

handles subtasks A, B, C, and D with a combination of fixed and configurable datapaths

Fig. 2.6 Extendable instruction set processor (simplified).

almost on the fly. Yet, the existence of reconfigurable logic and the business of coming up with a suitable hardware architecture are hidden from the system developer. The fact that overall program execution remains strictly sequential should further simplify the design process.

2.2.8 Digest

Program execution on a general-purpose processor and hardwired circuitry optimized for one specific flow of computation are two architectural antipodes. Luckily, many useful compromises exist in between, and this is reflected in figs.2.7 and 2.8. A general piece of advice is this:

Observation 2.4. *Rely on dedicated hardware only for those subfunctions that are called many times and are unlikely to change; keep the rest programmable via software, via reconfiguration, or both.*

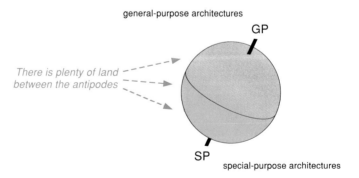

Fig. 2.7 The architectural solution space viewed as a globe.

Figure 2.8 gives rise to an interesting observation. While there are many ways to trade agility for computational efficiency and vice versa, the two seem to be mutually exclusive as we know of no architecture that would meet both goals at the same time.

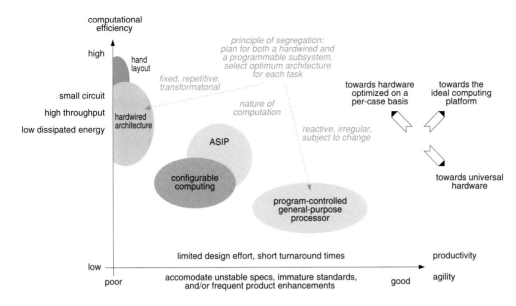

Fig. 2.8 The key options of architecture design.

2.3 | A transform approach to VLSI architecture design

Let us now turn our attention to the main topic of this chapter:

"How to decide on the necessary hardware resources for solving a given computational problem and how to best organize them."

Their conceptual differences notwithstanding, many techniques for obtaining high performance at low cost are the same for general- and special-purpose architectures. As a consequence, much of the material presented in this chapter applies to both of them. Yet, the emphasis is on dedicated architectures as the a priori knowledge of a computational problems offers room for a number of ideas that do not apply to instruction-set processor architectures.[14]

Observation 2.5. *Most data and signal processing algorithms would lead to grossly inefficient or even infeasible solutions if they were implemented in hardware as they are. Adapting processing algorithms to the technical and economic conditions of large-scale integration is one of the intellectual challenges in VLSI design.*

Basically, there is room for remodelling in two distinct domains, namely in the algorithmic domain and in the architectural domain.

---

[14] There exists an excellent and comprehensive literature on general-purpose architectures including [24] [25]. The historical evolution of the microprocessor is summarized in [26] [27] along with economic facts and trends. [28] [29] [30] emphasize the impact of deep-submicron technology on high-performance microprocessor architectures.

2.3.1 There is room for remodelling in the algorithmic domain...

In the algorithmic domain, the focus is on minimizing the number of computational operations weighted by the estimated costs of such operations. A given processing algorithm thus gets replaced by a different one better suited to hardware realization in VLSI. Data structures and number representation schemes are also subject to optimizations such as subsampling and/or changing from floating-point to fixed-point arithmetics. All this implies that alternative solutions are likely to slightly differ in their functionality as expressed by their input-to-output relations.

Six examples

When designing a digital filter, one is often prepared to tolerate a somewhat lower stopband suppression or a larger passband ripple in exchange for a reduced computational burden obtained, for instance, from substituting a lower order filter and/or from filling in zeros for the smaller coefficients. Conversely, a filter structure that necessitates a higher number of computations may sometimes prove acceptable in exchange for less stringent precision requirements imposed on the individual arithmetic operations and, hence, for narrower data words.

In a decoder for digital error-correction, one may be willing to sacrifice 0.1 dB or so of coding gain for the benefit of doing computations in a more economic way. Typical simplifications to the ideal Viterbi algorithm include using an approximation formula for branch metric computation, truncating the dynamic range of path metrics, rescaling them when necessary, and restricting traceback operations to some finite depth.

The autocorrelation function (ACF) has many applications in signal processing, yet it is not always needed in the form mathematically defined.

$$ACF_{xx}(k) = r_{xx}(k) = \sum_{n=-\infty}^{\infty} x(n) \cdot x(n+k) \tag{2.1}$$

Many applications offer an opportunity to relax the effort for multiplications because one is interested in just a small fragment of the entire ACF, because one can take advantage of symmetry, or because modest precision requirements allow for a rather coarse quantization of data values. It is sometimes even possible to substitute the average magnitude difference function (AMDF) that does away with costly multiplication altogether.

$$AMDF_{xx}(k) = r'_{xx}(k) = \sum_{n=0}^{N-1} |x(n) - x(n+k)| \tag{2.2}$$

Code-excited linear predictive (CELP) coding is a powerful technique for compressing speech signals, yet it has long been left aside in favor of regular pulse excitation because of its prohibitive computational burden. CELP requires that hundreds of candidate excitation sequences be passed through a cascade of two or three filters and be evaluated in order to pick the one that fits best. In addition, the process must be repeated every few milliseconds. Yet, experiments have revealed that the usage of sparse (up to 95% of samples replaced with zeros), of ternary (+1, 0, −1), or of overlapping excitation sequences has little negative impact on auditory perception while greatly simplifying computations and reducing memory requirements [31].

In designing computational hardware that makes use of trigonometric functions, look-up tables (LUTs) are likely to prove impractical because of size overruns. Executing a lengthy algorithm, on the other hand, may be just too slow, so a tradeoff among circuit size, speed, and precision must be found. The CORDIC (coordinate rotation digital computer) family of algorithms is one such compromise that was put to service in scientific pocket calculators in the 1960s and continues to find applications in DSP [32] [33]. Note that CORDIC can be made to compute hyperbolic and other transcendental functions too.

Computing the magnitude function $m = \sqrt{a^2 + b^2}$ is a rather costly proposition in terms of circuit hardware. Luckily, there exist at least two fairly precise approximations based on add, shift, and compare operations exclusively, see table 2.8 and problem 1. Better still, the performance of many optimization algorithms used in the context of demodulation, error correction, and related applications does not suffer much when the computationally expensive ℓ^2-norm gets replaced by the much simpler ℓ^1- or ℓ^∞-norm. See [34] for an example.

☐

The common theme is that the most obvious formulation of a processing algorithm is not normally the best starting point for VLSI design. Departures from some mathematically ideal algorithm are almost always necessary to arrive at a solution that offers the throughput and energy efficiency requested at economically feasible costs. Most algorithmic modifications alter the input-to-output mapping and so imply an **implementation loss**, that is a minor cut-back in signal-to-noise ratio, coding gain, bit-error-rate, mean time between errors, stopband suppression, passband ripple, phase response, false-positive and false-negative rates, data compression factor, fidelity of reproduction, total harmonic distortion, image and color definition, intelligibility of speech, or whatever figures of merit are most important for the application.

Experience tells us that enormous improvements in terms of throughput, energy efficiency, circuit size, design effort, and agility can be obtained by adapting an algorithm to the peculiarities and cost factors of hardware. Optimizations in the algorithmic domain are thus concerned with

"How to tailor an algorithm such as to cut the computational burden, to trim down memory requirements, and/or to speed up calculations without incurring unacceptable implementation losses."

What the trade-offs are and to what extent departures from the initial functionality are acceptable depends very much on the application. It is, therefore, crucial to have a good command of the theory and practice of the computational problems to be solved.

Observation 2.6. *Digital signal processing programs often come with floating-point arithmetics. Reimplementing them in fixed-point arithmetics, with limited computing resources, and with*

Table 2.8 | Approximations for computing magnitudes.

Name	aka	Formula
lesser	$\ell^{-\infty}$-norm	$l = \min(|a|, |b|)$
sum	ℓ^1-norm	$s = |a| + |b|$
magnitude (reference)	ℓ^2-norm	$m = \sqrt{a^2 + b^2}$
greater	ℓ^∞-norm	$g = \max(|a|, |b|)$
approximation 1		$m \approx m_1 = \frac{3}{8}s + \frac{5}{8}g$
approximation 2 [35]		$m \approx m_2 = \max(g, \frac{7}{8}g + \frac{1}{2}l)$

minimum memory results in an implementation loss. The effort for finding a good compromise between numerical accuracy and hardware efficiency is often underestimated.

The necessity to validate trimmed-down implementations for all numerical conditions that may occur further adds to the effort. It is not uncommon to spend as much time on issues of numerical precision as on all subsequent VLSI design phases together.

2.3.2 ...and there is room in the architectural domain

In the architectural domain, the focus is on meeting given performance targets for a specific data processing algorithm with a minimum of hardware resources. The key concern is

"How to organize datapaths, memories, controllers, and other hardware resources for implementing some given computation flow such as to optimize throughput, energy efficiency, circuit size, design effort, agility, overall costs, and similar figures of merit while leaving the original input-to-output relationship unchanged except, possibly, for latency."

As computations are just <u>reorganized</u>, not altered, there is no implementation loss at this point.

Given some data or signal processing algorithm, there exists a profusion of alternative architectures although the number of fundamental options available for reformulating it is rather limited. This is because each such option can be applied at various levels of detail and can be combined with others in many different ways. Our approach is based on reformulating algorithms with the aid of **equivalence transforms**. The remainder of this chapter gives a systematic view on all such transforms and shows how they can be applied to optimize VLSI architectures for distinct size, throughput, and energy targets.

2.3.3 Systems engineers and VLSI designers must collaborate

Systems theorists tend to think in purely mathematical terms, so a data or signal processing algorithm is not much more than a set of equations to them. To meet pressing deadlines or just for reasons of convenience, they tend to model signal processing algorithms in floating-point arithmetics, even when a fairly limited numeric range would amply suffice for the application. This is typically unacceptable in VLSI architecture design and establishing a lean **bit-true software model** is a first step towards a cost-effective circuit.

Generally speaking, it is always necessary to balance many contradicting requirements to arrive at a working and marketable embodiment of the mathematical or otherwise abstracted initial model of a system. A compromise will have to be found between the theoretically desirable and the economically feasible. So, there is more to VLSI design than just accepting a given algorithm and turning that into gates with the aid of some HDL synthesis tool.

Algorithm design is typically carried out by systems engineers whereas VLSI architecture is more the domain of hardware designers. The strong mutual interaction between algorithms and architectures mandates a close and early collaboration between the two groups, see fig.2.9.

Observation 2.7. *Finding a good tradeoff between the key characteristics of the final circuit and implementation losses requires an on-going collaboration between systems engineers and VLSI experts during the phases of specification, algorithm development, and architecture design.*

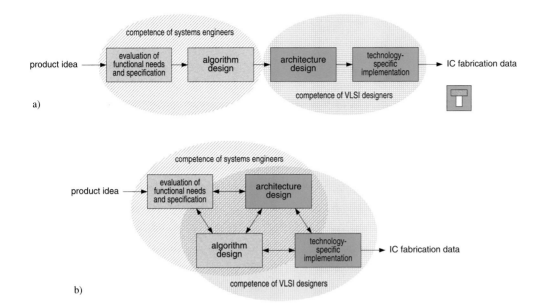

Fig. 2.9 Models of collaboration between systems engineers and hardware designers. Sequential thinking doomed to failure (a) versus a networked team more likely to come up with satisfactory results (b).

The fact that algorithm design is not covered in this text does not imply that it is of less importance to VLSI than architecture design. The opposite is probably true. A comprehensive textbook that covers the joint development of algorithms and architectures is [36]; anecdotal observations can be found in [37].

2.3.4 A graph-based formalism for describing processing algorithms

We will often find it useful to capture a data processing algorithm in a **data dependency graph** (DDG) as this graphical formalism is suggestive of possible hardware structures. A DDG is a directed graph where vertices and edges have non-negative weights, see fig.2.10. A vertex stands for a memoryless operation and its weight indicates the amount of time necessary to carry out that operation. The precedence of one operation over another one is represented as a directed edge. The weight of an edge indicates by how many computation cycles or sampling periods execution of the first operation must precede that of the second one.[15] Edge weight zero implies the two operations are scheduled to happen within the same computation or sampling period — one after the other, though. An edge may also be viewed as expressing the transport of data from one operation to another and its weight as indicating the number of registers included in that transport path.

To warrant consistent outcomes from computation, circular paths of total edge weight zero are disallowed in DDGs.[16] Put differently, any feedback loop shall include one or more latency registers.

---

[15] The term "computation cycle" is to be explained shortly in section 2.3.7.

[16] A **circular path** is a closed walk in which no vertex, except the initial and final one, appears more than once and that respects the orientation of all edges traversed. As the more customary terms "circuit" and "cycle" have

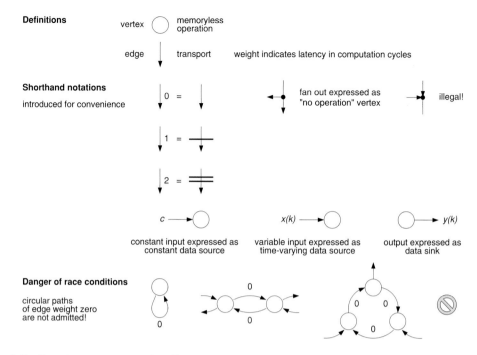

Fig. 2.10 Data dependency graph (DDG) notation.

2.3.5 The isomorphic architecture

No matter how one has arrived at some initial proposal, it always makes sense to search for a better hardware arrangement. Inspired VLSI architects will let themselves be guided by intuition and experience to come up with one or more tentative designs before looking for beneficial reorganizations. Yet, for the subsequent discussion and evaluation of the various equivalence transforms available, we need something to compare with. A natural candidate is the isomorphic architecture, see fig.2.11e for an example, where

- Each combinational operation in the DDG is carried out by a hardware unit of its own,
- Each hardware register stands for a latency of one in the DDG,
- There is no need for control because DDG and block diagram are isomorphic,[17] and
- Clock rate and data input/output rate are the same.

---

other meanings in the context of hardware design, we prefer "circular path" in spite of its clumsiness. For the same reason, let us use "vertex" when referring to graphs and "node" when referring to electrical networks.

A zero-weight circular path in a DDG implies immediate feedback and expresses a self-referencing combinational function. Such zero-latency feedback loops are known to expose the pertaining electronic circuits to unpredictable behavior and are, therefore, highly undesirable, see section 5.4.3 for details.

[17] Two directed graphs are said to be **isomorphic** if there exists a one-to-one correspondence between their vertices and between their edges such that all incidence relations and all edge orientations are preserved. More informally, two isomorphic graphs become indistinguishable when the labels and weights are removed from their vertices and edges. Remember that how a graph is drawn is of no importance for the theory of graphs.

Example

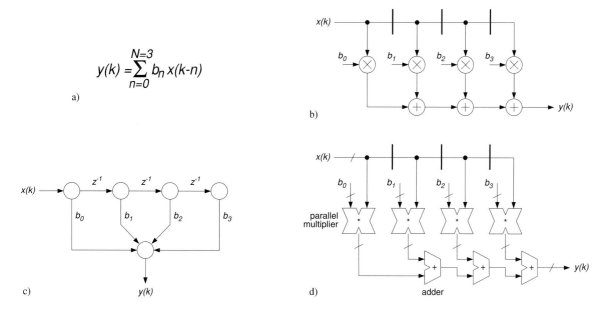

a)

$$y(k) = \sum_{n=0}^{N=3} b_n \, x(k-n)$$

b)

c)

d)

Fig. 2.11 Third order ($N = 3$) transversal filter expressed as a mathematical function (a), drawn as data dependency graph (DDG) (b), and implemented with the isomorphic hardware architecture (d). Signal flow graph shown for comparison (c).

☐

An architecture design as naive as this obviously cannot be expected utilize hardware efficiently, but it will serve as a reference for discussing both the welcome and the unfavorable effects of various architectural reorganizations. You may also think of the isomorphic architecture as a hypothetical starting point from which any more sophisticated architecture can be obtained by applying a sequence of equivalence transforms.[18]

2.3.6 Relative merits of architectural alternatives

Throughout our analysis, we will focus on the subsequent figures of merit.

Circuit size A. Depending on how actual hardware costs are best expressed, the designer is free to interpret size as area occupation (in mm^2 or lithographic squares F^2 for ASICs) or as circuit complexity (in terms of GE for ASICs and FPL).

Cycles per data item Γ denotes the number of computation cycles that separates the releasing of two consecutive data items, or — which is normally the same — the number of computation cycles between accepting two subsequent data items.

---

[18] See problem 2.10 for a more thorough exposure. Also observe that our transform approach to architecture design bears some resemblance to the theory of evolution.

Longest path delay t_{lp} indicates the lapse of time required for data to propagate along the longest combinational path through a given digital network. Path lengths are typically indicated in ns. What makes the maximum path length so important is that it limits the operating speed of a given architecture. For a circuit to function correctly, it must always be allowed to settle to a — typically new — steady state within a single computation period T_{cp}.[19] We thus obtain the requirement $t_{lp} \leq T_{cp}$, where the exact meaning of computation period is to be defined shortly in section 2.3.7.

Time per data item T indicates the time elapsed between releasing two subsequent data items. Depending on the application, T might be stated in µs/sample, ms/frame, or s/calculation, for instance. $T = \Gamma \cdot T_{cp} \geq \Gamma \cdot t_{lp}$ holds with equality if the circuit gets clocked at the fastest possible rate.

Data throughput $\Theta = \frac{1}{T}$ is the most meaningful measure of overall circuit performance. Throughput gets expressed in terms of data items processed per time unit; e.g. in pixel/s, sample/s, frame/s, data record/s, FFT/s, matrix inversion/s, and the like.[20] It is given by

$$\Theta = \frac{f_{cp}}{\Gamma} = \frac{1}{\Gamma \cdot T_{cp}} \leq \frac{1}{\Gamma \cdot t_{lp}} \tag{2.3}$$

for a circuit operated at computation rate f_{cp} or, which is the same, with a computation period T_{cp}.[21] Again, we are most interested in the maximum throughput where $T_{cp} = t_{lp}$.

Size–time product AT combines circuit size and computation time to indicate the hardware resources spent to obtain a given throughput. This is simply because $AT = \frac{A}{\Theta}$. The lower the AT-product, the more hardware-efficient a circuit.

Latency L indicates the number of computation cycles from a data item being entered into a circuit until the pertaining result becomes available at the output. Latency is zero when the result appears within the same clock cycle as that during which the input datum was fed in.

Energy per data item E is meant to quantify the amount of energy dissipated in carrying out some given computation on a data item. As examples consider indications in pJ/MAC, nJ/sample, µJ/datablock or mWs/videoframe.

The same quantity can also be viewed as the quotient $E = \frac{P}{\Theta}$ that relates power dissipation to throughput and is then be expressed in mW/$\frac{\text{Mbit}}{\text{s}}$, or W/GOPS (Giga operations per second), for instance. Using inverse term such as MOPS/mW and GOPS/W is more popular in the context of microprocessors.

Energy per data item is further related to the **power–delay product** (PDP) $pdp = P \cdot t_{lp}$, a quantity often used for comparing standard cells and other transistor-level circuits. The

---

[19] We do not consider multicycle paths, wave-pipelined operation, or asynchronous circuits here.

[20] Note that Mega Instructions Per Second (MIPS), a performance indicator most popular with IT specialists, neither reflects data throughput nor applies to architectures other than program-controlled processors.

[21] It is sometimes more adequate to express data throughput in terms of bits per time unit; (2.3) must then be restated as $\Theta = w \frac{f_{cp}}{\Gamma}$, where w indicates how many bits make up one data item.

difference is that our definition explicitly accounts for multicycle computations and for longer-than-necessary computation periods $E = PT = P \cdot \Gamma \cdot T_{cp} \geq P \cdot \Gamma \cdot t_{lp} = \Gamma \cdot pdp$.

Example

In the occurrence of the architecture shown in fig.2.11e, one easily finds the quantities below

$$A = 3A_{reg} + 4A_* + 3A_+ \tag{2.4}$$

$$\Gamma = 1 \tag{2.5}$$

$$t_{lp} = t_{reg} + t_* + 3t_+ \tag{2.6}$$

$$AT = (3A_{reg} + 4A_* + 3A_+)(t_{reg} + t_* + 3t_+) \tag{2.7}$$

$$L = 0 \tag{2.8}$$

$$E = 3E_{reg} + 4E_* + 3E_+ \tag{2.9}$$

where indices $*$, $+$, and *reg* refer to a multiplier, an adder, and a data register respectively.
□

A word of caution is due here. Our goal in using formulae to approximate architectural figures of merit is not so much to obtain numerical values for them but to explain roughly how they are going to be affected by the different equivalence transforms available to VLSI architects.[22]

2.3.7 Computation cycle versus clock period

So far, we have been using the term computation period without defining it. In synchronous digital circuits, a calculation is broken down into a series of shorter computation cycles the rhythm of which gets imposed by a periodic **clock** signal. During each computation cycle, fresh data emanate from a register, and propagate through combinational circuitry where they undergo various arithmetic, logic, and/or routing operations before the result gets stored in the next analogous register (same clock, same active edge).

Definition 2.1. *A computation period T_{cp} is the time span that separates two consecutive computation cycles.*

For the moment being, it is safe to assume that computation cycle, computation period, clock cycle, and clock period are all the same, $T_{cp} = T_{clk}$, which is indeed the case for all those circuits that adhere to single-edge-triggered one-phase clocking.[23] The inverse, that is the number of computation cycles per second, is referred to as **computation rate** $f_{cp} = \frac{1}{T_{cp}}$.

---

[22] As an example, calculation of the long path delay t_{lp} is grossly simplified in (2.6). For one thing, interconnect delays are neglected which is an overly optimistic assumption. For another thing, the propagation delays of the arithmetic operations are simply summed up which sometimes is a pessimistic assumption, particularly in cascades of multiple ripple carry adders where all operands arrive simultaneously. Synthesis followed by place and route often is the only way to determine overall path delays with sufficient accuracy.

[23] As an exception, consider dual-edge-triggering where each clock period comprises two consecutive computation periods so that $T_{cp} = \frac{1}{2}T_{clk}$. Details are to follow in section 6.2.3.

2.4 | Equivalence transforms for combinational computations

A computation that depends on the present arguments exclusively is termed combinational. A sufficient condition for combinational behavior is a DDG which is free of circular paths and where all edge weights equal zero.

Consider some fixed but otherwise arbitrary combinational function $y(k) = f(x(k))$. The DDG in fig.2.12a depicts such a situation. As suggested by the dashed edges, both input $x(k)$ and output $y(k)$ can include several subvectors. No assumptions are made about the complexity of f which could range from a two-bit addition, over an algebraic division, to the Fast Fourier Transform (FFT) operation of a data block, and beyond. In practice, designers would primarily be concerned with those operations that determine chip size, performance, power dissipation, etc. in some critical way.

Fig. 2.12 DDG for some combinational function f (a). A symbolic representation of the reference hardware configuration (b) with its key characteristics highlighted (c).

The isomorphic architecture simply amounts to a hardware unit that does nothing but evaluate function f, a rather expensive proposal if f is complex such as in the FFT example. Three options for reorganizing and improving this unsophisticated arrangement exist.[24]

1. **Decomposing** function f into a sequence of subfunctions that get executed one after the other in order to reuse the same hardware as much as possible.
2. **Pipelining** of the functional unit for f to improve computation rate by cutting down combinational depth and by working on multiple consecutive data items simultaneously.
3. **Replicating** the functional unit for f and having all units work concurrently.

It is intuitively clear that replication and pipelining both trade circuit size for performance while iterative decomposition does the opposite. This gives rise to questions such as

"Does it make sense to combine pipelining with iterative decomposition
  in spite of their antagonistic effects?" and
"Are there situations where replication should be preferred over pipelining?"

which we will try to answer in the following subsections.

---

[24] Of course, many circuit alternatives for implementing a given arithmetic or logic function also exist at the gate level. However, within the general context of architecture design, we do not address the problem of developing and evaluating such options as this involves lower-level considerations that strongly depend on the specific operations and on the target library. The reader is referred to the specialized literature on computer arithmetics and on logic design.

2.4.1 Common assumptions

The architectural arrangement that will serve as a reference for comparing various alternative designs is essentially identical to the isomorphic configuration of fig.2.12a with a register added at the output to allow for the cascading of architectural chunks without their longest path delays piling up. The characteristics of the reference architecture then are

$$A(0) = A_f + A_{reg} \tag{2.10}$$

$$\Gamma(0) = 1 \tag{2.11}$$

$$t_{lp}(0) = t_f + t_{reg} \tag{2.12}$$

$$AT(0) = (A_f + A_{reg})(t_f + t_{reg}) \tag{2.13}$$

$$L(0) = 1 \tag{2.14}$$

$$E(0) = E_f + E_{reg} \tag{2.15}$$

where subscript $_f$ stands for the datapath hardware that computes some given combinational function f and where subscript $_{reg}$ denotes a data register. For the sake of simplicity the word width w of the datapath is assumed to be constant throughout. For illustration purposes we use a graphical representation that suggests hardware organization, circuit size, longest path length, data throughput, and latency in a symbolic way, see figs.2.12b and c.

The quotient A_f/A_{reg} relates the size of the datapath hardware to that of a register, and t_f/t_{reg} does the same for their respective time requirements.[25] Their product $\frac{A_f}{A_{reg}} \frac{t_f}{t_{reg}}$ thus reflects the computational complexity of function f in some sense. (2.16) holds whenever logic function f is a fairly substantial computational operation. We will consider this the typical, although not the only possible, case.

$$A_{reg} t_{reg} \ll A_f t_f \tag{2.16}$$

Many of the architectural configurations to be discussed require extra circuitry for controlling datapath operation and for routing data items. Two additive terms A_{ctl} and E_{ctl} are introduced to account for this where necessary. As it is very difficult to estimate the extra hardware without detailed knowledge of the specific situation at hand, the only thing that can be said for sure is that A_{ctl} is on the order of A_{reg} or larger for most architectural transforms. Control overhead may in fact become significant or even dominant when complex control schemes are brought to bear as a result of combining multiple transforms.

As for energy, we will focus on the dynamic contribution that gets dissipated in charging and discharing electrical circuit nodes as a consequence of fresh data propagating through gate-level networks. Any dissipation due to static currents or due to idle switching is ignored, which is a reasonable assumption for comparing low-leakage static CMOS circuits that are fairly active.[26]

Throughout our architectural comparisons, we further assume all electrical and technological conditions to remain the same.[27] A comparison of architectural alternatives on equal grounds is otherwise

---

[25] Typical size A and delay figures t for a number of logic and arithmetic operations are given as illustrative material in appendix 2.12.

[26] Power dissipation, switching activities, leakage currents, and the like are the subjects of chapter 9.

[27] This includes supply voltage, cell library, transistor sizes, threshold voltages, fabrication process, and the gate-level structure of arithmetic units.

not possible as a shorter path delay or a lower energy figure would not necessarily point to a more efficient design alternative.

2.4.2 Iterative decomposition

The idea behind iterative decomposition — or decomposition, for short — is nothing else than **resource sharing** through *step-by-step execution*. The computation of function f is broken up into a sequence of d subtasks which are carried out one after the other. From a dataflow point of view, intermediate results are recycled until the final result becomes available at the output d computation cycles later, thereby making it possible to reuse a single hardware unit several times over. A configuration that reuses a multifunctional datapath in a time-multiplex fashion to carry out f in $d = 3$ subsequent steps is symbolically shown in fig.2.13. Note the addition of a control section that pilots the datapath on a per-cycle basis over a number of control lines.

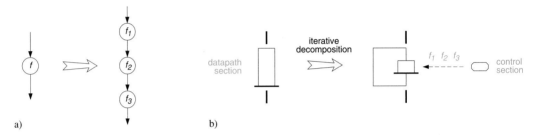

a)   b)

Fig. 2.13 Iterative decomposition. DDG (a) and hardware configuration for $d = 3$ (b).

PERFORMANCE AND COST ANALYSIS

Assumptions:

1. The total size requirement for implementing the various subfunctions into which f is decomposed ranges between $\frac{A_f}{d}$ and A_f.
2. The decomposition is lossless and balanced, i.e. it is always possible to break up f into d subfunctions the computations of which require a uniform amount of time $\frac{t_f}{d}$.

As a first-order approximation, iterative decomposition leads to the following figures of merit:

$$\frac{A_f}{d} + A_{reg} + A_{ctl} \leq A(d) \leq A_f + A_{reg} + A_{ctl} \quad (2.17)$$

$$\Gamma(d) = d \quad (2.18)$$

$$t_{lp}(d) \approx \frac{t_f}{d} + t_{reg} \quad (2.19)$$

$$d(A_{reg} + A_{ctl})t_{reg} + (A_{reg} + A_{ctl})t_f + A_f t_{reg} + \frac{1}{d}A_f t_f \leq AT(d) \leq$$
$$d(A_f + A_{reg} + A_{ctl})t_{reg} + (A_f + A_{reg} + A_{ctl})t_f \quad (2.20)$$

$$L(d) = d \quad (2.21)$$

$$E(d) \gtrsim E_f + E_{reg} \quad (2.22)$$

Let us confine our analysis to situations where the control overhead can be kept small so that $A_{reg} \approx A_{ctl} \ll A_f$ and $E_{reg} \approx E_{ctl} \ll E_f$. A key issue in interpreting the above results is whether size $A(d)$ tends more towards its lower or more towards its upper bound in (2.17). While iterative decomposition can, due to (2.16), significantly lower the AT-product in the former case, it does not help in the latter.

The lower bounds hold in (2.17) and (2.20) when the chunk's function f makes repetitive use of a single subfunction because the necessary datapath is then essentially obtained from cutting the one that computes f into d identical pieces only one of which is implemented in hardware. A monofunctional processing unit suffices in this case.

At the opposite end are situations where computing f asks for very disparate subfunctions that cannot be made to share much hardware resources in an efficient way. Iterative decomposition is not an attractive option in this case, especially if register delay, control overhead, and the difficulty of meeting assumption 2 are taken into consideration.

Operations that lend themselves well to being combined into a common computational unit include addition and subtraction in either fixed-point or floating-point arithmetics, and various shift and rotate operations. CORDIC units reuse essentially the same hardware for angle rotations and for trigonometric and hyperbolic functions.

As for energy efficiency, there are two mechanisms that counteract each other. On the one hand, iterative decomposition entails register activity not present in the original circuit. The extra control and data recycling logic necessary to implement step-by-step execution further inflate dissipation.

On the other hand, we will later find that long register-to-register signal propagation paths tend to foster transient node activities, aka glitches. Cutting such propagation paths often helps to mitigate glitching activities and the associated energy losses.[28] Such second-order effects are not accounted for in the simplistic unit-wise additive model introduced in (2.15), however, making it difficult to apprehend the impact of iterative decomposition on energy before specific circuit details become available.

Example

A secret-key block cipher operated in electronic code book (ECB) mode is a highly expensive combinational function. ECB implies a memoryless mapping $y(k) = c(x(k), u(k))$ where $x(k)$ denotes the plaintext, $y(k)$ the ciphertext, $u(k)$ the key, and k the block number or time index. What most block ciphers, such as the Data Encryption Standard (DES), the International Data Encryption Algorithm (IDEA), and the Advanced Encryption Standard (AES) Rijndael have in common is a cascade of several rounds, see fig.2.14 for the IDEA algorithm [38]. The only difference between the otherwise identical rounds is in the values of the subkeys used that get derived from $u(k)$. What is referred to as output transform is nothing else than a subfunction of the previous rounds.

If we opt for iterative decomposition, a natural choice consists in designing a datapath for one round and in recycling the data with changing subkeys until all rounds have been processed. As control is very simple, the circuit's overall size is likely to stay close to the lower bound in (2.17) after this first step of decomposition. On continuing in the same direction, however, benefits will diminish because the operations involved (bitwise addition modulo 2, addition modulo 2^{16}, and

---

[28] See chapter 9 for explanations.

multiplication modulo $(2^{16} + 1))$ are very disparate. In addition, the impact of control on the overall circuit size would be felt.

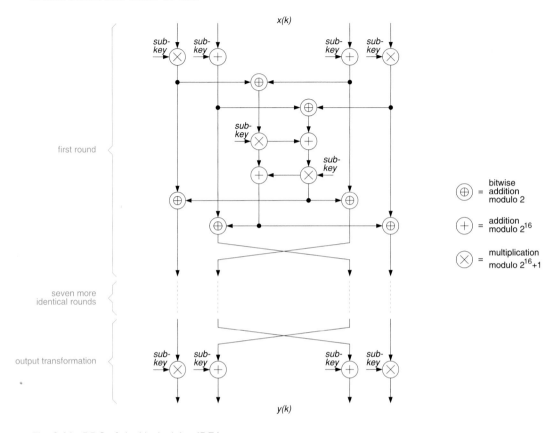

Fig. 2.14 DDG of the block cipher IDEA.

A more radical approach is to decompose arbitrary functions into sequences of arithmetic and/or logic operations from a small but extremely versatile set and to provide a single ALU instead. The datapath of any **microprocessor** is just a piece of universal hardware that arose from the general idea of step-by-step computation, and the **reduced instruction set computer** (RISC) can be viewed as yet another step in the same direction. While iterative decomposition together with programmability and time sharing, see section 2.4.5, explains the outstanding flexibility and hardware economy of this paradigm, it also accounts for its modest performance and poor energy efficiency relative to more focussed architecture designs.

Examples

Examples of ASICs the throughputs of which exceeded that of contemporary high-end general-purpose processors by orders of magnitude are given in sections 2.2 and 2.7.3.

2.4.3 Pipelining

Pipelining aims at increasing throughput by cutting combinational depth into several separate stages of approximately uniform computational delays by inserting registers in between.[29] The combinational logic between two subsequent **pipeline registers** is designed and optimized to compute one specific subfunction. As an ensemble, the various stages cooperate like specialist workers on an *assembly line*. Figure 2.15 sketches a functional unit for f subdivided into $p = 3$ pipeline stages by $p - 1$ extra registers. Note the absence of any control hardware.

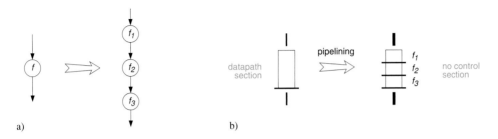

Fig. 2.15 Pipelining. DDG (a) and hardware configuration for $p = 3$ (b).

PERFORMANCE AND COST ANALYSIS

Assumptions:

1. The combinational logic for implementing function f is not affected by the number of pipeline stages introduced. Its overall size A_f, therefore, remains constant.
2. The pipeline is lossless and balanced, i.e. similarly to decomposition it is always possible to partition the logic into p stages such that all have identical delays $\frac{t_f}{p}$.
3. The size penalty of pipelining can be expressed by an additive term A_{reg} for each register accounting for the silicon area occupied by storage elements.
4. At each pipeline stage a performance penalty results from introducing a register delay t_{reg} which includes the delay caused by the storage element.

Pipelining changes performance and cost figures as follows:

$$A(p) = A_f + pA_{reg} \tag{2.23}$$

$$\Gamma(p) = 1 \tag{2.24}$$

$$t_{lp}(p) \approx \frac{t_f}{p} + t_{reg} \tag{2.25}$$

$$AT(p) \approx pA_{reg}t_{reg} + (A_{reg}t_f + A_f t_{reg}) + \frac{1}{p}A_f t_f \tag{2.26}$$

$$L(p) = p \tag{2.27}$$

$$E(p) \underset{\text{coarse grain}}{\overset{\text{fine grain}}{\gtrless}} E_f + E_{reg} \tag{2.28}$$

---

[29] For a more formal discussion see subsection 2.6.1.

Both performance and size grow monotonically with pipeline depth. The same holds true for latency. What is more interesting is that a modest number of pipeline stages each of which has a substantial depth dramatically lowers the AT-product due to (2.16). This regime is referred to as **coarse grain pipelining**.

Example

Equation (2.25) relates combinational delay to register delay. Another popular way to quantify the degree of pipelining is to express the delay on the longest path as a multiple of fanout-of-4 (FO4) inverter delays.[30]

CPU	year	clock freq. [MHz]	FO4 inverter delays per pipeline stage
Intel 80386	1989	33	$\approx$80
Intel Pentium 4	2003	3200	12–16
IBM Cell Processor	2006	3200	11

☐

Continuing along this line, one may want to insert more and more pipeline registers. However, (2.25) reveals that the benefit fades when the combinational delay per stage $\frac{t_l}{p}$ approaches the register delay t_{reg}. For large values of p the area–delay product is dominated by the register delay rather than by the payload function. A natural question for this type of deep or **fine grain pipelining** is to ask

"What is the maximum computation rate for which a pipeline can be built?"

The fastest logic gates from which useful data processing can be obtained are 2-input NAND or NOR gates.[31] Even if we are prepared to profoundly redesign a pipeline's logic circuitry in an attempt to minimize the longest path t_{lp}, we must leave room for at least one such gate between two subsequent registers. It thus is not possible to accelerate the computation rate beyond

$$T_{cp} \geq \min(t_{lp}) = \min(t_{gate}) + t_{reg} = \min(t_{nand}, t_{nor}) + t_{su\,ff} + t_{pd\,ff} \qquad (2.29)$$

which represents a lower bound for (2.25). Practical applications that come close to this theoretical minimum are limited to tiny subcircuits, however, mainly because of the disproportionate number of registers required, but also because meeting assumptions 1 and 2 is difficult with fine grained pipelines. Even in high-performance datapath logic, economic reasons typically preclude pipelining below 11 FO4 inverter delays per stage.

Equation (2.29) further indicates that register delay is critical in high speed design. In fact, a typical relation is $t_{reg} \approx$ 3–5 $\min(t_{gate})$. As a consequence, it takes twenty or so levels of logic between subsequent registers before flip-flop delays are relegated to insignificant proportions. A

---

[30] Comparing circuit alternatives in terms of FO4 inverters makes sense because fanout-of-4 inverters are common-place in buffer trees driving large loads and because the delays of other static CMOS gates have been found to track well those of FO4 inverters.

[31] This is because binary NAND and NOR operations (a) form a complete gate set each and (b) are efficiently implemented from MOSFETs, see sections A.2.10 and 8.1 respectively.

high-speed cell library must, therefore, not only include fast combinational gates but also provide bistables with minimum insertion delays.[32]

Example

Plugging into (2.29) typical numbers for a 2-input NOR gate and a D-type flip-flop with no reset from a 130 nm CMOS standard cell library, one obtains $T_{cp} \geq t_{\text{NOR2D1}} + t_{\text{DFFPB1}} = 18\,\text{ps} + 249\,\text{ps} \approx 267\,\text{ps}$ which corresponds to a maximum computation rate of about 3.7 GHz.
□

"How many stages yield optimum pipeline efficiency?"

Optimum hardware efficiency means minimum size–time product

$$AT(p) = \min \tag{2.30}$$

which is obtained for

$$p_0 = \sqrt{\frac{A_f t_f}{A_{reg} t_{reg}}} \tag{2.31}$$

Beyond this point, adding more pipeline registers causes the size–time product to deteriorate even though performance is still pushed further. It also becomes evident from (2.31) that, in search of an economic solution, the more complex a function, the more pipelining it supports. In practice, efficiency is likely to degrade before p_0 is reached because our initial assumptions 1 and 2 cannot be entirely satisfied. [39] indicates the optimal depth is 6 to 8 FO4 inverter delays per pipeline stage.

"How does pipelining affect energy efficiency?"

The additional registers suggest that any pipelined datapath dissipates more energy than the reference architecture does. This is certainly true for fine grain pipelines where the energy wasted by the switching of all those extra subcircuits becomes the dominant contribution.

For coarse grain designs, the situation is more fortunate. Experience shows that pipeline registers tend to reduce the unproductive switching activity associated with glitching in deep combinational networks, a beneficial side effect neglected in a simple additive model.

Interestingly, our finding that throughput is greatly increased makes it possible to take advantage of coarse grain pipelining for improving energy efficiency, albeit indirectly. Recall that the improved throughput is a result from cutting the longest path while preserving a processing rate of one data item per computation cycle. The throughput of the isomorphic architecture is thus readily matched by a pipelined datapath implemented in a slower yet more energy-efficient technology, e.g. by operating CMOS logic from a lower supply voltage or by using mostly minimum-size transistors. Our model cannot reflect this opportunity because we have decided to establish energy figures under the assumption of identical operating conditions and cell libraries. Another highly welcome property of pipelining is the absence of energy-dissipating control logic.

---

[32] Function latches where bistables and combinational logic get merged into a single library cell in search of better performance are to be discussed in sections 6.2.6 and 8.2.2.

PIPELINING IN THE PRESENCE OF MULTIPLE FEEDFORWARD PATHS

Although pipelining can be applied to arbitrary feedforward computations, there is a reservation of economic nature when a DDG includes many parallel paths. In order to preserve overall functionality, any latency introduced into one of the signal propagation paths must be balanced by inserting an extra register into each of its parallel paths. Unless those **shimming registers** help cut combinational depth there, they bring about substantial size and energy penalties, especially for deep pipelines where p is large.

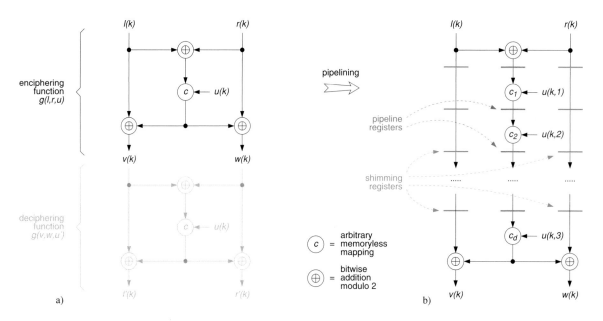

Fig. 2.16 Involutory cipher algorithm. DDG before (a) and after pipelining (b).

Example

With simplifications, fig.2.16a reproduces the block cipher IDEA. Variable k stands for the block index, $l(k)$ and $r(k)$ each denote half of a 64 bit plaintext block while $v(k)$ and $w(k)$ do the same for a 64 bit ciphertext block. $u(k)$ and $u'(k)$ stand for the keys used during enciphering and deciphering operations respectively. Provided the two keys are the same, i.e. $u'(k) = u(k)$, the net result is $l'(k) = l(k)$ and $r'(k) = r(k)$, which implies that the plaintext is recovered after calling g twice. Note that this involution property[33] is totally independent of function c which, therefore, can be designed so as to maximize cryptographic security.

Extensive pipelining seems a natural way to reconcile the computational complexity of c with ambitious performance goals. Yet, as a consequence of the two paths bypassing c, every pipeline

---

[33] A function g is said to be **involutory** iff $g(g(x)) \equiv x$, $\forall x$. As trivial examples, consider multiplication by -1 in classic algebra where we have $-(-x) \equiv x$, the complement function in Boolean algebra where $\overline{\overline{x}} \equiv x$, or a mirroring operation from geometry. Involution is a welcome property in cryptography since it makes it possible to use exactly the same equipment for both enciphering and deciphering.

register entails two shimming registers, effectively tripling the costs of pipelining, see fig.2.16b. This is the reason why pipeline depth had to be limited to eight stages per round in a VLSI implementation of the IDEA cipher in spite of stringent throughput requirements [40].

2.4.4 Replication

Replication is a brute-force approach to performance: If one functional unit does not suffice, allow for several of them. Concurrency is obtained from providing q instances of identical functional units for f and from having each of them process one out of q data items in a cyclic manner. To that end, two synchronous q-way switches distribute and recollect data at the chunk's input and output respectively. An arrangement where $q = 3$ is shown in fig.2.17.[34] Overall organization and operation is reminiscent of a *multi-piston pump*.

□

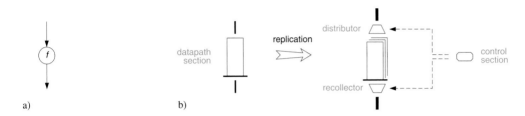

a)   b)

Fig. 2.17   Replication. DDG (a) and hardware configuration for $q = 3$ (b).

Performance and cost analysis

Assumptions:

1. Any size penalties associated with distributing data to replicated functional units
   and with recollecting them are neglected.
2. Any energy dissipated in data distribution and recollection is ignored.

The above assumptions hold fairly well provided the circuitry for computing f is much larger than that for data distribution and recollection. The key characteristics of replication then become

$$A(q) = q(A_f + A_{reg}) + A_{ctl} \tag{2.32}$$

$$\Gamma(q) = \frac{1}{q} \tag{2.33}$$

$$t_{lp}(q) \approx t_f + t_{reg} \tag{2.34}$$

$$AT(q) \approx \left(A_f + A_{reg} + \frac{1}{q} A_{ctl} \right)(t_f + t_{reg}) \approx (A_f + A_{reg})(t_f + t_{reg}) \tag{2.35}$$

$$L(q) = 1 \tag{2.36}$$

$$E(q) \approx E_f + E_{reg} + E_{ctl} \tag{2.37}$$

---

[34] Multiple processing units that work in parallel are also found in situations where the application naturally provides data in parallel streams, each of which is to undergo essentially the same processing. In spite of the apparent similarity, this must <u>not</u> be considered as the result of replication, however, because DDG and architecture are isomorphic. This is reflected by the fact that no data distribution and recollection mechanism is required in this case. Please refer to section 2.4.5 for the processing of multiple data streams.

As everyone would expect, replication essentially trades area for speed. Except for the control overhead, the *AT*-product remains the same. Pipelining, therefore, is clearly more attractive than replication as long as circuit size and performance do not become dominated by the pipeline registers, see fig.2.18 for a comparison.

Example

Consider a simple network processor that handles a stream of incoming data packets and does some address calculations before releasing the packets with a modified header. Let that processor be characterized by the subsequent cost figures: $A_f = 60w$ GE, $A_{reg} = 6w$ GE, where w is an integer relating to datapath width, $t_f = 12$ ns, $t_{reg} = 1.2$ ns, and $\min(t_{gate}) = 0.3$ ns.

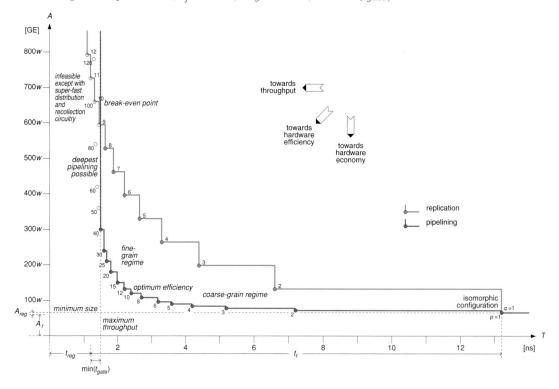

Fig. 2.18 *AT*-characteristics of pipelining and replication compared. Simplified by assuming perfect balancing, by not accounting for shimming registers in the occurrence of pipelining, and by abstracting from control, data distribution, and recollection associated with replication.

☐

Energywise, replication is indifferent except for the contributions for datapath control and data distribution/recollection. Also note, by the way, that replication does not shorten the computation period, which contrasts with iterative decomposition and pipelining.[35]

---

[35] Observe that the entirety of functional units must be fed with q data items per computation cycle and that processed data items emanate at the same rate. Only the data distribution and recollection subcircuits must

A more accurate evaluation of replication versus pipelining would certainly require revision of some of the assumptions made here and does depend to a large extent on the actual DDG and on implementation details. Nevertheless, it is safe to conclude that neither fine grain pipelining nor replication is as cost-effective as coarse grain pipelining.

Its penalizing impact on circuit size confines replication to rather exceptional situations in ASIC design. A megacell available in layout form exclusively represents such a need because adding pipeline registers to a finished layout would ask for a disproportionate effort. Replication is limited to high-performance circuits and always combined with generous pipelining.

Superscalar and multicore microprocessors are two related ideas from computer architecture.[36] Several factors have pushed the computer industry towards replication: CMOS technology offered more room for increasing circuit complexity than for pushing clock frequencies higher. The faster the clock, the smaller the region on a semiconductor die that can be reached within a single clock period.[37] Fine grain pipelines dissipate a lot of energy for relatively little computation. Reusing a well-tried subsystem benefits design productivity and lowers risks. A multicore processor can still be of commercial value even if one of its CPUs is found to be defective.

2.4.5 Time sharing

So far we have been concerned with the processing of a single data stream as depicted in fig.2.12. Now consider a situation where a number of parallel data streams undergo processing as illustrated in fig.2.19, for instance. Note that the processing functions f, g, and h may, but need not, be the same. The isomorphic architecture calls for a separate functional unit for each of the three operations in this case. This may be an option in applications such as image processing where a great number of dedicated but comparatively simple processing units are repeated along one or two dimensions, where data exchange is mainly local, and where performance requirements are very high.

More often, however, the costs of fully parallel processing are unaffordable and one seeks to cut overall circuit size. A natural idea is to pool hardware by having a single functional unit process the parallel data streams one after the other in a cyclic manner. Analogously to replication, a synchronous s-way switch at the input of that unit collects the data streams while a second one redistributes the processed data at the output. While the approach is known as time-sharing in computing, it is more often referred to as **multiplexing** or as **resource sharing** in the context of circuit design.[38] What it requires is that the circuitries for computing the various functions involved all be combined into a single datapath of possibly multifunctional nature. A *student sharing his time between various subjects* might serve as an analogy from everyday life.

---

be made to operate at a rate q times higher than the computational instances themselves. High data rates are obtained from configuring data distribution/recollection networks as heavily pipelined binary trees. Maximum speed is, again, determined by (2.29). Yet, circumstances permitting, it may be possible to implement data distribution and recollection using a faster technology than the one being used in the body of the processing chunk (superfast distribution and recollection). Also see [41] for further information on fast data distribution/recollection circuitry.

[36] A **superscalar** processor combines multiple execution units, such as integer ALUs, FPUs, load/store units, and the like, into one CPU so that superscalar CPU can fetch and process more than one instruction at a time. **Multicore** architectures go one step further in that they replicate entire CPUs on a single chip and so enable a processor to work on two or more threads of execution at a time.

[37] For a rationale, refer to section 6.3 that discusses delay in interconnect lines without and with repeaters.

[38] This is our second resource-sharing technique, after iterative decomposition introduced in section 2.4.2.

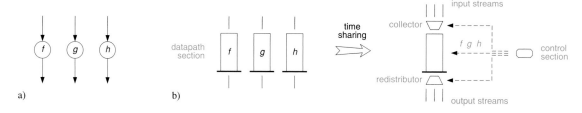

Fig. 2.19 Time sharing. DDG with parallel data streams (a) and hardware configuration for $s = 3$ (b).

PERFORMANCE AND COST ANALYSIS

Assumptions:

1. The size of a circuit capable of implementing functions f, g, and h with a single computational unit ranges between $\max\limits_{f,g,h}(A) = \max(A_f, A_g, A_h)$ and $\sum_{f,g,h} A = (A_f + A_g + A_h)$.
2. The time for the combined computational unit to evaluate any of the functions f, g, and h has a fixed value $\max\limits_{f,g,h}(t) = \max(t_f, t_g, t_h)$.
3. As for replication, any size and energy penalties associated with collecting and redistributing data are neglected.
4. The energy spent for carrying out functions f, g, and h (all together) with one shared unit is closer to $s \max\limits_{f,g,h}(E) = s\max(E_f, E_g, E_h)$ than to $\sum_{f,g,h} E = E_f + E_g + E_h$.

Time-sharing yields the following circuit characteristics:

$$\max_{f,g,h}(A) + A_{reg} + A_{ctl} \leq A(s) \leq \sum_{f,g,h} A + A_{reg} + A_{ctl} \tag{2.38}$$

$$\Gamma(s) = s \tag{2.39}$$

$$t_{lp}(s) \approx \max_{f,g,h}(t) + t_{reg} \tag{2.40}$$

$$s(\max_{f,g,h}(A) + A_{reg} + A_{ctl})(\max_{f,g,h}(t) + t_{reg}) \leq AT(s) \leq$$

$$s\left(\sum_{f,g,h} A + A_{reg} + A_{ctl}\right)(\max_{f,g,h}(t) + t_{reg}) \tag{2.41}$$

$$L(s) = s \tag{2.42}$$

$$E(s) \approx s\max_{f,g,h}(E) + E_{reg} + E_{ctl} \tag{2.43}$$

Similarly to what was found for iterative decomposition, whether size $A(s)$ tends more towards its lower or more towards its upper bound in (2.38) greatly depends on how similar or dissimilar the individual processing tasks are.

The most favorable situation occurs when one monofunctional datapath proves sufficient because all streams are to be processed in exactly the same way. In our example we then have $f \equiv g \equiv h$ from which $\max\limits_{f,g,h}(A) = A_f = A_g = A_h$ and $\max\limits_{f,g,h}(t) = t_f = t_g = t_h$ follow immediately. Apart from the overhead for control and data routing, $AT(s)$ equals the lower bound $s(A_f + A_{reg})(t_f + t_{reg})$ which is identical to the isomorphic architecture with its s separate computational units. It is in this best case exclusively that time-sharing leaves the size–time product unchanged and may, therefore, be viewed as complementary to replication.

The contrary condition occurs when f, g, and h are very dissimilar so that no substantial savings can be obtained from concentrating their processing into one multifunctional datapath. Time-sharing will then just lower throughput by a factor of s, thereby rendering it an unattractive option. Provided speed requirements are sufficiently low, a radical solution is to combine time-sharing with iterative composition and to adopt a processor style as already mentioned in subsection 2.4.2.

The energy situation is very similar. If the processing functions are all alike and if the computation rate is kept the same, then the energy spent for processing actual data also remains much the same.[39] Extra energy is then spent only for controlling the datapath and for collecting and redistributing data items. More energy is going to get dissipated in a combined datapath when the functions markedly differ from each other. As time-sharing has no beneficial impact on glitching activity either, we conclude that such an architecture necessarily dissipates more energy than a comparable non-time-shared one.

By processing s data streams with a single computational unit, time-sharing deliberately refrains from taking advantage of the parallelism inherent in the original problem. This is of little importance as long as performance goals are met with a given technology. When in search of more performance, however, a time-shared datapath will have to run at a much higher speed to rival the s concurrent units of the isomorphic architecture, which implies that data propagation along the longest path must be substantially accelerated. Most measures suitable to do so, such as higher supply voltage, generous transistor sizing, usage of high-speed cells and devices, adoption of faster but also more complex adder and multiplier schemes, etc., tend to augment the amount of energy spent for the processing of one data item even further.

Example

The Fast Fourier Transform (FFT) is a rather expensive combinational function, see fig.2.20. Luckily, due to its regularity, the FFT lends itself extremely well to various reorganizations that help reduce hardware size. A first iterative decomposition step cuts the FFT into $\log_2(n)$ rounds where n stands for the number of points. When an in-place computation scheme is adopted, those rounds become identical except for their coefficient values.[40] Each round so obtained consists of $\frac{n}{2}$ parallel computations referred to as **butterflies** because of their structure, see fig.2.20b. The idea of sharing

---

[39] Consider (2.27) and note that the equation simplifies to $sE_f + E_{reg} + E_{ctl}$ when f, g, and h are the same. The partial sum $sE_f + E_{reg}$ then becomes almost identical to $s(E_f + E_{reg})$ of the reference architecture. The apparent saving of $(s-1)E_{reg}$ obtained from making do with a single register does not materialize in practice because of the necessity to store data items from all streams.

[40] For a number of computational problems, it is a logical choice to have two memories that work in a **ping-pong** fashion. At any moment of time, one memory provides the datapath with input data while the other accommodates the partial results at present being computed. After the evaluation of one round is completed, their roles are swapped. Simple as it is, this approach unfortunately requires twice as much memory as needed to store one set of intermediate data. A more efficient technique is **in-place computation**, whereby some of the input data are immediately overwritten by the freshly computed values. In-place computation may cause data items to get scrambled in memory, though, which necessitates corrective action. Problems amenable to in-place computation combined with memory unscrambling include the FFT and the Viterbi algorithm [42].

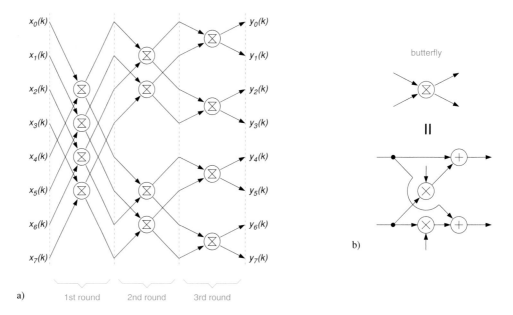

Fig. 2.20 DDG of 8-point FFT (a) and of butterfly operator (b) (reduced for simplicity).

one or two computational units between the butterfly operations of the same round is very obvious at this point.[41]

DDGs as regular as this offer ample room for devising a range of architectures that represent diverse compromises between a single-ALU microprocessor and a hardwired data pipeline maximized for throughput. Providing a limited degree of scalability to accommodate FFTs of various sizes is not overly difficult either. Favorable conditions similar to these are found in many more applications including, among others, transversal filters (repetitive multiply–accumulate operations), correlators (*idem*), lattice filters (identical stages), and block ciphers (cascaded rounds).
□

So far, we have come up with four equivalence transforms, namely

- Iterative decomposition,
- Pipelining,
- Replication, and
- Time-sharing.

Figure 2.21 puts them into perspective in a grossly simplified but also very telling way. More comments will follow in section 2.4.8.

---

[41] Alternative butterfly circuits and architectures have been evaluated in [43] with emphasis on energy efficiency. Also, in fig.2.20, we have assumed input samples to be available as eight concurrent data streams. FFT processors often have to interface to one continuous word-serial stream, however. Architectures that optimize hardware utilization for this situation are discussed in [44].

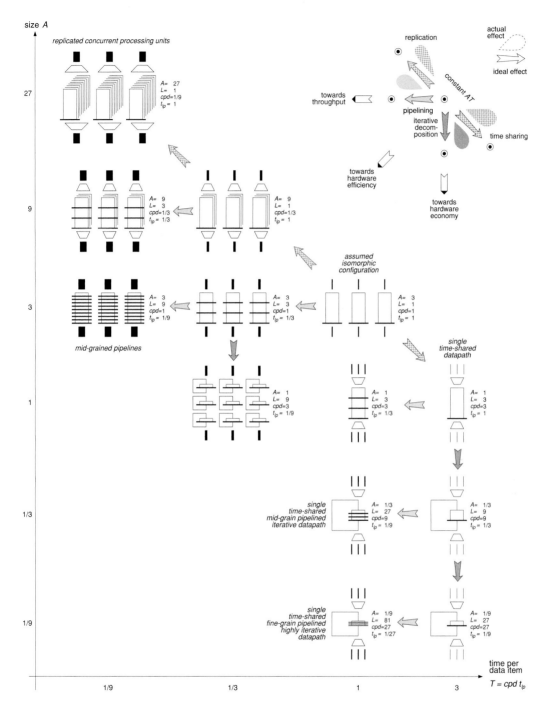

Fig. 2.21 A roadmap illustrating the four universal transforms for tailoring combinational hardware. Only a subset of all possible architectural configurations is shown, see problem 2.10. Greatly simplified by
- abstracting from register overhead ($A_{reg} = 0$, $t_{reg} = 0$), which also implies
- not making any difference between RAMs and flip-flops ($A_{RAM} = A_{ff} \cdot \#_{bits}$, $t_{RAM} = t_{ff}$),
- assuming ideal iterative decomposition and ideal time-sharing (lower bounds in (2.17) and (2.38)), and
- ignoring any overhead associated with control and/or data distribution and collection.

2.4.6 Associativity transform

All four architectural transforms discussed so far have one thing in common. Whether and how to apply them for maximum benefit can be decided from a DDG's connectivity and weights alone, no matter what operations the vertices stand for. In what follows, we will call any architectural reorganization that exhibits this property a **universal transform**.

The practical consequence is that any computational flow qualifies for reorganization by way of universal transforms. This also implies that any two computations the DDGs of which are isomorphic can be solved by the same architecture just with the vertices interpreted differently.

More on the negative side, universal transforms have a limited impact on the flow of computation because the number and precedence of operations are left unchanged.[42] As many computational problems ask for more specific and more profound forms of reorganization in order to take full advantage of the situation at hand, one cannot expect to get optimum results from universal transforms alone. Rather, one needs to bring in knowledge on the particular functions involved and on their algebraic properties. Architectural reorganizations that do so are referred to as operation-specific or **algebraic transforms**.

Probably the most valuable algebraic property from an architectural point of view is the associative law. Associativity can be capitalized on to

- o Convert a chain structure into a tree or vice versa, see example below,
- o Reorder operations so as to accommodate input data that arrive later than others do,
- o Reverse the order of execution in a chain as demonstrated in section 2.6, or
- o Relocate operations from within a recursive loop to outside the loop, see section 2.7.

This explains why the associativity transform is also known as **operator reordering** and as **chain/tree conversion**.

Example

Consider the problem of finding the minimum among I input values

$$y(k) = \min(x_i(k)) \quad \text{where} \ \ i = 0, 1, ..., (I - 1) \tag{2.44}$$

Assuming the availability of 2-way minimum operators, this immediately suggests a chain structure such as the one depicted in fig.2.22a for $I = 8$. The delay along the longest path is $(I - 1)t_{min}$ and increases linearly with the number of terms. As the 2-way minimum function is associative, the DDG lends itself to being rearranged into a balanced tree as shown in fig.2.22b. The longest path is thereby shortened from $I - 1$ to $\lceil \log_2 I \rceil$ operations, which makes the tree a much better choice, especially for large values of I. The number of operations and the circuit's size remain the same.

□

The conversion of a chain of operations into a tree, as in the above example, is specifically referred to as **tree-height minimization**. As a side effect, this architectural transform often has a welcome impact on energy efficiency. This is because glitches die out more rapidly and are more likely to

---

[42] While it is true that the number of DDG vertices may change, this is merely a consequence of viewing the original operations at a different level of detail.

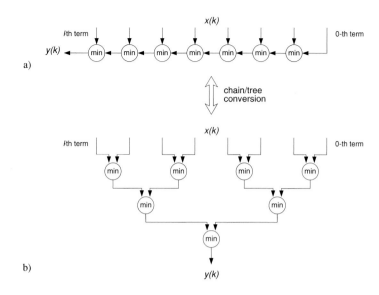

Fig. 2.22 8-way minimum function. Chain-type DDG (a), tree-type DDG (b).

neutralize when all data propagation paths are of similar lengths. In addition, we observe the same indirect benefit as with pipelining, in that a shortened longest path makes it possible to use a slower yet more energy-efficient circuit style or a reduced supply voltage if circumstances permit.

2.4.7 Other algebraic transforms

It goes without saying that many more algebraic laws can be put to use for improving dedicated architectures. **Distributivity** helps to replace the computation of $(a^2 - 2ax + x^2)$ by the more economic form of $(a - x)^2$, for instance, and is instrumental in exploiting known **symmetries** in coefficient sets. Together with **commutativity**, distributivity is also at the heart of distributed arithmetic, see subsection 2.8.3. **Horner's scheme** serves to evaluate polynomials with a minimum number of multiplications, the principle of **superposition** holds in linear systems, the **De Morgan theorem** helps in optimizing Boolean networks, and so on. See problem 5 for yet another operation-specific alteration. As a rule, always ask yourself what situation-specific properties might be capitalized on. The transforms discussed in this text just represent the more common ones and are by no means exhaustive.

2.4.8 Digest

- Iterative decomposition, pipelining, replication, and time-sharing are based on the DDG as a graph and make no assumptions on the nature of computations carried out in its vertices, which is why they are qualified as <u>universal</u>. The associativity transform, in contrast, is said to be an <u>algebraic</u> transform because it depends on the operations involved being identical and associative.

- Iterative decomposition, pipelining, replication, and a variety of algebraic transforms make it possible to tailor combinational computations on a single data stream to given size and performance constraints without affecting functionality. Time-sharing is another option in the presence of inherently parallel data streams and operations.

- For iterative decomposition to be effective, complex functions must be amenable to being broken into similar subfunctions so as to make it possible to reuse the same circuitry. Much the same reasoning also holds for time-sharing in that parallel functions must not be too diverse to share a single functional unit in a fairly efficient way.

- Pipelining is generally more efficient than replication, see fig.2.18. While coarse grain pipelining improves throughput dramatically, benefits decline as more and more stages are included. When in search of utmost performance, begin by designing a pipeline the depth of which yields close-to-optimum efficiency. Only then — if throughput is still insufficient — consider replicating the pipelined functional unit a few times; see problem 2 for an example. This approach also makes sense in view of design productivity because duplicating a moderately pipelined unit may be easier and quicker than pushing pipelining to the extreme.

- A theoretical upper bound on throughput, expressed as data items per time unit, that holds for any circuit technology and architecture is $\Theta \leq \frac{1}{\min(t_{gate})+t_{reg}}$.[43]

- Pipelining and iterative decomposition are complementary in that both can contribute to lowering the size–time product. While the former acts to improve performance, the latter cuts circuit size by sharing resources. Combining them indeed makes sense, within certain bounds, in order to obtain a high throughput from a small circuit.

- Starting from the isomorphic configuration, a great variety of architectures is obtained from applying equivalence transforms in different orders. Combining several of them is typical for VLSI architecture design. Figure 2.21 gives an idealized overview of the design space spanned by the four universal transforms.[44] Which configuration is best in practice cannot be decided without fairly detailed knowledge of the application at hand, of the performance requirements, and of the target cell library and technology.

- Program-controlled microprocessors follow an architectural concept that pushes iterative decomposition and time-sharing to the extreme and that combines them with pipelining, and often with replication too. Developers of general-purpose hardware cannot take advantage of algebraic transforms as their application requires detailed knowledge about the data processing algorithm.

- It can be observed from fig.2.21 that lowering the size–time product AT always implies cutting down the longest path t_{lp} in the circuit. This comes as no surprise as better hardware efficiency can be obtained only from keeping most hardware busy for most of the time by means of a higher computation rate f_{cp}.

---

[43] Further improvements are possible only by processing larger data chunks at a time i.e. by packing more bits, pixels, samples, characters, or whatever into one data item. Note this is tantamount to opting for a larger w in the sense of footnote 21.

[44] As a more philosophical remark, observe from fig.2.21 that there exists no single transform that leads towards optimum hardware efficiency. To move in that direction, designers always have to combine two or more transforms much as a yachtsman must tack back and forth to progress windward with his sailboat.

- Important power savings are obtained from operating CMOS logic with a supply voltage below its nominal value. Clever architecture design must compensate for the loss of speed that is due to the extended gate delays. Suggestions are given not only throughout this chapter, but also in the forthcoming material on energy efficiency.[45]

2.5 | Options for temporary storage of data

Except for trivial SSI/MSI circuits, any IC includes some form of memory. If the original data processing algorithm is of sequential nature and, therefore, mandates the temporary storage of data, we speak of <u>functional</u> memory. If storage gets introduced into a circuit as a consequence of architectural transformations, the memory is sometimes said to be of <u>nonfunctional</u> nature.

The major options for temporary storage of data are as follows:

- On-chip registers built from individual bistables (flip-flops or latches),
- On-chip memory (embedded SRAM or — possibly — DRAM macrocell), or
- Off-chip memory (SRAM or DRAM catalog part).[46]

There are important differences from an implementation point of view that matter from an architectural perspective and that impact high-level design decisions.

2.5.1 Data access patterns

Standard single-port RAMs provide access to data words one after the other.[47] This is fine in sequential architectures as obtained from iterative decomposition and time-sharing that process data in a step-by-step fashion. Program-controlled microprocessors with their "fetch, load, execute, store" processing paradigm are a perfect match for RAMs.

Fast architectures obtained from pipelining, retiming, and loop unfolding,[48] in contrast, keep data moving in every computation cycle, which mandates the usage of registers as only those allow for simultaneous access to all of the data words stored.

Incidentally, also keep in mind that the contents of DRAMs need to be periodically refreshed, which dissipates electrical power even when no data accesses take place.

2.5.2 Available memory configurations and area occupation

Next compare how much die area gets occupied by registers and by on-chip RAMs. While register files allow for any conceivable memory configuration in increments of one data word of depth and one data bit of width, their area efficiency is rather poor. In the occurrence of fig.2.23, registers occupy an order of magnitude more area than a single-port SRAM for capacities in excess of 5000 bits. This is because registers get assembled from individual flip-flops or latches with no sharing of hardware resources.

---

[45] In chapter 9.

[46] Please refer to section A.4 if totally unfamiliar with semiconductor memories.

[47] Dual-port RAMs can access two data words at a time, multi-port RAMs are rather uncommon.

[48] Retiming and loop unfolding are to be explained in sections 2.6 and 2.7 respectively.

Due to their simple and extremely compact bit cells, RAMs make much better use of area in spite of the shared auxiliary subcircuits they must accommodate. In a typical commodity DRAM, for instance, roughly 60% of the die is occupied by storage cells, the rest by address decoders, switches, precharge circuitry, sense amplifiers, internal sequencers, I/O buffers, and padframe. Yet, such circuit overheads tend to make RAMs less attractive for storing smaller data quantities, which is also evident from fig.2.23. A more serious limitation is that macrocells are available in a very limited number of configurations only. Adding or dropping an address bit alters memory capacity by a factor of two, and fractional cores are not always supported.

Always keep in mind that such effects have been ignored in the cost and performance analyses carried out in sections 2.4.2 through 2.4.5 where A_{reg} had been assumed to be fixed with no distinction between registers and RAMs. More specifically, this also applies to fig.2.21.

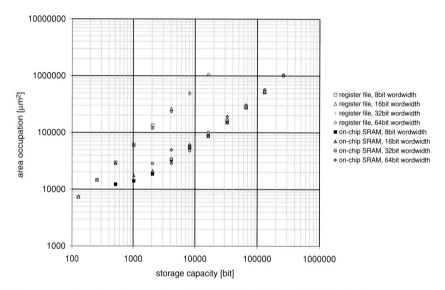

Fig. 2.23 Area occupation of registers and on-chip RAMs for a 130 nm CMOS technology.

2.5.3 Storage capacities

Embedded memories cannot rival the copious storage capacities offered by commodity RAMs. The densest memory chips available are DRAMs built from one-transistor cells whereas the macrocells intended for being embedded within ASICs typically get assembled from six-transistor SRAM storage cells, see table 2.9.[49]

The economic disparity between on-chip memories and cheap commodity DRAMs that are cost-optimized, fabricated in large quantities, and — all too often — subject to ruinous competition has

---

[49] DRAMs further take advantage of three-dimensional trench capacitors and other area-saving structures made possible by dedicated fabrication steps and process options unavailable to macrocells that are to be compatible with a baseline CMOS process. Also, processes for commodity memories are often ahead of ASIC processes in terms of feature size. Finally, the layout of memory chips is highly optimized by hand.

long discouraged the on-chip storage of very large amounts of data within ASICs. The design of true "systems on a chip" requires this gap to be closed. In fact, some integrated VLSI manufacturers are capable of embedding high-density DRAMs within their designs, but the approach has been slow to catch on with the ASIC community and, more particularly, with the providers of ASIC design kits.

Examples

Embedded DRAMs occupy a large part of the market for 3D graphics accelerator chips for laptops because higher performance and lower power dissipation are key value propositions [45]. The so-called "Emotion Engine" superscalar multimedia processor chip designed and produced by Toshiba for Sony's PlayStation 2 is another popular example.

☐

2.5.4 Wiring and the costs of going off-chip

On the negative side, off-chip memories add to pin count, package count, and board space. Communicating with them involves a profusion of parasitic capacitances and delays that cause major bottlenecks in view of operating speed, performance, and energy efficiency.

In addition, most commodity RAMs feature bidirectional data pins in an effort to keep costs and pin counts as low as possible. They thus impose the adoption of a bidirectional bus on any IC that is going to exchange data with them. Yet, note that bidirectional on-chip busses and even more so bidirectional pads require special attention during circuit design and test.

- Not only stationary but even transient drive conflicts must be avoided because of the strong drivers and important currents involved.
- Automated test equipment (ATE) must be made to alternate between read and write modes with no physical access to any control signal within the chip.
- Test patterns must be prepared for verifying bidirectional operation and high-impedance states during circuit simulation and testing.
- Electrical and timing measurements become more complicated.

2.5.5 Latency and timing

RAM-type memories further differ from registers in terms of latency, paging, and timing. Firstly, some RAMs have latency while others do not. In a read operation, we speak of latency zero if the content of a memory location becomes available at the RAM's data output in the very clock cycle during which its address has been applied to the RAM's address port. This is also the behavior of a register bank.

As opposed to this, we have a latency of one if the data word does not appear before an active clock edge has been applied. Latency is even longer for memories that operate in a pipelined manner internally. Latency may have a serious impact on architecture design and certainly affects HDL coding.[50]

---

[50] A trick that may help to conceal latency is to operate the RAM's memory address register on the opposite clock edge to the rest of the circuit. Note that this introduces extra timing conditions with respect to the supposedly inactive clock edge that do not exist in a single-edge triggered circuit, however.

Secondly, commodity DRAMs have their row and column addresses multiplexed over the same pins to cut down package size and board-level wiring. Latency then depends on whether a memory location shares the row address with the one accessed before, in which case the two are said to sit on the same page, or not. Paged memories obviously affect architecture design.

Thirdly, address decoding, precharging, the driving of long word and bit lines, and other internal suboperations inflate the access times of both SRAMs and DRAMs. RAMs thus impose a comparatively slow clock that encompasses many gate delays per computation period whereas registers are compatible with much higher clock frequencies.[51]

2.5.6 Digest

Table 2.9 | Key characteristics of register-based and RAM-based data storage compared.

architectural option	o n - c h i p				off-chip
	bistables		embedded		commodity
	flip-flop	latch	SRAM	DRAM	DRAM
fabrication process	compatible with logic				optimized
devices in each cell	20–30T	12–16T	6T|4T&2R	1T&1C	1T&1C
cell area per bit $(F^2)^a$	1700–2800	1100–1800	135–170	18–30	6–8
extra circuit overhead	none		$1.3 \leq$ factor ≤ 2		off-chip
memory refresh cycles	n o n e				y e s
extra package pins	none		none		address & data bus
nature of wiring	multitude of local lines		on-chip busses		package & board
bidirectional busses	none		optional		mandatory
access to data words	all at a time		one at a time		
available configurations	any		restricted		
energy efficiency	good[b]		fair	poor	very poor
latency and paging	none		no fixed rules		yes
impact on clock period	minor		substantial		severe

[a] Area of one bit cell in multiples of F^2, where F^2 stands for the area of one lithographic square.
[b] Depending on the register access scheme, conditional clocking may be an option.

Observation 2.8. *Most on-chip RAMs available for ASIC design*
± are of static nature (SRAMs),
+ have their views obtained automatically using a macrocell generator,
− offer a limited choice of configurations (in terms of $\#_{words}$ and w_{data}),
+ occupy much less area than flip-flops,
− but do so only above some minimum storage capacity,
+ greatly improve speed and energy efficiency over off-chip RAMs,

[51] It is sometimes possible to multiply maximum RAM access rate by resorting to **memory interleaving**, a technique whereby several memories operate in a circular fashion in such a way as to emulate a single storage array of shorter cycle time.

— *but cannot compete in terms of capacity,*
— *restrict data access to one read or write per cycle,*
— *impose rigid constraints on timing, minimum clock period, and latency.*

Examples

Cu-11 is the name of an ASIC technology by IBM that has a drawn gate length — and hence also a minimum feature size — of 110 nm and that operates at 1.2 V. The process combines copper interconnect with low-k interlevel dielectric materials. As part of the Cu-11 design library, IBM offers an SRAM macrocell generator for memories ranging from 128 bit to 1 Mibit as well as embedded DRAM megacells of trench capacitor type with up to 16 Mibit. A 1 Mibit eDRAM has a cycle time of 15 ns which is equivalent to 555 times the nominal delay of a 2-input NAND gate. eDRAM bit cell area is 0.31 μm^2, which corresponds to $25.6F^2$. A 1 Mibit eDRAM occupies an area of 2.09 mm^2 (with an overhead factor 1.84) and its 16 Mibit counterpart 14.1 mm^2 (overhead factor 1.63).

Actel's ProASIC^{PLUS} flash-based FPGA family makes embedded SRAMs available in chunks of 256×9 bit. The APA1000 part includes 88 such blocks, which corresponds to 198 kibit of embedded RAM if fully used.

□

Flash memories have not been addressed here as they do not qualify for short-time random-access storage. This is primarily because data must be erased in larger chunks before it becomes possible to rewrite individual words. The comparatively low speed and limited endurance are other limitations that make flash more suitable for longer-term storage applications such as retaining FPL configurations as explained in section 1.4.1.

Just for comparison, the physical bit cell area of flash technology is a mere 4 to $12F^2$ and, hence, comparable to DRAM rather than SRAM. What's more, by using four voltage levels intead of two, two bits can be stored per flash cell, bringing down the minimum area to just $2F^2$ per bit. Endurance is on the order of 100 000 write&erase cycles for flash cells that hold one bit (two states) and 10 000 cycles for those that hold two bits (four states). Still higher numbers are made possible by wear-leveling schemes implemented in special memory controllers.

As further details of the various memory technologies are of little importance here, the reader is referred to the literature [46] [47] [48] [49]. An excellent introduction to flash memory technology is given in [50] while [51] elaborates on improvements towards high-density storage and high-speed programming. According to the top-down approach adopted in this text, transistor-level circuits for bistables and RAMs will be discussed later, in sections 8.2 and 8.3 respectively.

2.6 | Equivalence transforms for nonrecursive computations

Unlike combinational computations, the outcome of sequential computations depends not only on present but also on past values of its arguments. Architectures for sequential computations must therefore include memory. In the DDG this gets reflected by the presence of edges with weights greater than zero. However, as nonrecursiveness implies the absence of feedback, the DDG remains free of circular paths. The storage capacity required by the isomorphic architecture is referred to as **memory bound** because no other configuration exists that could do with less.[52] Table 2.9 allows approximate translation from memory bits to chip area.

2.6.1 Retiming

The presence of registers in a circuit suggests a new type of reorganization known as retiming or as register balancing, whereby registers get relocated so as to allow for a higher computation rate without affecting functionality [52] [53]. The goal is to equalize computational delays between any two registers, thereby shortening the longest path that bounds the computation period from below. Referring to a DDG one must therefore know.

"In what way is it possible to modify edge weights without altering the original functionality?"

Fig. 2.24 Retiming. DDG (a) and a hardware configuration for $l = 1$ (b).

Let us follow an intuitive approach to find an answer.[53] Consider a DDG and pick a vertex, say h in fig.2.24, for instance. Now suppose the operation of vertex h is made to lag behind those of all others by adding latency to every edge pointing towards that vertex, and by removing the same amount of latency from every edge pointing away from that vertex. Conversely, any vertex could be made to lead the others by transferring latency from its incoming to its outgoing edges. Since any modifications made to the incoming edges are compensated for at the outgoing edges, nothing changes when the DDG is viewed from outside. The retimed DDG is, therefore, functionally equivalent to the initial one. As it is always possible (a) to think of an entire subgraph as one supervertex, and (b) to repeat the operation with changing vertices and supervertices, the idea paves the way for significant hardware reorganizations.

Not all DDGs obtained in this way are legal, though, because the general requirements for DDGs stated in subsection 2.3.4 impose a number of restrictions on edge weights or — which is the same — on latency registers. Any legal retiming must observe the rules below.

1. Neither data sinks (i.e. outputs) nor sources of time-varying data (i.e. variable inputs) may be part of a supervertex that is to be retimed. Sources of constant data (i.e. fixed inputs) do not change in any way when subjected to retiming.
2. When a vertex or a supervertex is assigned a lag (lead) by l computation cycles, the weights of all its incoming edges are incremented (decremented) by l and the weights of all its outgoing edges are decremented (incremented) by l.

---

[52] For combinational computations, the memory bound is obviously zero.

[53] A more formal treatise is given in [54].

3. No edge weight may be changed to assume a negative value.
4. Any circular path must always include at least one edge of strictly positive weight.[54]

Interestingly, rule 4 does not need to be taken into consideration explicitly — provided it was satisfied by the initial DDG — because the weight along a circular path will never change when rule 2 is observed. The proof is trivial.

As a direct consequence of rule 1, retiming does not affect latency. Retiming necessitates neither control logic nor extra data storage facilities but may alter the total number of registers in a circuit, depending on the structure of the DDG being subjected to the transformation.[55]

Energywise, it is difficult to anticipate the overall impact of retiming as switching activities, fanouts, and node capacitances all get altered. Yet, much as for pipelining, the reduced long path delay either allows for compensating a lower supply voltage or can be invested in using a slower but more energy-efficient circuit style or technology. The fact that retiming does not normally burden a circuit with much overhead renders it particularly attractive.

2.6.2 Pipelining revisited

Recall from section 2.4.3 that pipelining introduces extra registers into a circuit and necessarily increases its latency, which contrasts with what we have found for retiming. Pipelining can in fact be interpreted as a transformation of edge weights that differs from retiming in that rule 1 is turned into its opposite, namely

1. Any supervertex to be assigned a lag (lead) must include all data sinks (time-varying data sources).

What retiming and pipelining have in common is that they allow a circuit to operate at a higher computation rate. Most high-speed architectures therefore combine the two.

Example

Consider the problem of computing

$$y(k) = \sum_{n=0}^{N} h(c_n, x(k - n)) \tag{2.45}$$

i.e. a time-invariant Nth-order correlation where $h(c, x)$ stands for some unspecified — possibly nonlinear — function. Think of $h(c, x)$ as some distance metric between two DNA fragments c and x, for instance, in which case $y(k)$ stands for the overall dissimilarity between the DNA strings $(c_0, c_1, ..., c_N)$ and $(x(k), x(k - 1), ..., x(k - N))$ of length N.

---

[54] Although irrelevant here due to the absence of circular paths, this stipulation does apply in the context of recursive computations.

[55] For many applications it is important that a sequential circuit assumes its operation from a well-defined start state. (As a rule, initial state does matter for controllers but is often irrelevant for datapath circuitry.) If so, a mechanism must be provided for forcing the circuit into that state. Finding the start state for a retimed circuit is not always obvious. The problem of systematically computing the start state for retimed circuits is addressed in [55].

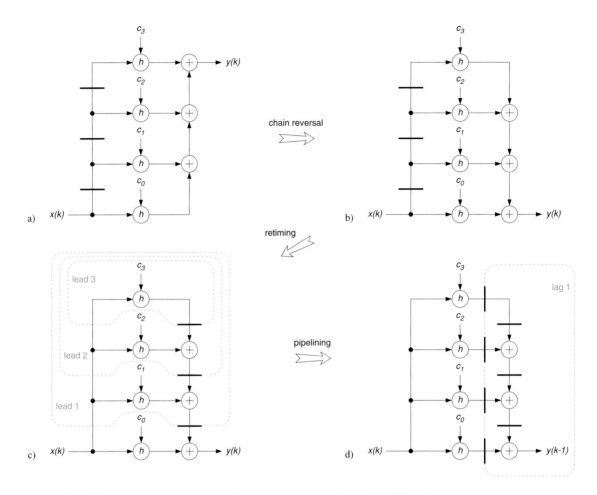

Fig. 2.25 Nonlinear time-invariant third-order correlator. Original DDG (a), with adder chain reversed by associativity transform (b), after retiming (c), with pipelining added on top so as to obtain a systolic architecture (d).

The DDG of a third-order correlator is shown in fig.2.25 together with its stepwise reorganization. For the sake of concreteness, let us assume that a register delay is $t_{reg} = 0.5$ ns, that computing one distance metric takes $t_h = 3$ ns, and that adding up two items costs $t_+ = 2$ ns.

(a) Original DDG as obtained from straight interpretation of (2.45). The delay along the longest path is stated in the table below, note that it grows with correlation order N. There is no retiming that would relocate the existing registers in a useful way. Although the configuration is amenable to pipelining, reformulating it first will eventually pay off.

(b) Same as (a) with the adder chain reversed by an associativity transform. Maximum delay and register count remain unchanged, but the computation has now become suitable for retiming. Also refer to problem 3.

(c) Functional registers transferred into the adder chain by retiming of (b). The three vertices and supervertices enclosed by dashed lines have been assigned leads of 1, 2, and 3 computation cycles respectively. Long path delay is substantially reduced with no registers added. Also notice that the maximum operating speed is no longer a function of correlation order N.

(d) Retiming complemented by pipelining. The supervertex enclosed by a dashed line, which includes the data sink, has been assigned a lag of 1. The longest path is further shortened at the price of extra registers and of an increased latency. Observe that it is not possible to improve performance any further unless one is willing to intervene into the adders on a lower level of circuit detail, also see section 2.8.1.

Key characteristics	Architectural variant			
	(a)	(b)	(c)	(d)
arithmetic units	$(N+1)A_h + NA_+$	*idem*	*idem*	*idem*
functional registers	NA_{reg}	*idem*	*idem*	*idem*
nonfunctional registers	0	*idem*	*idem*	$(N+1)A_{reg}$
cycles per data item Γ	1	*idem*	*idem*	*idem*
longest path delay t_{lp}	$t_{reg} + t_h + N t_+$	*idem*	$t_{reg} + t_h + t_+$	$t_{reg} + \max(t_h, t_+)$
for $N = 3$ (ns)	9.5	*idem*	5.5	3.5
for $N = 30$ (ns)	63.5	*idem*	5.5	3.5
latency L	0	*idem*	*idem*	1

☐

Make sure you understand there is a fundamental difference between the architectural transforms used in the above example. While retiming and pipelining are universal transforms that do not depend on the operations involved, changing the order of execution in the algebraic transform that leads from (a) to (b) insists on the operations concerned being identical and associative. The practical significance is that the sequence of reorganizations that has served to optimize the nonlinear correlator example also applies to transversal filters which are of linear nature, for instance, but not to DDGs of similar structure where addition is replaced by some non-associative operation.

2.6.3 Systolic conversion

Both pipelining and retiming aim at increasing computation rate by resorting and by equalizing register-to-register delays. For a given granularity, maximum speed is obtained when there is no more than one combinational operation between any two registers. This is the basic idea behind systolic computation.

A DDG is termed **systolic** if the edge weight between any two vertices is one or more. The architecture depicted in fig.2.25d is in fact systolic, and the ultimate pipeline in the sense of (2.29) is now also recognized as a circuit that is systolic at the gate level. Please refer to [56] for a more comprehensive discussion of systolic computation and to [57] for details on systolic conversion, that is on how to turn an arbitrary nonrecursive computation into a systolic one.

2.6.4 Iterative decomposition and time-sharing revisited

Applying the ideas of decomposition and time-sharing to sequential computations is straightforward. Clearly, only combinational circuitry can be multiplexed whereas functional memory requirements remain the same as in the isomorphic architecture.

Example

In the isomorphic architecture of a transversal filter, see fig.2.11e, each filter tap is being processed by its own multiplier. All calculations associated with one sample are thus carried out in parallel and completed within a single computation cycle. Nevertheless, the architecture is slow because the longest path traverses the entire adder chain, thereby mandating a long computation period. Also, hardware costs are immoderate, at least for higher filter orders N.

A more economic alternative that handles one filter tap after the other follows naturally, see fig.2.26. This architecture manages with a single multiplier that gets time-shared between taps. A single adder iteratively sums up the partial products until all taps that belong to one sample have been processed after $N + 1$ computation cycles. An accumulator stores the intermediate sums. Coefficients may be kept in a hardwired look-up table (LUT), in a ROM, or in some sort of writable memory. Datapath control is fairly simple. An index register that counts $n = N, N - 1, ..., 0$ addresses one coefficient at any time. The very same register also selects the samples from the input shift register, either by way of a multiplexer or a three-state bus, or by arranging the shift register in circular fashion and by having data there make a full cycle between any two subsequent samples. An output register maintains the end result while computation proceeds with the next sample.

For filters of higher order, one would want to substitute a RAM for the input shift register. While this requires some extra circuitry for addressing, it does not fundamentally change the overall architecture.

☐

2.6.5 Replication revisited

The concept of replication introduced in section 2.4.4 cannot normally be applied to sequential computations as the processing of one data item is dependent on previous data items. A notable exception exists in the form of so-called **multipath filters** (aka N-path filters) designed to implement sequential computations of linear time-invariant nature. With $H_1(z)$ denoting the transfer function of a single path, all of which are identical, and with q as replication factor, a composite transfer function

$$H(z) = H_1(z^q) \tag{2.46}$$

is obtained [58], which implies that the elemental transfer function $H_1(z)$ is compressed and repeated along the frequency axis by a scaling factor of q. Due to the resulting extra passbands the usefulness of this approach is very limited. An extended structure capable of implementing general FIR and IIR[56] transfer functions is proposed in [58] under the name of delayed multipath structures. Regrettably, the number of necessary operations is found to grow with q^2.

---

[56] FIR stands for finite impulse response, IIR for infinite impulse response.

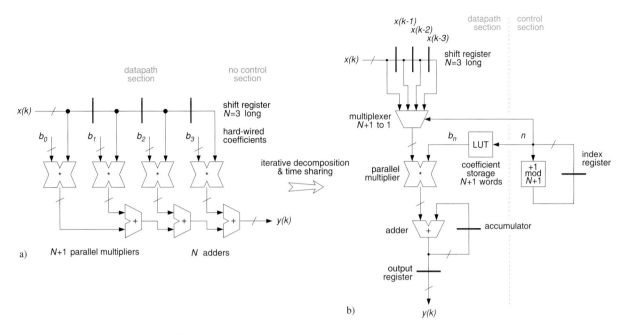

Fig. 2.26 Third-order transversal filter. Isomorphic architecture (a) and a more economic alternative obtained by combining time-sharing with iterative decomposition (b) (simplified).

2.6.6 Digest

- The throughput of arbitrary feedforward computations can be improved by way of retiming, by way of pipelining, and by combining the two. Replication is, in general, not a viable alternative.

- The associative law is often helpful in rearranging a DDG prior to iterative decomposition, pipelining, and especially retiming in order to take full advantage of these transforms.

- Much as for combinational computations, iterative decomposition and time-sharing are the two options available for reducing circuit size for feedforward computations. Highly time-multiplexed architectures are found to dissipate energy on a multitude of ancillary activities that do not directly contribute to data computation, however.

2.7 | Equivalence transforms for recursive computations

A computation is said to be timewise recursive — or recursive for short — if it somehow depends on an earlier outcome from that very computation itself, a circumstance that gets reflected in the DDG by the presence of a feedback loop. Yet, recall that circular paths of weight zero have been disallowed to exclude the risk of race conditions. Put differently, circular paths do exist but each of them includes at least one latency register.

This section examines equivalence transforms that apply specifically to recursive computations. We will find that such transforms are not universal. This is why we address linear time-invariant, linear time-variant, and nonlinear computations separately.

2.7.1 The feedback bottleneck

Consider a linear time-invariant first-order recursive function

$$y(k) = ay(k-1) + x(k) \tag{2.47}$$

which, in the z domain, corresponds to transfer function

$$H(z) = \frac{Y(z)}{X(z)} = \frac{1}{1 - az^{-1}} \tag{2.48}$$

The corresponding DDG is shown in fig.2.27a. Many examples for this and similar types of computations are found in IIR filters, DPCM[57] data encoders, servo loops, etc.

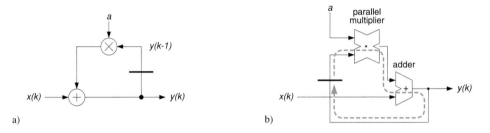

Fig. 2.27 Linear time-invariant first-order feedback loop. DDG (a) and isomorphic architecture with longest path highlighted (b).

Recursion demands that the result from the actual computation cycle be available no later than the next input sample. The longest path defined by all computations that are part of the loop must, therefore, not exceed one sampling interval. In the occurrence

$$\sum_{loop} t = t_{reg} + t_* + t_+ = t_{lp} \leq T_{cp} \tag{2.49}$$

As long as this **iteration bound** is satisfied by the isomorphic architecture of fig.2.27b implemented using some available and affordable technology, there is no out-of-the-ordinary design problem. As an example, we could easily provide a sustained computation rate of 100 MHz if the three delay figures for the actual word width were of 0.5 ns, 5 ns, and 2 ns respectively.

When in search of some higher throughput, say 200 MHz, recursiveness becomes a real bottleneck since there is no obvious way to make use of replication or to insert pipeline registers without altering the overall transfer function and input-to-output mapping. Retiming does not help either as the weight along a circular path is always preserved. So the problem is

"How to allow more time for those computations that are part of the recursion loop."

---

[57] DPCM is an acronym for differential pulse code modulation.

2.7.2 Unfolding of first-order loops

The key idea is to relax the timing constraint by inserting additional latency registers into the feedback loop while preserving the original transfer function. In other words, a tentative solution for a first-order loop must look like

$$H(z) = \frac{Y(z)}{X(z)} = \frac{N(z)}{1 - a^p z^{-p}} \tag{2.50}$$

where an unknown expression $N(z)$ is here to compensate for the changes that are due to the new denominator $1 - a^p z^{-p}$. Recalling the sum of geometric series we easily establish $N(z)$ as

$$N(z) = \frac{1 - a^p z^{-p}}{1 - az^{-1}} = \sum_{n=0}^{p-1} a^n z^{-n} \tag{2.51}$$

The new transfer function can then be completed to become

$$H(z) = \frac{\sum_{n=0}^{p-1} a^n z^{-n}}{1 - a^p z^{-p}} \tag{2.52}$$

and the new recursion in the time domain follows as

$$y(k) = a^p y(k - p) + \sum_{n=0}^{p-1} a^n x(k - n) \tag{2.53}$$

The modified equations correspond to a cascade of two sections. A first section, represented by the numerator of (2.52), is a DDG that includes feedforward branches only. This section is amenable to pipelining as discussed in section 2.4.3. The denominator stands for the second section, a simple feedback loop which has been widened to include p unit delays rather than one as in (2.47).

Using retiming, the corresponding latency registers can be redistributed into the combinational logic for computing the loop operations so as to serve as pipeline registers there. Neglecting, for a moment, the limitations to pipelining found in section 2.4.3, throughput can in fact be multiplied by an arbitrary positive integer p through this unfolding technique, several variations of which are dicussed in [59].

Unless p is prime, it is further possible to simplify the DDG — and hence the implementing circuit — by factoring the numerator into a product of simpler terms. Particularly elegant and efficient solutions exist when p is an integer power of two because of the lemma

$$\sum_{n=0}^{p-1} a^n z^{-n} = \prod_{m=0}^{\log_2 p - 1} (a^{2^m} z^{-2^m} + 1) \qquad p = 2, 4, 8, 16, ... \tag{2.54}$$

The feedforward section can then be realized by cascading $\log_2 p$ subsections each of which consists of just one multiplication and one addition.

Example

The linear time-invariant first-order recursive function (2.47) takes on the following form after unfolding by a factor of $p = 4$:

$$y(k) = a^4 y(k - 4) + a^3 x(k - 3) + a^2 x(k - 2) + ax(k - 1) + x(k) \tag{2.55}$$

which corresponds to transfer function

$$H(z) = \frac{1 + az^{-1} + a^2 z^{-2} + a^3 z^{-3}}{1 - a^4 z^{-4}} \tag{2.56}$$

Making use of (2.54), the numerator can be factorized to obtain

$$H(z) = \frac{(1 + az^{-1})(1 + a^2 z^{-2})}{1 - a^4 z^{-4}} \tag{2.57}$$

The DDG corresponding to this equation is shown in figure 2.28a. Note that the configuration is not only simple but also highly regular. Further improvements are obtained from pipelining in conjunction with retiming and shimming where necessary. The final architecture, shown in fig.2.28b, is equivalent except for latency. Incidentally, also note that threefold instantiation of one pipelined multiply–add building block favors further optimizations at lower levels of detail.

□

PERFORMANCE AND COST ANALYSIS

In the case of optimally efficient configurations, where p is an integer power of two, a lower bound for total circuit size can be given as follows:

$$A(p) \geq (\log_2 p + 1)A_f + p(\log_2 p + 1)A_{reg} \tag{2.58}$$

In the above example, we count three times the orginal arithmetic logic plus 14 extra (nonfunctional) registers. In return for an almost fourfold throughput, this is finally not too bad.

Analogously to what was found for pipelining in section 2.4.3, the speedup of loop unfolding tends to diminish while the difficulties of balancing delays within the combinational logic tend to grow when unfolding is pushed too far, $p \gg 1$.

A hidden cost factor associated with loop unfolding is due to finite precision arithmetics. For the sake of economy, datapaths are designed to make do with minimum acceptable word widths, which implies that output and intermediate results get rounded or truncated somewhere in the process. In the above example, for instance, addition would typically handle only part of the bits that emanate from multiplication. Now, the larger number of roundoff operations that participate in the unfolded loop with respect to the initial configuration leads to more quantization errors, a handicap which must be offset by using somewhat wider data words [60].

Loop unfolding greatly inflates the amount of energy dissipated in the processing of one data item because of the extra feedforward computations and the many latency registers added to the unfolded circuitry. More on the positive side, the shortened longest path may bring many recursive computations into the reach of a relatively slow but power-efficient technology or may allow a lower supply voltage.

The idea of loop unfolding demonstrated on a linear time-invariant first-order recursion can be extended in various directions, and this is the subject of the forthcoming three subsections.

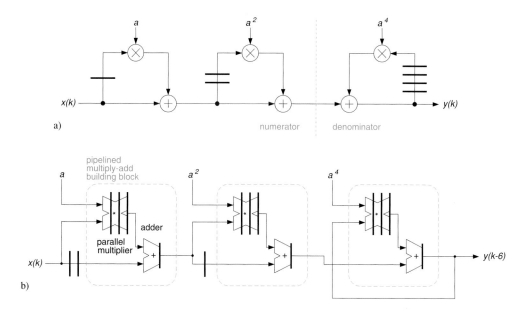

Fig. 2.28 Linear time-invariant first-order feedback loop. DDG after unfolding by a factor of $p = 4$ (a) and high-performance architecture with pipelining and retiming on top (b).

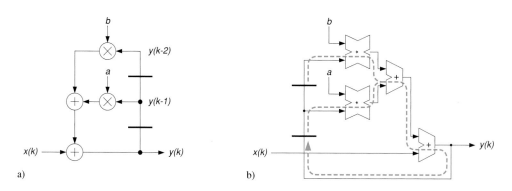

Fig. 2.29 Linear time-invariant second-order feedback loop. DDG (a) and isomorphic architecture with longest paths highlighted (b).

2.7.3 Higher-order loops

Instead of unfolding loops of arbitrary order directly, we make use of a common technique from digital filter design that consists in factoring a higher-order transfer function into a product of second- and first-order terms. The resulting DDG takes the form of cascaded second- and first-order sections. High-speed IIR filters of arbitrary order can be constructed by pipelining the cascade so obtained. As an added benefit, cascade structures are known to be less sensitive to quantization of coefficients and signals than direct forms. We will, therefore, limit the discussion to second-order recursive functions here,

$$y(k) = ay(k-1) + by(k-2) + x(k) \tag{2.59}$$

which correspond to the DDG depicted in fig.2.29. The equivalent in the z domain is

$$H(z) = \frac{Y(z)}{X(z)} = \frac{1}{1 - az^{-1} - bz^{-2}} \tag{2.60}$$

After multiplying numerator and denominator by a factor of $(1 + az^{-1} - bz^{-2})$ the transfer function becomes

$$H(z) = \frac{1 + az^{-1} - bz^{-2}}{1 - (a^2 + 2b)z^{-2} + b^2 z^{-4}} \tag{2.61}$$

which matches the requirements for loop unfolding by a factor of $p = 2$.
Analogously we obtain for $p = 4$

$$H(z) = \frac{(1 + az^{-1} - bz^{-2})(1 + (a^2 + 2b)z^{-2} + b^2 z^{-4})}{1 - ((a^2 + 2b)^2 - 2b^2)z^{-4} + b^4 z^{-8}} \tag{2.62}$$

Example

Figure 2.30 shows a DDG and a block diagram that implement the second-order recursion (2.62). Except for finite precision effects, the transfer function remains exactly the same as in (2.59), but the arithmetic operations inside the loop can now be carried out in four rather than one computation periods. The same pipelined hardware block is instantiated three times.

A high-speed fourth-order ARMA[58] filter chip that includes two sections similar to fig.2.30b has been reported in [61]. Pipelined multiply–add units have been designed as combinations of consecutive carry–save and carry–ripple adders. The circuit, fabricated in a standard 0.9 μm CMOS technology, has been measured to run at a clock frequency of 85 MHz and spits out one sample per clock cycle, so we have $\Gamma = 1$. Overall computation rate roughly is 1.5 GOPS[59], a performance that challenges more costly semiconductor technologies such as GaAs — or at least did so when it appeared in 1992. The authors write that one to two extra data bits had to be added in the unfolded datapath in order to maintain similar roundoff and quantization characteristics to those in the initial configuration. Circuit size is approximately 20 kGE, supply voltage 5 V, and power dissipation 2.2 W at full speed.
□

PERFORMANCE AND COST ANALYSIS

In comparison with the first-order case, the number of pipeline registers per subsection is doubled while the other figures remain unchanged. Hence, size estimation yields

$$A(p) \geq (\log_2 p + 1)A_f + (2p(\log_2 p + 1))A_{reg} \tag{2.63}$$

---

[58] ARMA is an acronym for "auto recursive moving average" used to characterize IIR filters that comprise both recursive (AR) and nonrecursive computations (MA).

[59] Multiply–add operations, in this case taking into account all of the filter's AR and MA sections.

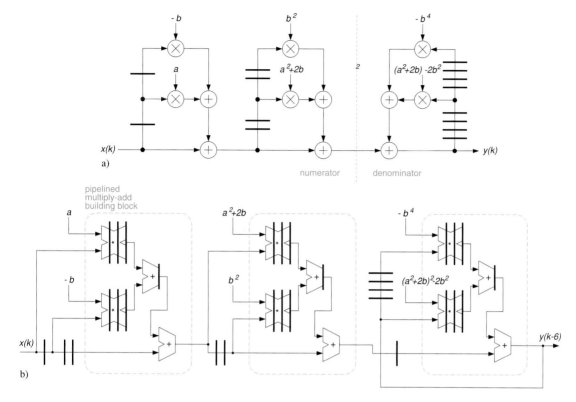

Fig. 2.30 Linear time-invariant second-order feedback loop. DDG after unfolding by a factor of $p = 4$ (a) and high-performance architecture with pipelining and retiming on top (b).

2.7.4 Time-variant loops

Here, the feedback coefficient a is no longer constant but varies as a function of time $a(k)$

$$y(k) = a(k)y(k-1) + x(k) \tag{2.64}$$

The unfolded recursions are derived in the time domain. Substituting $y(k-1)$ into (2.64) yields

$$y(k) = a(k)a(k-1)y(k-2) + a(k)x(k-1) + x(k) \tag{2.65}$$

which computes $y(k)$ from $y(k-2)$ directly, so the unfolding factor is $p = 2$. Repeating this operation leads to a configuration with $p = 3$ where

$$y(k) = a(k)a(k-1)a(k-2)y(k-3) + a(k)a(k-1)x(k-2) + a(k)x(k-1) + x(k) \tag{2.66}$$

and once more to $p = 4$

$$\begin{aligned}
y(k) = {}& a(k)a(k-1)a(k-2)a(k-3)y(k-4) \\
& + a(k)a(k-1)a(k-2)x(k-3) + a(k)a(k-1)x(k-2) + a(k)x(k-1) + x(k) \tag{2.67}
\end{aligned}$$

As for the time-invariant case, the process of unfolding can be continued to widen the recursive loop by an arbitrary positive integer p as expressed by

$$y(k) = \left(\prod_{n=0}^{p-1} a(k-n)\right) \cdot y(k-p) + \left(\sum_{n=1}^{p-1}\left(\prod_{m=0}^{n-1} a(k-m)\right) \cdot x(k-n)\right) + x(k) \qquad (2.68)$$

However, because precomputation is not applicable here, all necessary coefficient terms must be calculated on-line, which requires extra hardware. Depending on how the terms of (2.68) are combined, various DDGs can be obtained. One of them derived from (2.67) is depicted in fig.2.31.

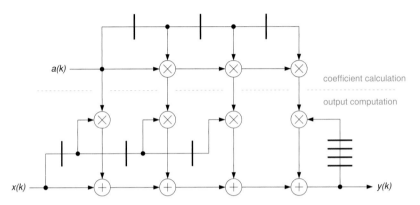

Fig. 2.31 Linear time-variant first-order feedback loop. DDG after unfolding by a factor of $p = 4$.

PERFORMANCE AND COST ANALYSIS

The count of adders and multipliers is proportional to the number of subsections p. Each subsection requires approximately $2p$ pipeline registers as both multipliers must be pipelined. Together with shimming registers, many of which are needed in this configuration due to the numerous parallel branches, roughly $4p^2$ registers are needed.

2.7.5 Nonlinear or general loops

A nonlinear difference equation implies that the principle of superposition does not hold. The most general case of a first-order recursion is described by

$$y(k) = f(y(k-1), x(k)) \qquad (2.69)$$

and can be unfolded an arbitrary number of times. For sake of simplicity we will limit our discussion to a single unfolding step, i.e. to $p = 2$ where

$$y(k) = f(f(y(k-2), x(k-1)), x(k)) \qquad (2.70)$$

The associated DDG of fig.2.32c shows that loop unfolding per se does not relax the original timing constraint, the only difference is that one can afford two cycles for two operations f instead of one

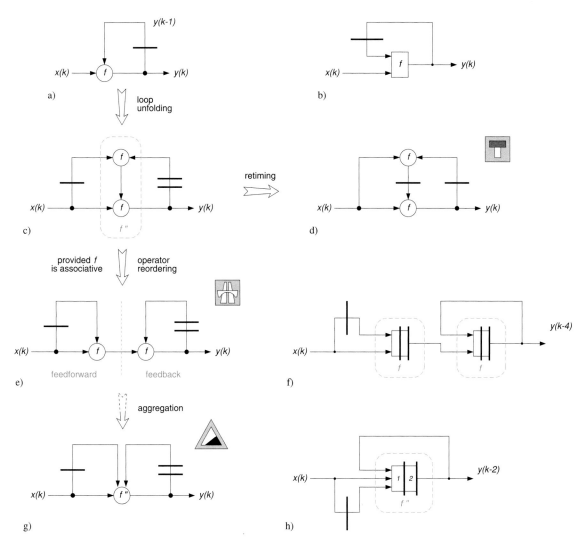

Fig. 2.32 Architectural alternatives for nonlinear time-variant first-order feedback loops. Original DDG (a) and isomorphic architecture (b), DDG after unfolding by a factor of $p = 2$ (c), same DDG with retiming added on top (d). DDG reorganized for an associative function f (e), pertaining architecture after pipelining and retiming (f), DDG with the two functional blocks for f combined into f'' (g), pertaining architecture after pipelining and retiming (h).

cycle for one operation. As confirmed by fig.2.32d, there is no room for any meaningful retiming in this case.

Yet, the unfolded recursion can serve as a starting point for more useful reorganizations. Assume function f is known to be associative. Following an associativity transform the DDG is redrawn as shown in fig.2.32e. The computation so becomes amenable to pipelining and retiming, see fig.2.32f, which cuts the longest path in half when compared with the original architecture of fig.2.32b. Even more speedup can be obtained from higher degrees of unfolding, the price to pay is multiplied circuit size and extra latency, though. In summary, architecture, performance, and cost figures resemble those found for linear computations.

The situation is definitely more difficult when f is not associative. Still, it is occasionally possible to relax the loop constraint to some extent by playing a trick. Reconsider fig.2.32c and think of the two occurrences of f being combined into an **aggregate computation**

$$y(k) = f"(y(k-2), x(k-1), x(k)) \tag{2.71}$$

as sketched in fig.2.32g. If that aggregate computation can be made to require less than twice as much time as the original computation, then the bottleneck gets somewhat alleviated. This is because it should then be possible to insert a pipeline register into the datapath unit for $f"$ so that the maximum path length in either of the two stages becomes shorter than the longest delay in a datapath that computes f alone.

$$t_{lp\,f"} = \max(t_{lp\,f"_1}, t_{lp\,f"_2}) < t_{lp\,f} \tag{2.72}$$

More methods for speeding up general time-variant first-order feedback loops are examined in [62]. One technique, referred to as **expansion** or **look-ahead**, is closely related to aggregate computation. The idea is to process two or more samples in each recursive step so that an integer multiple of the sampling interval becomes available for carrying out the necessary computations. In other terms, the recursive computation is carried out at a lower pace but on wider data words. This approach should be considered when the combinational logic is not amenable to pipelining, for example because it is implemented as table look-up in a ROM. The limiting factor is that the size of the look-up table (LUT) tends to increase dramatically.

Yet another approach, termed **concurrent block technique**, groups the incoming data stream into blocks of several samples and makes the processing of these blocks independent from each other. While data processing within the blocks remains sequential, it so becomes possible to process the different blocks concurrently.

The **unified algebraic transformation** approach promoted in [63] combines both universal and algebraic transforms to make the longest path independent of problem size in computations such as recursive filtering, recursive least squares algorithms, and singular value decomposition.

Any of the various architectural transforms that permit one to successfully introduce a higher degree of parallel processing into recursive computations takes advantage of algorithmic properties such as linearity, associativity, fixed coefficients, and limited word width, or of a small set of register states. If none of these applies, we can't help but agree with the authors of [62].

Observation 2.9. *When the state size is large <u>and</u> the recurrence is not a closed-form function of specific classes, our methods for generating a high degree of concurrency cannot be applied.*

Example

Cryptology provides us with a vivid example for the implications of nonlinear nonanalytical feedback loops. Consider a block cipher that works in **electronic code book** (ECB) mode as depicted in fig.2.34a. The algorithm implements a combinational function $y(k) = c(x(k), u(k))$, where $u(k)$ denotes the key and k the block number or time index. However complex function c, there is no fundamental obstacle to pipelining or to replication in the datapath.

Unfortunately, ECB is cryptologically weak as two identical blocks of plaintext result in two identical blocks of ciphertext because $y(k) = y(m)$ if $x(k) = x(m)$ and $u(k) = u(m)$. If a plaintext to be encrypted contains sufficient repetition, the ciphertext necessarily carries and betrays patterns from the original plaintext. Figure 2.33 nicely illustrates this phenomenon.

Fig. 2.33 A computer graphics image in clear text, ciphered in ECB mode, and ciphered in CBC-1 mode (from left to right, Tux by Larry Ewing).

To prevent this from happening, block ciphers are typically used with feedback. In **cipher block chaining** (CBC) mode, the ciphertext gets added to the plaintext before encryption takes place, see fig.2.34b. The improved cipher algorithm thus becomes $y(k) = c(x(k) \oplus y(k-1), u(k))$ and is sometimes referred to as CBC-1 mode because $y(k-1)$ is being used for feedback.

From an architectural point of view, however, this first-order recursion is awkward because it offers little room for reorganizing the computation. This is particularly true in ciphering applications where the nonlinear functions involved are chosen to be complex, labyrinthine, and certainly not analytical. The fact that word width (block size) is on the order of 64 or 128 bit makes everything worse. Inserting pipeline registers into the computational unit for c does not help since this would alter algorithm and ciphertext. Throughput in CBC mode is thus limited to a fraction of what is obtained in ECB mode.[60]

2.7.6 Pipeline interleaving is not an equivalence transform

It has repeatedly been noted in this section that any attempt to insert an extra register into a feedback loop with the idea of pipelining the datapath destroys the equivalence between original and pipelined computations unless its effect is somehow compensated. After all, circuits c and b of fig.2.34 behave differently. Although this may appear a futile question, let us ask

"What happens if we do just that to a first-order recursion?"

---

[60] Higher data rates must then be bought by measures on lower levels of abstraction, that is by optimizing the circuitry at the arithmetic/logic level, by resorting to transistor-level design in conjunction with full-custom layout, and/or by adopting a faster target technology, all of which measures ask for extra effort and come at extra cost. Only later have cryptologists come up with a better option known as **counter mode** (CTR) that does without feedback and still avoids the leakage of plaintext into ciphertext that plagues ECB.

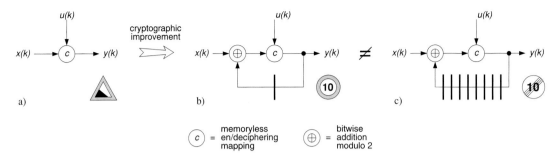

Fig. 2.34 DDGs for three block ciphering modes. Combinational operation in ECB mode (a) vs. time-variant nonlinear feedback loop in CBC mode (b), and CBC-8 operation b (c).

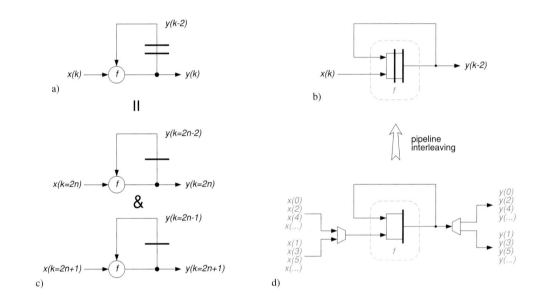

Fig. 2.35 Pipeline interleaving. DDG of nonlinear time-variant first-order feedback loop with one extra register inserted (a) and isomorphic architecture (b). Interpretation as two interleaved data streams each of which gets processed exactly as specified by the original nonlinear first-order recursion of fig.2.32a (c,d).

Adding an extra latency register to (2.69) results in the DDG of fig.2.35a and yields

$$y(k) = f(y(k-2), x(k)) \tag{2.73}$$

Observe that all indices are even in this equation. As k increments with time $k = 0, 1, 2, 3, \ldots$ indices do in fact alternate between even and odd values. It thus becomes possible to restate the ensuing input-to-output mapping as two separate recursions with no interaction between "even" data items

$x(k = 0, 2, 4, ..., 2n, ...)$ and "odd" items $x(k = 1, 3, 5, ..., 2n + 1, ...)$.

$$y(2n) = f(y(2n - 2), x(2n)) \tag{2.74}$$
$$y(2n + 1) = f(y(2n - 1), x(2n + 1)) \tag{2.75}$$

This pair of equations says that the original data processing recursion of (2.69) now gets applied to two separate data streams as depicted in fig.2.35c. From a more general perspective, it is indeed possible to cut the combinational delay in any first-order feedback loop down to $\frac{1}{p}$ by inserting $p - 1$ pipelining registers, yet the computation then falls apart into the processing of p interleaved but otherwise independent data streams. More often than not this is undesirable. However, practical applications exist where it is possible to take advantage of this effect.

Examples

Cipher block chaining (CBC) implements the recursion $y(k) = c(x(k) \oplus y(k - 1), u(k))$. What counts from a cryptographic point of view is that patterns from the plaintext do not show up in the ciphertext. Whether this is obtained from feeding back the immediately preceding block of ciphertext $y(k - 1)$ (CBC-1 mode) or some prior block $y(k - p)$ where $2 \le p \in \mathbb{N}$ (CBC-p mode) is of minor importance. Some cryptochips, therefore, provide a fast but nonstandard CBC-8 mode in addition to the regular CBC-1 mode, see fig.2.34c. In the case of the IDEA chip described in [64], maximum throughout is 176 Mbit/s both in pipelined ECB mode and in pipeline-interleaved CBC-8 mode as compared with just 22 Mbit/s in nonpipelined CBC-1.

For another example, take a subband coding or some other image processing algorithm where rows of pixels are dealt with independently from each other. Rather than scanning the image row by row, pixels from p successive rows are entered one by one in a cyclic manner before the process is repeated with the next column, and so on. It so becomes possible to deal with a single pipelined datapath of p stages [65].

□

Pipeline interleaving obviously does not qualify as an equivalence transform. Still, it yields useful architectures for any recursive computation — including nonlinear ones — provided that data items arrive as separate time-multiplexed streams that are to be processed independently from each other, or can be arranged to do so. From this perspective, pipeline interleaving is easily recognized as a clever and efficient combination of time-sharing with pipelining.

2.7.7 Digest

- When in search of high performance for recursive computations, reformulating a high-order system as a cascade of smaller-order sections in order to make the system amenable to coarse grain pipelining should be considered first. As a by-product, the reduced orders of the individual recursion loops offer additional speedup potential.

- Throughput of low-order recursive computations can be significantly improved by loop unfolding in combination with fine grain pipelining. This may bring computations into the reach of static CMOS technology that would otherwise ask for faster but also more expensive alternatives such as SiGe, BiCMOS, or GaAs.

- Whether the inflated latency is acceptable or not depends on the application. Also, the rapid growth of overall circuit size tends to limit the economically practical number of degrees of unfolding to fairly low values, say $p = 2$–8, especially when the system is a time-varying one.

- The larger number of roundoff operations resulting from loop unfolding must be compensated for by using longer word widths, which increases the cost of loop unfolding beyond solely the proliferation of computational units and intermediate registers.

- Nonlinear feedback loops are, in general, not amenable to throughput multiplication by applying unfolding techniques. A notable exception exists when the loop function is associative.

- Pipeline interleaving is highly efficient for accelerating recursive computations because it does not depend on any specific properties of the operations involved. Yet, as it modifies the input-to-output mapping, it is not an option unless the application admits that multiple data streams undergo the same processing independently from each other.

2.8 | Generalizations of the transform approach

2.8.1 Generalization to other levels of detail

As stated in section 2.3.4, DDGs are not concerned with the granularity of operations and data. Recall, for instance, figs.2.14 and 2.34a that describe the same block cipher at different levels of detail. As a consequence, the techniques of iterative decomposition, pipelining, replication, time-sharing, algebraic transform, retiming, loop unfolding, and pipeline interleaving are not limited to any specific level of abstraction although most examples so far have dealt with operations and data at the word level, see table 2.10.

Table 2.10 | An excerpt from the VLSI abstraction hierarchy.

Level of abstraction	Granularity	Relevant items	
		Operations	Data
Architecture	◯	subtasks, processes	time series, pictures, blocks
Word	○	arithmetic/logic operations	words, samples, pixels
Bit	·	gate-level operations	individual bits

ARCHITECTURE LEVEL

Things are pretty obvious at this higher level where granularity is coarse. As an example, fig.2.36 gives a schematic overview of a visual pattern recognition algorithm. Four subtasks cooperate in a cascade with no feedback, namely preprocessing, image segmentation, feature extraction, and object classification.

In a real-time application, one would definitely begin by introducing pipelining because four processing units are thereby made to work concurrently at negligible cost. In addition, each unit is thus dedicated to a specific subtask and can be optimized accordingly.

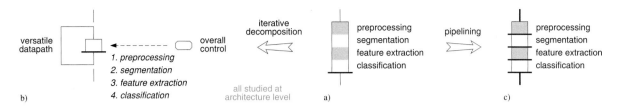

Fig. 2.36 Overall architectural alternatives for a pattern recognition system. Isomorphic architecture (a), iteratively decomposed computation flow (b), and pipelined operation (c).

The option of replicating the entire datapath would most likely get discarded at this point because it cannot compete with pipelining in terms of hardware efficiency. Replication of selected functional units could become an interesting alternative in later transforms, however, when the actual performance bottlenecks are known.

Iterative decomposition would be considered only if the desired throughput were so modest that it could be obtained from a single processing unit.

BIT LEVEL

Equivalence transformations can also be beneficial at low levels of abstraction. Take addition, for instance, which is an atomic operation when considered at the word level, see fig.2.37a. When viewed at the gate level, however, the same function appears as a composite operation that can be implemented from bit-level operations in many ways, the most simple of which is a ripple-carry adder shown in fig.2.37b. This detailed perspective clearly exposes the longest path and opens up new opportunities for reorganizing the DDG that remain hidden from a word or higher level of abstraction.

A gate-level pipelined version makes it possible to operate the circuit at a computation rate many times higher than with word-level pipelining alone, see fig.2.37c. As this amounts to fine-grain pipelining, the price to pay in terms of circuit size is likely to be excessive, however. In the above example, better solutions are obtained from more sophisticated arithmetic schemes such as carry-save, carry-select, or carry-lookahead adders [66] [67] [68] [69], possibly in combination with moderate pipelining. Incidentally, note that modern synthesis tools support automatic retiming of gate-level networks.

Conversely, the structure shown in fig.2.37d follows when the W-bit addition is decomposed into one-bit suboperations. Computation starts with the LSB. A flip-flop withholds the carry-out bit for the next computation period where it serves as carry-in to the next more significant bit. Obviously, the flip-flop must be properly initialized in order to process the LSB and any potential carry input in a correct way. Although this entails some extra control overhead, substantial hardware savings may nevertheless result when the words being processed are sufficiently wide.

2.8.2 Bit-serial architectures

The idea underlying fig.2.37d gives rise to an entirely different family of architectures known as bit-serial computation [70]. While most datapaths work in a bit-parallel fashion in that word-level

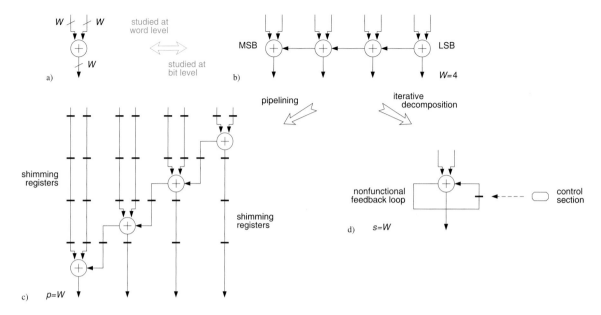

Fig. 2.37 4-bit addition at the register transfer level (a), broken up into a ripple-carry adder (b) before being pipelined (c) or iteratively decomposed (d).

operations are executed one after the other with all bits of a data word being processed simultaneously, organizing computations the other way round is also possible. Here, the overall structure remains isomorphic with the DDG, but the various word-level operations are decomposed into steps that are carried out one bit after the other.

Example

A bit-serial implementation of a third-order transversal filter is shown in fig.2.38, also see fig.2.11c for the DDG. The w-bit wide input samples $x(k)$ enter serially with the LSB first whereas coefficients $b_n(k)$ must be applied in a parallel format. The circuit is operated at a computation rate s times higher than the sampling frequency. Evaluation proceeds from the LSB to the MSB with computation periods numbered $w = s - 1, ..., 0$.

The first computation period ingests the LSB of the actual input sample $x(k)$, evaluates the LSBs of all samples $x(k), ..., x(k-3)$, and sees the LSB of the result $y(k)$ emerge at the output. The second period then handles the next significant bit and so on.[61] Shifts and carry-overs from one bit to the next higher position are obtained by using latency registers. As these registers may contain carries from previous additions after all W bits of the input have been processed, extra computation periods are required to bring out the MSB of $y(k)$, so that $s > W$.

---

[61] DSP applications frequently deal with input data scaled such that $|x_k| < 1$ and coded with a total of W bits in 2's-complement format, see (2.77). In this particular case, computation periods $w = s - 1, ..., 1$ process the input bits of weight 2^{-w} respectively, while the last computation period with $w = 0$ is in charge of the sign bit.

Note that iterative decomposition has led to <u>nonfunctional</u> feedback loops in the architecture of a transversal filter, although the DDG is free of circular paths by definition. As this kind of feedback is confined to within the multiply and add units, the filter as a whole remains amenable to pipelining, provided computations inside the loops are not affected.

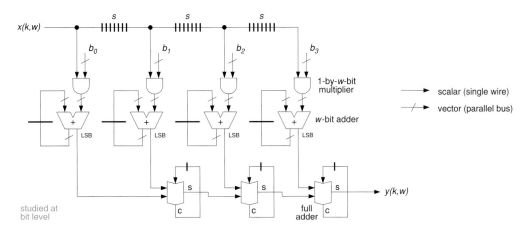

Fig. 2.38 Bit-serial implementation of a third-order transversal filter functionally equivalent to the bit-parallel architecture of fig.2.26 (simplified).

☐

Closer examination of bit-serial architectures reveals the following properties:

+ Control overhead is small when compared with bit-parallel alternatives because the isomorphism of DDG and architecture is maintained.
− As the DDG is hardwired into the datapath with no explicit control instance, changes to the processing algorithm, switching between different modes of operation, and exception handling are awkward to deal with.
+ The shallow combinational depth in conjunction with a high computation rate helps to keep all computational units busy for most of the time.
+ All global data communication is via serial links that operate at close to the maximum rate supported by the target technology, which cuts down on-chip wiring requirements.
+ Much of the data circulation is local, which contrasts favorably with the data items travelling back and forth between datapath and storage in bit-parallel architectures.
+ As FPL devices provide only limited wiring resources, the two previous assets tend to favor bit-serial architectures when designing for FPGAs or CPLDs.
− The word width must be the same throughout a serial architecture. Parts of the computation that do not make use of the full precision typically cause the hardware to idle for a number of computation periods.
+ Conversely, arbitrary precision can be accomodated with the same hardware when execution time is allowed to grow with word length.
− Bit-serial computation is incompatible with the storing of data in word-oriented RAMs and ROMs. Extra format conversion circuitry is required whenever such memories are preferred for their high storage densities.

- Division, data-dependent decisions, and many other functions are ill-suited to bitwise iterative decomposition and pipelining. While it is possible to incorporate bit-parallel functions, provided their interfaces are serialized and that they exhibit a fixed latency, the resulting hybrid structures often prove unsatisfactory and inefficient.
- Some algorithms based on successive approximation naturally operate with the MSB first. It is sometimes possible to reconcile LSB-first and MSB-first functions at the price of resorting to redundant number representation schemes.

In summary, bit-serial architectures are at their best for unvaried real-time computations that involve fixed and elementary operations such as addition and multiplication by a constant. The reader is referred to the specialized literature [70] [71] for case studies and for further information on bit-serial design techniques.

2.8.3 Distributed arithmetic

Bit-serial architectures have been obtained from breaking down costly word-level operations into bit-level manipulations followed by <u>universal</u> transforms such as iterative decomposition and pipelining. Another family of serial architectures results from making use of <u>algebraic</u> transforms at the bit level too. Consider the calculation of the following inner product:

$$y = \sum_{k=0}^{K-1} c_k x_k \tag{2.76}$$

where each c_k is a fixed coefficient and where each x_k stands for an input variable. Figure 2.39a shows the architecture that follows from routinely applying decomposition at the word level. Computation works by way of repeated multiply–accumulate operations, takes K computation periods per inner product, and essentially requires a hardware multiplier plus a look-up table for the coefficients.

Now assume that the inputs are scaled such that $|x_k| < 1$ and coded with a total of W bits in 2's-complement format.[62] We then have

$$x_k = -x_{k,0} + \sum_{w=1}^{W-1} x_{k,w} \, 2^{-w} \tag{2.77}$$

with $x_{k,0}$ denoting the sign bit and with $x_{k,w}$ standing for the bit of weight 2^{-w} in the input word x_k. By combining (2.76) and (2.77) the desired output y can be expressed as

$$y = \sum_{k=0}^{K-1} c_k \left(-x_{k,0} + \sum_{w=1}^{W-1} x_{k,w} \, 2^{-w} \right) \tag{2.78}$$

With the aid of the distributive law and the commutative law of addition, the computation now gets reorganized into the equivalent form below where the order of summation is reversed:

$$y = \sum_{k=0}^{K-1} c_k \left(-x_{k,0} \right) + \sum_{w=1}^{W-1} \left(\sum_{k=0}^{K-1} c_k x_{k,w} \right) 2^{-w} \tag{2.79}$$

---

[62] This is by no means a necessity. We simply assume $|x_k| < 1$ for the sake of convenience and 2's-complement format because it is the most common representation scheme for signed numbers in digital signal processing.

The pivotal observation refers to the term in parentheses

$$\sum_{k=0}^{K-1} c_k x_{k,w} = p(w) \tag{2.80}$$

For any given bit position w, calculating the sum of products takes one bit from each of the K data words x_k, which implies that the result $p(w)$ can take on no more than 2^K distinct values. Now, as coefficients c_k have been assumed to be constants, all those values can be precomputed and kept in a look-up table (LUT) instead of being calculated from scratch whenever a new data set x_k arrives at the input. A ROM is typically used to store the table. It must be programmed in such a way as to return the partial product $p(w)$ when presented with the address w, i.e. with a vector obtained from concatenating all bits of weight 2^{-w} across all input variables x_k. Playing this trick, and noting that $\sum_{k=0}^{K-1} c_k(-x_{k,0})$ is nothing else than $p(0)$ with the sign reversed, (2.79) takes on the utterly simple form

$$y = -p(0) + \sum_{w=1}^{W-1} p(w)\, 2^{-w} \tag{2.81}$$

While the isomorphic architecture calls for W LUTs with identical contents, a much smaller architecture can be obtained from decomposing the evaluation of (2.81) into a series of W consecutive steps. The new architecture manages with a single look-up table but requires a nonfunctional register for accumulating the partial products, see fig.2.39b. Calculation proceeds one bit position at a time, starting with the LSB in computation period $w = W - 1$ and processing the sign bit in the final cycle where $w = 0$.

A minor complication comes from the fact that the term $-p(0)$ has a sign opposite to all other contributions to y. A simple solution consists of using an adder–subtractor working under control of a "sign-bit cycle" signal from a modulo W counter that acts as a controller. The same counter is also in charge of releasing the fully completed result and of clearing the accumulator at the end of the last computation period (two details not shown in fig.2.39b). In addition, it guides the selection of the appropriate input bits unless the x_ks can be made to arrive in a bit-serial format LSB first.

The most striking difference between the two architectural alternatives of fig.2.39 is the absence of any multiplier in the second design. Rather than being concentrated in a single hardware unit, multiplication is spread over the circuit, which is why such architectures were given the name **distributed arithmetic**.

A limitation of distributed arithmetic is that memory size is proportional to 2^K, where K is the order of the inner product to be computed. Although a more impressive coding scheme makes it possible to introduce a symmetry into the look-up table which can then be exploited to halve its size [72], the required storage capacity continues to grow exponentially with K. More impressive memory savings are obtained from reformulating (2.80) in the following way:

$$\sum_{k=0}^{K-1} c_k x_{k,w} = \sum_{k=0}^{H-1} c_k x_{k,w} + \sum_{k=H}^{K-1} c_k x_{k,w} \tag{2.82}$$

where $0 < H < K$. Instead of having all K bits address a single ROM, they are split into two subsets of H and $K - H$ bits respectively, each of which drives its own LUT. The total storage requirement is so reduced from 2^K data words to $2^H + 2^{K-H}$, which amounts to $2^{\frac{K}{2}+1}$ when input bits are

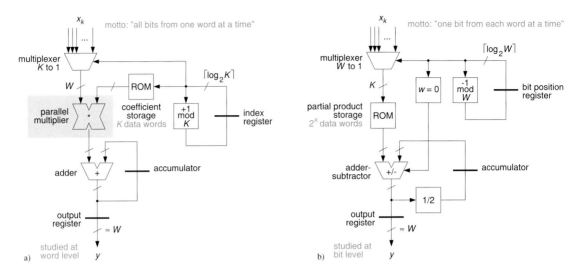

Fig. 2.39 Architectures for computing a sum of products by way of repeated multiply–accumulate operations (a) and with distributed arithmetic (b) (simplified).

split such that $H = \frac{1}{2}K$. The price to pay is an extra adder for combining the outputs from the two tables. Clearly, the idea can be extended to more than two tables.

Memory requirements may sometimes be slashed further by taking advantage of properties of the coefficient values at hand such as symmetries, repetitions, and relations between their binary codes, also see problem 2.10. Memory size and circuit structure thereby become highly dependent on the particular coefficient values, though, which makes it difficult to accomodate modifications.

In conclusion, distributed arithmetic should be considered when coefficients are fixed, when the number of distinct coefficient values is fairly small, and when look-up tables (LUT) are available at little cost compared with bit-parallel multipliers. This explains why this approach has recently regained popularity in the context of DSP applications with LUT-based FPGAs [73] [74]. Please refer to the literature for tutorials [72] and further VLSI circuit examples [75] if distributed arithmetic appears to be an option for your filtering problem.

2.8.4   Generalization to other algebraic structures

So far we have mostly been dealing with the infinite field[63] $(\mathbb{R}, +, \cdot)$ formed by the set of all real numbers together with addition and multiplication. Accordingly, most examples have been taken from digital signal processing, where this type of computation is commonplace. Now, as all algebraic fields share a common set of axioms, any algebraic transform that is valid for some computation in $(\mathbb{R}, +, \cdot)$ must necessarily hold for any other field.[64]

---

[63] See appendix 2.11 for a summary on algebraic structures.
[64] Universal transforms remain valid anyway as they do not depend on algebraic properties.

FINITE FIELDS

Galois fields[65] such as $\mathrm{GF}(2)$, $\mathrm{GF}(p)$, and $\mathrm{GF}(p^n)$ have numerous applications in data compression (source coding), error correction (channel coding), and information security (ciphering). Thus, when designing high-speed telecommunications or computer equipment, it sometimes proves useful to know that the loop unfolding techniques discussed for linear systems in $(\mathbb{R}, +, \cdot)$ directly apply to any linear computation in any Galois or other finite field too.

SEMIRINGS

The analysis of recursive computations in section 2.7 has revealed that almost all successful and efficient loop unfolding techniques are tied to linear systems over a field. That computation be performed in a field is a sufficient but not a necessary condition, however, as will become clear shortly. Recall how loop unfolding was derived for the time-variant linear case in (2.64) through (2.68). Substituting the generic operator symbols $\boxplus$ for $+$ and $\boxdot$ for $\cdot$ we can write

$$y(k) = a(k) \boxdot y(k-1) \boxplus x(k) \tag{2.83}$$

After the first unfolding step, i.e. for $p = 2$, one has

$$y(k) = a(k) \boxdot a(k-1) \boxdot y(k-2) \boxplus a(k) \boxdot x(k-1) \boxplus x(k) \tag{2.84}$$

and for arbitrary integer values of $p \geq 2$

$$y(k) = \left(\prod_{n=0}^{p-1} a(k-n) \right) \boxdot y(k-p) \boxplus \sum_{n=1}^{p-1} \left(\prod_{m=0}^{n-1} a(k-m) \right) \boxdot x(k-n) \boxplus x(k) \tag{2.85}$$

where $\sum$ and $\prod$ refer to operators $\boxplus$ and $\boxdot$ respectively. The algebraic axioms necessary for that derivation were closure under both operators, associativity of both operators, and the distributive law of $\boxdot$ over $\boxplus$. The existence of identity or inverse elements is not required. Also, we have never made use of commutativity of operator $\boxdot$, which implies (a) that the result also holds for other than commutative operators $\boxdot$, in which case (b) the above order of "multiplication" is indeed mandatory. The algebraic structure defined by these axioms is referred to as a semiring.

The practical benefit is that recursive computations of seemingly nonlinear nature when formulated in the field $(\mathbb{R}, +, \cdot)$ — or in some other field — become amenable to loop unfolding, provided it is possible to restate them as linear computations in a ring or semiring [76]. A number of problems related to finding specific paths through graphs are amenable to reformulation in this way. Suitable algebraic systems that satisfy the axioms of a semiring are listed in [77] under the name of path algebras and in appendix 2.11.

Example

Convolutional codes find applications in telecommunication systems for error recovery when data gets transmitted over noisy channels. While a convolutional coder is simple, the computational effort for decoding at the receiver end is much more substantial. The most popular decoding method is the **Viterbi algorithm** [78], a particular case of dynamic programming for finding the shortest path through a trellis graph.[65] Its sole recursive step includes an operation commonly referred to as **add–compare–select** (ACS) and goes[66]

$$y_1(k) = \min(a_{11}(k) + y_1(k-1), a_{12}(k) + y_2(k-1)) \qquad (2.86)$$

$$y_2(k) = \min(a_{21}(k) + y_1(k-1), a_{22}(k) + y_2(k-1)) \qquad (2.87)$$

As all other computations are of feedforward nature, the maximum achievable throughput of the decoding process is indeed limited by this nonlinear recursion in very high-speed applications, see fig.2.40a.

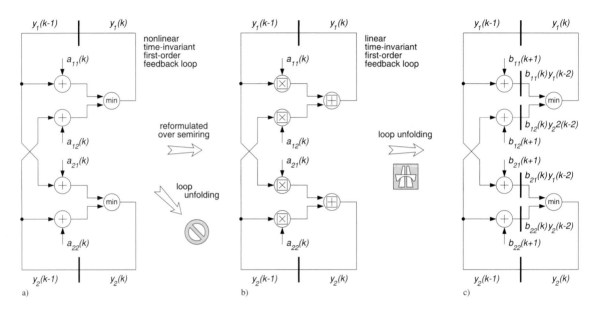

Fig. 2.40 The add–compare–select recursion in the Viterbi algorithm before (a) and after being reformulated over a semiring (b), and with loop unfolding on top (c) (simplified).

---

[65] **Dynamic programming** encompasses a broad class of optimization algorithms that decompose the search for a global optimum into a sequence of simpler decision problems at a local level. All decisions obey Bellman's principle of optimality, which states that the globally optimum solution includes no suboptimal local decisions. This is a very welcome property because it permits one to prune inferior candidates early during the search process. Dynamic programming finds applications in fields as diverse as telecommunications, speech processing, video coding, watermark detection, flight trajectory planning, and genome sequencing.

For an anecdotal introduction, think of the following situation. During the darkness of night, a group of four has to cross a fragile suspension bridge that can carry no more than two persons at a time. The four persons take 5, 10, 20 and 25 minutes respectively for traversing the bridge. A torch must be carried while on the bridge, the torch available will last for exactly one hour. The problem is how to organize the operation. Draw a graph where each state of affair gets represented by a vertex and each traversal of the bridge by an edge. By solving this quiz in a systematic way, you are bound to discover the ideas behind dynamic programming yourself.

[66] A 2-state convolutional code is assumed here. Codes with 32, 64, 128 or 256 states are more useful and hence also more widely used, but their discussion would unnecessarily complicate the argument.

Now consider a semiring where

- Set of elements: $S = \mathbb{R} \cup \{\infty\}$,
- Algebraic addition: $\boxplus = \min$, and
- Algebraic multiplication: $\boxdot = +$.

The new and linear — in the semiring — formulation of the ACS operation goes

$$y_1(k) = a_{11}(k) \boxdot y_1(k-1) \boxplus a_{12}(k) \boxdot y_2(k-1) \tag{2.88}$$
$$y_2(k) = a_{21}(k) \boxdot y_1(k-1) \boxplus a_{22}(k) \boxdot y_2(k-1) \tag{2.89}$$

which, making use of vector and matrix notation, can be rewritten as

$$\vec{y}(k) = A(k) \boxdot \vec{y}(k-1) \tag{2.90}$$

By replacing $\vec{y}(k-1)$ in (2.90) one gets the unfolded recursion for $p = 2$

$$\vec{y}(k) = A(k) \boxdot A(k-1) \boxdot \vec{y}(k-2) \tag{2.91}$$

To take advantage of this unfolded form, the product $B(k) = A(k) \boxdot A(k-1)$ must be computed outside the loop. Resubstituting the original operators and scalar variables, we finally obtain the recursion

$$y_1(k) = \min(b_{11}(k) + y_1(k-2), b_{12}(k) + y_2(k-2)) \tag{2.92}$$
$$y_2(k) = \min(b_{21}(k) + y_1(k-2), b_{22}(k) + y_2(k-2)) \tag{2.93}$$

which includes the same number and types of operations as the original formulation but allows for twice as much time. The price to pay is the extra hardware required to perform the nonrecursive computations below in a heavily pipelined way.

$$b_{11}(k) = \min(a_{11}(k) + a_{11}(k-1), a_{12}(k) + a_{21}(k-1)) \tag{2.94}$$
$$b_{12}(k) = \min(a_{11}(k) + a_{12}(k-1), a_{12}(k) + a_{22}(k-1)) \tag{2.95}$$
$$b_{21}(k) = \min(a_{21}(k) + a_{11}(k-1), a_{22}(k) + a_{21}(k-1)) \tag{2.96}$$
$$b_{22}(k) = \min(a_{21}(k) + a_{12}(k-1), a_{22}(k) + a_{22}(k-1)) \tag{2.97}$$

□

The remarkable hardware structure so obtained demonstrates that taking advantage of specific properties of an algorithm and of algebraic transforms has more potential to offer than universal transforms alone. Some computations can be accelerated by creating concurrencies that did not exist in the original formulation, which opens a door to solutions that would otherwise have remained off-limits.

2.8.5 Digest

- The transform approach to architecture design promoted in this text has been found to yield useful solutions at any level of granularity. Some of the resulting architectures are truly surprising.

- Both bit-serial architectures and distributed arithmetic follow quite naturally when arithmetic/logic operations are dissected into bit-level manipulations before the various equivalence transforms are applied. It is worthwhile to consider them when fixed and data-independent computations are to be carried out with limited hardware resources and moderate performance. After having sunk into oblivion for many years, the two techniques have had a comeback for filtering and other DSP applications with LUT-based FPGAs.

- All universal <u>and</u> algebraic transforms that apply to computations on the field of reals also apply to computations on Galois fields, of course.

- While loop unfolding is applicable to any linear computation in the field of reals, this is not a necessary condition. In case a recursion forms a bottleneck when in pursuit of higher performance, check whether it is possible to restate or modify the computations within the feedback loop in such a way as to make them linear over a semiring.

2.9 | Conclusions

2.9.1 Summary

We began this chapter by comparing instruction set processors with dedicated architectures. It was found that general-purpose computing asks for a high degree of flexibility that only program-controlled processors can provide. However, the ability to execute an arbitrary sequence of instructions on an unknown range of data types brings about numerous inefficiencies and largely precludes architectural optimizations. For well-defined computational tasks, much better performance and energy efficiency can be obtained from hardwired architectures with resources tailored to the specific computational needs of the target application. Segregation, weakly-programmable satellites, ASIPs, and configurable computing have been found to form useful compromises.

Next, we investigated a number of options for organizing datapath hardware. Our approach was based on reformulating a given data processing algorithm in such a way as to preserve its input-to-output relationship except, possibly, for latency, while improving on performance, circuit size, energy efficiency, and the like. Findings on how best to rearrange combinational, nonrecursive, and recursive computations were given in sections 2.4.8, 2.6.6, and 2.7.7 respectively. The approach was then generalized in terms of granularity and algebraic structure with the results summarized in section 2.8.5. The essence of these insights is collected in tables 2.12 and 2.11.

As energy efficiency depends on so many parameters, the pertaining entries of table 2.12 deserve further clarification. Assuming fixed energy costs per operation and ignoring any static currents, most architectural transforms discussed inflate the energy dissipated on a given calculation as conveyed by table entries E and "Extra hardware overhead". Put in other words, cutting circuit size and boosting throughput typically are at the expense of energy efficiency.

The picture is more favorable when there is room for cutting the energy spent per computational operation by playing with voltages, transistor sizes, circuit style, fabrication process, and the like. The most effective way to do so in CMOS is to lower the supply voltage since the energy dissipated per operation augments quadratically with voltage whereas a circuit's operating speed does not. The longer paths through a circuit are likely to become unacceptably slow, though. A suitable architecture transform may then serve to trim these paths in such a way as to compensate for

Table 2.11 | Options available for reorganizing datapath architectures. Upper-case letters denote transforms that are generally available whereas lower-case letters indicate some preconditions must be satisfied by the application and/or type of computation to make this a viable option.

		Type of computation		
		combinational or memoryless	sequential or memorizing	
			nonrecursive	recursive
	Data flow	feedforward	feedforward	feedback
	Memory	no	yes	yes
	Data dependency graph	DAG^a with all edge weights zero	DAG with some or all edge weights non-zero	Directed cyclic graph with no circular path of weight zero
	Response length	$M = 1$	$1 < M < \infty$	$M = \infty$
Nature of system	linear time-invariant	D,P,Q,S,a	D,P,q,S,a,R	D,S,a,R,i,U
	linear time-variant	D,P,Q,S,a	D,P,S,a,R	D,S,a,R,i,U
	nonlinear	D,P,Q,S,a	D,P,S,a,R	D,S,a,R,i,u
	Discussed in section	2.4	2.6	2.7

D : Iterative decomposition
P : Pipelining
Q : Replication
q : Multipath filtering as special case of replication
    provided the resulting repetitive transfer function is acceptable
S : Time-sharing
a : Associativity transform provided operations are identical and associative
R : Retiming
i : Pipeline interleaving, i.e. pipelining in conjunction with time-sharing,
    provided a number of data streams can be processed separately from each other
U : Loop unfolding
u : Loop unfolding provided computation is linear over a semiring

^{a} DAG is an acronym for directed acyclic graph, i.e. for a directed graph with no circular path.

the loss of speed incurred by opting for a more energy-efficient, yet also slower alternative. The attribute "Helpful for indirect energy saving" in table 2.12 refers to this option. Retiming, algebraic transforms, and coarse grain pipelining are the most promising candidates as they entail no or very little overhead. Whether such a potential for indirect energy optimization indeed materializes or not must be examined in detail on a per case basis.

2.9.2 The grand architectural alternatives from an energy point of view

Let us re-examine the fundamental architectural alternatives from an energy point of view. Program-controlled processors heavily rely on subcircuits and activities such as

Table 2.12 | Summary of the most important architectural transforms and their characteristics.

Impact on figure of merit below	Architectural transform						
	Decom-position	Pipe-lining	Repli-cation	Time-sharing	Associa-tivity	Retiming	Loop unfolding
A	$-... =$	$= ...+$	$+$	$-... =$	$=$	$=$	$+$
Γ	$+$	$=$	$-$	$+$	$=$	$=$	$=$
t_{lp}	$-$	$-$	$=$, mux $-$	$=$	$-...+$	$-$	$-$
$T = \Gamma \cdot t_{lp}$	$=$	$-$	$-$	$+$	$-...+$	$-$	$-$
AT	$-... =$	$-... =$	$=$	$= ...+$	$-...+$	$-$	$+$
L	$+$	$+$	$=$, mux $+$	$+$	$=$	$=$	$+$
E	$-...+$	$-...+$	$=$	$= ...+$	$-...+$	$=$	$+$
Extra hardware overhead	recy. and cntl.	none	distrib., recoll., and cntl.	collect., redist., and cntl.	none	none	extra word width
Helpful for indirect energy saving	no	coarse grain yes	possibly yes	no	yes	yes	possibly yes
Compatible storage type	any	register	register	any	any	register	register
Universal	yes	yes	yes	yes	no	yes	no
Discussed in subsection[s]	2.4.2 2.6.4	2.4.3 2.6.2	2.4.4	2.4.5 2.6.4	2.4.6	2.6.1 2.7.2	2.7.2

A :	circuit size	$=$:	approximately constant
Γ :	cycles per data item	$+$:	tends to increase
t_{lp} :	longest path length	$-$:	tends to decrease
T :	time per data item	$...$:	in less favorable situations
L :	latency in computation periods		
E :	energy per data item		

auxiliary circuitry for
recy.:   data recycling
cntl.:   datapath control
dist.:   data distribution
coll.:   data collection

- General-purpose multi-operation ALUs,
- Generic register files of generous capacity,
- Multi-driver busses, bus switches, multiplexers, and the like,
- Program and data memories along with address generation,

- Controllers or program sequencers,
- Instruction fetching and decoding,
- Stack operations and interrupt handling,
- Dynamic reordering of operations,
- Branch prediction and speculative execution,
- Data shuffling between main memory and multiple levels of cache, and
- Various mechanisms for maintaining cache and data consistency.

From a purely functional point of view, all of this is a tremendous waste of energy because none of it contributes to payload data processing. The welcome agility of instruction set processors is thus paid for with an overhead in terms of control operations and a formidable inflation of switching activity.[67]

The isomorphic architecture, in contrast, does not carry out any computations or data transfers unless mandated by the original data processing algorithm itself. There are no instructions to fetch and decode. There is no addressing and no accessing of memories either as all data are kept in registers. Data transfers are local, there is no data shuffling between registers, cache, and main memory. There are no busses with their important load capacitances to drive.

"Does this imply the isomorphic architecture is the most energy-efficient option then?"

Somewhat surprisingly, this is not so. The reason is glitching, a phenomenon observed in digital circuits that causes extra signal transients to occur on top of those stipulated by the computation flow. Glitch-induced switching is particularly intense when data recombine in combinational logic after having travelled along propagation paths of greatly disparate lengths because circuit nodes are then likely to rock back and forth several times before settling.[67] By cutting overlong propagation paths, moderate pipelining and iterative decomposition tend to abate glitching and so help to improve overall energy efficiency.

General-purpose processors further operate with data words of uniform and often oversized width throughout an entire algorithm.[68] As opposed to this, dedicated architectures make it possible to fine-tune the number of bits in every register and logic block to individual requirements as there is no compelling need to combine subfunctions with greatly different precision requirements into a single datapath sized for worst-case requirements. The overriding concern is to avoid switching activities that are not relevant to the final result. Turning off entire functional blocks whenever they sit idle naturally follows from applying this idea to higher levels of granularity.

Last but not least, the impressive throughputs of modern uniprocessors have been bought at the price of operating CMOS circuits under conditions that are far from optimal in view of energy efficiency (extremely fast clock, small MOSFET threshold voltages, large overdrive factors, and hence comparatively high supply voltage, significant leakage). An alternative design that takes advantage of concurrent processing to arrive at more favorable operating conditions may prove beneficial. Yet, as these issues are of electrical rather than architectural nature, their discussion will have to wait until chapter 9.[69]

---

[67] You may want to refer to appendix A.5 to learn more about the causes of glitching.

[68] Only the so-called multimedia instructions can provide programmers with an opportunity to process fewer bits per data item. Yet, not all instruction sets include them and not all algorithms lend themselves to taking advantage of sub-word parallelism.

[69] Which is also the place where a node's switching activity will be formally defined.

Energy considerations thus tend to give dedicated processing units new momentum over the unimaginative usage of general-purpose microcomputers. It is not unusual to find that a program-controlled processor dissipates two or three orders of magnitude as much energy as an application-specific architecture does for the same computation.[70]

Observation 2.10. *A key challenge of low-power design is to minimize all redundant switching activities by accommodating just as much flexibility as required by the application in an otherwise dedicated processing unit.*

2.9.3 A guide to evaluating architectural alternatives

In spite of our efforts to present a systematic overview on dedicated datapath architectures, we must admit that architecture design is more art than science. Many practical constraints and technical idiosyncrasies make it impossible to obtain a close-to-optimum solution by analytical means alone. The common procedure, therefore, is to come up with a variety of alternative ideas, to devise the corresponding architectures to a reasonable level of detail, and to evaluate their respective merits and drawbacks before decisions are taken. This approach — which is typical for many engineering activities — asks for creativity, methodology, and endeavor. It is nevertheless hoped that the material in this chapter gives some insight into the options available and some directions on when to prefer what option for tailoring VLSI architectures to specific technical requirements. What follows are some practical guidelines.

1. Begin by analyzing the algorithm. Section by section, identify the data flow and the nature of the essential computations. Estimate the necessary datapath resources by giving quantitative indications for
    - the word widths truly required (check [80] for references),
    - the data rates between all major building blocks,
    - the memory bounds, access rates, and access patterns in each building block, and
    - the computation rates for all major arithmetic operations.[71]

2. Identify the controllers that are required to govern the flow of computation along with its interplay with the external world. Analyze the control flow for data dependencies, overall complexity, and flexibility requirements. Find out where to go for a hard-wired dedicated architecture, where for a program-controlled processor, and where to look for a compromise.

3. Rather than starting from a hypothetical isomorphic architecture, let your intuition come up with a number of preliminary architectural concepts. Establish a rough **block diagram** for each of them. Make the boundaries between major subfunctions coincide with registers as you would otherwise have to trace path delays across circuit blocks during timing verification and optimization.

---

[70] [79] estimates the gap to be up to four orders of magnitude over direct-mapped architectures and growing.

[71] Watch out if you are given source code from some prior implementation, such as C code for a 32 bit DSP, for instance. You are likely to find items solely mandated by the resources available there or by software engineering considerations. Typical examples include operations related to (un)packing and (re)scaling, usage of computationally expensive data types, arithmetic operations substituted for bit-level manipulations, multitudes of nicely named variables that unnecessarily occupy distinct memory locations, and more.

4. It is always a good idea to prepare a comprehensive and, hence, fairly large table that opposes the different architectures under serious consideration.

   The rows serve to describe the hardware resources. Each major subfunction occupies a row of its own. Each such subfunction is then hierarchically decomposed into ever smaller subcircuits on a number of subsequent rows until it becomes possible to give numerical estimates for A and t_{lp}, and, possibly, for E as well. Once a subcircuit has been broken down to the RTL level, one can take advantage of HDL synthesis to obtain those figures with a good degree of precision. Finite state machines, in particular, are difficult to estimate otherwise.

   For each architectural variant, a few adjacent columns are reserved to capture A, t_{lp}, E and Γ. An extra column is set aside for n, a natural number that indicates how many times the hardware resource is meant to be instantiated for the architecture being considered. Γ stands for the number of cycles required to obtain one processed data item with the hardware resources available. Of course, this quantity tends to diminish with n but it is not possible to state the exact dependency in general terms.

5. Estimating the overall circuit size, cycles per data item, latency, and dissipated energy for each architecture now essentially becomes a matter of bookkeeping that can be carried out with the aid of spreadsheet software. Path delays are more tricky to deal with as logic and interconnect delays are subject to significant variation as a function of lower-level details.[72] It is thus quite common to code, synthesize, place, and route the most time-critical portions of a few competing architectures merely for the purpose of evaluating $\max(t_{lp})$ and of extrapolating clock frequency, overall computation rate, and overall throughput.

6. Analysis of the figures so obtained will identify performance bottlenecks and inacceptably burdensome subfunctions in need of more efficient implementations. This is the point where the architecture transforms discussed in this chapter come into play.

7. Compare the competing architectural concepts against the requirements. Narrow down your choice before proceeding to more detailed analyses and implementations.

Example

The table below shows results from exploring the design space for AES encryption with a key length of 128 bit [81]. The available options for trading datapath resources for computation time are evident. The narrower datapaths require extra circuitry for storing and routing intermediate results, which inflates complexity and adds to path delays. What all variants have in common is that the ten cipher rounds are carried out by a single datapath as a result from iterative decomposition of the AES algorithm. Also, none of the architectures makes use of pipelining, which results in latency and cycles per data item being the same. SubBytes refers to the cipher's most costly operation from a hardware point of view. While the figures include control logic and have been obtained from actual synthesis, simplifications have been made to obtain reasonably accurate estimates for the key figures of merit without having to establish the HDL code for each architectural alternative in full detail.

---

[72] Please refer to footnote 22 for a comment on the limitations of anticipating path delays.

Datapath width (bit)	8	16	32	64	128
Parallel SubBytes units	1	2	4	8	16
Circuit complexity (GE)	5 052	6 281	7 155	11 628	20 410
Area A (normalized)	1	1.27	1.47	2.43	4.27
Cycles per data item Γ	160	80	40	20	10
Longest path delay t_{lp} (normalized)	1.35	1.34	1.21	1.13	1
Time per data item T (normalized)	21.6	10.7	4.83	2.23	1
Size-time product AT	21.6	13.6	7.10	5.42	4.27

□

Designers of large VLSI chips running at elevated clock rates, such as high-performance uniprocessors, inevitably run into interconnect delay as another limiting factor. This is because it is no longer possible to transmit data from one corner of a chip to the opposite one within a single clock cycle. While this important aspect has been left aside in this chapter, more will be said in fig.6.18 and section 15.5.2.

As a concluding remark, we would like to recall once more that good solutions call for analysis and reorganization of data processing algorithms at all levels of details, including architecture (process/block), register transfer (arithmetic/word), and logic (gate/bit) levels. It has been shown on numerous occasions that viewing a problem from a totally different angle can pave the way to unexpected architectural solutions that feature uncommon characteristics. Also, the possibilities for replacing a given algorithm by a truly different suite of computations that is equivalent for any practical purpose of the application at hand, but better suited to VLSI, should always be investigated first.

2.10 | Problems

1. Computationally efficient approximations for the magnitude function $\sqrt{a^2 + b^2}$ have been presented in table 2.8. (a) Show that approximation 2 remains within ±3% of the correct result for any values of a and b. (b) Give three alternative architectures that implement the algorithm and compare them in terms of datapath resources, cycles per data item, longest path, and control overhead. Assume input data remain valid as long as you need them, but plan for a registered output. Begin by drawing the DDG.

2. Discuss the idea of combining replication with pipelining. Using fig.2.18 and the numbers that come along with it as a reference, take a pipelined datapath before duplicating it. Sketch the result in the AT-plane for various pipeline depths, e.g. for $p = 2, 3, 4, 5, 6, 8, 10$. Compare the results with those of competing architectures that achieve similar performance figures (a) by replicating the isomorphic configuration and (b) by extending the pipeline approach beyond the most efficient depth. How realistic are the various throughput figures when data distribution/recollection is to be implemented using the same technology and cell library as the datapath?

3. Reconsider the third-order correlator of fig.2.25a. (a) To boost performance, try to retime and pipeline the isomorphic architecture without prior reversal of the adder chain. How does the circuit so obtained compare with fig.2.25d. Give estimates for datapath resources, cycles per data item, longest path, latency, and control overhead. (b) Next assume your prime concern is area occupation. What architectures qualify?

4. Figure 2.26 shows a viable architecture for a transversal filter. Before this architecture can be coded using an HDL, one must work out the missing details about clocking, register clear, register enable, and multiplexed control signals. Establish a schedule that lists clock cycle by clock cycle what data items the various computational units are supposed to work on, what data items or states the various registers are supposed to hold, and what logic values the various control signals must assume to marshal the interplay of all those hardware items. Samples are to be processed as specified by fig.2.11a.

5. Arithmetic mean $\bar{x}$ and standard deviation σ are defined as

$$\bar{x} = \frac{1}{N} \sum_{n=1}^{N} x_n \qquad\qquad \sigma^2 = \frac{1}{N-1} \sum_{n=1}^{N} (x_n - \bar{x})^2 \qquad (2.98)$$

Assume samples x_n arrive sequentially one at a time. More specifically, each clock cycle sees a new w-bit data item appear. Find a dedicated architecture that computes $\bar{x}$ and σ^2 after N clock cycles and where N is some integer power of two, say 32. Definitions in (2.98) suggest one needs to store up to $N-1$ past values of x. Can you make do with less? What mathematical properties do you call on? What is the impact on datapath word width? This is actually an old problem the solution of which has been made popular by early scientific pocket calculators such as the HP-45, for instance. Yet, it nicely shows the difference between a crude and a more elaborate way of organizing a computation.

6. Most locations in the map of fig.2.21 can be reached from the isomorphic configuration on more than one route. Consider the location where $A = 1/3$ and $T = 1$, for instance. Possible routes include
   ○ (time share → decompose → pipeline) as shown on the map,
   ○ (time share → pipeline → decompose),
   ○ (pipeline → decompose → time share),
   ○ (pipeline → time share → decompose), and
   ○ (decompose → decompose).
   Architectures obtained when following distinct routes typically differ. Figure 2.21 indicates only one possible outcome per location and is, therefore, incomplete. Adding the missing routes and datapath configurations is left as a pastime to the reader. Purely out of academic interest, you may want to find out which transforms form commutative pairs.

7. Figure 2.21 shows a kind of compass that expresses the respective impact of iterative decomposition, pipelining, replication, and time-sharing. Include the impact of the associativity transform in a similar way.

8. Calculating the convolution of a two-dimensional array with a fixed two-dimensional operator is a frequent problem from image processing. The operator $c_{x,y}$ is moved over the entire original image $p(x,y)$ and centered over one pixel after the other. For each position X,Y the

pertaining pixel of the convoluted image $q(x,y)$ is obtained from evaluating the inner product

$$q(X,Y) = \sum_{y=-w}^{+w} \sum_{x=-w}^{+w} c_{x,y}\, p(X+x, Y+y) \tag{2.99}$$

Consider an application where $w = 2$. All pixels that contribute to (2.99) are then confined to a 5-by-5 square with the current location in its center. An uninspired implementation with distributed arithmetics would thus call for a look-up table with 2^{25} entries, which is exorbitant. A case study by an FPGA manufacturer explains how it is possible to cut this requirement down to one look-up table of a mere 16 words. Clearly, this remarkable achievement requires a couple of extra adders and flip-flops and is dependent on the particular set of coefficients given below. It combines putting together multiple occurrences of identical weights, splitting of the look-up table, taking advantage of nonoverlapping 1s across two coefficients, and clever usage of the carry input for the handling unit weights. Try to reconstruct the architecture. How close can you come to the manufacturer's result published in [73]?

$c_{x,y}$		x				
		-2	-1	0	1	2
	2	-16	-7	-13	-7	-16
	1	-7	-1	12	-1	-7
y	0	-13	12	160	12	-13
	-1	-7	-1	12	-1	-7
	-2	-16	-7	-13	-7	-16

2.11 | Appendix I: A brief glossary of algebraic structures

Any algebraic structure is defined by a set of elements S and by one or more operations. The nature of the operations involved determines which of the axioms below are satisfied.

Consider a first binary operation $\boxplus$

1. Closure wrt $\boxplus$: if a and b are in S then $a \boxplus b$ is also in S.
2. Associative law wrt $\boxplus$: $(a \boxplus b) \boxplus c = a \boxplus (b \boxplus c)$.
3. Identity element wrt $\boxplus$: There is a unique element e such that $a \boxplus e = e \boxplus a = a$ for any a, (e is often referred to as the "zero" element).
4. Inverse element wrt $\boxplus$: For every a in S there is an inverse $-a$ such that $a \boxplus -a = -a \boxplus a = e$.
5. Commutative law wrt $\boxplus$: $a \boxplus b = b \boxplus a$.

Consider a second binary operation $\boxdot$ that always takes precedence over operation $\boxplus$

6. Closure wrt $\boxdot$: if a and b are in S then $a \boxdot b$ is also in S.
7. Associative law wrt $\boxdot$: $(a \boxdot b) \boxdot c = a \boxdot (b \boxdot c)$.
8. Identity element wrt $\boxdot$: There is a unique element i such that $a \boxdot i = i \boxdot a = a$ for any a, (i is often referred to as "one" or the "unity" element).

9. Inverse element wrt $\boxdot$: For every a in S there is an inverse a^{-1} such that $a \boxdot a^{-1} = a^{-1} \boxdot a = i$, the only exception is e for which no inverse exists.

10. Commutative law wrt $\boxdot$: $a \boxdot b = b \boxdot a$.

11. Distributive law of $\boxdot$ over $\boxplus$: $a \boxdot (b \boxplus c) = a \boxdot b \boxplus a \boxdot c$ and $(a \boxplus b) \boxdot c = a \boxdot c \boxplus b \boxdot c$.

12. Distributive law of $\boxplus$ over $\boxdot$: $a \boxplus b \boxdot c = (a \boxplus b) \boxdot (a \boxplus c)$ and $a \boxdot b \boxplus c = (a \boxplus c) \boxdot (b \boxplus c)$.

13. Complement: For every a in S there is a complement $\bar{a}$ such that $a \boxplus \bar{a} = i$ and $a \boxdot \bar{a} = e$.

Name of algebraic structure	Opera-tions	Axioms satisfied												
		1	2	3	4	5	6	7	8	9	10	11	12	13
Set														
Semigroup	$\boxplus$	1	2											
Monoid	$\boxplus$	1	2	3										
Group	$\boxplus$	1	2	3	4									
Abelian or commutative group	$\boxplus$	1	2	3	4	5								
Abelian semigroup	$\boxplus$	1	2			5								
Abelian monoid	$\boxplus$	1	2	3		5								
Ring	$\boxplus \boxdot$	1	2	3	4	5	6	7				11		
Ring with unity	$\boxplus \boxdot$	1	2	3	4	5	6	7	8			11		
Division algebra aka skew field	$\boxplus \boxdot$	1	2	3	4	5	6	7	8	9		11		
Field	$\boxplus \boxdot$	1	2	3	4	5	6	7	8	9	10	11		
Commutative ring	$\boxplus \boxdot$	1	2	3	4	5	6	7			10	11		
Commutative ring with unity	$\boxplus \boxdot$	1	2	3	4	5	6	7	8		10	11		
Semiring	$\boxplus \boxdot$	1	2			5	6	7				11		
Commutative semiring	$\boxplus \boxdot$	1	2			5	6	7			10	11		
Boolean algebra	$\boxplus \boxdot$	1	2	3		5	6	7	8		10	11	12	13

EXAMPLES

Consider the set S_{DNA} of all possible DNA sequences of finite but non-zero length with characters taken from the alphabet $\{A, T, C, G\}$. This set together with the binary operation of string concatenation denoted as $\smile$[73] forms a **semigroup** $(S_{DNA}, \smile)$.

A **monoid** $(S_{DNA} \cup \{\epsilon\}, \smile)$ is obtained iff the empty sequence ϵ is also admitted.

All possible permutations of a given number of elements make up a **group** when combined with binary composition of functions[74] as sole operation. For a practical example, consider all six distinct rearrangements of three elements, shown below, and let us refer to them as set S_3.

1.	2.	3.	1.	2.	3.	1.	2.	3.	1.	2.	3.	1.	2.	3.	1.	2.	3.
↓	↓	↓	↓	↓	↓	↓	↓	↓	↓	↓	↓	↓	↓	↓	↓	↓	↓
1.	2.	3.	2.	3.	1.	3.	1.	2.	3.	2.	1.	1.	3.	2.	2.	1.	3.

---

[73] Alternative symbols for the concatenation operator include . (mathematics) and & (computer science).

[74] Binary composition of functions means that two functions are invoked one after the other $(...) \circ f \circ g \equiv g(f(...))$. Also keep in mind that a permutation is just a particular kind of function.

This set of permutations along with binary composition ∘ forms a group $(S_3, \circ)$.

The set of all positive integers $\mathbb{N}^+ = \{1, 2, 3, ...\}$ together with addition as sole operation constitutes an infinite **Abelian semigroup** $(\mathbb{N}^+, +)$.

When supplemented with 0, the above structure turns into an **Abelian monoid** $(\mathbb{N}, +)$.

The set of all integers $\mathbb{Z}$ together with addition forms an infinite **Abelian group** $(\mathbb{Z}, +)$.

A **commutative ring with unity** $(\mathbb{Z}, +, \cdot)$ results when multiplication is added as a second operation to the aforementioned Abelian group.

$(\mathbb{N}, +, \cdot)$, in contrast, is merely a **commutative semiring** because natural numbers have no additive inverses. Yet, in addition to the necessary requirements for a semiring, identity elements with respect to both operations also happen to exist in this example.

The set of all rational numbers $\mathbb{Q}$ together with addition as a first and multiplication as a second operation forms the **field**[75] $(\mathbb{Q}, +, \cdot)$. Other popular fields with infinitely many elements are $(\mathbb{R}, +, \cdot)$ over the real numbers, and $(\mathbb{C}, +, \cdot)$ over the complex numbers.

The set of all quotients $\frac{P(x)}{Q(x)}$ of two polynomials $P(x)$ and $Q(x)$ with real-valued coefficients together with addition and multiplication makes up yet another infinite field.

Any subset of integers $S = \{0, 1, ..., p-1\}$ forms a field together with addition modulo p and multiplication modulo p iff p is a prime number. Any such **finite field** is called a **Galois field** GF(p). The best-known finite field is the GF(2) ($\{0,1\}, \oplus, \wedge$). Observe that the additive inverse in GF(2) of a $-a$ is a itself and that the multiplicative inverse of 1 1^{-1} is 1 while 0 has no multiplicative inverse. As a second example, consider the GF(5) ($\{0, 1, 2, 3, 4\}, + \bmod 5, \cdot \bmod 5$).

Cardinalities of finite fields are not confined to prime numbers p but can take on any power p^n provided $n \in \mathbb{N}^+$. A Galois field where $n \geq 2$ is termed an **extension field** GF(p^n). All polynomials of degree $0, 1, ..., n-1$ with coefficients from GF(p) make up the set of elements. The first operation is addition modulo $M(x)$ and the second one multiplication modulo $M(x)$, where $M(x)$ is an irreducible polynomial of degree n with coefficients from GF(p). GF(3^2), for instance, is exemplified by ($\{0, 1, 2, x, x+1, x+2, 2x, 2x+1, 2x+2\}, + \bmod (x^2+1), \cdot \bmod (x^2+1)$).

For the set of all square matrices $\mathsf{M}_{n \times n}$ with coefficients taken from a field, there exist identity elements with respect to both addition and multiplication.[76] There is also an additive inverse for any element. As not every element has a multiplicative inverse, though, and as matrix multiplication is not commutative, the algebraic structure is an infinite **ring with unity**.

The factors of 30 together with operations least common multiple (lcm) and greatest common divisor (gcd) constitute a **Boolean algebra** of eight elements ($\{1, 2, 3, 5, 6, 10, 15, 30\}$, lcm, gcd). It necessarily follows that taking the complement $\bar{a}$ is tantamount to computing $\frac{30}{a}$ for any a.

$(S, \cup, \cap)$ is a Boolean algebra with union and intersection as binary operations iff S is a power set $\mathfrak{P}$. Consider a set of three elements $\Omega = \{a, b, c\}$, for instance. The set of all sets that can be

---

[75] The German term for a field is "(Zahlen)körper", the French "corps", and the Italian "campo".

[76] Incidentally note that all concepts of linear algebra (matrices, inverses, determinants, etc.) apply to matrices with coefficients from any field.

composed from these three elements, that is $\{\,\emptyset,\ \{a\},\ \{b\},\ \{c\},\ \{a,\,b\},\ \{a,\,c\},\ \{b,\,c\},\ \{a,\,b,\,c\}\,\}$, is called the power set of Ω and denoted as $\mathfrak{P}(\Omega)$. $(\mathfrak{P}(\Omega),\ \cup,\ \cap)$ then forms a Boolean algebra. Two particular elements of $\mathfrak{P}(\Omega)$, namely the empty set $\emptyset$ and the universal set Ω, act as identity elements e and i for the first and the second operation respectively. Each structure element $x \in \mathfrak{P}(\Omega)$ has a complement $\bar{x} = \Omega - x$.

The above structure is readily extended to an **infinite Boolean algebra** when elements and sets are chosen such that $|\mathfrak{P}(\Omega)| = \infty$, which, in turn, is obtained from making $|\Omega| = \infty$. As an example, assume Ω is made up of all DNA sequences of arbitrary length.

The well-known **switching algebra** $(\{0,1\},\ \vee,\ \wedge)$ is a Boolean algebra with just two elements. The complement of an element is its logic inverse and denoted with the negation operator $\neg$.

With no more than six axioms, the class of **semirings** is very broad. It includes but is not limited to the embodiments tabulated below:

Constituent	S	$\boxplus$	$\boxdot$
the commutative semiring of natural numbers	$\mathbb{N}$	$+$	$\cdot$
the commutative ring with unity of integers	$\mathbb{Z}$	$+$	$\cdot$
the "ordinary" fields	$\mathbb{Q}$	$+$	$\cdot$
	$\mathbb{R}$	$+$	$\cdot$
	$\mathbb{C}$	$+$	$\cdot$
all Galois fields, e.g.	$\{0,\,1\}$	$\oplus$	$\wedge$
all other fields, e.g.	$\frac{P(x)}{Q(x)}$	$+$	$\cdot$
the switching algebra	$\{0,\,1\}$	$\vee$	$\wedge$
other finite Boolean algebras, e.g.	$\{0,\,1\}$	$\wedge$	$\vee$
or	$\{1,\,2,\,3,\,4,\,6,\,12\}$	lcm	gcd
all other Boolean algebras, e.g.	$\mathfrak{P}(\Omega)$	$\cup$	$\cap$
the path algebras	$\{0,1\}$	max	min
	$\mathbb{R} \cup \{\infty\}$	min	$+$
	$\mathbb{R} \cup \{-\infty\}$	max	$+$
	$\{x \in \mathbb{R} \mid 0 \leq x \leq 1\}$	max	$\cdot$
	$\{x \in \mathbb{R} \mid x \geq 0\} \cup \{\infty\}$	max	min
the matrix algebras for every $n \in \mathbb{N}^+$	$\mathsf{M}_{n \times n}$	$+$	$\cdot$

2.12 | Appendix II: Area and delay figures of VLSI subfunctions

This appendix lists real-world numbers for common subfunctions such as logic gates, bistables, adders, and multipliers.[77] All data refer to commercial cell libraries in static CMOS technology

---

[77] Please refer back to section 2.5 for indications on the area occupation of register files and memories.

under typical operating conditions.[78] When talking about an individual cell, numbers relate to a version with simple output strength (1x drive).

Process generations and figures of merit compared

Process generation	M1 min. half-pitch F [nm]	Lithographic square F^2 [nm^2]	Number of metals	Supply [V]
250 nm	320	102 400	5	2.5
180 nm	240	57 600	6	1.8
130 nm	190	36 100	up to 8	1.2
90 nm	120	14 400	up to 9	1.0

A states the area occupied by the circuitry required to implement the target functionality. Except for individual standard cells, the intercell wiring has been completed and the associated overhead included. No provisions were made for global routing and I/O pads.

Parameter t_{id} denotes the insertion delay.[79] When referring to an individual library cell, a (purely capacitive) load of four standard inverters (FO4) is assumed. In the case of bistables, all delay figures refer to the non-inverting output.

Bistable storage functions

Table 2.13 | Selected flip-flops and latches (D flip-flops with no reset are found in pipelines).

D flip-flop with no reset	A [µm^2]	$[F^2]$	t_{id} [ps]
250 nm	97.9	956	267
180 nm	59.9	1040	203
130 nm	22.4	620	249
90 nm	14.3	993	160

D scan flip-flop with async. reset	A [µm^2]	$[F^2]$	t_{id} [ps]
250 nm	121.0	1181	279
180 nm	73.2	1271	202
130 nm	30.2	837	257
90 nm	19.8	1375	174

E (enable) flip-flop with async. reset	A [µm^2]	$[F^2]$	t_{id} [ps]
250 nm	126.7	1238	267
180 nm	76.5	1328	196
130 nm	32.5	900	245
90 nm	n.a.		

(transparent) latch with async. reset	A [µm^2]	$[F^2]$	t_{id} [ps]
250 nm	63.4	619	213
180 nm	36.6	635	151
130 nm	15.7	435	119
90 nm	11.0	764	251

---

[78] As a consequence of changes in the industry, it has not been possible to compile the table from datasheets of any single vendor; a horizontal line thus separates data from distinct companies. When comparing across process generations, be cautioned that cells are bound to differ significantly in their transistor-level circuits, MOSFET sizes, and threshold voltages due to divergent priorities (such as dense layout, high speed, low dynamic power, or low leakage). The lack of a universal standard for library characterization further contributes to distinctions.

[79] Insertion delay reflects the lapse of time that a subcircuit takes to pass on a data item from its input to the output and is defined in section A.6. As a reminder, $t_c = t_{id\,c} = \max(t_{pd\,c})$ for combinational functions, $t_{ff} = t_{id\,ff} = t_{su\,ff} + t_{pd\,ff}$ for flip-flops, and $t_{lc} = t_{id\,lc} = t_{su\,lc} + t_{pd\,lc}$ for latches.

Elementary logic functions

Table 2.14 | Selected logic gates.

Inverter	A		t_{id}	C_{inp}		Full adder	A		t_{id}
	$[\mu m^2]$	$[F^2]$	[ps]	[fF]			$[\mu m^2]$	$[F^2]$	[ps]
250 nm	11.5	113	69	5.6		250 nm	144.0	1410	230
180 nm	10.0	173	50	3.6		180 nm	76.5	1330	197
130 nm	3.4	93	17	1.9		130 nm	30.2	838	135
90 nm	3.3	229	18	2.2		90 nm	18.7	1300	92

2-input NAND	A		t_{id}		2-input NOR	A		t_{id}
	$[\mu m^2]$	$[F^2]$	[ps]			$[\mu m^2]$	$[F^2]$	[ps]
250 nm	17.3	169	64		250 nm	17.3	169	115
180 nm	10.0	173	63		180 nm	10.0	173	54
130 nm	4.5	124	25		130 nm	4.5	124	18
90 nm	4.4	305	34		90 nm	4.4	305	45

2-input XOR	A		t_{id}		2-to-1 MUX	A		t_{id}
	$[\mu m^2]$	$[F^2]$	[ps]			$[\mu m^2]$	$[F^2]$	[ps]
250 nm	51.8	506	214		250 nm	46.1	450	210
180 nm	26.6	462	178		180 nm	29.9	520	149
130 nm	10.1	279	50		130 nm	11.2	310	84
90 nm	8.8	610	61		90 nm	8.8	610	71

Arithmetic functions

Tables 2.15 and 2.16 refer to unpipelined adders and multipliers respectively. They include the approximate area for intercell wiring as estimated by Synopsys DesignCompiler. Synthesis results have been obtained by instantiating the appropriate DesignWare component followed by optimization with no timing constraint.

Table 2.15 | 2's complement adders with carry-in and carry-out.

ripple-carry adder DW01_add	A		t_{id}		carry-lookahead DW01_addsub	A		t_{id}
	$[\mu m^2]$	$[F^2]$	[ps]			$[\mu m^2]$	$[F^2]$	[ps]
90 nm 8 bit	149	10 400	720		90 nm 8 bit	237	16 500	800
90 nm 16 bit	299	20 700	1400		90 nm 16 bit	457	31 700	1480
90 nm 24 bit	448	31 100	2080		90 nm 24 bit	676	47 000	2150
90 nm 32 bit	598	41 500	2750		90 nm 32 bit	896	62 200	2830

Table 2.16 | 2's complement multipliers.

carry-save multiplier DW02_mult	A		t_{id}
	$[\mu m^2]$	$[F^2]$	[ps]
90 nm 8 bit × 8 bit	1 670	116 000	1350
90 nm 16 bit × 16 bit	5 670	394 000	2980
90 nm 24 bit × 24 bit	11 800	821 000	4200
90 nm 32 bit × 32 bit	20 530	1 430 000	5620

Chapter 3

Functional Verification

The ultimate goal of design verification is to avoid the manufacturing and deployment of flawed designs. Large sums of money are wasted and precious time to market is lost when a microchip does not perform as expected. Any design is, therefore, subject to detailed verification long before manufacturing begins and to thorough testing following fabrication. One can distinguish three motivations (after the late A. Richard Newton):

1. During specification: "Is what I am asking for what is really needed?"
2. During design: "Have I indeed designed what I have asked for?"
3. During testing: "Can I tell intact circuits from malfunctioning ones?"

In any of these cases, one can focus on different circuit properties.

Functionality describes what responses a system produces at the output when presented with given stimuli at the input. In the context of digital ICs, we tend to think of logic networks and of package pins but the concept of input-to-output mapping applies to information processing systems in general. Functionality gets expressed in terms of mathematical concepts such as algorithms, equations, impulse responses, tolerance bands for numerical inaccuracies, finite state machines (FSM), and the like, but often also informally.

Parametric properties, in contrast, relate to physical quantities measured in units such as Mbit/s, ns, V, µA, mW, pF, etc. that serve to express electrical and timing-related characteristics of an electronic circuit.

Observation 3.1. *Experience has shown that a design's functionality and its parametric properties are best checked separately since goals, methods, and tools are quite different.*

Our presentation is organized accordingly with section 3.1 discussing the options for specifying a design's functional behavior. Neither parametric issues nor the testing of physical parts will be addressed in this chapter.[1] After having exposed the puzzling limitations of functional verification

---

[1] The checking of timing-related quantities will be discussed in chapter 12 along with the verification of a circuit's inner layout. The testing of physical circuits is not part of this text.

in the first part of section 3.2, we will go on to suggest a couple of approaches that help to improve the likelihood of finding design flaws. Testbench design and other practical issues of how to organize simulation runs and simulation data are the subjects of section 3.3.

3.1 | How to establish valid functional specifications

Specifications available at the outset of a project are almost always inaccurate and incomplete. While parametric properties are relatively easy to state, expressing complex functionalities in precise yet concise terms is much more difficult. Functional specifications are, therefore, often stated verbally or graphically. There is a serious risk with doing so, however.

Warning example

An ASIC had to interface with an industry-standard microprocessor bus. Specifications made reference to official documents released by the CPU manufacturer, where bus read and write cycles were described in great detail along with precise timing diagrams. Although the ASIC was designed and tested with these requirements in mind, systems immediately crashed because of bus contentions when first prototypes were plugged into the target board.

What had gone wrong? It was found that the ASIC worked fine as long as its chip select line was active. When deselected, however, its pad drivers failed to release the bus by switching to a high-impedance state. This obvious necessity had been omitted in the original specifications and, as a consequence, also been ignored throughout the subsequent design and test phases.
□

In more general terms, the subsequent quote from [82] nicely summarizes an experience acquired by most designers of complex technical systems.

> Many computer systems fail in practice, not because they don't meet their specifications, but because the specifications left out some unanticipated circumstances or some unusual combination of events, so that when the unexpected occurred, the system was not able to deal with it. This is not necessarily due to sloppiness or stupidity on the part of the designer or to inadequate design methodology; it is a fundamental characteristic of the design process.

This leaves us with three important issues:

"How to have customers, marketing and engineers share the same understanding"
"How to ascertain specifications are precise, correct, and complete"
"How to make sure specifications describe the functionality that is really wanted and needed"

As natural language and informal sketches have been found to be inadequate, let us next discuss two approaches for arriving at more dependable specifications.

3.1.1 Formal specification

Ideally, all requirements for a circuit or system could be cast into a set of formal specifications which then would serve as a starting point for a rigorous mathematical proof of correctness. Over the years, a broad variety of formalisms has been devised for capturing behavioral aspects of numerous subsystems from many different fields, including truth tables, signal flow graphs, equations, state graphs, statecharts, Petri nets, and signal transition graphs (STGs).

A difficulty is the limited scope of each such formalism. While signal flow graphs, for instance, were developed for describing transformatorial systems, they are inadequate for modelling reactive systems. Although Petri nets and finite state machines can, in theory, describe any kind of computation, they tend to become unmanageable when applied to more complex processing algorithms due to combinatorial explosion. Yet, no real-world system of substantial size falls entirely into one of the above two categories. Most VLSI circuits include a great variety of subsystems, some of which are more of transformatorial nature (datapaths, look-up tables) and others more of reactive nature (controllers, interfaces). Relying on a single formal method for specifying the desired functionality of an entire chip or system is thus not normally practical.

A more mundane difficulty is that mathematical formalisms are unsuitable for communicating with customers and management. Also from a practical perspective, there must exist a straightforward and foolproof way to break down a system's specifications into specs for its various components in order to support collaborative development in a team, and to support products that comprise both hardware and software.

3.1.2 Rapid prototyping

Prototyping often is the only viable compromise between strictly formal and totally informal specification. By rapid or **virtual prototype** we understand an algorithmic model that emulates the functionality of the target circuit but not necessarily its architectural, electrical, and timing characteristics. A virtual prototype can be implemented

- As software code that runs on a general-purpose computer, microprocessor or DSP,
- With the aid of generic software tools for system-level simulations,[2] or
- By configuring FPGAs or other FPL devices.

The typical procedure goes

1. Apply formal methods (e.g. equations and statecharts) to capture specifications.
2. Use them as a starting point for developing a virtual prototype.
3. Make the prototype as widely available as possible for a thorough evaluation.
4. Refine specifications and prototype until satisfied before freezing them.

The pros and cons of rapid prototyping are as follows.

+ Demonstrations of the prospective functionality can be arranged at an early stage.
+ Shortcomings of the initial specifications are likely to get exposed in the process.

---

[2] Such as MATLAB/Simulink.

– Performance of the prototype does not normally come close to that of the final VLSI chip.

+ It is possible to submit a functional prototype to project management and/or customers for approval before investing time and money into actual hardware design.

+ The chances of discovering functional flaws early on are fairly good as functionality can be verified on the basis of real-world data.[3]

– There is no guarantee that all critical cases get covered during prototype testing.

+ A functional prototype is amenable to peer review and code inspection.

+ Design iterations and fine tuning are not penalized by the long turnaround times associated with IC design and manufacturing.

– Slips that are related to timing or electrical problems are unlikely to be found because they are not rendered correctly by a purely functional model.

Observation 3.2. *Rapid prototyping gives developers the opportunity to make and uncover more mistakes earlier and so to save precious cost and time.*

3.2 | Developing an adequate simulation strategy

Following fabrication, every physical part is subject to thorough tests. Most of those tests have **automated test equipment** (ATE) monitor the signal waveforms that result when predefined electrical waveforms are being applied to the **circuit under test** (CUT) over many, many clock cycles. Simulation essentially does the same, albeit with a virtual circuit model commonly referred to as the **model under test** (MUT).[4]

Both simulation and testing are said to be **dynamic verification** techniques as opposed to code inspection, formal verification, equivalence checking, timing analysis, and other **static verification** techniques that do not depend on signals, clocks, waveforms, or test patterns in any way. Simulation prevails when it comes to checking a design's functional behavior. This is mainly because of the limited capabilities of today's formal verification methods. However, simulation and testing both raise a couple of fundamental issues and bring about a variety of practical difficulties that we are going to address in the remainder of this chapter.

3.2.1 What does it take to uncover a design flaw during simulation?

Example

Consider a multiplexer buried within a large circuit. Assume that its two control inputs have been permuted by accident, see fig.3.1. A minor oversight during schematic entry or writing ... `to` ... rather than ... `downto` ... in the VHDL code suffices for this kind of mishap.

---

[3] It is even possible to operate a functional prototype within the target hardware environment, provided all surrounding equipment can be made to operate at a (reduced) clock rate that is consistent with the prototype's execution speed. This approach is particularly helpful for locating interface problems.

[4] A difference is that stimulation and observation of hardware are strictly confined to package pins, whereas all nodes can — at least in principle — be observed and controlled during simulation.

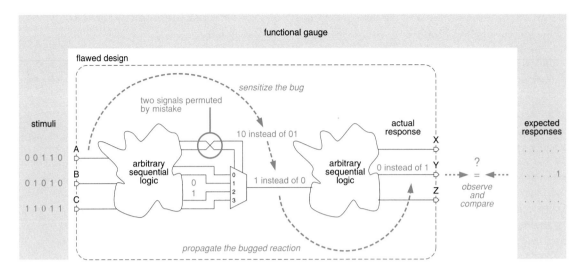

Fig. 3.1 A typical design mistake and the preconditions necessary for uncovering it during simulation.

Three preconditions must hold for that design flaw to become manifest during simulation.

Bug sensitization. The stimuli must drive the permuted nodes to opposite logic values. The data inputs to the multiplexer must further be adjusted such that a logic value opposed to the correct one indeed appears at the multiplexer output.

Bug propagation. The stimuli must permit the erroneous condition to propagate to observable nodes by causing a cascade of intermediate nodes to assume incorrect values.

Bug observation. The logic value observed on one or more of the nodes affected by the design error must get checked against the logic value that a correct design is expected to produce.

Unless all three conditions are met, the design flaw will have no consequence whatsoever during simulation, although circuits fabricated on the basis of the faulty netlist are almost sure to fail when put into service.

□

The same reasoning essentially applies to any other bug that affects functional behavior.

3.2.2 Stimulation and response checking must occur automatically

In the context of simulation, bug observation means checking the MUT's output. How to do so is dictated by the volume of data. Even a fairly modest subcircuit asks for keeping track of hundreds of waveforms over thousands of clock cycles. Digging through waveform plots, event lists, tabular printouts of logic values, and similar records from simulation runs is not practical for efficiency reasons. It is also unacceptable from a quality point of view because some incorrect data value hiding within myriads of correct items is very likely to be overlooked.

Observation 3.3. *Purely visual inspection of simulation data is not acceptable in VLSI design. Rather, designers must arrange for the simulator software to automatically check the actual responses from the model under test against the correct ones and to report any differences.*

The answer is to collect the **expected responses** along with the pertaining **stimuli** in a sequence of data patterns that we call a **functional gauge**. In mechanical engineering, gauges serve to verify the geometric conformity of manufactured parts so as to eliminate inaccurate parts before they are put together with other components. In some sense, gauges are specifications that have materialized. Similarly, a functional gauge serves to verify the functional correctness of some digital (sub)circuit. In its most simple expression, a functional gauge is a set of binary vectors listed cycle by cycle that specifies what kind of responses a correct design or circuit is supposed to provide when fed with certain stimuli.

Table 3.1 | Terms used in the context of dynamic verification.

In the	physical reality	world of simulation
a design exists as	fabricated circuit	HDL model or netlist
and is referred to as	circuit under test (CUT).	model under test (MUT).
As part of	prototype testing	functional verification
all those	stimuli and expected responses,	
collectively called	functional gauge,	
get administered by	automated test equipment	a software testbench
in search of	potential design flaws.	
As part of	production testing	fault simulation
all those	stimuli and expected responses,	
collectively called	the test vector set,	
get administered by	automated test equipment	a software testbench
in search of	potential fabrication defects.	

Test suite, test cases, and test patterns are often used as synonyms for functional gauge. The word **testbench** is also used in this context, but we reserve it for a somewhat different concept. A testbench is a piece of software used to pilot a simulation that provides the following services:

- Obtain stimuli vectors and apply them to the MUT at well-defined moments of time.
- Acquire the signal waveforms that emanate from the MUT as **actual response** vectors.
- Obtain expected response vectors and use them as a reference against which to compare.
- Establish a **simulation report** that points to problems (functional or timingwise), if any.
- Generate a periodic clock signal for driving simulation and clocked circuit models.

A testbench is to a MUT in the simulation world what ATE is to a fabricated circuit in the physical reality, see table 3.1, whereas functional gauge is a collective term for all those pairs of stimuli and expected responses being used to verify the functionality of a design, irrespective of whether that design is available as virtual HDL model or as tangible circuit.

Example

Consider a rising-edge-triggered Gray counter of word width $w = 4$ with enable and asynchronous reset. A suitable functional gauge is shown in fig.3.2a. Note that the gauge includes one stimulus/response pair per clock cycle and that both stimulus and response refer to the same cycle in each pair. Also observe that nothing else than the input-to-output mapping matters for the gauge. How states are being encoded inside the MUT and whether the counter is actually implemented as a Medvedev or as a full Moore machine is of no importance.[5]

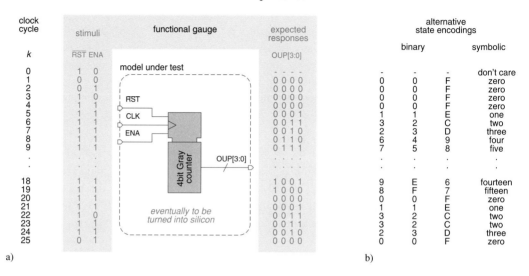

clock cycle k	stimuli RST ENA	functional gauge (model under test)	expected responses OUP[3:0]	alternative state encodings binary			symbolic
0	1 0		- - - -	-	-	-	don't care
1	0 0		0 0 0 0	0	0	F	zero
2	0 1		0 0 0 0	0	0	F	zero
3	1 0		0 0 0 0	0	0	F	zero
4	1 1		0 0 0 0	0	0	F	zero
5	1 1		0 0 0 1	1	1	E	one
6	1 1		0 0 1 1	3	2	C	two
7	1 1		0 0 1 0	2	3	D	three
8	1 1		0 1 1 0	6	4	9	four
9	1 1		0 1 1 1	7	5	8	five
.	. .		. . . .	.	.	.	.
18	1 1		1 0 0 1	9	E	6	fourteen
19	1 1		1 0 0 0	8	F	7	fifteen
20	1 1		0 0 0 0	0	0	F	zero
21	1 1		0 0 0 1	1	1	E	one
22	1 0		0 0 1 1	3	2	C	two
23	1 1		0 0 1 1	3	2	C	two
24	1 1		0 0 1 0	2	3	D	three
25	0 1		0 0 0 0	0	0	F	zero

a)   b)

Fig. 3.2 A functional gauge for a 4 bit Gray counter with enable and asynchronous reset (a). Three alternative state encoding schemes (b).

□

The functional gauge shown in fig.3.2 is very primitive and does not extend easily to more general situations, though. The concepts of functional gauge and testbench will thus have to be refined in numerous ways in the sections to come to make them more practically useful.[6] Let us begin with a fundamental problem.

3.2.3 Exhaustive verification remains an elusive goal

An unfailing way — in fact the sole one — to safeguard against any possible design flaw is to verify a design's functional behavior in perfect detail.[7] Let us see whether this is practical.

Exhaustive verification calls for traversing all edges in the design's state graph by exercising it with every possible input condition $i \in I$ in every possible state $s \in S$.[8] One might be tempted

---

[5] The difference is that a Medvedev machine has no output logic whereas a full Moore machine includes a non-trivial logic that translates each state into an output value. Refer to section B.1 for further explanations.

[6] This section focusses on devising functional gauges, testbench design is the subject of sections 3.3 and 4.4.

[7] Incidentally, note that the same argument also applies to the testing of physical parts for fabrication defects.

[8] Parallel edges are likely to exist, yet exhaustiveness indeed calls for checking the circuit's behavior for every single edge, that is for every state/condition pair, unless the presence of Mealy-type outputs can be ruled out.

to think that the product $|I||S|$ indicates the number of clock cycles c_{exh} necessary for exhaustive verification. In almost all practical applications, traversing every edge once will necessitate traversing others twice or more, however,[9] so that we must accept

$$c_{exh} \geq |I||S| \tag{3.1}$$

as a lower bound. Let w_i, w_s, and w_o denote the numbers of bits in the input, state, and output vectors respectively. The maximum number of possible input symbols $|I|$ then is 2^{w_i}, which figure must be discounted by the parasitic — i.e. unused — input codes. An analogous reasoning holds for $|S|$. Although it is not possible to accurately state $|I||S|$ in the general case, an upper bound can always be given as

$$c_{exh} \geq |I||S| \leq 2^{w_i + w_s} \tag{3.2}$$

The "less or equal" operator holds with equality in the absence of parasitic states and input symbols, i.e. when every combination of bits is being used for encoding some legal state and some input symbol respectively.

Let us plug in real figures to understand the practical significance. Consider the Intel 8080, an early microprocessor released in 1974 with an 8 bit datapath and almost trivial by today's standards. Abstracting from further details we find the following word widths:

input ports	8 bit data, 3 bit control, 1 bit reset	$w_i = 12$
registers	8 bit: A,B,C,D,E,H,L,IR; 16 bit: PC,SP; flags: 5	$w_s = 101$
output ports	8 bit data, 16 bit address, 6 bit status/control	$w_o = 30$

Assume there are no parasitic states and input symbols. The minimum number of clock cycles required for exhaustive simulation then is $2^{113} \approx 10^{34}$. Using test hardware running at 100 MHz, the process would run for more than $3 \cdot 10^{18}$ years. Software simulation would take orders of magnitude longer. To our regret, we must conclude that

Observation 3.4. *Exhaustive verification is not practical, even for relatively modest functions. Dynamic verification, therefore, almost always has to make do with a limited choice of test cases. The problem is to come up with a functional gauge of practical size <u>and</u> sufficient coverage.*

There is no cheap answer. We are thus going to discuss a number of more and less useful approaches to this problem that plagues both circuit simulation and IC testing.

3.2.4 All partial verification techniques have their pitfalls

TESTING DISTINCT FUNCTIONAL MECHANISMS SEPARATELY

The problem of exhaustive verification is **combinatorial explosion**. Exhaustive verification starts from a flat behavioral model obtained from combining the states of all data registers, counters, state machines, and the like into a single composite state. In addition, each possible state gets indiscriminately combined with each possible input. The Cartesian product so obtained describes all situations the circuit might conceivably encounter but, at the same time, causes the number of

---

[9] Fortunate exceptions are those cases where the state graph includes an Euler line. An (open) **Euler line** is a walk through a graph that runs through every edge exactly once.

test cases to explode as VLSI circuits actually include thousands of state registers and hundreds of inputs. A more pragmatic idea is to identify distinct subcircuits with fewer internal states and functional mechanisms, and to check them individually.

Example

Reconsider the Gray counter example. The functional gauge presented in fig.3.2 offers only partial coverage because it includes 26 cycles whereas (3.2) tells us 64 cycles is a lower bound for exhaustive verification.[10] Why is this nevertheless a reasonable compromise between functional coverage and verification costs?

Table 3.2 | Truth table of a w-bit binary Gray counter with enable and asynchronous reset.

$\overline{RST}$	CLK	ENA	OUP	
0	–	–	00...0	reset
1	↓	–	OUP	keep output unchanged
1	↑	0	OUP	*idem*
1	↑	1	$\mathrm{graycode}((\mathrm{bincode}(\mathtt{OUP}) + 1) \bmod 2^w)$	increment Gray-coded output

Begin with truth table 3.2 that specifies the behavior of Gray counters in more general and parametrized terms. Note that the desired functionality is made up of three mechanisms, namely a reset mechanism, an enable/disable mechanism, and the actual counting mechanism, see fig.3.3. The functional gauge of fig.3.2 addresses each of them separately.

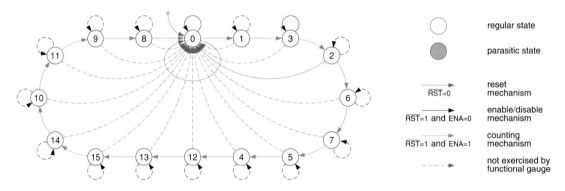

Fig. 3.3 The state graph of a 4 bit Gray counter with edges colored according to the functional mechanism they implement.

More precisely, the reset mechanism is being verified in cycles 0, 1, and 2, and the enable/disable mechanism in cycles 3, 4, and 5. The succession of output values is then being checked against the 4 bit Gray code in cycles 4 through 20. Provided the mechanisms involved function independently from each other, one can generalize from these partial checks and so obtain a high degree of

---

[10] Finding the exact minimum is left to the reader as an exercise, see problem 1.

confidence that the circuit will indeed work as intended. To be on the safe side, five extra vectors have been added to address a few more — and maybe more general — cases.

□

Though a workable and adequate solution in the above example, partial verification suffers from four limitations. Firstly, while refraining from traversing all edges (state transitions), the functional gauge of fig.3.2a does visit all nodes (states). The problem of combinatorial explosion persists and visiting all states rapidly becomes impractical on more substantial circuits.

Secondly, there exists a great variety of circuit models that comply with the gauge of fig.3.2a but, at the same time, contravene the functional specification of truth table 3.2. In the occurrence, one could easily come up with a modified design that allows counting to get disabled in some of the states but not in others. The example gauge would fail to uncover such a fault.

Thirdly, verifying each mechanism and subcircuit separately holds the risk of missing those problems that relate to the interaction of two (or more) of them.

Finally, identifying mechanisms and subcircuits for verification requires partial insight into its inner organization and working.[11]

Observation 3.5. *Unless functional coverage is exhaustive, a multitude of logic networks will necessarily exist that satisfy all verification steps deemed necessary and still fall short of meeting the original specifications for the overall circuit.*

Any simulation or test run that lacks exhaustiveness is tantamount to spot checking and, as such, perforce implies compromising among functional coverage, run time, and engineering effort. Ideally, an engineer would begin by enumerating all slips that might possibly occur during the design process before writing a functional gauge capable of sensitizing, propagating, and observing each of them. This is not possible in practice, though, because the number of potential bugs is virtually unlimited and our imagination insufficient to list them all.

Warning example

A tiny portion from an incorrect ASIC design is shown in fig.3.4. The designer's intention was to detect the zero state of a down counter by way of a 12-input NOR function.[12] Since no 12-input gate was available, he decided to compose the function from an 8-input and a 4-input NOR gate, but mistakenly instantiated a NAND gate during schematic entry. A simulation involving these four bits would have exposed the problem, but no such check was undertaken because the functional gauge never had the counter assume a state in excess of 18.

Why did the designer refrain from exercising the upper bits? Firstly, he wanted to keep simulation runs short, and exhausting a 12 bit counter with enable and reset would have required 16 384 cycles. More importantly, however, the designer was convinced that all input bits to a zero detector are

---

[11] A situation of limited knowledge is termed **gray box probing** and as opposed to the **black box** approach of exhaustive verification that makes no assumptions about the MUT whatsoever. A situation that assumes perfect knowledge of a circuit's inner details is referred to as **clear box probing**. The dilemma is this: Black box probing takes many vectors for a low probability of finding a problem. Clear box probing enables a test engineer to select such test cases as to address specific and likely problems, but may obstruct his view on other potential issues by contaminating his understanding with preconceptions from the circuit's design phase.

[12] Using the counter's carry/borrow bit instead would probably have been a more economic choice anyway.

interchangeable. He concluded it was sufficient to check the subcircuit's functioning by initializing the counter to 18 — or some other fairly low number — followed by having it count back to zero. Although he identified the end count mechanism and planned to check its functioning, the designer was just not prepared for a problem that would challenge his preconceptions.

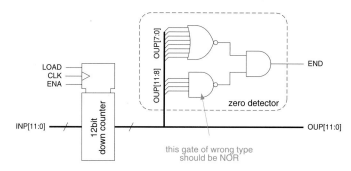

Fig. 3.4 A silly little oversight that managed to slip through simulation unnoticed because of poor coverage.

☐

What we learn from this example is that a critical difficulty of verification is to protect oneself against the unthinkable. Most examples of circuits and systems that have failed when put to service indeed confirm this.

MONITORING TOGGLE COUNTS IS OF LIMITED USE

A simple precautionary measure consists in collecting the toggle counts of all circuit nodes during simulation. Any node that never changes its logic value points to a weakness in the test suites chosen. In the example of fig.3.4, insisting on non-zero toggle counts would definitely have helped to recognize the stimuli as being inadequate. Yet, in spite of its utility and popularity, monitoring of node activities is far from solving all problems. A functional gauge may well toggle all nodes of a logic netlist back and forth and still be insufficient, see problem 3. Also, the concept of a circuit node is meaningless before a gate-level netlist has been established.

Observation 3.6. *The toggling of all nodes must be considered a desirable rather than a sufficient requirement for a good functional gauge.*

AUTOMATIC TEST PATTERN GENERATION DOES NOT HELP EITHER

Automatic test pattern generation (ATPG) is a technique that helps to tell intact ICs from defective ones following manufacturing. It is important to understand that ATPG does not normally help to uncover design flaws in nonmaterial circuit models such as HDL code or gate-level netlists. This is because ATPG starts from a presumably correct netlist and produces a set of test vectors for checking for the presence of predefined fabrication defects.[13]

---

[13] For the purpose of ATPG, fabrication defects are almost universally assumed to follow the so-called "single stuck-at fault" model whereby one circuit node at a time is assumed to be shorted to either logic 0 or 1. ATPG software attempts to cover close to 100% of all such faults with as few test vectors as possible.

Functional verification, in contrast, questions the correctness of a circuit model by setting its logic behavior against some kind of functional specification or reference model.

Monitoring code coverage is certainly useful but not sufficient

All decent HDL simulators can calculate code coverage figures by keeping track of how many times the individual statements in a MUT's source code are being exercised during a simulation run. 100% code coverage implies all executable statements have been executed once or more.

Observation 3.7. *However useful code coverage figures are, executing all statements in an RTL or behavioral circuit model implies neither that all states and transitions have been traversed nor that all conditions and subconditions for doing so have been checked.*

Also, code coverage relates to bug sensitization but neither to bug propagation nor to bug observation. Executing a flawed statement does not imply the bug must necessarily become manifest at an observable output. And whether it will be caught there or not depends on the expected responses, not on the stimuli.

The subsequent case study shows that functional coverage problems can take on much more subtle forms than in the above zero detector.

Warning example

Electronic dimmers for incandescent lamps work by varying the duty cycle of the load current. For every half wave of the 50 or 60 Hz mains voltage, a Triac connected in series with the lamp is turned on ("fired") at a phase angle adjustable between $0°$ and $180°$ and stays in the conducting state until the next zero crossing. A digital implementation is shown in fig.3.5.

The controller accepts commands from a touch key, converts the desired luminosity into a target phase angle, and fires the Triac via an optical coupler. The trigger impulse is initiated by a comparator when the actual phase angle matches the target value. The actual angle counter is clocked at 64 times the mains frequency so that in total 32 intensity levels are available. Synchronization with the mains is obtained through a zero-crossing detector that resets all 5 bits of the counter whenever a new half wave begins. Post-layout simulations and testing of fabricated samples on ATE confirmed circuit operation. Yet, the design of fig.3.5 is flawed.

When the first prototype was plugged into the target board, the dimmer was found to function o.k. except for a slight but disturbing oscillation of luminous intensity. The problem was quickly located in the synchronization mechanism. Since only the actual angle counter is reset, the clock divider proceeds from its current but otherwise indeterminate state whenever a new half wave begins. As a consequence, the next increment impulse for the actual angle counter can arrive anytime between a zero-crossing and $\frac{1}{32}$ half waves later. This, together with the fact that a free-running clock oscillator is being used, leads to a beat in firing angle and luminosity.

Why had this flaw passed unnoticed during circuit simulation and testing? The answer is that all simulations were carried out with the clock frequency an integer multiple of the mains frequency. It just never had occurred to the designers that non-integer frequency ratios might give rise to specific behavioral phenomena.

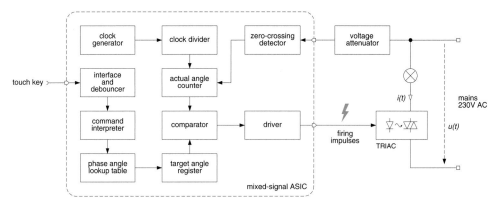

Fig. 3.5 Block diagram of flawed digital dimmer.

☐

ROUTINE IS THE DARK SIDE OF EXPERIENCE

The above examples have demonstrated the perils of unconcious, unspoken, and unjustified assumptions and misconceptions. We are now in a position to formulate what is considered to be the fundamental problem of any dynamic system verification approach, independently of whether the desired functionality is embodied in a piece of hardware, of software, or both.

Observation 3.8. *The innate difficulty with selecting critical test cases for dynamic verification is that human beings tend to check only for those problems that they expect.*

It is absolutely essential to select test cases with a clear and open mind, and section 3.2.5 will suggest how to get organized for this. Before doing so, however, let us point to another and particularly treacherous preconception.

Warning example

On its maiden flight on June 4, 1996, the Ariane 5 rocket had to be neutralized by its built-in self-destruct system at an altitude of 3500 m because excessive aerodynamic loads had ripped the solid boosters off the rocket shell after more than 30 s of seemingly normal flight [83]. Analysis of telemetry data revealed that there had been no structural failure but that the on-board computer had commanded the booster nozzles to maximum deflection two seconds before self-destruction occurred, and had so steered the rocket into an abnormal angle of attack.

Why did this happen? The flight control system of Ariane 5 depends on two computerized inertial reference platforms, one active and one for backup, that provide the on-board computer with velocity and attitude information. This flight control system was a proven design that had flown with Ariane 4 for years. On that fatal morning, however, both inertial reference platforms simultaneously ceased to deliver meaningful flight data and presented the on-board computer with diagnostic bit patterns instead. Misinterpreted as they were, these garbage data caused the on-board computer to initiate a sharp change in trajectory. The underlying reason was a numeric overflow that occurred when a parameter of minor importance was converted from a 64 bit floating-point number to a 16 bit signed integer combined with poor exception handling.

How was it possible for such a disastrous design flaw to remain undiscovered for so long? No numeric overflow had ever occurred in an Ariane 4 flight. Yet, the Ariane 5 trajectory implied considerably higher horizontal velocity values, which caused the critical parameter to accumulate beyond its habitual range. Tests for making sure that the navigational system would operate as intended in the new context had not been conducted. It was precisely the system's excellent reliability record that led the Ariane 5 design team to believe that everything worked fine and that no extra qualification steps were necessary.

☐

Observation 3.9. *How a system responds to the things going wrong shows how good it really is (quoted from Hans Stork, CTO of Texas Instruments).*

The fact that a subsystem has performed as expected when subjected to certain data sets, operating conditions, parameter configurations, and the like is no guarantee for its correct functioning in similar situations.

EVEN REAL-WORLD DATA SOMETIMES PROVE TOO FORGIVING

It is often argued that the functional coverage problem is best dealt with by using genuine data collected from real-world service instead of limiting dynamic verification to a small number of artificially prepared test suites. A test with data from the anticipated flight time sequence of Ariane 5 injected into the data processing section of the inertial reference system would indeed have disclosed its fatal limitation. Similarly, a software or a hardware prototype of the digital dimmer ASIC embedded within the remainder of the circuitry and operated with real-world waveforms would have led designers to recognize the oversight in their design.

Using actual data material is no panacea, however, because it may take an excessively large number of cycles before genuine stimuli might activate some rare but critical set of circumstances whereby a potential misbehavior of a design or its model could become apparent.

Warning example

A case that was given world-wide publicity in the fall of 1994 was the flaw in the floating-point division unit of early Pentium microprocessors [84] [85]. Due to a software problem, 5 out of 1066 table entries had been omitted from a PLA look-up table employed in the radix-4 SRT division algorithm.[14] Whether results from floating-point division came out wrong or not depended on the mantissa values involved. Intel scientists estimated the fraction of the total input number space that is prone to failure to be $1.14 \cdot 10^{-10}$ [86], which explains why it took several months before the user community eventually became aware of the problem.

☐

---

[14] SRT stands for Sweeney, Robertson, and Tocher, who independently invented the method in the 1950s.

RANDOM TESTING IS UNBIASED BUT ALSO UNFOCUSSED

The basic idea behind random testing is to get rid of human preconceptions and misconceptions by having an impartial random process select the test cases. Note that random testing does not simply mean random vectors, though. In order to exercise much of a circuit's state graph, it is the higher-level transactions that are to be chosen randomly, not the gate-level bit patterns. Except for simple combinational subcircuits, applying random bit patterns would be very inefficient.

3.2.5 Collecting test cases from multiple sources helps

We now know about the benefits and difficulties of simulation and testing. The trouble with dynamic verification is indeed knowing how to select a compact set of test cases that makes a mistaken design behave differently from a functionally correct one for any situation that might be relevant to the final circuit's operation. The fact that today's VLSI circuits comprise entire systems only exacerbates the problem. Some good advice is as follows.

Observation 3.10. *Except for the simplest subsystems where exhaustive testing is feasible, an acceptable selection of test cases shall comprise:*
- *A vast set of data that makes the MUT work in all regular regimes of operation, that exercises every functional mechanism, and that strains array and memory bounds.*
- *Particular numeric data that are likely to cause uncommon arithmetic conditions, including over-/underflow, division by small numbers, sign, carry, borrow, and not-a-number handling.*
- *Pathological cases that ask for exception handling and out-of-the-normal control flows.*
- *Genuine data sequences collected from real-world service.*
- *Randomly selected test cases.*

3.2.6 Assertion-based verification helps

Assertions are small pieces of simulation code embedded within a MUT that do not affect functionality. Instead, these Boolean tests are included solely to monitor the model's operation and to report any anomalous or unexpected condition that might occur, e.g.

- Memory addresses that point outside their legal range,
- State machines that assume parasitic, illegal or otherwise suspect states,
- Unforeseen input values and other out-of-the-ordinary conditions,
- Illegal instruction codes (opcodes) and unexpected status codes,
- Numeric over/underflows and other scaling problems,
- Event sequences unforeseen by the application or protocol,
- Resource conflicts and other situations of mutual lock-up,
- Excessive iteration counts or other unexpected state variables, and the like.

Assertions, aka in-code sanity checks, included in simulation models are effective at providing protection against design and coding errors. They nicely complement response checking because

- Feedback is immediate. There is no need for an abnormal condition to propagate to some distant node placed under constant monitoring.

- The link between a design flaw and its manifestation is short. There is no need for the engineer to trace back an output mismatch in an attempt to locate its place of origin.
- Assertions are a long-lasting investment. There is no need to repeatedly adjust them when submodels are being assembled to form larger design entities.

Especially the third feature is in sharp and welcome contrast to the stimulus/response pairs of a prerecorded functional gauge that must be modified whenever latency changes. Many assertions refer to the interface between a subcircuit and the embedding circuitry and between a subprogram and the calling code, but they are not limited to this.

Hint: While writing code for a MUT, enter an assertion wherever you explicitly or implicitly assume that a certain property would hold in real-life service.

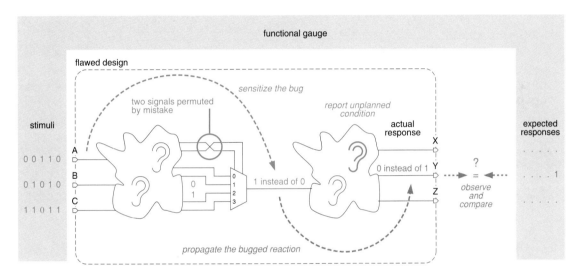

Fig. 3.6 Assertions resemble spies placed right into the MUT itself.

Assertions are not normally intended for synthesis. Unless they are ignored by synthesizers anyway — as VHDL assert statements are — assertions must get commented out or otherwise disabled prior to synthesis. As an alternative, a designer is free to implement part of the sanity checks in his HDL code in such a manner as to synthesize them into extra surveillance circuitry. By having that circuitry activate an alarm upon detection of an out-of-the-ordinary condition, he can take advantage of assertions to add **self-checking** capabilities to physical parts.

3.2.7 Separating test development from circuit design helps

The idea is to safeguard a design against oversights, misconceptions, and poor functional coverage by organizing manpower into two independent teams. A first team or person works towards designing the circuit while a second one prepares the functional gauge. Their respective interpretations are then crosschecked by verifying early behavioral models of the circuit against the gauge, see fig.3.7.

The goal of the circuit designers is to come up with a model that is functionally correct whereas the test engineers essentially try to prove them wrong.

The benefits and shortcomings of this healthy adversarial relationship are as follows.

+ Mistakes made by either team are likely to be uncovered by the crosschecking process.
+ The same holds for ambiguities in the initial specifications.
+ Having the design and the test teams work concurrently helps to cut down design time.
− A chance always remains that truly misleading specifications get interpreted in identical but erroneous ways by both teams.
− The difficult task of finding test cases of adequate coverage is left to humans.

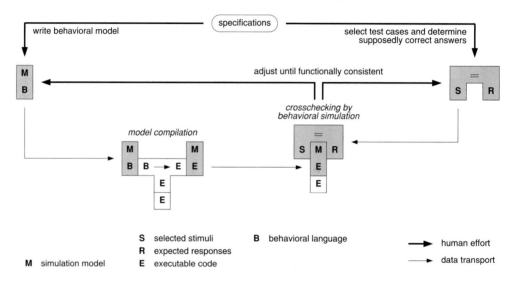

Fig. 3.7 Circuit models and functional gauge being prepared by separate teams.[15]

---

[15] Figures 3.7, 3.8, and 3.9 come as **T-diagrams**, a notation popular in the context of compiler engineering that serves to plan the porting of programs from one machine to another [87]. As an extension to the established T-diagram notation, two extra symbols have been added. One stands for a functional gauge and the other for a piece of information-processing hardware, such as a digital ASIC or an FPL device. Also note that synthesis tools are viewed here as compilers that turn behavioral models into gate-level netlists, and gate-level simulators as interpreters that translate between netlists and executable code.

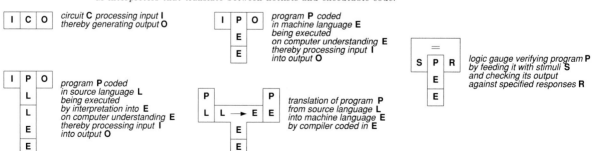

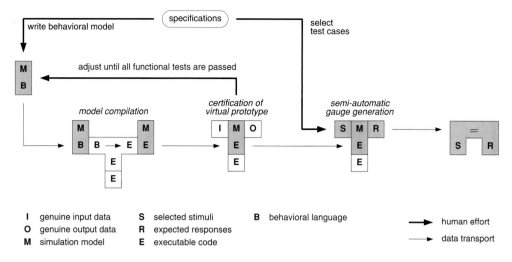

Fig. 3.8 Preparation of a functional gauge with the aid of a virtual prototype.

3.2.8 Virtual prototypes help to generate expected responses

Rapid prototyping has been introduced in section 3.1.2. Once a virtual prototype has been throughly certified in a multitude of test runs, it becomes a **golden model** which is assumed to be free of functional errors for the purpose of the upcoming design steps. As illustrated in fig.3.8, golden models come in handy for automatically computing expected responses from a selection of stimuli. Yet, note that the problem of selecting a set of relevant test cases is not addressed by this technique.

3.3 | Reusing the same functional gauge throughout the entire design cycle

Over the last few decades, functional verification has become ever more onerous because the overall set of potential circuit behaviors, correct and incorrect, has exploded with VLSI and ULSI circuit complexities. The concern is best expressed by a quote (from Walden Rhines):

> The question is whether the percentage for verification time tops out at 70%
> [of the total engineering effort in VLSI design] or it goes to 95% in the future.

Industry just cannot afford to rewrite functional gauges and testbenches over and over again as a design matures from a virtual prototype into synthesis code, a gate-level netlist, and — finally — into a physical part. Moreover, it is absolutely essential that a MUT be checked against the same specifications throughout the entire design cycle. Figure 3.9 again illustrates the design process but differs from fig.1.13 in that it emphasizes the reuse of stimuli and expected responses. How to do so is not immediately obvious because design views and tools greatly differ in their underlying assumptions, see table 3.3.

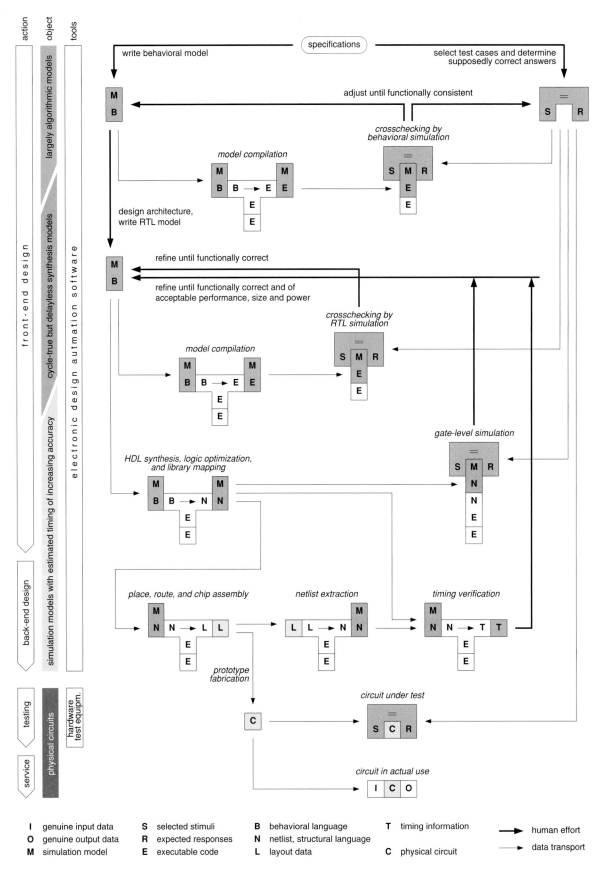

Fig. 3.9 Life cycle of a functional gauge during VLSI design and test in T-diagram notation.

Table 3.3 | Data, signal, and timing abstractions encountered during VLSI design and test.

level of abstraction	relevant data types and structures	numerical precision	modelling of electrical phenomena	relevant time scale for latency	timewise simulation resolution
immaterial circuit models					
algorithmic model	abstract and fairly free	essentially unlimited	not an issue	system-level transaction	not an issue
automata theory	discr. symbols	not an issue	not an issue	abstr. cycle	clock cycle
RTL synthesis model (VHDL)	numbers, bits, enumer. types	finite	optional	clock cycle	event sequence
gate-level netlist post-layout netlist	bits	finite	logic values	clock cycle	circuit delays
physical hardware					
autom. test equip.	bits	finite	discr. volt.	clock cycle	discr. strobes
physical circuit			cont. quant.		continuous

The difficulty is in

"How to make sure the same stimuli and expected responses can be reused across the full VLSI design cycle from purely algorithmic specifications to the testing of fabricated parts in spite of the various pieces of software and hardware being involved"

This is more than just a matter of format conversion. Ideally, a single functional gauge is reused with only minimal modifications to account for unavoidable differences in timewise and numerical resolution. Having to rewrite or to reschedule test patterns at each development step must be avoided for reasons of quality, trustworthiness, and engineering productivity.

From a hardware engineering point of view, a good simulation setup

- Is compatible with all formalisms and tools used during VLSI specification, design, and test (such as automata theory, MATLAB, HDLs, logic simulators, and ATE).
- Adheres to good software engineering practices (modular design, data abstraction, reuse, etc.).
- Translates stimuli and responses from bit-level manipulations to higher-level transactions.
- Consolidates simulation results in such a way a as to facilitate interpretation by humans.
- Is capable of handling situations where the timewise relationship between circuit input and output is unknown or difficult to predict.
- Manages with reasonable run times.

Next, we are going to discuss six measures that greatly contribute towards these goals.

3.3.1 Alternative ways to handle stimuli and expected responses

There is much liberty regarding how to organize a testbench. Even after one has selected the stimuli and response vectors that together form a functional gauge, there exist at least three conceptual alternatives for handling them.

Hardcoded testbench. Stimuli and expected responses are prepared before simulation begins and are included in the source code of the testbench itself, e.g. with the aid of repetitive instruction sequences, program loops, conditional branching, look-up tables, and the like.

File-based testbench. Stimuli and expected responses are prepared beforehand and are stored on disk files from where they are retrieved at run time under control of the testbench.

Golden-model-based testbench. Stimuli are generated at run time, e.g. by having the testbench invoke subprograms that act as test pattern generators. The expected responses are obtained from feeding some previously certified golden model with the same stimuli.

With the exception of trivial subcircuits, the first approach is too limited and too rigid to be of any practical value. We are thus going to elaborate on the latter two approaches, the most simple embodiments of which are depicted in figs.3.10a and b respectively.

3.3.2 Modular testbench design

Checking a MUT against a golden model often comes in handy, all the more so as modern simulators accept multilingual input.[16] As an example, you may want to check a synthesis model written in VHDL against a behavioral model developed earlier with SystemC. Yet, golden-model-based simulation implies recomputing the same sequence of expected responses for every single run. If debugging a MUT necessitates many iterations, a lot of computing resources get wasted in the process, and much the same applies to run-time stimuli generation. A better idea then is to compute stimulus/response pairs once and to store them on disk for subsequent file-based simulation runs as illustrated in figs.3.10f, g, and h.

The engineering effort to satisfy many different needs easily gets out of hand, however, unless one can identify a small number of versatile and reusable software modules from which all sorts of simulation setups can be readily assembled. This has been done in fig.3.10, while fig.3.14 shows the key modules in more detail. VHDL source code for a testbench that adheres to this concept will be given elsewhere in this text.[17]

Observation 3.11. *With testbenches being major pieces of software,*
it pays to have a look at them from a software engineering perspective.

3.3.3 A well-defined schedule for stimuli and responses

Another important choice refers to the timewise sequence of key events that repeat within every stimulus/response cycle, and to their relative timing. Poor timing may cause a gate-level model to report hundreds of hold-time violations per clock cycle during a simulation run, for instance, whereas a purely algorithmic model is simply not concerned with physical time. To complicate things further, engineers are often required to co-simulate an extracted gate-level netlist for one circuit block with a delayless model for some other part of the same design.

---

[16] ModelSim by Mentor Graphics is one such product, for instance.
[17] In section 4.9.4.

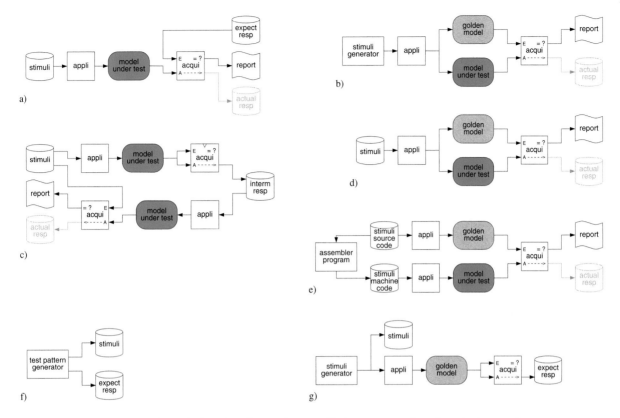

Fig. 3.10 Software modules from which testbenches can be assembled to serve a variety of needs. A setup that operates with a functional gauge previously stored on disk (a), a setup that generates stimuli and expected responses at run time (b), and an arrangement suitable for designs that implement involutory mappings (c). Alternatives for using a golden model as a reference (d,e); (e) addresses the special case where the stimuli exist as a piece of source code for a program-controlled processor. Options for preparing stimulus/response pairs (f,g).

Observation 3.12. *To be useful for comparing circuit models across multiple levels of abstraction, a testbench must schedule all major events in such a way as to respect the limitations imposed by all formalisms and tools involved in circuit specification, design, simulation, and test combined.*

Formalisms and tools are meant to include automata theory, HDLs, RTL models, gate-level netlists (whether delayless or backannotated with timing data), simulation software, and automated test equipment (ATE). In their choice of a schedule, many circuit designers and test engineers tend to be misled by the specific idiosyncrasies of one such instrument.

Key events

Consider some synchronous digital design.[18] The most important **events** that repeat in every clock cycle during both simulation and test then include

---

[18] We assume the popular single-phase edge-triggered clocking discipline where there is no difference between clock cycle and computation period, see section 6.2.2 for details.

- The application of a new stimulus denoted as △ (for Application),
- The acquisition and evaluation of the response denoted as ⊤ (for Test),
- The recording of a stimulus/response pair for further use denoted as □ (for storage),
- The active clock edge symbolically denoted as ↑, and
- The passive clock edge denoted as ↓ (mandatory but of subordinate significance).

When these events are ordered in an ill-advised way, the resulting schedules most often turn out to be incompatible. Exchanging functional gauges between software simulation and hardware testing then becomes very painful, if not impossible. The existence of a problem is most evident when supposedly identical simulation runs must be repeated many times over, fiddling around with the order and timing of these events just to make the schedule fit with the automated test equipment (ATE) at hand. A suspicion always remains that such belated manipulations of test vectors cannot be trusted because any change to the sequence of events or to their timing raises the question of whether the original and the modified simulation runs are equivalent, that is, whether they are indeed capable of uncovering exactly the same set of functional flaws.

A coherent stimulus/response schedule

The schedule of fig.3.11 has been found to be portable across the entire VLSI design and test cycle. Its formal derivation is postponed to section 3.7.

Observation 3.13. *At the RTL and lower levels, any consistent testbench shall*
- *provide a clock signal even if the MUT is of purely combinational nature,*
- *log one stimulus/response pair per clock cycle, and*
- *have all clock edges, all stimulus applications, and all response acquisitions occur in a strictly periodic fashion, symbolically denoted as* △ ↓ (⊤ = □) ↑.

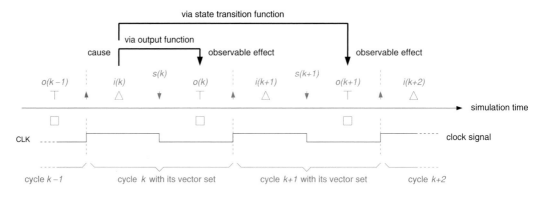

Fig. 3.11 A coherent schedule for simulation and test events with cause/effect relationships and single-phase edge-triggered clock signal waveform superimposed.

As becomes clear from fig.3.11, each computation period gets subdivided into four phases. A standard setup for a symmetric clock of 10 MHz is given below as an example. Of course, the numerical figures must be adapted to the situation at hand.

cycle	events with times of occurrence [ns]			
k	$\triangle$	$\downarrow$	$\top\square$	$\uparrow$
0	10	50	90	100
1	110	150	190	200
2	210	250	290	300
...	...	...	...	...

Observe that we have elected to assign the active clock edge to the end, rather than to the beginning, of a cycle because a finite state machine originates in some state — termed start state s_0 — without any intervention of a clock.

It is important to know when and how the effects of a specific stimulus $i(k)$ become observable. Note from fig.3.11 that $o(k) = g(i(k), s(k))$ is visible in the response acquired after applying $i(k)$, while its effect on the state $s(k)$ can be observed only in the response acquired one clock cycle later and only indirectly as $o(k + 1) = g(i(k + 1), f(i(k), s(k)))$.

3.3.4 Trimming run times by skipping redundant simulation sequences

Memories, counters, and state machines tend to inflate the number of stimulus/response pairs necessary to verify functionality. This is because stereotypical activities eat a lot of computation time without contributing much towards uncovering design flaws. Much of a simulation run then just reiterates the same state transitions many times over without moving on to fresh states and functional mechanisms for a long time. Examples are quite common in timers, filters, data acquisition equipment, and data transfer protocols, but the situation is notorious in image processing and man–machine interfaces.

For productivity reasons, designers seek to cut back cycles that feature little or highly recurrent computational activities in a design. What follows are suggestions of what they can do.

- Take advantage of the scan path facility to skip uninteresting portions of a simulation.
- Include auxiliary logic in the MUT that trims lengthy counting or waiting sequences and unacceptably large data quantities while in simulation mode.
- Do the same to synthesis models in view of the later testing of physical circuits.
- Model circuit operation on two different time scales (fine and coarse).

Example

Imagine you are designing a graphics accelerator chip. Instead of always simulating the processing of full-screen frames of 1280 pixels × 1024 pixels, make your MUT code capable of handling smaller graphics of, say, 40 pixels × 32 pixels as well. Use this thinned-out model for most of your simulation runs, but do not forget to run a couple of full-size simulations before proceeding to back-end design and prior to tapeout.

□

3.3.5 Abstracting to higher-level transactions on higher-level data

Abridged simulation techniques notwithstanding, humans are easily overwhelmed by the volume of bits and bytes when confronted with raw simulation data. Rather than drowning engineers with tons of 0s and 1s, functional gauges can be made to work at higher levels of abstraction. Table 3.4 makes a distinction among four grades.

Table 3.4 | Circuit models and testbenches can be made to cooperate in various ways.

Grade Operation steps	0 free style	1 cycle per cycle	2 word per word	3 trans. per trans.
Theoretical underpinning	none	automata theory		systems theory
Response checking	visually	by self-checking testbench		
Reporting is in terms of	waveforms	bit errors	word errors	figures of merit
Typical data items packaged in VHDL as	bit std_logic and ..._vector		string, sample, block, frame record	
I/O data formats	must match bit for bit		translated by protocol adapters	
High-level I/O processing	none		with MATLAB or similar tool	
Event sequence	arbitrary	periodic and locked to clock $\triangle \downarrow$ ($\top = \square$) $\uparrow$		
Simulation driven by	signal changes (VHDL events)	clock signal		transactions on high-level data
Clock generator	none (hardcoded)	is part of testbench		is part of MUT
Stimulus–response pair	n.a.	per clock	per data word	per transaction
Latency relationship between MUT and expected responses	n.a.	must match cycle for cycle	adjusted by start and compl. signals	adjusted by full handshake protocol
Overall quality	chaotic and much too limited	sound but low-level, matches ATE	offers welcome isolation from details	abstract, MUT to support handshaking
Best for	nothing	small subfunctions	larger circuits and systems	multi-clock systems

The functional gauges presented so far have worked with bit vectors locked to predetermined clock cycles. This straightforward concept is adequate for circuit blocks of modest complexity and will be referred to as **grade 1** simulation. A testbench organized in this way is shown in fig.3.14. As a matter of fact, cycle-true binary stimuli and responses are the only way to go when it comes to the testing of physical parts with automated test equipment (ATE).

Dealing with high-level stimuli and responses asks for protocol adapters

Grade 1 setups are inadequate for simulating larger circuits and systems. A first improvement is to collect stimuli and responses in composite data types such as records, data packets, audio fragments, or whatever is most appropriate for the application at hand.

Example

JPEG image compression in essence accepts an image frame, subdivides it into square blocks, and uses the Discrete Cosine Transform (DCT) to calculate a set of spectral coefficients for each block. Those coefficients are then quantized or outright replaced by zero when their impact on the perceived image quality is only minor. Image frames, blocks, and coefficient sets are the data items you would want to deal with when comparing the behavior of a JPEG MUT against a golden model. You would probably consider the compression of one block or, alternatively, of an entire frame to be a relevant transaction. Details such as the reading in of pixels or the toggling of individual data bits would just distract your attention (unless you were forced to debug a model at a very elementary level).
□

Observation 3.14. *A testbench serves not only to drive the MUT, its more noble duties are to translate stimuli and responses across levels of abstraction, and to consolidate simulation results so as to render interpretation by humans as convenient as possible.*

The difficulty is that the MUT — and possibly other system components as well — will undergo profound changes during the development process. What begins as a purely behavioral model is later refined into an RTL model, and ultimately becomes a gate-level netlist. The latter models will necessarily operate in terms of bits and clock cycles, however, exactly like the physical circuits they emulate. Any decent simulation setup must thus follow and support the process of successive refinement. A helpful aid for doing so are protocol adapters, aka bus-functional models (BFM), that translate stimuli and actual responses across levels of abstraction.

An input protocol adapter accepts a high-level stimulus (an image frame in the above example), breaks it down into smaller data items (e.g. blocks and pixels), and feeds those to the MUT word by word or bit by bit over a time span that may cover hundreds of clock cycles, see fig.3.12. Another adapter located downstream of the MUT does the opposite to consolidate output bits into a higher-level response (e.g. collecting bits into JPEG image data). The need to rework stimulus/response pairs each time a modification is made to the MUT can thus be avoided. Any change just affects the MUT itself and one or more of the protocol adapters but neither the testbench nor the functional gauge, thereby greatly simplifying maintenance.

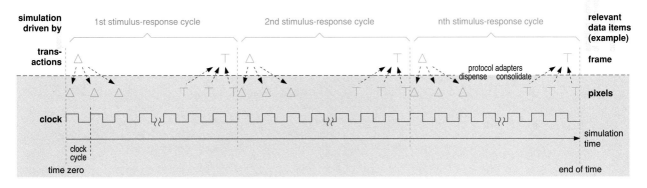

Fig. 3.12 Protocol adapters fill the gap between high-level transactions and cycle-true bit-level models (RTL or gate-level netlist).

Math packages are a great help for processing high-level data

Rather than just report the number of bit-level mismatches found, a high-level simulation setup ought to distill the essence of a simulation run into a few figures of merit relevant for the application. In the occurrence of lossy image compression, the decompressed and the original image will necessarily differ, and the system designer will be most interested in learning about the signal-to-quantization-noise ratio (SQNR), the maximum perceived color deviation, and similar overall ratings. It is a good idea to unburden the HDL testbench from such calculations by taking advantage of some standard mathematics tool such as MATLAB instead. Math packages not only provide high-level functions for data and signal processing and for statistics, but also offer superior means of visualization.

Simulation setups patterned after target system facilitate successive refinements

In practice, most simulation setups for large circuits follow the organization of the target system; see fig.3.13 for an example from wireless telecommunications where multiple antennas are being used at the transmitter and at the receiver end of a wireless channel to improve data rate and robustness.

In the setup of fig.3.13, a behavioral model is substituted for each RAM, IF (de)modulator, PCI interface, and other subfunction that collaborates with the MUT. The various HDL models for one design entity share the same interface so that they can serve as drop-in replacements for each other during the development process. The preparation of stimuli and the evaluation of responses is implemented in MATLAB so that the HDL testbench code remains essentially limited to configuring, controlling, and monitoring the MUT via the PCI interface. Protocol adapters take care of

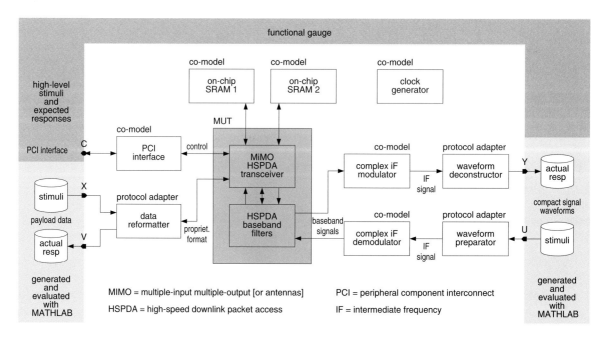

Fig. 3.13 Example of a sophisticated simulation setup patterned after a target system from wireless telecommunication (simplified).

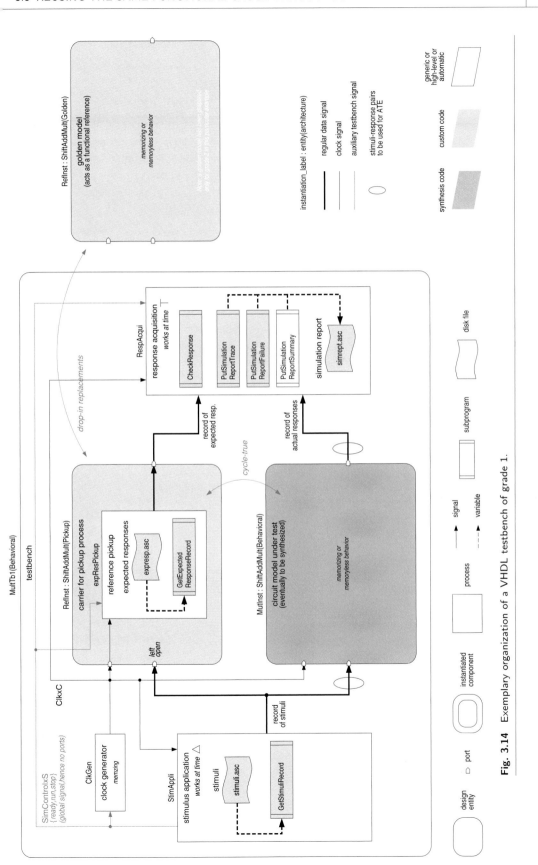

Fig. 3.14 Exemplary organization of a VHDL testbench of grade 1.

translating where necessary. All this together greatly simplifies the process of successive refinement and qualifies the setup as of **grade 2**.

3.3.6 Absorbing latency variations across multiple circuit models

Most RTL circuit models exhibit latencies not found in purely algorithmic models. This is a consequence of various optimization steps that designs undergo as part of architecture design. The necessity to accommodate RAM, parallel $\leftrightarrow$ serial conversions, and specific input/output protocols further contributes to latency in MUTs of actual circuits.

Example

From a mathematical point of view, JPEG decoding is a combinational function that can be computed by a delayless model, that is, with latency zero. As opposed to this, typical image decoding hardware ingests one set of quantized DCT coefficients at a time and takes many clock cycles before spitting out the pixels of a block or of the assembled image.
□

The point is that most architectural decisions have a dramatic impact on latency. For certain algorithms, the number of clock cycles required to complete a given computation may even depend on numerical data values. Probably the utmost tolerance towards latency variations is required when testing network processors because data packets do not necessarily emanate from such processors in the same order as that in which they were fed in. We must accept that
• latency is subject to change many times during a typical design cycle and that
• MUT and golden model might not exhibit identical latencies.
Locking stimuli and responses to specific clock cycles, as in the grade 1 setup of fig.3.2, for instance, does not offer the flexibility to handle such situations.

Observation 3.15. *To be truly reusable, a testbench must be capable of handling models where the timewise relationship between circuit input and output is unknown or difficult to predict.*

In a **grade 2** testbench, the comparison of responses and the calculation of high-level figures of merit can be made to absorb latency variations by delegating the task of knowing when to request and when to accept new data items to protocol adapters, see fig.3.15 for such a setup. Upstream adapters are designed so as to feed the MUT on request and, analogously, downstream adapters so as to wait for the next valid data item to emerge at the MUT's output. To make data-driven transfers a reality, each protocol adapter must either
○ interpret status flags from the MUT (such as "ready/$\overline{\text{busy}}$") or, in the absence thereof,
○ tacitly count clock cycles concurrently to the model's internal operation;
    emulating state machines that are part of the MUT may also be required.
While any substantial change to the MUT's architecture is likely to necessitate adjustments to the latency parameters coded into protocol adapters and co-models, the functional gauge, the testbench, and the golden model can hence remain the same.

A more radical solution is to impose handshaking for all data transfers between all subsystems and circuits involved. In a **grade 3** simulation, self-timed I/O transfers are supported not only by the testbench but also by the MUT and, hence, ultimately by the physical circuit itself. The MUT must

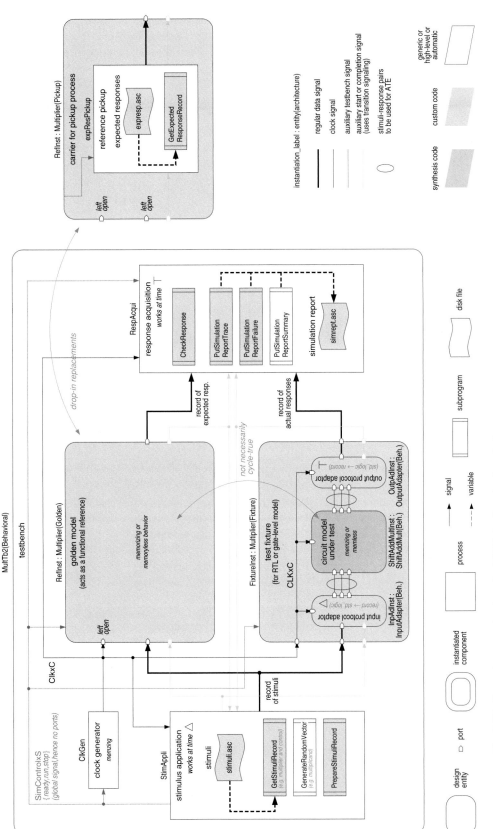

Fig. 3.15 Exemplary organization of a VHDL testbench of grade 2.

implement a full handshake protocol and must provide the necessary control signals on its input and output ports. The same applies to the golden model, if any. Clocking is viewed as a technicality internal to the MUT that is essentially unrelated to the high-level transactions that matter from a system perspective alone; see fig.3.16 for a sample setup.

An authoritative treatise on functional verification is [88]. Further ideas are discussed in [89] and other papers in the same journal issue. [90] discusses testbench reuse while [91] shows how to automate testbench generation with the *e* verification language.

3.4 | Conclusions

- More often than not, what is available at the onset of a VLSI project are intentions rather than specifications. Identifying the real needs and casting them into workable instructions always is the first step and a primary occupation in the design process. Rapid prototyping often is the only practical way to condense initial conceptions into detailed and unambiguous specs.

- In the absence of a more integral alternative such as formal verification, simulation remains the prevalent method for functional design verification. Being a dynamic technique, it gives only limited coverage against design flaws. This finding, that applies to both VLSI design and software engineering, reflects an unsolved puzzle of systems design in general.

- The challenge of dynamic verification is to safeguard oneself against all plausible design slips without attempting exhaustive software simulation or hardware testing. Coming up with a comprehensive collection of test cases simply requires foresight, care, precision, and a lot of work at the detail level. Albeit very general, the rules below give some guidance.

  - Cover all modes, situations, and conditions under which the system is to operate.
  - Have the test suites address uncommon situations and exceptional, if not pathological, inputs as well. To increase the likelihood of disclosing problems, blend genuine data with tests that focus on anomalous input, unusual states, numerical corners, and other exceptional conditions. Also consider adding test cases selected at random.
  - Identify distinct subsystems and functional mechanisms. You may address them separately provided they do indeed work independently from each other.
  - Make sure you understand what potential design flaws might pass undetected whenever a shortcut is taken.
  - Generously include in-code sanity checks (assertions) into simulation models.

- While establishing a verification plan, beware of preconceptions from the design process as to what situations and issues are to be considered uncritical. Insist on having persons other than the IC designers or HDL code writers select, or at least review, the test cases.

- Making the same functional gauge work across the entire VLSI design and test cycle is a necessity as functional consistency is otherwise lost. Doing so typically implies data abstraction and latency absorption.

- The desirable characteristics for hardware testbenches include:

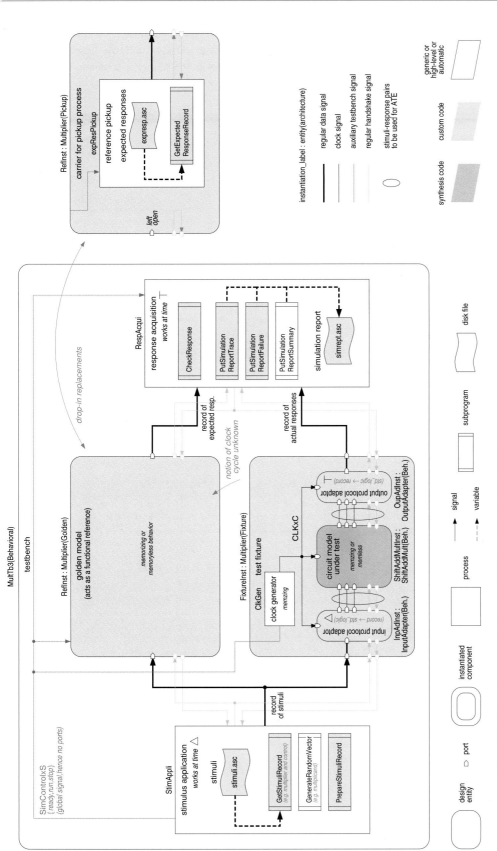

Fig. 3.16 Exemplary organization of a VHDL testbench of grade 3.

- Testbench design can bank on a library of reusable software modules.
- Code writing is confined to application-specific items as much as possible.
- Alternating between file-based and golden-model-based simulations is painless.
- Events are ordered so as to conform with modelling at all levels of abstraction.

- While established software engineering practices such as modular design, reuse, and data abstraction help to make testbench development more efficient, be prepared to spend more time on verifying functionality than on designing the circuit that implements it.

- There are virtually no limits to making simulation setups more sophisticated.

3.5 | Problems

1. Design a functional gauge for exhaustive verification of the Gray counter specified by truth table 3.2. With how many clock cycles can you manage? Indicate the formula for w-bit counters and the actual value for $w = 4$. How does this figure relate to the lower bound given in (3.2)?

2. Devise a functional gauge for a logic comparator function that tells whether two 6-bit vectors are the same or not. Go for a set of vectors that you consider a reasonable compromise between simulation time and functional coverage. Verify the VHDL architecture **correct** given below or the gate-level circuit obtained after synthesis against that gauge. No inconsistency must occur. Now check how your gauge performs on the flawed architectures given below. Note that the mistaken circuits named **flawedy** and **flawedu** are authentic outcomes from efforts by human designers who were using schematic entry tools. The deficiency of the fourth example **flawedz**, in contrast, has been built in on purpose in order to demonstrate the impact of an oversight during the editing of HDL code.

```
entity compara6 is
   port (
      INA: in Std_Logic_Vector(5 downto 0);
      INB: in Std_Logic_Vector(5 downto 0);
      EQ: out Std_Logic );
end compara6;


-----------------------------------------------------------------------

-- correct description of 6bit logic comparator function
architecture correct of compara6 is
begin
   EQ <= '1' when INA=INB else '0';
end correct;


-----------------------------------------------------------------------

-- flawed as one of the two arguments has its bits misordered
-- note: a wrong ordering of INB in the port list has the same effect
architecture flawedy of compara6 is
   signal INBM : Std_Logic_Vector(5 downto 0);
```

```
begin
   each_bit : for i in 5 downto 0 generate
      INBM(i) <= INB(5-i);
   end generate;
   EQ <= '1' when INA=INBM else '0';
end flawedy;


-------------------------------------------------------------------------


-- mistaken translation of desired function into boolean operations
architecture flawedu of compara6 is
   signal C1 : Std_Logic_Vector(5 downto 0);
   signal C2 : Std_Logic_Vector(2 downto 0);
begin
   first_level : for i in 5 downto 0 generate
      C1(i) <= not (INA(i) xor INB(i));
   end generate;
   second_level : for i in 2 downto 0 generate
      C2(i) <= not (C1(i) xor C1(i+3));
   end generate;
   EQ <= C2(2) xor C2(1) xor C2(0);
end flawedu;


-------------------------------------------------------------------------


-- corrupt due to a useless statement forgotten in the code
architecture flawedz of compara6 is
begin
   process (INA,INB)
   begin
      if INA=INB then EQ <= '1';
      else EQ <= '0';
      end if;
      if INA="110011" then EQ <= INA(0);
      end if;
   end process;
end flawedz;
```

3. The purpose of this problem is to show that a functional gauge may ensure full toggling of all nodes in a gate-level circuit and still be inadequate for functional verification. To that end, find two combinational networks together with a (nonexhaustive) gauge such that
   • all nodes get toggled back and forth,
   • both networks comply with the functional gauge, and
   • the two networks are functionally different.
   What are the simplest two such circuits you can think of? Generalizing to combinational n-input single-output functions, how does the number of test patterns necessary for full toggling relate to that required for exhaustive verification?

4. Consider a digital circuit that connects to a microprocessor bus, a situation sketched in fig.8.42 that mandates the usage of bidirectional pads on the data bus. A state machine inside the chip generates the enable signal for the pad drivers from its state and from $\mathrm{WR}/\overline{\mathrm{RD}}$ or some similar signal available at one of the chip's control pins. In order to stay clear of transient drive conflicts, the bus must not be driven from externally before the on-chip drivers have actually released the bus in reaction to the control pin asking them to do so. As a consequence, both

a testbench and physical test equipment must observe a brief delay between updating the control signal and imposing data on the bus. Extend the precedence graph and the schedule of figs.3.17 and 3.11 accordingly.

3.6 | Appendix I: Formal approaches to functional verification

Formal verification attempts to prove or disprove the correctness of some circuit representation by purely analytical means, i.e. without simulating the circuit's behavior over time. A successful proof gives the designer the ultimate confidence that his design will indeed function as previously specified at some higher level of abstraction, and this irrespective of the input as there is no need to apply stimuli. Most formal verification algorithms work by converting a given design representation into a state graph, an ordered binary decision diagram (OBDD), or some other graph-type data structure before analyzing that structure and/or comparing it against similar design representations. There are different degrees of ambition, though.

EQUIVALENCE CHECKING

Verifying the functional equivalence between two gate-level netlists or between a netlist and a piece of HDL code is not that difficult. Logic equations extracted from the gate-level netlist are compared against the reference set of logic equations using theorems from switching algebra. Software tools capable of doing so are typically used to check the consistency — in regular operation mode — of a gate-level netlist with the original RTL synthesis model after test structures have been added. Other relatively minor modifications such as clock tree insertion, logic reoptimization, and conditional clocking are covered as well.

While combinational subfunctions make up much of an RTL model, there are severe limitations when the checking is to be extended to sequential behavior. Automatic conformity checking of circuit models that are supposed to have equivalent external behavior but that differ in the number and/or location of registers, e.g. as a consequence of state reduction or architectural optimizations, remains a challenging research topic [92].

Last but not least, equivalence checking always presupposes the availability of a golden model.

MODEL CHECKING

As opposed to the above, model checking does not need any reference model, but aims at finding out whether a circuit model satisfies under all circumstances a set of specified criteria that any meaningful implementation must satisfy. A welcome property of model checking is that it provides a counterexample when some specification is violated by a design. A serious problem is the combinatorial explosion that confines the approach to subsystems with a fairly limited number of states. A detailed discussion is given in [93].

DEDUCTIVE VERIFICATION OR MODEL PROVING

Deductive verification is closely related to theorem proving. The goal is a mathematical proof that a given circuit model or protocol does indeed conform with its formal specifications. The answer essentially is of type "true" or "false" and thus provides few clues to developers as to what is wrong with their designs. Deductive verification further suffers from the problems mentioned in section 3.1.1, but remains an active research area. The reader is referred to [94] [95] for accounts on formal verification technology.

3.7 | Appendix II: Deriving a coherent schedule for simulation and test

This section serves to confirm that the simulation schedule presented in section 3.3.3 is indeed a well-founded one that conforms to the fundamental timing requirements of synchronous circuits without being unnecessarily constrained further. In order to do so, we approximate timing to a degree that makes it possible to describe how a circuit behaves when viewed from outside.[19]

EXTERNAL TIMING REQUIREMENTS IMPOSED BY A MODEL UNDER TEST (MUT)

Four sets of data propagation paths can be identified in any synchronous design that adheres to single-phase edge-triggered clocking.[20] These paths go
• from inputs to outputs with no intervening registers $(i \rightarrow o)$,
• from state-holding registers to outputs $(s \rightarrow o)$,
• from inputs to state-holding registers $(i \rightarrow s)$, and
• from state registers to state registers $(s \rightarrow s)$.

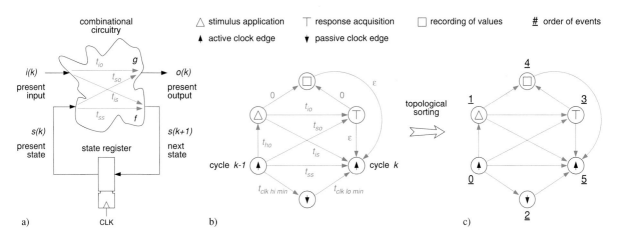

Fig. 3.17 Data propagation paths through single-phase edge-triggered synchronous circuits (a) along with the pertaining precedence graph for simulation events (b,c).

[19] We do not aim at modelling or even at understanding what exactly happens inside the circuit yet. A more accurate analysis will become feasible on the basis of a detailed timing model to be introduced in section 6.2.2.
[20] Any asynchronous reset input can be handled like an ordinary input in this context.

Not surprisingly, we find the same four sets of paths in a Mealy automaton. This is simply because the functionality of any synchronous circuit can be modelled as a Mealy-type state machine. In response to a new input or state, a physical circuit finishes re-evaluating g and f after the fresh data have propagated along all paths through the combinational logic. As input and state do not, in general, switch simultaneously, four delay parameters need to be introduced, namely t_{io}, t_{so}, t_{is}, and t_{ss}.[21] Parameter t_{io}, for instance, denotes the time required to compute a new output o after input i has changed.

The necessary precedence relations for the circuit to settle to a stationary state are[22]

$$t_{\top}(k) \geq \max(t_{\triangle}(k) + t_{io} , t_{\uparrow}(k-1) + t_{so}) \tag{3.3}$$

$$t_{\uparrow}(k) \geq \max(t_{\triangle}(k) + t_{is} , t_{\uparrow}(k-1) + t_{ss}) \tag{3.4}$$

PRECEDENCE RELATIONS CAPTURED IN A CONSTRAINT GRAPH

The above precedence relations can be expressed by the constraint graph of fig.3.17b where each node stands for a major event associated with clock period k. Note that there are two active clock nodes, namely one for the clock event immediately before the clock period under consideration and a second one for the clock event at its end. Each precedence relation is represented by a directed edge that runs from the earlier event to the later one. The minimum time span called for by the associated condition is indicated by the weight of that edge:

- Computational delays. The aforementioned data propagation paths t_{io}, t_{so}, t_{is}, and t_{ss} map to a first set of four edges.

- State register hold-time requirement. A fifth non-zero weight says that new stimuli must not be applied earlier than t_{ho} after the previous active clock event. Ignoring this constraint is likely to cause hold-time violations at some bistables or might otherwise interfere with the precedent state transition.[23]

- Clock minimum pulse widths. Two more edges of small but non-zero weight are labeled $t_{clk\,hi\,min}$ and $t_{clk\,lo\,min}$ and indicate the minimum time spans during which the driving clock signal must remain stable.

- Securing coherent vector sets. There are also four edges of weight zero or close to zero. One of them leads from $\top$ to $\uparrow$ and has an infinitesimally small weight ε. It reflects the requirement that response acquisition must occur before the circuit gets any chance to change its state in reaction to the next active clock event. Three more edges define the preconditions for recording a consistent stimulus/response pair for the current clock cycle.

---

[21] Most practical circuits have their set of input bits grouped into a number of vectors, each of which has its own delay parameters, and similarly for outputs. Extending our approach to cover such situations as well is left to the reader as an exercise, see problem 4.

[22] Of course, precedence relations may be simpler in a given particular case, say for a combinational circuit (automaton with no state where t_{so}, t_{is}, and t_{ss} are not defined) or for a counter (Medvedev machine where t_{io} is not defined and $t_{so} = t_{ss}$). However, by consistently sticking to a scheme that is suitable for the most general case, we can avoid having to reorder events whenever we must move from one circuit type to another.

[23] Note that t_{is} and t_{ss} are meant to include the setup times of the registers. This explains why t_{su} does not appear in the constraint graph, as opposed to t_{ho}, which cannot be subsumed anywhere else.

SOLVING THE CONSTRAINT GRAPH

Any desirable sequence must order the events in such a way as to satisfy the above precedence relations for positive but otherwise arbitrary values of the seven timing parameters involved. All solutions are obtained from topological sorting of the precedence graph.[24] One such ordering is indicated by the underlined numbers in fig.3.17c. It corresponds to a periodic repetition of stimulus application, response acquisition, and clocking, and is symbolically denoted as $\triangle \top \uparrow$.

The precedence graph also indicates minor liberties. While it is true that the recording of a stimulus/response pair may take place at any time between response acquisition and the subsequent active clock edge, there is nothing to be gained from defining an extra point in time for doing so. The events of response acquisition $\top$ and recording $\square$ may as well be tied together.

The passive clock edge, on the other hand, is free to float between two consecutive active edges as long as the two constraints $t_{clk\,hi\,min}$ and $t_{clk\,lo\,min}$ are respected. As a final result, the event order $\triangle \downarrow (\top = \square) \uparrow$ will almost always represent a workable solution. The recommended simulation schedule is depicted in fig.3.11.

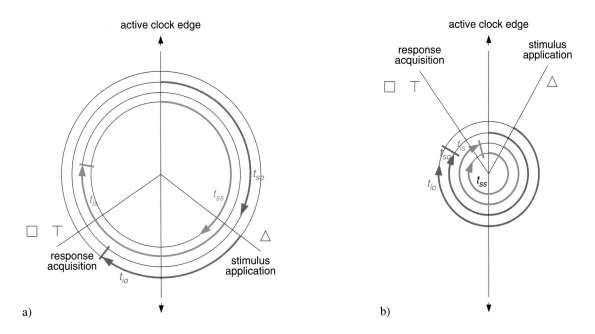

Fig. 3.18 Anceau diagram for a Mealy-type circuit operated at moderate speed (a) and close to maximum speed (b).

---

[24] The nodes of a graph are said to be in **topological order** if they are assigned integer numbers such that every edge leads from a smaller-numbered node to a larger-numbered one.

ANCEAU DIAGRAMS HELP VISUALIZE PERIODIC EVENTS AND TIMING

The Anceau diagram is very convenient for visualizing events and timewise relationships that repeat periodically. Each round trip corresponds to one clock cycle (or to one computation period). The example of fig.3.18a illustrates the simulation schedule just found.

Each arrow stands for the delay along one of the four signal propagation paths in a Mealy machine. The two short radial bars are just graphical representations of the max operators in (3.3) and (3.4) respectively. The outer one indicates when the output has settled to a new value and becomes available for acquisition. Similarly, the inner bar tells when the computation of the next state comes to an end and so determines the earliest point in time when an active clock edge can be applied.

If the operating speed is to be increased, the clock period grows shorter while delays and timing conditions remain the same. Beyond a certain point, stimuli must be applied earlier and/or responses will have to be acquired later in the clock cycle. This is shown in fig.3.18b, which corresponds to a circuit operating at close to its maximum speed.

Anceau diagrams come in handy for visualizing periodic event sequences and timing conditions. We will make use of them in later chapters in the context of clocking and input/output timing.

Chapter 4

Modelling Hardware with VHDL

4.1 | Motivation

4.1.1 Why hardware synthesis?

VLSI designers constantly find themselves in a difficult situation. On the one hand, buyers ask for microelectronic products that integrate more and more functions on a single chip. Following Moore's law, fabrication technology has always supported this aspiration by quadrupling the achievable circuit complexity every three years or so. Market pressure, on the other hand, vetoes a proportional dilation of product development times. Worse than this, time to market is even supposed to shrink. As a consequence, design productivity must constantly improve.

Hardware description languages (HDLs) and design automation come to the rescue in three ways: they

- Exonerate designers from having to deal with low-level details by moving design entry to more abstract levels,
- Allow designers to focus more strongly on functionality as synthesis tools construct the necessary circuits along with their structural and physical views automatically, and
- Facilitate design reuse by capturing a circuit description in a parametrized technology- and platform-independent form (as opposed to schematic diagrams, for instance).

Today, the transition from structural to physical is largely automated in digital VLSI design. The transition from purely behavioral to structural has not yet reached the same maturity, but HDL synthesis is routinely used for turning register transfer level (RTL) descriptions into gate-level networks that are then processed further with the aid of cell-based design automation software. A digital HDL essentially must be able to describe how subcircuits interconnect to form larger circuits and how those individual subcircuits behave functionally and timingwise.

Table 4.1 Languages commonly being used for modelling digital hardware.

Language	Originator/ Standard	Key characteristics
VHDL	DoD/ IEEE 1076	An HDL that supports not only structural and behavioral circuit models but also testbench models. A subset is synthesizable. Syntactically similar to Ada, see table 4.2 for more details.
Verilog	Gateway/ IEEE 1364	Conceptually very similar to VHDL, no type checking and more limited capabilities for design abstraction, though. Syntactically similar to C, see table 4.2 for more details.
System-Verilog	Accellera/ IEEE 1800	A superset of Verilog that includes many advanced features from VHDL and that may possibly supersede both of them. Supports object-oriented programming, not supported for synthesis yet.
SystemC (originally known as Scenic)	OSCI/ IEEE 1666	Extends C++ with class libraries and a simulation kernel. Makes it possible to add clocking information to C functions but does not support any timing finer than one clock cycle. Separates a block's behavior from communication details, synthesis path is via translation to RTL VHDL or Verilog with the aid of automatic allocation, scheduling, and binding.

4.1.2   What are the alternatives to VHDL?

As becomes evident from tables 4.1 and 4.2, **Verilog** [96] [97] shares most key concepts with VHDL. The same also applies to Verilog's recent and more advanced offspring **SystemVerilog**, making the differences between RTL synthesis models captured using those three languages largely a matter of syntax and coding style. As opposed to these, **SystemC** is not so much an HDL but more of a system description language targeted towards software/hardware co-design and co-simulation. It does not qualify for gate-level simulation and timing verification.

Here is why we have elected to go for VHDL in this text:

- The dissemination in the industry of HDLs other than those of table 4.1 is far too limited.
- Only VHDL and Verilog are widely supported by automatic synthesis tools.
- Strong typing, strict scoping, and stringent event ordering make VHDL a safer instrument than Verilog.
- VHDL has more sophisticated parametrization capabilities and is superior to Verilog when it comes to more abstract ways of modelling.

4.1.3   What are the origins and aspirations of the IEEE 1076 standard?

Providing spare parts over many years for industrial products that include ASICs and other non-standard state-of-the-art electronic components proves very difficult as technology evolves and as companies restructure. In search of a standard format for documenting digital ICs and for exchanging design data other than layout polygons, the US Department of Defense (DoD) in 1983 commissioned IBM, Intermetrics, and Texas Instruments to define an HDL. Ada was taken as a

Table 4.2 | Key features of VHDL and Verilog compared. See [98] for more.

Feature	VHDL	Verilog
Background and underlying concepts		
Industry standard	IEEE 1076	IEEE 1364
Initial acceptance / current revision	1987 / 2002	1995 / 2005
Roots	Ada	C
Overall character	dependable, verbose	concise, cryptic
Concurrent processes	yes	yes
Event-based concept of time	yes	yes
Circuit hierarchy and structure (netlist)	yes	yes
Discretization of electrical signals	adjunct package	part of language
Logic system	9-valued IEEE 1164	4 states, 8 strengths
Switch-level capability	no	yes
Language features and software engineering		
Interface declaration and implementation module	separate	no distinction made
Scoping consistent with module boundaries	yes	no
Strong typing	yes	no
Type conversion functions	adjunct package	none
Enumerated and other user-defined data types	supported	no
Data types acceptable at block boundaries	any	binary
Function arguments of variable word width	supported	no
Object classes with and without time attached	signals vs. variables	no distinction made
Timing and word size parametrization ("generics")	yes	yes
Conditional and repeated process generation and component instantiation ("generate")	yes	since 2001
Multiple models plus selection ("configuration")	yes	since 2001
Simulation		
Stringent order of events in the absence of delay	yes (via δ delay)	no
Event queue inspection (e.g. for timing checks)	part of language	via simulator calls
Text and file I/O	adjunct package	via simulator calls
Source code encryption mechanism	no	yes
Back-annotation from SDF files	VITAL IEEE 1076.4	yes
Acceleration of gate-level primitives	VITAL IEEE 1076.4	yes
Acceptance for sign-off simulation	yes	yes
Standard and macro cell models (for ASIC design)	commonly available	commonly available
3rd party components models (for PCB design)	scarce	commonly available
Synthesis		
Amenable to hardware synthesis	subset only	subset only
Timing constraints	not p.o.l. (SDF)	not p.o.l. (SDF)
Other synthesis directives	not part of language	not part of language
Model precomputation vs. hardware description	no distinction made	no distinction made
Analog and mixed-signal extension		
Designation	VHDL-AMS	Verilog-AMS
Industry standard	IEEE 1076.1	Accelera 2.2

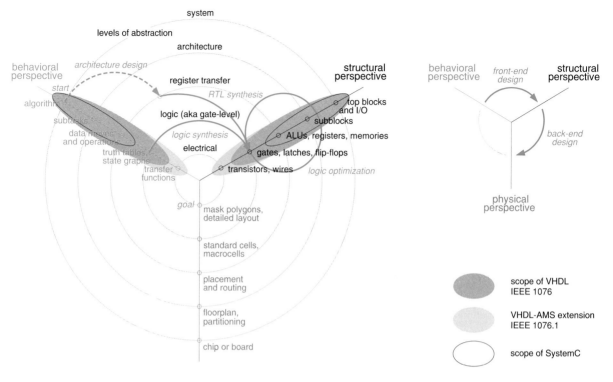

Fig. 4.1 VHDL and synthesis shown in the Y-chart.

starting point. As the project had originated from DoD's Very High Speed Integrated Circuits (VHSIC) program, the new language was given the acronym **VHDL**.

After a number of revisions and after military restrictions had been lifted, the proposal was eventually accepted as IEEE 1076(-87) standard in 1987. IEEE bylaws require any standard to be revised on a periodical basis and a first re-examination has led to the IEEE 1076-93 revision which has found wide adoption. Approximately at the same time IEEE also passed the IEEE 1164 standard, a nine-valued logic system used in conjunction with VHDL. Further re-examinations have been limited to relatively minor improvements and clarifications. The latest release of the standard is IEEE 1076-2002.

VHDL encompasses behavioral and structural views but not physical ones. The levels of abstraction covered range from purely algorithmic descriptions down to logic design. The omission of formalisms for describing time-continuous phenomena in terms of electrical quantities confines IEEE 1076 and 1164 to digital circuits. IEEE 1076.1, a more recent extension for analog and mixed-signal circuits, will be briefly touched upon in section 4.8.2.

The first software tools built around VHDL were compilers and simulators. The fact that a standard HDL did away with all those proprietary languages and products that had had a long tradition in logic simulation was a strong point that contributed to the popularity of VHDL. Only later did people want to come up with automatic synthesis tools that would accept behavioral specifications stated in VHDL and churn out gate-level netlists implementing them. As will become evident later in this chapter, VHDL continues to suffer from the fact that synthesis issues were ignored when the language was originally defined.

4.1.4 Why bother learning hardware description languages?

It seems a tempting idea to view HDLs as nothing more than intermediate formats for exchanging data between electronic system-level (ESL) tools and VLSI CAE/CAD suites.

"Why not skip learning hardware description languages and have electronic system-level tools generate HDL code from specifications automatically?"

There indeed exists a variety of software packages that deal with system design at levels of abstraction well above those addressed during actual VLSI design, see fig.1.12. Many of them deal with transformatorial systems as found in signal processing and telecommunications. With the aid of a tool similar to a schematic editor, a system is put together from high-level library functions such as oscillators, modulators, filters, phase shifters, delay lines, frequency dividers, phase locked loops (PLL), synchronizers, and the like. System behavior is then analyzed and optimized using specialized tools. Filter synthesis software capable of taking into account finite-word-length effects is also included in such DSP-oriented packages.

Other high-level tools model system behavior on the basis of condition/action pairs captured as state graphs, state charts, Petri nets, and the like. The graphical design capture and animation facilities provided are intuitive and very helpful for defining, checking, debugging, and improving the functionality of reactive systems.

More EDA packages are geared towards some field of application such as the analysis of communication channels, source and channel coders, data transfer networks, image processing, and optimization of instruction set computers.

The common theme is that EDA tools working above the HDL level tend to focus on fairly specific problem classes. Though probably unavoidable, this fragmentation is unfortunate in the context of VLSI design, where various building blocks of transformatorial and reactive nature coexist on a single chip.

ESL tools typically include code generators that produce software code for popular microcomputers or DSPs. Many of them are also capable of producing HDL code. All too often this code was nothing else than a translation of the processor code in the past. While this may be acceptable for simulation purposes under certain conditions, it is clearly not so for synthesis and the results so obtained remain unsatisfactory.

More recently, high-level synthesis tools have been developed specifically with computational hardware in mind. Most of them work on the basis of resource allocation, scheduling, and binding.[1] Coming up with a good overall solution implies exploring an immense solution space that involves both algorithmic and architectural issues. Yet, today's tool suites have limited optimization capabilities and are typically restricted to a few predefined hardware patterns.

---

[1] The tools essentially accept an algorithmic description in C and a series of pragmas or other human input that outlines the hardware resources to be made available. The output is RTL code that describes a VLIW ASIP operating under control of either a stored microprogram or a hardwired finite state machine. Final implementation is with FPGAs or as a cell-based ASIC. Catapult C by Mentor Graphics is a commercial example.

In conclusion, it will take a couple of years before true system-level synthesis pervades the industrial production environment. What's more, manual interventions in the source code are often indispensable to parametrize, adapt, interface or optimize circuit models. Finally, HDLs form the basis for virtual components used by many design tools and will, therefore, remain important for the development of model libraries.

Observation 4.1. *For the foreseeable future, hardware description languages such as VHDL, Verilog, and SystemVerilog are bound to remain prominent hubs for all digital design activities.*

4.1.5 Agenda

Section 4.2 introduces the key concepts that set HDLs apart from a programming language one by one. Sections 4.3 and 4.4 address issues specifically related to hardware synthesis (synthesis subset, FSMs, macrocells, timing constraints) and simulation (testbenches) respectively. VHDL textbooks and syntax descriptions are listed in appendix 4.7, language extensions in appendix 4.8. In addition to all that background-type material, code examples of selected subcircuits and of a testbench are given in appendix 4.9. Readers are strongly encouraged to go through that material to develop a better understanding of VHDL coding styles.

4.2 | Key concepts and constructs of VHDL

In this section, we shall give an overview on VHDL by asking ourselves

"What features are required to model digital electronic circuits for simulation and synthesis?"

In anticipation of our findings, we will identify a multitude of needs that can be collected into the six broad categories listed below. The language concepts addressing those needs will be introduced accordingly. As an exception, we have postponed the discussion of basic concepts that VHDL shares with modern programming languages to the end of our presentation, on the assumption that readers have had some exposure to software engineering.

Observation 4.2. *In a nutshell, VHDL can be characterized as follows:*

			standard	*subsection*
VHDL	=	*structured programming language*	*IEEE 1076*	*4.2.6*
	+	*circuit hierarchy and connectivity*	*idem*	*4.2.1*
	+	*concurrent processes and process interaction*	*idem*	*4.2.2*
	+	*a discrete replacement for electrical signals*	*IEEE 1164*	*4.2.3*
	+	*an event-based concept of time*	*IEEE 1076*	*4.2.4*
	+	*model parametrization facilities*	*idem*	*4.2.5*

A few more remarks are due before we start with our analysis.

- The entire chapter puts emphasis on the concepts behind VHDL and on applying the language to hardware modelling. There will be no comprehensive exposure to syntax or grammar. To become proficient in writing circuit models of your own, you will need a more detailed documentation on VHDL. An annotated bibliography is available in appendix 4.7.

- Our introduction of important VHDL concepts is accompanied by a series of illustrations that begins with fig.4.2 and ends with the full picture in fig.4.13. It might be a good idea to refer back to these synoptical drawings when in danger of getting lost in minor details.

- VHDL listings for a variety of subcircuits have been collected in appendix 4.9. You may want to refer to them while working through the more abstract material in this text. The same examples should also prove helpful as starting points when preparing your own models.

- Further keep in mind that this text discusses VHDL as defined by various international standards. Be aware of the fact that commercial EDA tools occasionally deviate in terminology and implementation.[2]

- Please observe the following linguistic ambiguity in the context of hardware modelling:

Meaning of "sequential" with reference to	Synonym	Antonyms
- program execution during simulation	step-by-step	concurrent, parallel
- nature of circuit being modelled	memorizing	combinational, memoryless

4.2.1 Circuit hierarchy and connectivity

The need for supporting modularity and hierarchical composition

Consider a motherboard from a personal computer, for instance. At the highest level of abstraction, you will discern a CPU chip, a graphics accelerator, all sorts of peripheral components, a ROM or two, several memory modules that themselves hold multiple RAM chips, plus a variety of passive components. When having a look into those ICs, you will discover datapaths, controllers, storage arrays, and the like. Each such subsystem in turn consists of many thousands of logic gates and bistables. Only at the bottom level of abstraction do we find transistor-level subcircuits that implement elementary logic and storage functions.

Electronic circuits and systems are organized into multiple layers of hierarchy because it long ago became entirely impractical to specify, understand, model, design, fabricate, test, and document electronic circuits as flat collections of transistors. The constant push to ever larger systems with hundreds of millions of gate equivalents has further accentuated this move.

Hierarchical composition essentially works by assembling larger entities from subordinate entities and by interconnecting them with the aid of busses and individual wires. Only by taking advantage of techniques such as abstraction, modular design, modular verification, and repetitive instantiation does it become possible to arrive at manageable descriptions of an overall circuit or system, see fig.4.2. Any HDL must, therefore, provide language elements for expressing hierarchical composition, and VHDL is no exception.

Design entity

The VHDL term design entity — or entity for short — refers to some clear-cut circuit or subcircuit. Clear-cut implies the (sub)circuit has not only an internal implementation but also an external interface. The benefits of information hiding have incited the originators of the VHDL language

---

[2] See footnotes 18, 21, 32, 33, 45, and appendix 4.8.7, for instance. Also see section 1.7 for a glossary of EDA terms that also includes Synopsys' vocabulary.

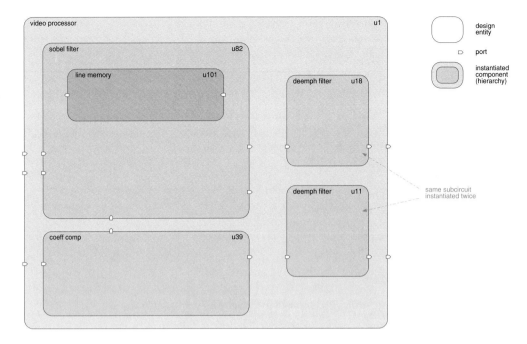

Fig. 4.2 Circuit modelling with VHDL I: Hierarchical composition.

to make a strict distinction between a subcircuit's external and internal views. The interface is specified in the entity declaration whereas the details of its implementation are captured in a language element referred to as the architecture body.

Entity declaration and ports

The entity declaration specifies the external interface of a design entity. VHDL requires an entity declaration for every piece of hardware that is going to be described by a model of its own. The most important part of an entity declaration is the **port clause** which lists all those nodes of the entity that are visible from outside. Put differently, every signal that appears in a port clause corresponds to a connector on the icon of that subcircuit as illustrated in fig.4.14. An example of an entity declaration is given below.

```
-- entity declaration
entity lfsr4 is
   port (
      CLKxC : in Std_Logic;
      RSTxRB : in Std_Logic;
      ENAxS : in Std_Logic;
      OUPxD : out Std_Logic );
end lfsr4;
```

Hint: Naming a signal or a port **IN** or **OUT** is all too tempting, yet these are reserved words in VHDL. We recommend the use of **INP** and **OUP** instead.

Architecture body (structural view)

An architecture body — colloquially often abbreviated architecture — is the place where the internal technicalities of a design entity are being described. Any design entity is permitted to contain instances of other design entities. A first code example is printed below. The circuit being modelled is a linear feedback shift register (LFSR). Although you are probably not yet in a position to understand everything, it should become clear that the circuit is composed of five logic gates and four flip-flops. As an exercise, draw a schematic diagram for the circuit.

```vhdl
-- architecture body
architecture structural of lfsr4 is

   -- component declarations
   component GTECH_FD2     -- D-type flip-flop with reset
      port (
         D, CP, CD : in  std_logic;
         Q : out std_logic );
   end component;
   component GTECH_FD4     -- D-type flip-flop with set
      port (
         D, CP, SD : in  std_logic;
         Q : out std_logic );
   end component;
   component GTECH_MUX2    -     input multiplexer
      port (
         A, B, S : in std_logic;
         Z : out std_logic );
   end component;
   component GTECH_XOR2    -- 2-input XOR gate
      port (
         A, B: in std_logic;
         Z : out std_logic );
   end component;

   -- signal declarations of internal nodes
   signal STATExDP : std_logic_vector(1 to 4);
   signal n11, n21, n31, n41, n42 : std_logic;

begin

   -- instantiate components and connect them by listing port maps
   u10 : GTECH_FD2
      port map( D => n11, CP => CLK, CD => RSTxRB, Q => STATExDP(1) );
   u20 : GTECH_FD2
      port map( D => n21, CP => CLK, CD => RSTxRB, Q => STATExDP(2) );
   u30 : GTECH_FD2
      port map( D => n31, CP => CLK, CD => RSTxRB, Q => STATExDP(3) );
   u40 : GTECH_FD4
      port map( D => n41, CP => CLK, SD => RSTxRB, Q => STATExDP(4) );
   u11 : GTECH_MUX2
      port map( A => STATExDP(1), B => n42, S => ENAxS, Z => n11 );
   u21 : GTECH_MUX2
      port map( A => STATExDP(2), B => STATExDP(1), S => ENAxS, Z => n21 );
   u31 : GTECH_MUX2
      port map( A => STATExDP(3), B => STATExDP(2), S => ENAxS, Z => n31 );
```

```
    u41 : GTECH_MUX2
       port map( A => STATExDP(4), B => STATExDP(3), S => ENAxS, Z => n41 );
    u42 : GTECH_XOR2
       port map( A => STATExDP(3), B => STATExDP(4), Z => n42 );

    -- connect state bit of rightmost flip-flop to output port
    OUPxD <= STATExDP(4);

end structural;
```

Component instantiation and port map

How do you proceed when asked to fit a circuit board with components? You think of the exact name of a part required, go and fetch a copy of it, and solder the terminals of that one copy in a well-defined manner to metal pads interconnected by narrow lines on a prefabricated circuit board. The component instantiation statement of VHDL does exactly this, albeit with virtual components and signals instead of physical parts and wires. How to connect instance terminals to circuit nodes gets specified in the port map clause.

In the above code example, nine components get instantiated following the keyword **begin**. As multiple copies of the same component must be told apart, each instance is assigned a unique identifier; u10, u20, ..., u42 in the occurrence. Further observe that the association operator => in the port maps does not indicate any assignment. Rather, it stands for an electrical connection made between the instance terminal to its left (formal part) and some node in the superordinate circuit the signal name of which is indicated to the right (actual part).

Component declaration

Each of the first four statements in the **lfsr4** architecture body specifies the name and the port list of a subcircuit that is going to be instantiated. VHDL requires that the names and external interfaces of all component models be known prior to instantiation.[3]

---

[3] There are essentially two ways for declaring a subcircuit model, yet the difference is a subtlety that can be skipped for a first reading. Assume you are describing a circuit by way of hierarchical composition in a **top-down** fashion, that is, beginning with the top-level design entity. In doing so, you must anticipate what subcircuits you will need. All that is really required for the moment are the complete port lists of those subcircuits-to-be that you are going to instantiate. Their implementations can wait until work proceeds to the next lower level of hierarchy. Declaring the external interfaces of such prospective subcircuits locally, that is within the current architecture body, is exactly what the component declaration statement is intended for.

Now consider the opposite **bottom-up** approach. You begin with the lowest-level subcircuits by capturing the interface in an entity declaration and the implementation in an architecture body for each subcircuit. These models are then instantiated at the next-higher level of the design hierarchy, and so on. Instantiation always refers to an existing design entity which explains why this type of instantiation is said to be direct. No component declarations are required in this case, yet the component instantiation statement is complemented with the extra keyword **entity** and with an optional architecture identifier as follows.

```
u6756 : entity lfsr4 (behavioral)
        port map( CLKxC => n18, RSTxRB => n8, ENAxS => n199, OUPxD => n4 );
```

For **direct instantiation** to work, design entities must be made visible with a **use work.all** clause. Use clauses and configuration specification statements are to be introduced in sections 4.2.6 and 4.2.5 respectively. Also note that direct instantiation is supported since the IEEE 1076-93 standard only.

VHDL further requires that the wires running back and forth between instances be declared. Those connecting to the outside world are automatically known from the port clause and need not be declared a second time. Internal nodes, in contrast, are defined in a series of supplemental signal declaration statements just before the keyword **begin**. More on signals is to follow shortly.

Let us conclude this section with a more general comment.

Observation 4.3. *VHDL can describe the hierarchical composition of a digital electronic circuit by instantiating components or entities and by interconnecting them with the aid of signals.*

A model that describes a (sub)circuit as a bunch of interconnected components is qualified as **structural**. Structural HDL models essentially hold the same information as circuit netlists do. Manually establishing entity declarations and structural architecture bodies in the style of the **lfsr4** example shown is not particularly attractive, though. Most structural models are in fact obtained from register-transfer level (RTL) and similar models with the aid of automatic synthesis software.

Still, situations exist where one needs to explicitly stipulate a circuit's connectivity, just think of how to embed a chip's core into the chip's padframe, for instance. In such a situation, the designer can either manually write structural VHDL code, or enter a schematic diagram into his CAD suite and have the schematic editor translate that into HDL code. Schematics have the advantage of being more suggestive of the circuit described than code listings. As will be explained in section 4.3.5, they lack the flexibility of a parametrized code model, however.

4.2.2 Concurrent processes and process interaction

The need for modelling concurrent activities

While we have learned how to capture a circuit's hierarchy in VHDL, our description remains devoid of life up to this point as we have no means for expressing circuit behavior. This is not only insufficient for simulating a circuit but also inadequate in view of dispensing designers from having to specify a circuit's composition in great detail. So there must be more to VHDL.

The most salient feature of any electronic system is the concurrent operation of its subcircuits; just think of all those ICs on a typical circuit board and of the many thousands of logic gates and storage devices within each such chip. This inherent parallelism contrasts sharply with the line-by-line execution of program code on a computer. Another innate trait is the extensive communication that permanently takes place between subcircuits and that is physically manifest in the multitude of wires that run across chips and boards. This is necessary simply because there can be no cooperation between multiple entities without on-going exchange of data.

Now assume you wanted to write a software model that imitates the behavior of a substantial piece of electronic hardware using some traditional programming language such as Pascal or C. You would soon get frustrated because of the absence of constructs and mechanisms to handle simultaneous operation and interaction. Hardware description languages such as VHDL and Verilog extend the expressive power of programming languages by supporting concurrent processes and means for

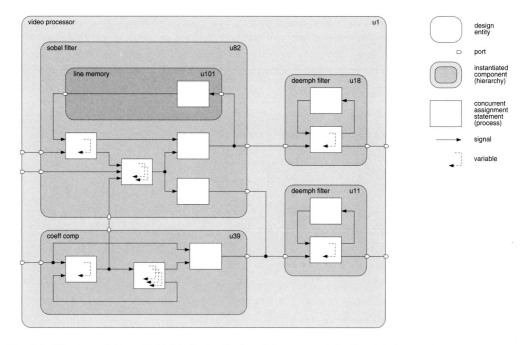

Fig. 4.3 Circuit modelling with VHDL II: A refined model capturing behavior with the aid of concurrent processes that communicate via signals. Also shown are variables that are confined to within processes by definition.

exchanging information between them, see fig.4.3 for a first idea. As this bears some resemblance to real-time languages, it comes as no surprise that Ada and its concept of concurrent tasks have been taken as a starting point for defining VHDL.

Signal

The vehicle for exchanging dynamic information between concurrent processes is the signal, there exists no other mechanism in VHDL for this purpose.[4] "Dynamic" means that the data being transferred as signals are free to evolve over time, that is to change their values during a simulation run. Any signal declared within an architecture body is strictly confined to that body and remains inaccessible from outside. A signal declaration must specify the name and the data type of the signal; an optional argument can be used to assign an initial value.

Example of a signal declaration   `signal THISMONTH : month;`

Second example   `signal ERROR, ACTUAL, WANTED : integer := 0;`

> Warning: An initial value assigned to a signal as part of its declaration is inadequate for modelling a hardware reset and gets ignored during VHDL synthesis.[5]

---

[4] The purpose of protected shared variables is a totally different one, see section 4.8.1 for explanations.

[5] Detailed explanations are to follow in observation 4.18 and in section 4.3.3.

Hint: VHDL code is easier to read when signals can be told from variables by their visual appearance. We make it a habit to use upper- and mixed-case names for **SIGNALS** and lower-case characters for **variables**.[6]

Concurrent processes (purpose)

Any of the four VHDL statements below provides us with the capability to alter signal values.

- Concurrent signal assignment (simplest).
- Selected signal assignment.
- Conditional signal assignment.
- Process statement (most powerful).

All four statements are subsumed as (concurrent) processes or as concurrent assignment statements, two generic terms being used in this text, not actual VHDL language constructs.[7] Unless the code is actually running on a multiprocessor, concurrent processes must of course be invoked one after the other during VHDL simulation but the observed effect is that they execute in parallel. How to obtain this effect makes up the major part of section 4.2.4.

Concurrent signal assignment (construct)

This is the most simple language construct that permits one to update the value of a signal. A concurrent signal assignment is used to describe a combinational operation with no need for branching; basic arithmetic and logic operations are typical examples. Concurrent signal assignments support function calls but no procedure calls.[8] The assignment operator for signals is **<=**, which choice is somewhat unfortunate because the same symbol also serves as relational operator (in lieu of $\leq$).

Example of a concurrent signal assignment   `THISMONTH <= august;`

Second example   `ERROR <= ACTUAL - WANTED;`

Selected signal assignment

This is a more elaborate form of a concurrent signal assignment reminiscent of a multiplexer (MUX) or data switch: One out of multiple possible values gets assigned to a signal under control of a selecting expression. An example follows.

```
with THISMONTH select
   QUARTER <= q1st when january | february | march,
              q2nd when april | may | june,
              q3rd when july | august | september,
              q4th when others;
```

---

[6] An elaborate naming convention for signals will be presented in section 5.7.

[7] The terms process, concurrent process, and parallel process are synonyms (in the sense of the German "nebenläufiger Prozess") whereas the term concurrent assignment statement stands for the subclass of active processes capable of altering signals. We will later learn about passive processes that do not assign any value to a signal. Also, do not equate the broad and conceptual notion of a process with the process statement, a VHDL syntax item. A process statement is just one particular case of a concurrent process.

[8] To be explained in section 4.2.6.

Conditional signal assignment

This language construct is very similar to the selected signal assignment but a bit more liberal in formulating the branching condition. An example follows.

```
SPRING <= true when (THISMONTH=march and THISDAY>=21) or
                     THISMONTH=april or THISMONTH=may or
                     (THISMONTH=june and THISDAY<20)
                     else false;
```

Process statement (construct)

The process statement is an even more powerful — but also more tricky and more verbose — VHDL construct for expressing a concurrent process. What sets it apart from the signal assignment statements discussed before are essentially

- Its capability to update two or more signals at a time,
- The fact that the instructions for doing so are captured in a sequence of statements that are going to be carried out one after the other,
- The liberty to make use of variables for temporary storage, and
- A more detailed control over the conditions for activating the process.[9]

Process statements cannot be nested but may call subprograms. Conditional execution and branching are supported, of course. The process statement is best summed up as being concurrent outside and sequential inside.[10] The example given next is semantically and functionally identical to the conditional signal assignment given before.

```
memless1: process (THISMONTH, THISDAY)
-- an event on any signal listed activates the process
begin
    SPRING <= false;    -- execution begins here
  if THISMONTH=march and THISDAY>=21 then SPRING <= true; end if;
  if THISMONTH=april              then SPRING <= true; end if;
  if THISMONTH=may                then SPRING <= true; end if;
  if THISMONTH=june  and THISDAY<=20 then SPRING <= true; end if;
end process memless1;    -- process suspends here
```

A process statement can be made to capture almost anything from a humble piece of wire up to an entire image compression circuit, for instance. The decision is left to the discretion of the VHDL programmer. More particularly, a process statement can model a combinational function, a data storage operation, or any combination of the two. This depends on how the code is written; guidelines are to follow in observation 4.14.

---

[9] Several of these items will be clarified in section 4.2.4.

[10] Make sure you understand that "sequential" refers to code execution during VHDL simulation here and not to the nature of the hardware being modelled, which may be either combinational or sequential depending on how the code is organized, see observation 4.14. After all, it is perfectly natural to fill the truth table of a complex function by way of a sequential algorithm. Also note that the identifier memless1 in the code example is just an optional free-choice label that has no impact on simulation and synthesis whatsoever.

Hint: For the sake of modularity and legibility, do not cram too much functionality into a concurrent process. As a rule, concurrent, selected, and conditional signal assignment statements serve to describe combinational operations whereas process statements are primarily used for modelling all sorts of data storage registers.[11]

Architecture body (behavioral view)

Most architecture bodies include a collection of concurrent processes that together make up the entity's overall functionality. Such models are called **behavioral** because they specify how the entity is to react in response to changing input signals. Potential reactions include the updating of output signals, the updating of the entity's current state, the checking of compliance with some predefined timing conditions, or simply ignoring the new input.

The architecture body given below matches the entity declaration given in the previous section. In fact, the functionality being modelled is again that of an LFSR of length four with enable and asynchronous reset. Taking fig.4.3 as a pattern, make a small drawing that illustrates the processes and the signals that are being exchanged. Find out what hardware item each concurrent process stands for and compare the drawing with that established earlier. What liberties do you have in coming up with a schematic diagram?

```
-- architecture body
architecture behavioral of lfsr4 is
   signal STATExDP, STATExDN : std_logic_vector(1 to 4);
   -- for present and next state respectively
begin

   -- computation of next state
   STATExDN <= (STATExDP(3) xor STATExDP(4)) & STATExDP(1 to 3);

   -- updating of state
   process (CLKxC,RSTxRB)
   begin
      -- activities triggered by asynchronous reset
      if RSTxRB='0' then
         STATExDP <= "0001";
      -- activities triggered by rising edge of clock
      elsif CLKxC'event and CLKxC='1' then
         if ENAxS='1' then
            STATExDP <= STATExDN;
         end if;
      end if;
   end process;

   -- updating of output
   OUPxD <= STATExDP(4);

end behavioral;
```

Observation 4.4. *In VHDL, the behavior of a digital electronic circuit typically gets described by a collection of concurrent processes that execute simultaneously and that communicate via signals, and where each such process represents some subfunction.*

---

[11] This is particularly true for RTL synthesis models. More detailed advice is to follow in section 4.3.

VHDL supports a great variety of modelling styles

As we have learned so far, VHDL covers both behavioral and structural circuit descriptions, but no physical ones. Within this limitation, VHDL supports a great variety of modelling styles.

A **procedural model** essentially describes functionality in a sequence of steps much as a piece of conventional program code does. A design entity is captured in one process statement and its behavior gets implemented with the aid of sequential statements there.[12]

A **dataflow model** describes the behavior as a collection of concurrent signal assignments the respective operations of which get coordinated by the signals exchanged. As an extension thereof, block statements might also be used.

A **structural model** describes the inner composition of a design entity as a set of instantiated components together with their interconnections and is equivalent to a netlist. Component declaration and instantiation statements are typical for structural models.

Listing 4.1 below juxtaposes three architecture bodies that describe the same function from all three perspectives while fig.4.5 illustrates the differences and commonalities.[13] Make sure you understand the conceptual difference between the procedural and dataflow models in spite of the apparent similarity of their codes, also problem 3.

Observation 4.5. *VHDL allows procedural, dataflow, and structural modelling styles to be freely combined in a single model.*

Except for the most simple subcircuits, a typical VHDL model includes a mix of elements from all three styles as suggested by fig.4.4.[14] Experimental results on how VHDL coding style affects simulation performance are reported in [99].

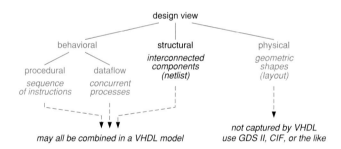

Fig. 4.4 Modelling styles and their relationships to VHDL and other EDA languages.

---

[12] The VHDL community, which has its own vocabulary, would typically call this a behavioral model. However, in order to stay in accordance with the universally accepted Y-chart of fig.4.1, we prefer to use the more precise designation procedural model and to reserve the term behavioral for the superclass of procedural and dataflow models, see fig.4.4. Incidentally, note that behavioral, procedural, dataflow, and structural are not reserved words of VHDL. These are just user-defined terms that serve to convey information about modelling style.

[13] A full-adder has been chosen because of its simplicity and commonplace nature. Clearly, none of the three architecture bodies reflects how one would normally model an adder as VHDL supports the arithmetic operator **+**.

[14] Independently from whether a VHDL model is of procedural, dataflow or structural nature, or mixes all of them, some behavioral model must ultimately be given for every elemental component for simulation to work.

Listing 4.1 Procedural, dataflow, and structural styles compared. Note: Adders are normally synthesized from algebraic expressions, a full-adder has been chosen here for its simplicity.

```vhdl
entity fulladd is
   port (  INPA, INPB, INPC : in std_logic;
      OUPS, OUPC : out std_logic );
end fulladd;
--------------------------------------------------------------------------------
-- compute results in a series of sequential steps
architecture procedural1 of fulladd is
begin
   process (INPA,INPB,INPC)
      variable loc1, loc3, loc4 : std_logic;
   begin
      loc1 := INPA xor INPB;
      OUPS <= INPC xor loc1;
      loc3 := INPC nand loc1;
      loc4 := INPA nand INPB;
      OUPC <= loc3 nand loc4;
   end process;
end procedural1;
--------------------------------------------------------------------------------
-- spawn a concurrent signal assignment for each logic gate
architecture dataflow1 of fulladd is
   signal LOC1, LOC3, LOC4 : std_logic;
begin
   LOC1 <= INPA xor INPB;
   OUPS <= INPC xor LOC1;
   LOC3 <= INPC nand LOC1;
   LOC4 <= INPA nand INPB;
   OUPC <= LOC3 nand LOC4;
end dataflow1;
--------------------------------------------------------------------------------
-- describe logic network as a bunch of interconnected logic gates
architecture structuralgtech of fulladd is

   -- list cells from Synopsys' generic cell library to be used
   component GTECH_XOR2
      port ( A, B : in std_logic;
             Z : out std_logic );
   end component;
   component GTECH_NAND2
      port ( A, B : in std_logic;
             Z : out std_logic );
   end component;

   -- declare internal signals
   signal LOC1, LOC3, LOC4 : std_logic;

begin
   -- instantiate cells and connect them by listing port maps
   U1: GTECH_XOR2
      port map ( A=>INPB, B=>INPA, Z=>LOC1 );
   U2: GTECH_XOR2
      port map ( A=>INPC, B=>LOC1, Z=>OUPS );
```

```
    U3: GTECH_NAND2
        port map ( A=>INPC, B=>LOC1, Z=>LOC3 );
    U4: GTECH_NAND2
        port map ( A=>INPA, B=>INPB, Z=>LOC4 );
    U5: GTECH_NAND2
        port map ( A=>LOC3, B=>LOC4, Z=>OUPC );
end structuralgtech;
```

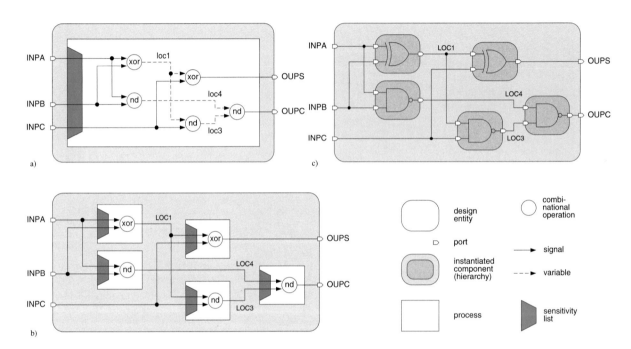

Fig. 4.5 Procedural (a), dataflow (b), and structural (c) models of a full-adder function.

Another example where a memoryless input-to-output mapping function is captured in accordance with very different coding styles is reproduced in appendix 4.9.1.

4.2.3 A discrete replacement for electrical signals

The need for representing multiple logic values

An innocent approach to hardware modelling would be to use one binary digit per circuit node. VHDL actually provides two predefined data types for describing two-valued data:

bit   which can take on value 0 or 1.

boolean   which can take on value **false** or **true**.

Yet, even digital hardware exhibits a number of characteristics and phenomena such as transients, three-state outputs, multiple buffers driving a common node with an inherent potential for conflicts, indeterminate circuit state following power-up, and the like. None of these circumstances can be captured with a two-valued logic abstraction.

Observation 4.6. *Distinguishing between logic 0 and 1 is inadequate for modelling the binary signals found in digital electronic circuitry. A more elaborate multi-valued logic system must be sought that is capable of capturing the effects of both node voltage and source impedance.*

The idea behind a **logic system** is to discretize the continuous-valued node voltage and source impedance separately. Voltage gets quantized into a number of **logic states** while the amount of current that a subcircuit can sink or source is mapped onto discrete **drive strengths**, or strengths for short. It thus becomes possible to condense the electrical condition of a circuit node into one logic value at any time.

A standard multi-valued logic system

A fairly simple and universal logic system to be used in conjunction with VHDL is available from a package named **ieee.std_logic_1164**.[15]

The **logic states** implemented are:

low	logic low, that is below U_l.
high	logic high, that is above U_h.
unknown	may be "low", "high", or anywhere in the forbidden interval
	in between, e.g. as a result from a short between two conflicting drivers.

Note, by the way, that no distinction is made between a drive conflict, the outcome of which is truly unknown, and a ramping node, the voltage of which is known to assume values between thresholds U_l and U_h for some time. Either one is modelled as "unknown".

The **drive strengths** being used are:

strong	the low impedance value commonly exhibited by a driving output.
high-impedance	the almost infinite impedance exhibited by a disabled three-state output.
weak	an impedance somewhere between "strong" and "high-impedance",
	e.g. as exhibited by a passive pull-up/-down resistor or a snapper.

A regular matrix with nine logic values should result when three logic states are combined with three drive strengths. The 1164 standard committee has, however, refrained from differentiating between "charged high" and "charged low" by collapsing all high-impedance conditions to a single

---

[15] The originators of VHDL have deliberately chosen not to incorporate any logic system into the IEEE 1076 standard itself as this would have biased the language towards some circuit technology such as CMOS, ECL or GaAs, for instance, and would preclude its evolution towards unforeseen technologies in the future. Instead, a logic system has been defined as separate standard IEEE 1164 and made available in the said package. It is thus possible to replace it by some user-defined logic system at any time should the necessity occur.

value of undetermined state. This conservative choice avoids the difficulties of fixing realistic charge decay times.[16]

On the other hand, two extra values have been added, namely:

uninitialized   has never been assigned any value, e.g. the internal state of a storage element immediately after power-up, distinguished from "unknown" as the latter can arise from causes other than failed initialization (applicable to simulation only).

don't care   whether the node is "low" or "high" is considered immaterial, used by designers to leave the choice to the logic optimization tool (applicable to synthesis only).

The matrix of table 4.3 summarizes the IEEE 1164 standard 9-valued logic system which is sometimes also referred to as MVL-9 (multiple-value logic). The data type implementing this logic system is called std_ulogic.

Table 4.3   IEEE 1164 standard 9-valued logic system.

logic value → ↓	logic state			acceptable for	
	low	unknown	high	simulation	synthesis
uninitialized		U		U	
strong	0	X	1	0 X 1	0 1
strength weak	L	W	H	L W H	
high-impedance	Z	Z	Z	Z	Z
don't care		–			–

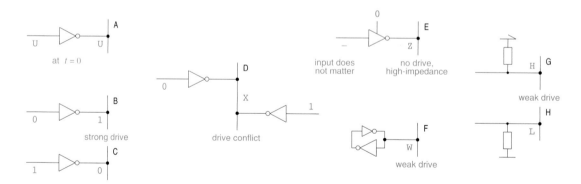

Fig. 4.6   The nine logic values of the IEEE 1164 standard illustrated.

---

[16] Short-term charge retention on circuit nodes is a typical trait of CMOS not found in TTL or ECL circuits. Another item that sets IEEE 1164 apart from other logic systems is the absence of an extra drive strength "forced", which is sometimes introduced to model a node driven with impedance zero or almost so, e.g. when strapped to ground by some metal wire.

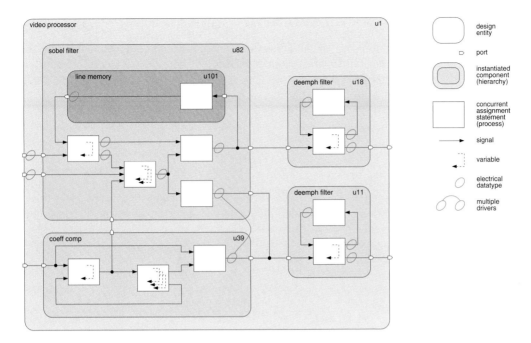

	design entity
	port
	instantiated component (hierarchy)
	concurrent assignment statement (process)
	signal
	variable
	electrical datatype
	multiple drivers

Fig. 4.7 Circuit modelling with VHDL III: Refined model that uses special data types to model electrical phenomena.

Modelling of three-state outputs and busses

Section 4.2.2 has taught us how to assign some value to a signal, but how do we tell we want a node to be released or — which is the same — reverted to an undriven condition? With the aid of the IEEE 1164 logic system, the answer is straightforward: just assign logic value Z.

Example
```
OUP <= not INP when ENA='1' else 'Z';
```

As illustrated in fig.4.7, many digital circuits include **busses** and other **multi-driver nodes** that operate under control of multiple processes. How to model them is now obvious. In the code fragment below, the common node COM is left floating, that is in a high-impedance condition, when neither of the two drivers is enabled.

```
.....
COM <= not INPA when SELA='1' else 'Z';
.....
COM <= not INPB when SELB='1' else 'Z';
.....
```

Multiple drivers and conflict resolution during simulation

Let us now find out how multi-driver signals are actually handled during simulation. What if a drive conflict occurs?[17] How is such an electrical issue modelled in VHDL? Built on top of std_ulogic sits a subtype called std_logic that shares the same set of nine values. As shown in fig.4.8, the difference matters in the presence of multiple drivers.

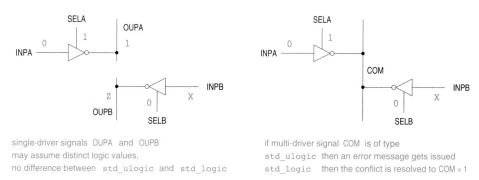

single-driver signals OUPA and OUPB
may assume distinct logic values,
no difference between std_ulogic and std_logic

if multi-driver signal COM is of type
std_ulogic then an error message gets issued
std_logic then the conflict is resolved to COM = 1

Fig. 4.8 Handling of a multi-driver node with types std_ulogic and std_logic.

If the signal is of type std_ulogic, some form of error message will be generated during compilation or simulation.[18] On a std_logic type signal, in contrast, any conflict between diverging values is tacitly solved at simulation time by calling a **resolution function** that determines the most plausible outcome. One such function is part of the IEEE 1164 standard, its name is **resolved**. This function defines, among many other things, that if an attempt is made to drive a node to Z, W, and 0 at the same time, simulation is to continue with a 0, see listing 4.2.

Listing 4.2 | IEEE 1164 standard resolution function.

```
--------------------------------------------------------------------------
-- resolution function "resolved"
--------------------------------------------------------------------------
constant resolution_table : stdlogic_table := (
--    ------------------------------------------------------------------
--    |  U    X    0    1    Z    W    L    H    -        |  |
--    ------------------------------------------------------------------
      ( 'U', 'U', 'U', 'U', 'U', 'U', 'U', 'U', 'U' ), -- | U |
      ( 'U', 'X', 'X', 'X', 'X', 'X', 'X', 'X', 'X' ), -- | X |
      ( 'U', 'X', '0', 'X', '0', '0', '0', '0', 'X' ), -- | 0 |
      ( 'U', 'X', 'X', '1', '1', '1', '1', '1', 'X' ), -- | 1 |
      ( 'U', 'X', '0', '1', 'Z', 'W', 'L', 'H', 'X' ), -- | Z |
      ( 'U', 'X', '0', '1', 'W', 'W', 'W', 'W', 'X' ), -- | W |
      ( 'U', 'X', '0', '1', 'L', 'W', 'L', 'W', 'X' ), -- | L |
      ( 'U', 'X', '0', '1', 'H', 'W', 'W', 'H', 'X' ), -- | H |
      ( 'U', 'X', 'X', 'X', 'X', 'X', 'X', 'X', 'X' )  -- | - |
   );
```

---

[17] A **drive conflict** implies that two or more processes attempt to drive a signal to incompatible logic values.

[18] The details depend on the product being used. ModelSim categorically rejects any unresolved signals that are driven from multiple processes whereas Synopsys accepts them with warnings issued at elaboration time.

Observation 4.7. *Signals of type* std_logic *can accommodate multiple drivers whereas those of type* std_ulogic *cannot. In more general terms, a signal is allowed to be driven from multiple processes iff a resolution function is defined that determines the outcome.*

The IEEE 1076 standard insists that a resolution function be available for any signal that is being driven from multiple sources but places the details under the designer's control. Function **resolved** is fine for fully complementary static CMOS logic. By programming his own resolution functions, the designer can indicate how to solve driver conflicts in other situations.[19] Make sure you understand that there can be no such thing as a resolution function for variables. The same applies to bits, bit vectors, integers, reals, and similar data types.

Collapsing of logic values for the purpose of synthesis

Not all nine values of the IEEE 1164 logic system make sense from a synthesis point of view. The semantics of 0 and 1 are obvious. A don't care symbol – on the right-hand side of an assignment implies logic value is of no importance, in which case logic synthesis is allowed to select either a 0 or a 1 so as to minimize gate count. Z also has a well-defined meaning because it calls for a driver with built-in three-state capability as discussed before.

Values U, X, and W, in contrast, capture specific situations that occur during simulation, but have no sensible interpretation for synthesis. As far as L and H are concerned, one might imagine an EDA tool that would insert pull-down/-up devices or otherwise play with weak drivers. However, as this would entail static currents unpopular with CMOS circuit designers, L and H are not normally honored by today's synthesis software. Most synthesis tools collapse meaningless (to them) values to more sensible ones, e.g. L to 0, H to 1, and X or W to –.

Hint: For the sake of clarity and portability, do not use logic values other than 0, 1, Z, and – in VHDL source code that is intended for synthesis.

Data types for modelling of scalar (single-bit) signals

IEEE 1164 types std_ulogic and std_logic have been introduced to emulate the electrical behavior of circuit nodes in a more realistic way than IEEE 1076 type **bit** does. Using them for simulation purposes is not without cost, however. After all, multiple values occupy more storage capacity than a two-valued data type does, and their processing asks for a higher computational effort. The latter is particularly true when a resolution function is being called. Thus, before opting for a type for a VHDL signal or variable, find out what features you need to model, and what effects you can afford to neglect. Then refer to the selection guide below.

---

[19] As an example, it thus becomes possible to handle open-collector outputs and open-drain when constructing wired-AND operations.

data type	bit	std_ulogic	std_logic
defined in	VHDL	ieee.std_logic_1164	
for simulation purposes			
modelling of power-up phase	no	yes	yes
modelling of weakly driven nodes	no	yes	yes
modelling of multi-driver nodes	no	yes	yes
handling of drive conflicts	n.a.	reported	resolved
storage requirements	minimal	substantial	
computational effort	minimal	substantial	
for synthesis purposes			
three-state drivers	no	yes	yes
don't care conditions	no	yes	yes

As an example, assume you want to synthesize a circuit node with a single driver. If the code is intended for synthesis exclusively, type **bit** will do. If you further want to simulate your code, you will want to learn whether the node has ever been initialized or not. Also, you will want to make sure you get a message from the simulator, should a short circuit between this and some other node inadvertently creep into your design. Type **std_ulogic** would then be the safest choice, although most designers tend to use **std_logic** throughout.

Data types for modelling vectored (multi-bit) signals

Both **std_ulogic** and **std_logic** represent a single bit, whereas most digital circuits operate on several bits at a time. One data type that gets coded using multiple bits in computers is **integer**. There are limitations for describing circuit hardware at lower levels of detail using integers, though. Word width is fixed — to 32 bit in the case of VHDL — and there is no way to access just a portion of a data word. Integers also suffer from a lack of expressiveness for describing electrical phenomena much as type **bit** does.

VHDL further supports the collection of multiple bits into a vector such as in **bit_vector**, **std_ulogic_vector**, and **std_logic_vector**, all of which imply a one-dimensional array built from their scalar counterparts. The problem here is the absence of arithmetic operations for those data types in the IEEE 1076 and 1164 standards.

As the existing standards offered no solution, two new packages were developed and accepted as IEEE standard 1076.3 in 1997. Both packages define two extra data types called **signed** and **unsigned** that are overloaded for standard VHDL arithmetic operators as much as possible. Objects of type **unsigned** are interpreted as unsigned integer binary numbers, and objects of type **signed** as signed integer binary numbers coded in 2's complement (2'C) format. No provisions are made to support other number representation schemes such as 1's complement (1'C), sign-and-magnitude (S&M), or any floating-point format.[20] The programmer is free to specify how many bits shall be set aside for coding an **unsigned** or a **signed** when declaring a constant, variable or signal.

---

[20] A floating-point standard is currently in preparation. Please check section A.1.1 if you are not familiar with binary number representation schemes.

The difference between the two packages is that `ieee.numeric_bit` is composed of `bit` type elements, whereas `ieee.numeric_std` operates on `std_logic` elements. As they otherwise define identical data types and functions, only one of the two packages can be used at a time. Clearly, what has been said about the costs of simulating with multi-valued data types in the context of single-bit nodes also applies to multi-bit nodes.

data type(s)	integer, natural, positive	bit_ vector	std_logic _vector	signed, unsigned	signed, unsigned
defined in	VHDL	VHDL	ieee.std_ logic_1164	ieee.nu- meric_bit	ieee.nu- meric_std
word width	32 bit	at the programmer's discretion			
arithmetic operations	yes	no[21]	no[21]	yes	yes
logic operations	no	yes	yes	yes	yes
access to subwords or bits	no	yes	yes	yes	yes
modelling of electrical effects	no	no	yes	no	yes
simulation costs	low	moderate	high	moderate	high

Orientation of binary vectors

Whenever a positional number system is used to encode a numeric value, there is a choice whether to spell the data word with the MSB or the LSB first. In VHDL, this applies to data types **signed**, **unsigned**, **bit_vector**, **std_logic_vector**, and **std_ulogic_vector**. Any misinterpretation is likely to cause serious problems for a circuit's simulation and functioning.

> Hint: Any vector that contains a data item coded in some positional number system should consistently be declared as $(i_{MSB}$ **downto** $i_{LSB})$, where 2^i is the weight of the binary digit with index i. The MSB will thus have the highest index referring to it and will appear in the customary leftmost position because $i_{MSB} \geq i_{LSB}$.

Example   `signal HOUR : unsigned(4 downto 0) := "10111";`

Most designers go for resolved data types

Simulating with unresolved **std_ulogic** and **std_ulogic_vector** types is definitely more conservative than simulating with their resolved counterparts because an error message will tell you, should any of those accidentally get involved in a drive or naming conflict. Yet, the IEEE 1164 standard recommends that "For scalar ports and signals, the developer may use either **std_ulogic** or **std_logic**

---

[21] A historical note is due here. In anticipation of the IEEE 1076.3 standard, most vendors of VHDL software tools had introduced extensions of their own, thereby turning a "no" into a "yes" where indicated. Yet, as all such efforts were made on a proprietary basis, relying on them is detrimental to code portability. While the interpretation of arithmetic operators was unlikely to differ, the names and coding schemes of the extra data types were not always the same. Unofficial extensions, such as the former `ieee.std_logic_arith` package, must be viewed as obsolete temporary fixes that must no longer be used, now that the **numeric** packages are available.

type. For vector ports and signals, the developer should use **std_logic_vector** type."[22] In practice, the types **std_logic** and **std_logic_vector** prevail.

4.2.4 An event-based concept of time for governing simulation

The need for a mechanism that schedules process execution

It has been observed in section 4.2.2 that VHDL simulation is to yield the same result as if the many processes present in a circuit model were operating simultaneously, although no more than a few processors are normally available for running the simulation code. What is obviously required then is a mechanism that schedules processes for sequential execution and that combines their effects so as to perfectly mimic concurrency. This mechanism that always sits in the background of VHDL models is the central theme of this section.

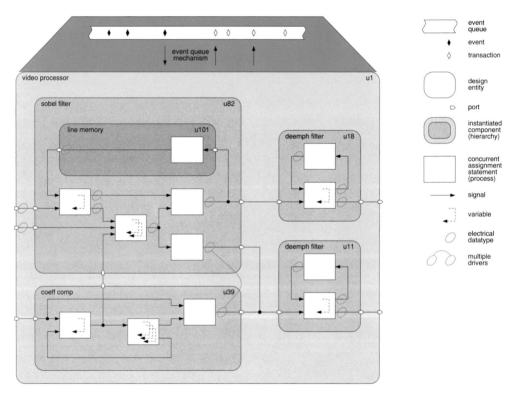

Fig. 4.9 Circuit modelling with VHDL IV: Circuit model augmented with an event queue mechanism that governs process activation.

---

[22] Two reasons are given for this surprising advice: Concerns expressed by EDA vendors that they might not be able to optimize simulator performance for both data types, and interoperability of circuit and testbench models from different sources. It is in fact a bizarre quirk of VHDL that **std_logic** is a subtype of **std_ulogic** which allows for cross assignments without type conversion, whereas **std_logic_vector** and **std_ulogic_vector** are two distinct types and, hence, make type conversion compulsory when assigning one type to the other.

Observation 4.8. *A perfect comprehension of how a model's concurrent processes are being scheduled during simulation is essential for hardware modelling. Understanding and writing code for synthesis is no exception.*

Simulation time versus execution time

First of all, we must distinguish between simulation time and execution time. Simulation time is to a VHDL model what physical time is to the hardware described by that model. The simulator software maintains a counter that is set to zero when a new simulation run begins and that registers the progress of simulation time from then on. This counter can be likened to a stopwatch and any event that occurs during simulation can be thought of as being stamped with the time currently displayed by that clock.

Execution time, in contrast, refers to the time a computer takes to execute statements from the VHDL code during simulation. It is of little interest to circuit designers as long as their simulation runs do complete within an acceptable lapse of time.

The benefits of a discretized model of time

Assume you wanted to model a digital circuit using some conventional programming language. Capturing the logic behavior of its gates and registers poses no major problem, but how about taking into account their respective propagation delays? How would you organize a simulation run? You would find that no computations are required unless a node switches. For the sake of efficiency, you would thus decide to consider time as being discrete and would devise some data structure that activates the relevant circuit models when they have to (re)evaluate their inputs. These are precisely the ideas underlying event-driven simulation.

Observation 4.9. *In VHDL simulation, the continuum of time gets subdivided by events each of which occurs at a precise moment of simulation time. An **event** is said to happen whenever the value of a signal changes.*

Event-driven simulation

The key element that handles events and that invokes processes is called **event queue** and can be thought of as a list where entries are arranged according to their time of occurrence, see fig.4.10. An entry is referred to as a **transaction**.

Event-driven simulation works in cycles where three stages alternate:

1. Advance simulation time to the next transaction in the event queue, thereby making it the current one.[23]
2. Set all signals that are to be updated at the present moment of time to the value associated with the current transaction.
3. Invoke all processes that need to respond to the new situation and have them (re)evaluate their inputs. Every signal assignment supposed to modify a signal's value causes a transaction to be entered into the event queue at that point in the future when the signal is anticipated to take

---

[23] Multiple entries may be present for the same moment of time, but the general procedure remains the same.

on its new value. This stage comes to an end when all processes invoked suspend after having finished to schedule signal updates in response to their current input changes.

After completing the third stage, a new simulation cycle is started. Simulation stops when the event queue becomes empty or when simulation time reaches some predefined final value.

As nothing happens between transactions, an event-driven simulator essentially skips from one transaction to the next. No computational resources are wasted while models sit idle. Parallel processes and event queue together form a powerful mechanism for modelling the behavior of discrete-time systems.[24] Refer to fig.4.15 for a wider perspective on the simulation cycle.

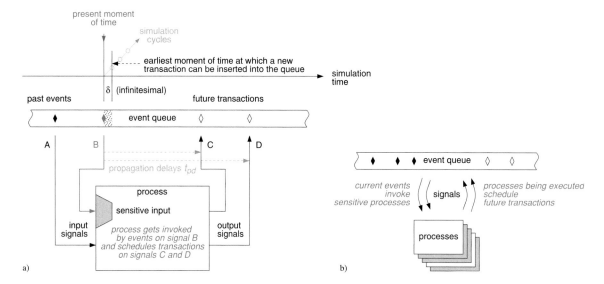

Fig. 4.10 Event-driven simulation. Interactions between the event queue and a VHDL process (a), actions that repeat during every simulation cycle (b).

Transaction versus event

It is important to note that any signal update actually occurs in two steps. Execution of a signal assignment causes a transaction to be entered into the event queue but has no immediate impact. The update is to become effective only later when simulation time has reached the scheduled time for that transaction.

By the same token, not every signal assignment that is being carried out necessarily causes a signal to toggle. All too often, a process gets evaluated in response to some event on one of the wake-up signals just to find out that the result is the same as before. Consider a four-input NAND gate or an E-type flip-flop, for instance. Also, the effect of a first transaction may get nullified by a second

---

[24] Incidentally, note that the event queue mechanism is by no means confined to electronic hardware but is also being used for simulating land, air, and data traffic, for evaluating communication protocols, for planning fabrication and logistic processes, and in many other discrete-time applications.

transaction inserted into the event queue afterwards. This is why transactions and events are not the same. A transaction that does not alter the value of a signal is still a transaction but it does not give rise to an event.[25]

Observation 4.10. *Events are observable from the past evolution of a signal's value up to the present moment of simulation time whereas transactions merely reflect future plans that may, or might not, materialize.*

Delay modelling

The lapse of time between an event at the input of a process and the ensuing transaction at its output reflects the delay of the piece of hardware being modelled. In VHDL, delay figures are typically conveyed by the after clause which forms an optional part of the signal assignment statement.[26] The statement below, for instance, models the propagation delay of an adder by scheduling a transaction on its output t_{pd} after an event at either input.

Example   `OUP <= INA + INB after propdelay;`

The δ delay

For obvious reasons, a process cannot be allowed to schedule signal updates for past or present moments of time. It is, therefore, natural to ask

"What is the earliest point in time at which a new transaction can be entered into the queue?"

In the occurrence of VHDL, the answer is δ time later, where δ does not advance simulation time but requires going through another simulation cycle. Put differently, δ can be thought of as an infinitesimally small lapse of time greater than zero. This refinement to the basic event queue mechanism serves to order transactions when the simulation involves models that are supposed to respond with delay zero. Without the δ time step, there would be no way to order zero-delay transactions and simulation could, therefore, not be guaranteed to yield consistent and reproducible results. Although simulation time does not progress in regular intervals, δ may, in some sense, be interpreted as the timewise resolution of the simulator.

"How does a simulator handle signal assignments with no after clause?"

The answer is that delay is assumed to be zero exactly as if the code read   ... **after** 0 ns ... . The transaction is then scheduled for the next simulation cycle or, which is the same, one δ delay

---

[25] An event queue resembles very much an agenda in everyday life. Transactions are analogous to entries there. Signals reflect the evolution of the state of our affairs such as current location and occupation, health condition, social relations, material possessions, and much more. An entry in the agenda stands for some specific intention as anticipated today. At any time, an event, such as a phone call, may force us to alter our plans, i.e. to add, cancel or modify intended activities to adapt to a new situation. Some of our activities remain in vain and do not advance the state of affairs, very much as some of the transactions do not turn into events. Finally, in retrospect, an agenda also serves as a record of past events and bygone states.

[26] Related language constructs that also express time intervals are **wait for** and **reject**. A wait for statement causes a process to suspend for the time indicated before being reactivated. The reject clause, a feature added in the 1993 standard update, helps to describe rejection phenomena on narrow pulses in a more concise way. You may also want to refer to section 12.2.1 for a comment on how to model transients in VHDL more precisely.

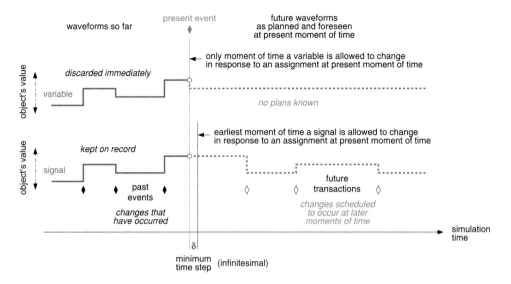

Fig. 4.11 The past, present, and future of VHDL variables and signals.

later. Omitting the after clauses is typical in RTL synthesis models because physically meaningful delay data are unavailable at the time when such models are being established. Much the same applies to behavioral models at the algorithmic level.

Hint: When simulating models with no delays other than infinitesimally small δ delays, it becomes difficult to distinguish between cause and effect from graphical simulation output because the pertaining events appear to coincide. It then helps to artificially postpone transactions by a tiny amount of time in otherwise delayless signal assignments.

To allow for quick adjustments, a constant of type time is best declared in a package and referenced throughout a model hierarchy. Note that the largest sum of fake delays must not exceed one clock period, though.

Example   OUP <= INA + INB after fakedelay;

Signal versus variable

We now are in a good position to understand those features that separate signals from variables. While the difference in terms of scope has already been illustrated in fig.4.3, fig.4.11 exposes those particularities that relate to time.

A variable has no time dimension attached, which is to say that it merely holds a present value. Neither transactions nor events are involved. The effect of a variable assignment is thus felt immediately, that is, in the next statement exactly as in any traditional programming language.

A signal, in contrast, is a typical element of an HDL. It is defined over time, which implies that a signal holds not only a present value, but also past values, plus those values that are anticipated to become manifest in the future. The effect of a signal assignment is not felt before the delay specified in the after clause has expired. The minimum delay, and default value in the absence of an after clause, is δ. As another particularity, it is possible to schedule multiple transactions in a single signal assignment statement.

Example   `THISMONTH <= august, september after 744 hr, october after 1464 hr;`

Observation 4.11. *VHDL signals convey time-varying information between processes via the event queue. They are instrumental in process invocation, which is directed by the same mechanism. Variables, in contrast, are confined to within a process statement or subprogram and do not interact with the event queue in any way.*

Be sure to understand the observation below as ignoring it gives rise to frequent misconceptions.

Observation 4.12. *A signal assignment (<=) does not become effective before the delay specified in the after clause has expired. In the absence of an explicit indication, there is a delay of one simulation cycle, so the effect can never be felt in the next statement. This sharply contrasts with a variable assignment (:=), the effect of which is felt immediately, that is, in the next statement exactly as in any programming language.*

Concurrent processes (order of execution)

A process is either active or suspended at any time. Simulation time is stopped while the code of the processes currently active is being carried out, which implies that
(a) all active processes are executed concurrently with respect to simulation time, and
(b) all sequential statements inside a process statement are executed in zero simulation time. The order of process invocation with respect to execution time is undetermined.

Observation 4.13. *As opposed to conventional programming languages where the thread of execution is strictly defined by the order of statements in the source code, there is no fixed ordering for carrying out processes (including concurrent signal assignments and assertion statements) in VHDL. When to invoke a process gets determined solely by events on the signals that run back and forth between processes.*

Sensitivity list

Each process has its own set of signals that cause it to get (re)activated whenever an event occurs on one or more of them. The process is said to be sensitive to those signals. The entirety of such signals are aptly qualified as its wake-up or trigger signals, although this is not official VHDL terminology. What are the wake-up signals of a given process? The answer depends on the type of process.

Concurrent, selected, and conditional signal assignments (activation)

Specifying wake-up signals is neither necessary nor legal as the process is simply sensitive to any signal that appears anywhere on the right-hand side of the assignment operator <= . In the example below, THISMONTH and THISDAY act as wake-up signals.

```
SPRING <= true when (THISMONTH=march and THISDAY>=21) or
                    THISMONTH=april or THISMONTH=may or
                    (THISMONTH=june and THISDAY<20)
                    else false;
```

Process statement (activation)

The process statement is much more liberal in that it provides a special clause, termed sensitivity list, where all wake-up signals must be indicated explicitly. This feature gives the engineer more freedom but also more responsibility. As we will see shortly, including or omitting a signal from a sensitivity list profoundly modifies process behavior. Upon activation by a wake-up signal, instructions get executed one after the other until the **end process** statement is reached. The process then reverts to its suspended state.

The example below is semantically identical to the conditional signal assignment above. The sensitivity list is included within parentheses to the right of the keyword **process**.

```
memless1: process (THISMONTH , THISDAY)
-- an event on any signal listed activates the process
begin
   SPRING <= false;   -- execution begins here
   if THISMONTH=march and THISDAY>=21 then SPRING <= true; end if;
   if THISMONTH=april                 then SPRING <= true; end if;
   if THISMONTH=may                   then SPRING <= true; end if;
   if THISMONTH=june  and THISDAY<=20 then SPRING <= true; end if;
end process memless1;   -- process suspends here
```

Wait statement

Another option for indicating where execution of a process statement is to suspend and when it is to resume, is to include a wait statement. Note that the two forms are mutually exclusive. That is, no process statement is allowed to include both a sensitivity list and **wait**s.

As the name suggests, process execution suspends when it reaches a wait statement. It resumes with the subsequent instruction as soon as a condition specified is met and continues until the next wait is encountered, and so on. The wait statement comes in four flavors that differ in the nature of the condition for process reactivation:

statement	wake-up condition
wait on ...	an event (signal change) on any of the signals listed here
wait until ...	*idem* plus the logic conditions specified here
wait for ...	a predetermined lapse of time as specified here
wait	none, sleep forever as no wake-up condition is given

The code below is functionally interchangeable with process **memless1** shown above but uses a **wait on** statement instead of a sensitivity list.

```
memless2: process    -- no sensitivity list because a wait statement is used
begin
   SPRING <= false;   -- execution begins here
   if THISMONTH=march and THISDAY>=21 then SPRING <= true end if;
   if THISMONTH=april              then SPRING <= true end if;
   if THISMONTH=may                then SPRING <= true end if;
   if THISMONTH=june  and THISDAY<=20 then SPRING <= true end if;
   wait on THISMONTH, THISDAY;   -- process suspends here until reactivated
                                 -- by an event on any of these signals
end process memless2;   -- execution continues with first statement
```

Note that process execution does not terminate with the **end process** statement but resumes at the top of the process body. In a process statement with a single wait statement, execution thus necessarily makes a full turn through the process code every time the process gets (re)activated.[27] As opposed to this, only a fragment of the code gets executed in a process with multiple waits, which also implies that there can be no equivalent process with a sensitivity list in this case.

A process statement may, but need not, exhibit sequential behavior

Reconsider process **memless1** and imagine **THISDAY** is omitted from the process sensitivity list. What does that change? Events on **THISDAY** are unable to activate the process and, hence, no longer update signal **SPRING**. Only an event on signal **THISMONTH** causes the logic value of **THISDAY** to get (re)-evaluated. The state of **SPRING** thus depends on past values of **THISDAY**, rather than just on the present one, which implies memory and sequential behavior.

"What exactly is it that makes a process statement exhibit sequential behavior?"

The difference is in the organization of the source code and the criteria are as follows.

Observation 4.14. *A process statement implies memory whenever one or more of the conditions below apply. Conversely, memoryless behavior is being modeled iff none of them holds.*
o *The process statement includes multiple* **wait on** *or* **wait until** *statements.*
o *The process statement evaluates input signals that have no wake-up capability.*
o *The process statement includes variables that get assigned no value before being used.*
o *The process statement fails to assign a value to its output signals for every possible combination of values of its inputs.*

Process statements with multiple waits are not supported for synthesis. How to capture sequential subcircuits in VHDL synthesis models will be the subject of section 4.3.3.

Observation 4.15. *VHDL knows of no specific language construct that could distinguish a sequential model from a combinational one. Similarly, there are no reserved words to indicate whether a piece of code is intended to model a synchronous or an asynchronous circuit, or whether a finite state machine is of Mealy, Moore or Medvedev type. What makes the difference is the detailed construction of the source code.*

---

[27] The reason why the **wait** is placed at the end — rather than at the beginning — in the **memless2** code is that all processes get activated once until they suspend as part of the initialization phase at simulation time zero.

Hint: As it is not immediately obvious whether a given process statement actually models memorizing or memoryless behavior, it is a good idea to make this information explicit in the source code, either by adding a comment or by choosing some meaningful name for the optional process label.

This habit not only makes code easier to understand, but also helps to check the presence, nature, and number of bistables that are obtained from synthesis against the code writer's intentions. This is important because a minor oversight during VHDL coding may turn a memoryless process statement into a memorizing one, and hence the associated circuit as well.

The need for detecting events on certain signals

Many subcircuits respond to specific input waveforms, and so must their models. As an example, a positive-edge-triggered register updates its state in response to a rising clock. An HDL must therefore provide a language construct to detect signal transitions. In VHDL, this role is assumed by the signal attribute `'event`, a boolean that is true iff an event has occurred during the current simulation cycle. It is frequently used in conjunction with an **if** clause to detect the completion of a clock edge and to qualify the actions listed in the **then** clause. During automatic circuit synthesis, an **if** `FOO'event` ... clause instructs the software to instantiate flip-flops triggered by events on signal `FOO` when the netlist is being built.

Example of clock edge detection   `if CLK'event and CLK='1' then ... endif;`

Alternatively, you may prefer to call upon function `rising_edge()` defined in the IEEE 1164 standard along with its counterpart `falling_edge()`. Both functions work for signals of type `std_logic` and `std_ulogic` and make use of the `'event` attribute internally.

Alternative form of edge detection   `if rising_edge(CLK) then ... endif;`

Signal attributes

Broadly speaking, a signal attribute is a named characteristic of a signal. In addition to `'event`, VHDL knows of `'transaction`, `'stable()`, `'driving`, `'last_value`, `'delayed()`, and five more. Users are free to declare their own signal attributes on top of those predefined by IEEE 1076. Not all signal attributes are supported by synthesis, though; `'event` in fact often is the only one.

The need for monitoring signal waveforms

Latches, flip-flops, RAMs, and other subcircuits with memory impose specific timing requirements such as setup and hold times on data, and minimum pulse widths on clock inputs. Should any of these timing conditions get violated, their behavior becomes unpredictable. Checking for compliance is thus absolutely essential for meaningful simulations.

Observation 4.16. *Any (sub)circuit's simulation model is in charge of two things:*
- *Evaluate input data and (re)compute the model's various equations to update its outputs and — in the case of a memorizing subcircuit — also its state.*
- *Check whether input waveforms do indeed conform with the timing requirements imposed by the subcircuit being modelled.*

Actually, timing checks must be completed before logic evaluation can begin. They are essentially carried out by having the simulator inspecting the event queue for events on the relevant signals. A timing condition is considered as respected if no events are found within the time span during which a given signal is required to remain stable. VHDL supports this idea with signal attributes such as `'stable` and with concurrent assertion statements.

Concurrent assertion statements

Above all, a concurrent assertion statement is a process that gets (re)activated by an event on any signal present in the assert expression much like a concurrent signal assignment. The difference is that a concurrent assertion statement is neither intended to schedule any new transaction nor capable of doing so. It is, therefore, qualified as a **passive process** as opposed to the process statement and to concurrent/selected/conditional signal assignments that specifically serve to update signal values. Passive processes are typically used to monitor user-defined conditions and for collecting statistical data during simulation runs.

The assertion statement checks the value of a boolean expression and takes no further action if the expression evaluates to **true**. If not so, the string following the keyword **report** is sent to the output device. The severity level is reported to the simulator, which then decides on whether to proceed or to abort the current simulation run. Though the example below works at the cell level, assertions are allowed at any level of abstraction in a design hierarchy.

A D-type flip-flop model that includes setup and hold-time checks is shown next. In this example, two concurrent assertion statements are placed between the keywords **begin** and **end** in the entity declaration. Incidentally, note that the assertion statement is also available in a sequential form for inclusion in process statements and subprograms, for instance.

```
entity setff is
    .....
begin
    assert (not (CLK'event and CLK='1' and not INPD'stable(1.09 ns)))
        report "setup time violation" severity warning;
    assert (not (INPD'event and CLK='1' and not CLK'stable(0.60 ns)))
        report "hold time violation" severity warning;
end setff;

architecture behavioral of setff is
    signal PREST : bit;    -- state signal
begin
    memzing: process (CLK,RST)
    begin
        -- activities triggered by asynchronous active-low reset
        if RST='1' then
            PREST <= '0';
```

```
          -- activities triggered by rising edge of clock
          elsif CLK'event and CLK='1' then
             PREST <= INPD;
          end if;
       end process memzing;
       .....
end behavioral;
```

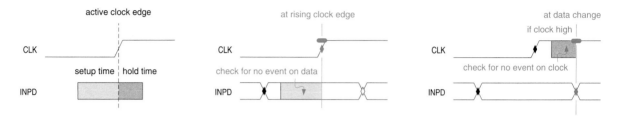

Fig. 4.12 Checking setup and hold conditions by searching the event queue for past events.[28]

Observation 4.17. *Any inspection of the event queue for compliance with timing requirements must necessarily look backward in time as forthcoming transactions might be added at any time and change the anticipated future evolution of the signal being examined.*

Initial values cannot replace a reset mechanism

As discussed earlier, the VHDL syntax supports assigning an initial value to a signal as part of the signal's declaration statement, and the same applies for variables.

Observation 4.18. *The initial value given to a signal or variable defines the object's state at $t = 0$, just before the simulator enters the first simulation cycle. A hardware reset mechanism, in contrast, remains ready to reconduct the circuit into a predetermined start state at any time $t \geq 0$. Modelling such a mechanism with an HDL requires the distribution of a special reset signal to all bistables concerned.*

Make sure you understand these are two totally different things. An initialized signal or variable does not model a hardware reset facility and will, therefore, not synthesize into one. A working example for a hardware reset mechanism has been given in the setff code above.

---

[28] Setup and hold times are assumed to be positive in the figure. The modelling of sequential subcircuits that feature a negative timing constraint becomes possible by adding fictitious input delays and by adjusting their values such as to make both setup and hold times positive. In order to preserve the original delay figures, all delays added at the input must be compensated for at the output. This is how negative timing constraints are to be handled according to the VITAL standard, at any rate.

4.2.5 Facilities for model parametrization

The need for supporting parametrized circuit models

Imagine you have devised a behavioral HDL model for some datapath unit that includes 16 data registers and that is capable of carrying out seventeen distinct arithmetic and logic operations on data words 32 bits wide. While continuing on your project, you find out that you need a similar unit for address computations. There are significant differences, however. Addresses are just 24 bits wide, no more than five registers are required, and a subset of eight ALU operations suffices for the purpose. How do you handle such a situation?

It would be fairly easy to derive a separate model by pruning the existing code and by downsizing certain index ranges there. But what if you later needed a third and a fourth model? What if new requirements asked for substantial extensions to the existing model? The problem lies not so much in the initial effort of creating yet another model. Rather, it is the maintenance of a multitude of largely identical source codes that renders this approach so onerous.

A truly reusable model, in contrast, should be written in a parametrized form so as to accommodate distinct choices and parameter settings within a single piece of code. The most salient features of VHDL for this goal are generic parameters, configurations, conditional spawning of processes, and conditional instantiation, all of which items are going to be introduced in this section. Please refer to fig.4.13 to see how these fit into the general picture.

Generics

As stated earlier, signals carry dynamic, i.e. time-varying, information between processes and indirectly also between design entities. Generics, in contrast, serve to disseminate static, i.e. time-invariant, indications, to design entities such as

- Word width,
- Active-low or -high signaling on inputs and outputs,
- Output drive capability,
- Functional options (e.g. details about an instruction set),
- Timing quantities (propagation and contamination delay, setup and hold time), and
- Capacitive load figures.

For a generic to become visible within a given entity, the generic's name must be included in the entity declaration in a so-called **generic clause** in much the same way as signals are made accessible with a port clause. Yet, as opposed to ports, generics do not have any direct hardware counterpart. Also, it is possible to indicate a default for the case that no value is specified when the component gets instantiated. An example follows.

```
component parityoddw    -- w-input odd parity gate
   generic (
      width : natural range 2 to 8;    -- number of inputs with supported range
      tpd : time := 1.0 ns );    -- propagation delay with default value
   port (
      INP : in  std_logic_vector(width-1 downto 0);
      OUP : out std_logic );
end component;
```

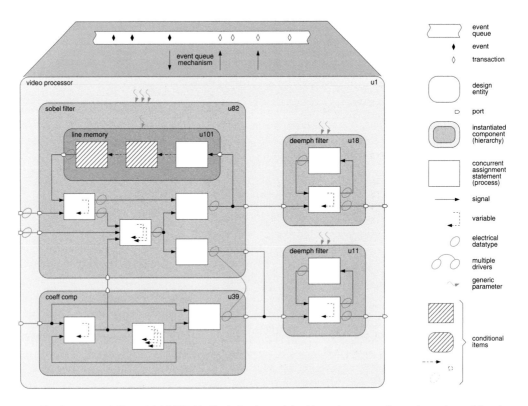

Fig. 4.13 Circuit modelling with VHDL V: Final circuit model taking advantage of generics and conditional items for parametrization.

Conditional spawning of processes with the generate statement

Situations exist where it is impossible to know and freeze the number of processes at the time when the source code is being written. Cosider array-type circuits where some elemental subfunction is repeated along one or more dimensions and where each subfunction interacts with its neighbors. As it comes naturally to describe the behavior of each such subfunction in a concurrent process of its own, the total number of processes is bound to vary with array size. Yet, writing a hardware model that suits arbitrary array sizes is not possible with the customary computer language constructs that govern code execution at run time.

With its generate statement, VHDL provides a mechanism for producing processes under control of constants and generics. An example is given in the code fragment below that implements the Game of Life.[29]

---

[29] The Game of Life by John Horton Conway dates back to 1970. It is not really a game, in fact it is a 2-dimensional cellular automaton where each cell has eight neighbors. Each cell is either dead or alive, and its birth, survival or death depends on how many living neighbors the cell has. In the case of the Game of Life, the rule is often abbreviated as "S23/B3", which means that a living cell survives if it has two or three live neighbors, and that a dead cell is turned into a living one if exactly three of its neighbors are alive.

```
.....
-- spawn a process for each cell in the array
row : for ih in height-1 downto 0 generate    -- repetitive generation
   cell : for iw in width-1 downto 0 generate    -- repetitive generation
      memzing: process(CLK)
         subtype live_neighbors_type is integer range 0 to 8;
         variable live_neighbors : live_neighbors_type;
      begin
         if CLK'event and CLK='1' then
            live_neighbors := live_neighbors_at(ih,iw);
            if PREST(ih,iw)='0' and live_neighbors=3   then
               PREST(ih,iw) <= '1';    -- birth
            elsif PREST(ih,iw)='1' and live_neighbors<=1 then
               PREST(ih,iw) <= '0';    -- death from isolation
            elsif PREST(ih,iw)='1' and live_neighbors>=4 then
               PREST(ih,iw) <= '0';    -- death from overcrowding
         end if;
      end process memzing;
   end generate;
end generate;
.....
```

Incidentally, there exists not only a replicative form of the generate statement but also a conditional form that is discussed next, along with a somewhat different usage of this statement.

Conditional instantiatiation with the generate statement

Many circuits are obtained by replicating subcircuits in a more or less regular way. It thus makes sense to describe the repetitive subcircuits only once and to indicate how many times each is going to be instantiated. Similarly, when writing parametrized netlist generators, one often finds that circuit details depend on the exact word width, the actual bit position, or the like. Straight component instantiation statements cannot cope with such situations, what is needed is a controllable instantiation mechanism. As this bears much resemblance to the conditional spawning of processes discussed before, it comes as no surprise that the generate statement is used to handle such tasks as well.

Consider the source code below and note that the generate comes in two flavors. A replicative **for...generate** serves to iteratively compose the circuit from bit slices whereas conditional **if...generate** statements take care to implement different logic subcircuits depending on whether the most significant bit or one of the other bits is being handled.

```
-- architecture body
architecture structural of binary2gray is

   -- component declarations
   component xnor2_gate
      port (A1, A2 : in bit;
            ZN : out bit);
   end component;
   component inverter_gate
      port (I : in bit;
            ZN : out bit);
```

```
   end component;

   -- signal declarations
   signal INODE : bit_vector(width-1 downto 0);

begin

   -- assemble logic network by instantiating and interconnecting components
   any_slice : for i in 0 to width-1 generate  -- repetitive generation
      -- a row of eqvs except for the MSB which requires an inverter
      less_significant : if i<width-1 generate  -- conditional generation
         uxn : xnor2_gate  -- component instantiation
            port map (A1=>INP(i), A2=>INP(i+1), ZN=>INODE(i));
      end generate;
      most_significant : if i=width-1 generate  -- conditional generation
         uin : inverter_gate  -- component instantiation
            port map (I=>INP(i), ZN=>INODE(i));
      end generate;
      -- a final row of inverters
      uin : inverter_gate  -- component instantiation
         port map (I=>INODE(i), ZN=>OUP(i));
   end generate;

end structural;
```

Observation 4.19. *Generate statements get interpreted before actual circuit simulation and synthesis can begin.*[30] *Variables and signals are not acceptable as conditions or as loop boundaries since they are subject to vary at run time, only constants and generics are.*

The need for multiple architecture bodies

Consider the role of VHDL simulation in a typical VLSI design project. A first model established early on in the design cycle serves to validate the intended functionality. Such an early model is relatively compact because it can be written in a purely behavioral style with little attention paid to architectural issues or to current limitations of synthesis technology.

Later in the design cycle, a more detailed model is written — probably at the RTL level — as a starting point for synthesis. Here, simulation serves to make sure this second and much more voluminous piece of code is correct before it gets implemented by myriads of gates.

A third simulation round is then carried out on the gate-level netlist obtained from synthesis. In the end, three models of distinct levels of detail have been used to capture the same functionality.

A similar situation occurs when one wants to evaluate different approaches for implementing a given functionality in hardware with respect to their relative merits for synthesis and simulation (gate count, longest path delay, energy efficiency; run time, memory requirements, etc.).

VHDL accommodates multiple circuit models by allowing a design entity to have more than one architecture body, see fig.4.14.

---

[30] As part of the elaboration phase to be explained shortly.

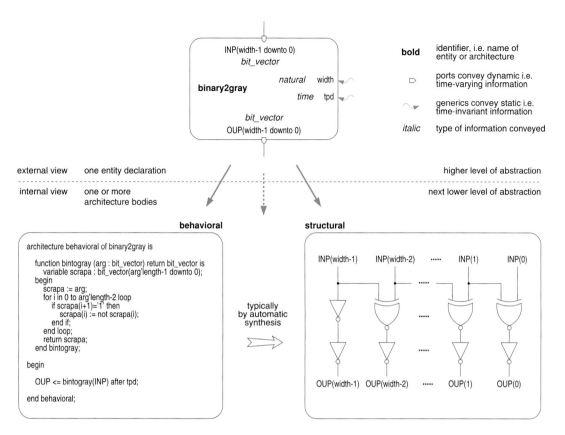

Fig. 4.14 Entity declaration versus architecture body and port versus generic.

Configuration specification and binding

In the presence of multiple architecture bodies for one entity declaration, designers must be given an instrument to indicate which architecture to consider during simulation and synthesis. This is the configuration specification statement. The mechanism is more general, however, in that it makes it possible to specify a binding between any one instance of a component and some entity–architecture pair. Put differently, a component instantiated under one name can be bound to an entity with a different name, and this binding does not need to be the same for all instances drawn from that component. Configuration specification statements must appear in the design unit where the components concerned are being instantiated.

Prior to simulation and synthesis, the VHDL analyzer software must follow all the configuration specification statements present in a circuit model and must then bring to bear the architecture bodies indicated. This preparatory step is also referred to as binding.

The code fragment below specifies that instance u113 of entity or component **binary2gray** is to be implemented by the architecture body **behavioral** of entity **binary2gray** whereas the body **structural** is to be used for u188.

```
.....
for u113: binary2gray use entity binary2gray(behavioral);
for u188: binary2gray use entity binary2gray(structural);
.....
```

The code fragment below specifies that all instances of EQV gates and inverters that have been instantiated in the previous example are to be modelled by the respective entity–architecture pairs indicated.

```
.....
for all: xnor2_gate use entity GTECH_XNOR2(behavioral);
for all: inverter_gate use entity GTECH_NOT(behavioral);
.....
```

Observation 4.20. *VHDL provides a range of complementary constructs that are instrumental in writing parametrized circuit models. More particularly, it is possible to establish a model without committing the code to any specific number of processes and/or instantiated components.*

Elaboration

The fact that the processes, and signals in a VHDL model are not a priori fixed but obtained under control of the VHDL code itself necessitates a preparatory step before simulation or synthesis can begin, see fig.4.15. During that phase, termed elaboration, multiple instantiations get expanded and generate statements unrolled so that the final inventory of instances, processes and signals can be established. No new instances, processes or signals can be created after elaboration is completed.

In preparation to simulating a design, all lowest-level design entities (leaf cells) must become available as behavioral models, either as part of the current VHDL design file itself or from a separate components library. Once elaboration is completed, the memory space necessary to hold the entirety of signals can be reserved. As processes work concurrently, memory space also needs to be set aside for the variables associated with every single process.

In preparation to synthesizing a design, the software has to find out where to resort to actual synthesis and where to simply assemble a netlist from library components. Any elemental component instantiated as part of a structural model must be available from the cell library targeted. The numbers of occurrences for all elemental and non-elemental components are frozen when elaboration completes.

4.2.6 Concepts borrowed from programming languages

Structured flow control statements

Structured programming is universally accepted by the programming community. The IEEE 1076 standard defines a set of flow control statements that is consistent with this discipline and that supports exception handling in nested loops. Constructs include `if...then...else`, `case`, `loop`, `exit`, `next` the semantics of which are self-explanatory.

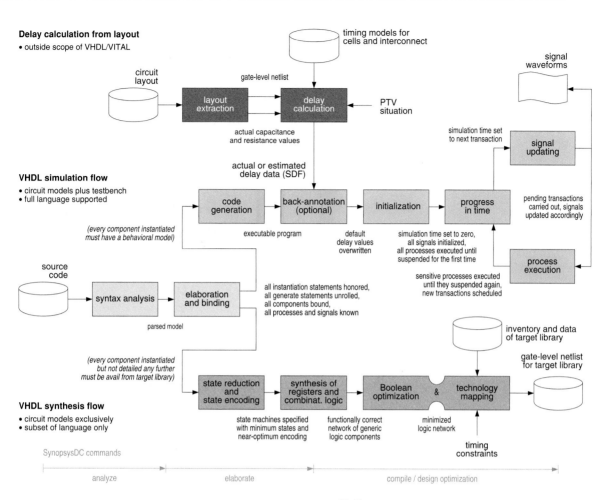

Fig. 4.15 Major steps in VHDL simulation and synthesis (simplified).[31,32]

Object

A data object, or simply an object, is a constant, a variable, a signal, or a file.[33]

---

[31] It may seem counterintuitive that stimuli and expected responses appear nowhere in the simulation part of the drawing. The reason is that the testbench which feeds the event queue with the necessary transactions is itself being described in VHDL and thus enters the processing chain from the left much as the source code of the model under test (MUT) does.

[32] Organization and vocabulary of commercial products may deviate considerably from our generic figure, take DesignCompiler by Synopsys, for instance. While the "analyze" command does what its name suggests, the subsequent "elaborate" encompasses elaboration, binding, parts of synthesis, plus a preliminary mapping to logic gates from the GTECH generic library. Completing synthesis, timing-driven logic optimization, and technology mapping are then handled by another command named "compile". Also, FSMs do not get processed and optimized before logic synthesis occurs but are re-extracted later from the gate-level netlist where necessary.

[33] Files were not considered as objects before the VHDL'93 standard update. In addition, file handling operations differ between VHDL'87 and '93, so porting legacy code that reads from or writes to files necessitates adaptations of the source.

Constant

A constant holds a fixed value that gets assigned where the constant is declared.

Example of a constant declaration `constant Fermat_prime_4 : integer := 65537;`

Variable

A variable holds a changeable value. The scope is limited to the subprogram or process where the variable is declared; global variables do not exist.[34] As an option, an initial value may be indicated where the variable is declared. Variables declared in a subprogram are (re)initialized whenever the subprogram is called, whereas variables declared in a process retain their value from one process activation to the next, that is until they are assigned a new value. The assignment operator for variables is := whereas <= must be used when assigning to a signal.[35]

Example of a variable declaration `variable brd : real := 2.48678E5;`
Example of a variable assignment `brd := brd + ddr;`

Data types

In VHDL, type is synonym for data type. Basically, a data type defines a set of values and a set of operations that can be performed on them. VHDL is a strongly typed language, which implies that every object has a type and that extensive type checking is performed. Types must be converted before an assignment or a comparison becomes possible across distinct types, also see section 4.8.7 for conversion functions.

Users may declare their own data types or use predefined ones. The types **real**, **integer**, and **time** are part of the IEEE 1076 standard. Enumerated types are also supported, and the predefined types **character**, **boolean**, and **bit** are in fact such types. More data types added on top of the VHDL language for the modelling of various electrical phenomena have been discussed in section 4.2.3.

Example of a type declaration `type month is (january, february,..., december);`

Subtypes

A subtype shares the operations with its parent type, but differs in that it takes on a subset of data values only.[36]

Example of a subtype declaration `subtype day is integer range 1 to 31;`

---

[34] So-called protected shared variables were added later, please refer to section 4.8.1 for their usage.

[35] Further insight on how variables relate to signals and time can be obtained from section 4.2.4.

[36] It is perfectly legal to declare a subtype with an improper subset, i.e. with a data set identical to that of its parent type.

Hint: It is good engineering practice to indicate an upper and a lower bound when using integers for the purpose of hardware modelling. A simulation tool can then incorporate on-line checks to ascertain that data values do indeed remain within their legal range, while synthesis is put in a position to selectively cut down the number of hardware bits rather than using the default word width of 32 bit.

Arrays and records

Scalar, i.e. atomic, data types can be lumped together to form composite data types. An array is a collection of elements all of which are of the same type.

Example of an array type declaration   `type byte is bit_vector(7 downto 0);`

As opposed to the homogeneous composition of an array, a record may be assembled from elements of different types, which is why it is qualified as heterogeneous. Another difference is that each element is identified by a name unique within the record. An example follows.

```
type date is record
   date_year : integer;
   date_month : month;
   date_day : day;
end record;
```

Hint: Records come in handy when a set of signals connects to many entities or traverses multiple levels of hierarchy. Keeping clock and reset out, use a record to collect the entire set into one wholesale quantity that can be referenced as such. Adding or dropping an information signal or changing the cardinality of a vector then becomes just a matter of modifying one record declaration in one package. Designers can thus dispense with rewriting the port clauses in all entity declarations.

Type attributes and array attributes

A type attribute is a named characteristic of a data type or data value. Type attributes make it possible to recover information such as the range of a type or subtype, the position number, the successor, or the predecessor of a given value in an enumerated type, and the like. An example for the usage of type attributes `'left` and `'right` is given further down in function **NextMonth**.

Array attributes operate in a similar way on array types and array objects to obtain their bounds and cardinalities. In the example below, **dateandtime'range** returns 0 to 5 whereas asking for **dateandtime'length** yields a value of 6.

```
type sixtupel is array (0 to 5) of integer;   -- type declaration
variable dateandtime : sixtupel;              -- variable declaration
```

Subprogram, function, and procedure

A subprogram is either a function or a procedure. A function returns a value and has no side effects, whereas the opposite is true for a procedure. Any subprogram is dynamic, which is to say that it does not persist beyond its current invocation. An example of a function, named `NextMonth`, is given below as part of a package body.

Package

A package is a named collection of widely used constants, types, subprograms, and/or component declarations that becomes visible within some other design unit when being referred to it in a **use** clause. Packages make it possible to circumvent the waste and perils of repeating supposedly identical declarations at multiple places. VHDL applies the principles of **information hiding** to packages by separating package declaration from package body.[37]

```
-- package declaration
package calendar is
   type month is (january, february, march, april, may, june, july,
                  august, september, october, november, december);
   subtype day is integer range 1 to 31;
   function NextMonth (given_month : month) return month;
   function NextDay (given_day : day) return day;
end calendar;

-- package body
package body calendar is

   function NextMonth (given_month : month) return month is
   begin
      if given_month=month'right then return month'left;
      else return month'rightof(given_month);
      end if;
   end NextMonth;

   function NextDay (given_day : day) return day is
   .....
   end NextDay;

end calendar;
```

Predefined package standard

All data types and subtypes of VHDL are actually defined in this package along with the pertaining logic and arithmetic operations and a few more features. As a package **standard** comes with the language, always gets precompiled into design library **std**, and is made available there by default; users do not normally need to care much about it.

---

[37] Information hiding is a well-established principle in software engineering whereby a piece of software is divided into a declaration module or interface which must be made accessible to the caller, and a separate implementation module or body which is deliberately withheld. It is based on the observation that the interactions between software entities remain the same regardless of their implementation details, as long as their interfaces and overall functionalities do not change. Note the similarity to the **black box** concept from electrical engineering.

Predefined package `textio`

Type declarations and subprograms related to the reading and writing of ASCII files are collected in a special package `textio`, which greatly facilitates the coding of testbenches and other programs that do interface with text files. For obvious reasons, file I/O code is not intended for synthesis. Package `textio` routinely gets precompiled into library **std** as well. Yet, to make its definitions immediately available within a design unit, the pertaining source code must include the line **use** `std.textio.all;`.

Design unit and design file

Incremental compilation is based on the separate processing of individual program modules. The VHDL term for a language construct that is amenable to successful compilation[38] independently from others is design unit. VHDL provides five kinds of design units, namely

- Package declaration,
- Package body,
- Entity declaration (see section 4.2.2),
- Architecture body (see sections 4.2.2 and 4.2.1), and
- Configuration declaration (see section 4.2.5).

One or more design units are stored in a design file.[39] VHDL compilers, aka VHDL analyzers, accept one design file at a time and store the output in a design library, see fig.4.16.

Observation 4.21. *VHDL supports information hiding and incremental compilation.*

Design library

A design library is a named repository for a collection of design units after compilation on a host computer, which has two major implications. Firstly a design library is normally not portable but specific for a platform. That is, the result depends on the computer for which it has been compiled and on the software tool being used (manufacturer, product, simulator or synthesizer). Secondly, a design library can accommodate many design files and design units. As a consequence, a library is typically referenced under a logical name that differs from the original name(s) of the design file(s) included. Also note that a VHDL design library can hold a program library, a component library, or both.

---

[38] Compilation here loosely refers to the early processing steps that are necessary to simulate and/or synthesize a circuit from VHDL source code, also see fig.4.15. More precisely, one can distinguish between four operating principles in VHDL simulation. A first category translates the original VHDL source into some **pseudo code** which then gets interpreted by a simulation kernel. A second category begins by translating VHDL into C. This C code is then translated into the host's machine code by the local C compiler before being linked with the simulation kernel. Simulation is done by executing this machine code. The third category avoids the detour and compiles from VHDL to machine code directly, which is why the approach is termed **native compiled code**. All three operating principles can be found among commercial VHDL simulators; a discussion of their relative merits is given in [100] [101]. Yet another approach is **hardware acceleration** by which a design gets downloaded to multiple processors tailored to the purpose after elaboration. As we focus on the language itself, we will not differentiate between such implementation alternatives in this text.

[39] Incidentally, note that `design_file` serves as start symbol in formal definitions of VHDL such as syntax diagrams or the Extended Backus Naur Form (EBNF).

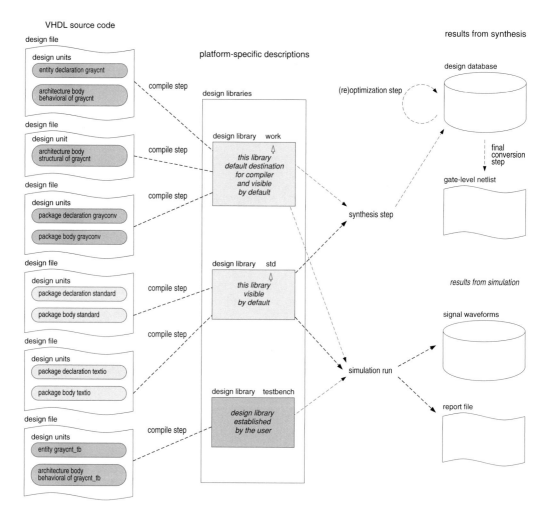

Fig. 4.16 Organization of VHDL source code and intermediate data (simplified).

Library and use clauses

As most designs make use of VHDL code that is distributed over several design libraries, a mechanism is needed for importing outside subprograms, types, entities, and components. This is exactly the *raison d'être* behind the VHDL library and the **use clause**.

In the code fragment below, the first statement makes the design library **testbench** visible. That is, subprogram **PutSimulationReportLine** from package **graycnt_tb**, for instance, could then be referenced as **testbench.graycnt_tb.PutSimulationReportLine**, provided the package had previously been compiled into that very design library. Much the same holds for other items declared in the package.

The subsequent use clause goes one step further in that it dispenses with the need to write the full library and package names each time an item from package **graycnt_tb** is being referenced.

In the occurrence, a simple `PutSimulationReportLine` suffices.[40] Another example is presented in section 4.9.2; you may want to have a look at it.

```
-- library and use clauses
library testbench;
use testbench.graycnt_tb.all;
.....
```

Special libraries `work` and `std`

The design library named `work` differs from all other libraries in that VHDL compilers direct their output to that particular library unless explicitly instructed to do otherwise. Another special library termed `std` accommodates the compilation results from the two packages `standard` and `textio` that come with VHDL. In addition, libraries `work` and `std` are universally visible because the two statements below are tacitly included in any design unit.

```
library std, work;
use std.standard.all;
```

Note that `standard` is the only package that is immediately available by default; all others ask for an explicit use clause. In the example of fig.4.16, the programmer has to include a statement such as `use work.grayconv.all;` in the source code of `graycnt` in order to make the code of package `grayconv` available for reference there.

4.3 | Putting VHDL to service for hardware synthesis

4.3.1 Synthesis overview

Automatic synthesis aims at turning some sort of behavioral description (RTL or other) into a gate-level netlist with as little human intervention as possible. Using the standard cells available from a target library, VHDL synthesizers attempt to come up with a gate-level circuit that meets all user-defined performance targets at the lowest possible hardware costs, see fig.4.17.

A basic overview of the synthesis process is available from fig.4.15 in section 4.2.5 where elaboration and binding have been discussed. State reduction eliminates redundant states, if any, while state encoding assigns a unique binary code to each state.[41] The subsequent **synthesis** step builds all necessary state registers and specifies the combinational subfunctions in between. The result is a preliminary network described at an intermediate level of detail in terms of logic equations and generic components rather than actual logic gates.

Boolean optimization attempts to rework the logic networks in such a way as to bring their longest signal propagation paths below the relevant user-defined timing constraint while, at the

---

[40] The improved convenience also opens the door for ambiguous references, however, if the same name happens to exist more than once across the various design libraries being made visible in this way.

[41] Please refer to appendix B for more details.

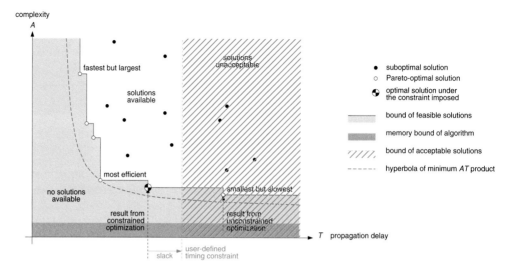

Fig. 4.17 Tradeoffs between size and performance for a hypothetical circuit.

same time, minimizing hardware complexity. Finding an optimal or near-optimal circuit depends on numerous characteristics of the library cells available, so simplifying and reorganizing logic equations and networks is closely intertwined with the subsequent **technology mapping** [102] phase where the generic gates in the netlist get replaced by components that are actually available from the target library.[42] More sophisticated tools also address energy efficiency.

As stated in section 4.1.3, VHDL was not originally intended for synthesis. Existing synthesis tools handle only a subset of the VHDL grammar. Much of this section is devoted to discussing the existing limitations and to presenting workable solutions for synthesis code.

Observation 4.22. *While almost all VHDL simulators support the full IEEE 1076 standard, only a subset of the legal language constructs is amenable to synthesis. As good VHDL code must be portable across platforms and synthesis tools, model writers must confine themselves to safe, unambiguous, and universally accepted constructs.*

4.3.2 Data types

Only a subset of the VHDL types is amenable to hardware synthesis. Support covers
+ `integer`,
+ `boolean` and `bit`,
+ `std_logic` and `std_ulogic` (see section 4.2.3),
+ `signed` and `unsigned` (see section 4.2.3),
+ user-defined enumerated types,
along with the pertaining array and record data types. Unlike the above items,

---

[42] Much of today's logic optimization software descends from programs such as ESPRESSO (two-level logic) and MIS (multilevel logic). A major challenge in coming up with adequate algorithms is to achieve a low asymptotic complexity in order to cope with complex functions and large networks.

— `real`,

— `time`,

— `character`, and

— `file`

are not normally supported.

4.3.3 Registers, finite state machines, and other sequential subcircuits

With respect to sequential subcircuits, an important limitation of synthesis technology is this.

Observation 4.23. *While VHDL allows the modelling of arbitrary behavior (as long as it is causal, discrete in value, and discrete in time), automatic synthesis supports synchronous clock-driven subcircuits and — at a higher level — conglomerates of such subcircuits.*

Synchronous circuit operation means that state transitions are restricted to occuring exclusively at precise moments of time as defined by a clock signal.[43] Not all code that is syntactically correct VHDL and that works during simulation is thus acceptable for synthesis. More particularly:

Observation 4.24. *Synthesis tools do not support multiple waits in a process statement.*

Only process statements with a sensitivity list or with a single wait are amenable to synthesis. The reason is that each wait statement is allowed to carry its own condition as to when process execution is to resume. Depending on the details postulated in those conditions, the source code may imply synchronous or asynchronous behavior. While the former readily maps to a clock-driven circuit assembled from flip-flops and combinational gates, the latter is likely to express a behavior that depends on a haphazard collection of events on various signals in a delicate way. Coming up with a physical circuit that is safe and functionally equivalent to the source code may then prove extremely difficult, if not impossible.

HOW TO MODEL A REGISTER IN VHDL

To make sure synthesis code will be met with universal acceptance, any process statement that is supposed to exhibit sequential behavior should be written in such a way as to conform with the skeleton shown in listing 4.3. A clock signal is mandatory. One optional signal is accepted for implementing an asynchronous reset function on the state register. No other signals are permitted to (re)activate the process. The reason for marking various items as disallowed is that their presence in the code would render the model's behavior inconsistent with a single-edge-triggered register

Listing 4.3 | Skeleton of a process statement with memory that is safe and universally accepted for synthesis.

```
---- updating of state
------------------------------------------------------------------------
process (CLOCK,RESET) <--------- sensitivity list, no more signals accepted!
begin
    <--------- no other statement allowed here!
```

---

[43] An in-depth discussion is to follow in section 5.2.1.

```
-- activities triggered by asynchronous active-high reset
if RESET='1' then
   PRESENTSTATE <= STARTSTATE;
   .....
-- activities triggered by rising edge of clock
elsif CLOCK'event and CLOCK='1' then <--------- no more term allowed here!
   <--------- extra subconditions, if any, accepted here.
   PRESENTSTATE <= NEXTSTATE;
   .....
<--------- no further elsif or else clause allowed here!
end if;
<--------- no more statement allowed here!
end process;
```

Portability is not the only merit of this restricted coding scheme, it also ensures that subcircuits so described behave safely and predictably. A process statement for a register is given in the code example of listing 4.4.

Listing 4.4 | Code example for a register that features an asynchronous reset, a synchronous load, and an enable. Actual designs are unlikely to combine all three mechanisms in a single register, so a subset of the clauses shown will typically do.

```
---- updating of state
--------------------------------------------------------------------------
p_memzing : process (CLKxC,RSTxRB)
begin
   -- activities triggered by asynchronous reset
   if RSTxRB='0' then
      STATEVECTORxDP <= (others => '0')   -- shorthand for all zeros
   -- activities triggered by rising edge of clock
   elsif CLKxC'event and CLKxC='1' then
      -- when synchronous load is asserted
      if LODxL='1' then
         STATEVECTORxDP <= (others => '1')   -- shorthand for all ones
      -- else assume new value iff enable is asserted
      elsif ENAxS='1' then
         STATEVECTORxDP <= STATEVECTORxDN;
      end if;
   end if;
end process p_memzing;
```

EXPLICIT VERSUS IMPLICIT STATE MODELS

These two classes are easily told apart by counting the wait statements per process statement. In an **explicit state model**, there is no process statement with more than one **wait on** or **wait until**; process statements with a sensitivity list (and hence no wait) are more typical, however. In either case, program execution always returns to the same line of code and suspends there after having completed one full turn. One or more VHDL signals — or variables — serve to preserve the current state from one process activation to the next. Explicit state models come most naturally to

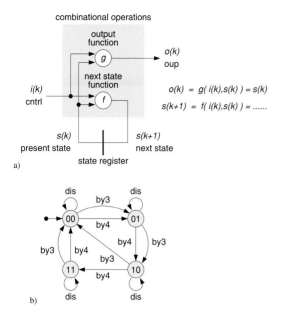

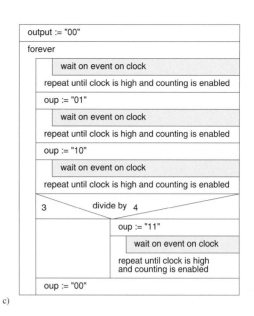

Fig. 4.18 Three formalisms that affect the writing of software models for sequential behavior. Data dependency graph (a), state chart (b), Nassi–Shneiderman diagram (c). Note that the programmable divide-by-3/divide-by-4 counter chosen for illustration is a Medvedev machine.

hardware designers who are accustomed to thinking in terms of finite state machines (FSMs) along with visual formalisms such as schematics, state graphs, and the like.

An **implicit state model**, in contrast, is immediately recognized by the presence of multiple waits in process statements. Upon activation, the simulator executes instructions until the next wait statement encountered causes it to suspend. Suspension may thus occur at distinct lines of code, and each wait statement represents a specific state of the model. There is no tangible state variable. Rather, it is the return address to the suspended process that assumes this role during simulation, hence the name implicit state. Implicit state models resemble the way how software designers code algorithms, and are related to Nassi–Shneiderman diagrams, aka structograms, and flowcharts. See fig.4.18 and table 4.4 for a comparison.

Implicit state models occasionally find applications in the context of purely behavioral simulation, that is above the RTL level in the abstraction hierarchy. They must be translated into an explicit state model at the RTL level before synthesis can begin. Code examples in this text refer to explicit state models exclusively.

How to capture a finite state machine in VHDL

The restriction to explicit state models notwithstanding, one still has the choice of putting an FSM into one VHDL process statement or of distributing it over two or more concurrent processes.

Table 4.4 | Explicit and implicit state models compared.

modelling style	explicit state		implicit state
	computed state	enumerated state	
inspired by	data dependency graph or schematic diagram	state chart, state graph, or state table	Nassi–Shneiderman diagram
synchronization mechanism	sensitivity list or single wait statement (semantically equivalent)		multiple wait statements
state variable	declared explicitly as signal or variable and thus of user-defined type		hidden in pointer to current statement
states captured by	(sub)range of integer or of bit vector type	range of enumerated type	multiple wait statements
state transitions captured by	arithmetic and/or logic operations	one-to-one translation from state table	control flow
output function captured by	arithmetic and/or logic operations	one-to-one translation from state table	assignment statements
immediate hardware equivalent	yes		depending on wait conditions
synchronous	yes		*idem*
synthesizable	yes		no

Packing an entire FSM into a single process statement

The code is organized as illustrated by fig.4.19a through c for Mealy, Moore, and Medvedev automata respectively.[44] Evaluation begins with the asynchronous reset and clock inputs to find out whether the machine's state must be updated. The remaining inputs are processed further down in the process statement. The state must be stored as a VHDL <u>variable</u> because any state change must become visible to the output function within the same process activation. The process statement must be made sensitive to events on clock and reset and, in the case of a Mealy machine, to events on other input signals as well.

Although this coding style is legal VHDL and perfectly acceptable for simulation, its general adoption is discouraged because it is not supported by many synthesis tools and because it may result in inefficient gate-level networks.

Distributing an FSM over two (or more) concurrent processes

As depicted in fig.4.19d through f, a memorizing process is essentially in charge of maintaining the current state from one activation to the next. A second process statement of memoryless nature computes the next state and the present output value. This combinational part might just as well be implemented using concurrent/selected/conditional signal assignments. Both present state and next state are being modelled as VHDL <u>signals</u> that go back and forth between the two (or more) processes involved.

---

[44] The three classes of automata essentially differ in the nature of their output functions. Please refer to sections B.1 and B.2 if you have doubts about the characteristics and equivalence relations among those classes.

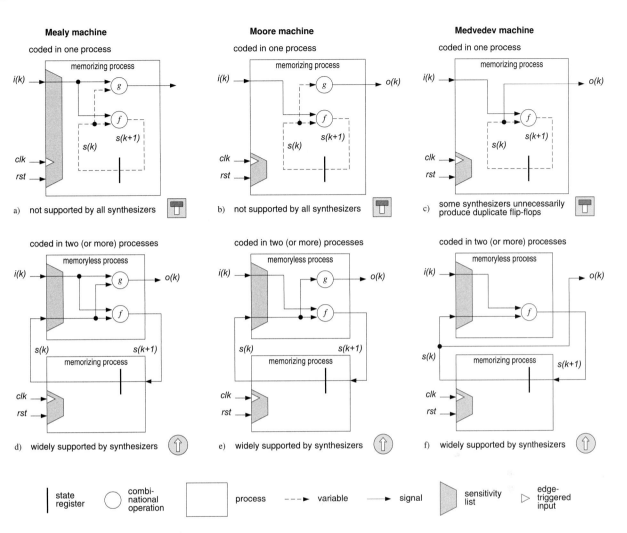

Fig. 4.19 Coding schemes for synchronous Mealy, Moore, and Medvedev machines. Note that nothing prevents a designer from distributing the combinational operations subsumed here as "memoryless process" over two or more concurrent processes in his code. Also, it is left to the discretion of the programmer whether to capture them as concurrent signal assignments or with the aid of process statements.

Capturing sequential and combinational behavior in separate processes is universally accepted and tends to lead to more economic circuit structures during synthesis. As an additional benefit, the type of automaton (Mealy, Moore, Medvedev) can be changed at any time without having to reorganize the source code too much.

Observation 4.25. *For the sake of portability, VHDL code intended for synthesis shall*
a) model circuits at the register transfer level (RTL) throughout,
b) collect combinational and sequential logic in separate processes, and
c) have all memorizing process statements conform with the skeleton of program 4.3.

Hint: If a process statement is to model

o Combinational logic, exhaustively enumerate all inputs in the sensitivity list because any one of them must act as a wake-up signal. Ensure that you assign a value to each output for all possible combinations of input values.

o A flip-flop, register, counter, or a similar sequential subcircuit, include the driving clock plus an optional asynchronous reset signal in the sensitivity list, but nothing else. Do not forget to initialize all state-preserving objects in the reset clause, if any.

These are the only two code arrangements compatible with the synthesis of synchronous edge-triggered circuits. Forgetting signals in sensitivity lists is a frequent mistake.[45] Failure to assign a don't care to a signal — or to a variable declared within a process statement — in situations where the object's value does not seem to matter is another common oversight. Keep in mind that not assigning anything means that the object's present value shall be maintained and, as a consequence, implies memory.

Hint: The emergence during synthesis of latches or other bistables that are unplanned for must alert the VHDL designer that his code is badly wrong. For each subcircuit, check the number of flip-flops and latches obtained from synthesis against your expectations. As a general rule, do not ignore warnings and error messages from the synthesizer unless you understand what they mean.

Hint: In applications where state transitions depend on an enable signal, you have two options: Either include a subcondition in the memorizing process, as in listing 4.4, or add an extra conditional branch in the memoryless process in charge of determining the next state. Though the two approaches are functionally identical, we recommend the first because the synthesizer may otherwise be unable to map to E-type flip-flops and, hence, to implement clock gating.

VHDL code examples for Mealy, Moore, and Medvedev machines amenable to synthesis are given in appendix 4.9.2. Please note that the same examples also serve to illustrate how to model an asynchronous reset, a synchronous clear, and an enable.

FSM optimization ignored in the language standard

Another concern arises when one wants to take advantage of automatic state reduction and/or automatic state encoding in an FSM that uses an enumerated data type to capture its set of states. Some FSM optimization tools have been designed to recognize automata from the code alone, others need some form of guidance to identify the signals and/or variables that act as a repository for the current state. The problem is that VHDL provides no such construct.

---

[45] Incidentally, note that not all EDA tools handle sensitivity lists alike. DesignCompiler by Synopsys, for instance, checks a sensitivity list for consistency with the process body, but essentially synthesizes from the sequential statements in the process body alone. While this is no problem for correct models, a netlist obtained from an RTL model with an incomplete sensitivity list may behave and simulate differently from the original model.

Vendors of EDA tools have come up with proprietary extensions to fill the gap. Perhaps the most elegant solution is the one adopted by Synopsys because it neatly extends the VHDL IEEE 1076 standard syntax by one user-defined attribute.

```
architecture enumerated_state of ..... is
   type state is (st1, st2, ..... );   -- declaration of enumerated state type
   signal PRESENTSTATE, NEXTSTATE : state;   -- declaration of state signals
   attribute state_vector : string;   -- this attribute is not part of IEEE 1076
   attribute state_vector of enumerated_state : architecture is "PRESENTSTATE";
   .....
begin
   .....
```

Please refer to section 4.9.3 for an unabridged example.

4.3.4 RAMs, ROMs, and other macrocells

On-chip RAMs are used to temporarily store all sorts of intermediate data whereas on-chip ROMs serve as repositories for program code, look-up tables (LUT), and other permanent information. This section will discuss how to incorporate RAM and ROM macrocells into VHDL models in a way that is acceptable for synthesis.

The most innocent approach is to declare a storage array as if the code were intended for simulation purposes and to assume the synthesizer will take care of all the rest with no human interaction. As an example, consider a 4 bit-binary-to-seven-segment display decoder. The content of an adequate LUT can be captured as an array of constants as shown below. While a workable solution, this piece of code will not synthesize into a ROM, but into random logic as would any other RTL model of combinational nature.

```
.....
-- address of array must be of type integer or natural
p_memless : process (BIN4xD)
   variable address : natural range 0 to 15;
   type array16by7 is array(0 to 15) of std_logic_vector(1 to 7)
   constant segment_lookup_table : array16by7 :=   -- segments ordered a...g
      ("1111110","0110000","1101101","1111001",   -- digits 0,1,2,3,
       "0110011","1011011","0011111","1110000",   --         4,5,6,7,
       "1111111","1110011","1110111","0011111",   --         8,9,A,b,
       "1001110","0111101","1001111","1000111");   --         C,d,E,F;
begin
   -- use binary input as index, look up in table, and assign to segment output
   address := to_integer(unsigned(BIN4xD));
   SEG7xD <= segment_lookup_table(address);
end process p_memless;
.....
```

Trying to do the same in the case of a 64 byte RAM, for instance, would mean having to include the code fragment below in the declaration section of the superordinate architecture body. Reading and writing one byte at a time would involve assignment statements and an address pointer that selects one out of the 64 storage vectors from the array of signals.

```
.....
   type array64by8 is array(0 to 63) of std_logic_vector(7 downto 0);
   signal STORAGExD : array64by8;
.....
```

The idea is impractical, however, because the behavior so defined is a far cry from actual RAM macrocells and their interface specifications. Worse than this, automated synthesis would hardly churn out a safe and synchronous gate-level circuit either.

From a more general perspective, whether to implement a storage array in a RAM, from flip-flops or otherwise is a decision that has far-reaching consequences for circuit architecture, performance, energy efficiency, design effort, and tool costs.

Observation 4.26. *Spontaneous incorporation of macrocells is neither a practical nor really a desirable proposition for RTL synthesis because it would deprive designers of control over a circuit's architecture.*

A more realistic approach would be to instantiate a RAM stating the macrocell generator to be used and to pass on all further specifications in a generic map. With `cmosram01` the name of some fictive generator for clocked SRAMs, for instance, this would require a code fragment similar to the one below.

```
u39: cmosram01
   generic map ( number_of_words => 64, word_width => 8,
                 data_input_output_separate => false )
   port map ( CLOCK => CLKxC, WRITE_ENABLE => RAMWRxS,
              ADDRESS => RAMADRxS, DATA_IO => RAMDATxD );
```

Regrettably, this approach is not currently feasible due to the lack of standardization and the absence of interfaces between VHDL synthesis and all those proprietary macrocell generators in existence. For the time being, a macrocell must get instantiated like any other component.

```
u39: myram64by8
   port map ( CLOCK => CLKxC, WRITE_ENABLE => RAMWRxS,
              ADDRESS => RAMADRxS, DATA_IO => RAMDATxD );
```

Observation 4.27. *The necessary design views of a macrocell (simulation model, schematic icon, detailed layout, etc.) must all be obtained from outside the VHDL environment.*

To that end, either the IC designer must gain access to the process-specific macrocell generator software and run it with an appropriate parameter setting, or he must commission the silicon foundry to do so for him. The choice is typically determined by commercial considerations.

Table 4.5 shows how a VHDL synthesis model must be organized in order to obtain various read-only and read–write storage functions.[46]

---

[46] Note that the market offers functional replacements for RAMs that are built on the basis of individual bistables. Such models are amenable to VHDL synthesis in the normal way, that is they ultimately map to standard cells much as does the designer's own code. Of course, such cell-based implementations cannot compete with true RAMs in terms of layout density. Also, many of them combine a mix of gates, latches, and flip-flops into a circuit that does not adhere to a pure and unconditionally safe synchronous clocking discipline.

Table 4.5 The desired circuit type determines how its synthesis model must be organized.

Look-up table (LUT) (memoryless)		
desired hardware organization	random logic	ROM (tiled layout)
function must be modelled	as an array-type constant or with logic equations	by instantiating a ROM macrocell as a component
Storage array (memorizing)		
desired hardware organization	register file built from flip-flops or latches	RAM (tiled layout)
function must be modelled	as an array of (clocked) storage registers	by instantiating a RAM macrocell as a component
Common characteristics of implemented circuit		
area-efficient when data quantity is	small	large
technology-specific software required	no	macrocell generator
VHDL code amenable to retargeting	yes	manual rework needed
pre-synthesis simulation works from	RTL source code	extra behavioral model
post-synthesis simulation works from	gate-level model	*idem*

4.3.5 Circuits that must be controlled at the netlist level

ARITHMETIC UNITS AND OTHER PARAMETRIZED STRUCTURAL SYNTHESIS MODELS

Designers cannot always afford to leave decisions on a circuit's organization to the discretion of automatic synthesis, they sometimes need to exactingly control the outcome at the gate level. As an example, assume you had to implement a high-performance multiplier for sign–magnitude numbers, a format not really supported by VHDL synthesis tools. While schematic entry offers full control over a circuit's structure, it is always tied to specific components and to particular circumstances in terms of word widths, pipeline depth, output format, and the like. The chance of ever reusing such a rigid circuit model is extremely low.

HDLs make it possible to write synthesis models that are structural <u>and</u> parametrized at the same time. These differ from captive structural models in that they make extensive use of generics and of conditional component instantiation statements, see architecture **structural** of `binary2gray` in section 4.2.5 for a simple example. The role of synthesis in the processing of VHDL source code of this kind is essentially limited to elaboration, technology mapping, and timing optimization with the overall organization of the original network being preserved. The procedure as a whole can be viewed as HDL-controlled netlist generation.[47]

CLOCKS, SYNCHRONIZERS, SCAN PATHS, AND THE LIKE

Every VLSI chip includes subcircuits that must conform to predefined structural patterns if the circuit as a whole is to function as intended. Clock distribution networks, clock gating circuitry,

---

[47] Predesigned, preverified, and optimized, yet configurable and technology-independent circuit models are being marketed by Synopsys under the product name **DesignWare**. They range from fairly simple arithmetic units and register files to an entire video decoder. VHDL models of adders, multipliers, dividers, and square root and trigonometric functions are available from [68] along with extraordinarily vivid explanations.

synchronizers, scan paths, and leakage suppression circuits are common examples. While their functionality is trivial, their structural, electrical and/or timing characteristics must conform to precise specifications. A loose collection of inverters is no valid substitute for a clock tree, for instance; nor does a simple AND operation qualify as a clock gate. Similarly, scan testing implies the presence of a shift register in the actual circuit hardware as illustrated in fig.6.6, not just another way of transiting from one state to the next.

Observation 4.28. *Boolean optimization algorithms and general-purpose logic synthesis tools are not designed to handle clock gates, synchronizers, clock distribution networks, scan paths, high-performance arithmetic units, and other portions of a design that must comply with structural rather than just with behavioral specifications.*

> Hint: Use dedicated tools or fall back on structural HDL code or on schematic entry wherever tight control over a subcircuit's gate-level netlist is sought. Take advantage of proven synthesis models for arithmetic units (DesignWare). Do not reoptimize subcircuits so obtained as critical properties may deteriorate. Most synthesis tools accept "don't touch" commands to prevent them from altering critical subcircuits while attempting to optimize the main body of a design.

Generating balanced clock trees, for instance, is typically postponed to the physical design phase and handled by specialized EDA software there.

4.3.6 Timing constraints

Synthesis constraints are not part of the VHDL standard

The timing-related constructs in the IEEE 1076 standard have been defined exclusively with simulation in mind. Inertial effects of physical circuits are typically modelled with after clauses, sometimes with the help of extra reject clauses. These language constructs are meaningless in the context of synthesis, however. After all, it is not possible to stipulate some arbitrary timing data and then to come up with a circuit that exactly meets those predefined numbers.[48] Synthesis tools thus simply ignore after and reject clauses. The same applies to wait for statements and to timing-related assertion statements.

Example of timing data being ignored by synthesis   OUP <= INA + INB after 1.7 ns;

Observation 4.29. *Timing-related VHDL constructs, such as* **after** *... and* **wait for** *..., are for simulation purposes exclusively and get ignored during synthesis. They serve to model the behavior of existing circuits, not to impose target requirements for the synthesis process.*

A more sensible goal is to define bounds that could guide synthesis and logic optimization by telling apart acceptable from unacceptable timing characteristics. Such timing constraints have

---

[48] This is because the exact timing of a circuit depends not only on the gate-level network, but also on load capacitances, wiring parasitics, PTV and OCV, cross-coupling effects, and more. Also, the timing of library cells is not continuously adjustable. In the case of a 2-input AND function, the target library would provide a few standard cells that differ in terms of transistor sizing, drive strength, and layout. Each such cell has its proper delay vs. load characteristics. Synthesis software just picks one or the other depending on the current requirements, but there is no sensible way to fashion an arbitrary delay at will.

never been adopted in the IEEE 1076 language standard, though. VHDL is, therefore, not capable of expressing any upper bound for a long path such as the one illustrated in fig.4.17.

Unsupported construct: `OUP <= INA + INB with_propdelay_no_more_than 1.7 ns;`

As a workaround, timing and other synthesis directives must be expressed with the aid of proprietary language extensions or with scripting languages such as Tcl. Portable formulations are important as the same timing constraints are to be reused during timing verification to check whether a design does indeed meet its specifications before being sent for fabrication.

HOW TO FORMULATE TIMING CONSTRAINTS

In single-edge-triggered one-phase circuits, an upper bound for the delay from one register to the next gets imposed by the clock period.[49] Indicating a target clock period is thus mandatory and straightforward, see fig.4.20a. Formulating constraints for input and output paths for commercial synthesis and timing verification tools is more tricky because of ambiguous naming habits. In fact, it is possible to define input/output timing from either of two perspectives:

o Indicate how much time is available to the circuit under construction.
o Quantify the amount of time that must be set aside for the surrounding circuitry.

As specifying the circuit under construction itself comes most naturally, the first perspective is normally adopted in this text. Yet, most EDA tools adopt the second view because it permits one to alter the target clock period without having to numerically readjust all I/O timing constraints. Unfortunately, the names being used, input and output delay, give rise to confusion. The material below, including table 4.6 and fig.4.20, attempts to reconcile the two views.

You will typically want to give upper bounds for the long path delays by specifying:

$$t_{idel\,max} = t_{pd\,upst} = t_{pd\,ff\,upst} + t_{pd\,a} - t_{di} \Leftarrow t_{pd\,b} + t_{su\,ff} \leq T_{clk} - t_{pd\,ff\,upst} - t_{pd\,a} + t_{di} \quad (4.1)$$

$$t_{odel\,max} = t_{su\,dnst} = t_{pd\,e} + t_{su\,ff\,dnst} + t_{di} \Leftarrow t_{pd\,ff} + t_{pd\,d} \leq T_{clk} - t_{pd\,e} - t_{su\,ff\,dnst} - t_{di} \quad (4.2)$$

If you make no distinction between $t_{idel\,max}$ and $t_{idel\,min}$, EDA tools will assume they are the same, which implies that input data get updated once per clock period at time ③ and then remain valid for one entire period. If this is not so, the short path delay must be constrained as well. Use a separate Tcl statement where you indicate a lower bound of

$$t_{idel\,min} = t_{cd\,upst} = t_{cd\,ff\,upst} + t_{cd\,a} - t_{di} \Leftarrow t_{cd\,b} - t_{ho\,ff} \geq -t_{cd\,ff\,upst} - t_{cd\,a} + t_{di} \quad (4.3)$$

while observing that any physical circuit must satisfy

$$t_{idel\,min} < t_{idel\,max} \Leftarrow t_{valid\,upst} = t_{pd\,upst} - t_{cd\,upst} = t_{idel\,max} - t_{idel\,min} > 0 \quad (4.4)$$

---

[49] An in-depth analysis of the operation and timing of synchronous circuits is to follow in sections 6.2 and 6.4. You may want to read through that material before coming back to this section when preparing timing constraints for synthesis and/or timing verification.

Table 4.6 | Cross reference for input and output timing constraints.

Event	Symbol	Quantity	Synopsys term
relating to the interface with the upstream circuitry			
③ data valid window begins	$t_{su\,inp}$ $\leq T_{clk} - t_{pd\,upst}$ $t_{pd\,upst} = t_{idel\,max}$	setup time of circuit under construction clock-to-output propagation delay of upstream circuitry	maximum input delay
② data valid window ends	$t_{ho\,inp}$ $\leq t_{cd\,upst}$ $t_{cd\,upst} = t_{idel\,min}$	hold time of circuit under construction clock-to-output contamination delay of upstream circuitry	minimum input delay
relating to the interface with the downstream circuitry			
④ data call window begins	$t_{pd\,oup}$ $\leq T_{clk} - t_{su\,dnst}$ $t_{su\,dnst} = t_{odel\,max}$	clock-to-output propagation delay of circuit under construction setup time of downstream circuitry	maximum output delay
① data call window ends	$t_{cd\,oup}$ $\geq t_{ho\,dnst}$ $t_{ho\,dnst} = -t_{odel\,min}$	clock-to-output contamination delay of circuit under construction hold time of downstream circuitry	minus minimum output delay

Similarly, if you do not distinguish between $t_{odel\,max}$ and $t_{odel\,min}$, the synthesizer will try its best to meet the setup condition of the downstream circuitry, but will do nothing particular about the hold condition there. In the extreme case, a circuit that just flashes valid output data at time ④ might pass as acceptable because $t_{odel\,max} = t_{odel\,min}$ is the same as $t_{su\,dnst} = -t_{ho\,dnst}$, which indeed stands for a downstream circuit that is capable of picking up data in zero time. To prevent this from happening, you can either bank on automatic hold-time fixing or you can explicitly constrain the short path from below by specifying

$$t_{odel\,min} = -t_{ho\,dnst} = t_{cd\,e} - t_{ho\,ff\,dnst} + t_{di} \Leftarrow t_{cd\,ff} + t_{cd\,d} \geq t_{ho\,ff\,dnst} - t_{cd\,e} - t_{di} \qquad (4.5)$$

while observing

$$t_{odel\,min} < t_{odel\,max} \Leftarrow t_{call\,dnst} = t_{su\,dnst} + t_{ho\,dnst} = t_{odel\,max} - t_{odel\,min} > 0 \qquad (4.6)$$

CIRCUIT PARTITIONING IN VIEW OF SYNTHESIS AND OPTIMIZATION

Timing constraints on propagation paths that extend across multiple circuit blocks render synthesis unnecessarily difficult and are likely to result in suboptimal circuits. This is because logic optimization and technology mapping are carried out in chunks to avoid excessive computer run times and memory requirements on large designs. Most tools accept proprietary directives for merging and segregating circuit logic into **synthesis chunks**. However, the better the initial architecture and

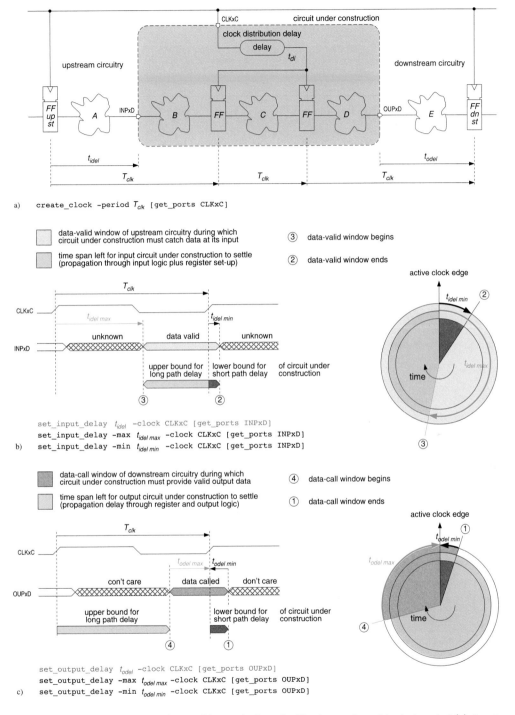

Fig. 4.20 Timing constraints as understood by synthesis tools. Circuit overview with clock period (a), input timing (b), and output timing (c).

the various design entities in the VHDL source code reflect a sensible hardware organization, the less effort will have to be wasted in repartitioning at synthesis time.

> Hint: Synthesis and optimization work much better if a design is organized such that
>
> - related or tightly connected subcircuits belong to the same design entity,
> - all outputs from a synthesis chunk are registered, and
> - critical paths are confined to within one synthesis chunk.

Registered outputs further preclude the unwanted emergence of zero-latency loops and hazards.

4.3.7 Limitations and caveats for synthesis

In conclusion, it is important to understand that only a subset of the VHDL language defined in the IEEE 1076 standard is supported for synthesis and that subset is not exactly the same for all commercial products. Also, tool builders have added proprietary directives, constraints, and even extra data types to fill gaps that were left open when the VHDL standard was established. All these factors are at the detriment expense of compatibility. In addition to this, there is a number of language and circuit constructs that require special attention because unsafe or inefficient designs may otherwise result from synthesis. Tables 4.7 and 4.8 attempt to collect the most severe limitations and caveats.

4.3.8 How to establish a register transfer-level model step by step

While VHDL is perfectly suitable for coding a data processing algorithm, do not expect an EDA tool to accept a purely behavioral model and to turn that into a circuit design of acceptable performance, size, and energy efficiency. Exceptions are limited to circuits of fairly modest or fairly specific functionality. Rather, the fun and the burden of architecture design rests with the hardware developer. Only after an architecture has been worked out by human engineers does it make sense to describe the hardware organization at an intermediate level of detail, typically RTL, and to submit the HDL code so obtained to a synthesis tool.

RTL modelling is best carried out in a procedure of successive refinement:

1. Begin by drawing a fairly detailed block diagram of the architecture to be implemented.

2. Check where you can take advantage of off-the-shelf synthesis models (DesignWare).

3. Organize the circuit in such a way as confine critical propagation paths to within circuit blocks. Make your design entities match with those circuit blocks.

4. Identify macrocells such as RAMs and ROMs and prepare for generating the necessary design views outside the HDL environment.

5. Identify <u>all</u> registers (data, I/O, pipeline, address, control, status, mode, test etc.) and loosely collect the combinational operations in between into clouds.

6. For each combinational cloud, specify the operations in mathematical terms (equations, truth tables, structograms, pseudo code, etc.) and figure out how to compute the desired outputs in an efficient and — where meaningful — also parametrizable way.

Table 4.7 | VHDL constructs that are unacceptable for synthesis.

item	reason	remedy
constructs that will never synthesize		
text and file I/O	no hardware equivalent	do not use
access data type	no hardware equivalent	do not use
initialized variables	hardware needs explicit reset mechanism	do not use
after, reject, wait for	no way to synthesize precise timings	do not use
assert	no hardware equivalent	expect no effect
weakly driven nodes **L, H**	resistors incompatible with CMOS	do not use
capacitively charged nodes	not modelled in IEEE 1164	do not use
constructs that should not be left to general VHDL synthesis		
scan path	needs specific network structure	add after synthesis
clock distribution network	needs specific network and exact timing	add after synthesis
synchronizer	asks for specific network structure	structural code
clock gating	perilous unless network and timing	structural code
	are tightly controlled	or qualified tool
sleep mode	asks for specific network structure	add after synthesis
padframe	purely structural	structural code
constructs that are not (yet) synthesized (properly) today		
process with multiple waits (implicit state) reactivated by		
- same event throughout	not supported by all synthesizers	use explicit state
- disparate events	models asynchronous circuit behavior	avoid anyway
macrocell (RAM, ROM)	insufficient tool integration	instantiate
transmission gate	not a logic function	instantiate
snapper	not a logic function	instantiate

7. Establish a **schedule** that specifies what is to happen during each clock cycle. This is a table with one line per computation period and an entry for each relevant building block that expresses the following items:
   - ALU or arithmetic unit: operation being carried out, data set being processed.
   - Other major combinational block: data set being processed.
   - Finite state machine: present state, present output.
   - Register: present datum, being cleared or not, being enabled or not.
   - Important signal: present datum.
   - Output pin or connector: present datum, being driven or not.
   - Input pin or connector: datum that must be available.
   - Bidirectional pin or on-chip bus: present datum, being driven or not.

8. Identify all finite state machines and find out what type is most appropriate.[50]

9. Capture each register in a memorizing process statement.

---

[50] Table B.5 may help in doing so.

Table 4.8 | VHDL constructs that require particular attention if intended for synthesis.

item	potential problem if neglected	remedy
in general		
δ delay in signal assignments	unwanted behavior	use variables
integer-type signals and variables	inefficient circuit	indicate ranges
circuit partitioning	inefficient circuit and/or excessive run times	careful coding (or repartitioning at synthesis time)
in combinational functions		
exhaustive specification	unwanted sequential behavior	careful coding
exclusive specification	unexpected behavior	careful coding
operator precedence	unexpected behavior	careful coding
hardware costs for the processing of arbitrary numbers (e.g. $x : 2^n$ vs. $x : y$, same for $\cdot$ and mod)	inefficient circuit or rejection	adequate algorithm design
hardware costs of operators (e.g. $=$ vs. $\leq$, $\sqrt{x}$ vs. x^2)	inefficient circuit	adequate algorithm and architecture design
reuse of hardware for related operators (e.g. $+$ and $-$)	inefficient circuit	apply algebraic transforms beforehand
alternative hardware structures for arithmetic operations (e.g. RCA, CLA, CSLA, or PPA-BK)[a]	inefficient circuit	use specialized netlist generator (e.g. DesignWare)
in sequential functions		
code organization	inefficient circuit	comb. and seq. logic coded in separate processes
location of output assignments in FSMs	unexpected type of automaton	careful coding
parasitic inputs and states	lock-up or unexpected behavior	tie down
state reduction and state encoding	inefficient circuit	use directives and explore alternatives

[a] The abbreviations stand for ripple–carry adder, carry–lookahead adder, carry–select adder, and parallel-prefix algorithm Brent-Kung adder respectively [67].

10. For each combinational cloud, decide on the number of processes you want to use. Prefer concurrent, selected, and conditional signal assignments for simpler operations; plan to use process statements to capture lengthy computation sequences. Give a meaningful name to each process.

11. Note that all data items that run back and forth between the various processes must be declared as signals and decide on the most appropriate data type for each.

12. Only now begin with translating your draft into actual HDL code.
    • Organize finite state machines as suggested by fig.4.19 and pattern the code of registers after program 4.4. Be careful to handle special signals such as clock, asynchronous reset, synchronous initialization, and enable properly.

- Use the schedule previously established to obtain the various subfunctions in full detail.
- Specify don't care entries wherever possible.
- For each memoryless process statement, do not forget to include all of its inputs on the sensitivity list. If you make use of auxiliary variables, make sure they are assigned a value before being used. Ensure that you assign a value to each output for all mathematically possible combinations of input values; default assignments may help.

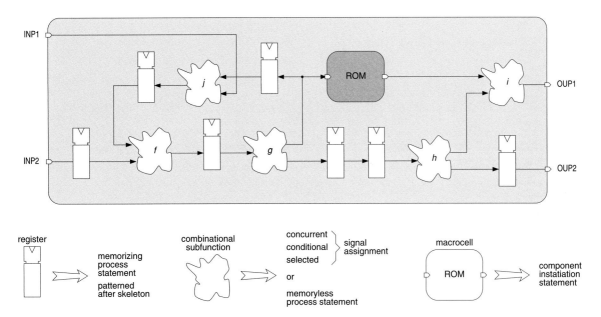

Fig. 4.21 Translating an RTL diagram into VHDL code.

Observation 4.30. *Writing code for VHDL synthesis is not the same as writing software for a program-controlled computer. Always think in terms of circuit blocks (i.e. entities) and concurrent activities (i.e. processes) rather than in terms of instruction sequences.*

Golden rule:
Establish a block diagram of your architecture first,
then code what you see!

Hint: Most EDA tool suites include a schematic editor that can translate back and forth between a schematic diagram and structural VHDL code. Take advantage of schematic entry to assist you in establishing the VHDL model of your architecture.

4.4 | Putting VHDL to service for hardware simulation

4.4.1 Ingredients of digital simulation

This section gives an example for a complete VHDL simulation setup that is based on the general principles of functional verification and testbench design established earlier in this text.[51] As discussed there, digital circuit simulation essentially involves

- A model under test (MUT),
- Stimuli, i.e. signal waveforms coded as data vectors, for exercising the MUT,
- Expected responses that are presumed to be correct, and
- A testbench, that is a facility that applies stimuli, that acquires responses, and that compares the actual against the expected responses.

The event queue mechanism of VHDL enables designers to specify stimuli of almost arbitrary waveforms and to compress those waveforms into compact data vectors. The assertion statement further makes it possible to compare responses and to generate error messages automatically. VHDL can thus capture both circuit models and their testbenches, which dispenses with product-specific simulation control languages. In addition, verification engineers are not restricted any particular subset of the HDL as testbenches are not intended for synthesis.

Observation 4.31. *As opposed to IC designers, who are concerned with circuit details down to the RTL level and below, verification engineers must think in terms of behavior and functionality, which requires a different mindset.*

4.4.2 Anatomy of a generic testbench

Our focus here is on the overall setup for a grade 1 simulation.[51] VHDL source code organized along the lines shown in fig.3.14 can be found in appendix 4.9.4. Obtaining a good understanding requires working through that code. The comments below are intended as a kind of travel guide that points to major attractions there.

As a circuit example, we have chosen a simple shift-and-add multiplier for three reasons. Firstly, its functionality is obvious enough not to distract the reader's attention from testbench design. Secondly, a number of clock cycles are required to process one set or arguments because calculation occurs sequentially. Thirdly, with multiplication being a combinational function, this example offers room for upgrading to a grade 2 testbench later.

All disk files handled by separate processes and stored in ASCII format

A simulation run involves files of different kinds to accommodate

- The stimuli (as they seldom get obtained entirely from a random generator),
- The expected responses (unless they get computed online by a golden model), and
- A simulation report (mandatory).

---

[51] How to uncover design flaws with the aid of simulation is the subject of chapter 3 What is meant by a grade 1, 2 or 3 simulation is also explained there.

For each file, there is one VHDL process that is in charge of opening, reading or writing, and closing it. They get notified of the pending end of a simulation run via an auxiliary two-valued signal named `EndOfSimxS` so that they can carry out the necessary activities.

ASCII text files are definitely preferred over binary data because the former are human-readable and platform-independent, whereas the latter are not. What's more, the usage of ASCII characters makes it possible to take advantage of the full IEEE 1164 logic system for specifying stimuli and expected responses. Designers are thereby put in a position to check for a high-impedance condition by entering a `Z` as expected response, for instance, or to neutralize a response by entering a – (don't care) as reference value.

No provisions have been made to log stimulus/response pairs in view of the testing of physical parts with ATE. As most simulators can be set up to carry out this function at user-defined points in time (□), there is no need to burden the testbench code with that.

Separate processes for stimulus application and for response acquisition

The scheduling of the simulation events listed below has been identified as crucial.

- The application of a new stimulus denoted as △ (for Application),
- The acquisition and evaluation of the response denoted as ⊤ (for Test),
- The two clock edges, symbolically denoted as ↑ and ↓.

It would be naive to include the time of occurrence of such key events hardcoded into a multitude of `wait for` statements or `after` clauses dispersed throughout the testbench code. A much better solution is to assign stimulus application and response acquisition to separate processes that get periodically activated at times △ and ⊤ respectively. All relevant timing parameters are expressed as constants or as generics, thereby making it possible to adjust them from a single place in the testbench code.

Testbench supports both file-based and golden-model-based simulation

The expected responses are either

○ computed on the fly from the stimuli by a golden model of behavioral nature or
○ read from a previously prepared file with the aid of a pickup process.

Basically, response acquisition and evaluation are the same no matter whether the expected responses are read from a file or calculated by a golden model. By designing MUT, golden model, and reference pickup to share exactly the same interface and to be fully interchangeable, the response acquisition process can be kept unchanged. Technically, switching from file-based to golden-model-based simulation or back is just a matter of exchanging two architecture bodies in a configuration statement.

An implication is that the reference pickup must turn the expected response read from disk into a VHDL signal of the same type as the actual responses from the MUT. As an extra benefit, stimuli, expected responses, and actual responses must be declared as top-level signals within the testbench entity. This in turn renders those key signals directly accessible and tends to facilitate monitoring the course of action during a simulation run.

Stimuli and responses collected in records

As stated earlier, keeping the testbench code modular and reusable is highly desirable. Two measures contribute towards rendering the VHDL processes that apply the stimuli and that acquire the responses independent of the MUT.

a) All input signals are collected into one stimulus record and, analogously,
   all output signals into one response record.[52]

b) The subsequent operations are delegated to specialized subprograms:

   - All file read and write operations,
   - Unstuffing of stimuli (if read from file),
   - Preparation of stimuli (if generated at run time),
   - Unstuffing of expected responses (if read from file),
   - Stuffing of actual responses (if written to file),
   - Response checking (actual against expected), and
   - The compilation of a simulation report.

The main processes that constitute the testbench are thus put in a position to handle stimulus and response records as wholesale quantities without having to know about their detailed composition. The writing of custom code is hence confined to a handful of subprograms.[53]

Simulation to proceed even after expected responses have been exhausted

Occasionally, a designer may want to run a simulation before having prepared a complete set of expected responses. There will be more stimuli vectors than expected responses in this situation. To support this policy, the testbench has been designed so as to continue until the end of the stimuli file, albeit with no automatic checking of responses. Simulation in the absence of any reference file is not accepted, though.

Stoppable clock generator

A simulation run comes to its end when the processing of the last stimulus record has been completed and the pertaining response record has been acquired. A mundane difficulty is to halt the VHDL simulator. Basically, there exist three alternatives for doing so, namely

---

[52] Bidirectional input/output signals must appear both as stimuli and as expected responses. Also note that an asynchronous reset signal is treated as an ordinary input signal and thus included in the stimuli record.

[53] As an example, the subprograms that check the responses against each other and that handle the reporting need to be reworked when a MUT connector gets added, dropped or renamed. Ideally, one would prefer to do away with this dependency by having those subprograms find out themselves how to handle the response records. This is, unfortunately, not possible as VHDL lacks a mechanism for inquiring about a record's composition at run time in analogy to the array attributes `'left`, `'right`, `'range`, and `'length`.

a) Have the simulator stop after a predetermined amount of time,
b) Cause a failure in an assert or report statement to abort the run, or
c) Starve the event queue, in which case simulation comes to a natural end (see section 4.2.4).

Alternative a) is as restrictive as b) is ugly, so c) is the choice to retain. A clock generator that can be shut down is implemented as concurrent procedure call, essentially a shorthand notation for a procedure call embedded in a VHDL process. The reason for using this construct rather than a regular process statement is that the clock generator is reusable, and that we want to make it available in a package. This is not otherwise possible, as the VHDL syntax does not allow a package to include process statements.

Note: Make sure you understand that a clock signal is needed to drive simulation even if the circuit being modelled is of purely combinational nature.

Reset considered an ordinary stimulus bit

Timingwise, the reset signal, whether synchronous or asynchronous, is handled like any other stimulus bit and gets updated at time $\triangle$.

Fake delays help visualize cause/effect relationship

Interpreting simulation waveforms is often confusing when too many signals get assigned new values with zero delay within a MUT or a golden model. Fake delays in otherwise delayless signal assignments greatly help to make the cause $\rightarrow$ effect relationships evident.

4.4.3 Adapting to a design problem at hand

In the context of a specific design project, one has to write new or to adapt existing VHDL source code for

- Declaring the data types involved in the various interfaces,
- Reading and writing those data types from and to files,
- Translating between those data types where necessary,
- Generating stimuli vectors (if working with random data), and
- Comparing the actual and expected responses against each other.

The rest of the testbench code and the general simulation setup given in fig.3.14 are reusable.

4.4.4 The VITAL modelling standard IEEE 1076.4

VITAL is an acronym for "VHDL initiative towards ASIC libraries", a move borne jointly by electronic design automation (EDA) vendors and the ASIC industry that aimed at making VHDL simulation a viable option for sign-off. Sign-off quality simulation asks for accuracy, capacity, and speed.[54] A working group established in 1992 came up with two VHDL packages that were accepted as the IEEE 1076.4 standard in 1995 and revised in 2000.

Back-annotation

Back-annotation refers to the process of complementing or of adjusting HDL circuit models with extraneous timing data illustrated in fig.4.15. VITAL provides the necessary hooks for doing so in a package named `ieee.Vital_Timing` that defines standard names and types for ports and also for the generics that convey the pertaining timing parameters. Adhering to those naming conventions enables a VITAL-compliant simulator to read numerical timing data from a special file established in standard delay format (SDF) and to substitute them for the default values given in the entity declaration of the VHDL model. Note that VITAL provides no particular construct for wires. Instead, interconnect delays are lumped to the inputs of those gates that are being driven from a wire.

Delay modelling, that is the problem of describing how the various timing parameters of gates and interconnect depend on fanout, input ramps, layout parasitics, PTV conditions, and the like, is outside the scope of VITAL. SDF files must be obtained from some extra delay calculator prior to being transferred to the VHDL environment.

Acceleration

Making VHDL simulation comparable to logic simulation in terms of capacity and speed was another key goal. It is addressed by imposing the usage of specific VITAL subprograms for the modelling of gate-level subcircuits. Subprograms included in package `ieee.Vital_Timing` have been prepared to help with the modelling of delays and with the monitoring of timing conditions whereas others included in `ieee.Vital_Primitives` have been defined for the functional modelling of both combinational and sequential subcircuits. What they have in common is that the number of expensive processes and signals is kept to a miminum. Also, the fact that such functions and procedures are standardized permits further optimization such as assembly coding and inclusion in the simulator's kernel. VITAL-accelerated simulation has been reported to speed up program execution by a factor of 25 over gate-level models written in plain VHDL.

A VHDL model is said to comply with **VITAL level 1** if its entity declaration follows the VITAL naming conventions and if its architecture body is built from VITAL functions and procedures exclusively. If so, the model supports both back-annotation <u>and</u> acceleration. To be of any practical value, VHDL models for gate-level library elements must be written according to VITAL level 1 and most cell libraries available today actually comprise compliant simulation models. About the only persons concerned with writing such code are library developers.

A simulation model where VITAL compliance is limited to the entity declaration while the architecture body follows some arbitrary VHDL coding style is said to comply with **VITAL level 0**. As a consequence, such a piece of code can support back-annotation but no acceleration. VITAL

---

[54] Accuracy is concerned with anticipating the timing characteristics of the fabricated parts from parameters that are available during simulation. Capacity quantifies how large a network a simulator can reasonably handle with a given amount of computer memory. Speed refers to the ratio between simulation time and execution time.

level 0 models are appropriate for the behavioral models of larger building blocks that most electronic engineers happen to write as part of the VLSI design cycle.

Memory modelling

A new package `ieee.Vital_Memory` with types, constants, and subprograms for use in ROM and SRAM memory models has been added in the 1076.4-2000 standard revision. A table-based modelling style extends the benefits of VITAL level 1 to memories by encouraging more uniform coding practices, by supporting back-annotation via the standard delay format (SDF), and by contributing to better simulator performance.

4.5 | Conclusions

- The practical benefits of using VHDL in digital hardware design are as follows.

  - VHDL simulation supports a top-down design methodology of successive refinements from behavioral level down to gate-level models using a single standard language.
  - RTL synthesis makes it possible to do away with what used to be the two to four bottom levels of schematic drawings in a typical VLSI design hierarchy.
  - Automatic technology mapping makes it unnecessary to commit a VHDL-based design to some specific cell library or fabrication process until late in the design cycle, even allowing for retargetting between field- and mask-programmable ASICs.
  - VHDL enables sharing, reusing, and porting of subfunctions and subcircuits in a parametrized and therefore more useful form than traditional schematics.
  - VHDL simulation also dispenses with proprietary simulation control languages.

- While the IEEE 1076 and 1164 standards are fully supported for simulation, only a subset of VHDL is amenable to synthesis because the language was not originally developed with that application in mind.

- The most important engineering decisions that set efficient designs apart from inefficient ones do not relate to VHDL, but to architectural issues. Algorithmic and architectural questions must be answered <u>before</u> the first line of synthesis code is written.

- The impact of coding style on random logic tends to be overstated. Also, do not expect timing-wise synthesis constraints to do away with architectural bottlenecks. All too often, their effects are limited to buying moderate performance gains at the expense of using substantially larger circuits.

- For the time being, there remains a gap between system design, which focusses on over-all circuit behavior, and actual hardware design, which involves many structural and implementation-specific issues. The necessary manual translation from a purely behavioral model to an RTL description amenable to synthesis and the ensuing re-entry of design data tend to lead to errors and misinterpretations. Progress towards high-level synthesis is expected to gradually narrow the gap.

- HDL synthesis does not do away with architecture design!

4.6 | Problems

1. Section 4.2.2 includes examples of conditional and non-conditional signal assignments. For each such example, state the conditions that cause the code to get re-evaluated during simulation.

2. Consider process statement `memless1` from section 4.2.2 and note that signal `SPRING` is unconditionally set `false` before being assigned its actual value in a series of conditional statements. That value will thus evolve from `true` to `false` and back again when the process is invoked during springtime, e.g. at midnight of May 18. Do you think this trait will become visible as a brief transient during VHDL simulation? Would a circuit synthesized from this model exhibit a hazard? Explain your reasoning.

3. Listing 4.1 includes a procedural model, a dataflow model, and a structural model for a small subcircuit.
   (a) For each of the three architecture bodies, find out in what way it is possible to reorder the VHDL statements without affecting the model's functionality.
   (b) Although the three bodies describe exactly the same functionality at exactly the same level of abstraction, they greatly differ in the total count of signals, variables, design entities, instances, processes, and statements involved. Determine the respective numbers.
   (c) Establish three schedules that list what is happening simulation cycle after simulation cycle in response to an event on any of the inputs. Think about the impact on computational efficiency when the entity gets simulated.

4. Consider the VHDL source code given in appendix 4.9.1. Notice that the model assumes fixed input and output widths of 4 and 3 bits respectively. Examine how the various architecture bodies would scale if the model were to be parametrized in order to handle an arbitrary word width at its input. Add the necessary generic interface constant(s) in the entity declaration and rewrite a few architectures so as to obtain a scalable model. You may further want to synthesize and to compare the resulting networks.

5. Explain the differences between the conditional constructs `if ... then ... end if;` and `if ... generate ... end generate;`. Are there any other VHDL statements that are related to each other in the same way?

6. Write a VHDL model for a binary-coded-decimal (BCD) counter that is amenable to both simulation and synthesis. A control input of two bits is to decide between count-up (01), count-down (10), and hold (00) condition. Input-to-output latency shall not exceed one clock cycle. Under no circumstance are hazards tolerated on any of the output signals. Also, do not forget to address parasitic inputs and parasitic states.

7. Consider the code below, where the output of a linear feedback state register (LFSR) and of a counter are combined into a four-bit random output. What's wrong with this design? Hint: Actually, there is one obvious and one more subtle problem. They are related to the same clause in the process statement.

```
entity partres is
   port (
       CLKxC : in std_logic;
       RSTxRB : in std_logic;
       OUPxD : out std_logic_vector(3 downto 0) );
end partres;
--------------------------------------------------------------------
architecture behavioral of partres is
   signal STATEAxDP, STATEAxDN : std_logic_vector(1 to 4);
   signal STATEBxDP, STATEBxDN : unsigned(3 downto 0);
begin

   -- computation of next state
   STATEAxDN <= (STATEAxDP(3) xor STATEAxDP(4)) & STATEAxDP(1 to 3);
   STATEBxDN <= STATEBxDP + "1001";

   -- updating of state
   process (CLKxC,RSTxRB)
   begin
      -- activities triggered by asynchronous reset
      if RSTxRB='0' then
         STATEAxDP <= "0001";
      -- activities triggered by rising edge of clock
      elsif CLKxC'event and CLKxC='1' then
         STATEAxDP <= STATEAxDN;
         STATEBxDP <= STATEBxDN;
      end if;
   end process;

   -- updating of output
   combine: for i in 3 downto 0 generate
      OUPxD(i) <= STATEAxDP(4-i) xor STATEBxDP(i);
   end generate combine;

end behavioral;
```

8. Consider a shaft equipped with an angle encoder that indicates the shaft's current position using a two-bit unit distance code, see fig.4.22. Design a state machine that accepts this code and tells you whether the shaft is currently rotating clockwise or counterclockwise. The former sense of rotation shall be indicated when the shaft is at a standstill. Establish a VHDL synthesis model. You may want to extend the functionality in such a way as to indicate the angular position of the shaft and the number of turns it has made since time zero.

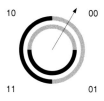

Fig. 4.22 Unit distance encoding of shaft angle.

4.7 | Appendix I: Books and Web Pages on VHDL

Reference	Year	IEEE 1076	IEEE 1164	model-ling	test-bench	syn-thesis	Comments and special topics
Specifications of language and syntax							
IEEE [103]	'02	-02	no	no	no	no	language reference manual
IEEE [104]	'94	-93	no	no	no	no	language reference manual
Zimmermann [105]	'02	.1	no	no	no	no	AMS syntax on www
Zimmermann [106]	'97	-93	no	no	no	no	syntax in EBNF on www
Bhasker [107]	'95	-93	no	no	no	no	syntax diagrams, +
Bergé et al. [108]	'93	-93	no	(yes)	no	(yes)	standard revision report
Textbooks							
Volnei Pedroni [109]	'04	-93	yes	yes	no	yes	many examples, FPL
Molitor & Ritter [110]	'04	-93	yes	yes	yes	(yes)	examples, in German
Ashenden et al. [111]	'03	-93	yes	yes	yes	no	VHDL-AMS
Ashenden [112]	'02	-02	yes	yes	yes	yes	pointers, std versions, +
Yalamanchili [113]	'01	-93	yes	yes	(yes)	yes	FPGAs, memory model
Armstrong & Gray [114]	'00	-87	yes	yes	yes	yes	high-level synthesis
Heinkel [115]	'00	-93	yes	yes	(yes)	yes	VHDL-AMS
Zwolinski [116]	'00	-93	yes	yes	(yes)	yes	asychronous circuits
Chang [117]	'99	-93	no	yes	no	yes	based on [118], examples
Cohen [119]	'99	-93	yes	yes	yes	yes	coding style, testbenches
Ashenden [120]	'98	-93	yes	yes	(yes)	no	subset of full edition
Navabi [121]	'98	-93	yes	yes	yes	(yes)	small CPU example
Sjoholm & Lindh [122]	'97	-93	yes	yes	yes	yes	behavioral synthesis
Chang [118]	'97	-93	yes	yes	yes	yes	testbenches, project, +
Cohen [123]	'97	-93	yes	yes	yes	yes	coding style
Bhasker [124]	'96	-93	yes	yes	(yes)	yes	code to circuit mappings
Hsu et al. [125]	'95	-87	(yes)	no	(yes)	yes	numeric packages, +
Ott & Wilderotter [126]	'94	-87	(yes)	(yes)	yes	yes	management issues
Airiau et al. [127]	'94	-93	yes	no	no	yes	numeric packages, +
Bhasker [128]	'94	-93	no	yes	yes	no	-87 to -93 portability, +
References with a specific focus							
Bergeron [88]	'00	-93	no	yes	yes	no	functional verification +
Bhatnagar [129]	'99	-93	no	no	no	yes	synthesis with Synopsys
Other resources							
Hamburg Archive [130]	'06			n.a.			free models, FAQ, links

( ) = light coverage only, + = personal preference.

4.8 | Appendix II: Related extensions and standards

4.8.1 Protected shared variables IEEE 1076a

The normal way of exchanging time-varying data between processes in VHDL is via signals. Signals essentially stand for electrical wires running between subcircuits and the non-zero delays of those subcircuits are expressed as part of signal assignment statements.

As opposed to this, shared variables are intended to support inter-process communication for bookkeeping and supervision tasks during simulation runs, e.g. counting the number of process invocations, keeping track of exceptions or other special events, coordinating activities among the different processes in a testbench, or collecting statistical data. Zero-delay communication is fine in this context. Were it not for shared variables, programmers would be forced to employ signals, thereby obscuring their original intention and unnecessarily inflating execution time.

To stay clear of problems that might result from simultaneous read or write operations to a global variable by distinct processes, the access must be controlled. Protected shared variables provide a means for synchronization. Access to a protected shared variable is exclusive and must always be made to occur by calling one of the functions or procedures written for that purpose. That is, when a first process gains access to a protected variable by calling one such subprogram to work on it, any further process attempting to read or modify the same variable by invoking the same or another subprogram must wait until the current access has terminated.

Declaring a protected shared variable of an existing data type could hardly be simpler.

Example
```
shared variable event_counter : shared_counter;
```

Obviously, the variable's type is to be declared beforehand, along with all functions and/or procedures necessary for access. Consistently with VHDL's guiding principles, the declaration of a protected type is separated from its implementation as shown in the subsequent example [131].

```
-- protected type declaration
type shared_counter is protected
   procedure : reset;
   procedure : increment ( by : integer := 1 );
   impure function value return integer;
end protected;
```

```
-- protected type body
type shared_counter is protected body

   variable count : integer := 0;

   procedure reset is
   begin
      count := 0;
   end procedure reset;

   procedure increment ( by : integer := 1 ) is
```

```
    begin
       count := count+by;
    end procedure increment;

    impure function value return integer is
    begin
       return count;
    end function value;

end protected body shared_counter;
```

Another code fragment shows that subprograms for reading or modifying a protected shared variable get invoked by prefixing their names with that of the variable meant to be accessed.

```
...
event_counter.reset;
event_counter.increment (3);
assert event_counter.value > 0;
...
```

Protected variables are part of the 2002 revision of the IEEE 1076 standard. Shared but unprotected variables had been introduced in VHDL'93 as a result of controversial debates in the standard committee; using them is discouraged as no access control mechanism was provided.

4.8.2 The analog and mixed-signal extension IEEE 1076.1

This standard, which was accepted in 1999, is informally known as **VHDL-AMS** and aims at augmenting the capabilities of the original language towards describing and simulating lumped analog and mixed-signal circuits. It has been a guiding principle to augment the existing VHDL constructs and to add new ones so as to make the new IEEE 1076.1 language a proper superset of VHDL, which has obvious benefits.

To capture continuous quantities such as voltages, currents, and charges, a new kind of object has been introduced that complements the constants, variables, and signals defined in the IEEE 1076 standard. This class is termed **quantity** and any object that belongs to it takes on floating-point values exclusively.

The supplemented language supports the modelling of time-continuous behavior by accommodating (possibly nonlinear) differential and algebraic equations in the time domain such as $F(\dot{x}(t), x(t), t) = 0$. So-called implicit quantities have been included to denote derivatives and integrals over time. If x has been declared as a quantity, for instance, then x'dot automatically refers to $\frac{d}{dt}x$.

Elaboration of a VHDL-AMS model yields a digital part (made up of signals and processes) and an analog part (consisting of quantities and differential algebraic equations). Simulation begins with determining the model's initial condition at time zero. The standard further defines how to synchronize the traditional event-driven simulation cycle with a solver for a system of simultaneous differential and algebraic equations.

Observation 4.32. *In a nutshell, VHDL's analog and mixed-signal extensions are as follows:*

			standard
VHDL-AMS	=	*VHDL*	*IEEE 1076*
	+	*continuous-value objects*	*IEEE 1076.1*
	+	*simultaneous differential algebraic equations*	*idem*
	+	*coupled continuous and discrete models of time*	*idem*
	+	*standard transistor models*	*(planned)*

It is worthwhile to note that VHDL-AMS extends the modelling capabilities in many ways. The significance of nonlinear and/or differential and algebraic equations in stating the static and continuous-time characteristics of electrical components and circuits is immediately evident. Entire subsystems from data transmission, signal processing, control systems, etc. can be condensed into abstract high-level mathematical models.

What further sets VHDL-AMS apart from SPICE is the absence of built-in transistor models in the simulator kernel. Model writers are no longer confined to a structural view that describes how more complex (sub)circuits are pieced together from a few built-in primitives (such as resistors, capacitors, and transistors). Rather, they are put in a position to describe opamps, active filters, phase locked loops (PLLs), etc. from a purely behavioral perspective using mathematical equations as building blocks and combining them with event-driven submodels where appropriate. This also enables them to create their own primitive models and to include them in circuit simulation at any time with no need for assistance from the software vendor.

Lastly, there is nothing that would limit quantities to being of electrical nature, which opens the door for modelling thermal, micromechanical, optoelectronic, magnetic, and other effects. Yet, we will not elaborate on VHDL-AMS as analog, mixed-signal, and multi-domain models are beyond the scope of this text. Please refer to [132] [111] [133].

4.8.3 Mathematical packages for real and complex numbers IEEE 1076.2

While VHDL provides the floating-point data type **real**, it does not support operations other than basic arithmetic operators.[55] To overcome this limitation, two new packages were defined and accepted as IEEE standard 1076.2 in 1996.

Package **math_real** includes

- Definitions of constants including e and π.
- Sign, floor, ceiling, round, truncate, min, and max functions.
- Square root, cubic root, power (x^y), exponential (e^x), and logarithm ($\ln(x)$) functions.
- Trigonometric functions (sin, cos, tan, arcsin, etc.).
- Hyperbolic functions (sinh, cosh, tanh, arcsinh, etc.).
- A pseudo-random number generator for reals uniformly distributed in the interval [0, 1].

Package **math_complex** includes

- Definitions of complex number types (in Cartesian and polar form).

---

[55] Addition +, subtraction −, sign inversion −, multiplication *, division /, integer power (x^n, $n \in \mathbb{N}$) **, and absolute value **abs**.

- Absolute value, angle, (argument), negate, and conjugate functions.
- Square and exponential (e^z) functions.
- Overloaded versions of basic arithmetic operators (for Cartesian and polar operands).
- Type conversion functions.

Both packages have been added to the existing design library **ieee**. Clearly, they are intended for modelling system behavior at higher levels of abstraction and for auxiliary functions in testbenches, not for synthesis.

4.8.4 The arithmetic packages IEEE 1076.3

As no single IEEE 1076 or IEEE 1164 data type supports computer arithmetics with adequate precision and convenience, more data types were needed. The electronic design automation community quickly filled the gap with add-on packages, but their proprietary nature jeopardized the portability of VHDL models that made use of such unofficial extensions. As a consequence, two new packages called **numeric_bit** and **numeric_std** were developed and accepted as IEEE standard 1076.3 early in 1997. They include

- The definition of data types **unsigned** and **signed** (discussed in section 4.2.3) for unsigned and 2's-complement arithmetic respectively,
- Overloaded versions of IEEE 1076 arithmetic, logical, and relational operators.
- Arithmetic shift and rotate functions.
- Resizing functions with sign extension and reduction.
- Type conversion functions (see appendix 4.8.7).

In addition, the IEEE 1076.3 standard indicates how to interpret for the purpose of VHDL synthesis logical values, such as "L", "X", and "U", that have a physical meaning as outcomes from simulation but not as specifications for a circuit to be. This is also why the two packages are sometimes referred to as **synthesis packages**.

4.8.5 A language subset earmarked for synthesis IEEE 1076.6

Accepted as a standard in 2004 and superseding a 1999 edition, this standard defines a subset of the IEEE 1076, 1076.3, and 1164 standards that is suitable for RTL synthesis. Constructs not amenable to synthesis are identified. In addition, the semantics are unambiguously defined such as to ensure uniformity accross synthesis tools. That is, given any RTL model strictly limited to VHDL constructs from the subset defined, the functional characteristics of the circuits obtained from synthesis must be the same for all VHDL synthesizers that comply with IEEE 1076.6.

4.8.6 The standard delay format (SDF) IEEE 1497

The SDF was originally developed by Open Verilog International (OVI) and later modified to become SDF version 4.0, which was accepted as IEEE 1497 standard in 2001. SDF files are written in ASCII-readable form and store timing data in a non-proprietary format for later use during the VLSI design and verification process.

SDF makes it possible to share gate delays, timing conditions, and interconnect delays between cell libraries, delay calculators, HDL simulators, and static timing analysis software. More particularly,

SDF supports back-annotating existing netlists with numerical timing data obtained from layout extraction as illustrated in figs.4.15 and 12.2.

The language also includes constructs for forward-annotation, that is for specifying timing constraints that are to guide the synthesis process of prospective circuits shown in figs.4.15 and 4.17. Further provisions allow for documenting the PTV conditions for which the timing data stored in an SDF file apply.

4.8.7 A handy compilation of type conversion functions

with contributions by R. Zimmermann

The table below summarizes type conversion functions between the most important VHDL data types. Please note that proprietary types and functions render source code awkward to port from one EDA platform to another. Proprietary packages should, therefore, be dismissed in favor of vendor-independent international standards.

Table 4.9 | Type conversion functions.

according to	IEEE 1076	
as defined in package	`std.standard`	
Conversion		
real ▷ integer	integer(arg)	
integer ▷ real	real(arg)	

according to	IEEE 1076.3	Synopsys proprietary
as defined in package	`ieee.numeric_std`	`ieee.std_logic_arith`
Conversion		
std_logic_vector ▷ unsigned	unsigned(arg)	unsigned(arg)
std_logic_vector ▷ signed	signed(arg)	signed(arg)
unsigned ▷ std_logic_vector	std_logic_vector(arg)	std_logic_vector(arg)
signed ▷ std_logic_vector	std_logic_vector(arg)	std_logic_vector(arg)
integer ▷ unsigned	to_unsigned(arg,size)	conv_unsigned(arg,size)
integer ▷ signed	to_signed(arg,size)	conv_signed(arg,size)
unsigned ▷ integer	to_integer(arg)	conv_integer(arg)
signed ▷ integer	to_integer(arg)	conv_integer(arg)
integer ▷ std_logic_vector	integer ▷ unsigned|signed ▷ std_logic_vector	
std_logic_vector ▷ integer	std_logic_vector ▷ unsigned|signed ▷ integer	
Resizing		
unsigned	resize(arg,size)	conv_unsigned(arg,size)
signed	resize(arg,size)	conv_signed(arg,size)

4.9 | Appendix III: Examples of VHDL models

The subsequent pages include listings of VHDL code for a variety of applications. Albeit rather small, these models complement the explanations of individual language features given in previous sections with self-contained examples that are fully functional. They also provide material from which newcomers can get inspiration when developing their own models.

To these ends, examples have been selected such as to

- Include both circuit and testbench models,
- Cover a variety of circuit features and coding styles,
- Address both combinational (memoryless) and sequential (memorizing) behavior,
- Illustrate numerous language constructs including text and file I/O, and to
- Show circuit models that are amenable to both simulation and synthesis.

All code examples mentioned below follow the naming convention to be presented in section 5.7 and are available for download from the author's website
`http://dz.ee.ethz.ch/support/ic/vhdl/vhdlexamples.en.html`

4.9.1 Combinational circuit models

A LOGIC NETWORK THAT TELLS HOW MANY OF ITS INPUTS ARE AT LOGIC 1

```
-- Mission: Illustrate numerous ways for capturing combinational logic.
--     Example includes architecture bodies that follow behavioral, dataflow,
--     and structural coding styles. In an attempt to demonstrate the richness
--     of VHDL, many diverse varieties coding styles have been covered. Note that
--     most of them are inappropriate if data word width were allowed to vary.
-      In fact, iterative* are the only architecture bodies that do scale well.
--                     interesting to study the impact of coding style on the
--     network obtained from automatic synthesis.
-- Functionality: so-called (4,3)-counter, see below for a truth table.
-- Findings in terms of hardware costs after unconstrained optimization
--     for UMCL250 library with SynopsysDC:
--     architecture      cells types nets area  attr_setting  version
--     selectassign      8     7     12   324   default       2004.12
--     concurassign      7     5     11   348   "             "
--     caseselect        8     7     12   324   "             "
--     ifelsifelse       8     7     12   324   "             "
--     lookup            8     7     12   324   "             "
--     iterativecount    8     7     12   324   "             "
--     iterativelogic    7     5     11   348   "             "
--     iterativepad      7     5     11   348   "             "
--     dataflow          8     7     12   324   "             "
--     generated         7     5     11   348   "             "
--     structuralgtech   8     7     12   324   "             "
-- Author: H.Kaeslin.
---------------------------------------------------------------------
library ieee;
use ieee.std_logic_1164.all;
use ieee.numeric_std.all;
```

```vhdl
library gtech;                    -- generic component library by Synopsys
use gtech.gtech_components.all;
-------------------------------------------------------------------------------

entity onescnt is
   port ( INAxDI : in  std_logic_vector(3 downto 0);
          CNTxDO : out std_logic_vector(2 downto 0) );
end onescnt;


-------------------------------------------------------------------------------
--                    CNTxDO       CNTxDO(2)      CNTxDO(1)      CNTxDO(0)
--    INAxDI(1:0)  00 01 11 10    00 01 11 10    00 01 11 10    00 01 11 10
-- INAxDI(3:2) 00   0  1  2  1     0  0  0  0      0  0  1  0     0  1  0  1
--     "      "  01   1  2  3  2     0  0  0  0      0  1  1  1     1  0  1  0
--     "      "  11   2  3  4  3     0  0  1  0      1  1  0  1     0  1  0  1
--     "      "  10   1  2  3  2     0  0  0  0      0  1  1  1     1  0  1  0
-------------------------------------------------------------------------------


-- ============================================================================

-- mapping specified by exhaustive enumeration in a selected signal assignment
architecture selectassign of onescnt is
begin
   -- assign the output signal the values listed below
   with INAxDI select
      CNTxDO <= "000" when "0000",
                "001" when "0001",
                "001" when "0010",
                "001" when "0100",
                "001" when "1000",
                "011" when "0111",
                "011" when "1011",
                "011" when "1101",
                "011" when "1110",
                "100" when "1111",
                "010" when others;
end selectassign;


-- ============================================================================

-- mapping specified by arithmetic computation in a concurrent signal assignment
architecture concurassign of onescnt is
begin
   -- extend input bits to final word width before adding them

   -- note: [type-]qualified expressions make it possible to interpret
   -- the result of concatenation as unsigned without explicit conversion;

   CNTxDO <= std_logic_vector( unsigned'("00" & INAxDI(3)) +
                               unsigned'("00" & INAxDI(2)) +
                               unsigned'("00" & INAxDI(1)) +
                               unsigned'("00" & INAxDI(0)) );
end concurassign;


-- ============================================================================
```

```vhdl
-- mapping specified by exhaustive enumeration in a process statement
architecture caseselect of onescnt is
begin
   p_memless : process (INAxDI)
   begin
      -- assign the output signal the values listed below
      case INAxDI is
         when "0000" => CNTxDO <= "000";
         when "0001" => CNTxDO <= "001";
         when "0010" => CNTxDO <= "001";
         when "0011" => CNTxDO <= "010";
         when "0100" => CNTxDO <= "001";
         when "0101" => CNTxDO <= "010";
         when "0110" => CNTxDO <= "010";
         when "0111" => CNTxDO <= "011";
         when "1000" => CNTxDO <= "001";
         when "1001" => CNTxDO <= "010";
         when "1010" => CNTxDO <= "010";
         when "1011" => CNTxDO <= "011";
         when "1100" => CNTxDO <= "010";
         when "1101" => CNTxDO <= "011";
         when "1110" => CNTxDO <= "011";
         when "1111" => CNTxDO <= "100";
         when others => CNTxDO <= "111";  -- SynopsysDC insists on this
      end case;
   end process p_memless;
end caseselect;

--================================================================================

-- mapping specified by telescoped branching in a process statement
architecture ifelsifelse of onescnt is
begin
   p_memless : process (INAxDI)
   begin
      -- assign the output signals the values listed below
      if INAxDI="0000" then
         CNTxDO <= "000";
      elsif (INAxDI="0001" or INAxDI="0010" or INAxDI="0100" or INAxDI="1000")
      then
         CNTxDO <= "001";
      elsif (INAxDI="0111" or INAxDI="1011" or INAxDI="1101" or INAxDI="1110")
      then
         CNTxDO <= "011";
      elsif INAxDI="1111" then
         CNTxDO <= "100";
      else
         CNTxDO <= "010";
      end if;
   end process p_memless;
end ifelsifelse;

--================================================================================

-- mapping specified by indexing a look-up table in a process statement
-- reminder: check when a ROM can be obtained via synthesizer directives
```

```vhdl
architecture lookup of onescnt is
begin
   -- address of array is requested to be of type integer or natural
   p_memless : process (INAxDI)
      variable num : natural range 0 to 4;
      variable address : natural range 0 to 15;
      type table_16 is array(0 to 15) of natural range 0 to 4;
      constant LOOKUP_TABLE: table_16 := (0,1,1,2,1,2,2,3,1,2,2,3,2,3,3,4);
   begin
      -- use input as index, look up in table, and assign to output
      address := to_integer(unsigned(INAxDI));
      num := LOOKUP_TABLE(address);
      CNTxDO <= std_logic_vector(to_unsigned(num,3));
   end process p_memless;
end lookup;

--================================================================================

-- mapping specified by iterative counting in a process statement
architecture iterativecount of onescnt is
begin
   p_memless : process (INAxDI)
      variable num : natural range 0 to 4;
   begin
      num := 0;
      -- count those bits that equal one in a sequential loop
      for idx in 0 to 3 loop
         if INAxDI(idx)='1' then num := num+1; end if;
      end loop;
      -- convert the resulting number into a binary coded signal
      CNTxDO <= std_logic_vector(to_unsigned(num,3));
   end process p_memless;
end iterativecount;

--================================================================================

-- mapping specified by iterative logic operations in a process statement
-- after K.C. Chang, Digital Systems Design with VHDL and Synthesis, p121
architecture iterativelogic of onescnt is
begin
   p_memless : process (INAxDI)
      variable tmp : std_logic_vector(2 downto 0);
      variable cin, cout : std_logic;
   begin
      tmp := "000";
      -- increment binary sum bit by bit in a sequential loop
      for idx in INAxDI'reverse_range loop
         if INAxDI(idx)='1' then
            cin := '1';
            -- propagate carries
            for idk in tmp'reverse_range loop
               cout := tmp(idk) and cin;
               tmp(idk) := tmp(idk) xor cin;
               cin := cout;
            end loop;
         end if;
```

```
         end loop;
      CNTxDO <= tmp;
   end process p_memless;
end iterativelogic;


-- ============================================================================

-- mapping specified by iterative arithmetic computation in a process statement

-- note: the basic approach is the same as in architecture body concurassign,
-- yet the embodiment as a process permits the usage of an iterative loop
-- and of variables.

architecture iterativepad of onescnt is
begin
   p_memless : process (INAxDI)
      variable pad, sum : unsigned(2 downto 0);
   begin
      sum := (others => '0');    -- with-independent shorthand
      -- pad input bits with zeros before adding them in a sequential loop
      for idx in INAxDI'range loop
         pad := "00" & INAxDI(idx);
         sum := sum + pad;
      end loop;
      CNTxDO <= std_logic_vector(sum);
   end process p_memless;
end iterativepad;


-- ============================================================================

-- mapping specified by concurrent signal assignments at the Boolean level
architecture dataflow of onescnt is
   signal ROW0xD, ROW3xD, ROW1OR2xD, COL3xD, COL1THRU3xD: std_logic;
begin
   -- indicate a combinational function for each output bit
   -- by exploiting the similarities of rows in the truth table
   ROW0xD <= INAxDI(3) nor INAxDI(2);
   ROW3xD <= INAxDI(3) and INAxDI(2);
   ROW1OR2xD <= (not ROW0xD) and (not ROW3xD);
   COL3xD <= INAxDI(1) and INAxDI(0);
   COL1THRU3xD <= INAxDI(1) or INAxDI(0);
   CNTxDO(2) <= ROW3xD and COL3xD;
   CNTxDO(1) <= (ROW0xD and COL3xD) or (ROW3xD and (not COL3xD)) or
                (ROW1OR2xD and COL1THRU3xD);
   CNTxDO(0) <= INAxDI(3) xor INAxDI(2) xor INAxDI(1) xor INAxDI(0);
end dataflow;


-- ============================================================================

-- mapping specified by arithmetic computation at the bit level and by
-- generating a series of cascaded concurrent signal assignments there
architecture generated of onescnt is
   type support4x3 is array(1 to 4) of unsigned(2 downto 0);
   signal INNERCNTxD : support4x3;
begin
   -- pad input bits with zeros, embed additions between support nodes
```

```
  G1 : for i in 3 downto 1 generate
      INNERCNTxD(i+1) <= INNERCNTxD(i) + ("00" & INAxDI(i));
  end generate;
  INNERCNTxD(1) <= "00" & INAxDI(0);
  CNTxDO <= std_logic_vector(INNERCNTxD(4));
end generated;


--===============================================================================


-- logic network described as a bunch of interconnected logic gates
architecture structuralgtech of onescnt is

  -- note: a list of explicit component declarations is redundant in this
  -- context because the GTECH generic component library has been made
  -- available by the library and use statements early in this program.

  -- declare internal signals
  signal ORANGExD, YELLOWxD, GREENxD, BLUExD, VIOLETxD: std_logic;

begin

  -- instantiate components and connect them by listing port maps
  U1 : gtech_OA21
      port map ( A=>INAxDI(3), B=>BLUExD, C=>GREENxD, Z=>CNTxDO(0) );
  U2 : gtech_NOR2
      port map ( A=>GREENxD, B=>VIOLETxD, Z=>CNTxDO(2) );
  U3 : gtech_XOR2
      port map ( A=>INAxDI(1), B=>INAxDI(0), Z=>YELLOWxD );
  U4 : gtech_XOR2
      port map ( A=>YELLOWxD, B=>INAxDI(2), Z=>BLUExD );
  U5 : gtech_XOR2
      port map ( A=>GREENxD, B=>VIOLETxD, Z=>CNTxDO(1) );
  U6 : gtech_NAND2
      port map ( A=>INAxDI(3), B=>BLUExD, Z=>GREENxD );
  U7 : gtech_AOI22
      port map ( A=>INAxDI(0), B=>INAxDI(1), C=>YELLOWxD, D=>INAxDI(2),
                 Z=>VIOLETxD );

end structuralgtech;
```

4.9.2   Mealy, Moore, and Medvedev machines

What follows is an example for each of the three classes of finite state machines. Refer to section 4.3 and fig.4.19 if in doubt on how to cast finite state machines into VHDL processes.

A Mealy-type state machine

```
-- Mission: Illustrate how to model a Mealy machine with two processes.
--     Example designed to include an asynchronous reset, three-state outputs,
--     self loops, symbolic encoding of states and outputs, plus handling of
--     parasitic states and inputs so as to eliminate any chance of lock-up.
-- Functionality: See state graph below.
-- Author: H.Kaeslin.
```

```
--------------------------------------------------------------------------------
library ieee;
use ieee.std_logic_1164.all;
--------------------------------------------------------------------------------

entity mealy5st is
   port (CLKxCI  : in   std_logic;
         RSTxRBI : in   std_logic;
         INPxDI  : in   std_logic_vector(1 downto 0);
         OUPxDO  : out  std_logic_vector(1 downto 0) );  -- ternary, 11 not used
end mealy5st;


--------------------------------------------------------------------------------
-- INPxDI/OUPxDO                                                       (STATExDP)
--
--            00/e          00/i          00/i          00/i          00/f
--                 -             -             -             -
--           | |  01/i    | |  01/i    | |  01/i    | |  01/f    | |
--        ---> v / ------> v / ------> v / ------> v / ------> v / ----
--  10/e |     (0)          (1)          (2)          (3)          (4)    | 01/f
--        -----   <-------- <-------- ^ <-------- <-------- <----
--                10/e         10/i   |   10/i         10/i
--                                    o
--
--------------------------------------------------------------------------------

architecture enumerated_state of mealy5st is

   type state is (st0, st1, st2, st3, st4);  -- enumerated state type
   signal STATExDP, STATExDN : state;        -- present state and next state
   -- symbolic encodings of output
   constant syme : std_logic_vector(1 downto 0) := "10";   -- empty
   constant symi : std_logic_vector(1 downto 0) := "00";   -- in between
   constant symf : std_logic_vector(1 downto 0) := "01";   -- full
   constant symz : std_logic_vector(1 downto 0) := "ZZ";   -- high-impedance

begin

   ---- computation of next state and present outputs
   --------------------------------------------------------------------------
   p_memless : process (INPxDI, STATExDP)
   begin
      -- default assignments
      STATExDN <= STATExDP;    -- remain in present state
      OUPxDO <= symi;          -- output "in between"
      -- nondefault transitions and outputs
      case STATExDP is
         when st0 =>   -- formulated by handling each input code separately
            if    INPxDI="00" then OUPxDO <= syme;
            elsif INPxDI="10" then OUPxDO <= syme;
            elsif INPxDI="01" then STATExDN <= st1;
            else OUPxDO <= syme;   -- parasitic input 11, treat as 00
            end if;
         when st1 =>
            if    INPxDI="00" then null;   -- defaults suffice, may be omitted
            elsif INPxDI="10" then OUPxDO <= syme; STATExDN <= st0;
```

```vhdl
            elsif INPxDI="01" then STATExDN <= st2;
            else null;   -- parasitic input 11, treat as 00
            end if;
        when st2 =>   -- adopting more concise formulations from now on
            if    INPxDI="10" then STATExDN <= st1;
            elsif INPxDI="01" then STATExDN <= st3;
            end if;
        when st3 =>
            if    INPxDI="10" then STATExDN <= st2;
            elsif INPxDI="01" then OUPxDO <= symf; STATExDN <= st4;
            end if;
        when st4 =>
            if    INPxDI="10" then STATExDN <= st3;
            else  OUPxDO <= symf;   -          r parasitic input 11
            end if;
        when others =>   -- tie up parasitic states for synthesis
            OUPxDO <= symz; STATExDN <= st2;
      end case;
   end process p_memless;

   ---- updating of state
   --------------------------------------------------------------------------
   p_memzing : process (CLKxCI, RSTxRBI)
   begin
      -- activities triggered by asynchronous reset (active low)
      if RSTxRBI = '0' then
         STATExDP <= st2;
      -- activities triggered by rising edge of clock
      elsif CLKxCI'event and CLKxCI = '1' then
         STATExDP <= STATExDN;
      end if;
   end process p_memzing;

end enumerated_state;
```

A Moore-type state machine

```vhdl
-- Mission: Illustrate how to model a Moore machine with two processes. The
--     example, taken from the author's lecture notes "Finite State Machines",
--     includes a synchronous clear and also serves to demonstrate how to use
--     symbolic state names in conjunction with a specific state assignment.
-- Note: State encoding could just as well be controlled by synthesizer directives.
-- Functionality: See state table below.
-- Author: H.Kaeslin.
--------------------------------------------------------------------------

library ieee;
use ieee.std_logic_1164.all;
--------------------------------------------------------------------------

entity moore6st is
   port ( CLKxCI : in  std_logic;
          CLRxSI : in  std_logic;          -- synchronous clear
          INPxDI : in  std_logic;          -- present input
```

```
             OUPxDO : out std_logic );     -- present output
end moore6st;


----------------------------------------------------------------------------------
--                                       present input
--                                       a          b
--              present     present      next       next
--              state       output       state      state
--   start state    stu         1            stz        stw
--                  stv         0            stz        stw
--                  stw         0            stx        stu
--                  stx         0            sty        stx
--                  sty         1            stv        sty
--                  stz         1            sty        stx
----------------------------------------------------------------------------------


architecture symbolic_state of moore6st is

   -- symbolic encodings of state
   subtype state is std_logic_vector(5 downto 0);  -- vector state type
   constant stu : state := "000001";    -- force one-hot state assignment
   constant stv : state := "000010";
   constant stw : state := "000100";
   constant stx : state := "001000";
   constant sty : state := "010000";
   constant stz : state := "100000";
   signal STATExDP, STATExDN : state;   -- present state and next state
   -- symbolic encodings of input
   constant syma : std_logic := '0';
   constant symb : std_logic := '1';

begin

   ---- computation of next state and present outputs
   ------------------------------------------------------------------------------
   p_memless : process (INPxDI,STATExDP)
   begin
      -- default assignments
      STATExDN <= stu;  -- jump to start state
      OUPxDO <= '1';     -- output a logic 1
      -- nondefault transitions and outputs
      case STATExDP is
      when stu =>
         if INPxDI=syma then STATExDN <= stz; else STATExDN <= stw; end if;
      when stv =>
         if INPxDI=syma then STATExDN <= stz; else STATExDN <= stw; end if;
         OUPxDO <= '0';   -- output assignment placed outside input branching
      when stw =>
         if INPxDI=syma then STATExDN <= stx; end if;
         OUPxDO <= '0';
      when stx =>
         OUPxDO <= '0';   -- note that order of statements is immaterial
         if INPxDI=syma then STATExDN <= sty; else STATExDN <= stx; end if;
      when sty =>
         if INPxDI=syma then STATExDN <= stv; else STATExDN <= sty; end if;
      when stz =>
```

```
        if INPxDI=syma then STATExDN <= sty; else STATExDN <= stx; end if;
     -- the remaining 58 cases need not be handled explicitly because
     -- parasitic states are properly tied up by the default assignments
     when others => null;
     end case;
  end process p_memless;


  ---- updating of state
  ----------------------------------------------------------------
  p_memzing : process (CLKxCI)
  begin
     -- activities triggered by rising edge of clock
     if CLKxCI'event and CLKxCI='1' then
        -- when synchronous clear is asserted
        if CLRxSI='1' then
           STATExDP <= stu;
        -- otherwise
        else
           STATExDP <= STATExDN;
        end if;
     end if;
  end process p_memzing;

end symbolic_state;
```

A MEDVEDEV-TYPE STATE MACHINE

In this example, the source code has been distributed over two design units to illustrate the concept
of separate compilation and the usage of **library** and **use** statements. The first design unit describes
the circuit itself; the second one is a package of code conversion functions.

```
-- Mission: Illustrate how to model a Medvedev machine in two processes.
--    Also serves to study more subtle issues of type conversion.
--    Example designed to include an asynchronous reset,
--    a subcondition in the clock clause, and a parametrized word width,
--    and to make light usage of the IEEE numeric_std package.
-- Functionality: w-bit Gray counter with enable and asynchronous reset.
-- Author: H.Kaeslin.
----------------------------------------------------------------

library ieee;
use ieee.std_logic_1164.all;
use ieee.numeric_std.all;
use work.grayconv.all;             -- my own set of Gray code converter functions
----------------------------------------------------------------

entity graycnt is
   generic (
      width : integer := 5 );    -- default value for number of state bits
   port (
      CLKxCI   : in   std_logic;
      RSTxRBI  : in   std_logic;
      ENAxSI   : in   std_logic;
      COUNTxDO : out  std_logic_vector((width-1) downto 0) );
```

```
end graycnt;

---------------------------------------------------------------------------
-- Approach is with double conversion: Gray->binary, increment, binary->Gray.
-- Note this is not necessarily the most economic or the fastest solution.
-- Declaring COUNTxD? of type unsigned would help to slightly simplify code
-- but requires overloading functions bintogray and graytobin for this type.
---------------------------------------------------------------------------

architecture computed_state of graycnt is

   -- present state and next state
   signal COUNTxDP, COUNTxDN : std_logic_vector(width-1 downto 0);

begin

   ---- computation of next state
   ------------------------------------------------------------------------
   COUNTxDN <= bintogray(std_logic_vector(
      unsigned(graytobin(COUNTxDP)) + 1 ));

   ---- updating of state
   ------------------------------------------------------------------------
   p_memzing : process (CLKxCI,RSTxRBI)
   begin
      -- activities triggered by asynchronous reset
      if RSTxRBI='0' then
         COUNTxDP <= (others => '0');   -- width-independent shorthand
      -- activities triggered by rising edge of clock
      elsif CLKxCI'event and CLKxCI='1' then
         -- proceed to next state only if enable is asserted
         if ENAxSI='1' then
            COUNTxDP <= COUNTxDN;
         end if;
      end if;
   end process p_memzing;

   ---- assignment to output only signal
   ------------------------------------------------------------------------
   COUNTxDO <= COUNTxDP;

end computed_state;
```

```
-- Mission: illustrate the use of a package in VHDL.
-- Functionality: Gray code <-> binary code conversion functions.
-- Author: H.Kaeslin.
---------------------------------------------------------------------------

library ieee;
use ieee.std_logic_1164.all;
---------------------------------------------------------------------------

-- package declaration
package grayconv is
   function bintogray (arg : std_logic_vector) return std_logic_vector;
```

```vhdl
  function graytobin (arg : std_logic_vector) return std_logic_vector;
end grayconv;

-------------------------------------------------------------------------------
--
--                         |     |     |     |     |
-- binary to Gray          |--.  |--.  |--.  |--.  |          # of X-ops on
-- conversion              |  '--X  '--X  '--X  '--X          longest paths
-- for width=5             |     |     |     |     |          = 1
--                         v     v     v     v     v
--
-- bit positions           4     3     2     1     0          X = XOR
--
--                         |     |     |     |     |
-- Gray to binary          |  ,--X  ,--X  ,--X  ,--X          # of X-ops on
-- conversion              |--'  |--'  |--'  |--'  |          longest path
-- for width=5             |     |     |     |     |          = width-1
--                         v     v     v     v     v
--
-------------------------------------------------------------------------------

-- package body
package body grayconv is

  -- purpose: converts binary code into Gray code
  -            place computation on a scratchpad variable
  function bintogray (arg : std_logic_vector) return std_logic_vector is
     variable scrapa : std_logic_vector(arg'length-1 downto 0);
  begin
     scrapa := arg;
     for i in 0 to arg'length-2 loop       -- note MSB remains unchanged
       if scrapa(i+1)='1' then
          scrapa(i) := not scrapa(i);
       end if;
     end loop;
     return scrapa;
  end bintogray;

  -- purpose: converts Gray code into binary code
  -            place computation on a scratchpad variable
  function graytobin (arg : std_logic_vector) return std_logic_vector is
     variable scrapa : std_logic_vector(arg'length-1 downto 0);
  begin
     scrapa := arg;
     for i in arg'length-2 downto 0 loop   -- note MSB remains unchanged
       if scrapa(i+1)='1' then
          scrapa(i) := not scrapa(i);
       end if;
     end loop;
     return scrapa;
  end graytobin;

end grayconv;
```

4.9.3 State reduction and state encoding

A STATE MACHINE THAT FEATURES EQUIVALENT STATES

The state graphs before and after reduction are shown in figs.B.8 and B.9 respectively.

```vhdl
-- Mission: Demonstrate state reduction and state encoding techniques,
--    serve as a benchmark for FSM optimization software,
--    and show how to identify state vectors for Synopsys Design Compiler.
-- Functionality: fictive, 5 of the 11 states can be collapsed with others.
--    For state graph see chapter "Finite State Machines" in author's book.
-- Author: H.Kaeslin
-------------------------------------------------------------------------------
-- The state reduction can be performed with the following instructions
-- in Synopsys Design Compiler (version W-2004.12):
--    read_file -format vhdl {your_path/phantasy.vhd}
--    set_ultra_optimization true
--    set fsm_auto_inferring true
--    set fsm_enable_state_minimization true
--    set_fsm_encoding_style binary
--    compile
--    report_fsm
-------------------------------------------------------------------------------

library ieee;
use ieee.std_logic_1164.all;
-------------------------------------------------------------------------------

entity phantasy is
   port (
      CLKxCI  : in  std_logic;
      RSTxRBI : in  std_logic;
      INPxDI  : in  std_logic_vector(1 downto 0);
      OUPxDO  : out std_logic );
end phantasy;


-------------------------------------------------------------------------------


architecture enumerated_state of phantasy is

   type state is (st1, st2, st3, st4, st5, st6, st7, st8, st9, st10, st11);
   signal STATExDP, STATExDN : state;
   -- identify state vector for automatic state reduction
   attribute state_vector : string;   -- proprietary attribute of Synopsys
   attribute state_vector of enumerated_state : architecture is "STATExDP";

begin

   ---- computation of next state and present outputs
   -------------------------------------------------------------------------------
   p_memless: process (INPxDI, STATExDP)
   begin
      -- default assignments
      STATExDN <= STATExDP;   -- remain in present state
      OUPxDO <= '0';          -- output a logic 0
      -- nondefault transitions and outputs
```

```vhdl
case STATExDP is
when st1 =>
   OUPxDO <= '1';
   if    INPxDI="00" then STATExDN <= st1;
   elsif INPxDI="11" then STATExDN <= st2;
   else                   STATExDN <= st1;
   end if;
--
when st2 =>
   if    INPxDI="00" then STATExDN <= st2;
   elsif INPxDI="11" then STATExDN <= st3;
   else                   STATExDN <= st8;
   end if;
when st3 =>
   if    INPxDI="00" then STATExDN <= st3;
   elsif INPxDI="11" then STATExDN <= st4;
   else                   STATExDN <= st5;
   end if;
when st4 =>
   if    INPxDI="00" then STATExDN <= st4;
   elsif INPxDI="11" then STATExDN <= st2;
   else                   STATExDN <= st5;
   end if;
--
when st5 =>
   if    INPxDI="00" then STATExDN <= st6;
   elsif INPxDI="11" then STATExDN <= st9;
   else                   STATExDN <= st5;
   end if;
when st6 =>
   if    INPxDI="00" then STATExDN <= st7;
   elsif INPxDI="11" then STATExDN <= st3;
   else                   STATExDN <= st11;
   end if;
when st7 =>
   if    INPxDI="00" then OUPxDO <= '1';
   elsif INPxDI="11" then OUPxDO <= '1';
   end if;
   STATExDN <= st1;
--
when st8 =>
   if    INPxDI="00" then STATExDN <= st9;
   elsif INPxDI="11" then STATExDN <= st6;
   else                   STATExDN <= st8;
   end if;
when st9 =>
   if    INPxDI="00" then STATExDN <= st10;
   elsif INPxDI="11" then STATExDN <= st4;
   else                   STATExDN <= st11;
   end if;
when st10 =>
   if    INPxDI="00" then OUPxDO <= '1';
   elsif INPxDI="11" then OUPxDO <= '1';
   end if;
   STATExDN <= st1;
--
```

```
      when st11 =>
         STATExDN <= st1;
      end case;
   end process p_memless;


   ---- updating of state
   -------------------------------------------------------------------------
   p_memzing: process (CLKxCI, RSTxRBI)
   begin
      -- activities triggered by asynchronous reset
      if RSTxRBI='0' then
         STATExDP <= st1;
      -- activities triggered by rising edge of clock
      elsif CLKxCI'event and CLKxCI='1' then
         STATExDP <= STATExDN;
      end if;
   end process p_memzing;

end enumerated_state;
```

4.9.4 Simulation testbenches

The testbenches given below have been written along the lines discussed in section 4.4.2 and in chapter 3. This implies, among many other things, that the events "$\triangle \downarrow \top \uparrow$" repeat periodically. The MUT is a sequential circuit that takes multiple clock cycles to carry out one unsigned n-by-n multiplication. Simulation is a for a 4-by-4 bit configuration. Make sure you understand that none of the code below is intended for synthesis.

Model-independent declarations and subprograms

simulstuff.vhd contains a collection of type declarations and subprograms that do not depend on the MUT and can, therefore, be reused in almost any VHDL simulation. The package includes

- An adjustable clock generator.
- Procedures for file handling.
- Procedures for generating random stimuli.
- A procedure for checking actual against expected responses.
- Procedures for evaluating and reporting mismatches.
- Procedures for compiling a simulation report.

For sake of brevity, the code is not reprinted here but available for download.

A testbench of grade 1

In a grade 1 simulation, stimulus/response pairs are expressed as bits exclusively and locked to predetermined clock cycles. Overall organization follows fig.3.14. Source code is distributed over five design files.

- multtb1.vhd: Testbench including clock generator and two VHDL processes
  for stimulus application and response acquisition.

- `multtb1pkg.vhd`: A collection of model-specific declarations and subprograms.
- `shiftaddmult.vhd`: The MUT, a sequential shift-and-add multiplier.
- `multref.vhd`: An entity that reads the expected responses from a file.
- `simulstuff.vhd`: Same as always, see above.

```vhdl
-- Mission: Provide a code example for a file-based testbench.
-- Functionality: A simple testbench for a 4-bit shift add multiplier. The
--    source code of the multiplier is located in the file shiftaddmult.vhd.
--    The testbench reads stimuli and expected responses from separate
--    ASCII files and writes a simulation report to a third ASCII file.
--    The testbench applies stimuli and acquires actual responses every clock
--    cycle.
--    When stimuli vectors outnumber expected responses, the simulation run
--    continues with "don't cares".
-- Companion files: simulstuff.vhd, multtb1pkg.vhd, shiftaddmult.vhd,
--    multref.vhd
-- Platform: This testbench was written and tested with Modelsim 6.0.
-- Company: Microelectronics Design Center, ETH Zurich.
-- Authors: Hubert Kaeslin, Thomas Kuch
------------------------------------------------------------------------------

use std.textio.all;
library ieee;
use ieee.std_logic_textio.all;   -- read and write overloaded for std_logic
use ieee.std_logic_1164.all;
use work.simulstuff.all;
use work.multTb1Pkg.all;


------------------------------------------------------------------------------


entity MultTb1 is
  -- a testbench does not connect to any higher level of hierarchy
end MultTb1;


------------------------------------------------------------------------------


architecture Behavioral of MultTb1 is

  -- declaration of model under test (MUT) and functional
  -- reference (expected response pickup)
  component ShiftAddMult is
    generic (
      width         : natural );
    port (
      ClkxCI        : in  std_logic;
      RstxRBI       : in  std_logic;
      StartCalcxSI  : in  std_logic;
      InputAxDI     : in  std_logic_vector(width-1 downto 0);
      InputBxDI     : in  std_logic_vector(width-1 downto 0);
      OutputxDO     : out std_logic;
      OutputValidxSO : out std_logic );
  end component;

  for RefInst : ShiftAddMult use entity work.ShiftAddMult(Pickup);
  for MutInst : ShiftAddMult use entity work.ShiftAddMult(Behavioral);
```

```
begin

  -- instantiate MUT and functional reference and connect them to the
  -- testbench signals
  -- note: any bidirectional must connect to both stimuli and responses
  ----------------------------------------------------------------------------

  MutInst : ShiftAddMult
    generic map (
      width          => 4 )
    port map (
      ClkxCI          => ClkxC,
      RstxRBI         => StimuliRecxD.RstxRB,
      StartCalcxSI    => StimuliRecxD.StartCalcxS,
      InputAxDI       => StimuliRecxD.InputAxD,
      InputBxDI       => StimuliRecxD.InputBxD,
      OutputxDO       => ActResponseRecxD.OutputxD,
      OutputValidxSO  => ActResponseRecxD.OutputValidxS );

  RefInst : ShiftAddMult
    generic map (
      width          => 4 )
    port map (
      ClkxCI          => ClkxC,
      RstxRBI         => StimuliRecxD.RstxRB,
      StartCalcxSI    => StimuliRecxD.StartCalcxS,
      InputAxDI       => StimuliRecxD.InputAxD,
      InputBxDI       => StimuliRecxD.InputBxD,
      OutputxDO       => ExpResponseRecxD.OutputxD,
      OutputValidxSO  => ExpResponseRecxD.OutputValidxS );


  -- pausable clock generator with programmable mark and space widths
  ----------------------------------------------------------------------------
  -- The procedure ClockGenerator is defined in the package simulstuff.
  -- This concurrent procedure call is a process that calls the procedure,
  -- with a syntax that looks like a "process instance".

  ClkGen : ClockGenerator(
    ClkxC        => ClkxC,
    clkphaselow  => clk_phase_low,
    clkphasehigh => clk_phase_high );


  -- obtain stimuli and apply it to MUT
  ----------------------------------------------------------------------------
  StimAppli : process
  begin

    AppliLoop : while not (endfile(stimulifile)) loop
      wait until ClkxC'event and ClkxC = '1';

      -- wait until time has come for stimulus application
      wait for stimuli_application_time;
      -- apply stimulus to MUT
```

```vhdl
    StimuliRecxD <= GetStimuliRecord(stimulifile);
  end loop AppliLoop;

  -- tell clock generator to stop at the end of current cycle
  -- because stimuli have been exhausted
  EndOfSimxS <= true;
  -- close the file
  file_close(stimulifile);
  wait;
end process StimAppli;


-- acquire actual response from MUT and have it checked
-------------------------------------------------------------------------
RespAcqui : process

  -- variables for accounting of mismatching responses
  variable respmatch : respMatchArray;
  variable respaccount : respaccounttype := (0, 0, 0, 0, 0, 0);
  -- variable for counting the lines written to the simulation report
  variable simRepLineCount : natural := 0;

begin
  -- This wait statement is useful only if the stimuli file is empty. In that
  -- case, EndOfSimxS gets true after one delta delay. Without the wait, the
  -- exit statement below would be executed before EndOfSimxS gets true.
  wait until ClkxC'event and ClkxC = '0';

  AcquiLoop : loop
    -- leave the loop if there are no more stimuli left
    exit AcquiLoop when EndOfSimxS = true;

    wait until ClkxC'event and ClkxC = '1';

    -- wait until time has come for response acquisition
    wait for response_acquisition_time;

    -- compare the actual with the expected responses
    CheckResponse(ActResponseRecxD, ExpResponseRecxD,
                  respmatch, respaccount);

    -- add a trace line to report file
    PutSimulationReportTrace(simreptfile, StimuliRecxD, ActResponseRecxD,
                             respmatch, respaccount, simRepLineCount);

    -- add extra failure message to report file if necessary
    PutSimulationReportFailure(simreptfile, ExpResponseRecxD, respmatch);

  end loop AcquiLoop;

  -- when the present clock cycle is the final one of this run
  -- then establish a simulation report summary and write it to file
  PutSimulationReportSummary(simreptfile, respaccount);
  -- close the file
  file_close(simreptfile);
  report "Simulation run completed!";
```

```
    wait;
  end process RespAcqui;

end architecture Behavioral;  -- of MultTb1
```

```
-- Mission: See associated testbench file.
-- Functionality: This package contains type declarations, signals, and
--   constants for the testbench. It also contains some MUT-specific functions
--   and procedures.
-- Platform: Modelsim 6.0.
-- Company: Microelectronics Design Center, ETH Zurich.
-- Authors: Hubert Kaeslin, Thomas Kuch
-------------------------------------------------------------------------------

use std.textio.all;
library ieee;
use ieee.std_logic_1164.all;
use ieee.std_logic_textio.all;
use work.simulstuff.all;


--=============================================================================


package multTb1Pkg is

  -- declarations of all those signals that do connect to the MUT
  -- most of them are collected in records to facilitate data handling
  -- note: any bidirectional must be made part of both stimuli and responses
  type stimuliRecordType is record
    RstxRB      : std_logic;
    StartCalcxS : std_logic;
    InputAxD    : std_logic_vector(3 downto 0);
    InputBxD    : std_logic_vector(3 downto 0);
  end record;

  -- same for actual and expected response
  type responseRecordType is record
    OutputxD     : std_logic;
    OutputValidxS : std_logic;
  end record;

  -- as there are two elements in the response record, an array with two
  -- elements of type respmatchtype is needed
  type respMatchArray is array (1 to 2) of respmatchtype;

  signal ClkxC             : std_logic := '1';    -- driving clock
  signal StimuliRecxD      : stimuliRecordType;   -- record of stimuli
  signal ActResponseRecxD : responseRecordType;   -- record of actual responses
  signal ExpResponseRecxD : responseRecordType;   -- record of expected responses

  -- timing of clock and simulation events
  constant clk_phase_high         : time := 50 ns;
  constant clk_phase_low          : time := 50 ns;
  constant response_acquisition_time : time := 90 ns;
  constant stimuli_application_time  : time := 10 ns;
```

```vhdl
  -- declaration of stimuli, expected responses, and simulation report files
  constant stimuli_filename : string := "../simvectors/stimuli.asc";
  constant expresp_filename : string := "../simvectors/expresp.asc";
  constant simrept_filename : string := "../simvectors/simrept.asc";

  -- the files are opened implicitly right here
  file stimulifile : text open read_mode  is stimuli_filename;
  file simreptfile : text open write_mode is simrept_filename;

------------------------------------------------------------------------------

  -- function for reading stimuli data from the stimuli file
  function GetStimuliRecord (file stimulifile : text)
    return stimuliRecordType;

  -- function for reading expected responses from the expected response file
  function GetExpectedResponseRecord (file exprespfile : text)
    return responseRecordType;

  -- procedure for comparing actual and expected response
  procedure CheckResponse
    (actRespRecord : in    responseRecordType;
     expRespRecord : in    responseRecordType;
     respmatch     : inout respMatchArray;
     respaccount   : inout respaccounttype);

  -- procedure for writing stimuli and actual responses to the report file
  procedure PutSimulationReportTrace
    (file simreptfile :    text;
     stimuliRecord   : in stimuliRecordType;
     actRespRecord   : in responseRecordType;
     respmatch       : in respMatchArray;
     respaccount     : in respaccounttype;
     simRepLineCount : inout natural);

  -- compose a failure message line and write it to the report file
  procedure PutSimulationReportFailure
    (file simreptfile :    text;
     expRespRecord    : in responseRecordType;
     respmatch        : in respMatchArray);

end package MultTb1Pkg;

--==============================================================================

package body MultTb1Pkg is

    -- purpose: get one record worth of stimuli from file.
  function GetStimuliRecord
    (file stimulifile : text)
    return stimuliRecordType
  is
    variable in_line, in_line_tmp : line;
    -- stimuli to default to unknown in case no value is obtained from file
    variable stimulirecord : stimuliRecordType :=
      (RstxRB => 'X', StartCalcxS => 'X',
```

```vhdl
      InputAxD => (others => 'X'), InputBxD => (others => 'X') );
  begin
    -- read a line from the stimuli file
    -- skipping any empty and comment lines encountered
    loop
      readline(stimulifile, in_line);
      -- copy line read to enable meaningful error messages later
      in_line_tmp := new string'(in_line(in_line'low to in_line'high));
      if in_line_tmp'length >= 1 then
        exit when in_line_tmp(1) /= '%';
      end if;
      deallocate(in_line_tmp);
    end loop;
    -- extract all values of a record of stimuli
    GetFileEntry(stimulirecord.RstxRB, in_line, in_line_tmp, stimuli_filename);
    GetFileEntry(stimulirecord.StartCalcxS, in_line, in_line_tmp,
      stimuli_filename);
    GetFileEntry(stimulirecord.InputAxD, in_line, in_line_tmp,
      stimuli_filename);
    GetFileEntry(stimulirecord.InputBxD, in_line, in_line_tmp,
      stimuli_filename);
    -- deallocate line copy now that all entries have been read
    deallocate(in_line_tmp);
    return stimulirecord;
  end GetStimuliRecord;

----------------------------------------------------------------------------------

  -- purpose: get one record worth of expected responses from file.
  function GetExpectedResponseRecord (file exprespfile : text)
    return responseRecordType
  is
    variable in_line, in_line_tmp : line;
    -- expected responses to default to don't care
    -- in case no value is obtained from file
    variable expresprecord : responseRecordType :=
      (OutputxD => '-', OutputValidxS => '-');
  begin
    -- read a line from the expected response file as long as there are any
    -- skipping any empty and comment lines encountered
    if not(endfile(exprespfile)) then
      loop
        readline(exprespfile, in_line);
        -- copy line read to enable meaningful error messages later
        in_line_tmp := new string'(in_line(in_line'low to in_line'high));
        if in_line_tmp'length >= 1 then
          exit when in_line_tmp(1) /= '%';
        end if;
        deallocate(in_line_tmp);
      end loop;
      -- extract all values of a record of expected responses
      GetFileEntry(expresprecord.OutputxD, in_line, in_line_tmp,
        expresp_filename);
      GetFileEntry(expresprecord.OutputValidxS, in_line, in_line_tmp,
        expresp_filename);
      -- deallocate line copy now that all entries have been read
```

```vhdl
      deallocate(in_line_tmp);
      -- return default value in case EOF is overrun, no else clause needed
    end if;
    return expresprecord;
  end GetExpectedResponseRecord;


  ------------------------------------------------------------------------------


  -- purpose: procedure for comparing actual and expected response
  procedure CheckResponse
    (actRespRecord : in    responseRecordType;
     expRespRecord : in    responseRecordType;
     respmatch     : inout respMatchArray;
     respaccount   : inout respaccounttype)
  is
  begin

    CheckValue(ActResponseRecxD.OutputxD, ExpResponseRecxD.OutputxD,
               respmatch(1), respaccount);
    CheckValue(ActResponseRecxD.OutputValidxS, ExpResponseRecxD.OutputValidxS,
               respmatch(2), respaccount);

  end CheckResponse;


  ------------------------------------------------------------------------------


  -- purpose: writing stimuli and actual responses to the report file.
  procedure PutSimulationReportTrace
    (file simreptfile :    text;
     stimuliRecord     : in stimuliRecordType;
     actRespRecord     : in responseRecordType;
     respmatch         : in respMatchArray;
     respaccount       : in respaccounttype;
     simRepLineCount   : inout natural)
  is
    constant N        : natural := 60;
    variable out_line : line;
  begin
    -- every Nth line, (re)write the signal caption to the simulation report
    if simRepLineCount mod N = 0 then
      write(out_line,
        string'(" "));
      writeline(simreptfile, out_line);
      write(out_line,
        string'("Time            RstxRB                  OutputxD"));
      writeline(simreptfile, out_line);
      write(out_line,
        string'("|                |    StartCalcxSI        |  OutputValidxS"));
      writeline(simreptfile, out_line);
      write(out_line,
        string'("|                |  |    InputAxD          |     |"));
      writeline(simreptfile, out_line);
      write(out_line,
        string'("|                |  |   3210   InputBxD    |     |"));
      writeline(simreptfile, out_line);
      write(out_line,
```

```vhdl
          string'("|                 |   |   |      3210      |      |"));
      writeline(simreptfile, out_line);
      write(out_line,
          string'("|                 |   |   |      |          |      |"));
      writeline(simreptfile, out_line);
    end if;
    simRepLineCount := simRepLineCount + 1;

    -- begin with simulation time
    write(out_line, string'("at "));
    write(out_line, now);
    -- add stimuli
    write(out_line, ht);
    write(out_line, stimuliRecord.RstxRB);
    write(out_line, string'("   "));
    write(out_line, stimuliRecord.StartCalcxS);
    write(out_line, string'("   "));
    write(out_line, stimuliRecord.InputAxD);
    write(out_line, string'("   "));
    write(out_line, stimuliRecord.InputBxD);
    -- add actual response 1
    write(out_line, string'("         "));
    write(out_line, actRespRecord.OutputxD);

    case respmatch(1) is
      when mok =>
        -- if the actual response matches with the expected one, append nothing
        write(out_line, string'("     "));
      when mne =>
        -- if there was no expected response for the actual one, append a '-'
        write(out_line, string'(" -   "));
      when mlf =>
        -- if the actual response doesn't match logically, append an 'l'
        write(out_line, string'(" l   "));
      when msf =>
        -- if the actual doesn't match in strength, append an 's'
        write(out_line, string'(" s   "));
      when others =>   -- when mil
        -- if the actual response is "don't care", append an 'i'
        write(out_line, string'(" i   "));   --
    end case;

    -- add actual response 2
    write(out_line, actRespRecord.OutputValidxS);

    case respmatch(2) is
      when mok => null;
      when mne =>
        write(out_line, string'(" -"));
      when mlf =>
        write(out_line, string'(" l"));
      when msf =>
        write(out_line, string'(" s"));
      when others =>   -- when mil
        write(out_line, string'(" i"));
    end case;
```

```vhdl
    -- write the output line to the report file
    writeline(simreptfile, out_line);
  end PutSimulationReportTrace;


--------------------------------------------------------------------------------

  -- purpose: compose a failure message line and write it to the report file.
  procedure PutSimulationReportFailure
    (file simreptfile :    text;
     expRespRecord    : in responseRecordType;
     respmatch        : in respMatchArray)
  is
    variable out_line : line;
  begin

    -- if at least one actual doesn't match with its expected response
    if (respmatch(1) /= mok and respmatch(1) /= mne) or
       (respmatch(2) /= mok and respmatch(2) /= mne) then

      write(out_line, string'("^^ Failure! Expected was :"));

      -- if actual response 1 doesn't match with its expected response
      if respmatch(1) /= mok and respmatch(1) /= mne then
        -- add expected response
        write(out_line, string'("                "));
        write(out_line, expRespRecord.OutputxD);

      else
        write(out_line, string'("                "));
      end if;

      -- if actual response 2 doesn't match with its expected response
      if respmatch(2) /= mok and respmatch(2) /= mne then
        -- add expected response
        write(out_line, string'("      "));
        write(out_line, expRespRecord.OutputValidxS);
      end if;

      writeline(simreptfile, out_line);
    end if;
  end PutSimulationReportFailure;

end package body MultTb1Pkg;
```

```vhdl
-- Mission: See associated testbench file.
-- Functionality: The following architecture describes a shift add multiplier
--   for unsigned numbers. The multiplier takes two W-bit input values. The
--   2W-bit wide result is put out serially (LSB first).
-- Platform: Modelsim 6.0
-- Company: Microelectronics Design Center, ETH Zurich
-- Author: Thomas Kuch
--------------------------------------------------------------------------------

library ieee;
```

```vhdl
use ieee.std_logic_1164.all;
use ieee.numeric_std.all;
--------------------------------------------------------------------------------

entity ShiftAddMult is
  generic (
    width          : natural := 8 );   -- word width of the two inputs
  port (
    ClkxCI         : in  std_logic;
    RstxRBI        : in  std_logic;
    StartCalcxSI   : in  std_logic;
    InputAxDI      : in  std_logic_vector(width-1 downto 0);
    InputBxDI      : in  std_logic_vector(width-1 downto 0);
    OutputxDO      : out std_logic;
    OutputValidxSO : out std_logic );
end ShiftAddMult;


--------------------------------------------------------------------------------

architecture Behavioral of ShiftAddMult is

  signal ClearRegxS  : std_logic;
  signal InputRegEnxS : std_logic;
  signal ShiftEnxS    : std_logic;

  signal InputAxDP    : unsigned(2*width-1 downto 0);
  signal InputBxDP    : unsigned(width-1 downto 0);

  signal Summand1xD : unsigned(width-1 downto 0);
  signal Summand2xD : unsigned(width-1 downto 0);
  signal SumxD      : unsigned(width downto 0);

  type statetype is (idle, newData, calculate);
  signal StatexDP, StatexDN : statetype;

  signal CounterxDP, CounterxDN : natural range 0 to 2*width -1;


begin  -- Behavioral

  -- Input registers
  --------------------------------------------------------------------------------

  -- Shift register for input A
  InpAReg: process (ClkxCI, RstxRBI)
  begin
    if RstxRBI = '0' then                -- asynchronous reset (active low)
      InputAxDP <= (others => '0');
    elsif ClkxCI'event and ClkxCI = '1' then  -- rising clock edge
      if InputRegEnxS = '1' then
        InputAxDP(width-1 downto 0) <= unsigned(InputAxDI);
      elsif ClearRegxS = '1' then
        InputAxDP <= (others => '0');
      elsif ShiftEnxS = '1' then
        InputAxDP <= InputAxDP srl 1;  -- the MSB is padded with 0
      end if;
```

```vhdl
    end if;
end process InpAReg;


-- Register for input B
InpBReg: process (ClkxCI, RstxRBI)
begin
  if RstxRBI = '0' then            -- asynchronous reset (active low)
    InputBxDP <= (others => '0');
  elsif ClkxCI'event and ClkxCI = '1' then  -- rising clock edge
    if InputRegEnxS = '1' then
      InputBxDP <= unsigned(InputBxDI);
    elsif ClearRegxS = '1' then
      InputBxDP <= (others => '0');
    end if;
  end if;
end process InpBReg;



-- Bitwise multiplication of one bit of input A with the whole input B
------------------------------------------------------------------------------

AndGate: process (InputAxDP, InputBxDP)
begin
  for i in width-1 downto 0 loop
    Summand1xD(i) <=  InputAxDP(0) and InputBxDP(i);
  end loop;
end process AndGate;



-- Adder
------------------------------------------------------------------------------
SumxD <= resize(Summand1xD, width+1) + resize(Summand2xD, width+1);



-- Register for summation
------------------------------------------------------------------------------

SumReg: process (ClkxCI, RstxRBI)
begin
  if RstxRBI = '0' then            -- asynchronous reset (active low)
    Summand2xD <= (others => '0');
  elsif ClkxCI'event and ClkxCI = '1' then  -- rising clock edge
    if ClearRegxS = '1' then
      Summand2xD <= (others => '0');
    else
      Summand2xD <= SumxD(width downto 1);
    end if;
  end if;
end process SumReg;



-- Output assignment
------------------------------------------------------------------------------
OutputxDO <= std_logic(SumxD(0));
```

```vhdl
-- Controller FSM
--------------------------------------------------------------------------------

-- Combinational process
FSMcomb: process (StatexDP, StartCalcxSI, CounterxDP)
begin
  -- default assignments
  InputRegEnxS   <= '0';
  ShiftEnxS      <= '0';
  OutputValidxSO <= '0';
  ClearRegxS     <= '0';
  StatexDN       <= StatexDP;
  CounterxDN     <= CounterxDP;

  -- nondefault transitions and assignments
  case StatexDP is
    when idle =>
      ClearRegxS <= '1';
      if StartCalcxSI = '1' then
        StatexDN <= newData;
      end if;

    when newData =>
      InputRegEnxS <= '1';
      StatexDN     <= calculate;

    when calculate =>
      ShiftEnxS      <= '1';
      OutputValidxSO <= '1';
      if CounterxDP = 2*width - 1 then
        CounterxDN <= 0;
        StatexDN   <= idle;
      else
        CounterxDN <= CounterxDP + 1;
      end if;

    when others =>
      StatexDN <= idle;
  end case;
end process FSMcomb;

-- State register
FSMseq: process (ClkxCI, RstxRBI)
begin
  if RstxRBI = '0' then              -- asynchronous reset (active low)
    StatexDP   <= idle;
    CounterxDP <= 0;
  elsif ClkxCI'event and ClkxCI = '1' then  -- rising clock edge
    StatexDP   <= StatexDN;
    CounterxDP <= CounterxDN;
  end if;
end process FSMseq;

end architecture Behavioral; -- of ShiftAddMult
```

```vhdl
-- Mission: See associated testbench file.
-- Functionality: This entity contains the functional reference of the
-- testbench. The reference picks up expected responses from an ASCII file.
-- Platform: Modelsim 6.0
-- Company: Microelectronics Design Center, ETH Zurich.
-- Authors: Thomas Kuch
--------------------------------------------------------------------------

use std.textio.all;
library ieee;
use ieee.std_logic_1164.all;
use ieee.std_logic_textio.all;
use ieee.numeric_std.all;
use work.simulstuff.all;
use work.multTb1Pkg.all;

-- This architecture picks up the expected responses from a file
-- (the architecture uses the same entity as the MUT)
--------------------------------------------------------------------------
architecture Pickup of ShiftAddMult is

  -- the file is opened implicitly right here
  file exprespfile : text open read_mode is expresp_filename;

begin
  ExpResPickup : process
    variable ResponsexD : responseRecordType;
  begin

    PickupLoop : loop
      wait until (ClkxCI'event and ClkxCI = '1') or EndOfSimxS = true;

      -- leave the loop if there are no more stimuli left
      exit PickupLoop when EndOfSimxS = true;

      -- update expected response from file
      ResponsexD := GetExpectedResponseRecord(exprespfile);
      OutputxDO       <= ResponsexD.OutputxD;
      OutputValidxSO <= ResponsexD.OutputValidxS;

    end loop PickupLoop;

    file_close(exprespfile);  -- close the file
    wait;
  end process ExpResPickup;

end architecture Pickup;  -- of ShiftAddMult
```

Admittedly, the above testbench code is fairly large and complex for the modest MUT. This is because the code has been designed to scale to more complex MUTs and because it includes all necessary facilities for file handling, clocking, response checking, and error reporting, and for producing a summary at the end of a simulation run.

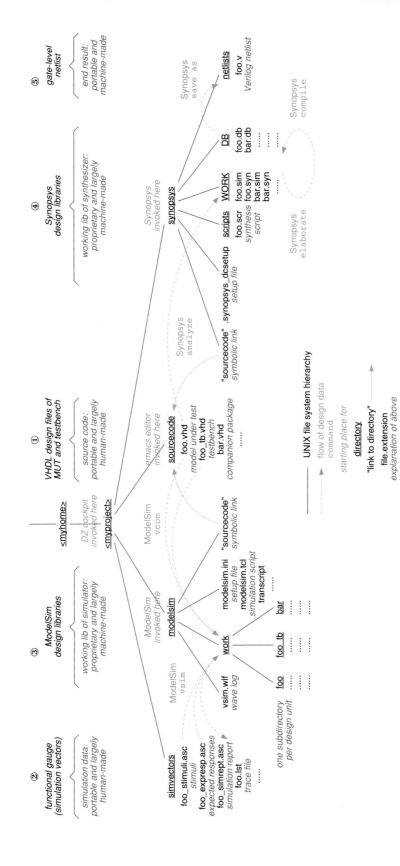

Fig. 4.23 A practical directory organization.

A TESTBENCH OF GRADE 2

In a grade 2 simulation, the testbench includes extra protocol adapters to translate between word-level stimulus/response pairs and cycle-true bit-level data. Overall organization follows fig.3.15. Source code is distributed over six design files.

- `multtb2.vhd`: Testbench including two VHDL processes for stimulus application and response acquisition; also configures simulation as file-based or golden-model-based.
- `multtb2pkg.vhd`: A collection of model-specific declarations and subprograms.
- `protocoladapter.vhd`: The two protocol adapters for the MUT.
- `shiftaddmult.vhd`: The MUT, a sequential shift-and-add multiplier (same as for grade 1).
- `mult.vhd`: Three entities that respectively
  - instantiate the MUT along with its two protocol adapters,
  - serve as a golden model (of purely combinational nature!),
  - read the expected responses from a file.
- `simulstuff.vhd`: Same as always, see above.

Again, the code is made available for download instead of being reprinted here.

4.9.5 Working with VHDL tools from different vendors

Making VHDL tools from different EDA vendors cooperate smoothly can be difficult because each tool has its own preferences on where and in what format design data should be stored. A well-defined directory organization greatly simplifies data management during design iterations, engineering change orders, tapeout, backup, reuse, and the like.

> Repositories for sourcecode, for simulation input data, for simulation output, for tool-specific intermediate data, and for finished netlists are to be kept apart.

Fig.4.23 suggests a directory structure for those who want or need to use ModelSim for simulation and Synopsys Design Compiler for synthesis.

Chapter 5

The Case for Synchronous Design

5.1 | Introduction

Experience tells us that malfunctioning digital circuits and systems often suffer from **timing problems**. Symptoms include

- Bogus output data,
- Erratic operation, typically combined with a
- Pronounced sensitivity to all sorts of variabilities such as PTV and OCV.

Erratic operation often indicates that the circuit operates at the borderline of a timing violation. Searching for the underlying causes not only is a nightmare to engineers but also causes delays in delivery and undermines the manufacturer's credibility.[1]

Observation 5.1. *To warrant correct and strictly deterministic circuit operation, it is absolutely essential that all signals have settled to a valid state before they are admitted into a storage element (such as a flip-flop, latch or RAM).*

This truism implies that all combinational operations and all propagation phenomena involved in computing and transporting some data item must have come to an end before that data item is being locked in a memory element. Data that are free to change their values at any time are dangerous because they may give rise to bogus results and/or may violate timing requirements imposed by the electronic components involved.[2] Hence the need for regulating all state changes and data storage operations.

---

[1] Malfunctioning that occurs intermittently or that depends on minor variations of temperature, voltage, signal waveforms, and similar circumstances makes debugging extremely painful. The fact that one can never be sure whether simulation accuracy suffices to predict transient waveforms, a subcircuit's reactions to a marginal triggering condition, and other details with sufficient precision does not help either.

[2] Timing requirements are meant to include setup and hold conditions, minimum clock high and low times, and maximum clock rise and fall times. All these quantities are explained in section A.6.

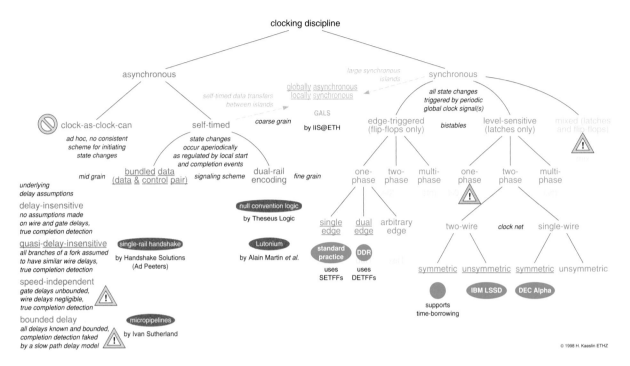

Fig. 5.1 Family tree of clocking disciplines (simplified). Underlined are those options that this author considers to be safe and economically viable in the context of VLSI, provided they are implemented correctly.

Many schemes for doing so have been devised over the years, see fig.5.1. From a conceptual perspective, we must distinguish between two diametrically opposed alternatives, namely synchronous clocking and self-timed operation. A third category that includes all unstructured ad hoc clocking styles — occasionally referred to as "clock-as-clock-can" in this text — is not practical except for the smallest subcircuits perhaps.

The present chapter first introduces and compares these approaches before commenting on why synchronous clocking is considered to be the best choice for staying clear of timing problems in the context of digital VLSI. Several general design rules are explained. Further details such as what bistables to use, how to clock them, and how to distribute the clock signal(s) over a circuit (or a clock domain) will be the subject of chapter 6.

5.2 | The grand alternatives for regulating state changes

5.2.1 Synchronous clocking

Definition 5.1. *A circuit or subcircuit is said to operate synchronously iff __all__ data storage operations and, hence, __all__ state transitions, are restricted to occurring periodically at precise moments of time that are determined by a special signal referred to as the* **clock***.*

A single clock that drives a set of flip-flops is the most straightforward pattern, but the above definition is meant to include circuits where bistables are being driven by a pair of complementary clock signals, e.g. CLK and $\overline{\text{CLK}}$. The same also applies to multi-phase clocks and even to multiple clocks of distinct frequencies, provided those subclocks are all locked to one primary clock signal. State changes are effectively restricted to occurring synchronously to the primary clock as long as all driving clocks maintain fixed frequency and phase relationships.[3]

Definition 5.2. *A **clock domain** or, which is the same thing, a **synchronous island** is a (sub)circuit where all clock signals maintain fixed frequency and phase relationships because they are derived from a common source.*

A clock domain may be confined to one subblock on a chip or may extend over several chips or even printed circuit boards. Any line that separates two distinct clock domains is referred to as a **clock boundary**. As stated before, most designs make do with a single clock signal per domain, but there are exceptions to this rule.

5.2.2 Asynchronous clocking

Definition 5.3. *A circuit or subcircuit works **asynchronously** when some or all of the memory elements therein are permitted to change their states independently from a global reference.*

Asynchronous circuits are easily identified by the presence of

- Unclocked bistables (SR-seesaw, Muller-C, MUTEX, etc.),
- Zero-latency feedback loops, e.g. as part of
- Asynchronous state machines (ASM),
- Logic gates other than buffers and inverters in clock nets,[4]
- Gated asynchronous (re)set signals,
- Logically redundant circuitry for hazard suppression,
- Imposed delays, ring oscillators,
- One-shots, pulse shapers, and similar subcircuits.

Examples of such constructs that never appear in synchronous designs are given in fig.5.2b and discussed in more detail in section 5.4.3. A multitude of clock signals, clock domains, and clock boundaries is another indication of asynchronous circuit operation.

5.2.3 Self-timed clocking

A more recent addition to the category of asynchronous circuits relies on self-timing throughout.

Definition 5.4. *An asynchronous circuit or subcircuit is said to operate in a **self-timed** manner when all state transitions get regulated by local start and completion events on a per case basis.*

---

[3] This is the case when some primary clock is being subdivided by way of some counter circuitry in order to derive one or more slower clocks. The converse, that is clock frequency multiplication with the aid of a phase locked loop (PLL), also results in multiple signals with fixed frequency and phase relationships.

[4] Gated clocks are an exception, please refer to section 6.5 for advice.

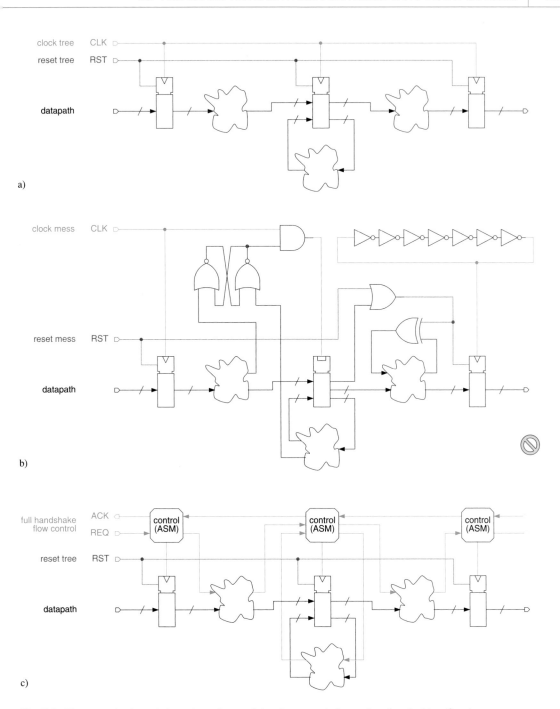

Fig. 5.2 Toy examples intended to give a flavor of the three grand alternatives for clocking. Synchronous clocking (a), clock-as-clock-can (b), and self-timed clocking (c).

The guiding principle is that no storage register is allowed to accept any new data from its predecessor before its successor has safely acquired the data item currently stored in the register. Whether the data items are subject to combinational operations while travelling from one register to the next or not is immaterial.

As opposed to synchronous clocking, there is no global signal that would trigger operations, which explains why this scheme is also known as clockless logic. Instead, the various building blocks in the circuit coordinate their activities by way of handshake protocols, which necessitates an on-going mutual exchange of status information. There is no way for an external observer to tell when a given computational step or data transfer is to happen because everything occurs in an entirely data-driven and, therefore, aperiodic way. The operation of a self-timed circuit proceeds at its own speed and automatically adjusts to PTV conditions.

Self-timed schemes come in numerous varieties [134] [135] [136] that essentially differ in how status information is actually communicated and in the assumptions on circuit delays that underlie protocol and circuit design.

5.3 | Why a rigorous approach to clocking is essential in VLSI

The propagation of new data values through a digital network gives rise to transient phenomena within the network itself and at the output. Let us briefly review the impact of spurious events on the functioning of digital circuits to prepare the ground for a subsequent discussion of clocking schemes and circuit design styles.

5.3.1 The perils of hazards

Hazard and glitch are two names for unwanted transients on binary signals. How they originate is examined in appendix A.5, key findings are as follows:

- Be prepared to observe glitching at the output of any combinational circuit with the sole exception of fanout trees. This is because hazards can originate
  (a) if two or more inputs change at the same time — or almost so —, or
  (b) if the combinational logic includes reconvergent fanout paths.

- Glitches are difficult to predict as their manifestation — in terms of waveform, amplitude, and duration — depends on many low-level details unknown at synthesis time, including placement, wiring, detailed layout, gate and interconnect delays, crosstalk, PTV conditions, and on-chip variations (OCVs). A minute departure may determine whether a hazard manifests itself as a rail-to-rail excursion, becomes visible as a runt pulse, or results in no tangible manifestation at all.

It is prudent to assume that any combinational network gives rise to hazards unless one has proof to the contrary. To determine whether a network actually develops glitches or not requires a detailed analysis of its transistor-level circuitry, timing parameters, layout parasitics, input waveforms, and the like. Manual elimination of hazards is a painful and unprofitable exercise as only small networks are amenable to analysis. Imposing narrow delay bounds on signal propagation paths is not compatible with automated synthesis and place-and-route (P&R) of VLSI circuits where interconnect delays often dominate over gate delays.

Observation 5.2. *All clocks, all asynchronous reset and presets, all write lines of asynchronous RAMs plus any other signal that might trigger a state transition in a memorizing subcircuit must be kept free of hazards under all circumstances.*[5]

There is an abundance of catastrophic failures that could result if this rule were violated:

- Unwanted state transitions in any kind of finite state machine.
- Multiple registration of a single event by a counter.
- Erroneous activation of an edge-triggered interrupt request line of a microprocessor.
- Unplanned-for return of a sequential (sub)circuit to its start state.
- Storage of bogus data in a register or a RAM.
- Data losses or duplications during data transfer operations.
- Deadlocks in asynchronous communication protocols (such as handshaking).
- Marginal triggering and, hence, metastable behavior of bistables.[6]

Data, address, status, control, and other signals not capable of sparking off a state change without the intervention of a clock are not affected. Similarly, combinational subcircuits are not normally sensitive to hazards because all transient effects are reversible and eventually die out.[7] Glitches also cause no problem when driving sluggish peripheral equipment such as indicator lamps, electromechanical relays, teletypes, and the like.

5.3.2 The pros and cons of synchronous clocking

There are ten essential benefits that are shared by all synchronous clocking disciplines.

1. Hazards do not compromise functionality. Clock and asynchronous reset are the only two signals that must be kept free of hazards under all circumstances. Doing so is easy, strictly limiting distribution networks to fanout trees suffices.

2. As no timing violations ever occur within a properly designed synchronous circuit, there is no chance for inconsistent data, marginal triggering, and metastability to develop.

3. Immunity to noise and coupling effects is maximum because all nodes are allowed to settle before any storage operations and state changes occur.

---

[5] Be warned that state-change-triggering inputs are not always identified as such in the documentation of commercially available components or library elements. They sometimes hide under unconspicuous names such as "chip enable" (CEB), "write enable" (WEB), "interrupt request" (IRQ), and "strobe", to name just a few.

[6] The term "metastability" refers to an unpredictable behavior of a memory element that may, or might not, result from violating its timing conditions, see chapter 7 for details.

[7] There are a few exceptions, however, where hazards are unacceptable in spite of the memoryless nature of the subsystems involved. These include:
- Digital modulators and other circuits where signals are required to follow a well-defined waveform or spectrum.
- Output enable signals where hazards could occasion transient drive conflicts, thereby leading to exaggerated crossover currents, needless power dissipation, and excessive ground bounce.
- Electronically controlled power drives and power converters where unforeseen current spikes are likely to cause permanent damage.

4. All timing constraints are one-sided. For a circuit to function correctly, any timing quantity is bounded either from above (such as the longest propagation delay, for instance) or from below (such as the contamination delays). Two-sided constraints do not exist.[8]

5. Together, the four above properties warrant deterministic behavior of circuits independently from low-level details.[9] Synchronous designs do not rely on delay tuning in any way, what matters for functional correctness are the data operations at the RTL level exclusively. This argument cannot be overestimated in view of

   - Automatic placement, routing, and physical design verification,
   - Automatic HDL synthesis, logic optimization, clock tree generation, and rebuffering,
   - Automatic insertion of test structures,
   - Reusing a HDL model or a netlist in multiple designs, and
   - Retargetting a design from one cell library and/or fabrication process to another (e.g. from FPL to a mask-programmed IC, or vice versa).

6. A systematic, modular, and efficient approach to design, test, and troubleshooting is impossible unless history-dependent behavior is strictly confined to clocked storage elements. More specifically, synchronous operation makes it possible to separate functional verification from timing analysis and to take advantage of automata theory and related concepts.

7. There is no need for any redundant circuitry to suppress hazards. Standard synthesis tools are geared towards minimizing circuit complexity while meeting performance constraints; they are not concerned with transients and their elimination.

8. The operations that are to be carried out in each clock cycle can be stated and collected at compile time, thereby opening a door for cycle-based simulation techniques that are more efficient when circuits grow large. Asynchronous circuits, in contrast, are entirely dependent on event-driven simulation.

9. Established methods for circuit testing (such as fault grading, test vector generation, and the insertion of test structures) start from the assumption of synchronous operation. What's more, almost all test equipment is designed accordingly.

10. Synchronous clocking makes it possible to slow down and even to suspend circuit operation in any state and for an arbitrary lapse of time,[10] which greatly facilitates the tracing of state transitions, data transfers, protocol sequences, and computation flow when a malfunctioning circuit must be debugged. The ability to operate synchronous circuits in speed-limited environments is often welcome for prototyping purposes.

Undeniably, synchronous circuit operation also has its drawbacks.

1. Performance is determined by the worst rather than by the average delay over all data.[11]

---

[8] With the exception of one-phase level-sensitive operation, which is considered impractical, see section 6.2.7.

[9] This is to say that buffer sizing, library changes, parameter variations (PTV and OCV), physical arrangement, layout parasitics, etc. are likely to impact maximum clock rate, I/O timing, power dissipation, and other quantitative figures of merit, but not a circuit's functionality.

[10] Unless capacitive data storage is involved such as in DRAMs or in dynamic CMOS logic.

[11] A workaround is to be presented in footnote 18.

2. There may be unnecessary power dissipation because each register dissipates energy in each clock cycle regardless of the extent of state change. Yet, there exists a broad variety of techniques for lowering clock-induced power dissipation while maintaining overall synchronous circuit operation.[12]

3. Synchronous operation causes periodic surges in supply currents. This not only strains the power and ground nets but also entails electromagnetic radiation at the clock frequency and at higher harmonics.

4. Synchronization problems are unavoidable at the interface between any two clock domains.[13] However, similar problems arise wherever an asynchronous subsystem interfaces with a clock-driven environment such as a sampled data source or data sink.

5. Most synchronous clocking disciplines insist on tightly controlled delays within the clock distribution network. Special software tools that address this need during physical design make up part of all major VLSI CAD suites.

5.3.3 Clock-as-clock-can is not an option in VLSI

Unsafe circuits often emanate from obsolete or perfunctory design methodologies. This is particularly true for asynchronous circuits, the design of which is very demanding, both in terms of profound understanding and in terms of engineering effort. Sporadic timing violations and a pronounced sensitivity to delay variations are typical consequences. Although popular with digital pioneers, ad-hoc clocking schemes have become unacceptable because VLSI technology has changed the picture in the following way.

- Most asynchronous circuits must be considered fragile because their functional behavior critically depends on certain delay figures and, therefore, also on their layout arrangements. This makes it difficult to anticipate whether fabricated circuits will indeed behave as simulated as no simulation model is capable of rendering all effects that contribute to timing variations with perfect precision.

- Finding and correcting timing problems is difficult enough on a board in spite of the fact that designers have access to almost all circuit nodes and can add extra components for tuning delays. It is next to impossible on a monolithic chip that offers no such possibilities.

- Historically, emphasis was on making do with as few SSI/MSI packages as possible. The prime challenge today is first-time-right design, a few more logic gates do not normally matter.

- Early logic MOS and TTL devices were so slow that wiring-induced delays could be neglected altogether. As opposed to this, interconnect delays due to wiring parasitics tend to dominate over gate delays in VLSI, thereby making it impossible to predict delays from circuit diagrams.

- VLSI designers cannot afford delay tuning of a vast collection of signals. For the sake of productivity, they must rely on design automation as much as possible for logic synthesis, optimization, placement, routing, and verification. The higher productivity so obtained is at the expense of control over most implementation details.

---

[12] You may want to refer to sections 6.5, 8.2, and 9.2.2 for energy-conserving circuit design.

[13] This is the subject of chapter 7.

In conclusion, an industrial circuit designer concerned with design productivity, first-time-right design, and fabrication yield is well advised to follow the recommendations below.

Observation 5.3. *Ad-hoc approaches to clocking are no more than unfortunate leftovers from the early days of digital design that are incompatible with the requirements of VLSI. Instead, strive to make do with as few clock domains as possible and strictly adhere to one synchronous clocking discipline within each such domain.*

Let us bring this matter to an end with two quotes from experts

"Just say NO to asynchronous design!" [137]
"KISS those asynchronous-logic problems good-bye, Keep It Strictly Synchronous!"[14] [138]

5.3.4 Fully self-timed clocking is not normally an option either

As opposed to unsophisticated asynchronous clocking schemes, self-timed clocking follows a strict discipline, holds the promise of achieving better performance, and provides valuable hooks for improving on energy efficiency [139]. In comparison with synchronous clocking schemes, the notorious difficulty of domain-wide clock distribution is replaced by a multitude of local synchronization or arbitration problems, which in turn calls for a specific design methodology.

In spite of its theoretical benefits,

- The hardware and energy overheads associated with implementing handshaking or related request–acknowledge protocols all the way down to the level of logic gates,
- The difficulties of interfacing with clock-driven peripheries and test equipment,
- The absence of self-timed components in commercial cell libraries,[15]
- The shortage of adequate EDA support,
- The excruciating subtleties of the design process, and
- The lack of widespread know-how, together with
- The ensuing time-to-market penalty

have — so far — prevented fully self-timed logic from becoming a practical alternative.[16]

5.3.5 Hybrid approaches to system clocking

Current efforts are attempting to combine the best of both worlds into **globally asynchronous locally synchronous** (GALS) circuits where synchronous islands communicate via self-timed data exchange protocols. The usage of arbiters and pausable clocks is typical for GALS circuits. Such **heterochronous** architectures are being investigated as an alternative to overly large synchronous systems in search of improved energy efficiency, better performance, more manageable clock distribution, and facilitated design reuse [141] [142].

---

[14] K.I.S.S. originally was a slogan to improve the success rate of complex operations, "Keep It Simple, Stupid".

[15] Such as the Muller-C and the mutual exclusion elements (MUTEX) explained in appendix A.4. Dual-rail encoding further necessitates special circuits for logic gates and bistables that are not found in regular cell libraries.

[16] Industrial interest is documented by startups such as Theseus Logic Inc. and Handshake Solutions. The former has patented NULL Convention Logic (NCL), which, however, is penalized by an important overhead factor of 2 to 2.5 over traditional synchronous logic [140].

Table 5.1 | The grand alternatives for clocking compared.

| | Clocking discipline | | |
| | asynchronous | | synchronous |
Desirable characteristics	ad hoc	self-timed	
Fundamentals			
Immune to hazards	no	yes except for protocol signals	yes except for clock and reset
No need for hazard-suppression logic	no	yes	yes
One-sided timing constraints only	no	yes	yes
No marginal triggering during circuit operation	maybe, maybe not	yes	yes within synchr. island
Avoids timing problems at interfaces	no	no	no
Design process			
No particular library cells needed	yes	no	yes
Does without tightly controlled delays in logic and interconnect	no	yes except for local subcircuits	yes except for clock[a]
Circuit to function irrespective of logic and layout details	no	yes	yes
Functionality and timing separable	no	mostly yes	yes
Systematic and modular design methodology, reuse facilitated	no	yes	yes
Matches with prevalent flows and tools	no	no	yes
Arbitrarily slow operation supported for debugging purposes (step-by-step)	no	no	yes[b]
Periodicity of circuit operation			
All signals to settle before clocking	no	no	yes
Non-periodic "random" supply current	more or less	yes	no
Supports cycle-based simulation	no	no	yes
Works with existing test equipment	no	no	yes
Figures of merit of final circuit			
Good area efficiency	sometimes yes, more often no	only if overhead remains modest	yes
Better than worst-case performance	more or less	yes	no
Good performance in practice	in particular applications	debatable	yes
Good energy efficiency	in particular applications	yes if overhead remains modest	yes if designed accordingly

[a] Skew-tolerant schemes are available, see section 6.2.
[b] Unless DRAMs or dynamic CMOS logic is being used.

A different strategy termed **mesochronous clocking** is to distribute a global clock signal without much concern for skew. Specially designed local synchronizer circuits are then used to sample data at multiple points in time, to detect synchronization failures, and to retain valid data only [143]. A related idea is to insert tunable delay lines within the clock distribution network and to calibrate them automatically at startup time such as to make all blocks work synchronously together [144].

For the time being, most this must be considered research, however, as industry is reluctant to embrace unproven concepts and design flows.

5.4 | The dos and don'ts of synchronous circuit design

Synchronous operation essentially rests on the two guiding principles to be presented next.

5.4.1 First guiding principle: Dissociate signal classes!

While digital VLSI designers must devise circuits that function in a predictable and dependable way, they cannot afford to study the transient behavior of every single circuit node in detail. What is needed is a robust and well-understood clocking discipline that, when properly implemented, warrants correct circuit timing under all operating conditions.

From the background of our findings on transients in digital circuits in section 5.3, there is a fairly obvious solution. Rather than attempting to suppress hazards here and there — which is symptomatic for clock-as-clock-can design — the set of acceptable circuit structures is voluntarily and consistently restricted to those that
• do not let hazards originate in clock and in asynchronous (re)set nets, and that
• tolerate hazards on all other signals with no impact on functionality whatsoever.

The most efficient way to prevent dangerous hazards from coming into existence is to shut out all signals that might possibly prompt a state change from participating in logic operations. Table 5.2 begins by distinguishing between hazard-sensitive and hazard-tolerant signals before classifying signals further as a function of their respective roles in a circuit.

Reset signals cause a sequential circuit to fall into some predetermined start state without the intervention of a clock. Their effect is immediate, unconditional, and always the same. Practically speaking, this includes all **asynchronous reset** and set inputs present on many bistables, but none of the synchronous initialization signals.

Clock signals are in charge of sparking off all regular transitions of a sequential circuit from one state to the next, but have no influence on what that next state will be.

Information signals is a collective term for all those signals that contribute to deciding what state a circuit is to assume in response to an active clock edge and/or what output that circuit shall produce.

Table 5.2 | Taxonomy of signals within a synchronous island. The bottom part suggests a naming convention for signals in HDL models, schematics, and netlists detailed further in appendix 5.7.

Class	Electrical signals				
	Reset signal	Clock signal(s)	Information signals		
determines	when to (re)enter the start state	when to move from one state to the next	what state to enter next and/or what output to produce		
Hazards	inadmissible	inadmissible	harmless		
Role during simulation	general model wake up	general model wake up	evaluated at model activation time, no wake up of memorizing models		
Subclass	—	—	Functional signals		Test signals
serves to			implement the desired functionality		improve observability and controllability
Members	Asynchronous (re)set	Clock(s)	Status, control	Data, address	Block isolation, scan path(s)
Switching is	Examples and their identification by way of naming and color code				
synchronous to local clock	—	CLK	many	many	TST, SCM, SCI, SCO
Class_char Color		C green	S blue	D black	T yellow
asynchronous to local clock	RST, SET	—	any input prior to synchronization		
Class_char Color	R red		A orange	A orange	A orange

We further distinguish between functional signals and test signals, with the first subclass largely outnumbering the second. Functional signals comprise data, address, control, and status signals, which together implement the desired functionality. Test signals get added on top during the design process to improve circuit testability. From this perspective, **synchronous clear** and load inputs are nothing else than particular control signals.

We now stipulate

Observation 5.4. *Synchronous circuits boast a clear-cut separation into signals that decide on <u>when</u> state transitions and output changes are to take place, and others that determine <u>what</u> data values to output and what state transitions to carry out, if any. As a consequence, (asynchronous) reset signals, clock signals, and information signals never mix. Combinational operations (other than unary negation) are strictly confined to information signals.*

5.4.2 Second guiding principle: Allow circuits to settle before clocking!

Another essential precondition for safe operation follows immediately from observation 5.1, which implies that any combinational network shall be allowed to settle to its steady-state condition before the emanating output signals are clocked into some memory device. In the context of a (sub)circuit driven by a single clock, this amounts to the following requirement.

Observation 5.5. *Synchronous designs must be operated with a clock period long enough to make sure that all transient effects have died out before the next active clock edge instructs registers and other storage devices to accept new data.*[17]

While this indeed prevents hazards on information signals from having any effect on the circuit's (next) state, it comes at a cost. The length of the computation period, and hence also the clock period, are bounded from below by the slowest signal that travels between any two consecutive registers (longest propagation path, most penalizing set of data, slowest operating condition). The clock rate must be chosen to be such as to conform with the worst-case timing, thereby denying the possibility of taking advantage of more favorable situations, no matter how frequently these might occur.[18]

5.4.3 Synchronous design rules at a more detailed level

In a certain sense, observations 5.4 and 5.5 form the "constitution" of synchronous circuit design from which many "laws" of more specific nature can be easily derived given some particular situation. A couple of them will be explained and illustrated next, but we do not intend to present all such rules here as many of them have a rather narrow focus.[19]

HDL synthesis has greatly simplified things in that designers no longer need to manually assemble subfunctions from primitive gates and bistables. Yet, the responsibility for writing circuit models that properly synthesize to robust synchronous circuits is still that of the engineer. To do so, he must be capable of recognizing and correcting dangerous constructs.

UNCLOCKED BISTABLES PROHIBITED

Observation 5.6. *As opposed to flip-flops and latches, unclocked bistables such as seesaws, snappers, and Muller-C elements do not qualify as storage elements in synchronous designs.*

---

[17] A more precise, quantitative formulation of this and other constraints will be given in section 6.2.

[18] This is precisely the starting point for **speculative completion** that borrows from self-timed execution while preserving strictly synchronous circuit operation. The idea is based on the statistical observation that only few data vectors do indeed exercise the longest path in ripple-carry adders and related arithmetic circuits. The clock frequency is chosen such that the worst-case delay fits into two clock periods instead of one. A fast auxiliary circuit monitors carry generation and carry propagation signals in order to determine whether the current calculation involves a short or a long ripple-carry path. System operation is made to continue immediately if the adder is found to settle before the end of the first cycle. If not so, system operation is stalled for one extra clock cycle. Please refer to [145] for a more detailed account of this out-of-the-ordinary technique.

[19] More rules are included in appendix C. [146] is a valuable reference on safe digital design; the textbook gives a list of nine detailed rules but implicitly excludes all clocking disciplines other than edge-triggered one-phase clocking, which is overly restrictive.

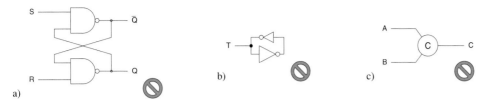

Fig. 5.3 Examples of bistables that do not qualify for data storage in synchronous designs. Seesaw (a), snapper (b), and Muller-C element (c).

Unclocked bistables do not differentiate among reset, clock, and information signals, which renders them vulnerable to hazards and is contrary to the postulates of observation 5.4. The usage of naked SR-seesaws is strongly discouraged in spite of the fact that certain latch and flip-flop designs include them as subcircuits.[20] Snappers are not for storing data; their sole legal usage is to prevent the voltage from drifting away while a three-state node waits in a high-impedance condition. The Muller-C element is a building block of self-timed circuits.

ZERO-LATENCY LOOPS PROHIBITED

Zero-latency loop is just another name for a circular signal propagation path in a network of combinational subcircuits, see fig.5.4 for examples. The problem with such circuits is that it is not possible to determine a logic value for the output even if all input values are known. The underlying reason is that some combinational function makes reference to its own result.

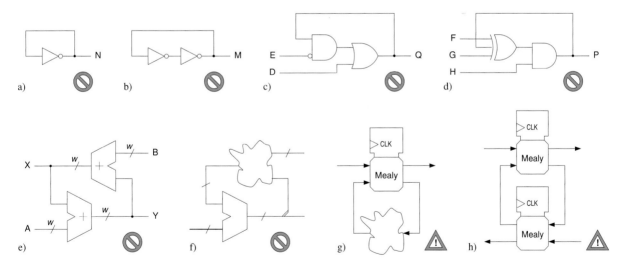

Fig. 5.4 A few examples of undesirable zero-latency loops.

---

[20] Refer to section 8.2 for detailed circuit diagrams.

Variable N in fig.5.4a, for instance, can settle neither to 0 nor to 1 without causing a logic contradiction. A physical circuit may then oscillate or assume a precarious equilibrium. The circuit of fig.5.4b is free of contradictions and quickly locks into one of two stable states of equilibrium, but there is no way to predict M without knowledge of its original value.

For a more substantial example consider the cross-coupled adders of fig.5.4e. Inputs A, B and outputs X, Y all have the same word width $w \geq 2$ (carry bits are left aside for simplicity).

$$X = B + Y \qquad\qquad Y = A + X \qquad\qquad\qquad (5.1)$$

Behavior depends on the input values applied. If $A + B = 0$, equations reduce to $X = X$ and $Y = Y$, which implies that X and Y preserve their values. Although built from combinational gates exclusively, a physical circuit would lock into one of many possible states and remain there as long as the inputs are kept the same. Yet, a circuit with memory but no distinction between <u>when</u> and <u>what</u> signals does not conform with the dissociation principle of synchronous design.

Conversely, if $A + B \neq 0$, one has $X = (B + A) + X$ and $Y = (A + B) + Y$. A physical circuit would be bound to add and switch at a frenzied pace forever. Failure to settle to a steady-state is against the second guiding principle of synchronous operation, however. What's more, none of the circuits shown behaves as one would expect from logic and arithmetic subfunctions.

HDL code can include circular references without the author being aware of them. Examples often involve a Mealy machine plus a datapath, glue logic, or some other surrounding circuitry, see fig.5.4g. A feedback path in the logic can then combine with a through path in the finite state machine (FSM) to form a circular signal propagation path. Note that not all FSM input and output bits need to be part of the loop, one of each suffices. Another particularity is that such paths may open and close as a function of state and input value.[21] While this is likely to confine malfunction to just a few situations, it also renders debugging particularly difficult.

Observation 5.7. *By prohibiting zero-latency loops, gate-level circuits are effectively prevented from oscillating, from locking into unwanted states, and from otherwise behaving in an unpredictable way. In synchronous designs, each circular path must include one register at least.*[22]

MONOFLOPS, ONE-SHOTS, EDGE DETECTORS, AND CLOCK CHOPPING PROHIBITED

Monoflop, monostable multivibrator, one-shot, and edge detector are names for a variety of subcircuits that have one thing in common: they output multiple edges or even multiple pulses for a single transition at their input. A clock chopper does the same with a clock signal. Any of these can be built from a delay line in conjunction with reconvergent fanout, though other implementations also exist. See fig.5.5 for an example and observe that the exact nature of the contraption gets determined by the logic operation at the point of reconvergence.

---

[21] A perfidious example is given in appendix B.2.3.

[22] Exceptions exist as the absence of zero-latency loops is a sufficient but not a necessary condition for memoryless behavior. A notable example of a circuit that settles to well-defined steady states for arbitrary stimuli in spite of a zero-latency loop is the end-around-carry used in 1's complement and sign–magnitude adders [66]. Proving that a feedback circuit predictably exhibits combinational behavior under all circumstances may be a major effort, though.

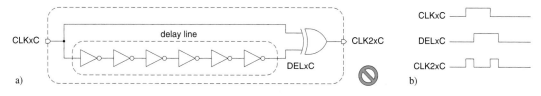

Fig. 5.5 Clock chopper circuit (a) with waveforms (b).

Unwary designers are sometimes tempted to resort to clock chopping when they have to incorporate blocks, e.g. unclocked RAMs, that impose timing constraints incompatible with normal synchronous circuit operation. Frequency multiplication is another usage. Once again, the requirements of observation 5.4 are obviously violated. Major problems are testability, ill-defined pulse width, critical place and route, timing verification, and vulnerability to delay variations.

Observation 5.8. *Monoflops, one-shots, edge detectors, and especially clock choppers are absolute no-nos in synchronous design.*

Clock and reset signals to be distributed by fanout trees

Clock and asynchronous (re)set signals can trigger a state transition at any time. A hazard on a signal of either of these two classes is, therefore, very likely to lead to catastrophic failure.

Observation 5.9. *No clock and asynchronous (re)set signals shall ever participate in any logic operation other than the unary operation of taking the complement. Instead, any signal capable of inducing a state change must be distributed by a fanout tree exclusively.*

This is because fanout trees are a priori known not to generate hazards.[23] Binary logic operations such as NOR, AND, and XOR, are strictly prohibited. Buffers and inverters are the sole logic gates that are acceptable in clock and in reset networks.

Beware of unsafe clock gates

Clock gating implies enabling and disabling state transitions by suppressing part of the clock edges depending on the present value of some control signal. Doing so with the aid of an AND or some other simple gate spliced into the clock net as shown in fig.5.6 has always been a poor technique because it is unsafe and against the principle of dissociation. However, as conditional clocking has seen a renaissance and has indeed become a necessity in the context of low-power design, safer ways of doing so will be studied in great detail in section 6.5.

No gating of reset signals

Very much like a gated clock, we speak of a **gated reset** when the asynchronous reset or preset input of a latch, flip-flop, register, counter, or some other state-preserving subcircuit participated in

---

[23] No hazard can arise in a single-input network that is free of reconvergence. Please refer back to section 5.3 or see section A.5 for a more complete rationale.

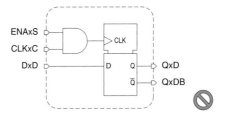

Fig. 5.6 Unsafe D-type flip-flop with enable resulting from malformed clock gating.

combinational operation with some other signal. This anachronistic practice of using a reset signal for any other purpose than for overall circuit initialization is against the principles of synchronous design and holds serious dangers.

Warning example

Specifications had asked for a modulo-15 counter, i.e. a counter that steps from state 0 to 14 before returning to 0. Counter slices were available from the standard cell library. The designer elected to start with a modulo-16 counter and to skip the unutilized state 15 by activating the asynchronous reset mechanism whenever the counter would enter that state. He devised the circuit of fig.5.7a, which behaved as expected during logic simulation. When the IC was put into service, however, the circuit missed out state 14 and went from state 13 to 0 directly. Why?

What the designer had overlooked was the transient behavior of the decode logic he had created. He implicitly started from the assumption that all inputs to the 4-input NAND would switch at the same time. In reality, all sorts of imbalances contributed to making them arrive staggered in time. In the occurrence above, the LSB was slightly delayed with respect to the other bits, which made the 4-input NAND temporarily see a 1111 during the transition from 1101 to 1110, see fig.5.7b.[24] Further note that other delay patterns could equally well have caused the counter to return to state 0 from states 7 or 11.

The design might have developed yet another failure because the intended clearing of the counter slices in state 15 is by way of a feedback loop with latency zero. The reset condition comes to an end whenever the output of the fastest slice begins to flip. There is no guarantee that the impulse so generated is sufficiently long to clear all other bistables too. Some of the flip-flops might also be subject to marginal triggering and eventually fall back to their previous states.

☐

Observation 5.10. *The sole purpose of asynchronous (re)set inputs is to bring an entire circuit into its predefined start state. Do not gate them with information signals.*

---

[24] The phenomenon can be recognized as a function hazard.

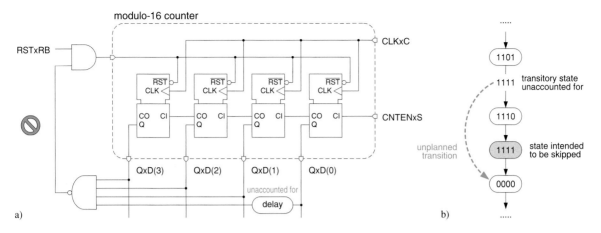

Fig. 5.7 Unsafe modulo-15 counter subcircuit. Schematic (a) and state graph (b).

It is interesting to study the motivations that lead people to expose themselves to the hazards of unsafe circuits such as the ones of figs 5.6 and 5.7. Two rationales are often heard:

- Some desired functionality was unavailable in a synchronous implementation from the target cell library. In the case of the above modulo-15 counter, the designer flatly preferred to misuse the asynchronous reset instead of figuring out how to add a synchronous clear/load to an elementary D-type flip-flop. The need for an enable/disable mechanism, for conditional clocking, and for data transfers across clock boundaries are further situations that tend to expose designers to the temptations of asynchronous design. HDL synthesis certainly has helped designers to stay away from such practices.

- The asynchronous implementation was believed to result in a smaller circuit, faster speed, and/or better energy efficiency than a synchronous alternative. While this may be true in some cases, the contrary has been demonstrated in many others. Keep in mind that any redundant circuitry necessary to generate multiple auxiliary clocks, to stretch impulses, to suppress unwanted glitches, or to make sure local delay constraints are satisfied does not come without its price either. In view of all the limitations cited in section 5.3.3, overall cost-effectiveness remains questionable to say the least.[25]

BISTABLES WITH BOTH ASYNCHRONOUS RESET AND PRESET INPUTS PROHIBITED

Some component and cell libraries include flip-flops that feature both an asynchronous set and an asynchronous reset. Synchronous design knows of no useful application for such subcircuits. Also, behavior is unpredictable and resembles that of a seesaw when S and R are deasserted simultaneously.

---

[25] You may want to consult section A.10, where more details on how to construct safe flip-flops with enable, with synchronous clear, with scan facility, and the like are given.

RESET SIGNALS TO BE PROPERLY CONDITIONED

Please note, to begin with, that the total load controlled by the global reset is on the same order of magnitude as that driven by the circuit's clock signal. It typically takes a large buffer or a buffer tree to drive that net with acceptable rise and fall times.

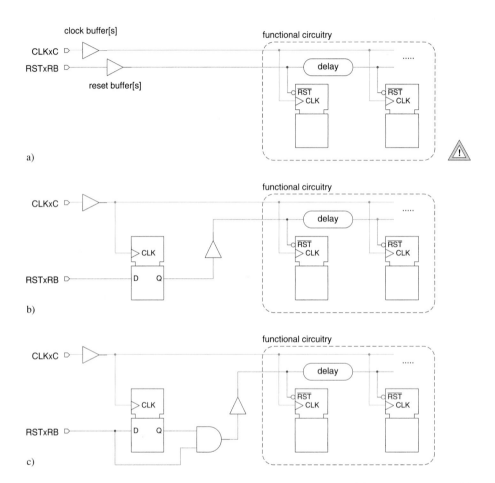

Fig. 5.8 Conditioning of global asynchronous reset signal. Fully asynchronous (a), fully synchronized (b), and unconditional application combined with synchronized removal (c).

A more specific peril associated with the usage of asynchronous (re)sets exists in applications with a **free-running clock**, that is where a reset impulse may occur at any time with respect to the driving clock, see fig.5.8a. If the circuit's asynchronous reset input gets deactivated near the active clock edge, then part of the bistables may be allowed to take up normal operation before the end of one clock cycle while others would remain locked until early in the next cycle. Going through the same efforts as with the clock distribution network to minimize **reset skew**, i.e. local differences in delays within the reset distribution network, is a rather costly option.

Incorporating the auxiliary subcircuit depicted in fig.5.8b solves the problem by synchronizing the reset signal to the on-chip clock before it gets distributed to the functional bistables. All bistables are thus guaranteed to get properly (re)set during the same clock cycle. Any transitions on the internal reset occur shortly after the active clock edge, which leaves the designer with ample time, i.e. almost one clock period, for distributing the signal across the chip. As a side effect, the circuit responds to the external reset signal with a latency of one clock cycle.

Yet, a new difficulty has been introduced because the reset operation now depends on the functioning of the clock subsystem. In the absence of a clock or in the presence of specific hardware faults, the circuit would fail to settle to a defined start state. Drive conflicts, static power dissipation, inconsistent output data, destructive overwrite of nonvolatile memories, and other operating conditions unsafe for the surrounding system may be the undesirable consequences. In the design of fig.5.8c, a simple combinational bypass makes sure that a reset is immediately and unconditionally brought to bear. Deactivation, in contrast, takes place at well-defined moments of time. An equivalent circuit alternative does away with the AND-gate by connecting the external reset input to a synchronizer flip-flop with an asynchronous reset terminal.

Keep in mind that all three designs impose conditions on the relative timing of RSTxRB and CLKxC. Violating them may cause marginal triggering in one or more of the flip-flops. Also remember that asynchronous (re)set inputs are very sensitive to glitches and noise.

PAY ATTENTION TO PORTABLE DESIGN

IC designers time and again face the problem of porting a design from one implementation platform to another, e.g. when they

- Port a design from an FPGA to a cell-based ASIC, or vice versa,
- Upgrade to a more up-to-date target process or library,
- Incorporate a third-party circuit block into a larger design, or
- Accept a heritage design for integration.

While it is possible to fine-tune almost any design for some given technology, employing asynchronous techniques can be disastrous when it comes to porting a design from one target platform to another. Think ahead and design for portability in the first place.

Observation 5.11. *Accept that almost all VLSI designs are subject to porting during their lifetime and stick to the established rules of safe design and synchronous operation to render the porting smooth and cost-effective.*

Design rules other than those given in this chapter are to follow in sections 6.5, 8.5, and 13.4.2.

Conversely, when being proposed an existing design, do not accept it unseen. Be prepared to distillate the necessary functionality and reimplement it using a clocking discipline that is compatible with VLSI or your target PLD. Heritage designs on the basis of standard parts such as microprocessors and LSI/MSI circuits notoriously include the worst examples of asynchronous design tricks. Also note that on-chip memories often are at the origin of portability problems because of the many varieties being offered.

5.5 | Conclusions

- Among the three grand alternatives for clocking digital circuits, ad-hoc asynchronous operation has been found to be unsafe and inefficient in VLSI.

- While safe if implemented correctly, self-timed operation at the gate level entails an unacceptable overhead in terms of hardware and energy, and necessitates out-of-the-normal design methodologies, software tools, and library cells.

- Synchronous operation of large system chunks does away with almost all timing problems, results in efficient circuits, and is compatible with today's design automation flows and cell libraries. There hardly is a better choice for VLSI, especially when there is high pressure on tight schedules and high design productivity.

- Synchronous circuits exhibit a strict dissociation of signals into

  - One clock signal (possibly more of fixed frequency and phase relationship),
  - One asynchronous reset signal (optional),
  - An arbitrary number of information signals.

- HDL synthesis does not relieve designers of deciding about clocking disciplines and clock domains as it is possible to express any clocking discipline in an RTL circuit model.

5.6 | Problems

1. Fig.5.4c and d depict two feedback circuits, each built from a few logic gates. Establish their respective truth tables and discuss your findings.

2. The circuit of fig.5.4c behaves much like a latch where D acts as data input and E as enable input. Explain why this construct does not qualify as a latch in synchronous designs.

3. Come up with a synchronous implementation for the modulo-15 counter of fig.5.7. Compare the relative sizes of the two alternatives on the basis of the assumption that both counters are built from D-type flip-flops.

4. FireWire is the name of a serial bus for interconnecting computer and multimedia equipment. To facilitate the delimiting and recovering of individual bits from the data stream at the receiver end, data get conveyed one bit after the other coded using two peer signals (the fact that two differential signal pairs are actually being used does not matter in this context). The first signal termed "data" simply corresponds to the incoming data whereas its "companion" is to feature a transition at the boundary between any two adjacent bits iff the first signal does not. Thus, at either end of a FireWire link, a modulator circuit in the transmitter converts the incoming serial data stream into a data-plus-companion pair, while a demodulator on the receiving side is in charge of recovering the initial data from that signal pair. Design both a modulating and a demodulating circuit. What class is the automaton you must use?

5.7 | Appendix: On identifying signals

with contributions by N. Felber, R. Zimmermann, and M. Brändli

Many mistakes in hardware design can be traced down to minor oversights in specifications, datasheets, truth tables, HDL code, schematics, and the like. Misinterpretations are particularly likely to occur at the interfaces between different subsystems, subcircuits, and clock domains because correct operation implies mutual agreement on the meaning behind data formats, signal waveforms, and transmission protocols. A clear and unambiguous, yet simple, scheme for naming signals and for drawing diagrams greatly helps to avoid such problems, especially when working in a team. To be helpful, a notational convention must keep track of

- Signal class.
- Active level.
- Signaling waveform.
- Input, output or bidirectional.
- Present state vs. next state.
- Clock domain.

5.7.1 Signal class

A total of six signal subclasses have been identified in section 5.4 and catalogued in table 5.2. The same table suggests appending one of the characters below to make a signal's role evident from its name. Please refer to section 5.7.7 for syntactical details.

- ○ R asynchronous (re)set,
- ○ A any other signal subject to asynchronous switching,
- ○ C clock,
- ○ S status or control,
- ○ D data, address or the like,
- ○ T test.

Colors in schematic diagrams and HDL source code

Wires colored in a meaningful way greatly expedite the understanding of large schematic diagrams and render many potential problems immediately visible.[26] Let us cite five examples on the basis of the coloring scheme of table 5.2:

- Any clock signal (green) routed through a logic gate or other combinational subcircuit with two or more inputs draws attention to a potentially unsafe clock gating practice.

---

[26] The coloring scheme proposed in fig.5.2 is intended for usage in schematic diagrams that are drawn on light backgrounds. On dark backgrounds, white must be substituted for black. Also, swapping blue and yellow reestablishes the original idea of fading all test-related signals, since they tend to distract from a circuit's functionality.

- A register with neither an asynchronous (re)set wire (red) nor a synchronous clear (blue) attached indicates the subcircuit is in need of a special homing sequence for initialization.
- Any combinational logic that drives an asynchronous (re)set line (red) indicates (re)set inputs are exposed to hazards as a consequence of their being misused for functional purposes.
- Any signal emanating from a foreign clock domain (orange) that drives combinational logic points to a lack of synchronization.
- The absence of any test signals (yellow) makes it obvious that no test structures have been incorporated in a functional block.

Coloring signal names in HDL source code facilitates the interpretation of hardware models in much the same way as colored wires used to do in schematics. The Emacs editor has been extended by a special VHDL mode that checks and supports syntactical correctness [147]. Automatic coloring of signal names further assists authors and readers of code.

Clock symbols and clock domains in schematic diagrams

It is good engineering practice to identify the clock inputs by way of graphical symbols attached to icons, see fig.5.9a, b, and c. Standard single-edge-triggering clocks are marked by a small triangle, the more unusual double-edge-triggering clocks by two dovetailed triangles, and level-sensitive clocks by a small rectangle. This rule is by no means restricted to gate-level diagrams as its application is beneficial to the clarity of schematics at any level of hierarchy.

Any signal that crosses over from one clock domain to another deserves particular attention as it may switch at any time irrespective of the receiver's clock. Making clock domains explicit in block diagrams and schematics is very important. All subcircuits that belong to one clock domain shall be placed close together and all clock boundaries shall be made evident.

5.7.2 Active level

Naming of complementary signals

As active-high and active-low signals coexist within the same circuit, it is important to make it clear in the identifier whether a given signal is to be understood in positive or in negative logic. Overlining the name is how negated terms are identified in mathematics, e.g. if FOO stands for an active-high signal, then its active-low counterpart is named $\overline{\text{FOO}}$ (pronounced "foo bar").

Since this notation does not yield character strings acceptable for processing by computers, many engineers prefix or postfix the original identifier, e.g. they might resort to something like ˜FOO, /FOO or FOO\$. However, the problem with most non-alphanumerical characters is that today's EDA tools, HDLs, and related standards disagree on which such symbols they support within node names and on where they are allowed to occur in the name's character string.

For reasons of cross-platform compatibility, we prefer to use the attribute B (for "bar") in conjunction with the syntactical conventions given in section 5.7.7. In the occurrence, we do write FOOxB, also see fig.5.9d and e. The popular practice of appending a simple B or N is not recommended because it may give rise to confusion in names such as SELB or OEN.

The same argument also holds for conditions and actions related to edges rather than to logic levels. By default, the rising edges are considered to be the active ones for a non-inverted signal.

Inversion symbols in schematic diagrams

Signal polarities must also become clear from schematics and icons. In fact, cell icons often include small circles that are supposed to stand as a shorthand notation for regular gate-type inverter symbols. One would thus naturally expect to find such an inversion circle wherever a signal gets translated from active-high to active-low or vice versa, and wherever polarity changes from rising-edge- to falling-edge-triggering or back. This is the mindset behind fig.5.9d.

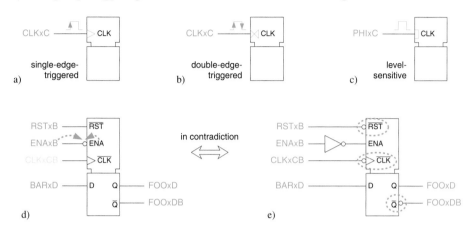

Fig. 5.9 Notational conventions in schematic diagrams. Identification of clock inputs (a,b,c) and two competing mindsets for reflecting signal polarities (d,e).

Most industrial cell libraries follow the notation depicted in fig.5.9e, though. Please observe that signals RSTxB and CLKxB are <u>not</u> subject to any negation upon entering the flip-flop here; the small circles just reconfirm that the cell's inputs $\overline{\text{RST}}$ and $\overline{\text{CLK}}$ are active-low and falling-edge-triggered respectively. Much the same observation applies to output $\overline{\text{Q}}$. The reason for this departure from mathematical rigor is that a cell can thus keep the same graphical icon independently from the context, a quality not shared by the notational convention of fig.5.9d.

> Hint: As there are no generally adhered-to standards, designers are well advised to double-check the exact meaning of signal identifiers, terminal names, and inversion circles when working with third-party components, cell libraries, or HDL code.

5.7.3 Signaling waveforms

In distinguishing between active-high and active-low signals, we have tacitly assumed that it was a signal's level that conveyed information from a transmitter to the receiver(s). While this is often true, this is not the only way to code information into a signal waveform. Figure 5.10 shows a data signal along with three different interpretations of its waveform.[27]

---

[27] Three is by no means exhaustive. Calling upon pairs of complementary signals to improve on noise immunity (differential signaling) comes as a natural extension. Other schemes map data onto waveforms in a more elaborate way to allow safe recovery of individual bits from a serial data stream at the receiving end (self-clocking),

Clock-qualified signaling means that the logic level, 1 or 0, of the signal at the instant of each active clock edge is what matters. Key characteristics of this scheme are as follows.

- Qualifier-type signals are always meant relative to a specific clock.
- They must remain stable throughout the data-call window (i.e. setup/hold interval) of the receiving circuit.
- Each timewise coincidence gets interpreted as a relevant event of its own even if the signal does not change in between.
- Hazards between active clock edges have no effect.

Examples: The enable input of a synchronous counter, the write line of a clocked (synchronous) RAM, the zero flag from an ALU when followed by a flip-flop.

Impulse signaling relies on the presence of an impulse, mark or pause, which implies that

- Each rising edge <u>or</u>, alternatively, each falling edge is counted as a relevant event.
- The signal must return to its initial, passive level before it can possibly become active for a second time.
- The durations of marks and pauses are immaterial.
- Static and dynamic hazards are unacceptable as they would cause erroneous registration of events where there are none.

Examples: Write line of an unclocked (asynchronous) RAM, request and acknowledge signals in a four-phase handshake protocol.

Transition signaling does the signaling with the presence of an edge, which means that

- Each rising <u>and</u> falling edge is counted as a relevant event.
- The signal must not restore its initial value between two consecutive events.
- The durations of marks and pauses are immaterial.
- Static and dynamic hazards are unacceptable as they would cause erroneous registration of events where there are none.

Example: Request and acknowledge signals in a two-phase handshake protocol.

Note from fig.5.10 that the number of relevant events and their respective instants of registration are not the same. It is thus absolutely essential that transmitter and receiver agree on the signaling waveform if a system is to function as intended. This is not normally a problem within a clock domain where information signals are clock-qualified as a consequence of how they are implemented in hardware. For the sake of simplicity, we can omit an attribute in this case. However, care must be exercised where a signal traverses the boundary of a subsystem or clock domain. The signaling waveform shall thus be stated as follows:

---

whereas yet others transmit two peer signals — neither of which can be interpreted as a clock — for the same purpose (companion signaling). The FireWire bus, for instance, uses four wires and a waveform referred to as **non-return to zero with data strobe** (NRZ-DS) which is a combination of differential with companion signaling. We refrain from discussing such sophisticated signaling schemes here as they are mainly intended for communication between more distant subsystems.

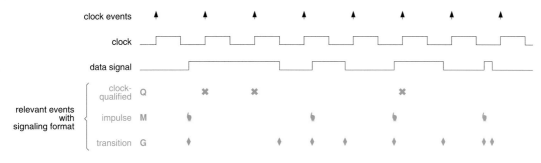

Fig. 5.10 Three signaling waveforms in common use.

- ○ Q clock-qualified signaling,
- ○ M impulse signaling (for "mark"),
- ○ G transition signaling (for "grade").

5.7.4 Three-state capability

Signals with high-impedance capability shall be identified by a Z character. This is to remind designers of the various issues associated with a three-state circuit node such as resolution function, drive conflicts, floating potential, and testability.

5.7.5 Inputs, outputs, and bidirectional terminals

Fig. 5.11 Terminals in schematic diagrams.

HDL code is easier to understand when inputs and outputs are immediately discernible from an appended I or O.[28] IO then obviously stands for a bidirectional signal. The same information is often conveyed by way of graphical symbols in schematic diagrams, see fig.5.11.

---

[28] As an added benefit, this also circumvents a quirk of VHDL in that the language disallows reading back from an output terminal. That is, an interface signal declared as out in the port clause of some entity declaration is permitted neither on the right-hand side of a signal assignment in the pertaining architecture body nor in a relation operation (such as =, /=, <, etc.). Distinguishing between two signals does away with the problem: while FOOxD can be freely used within the architecture body, a copy named FOOxDO is to drive the output port.

Some programmers prefer to declare the port to be of mode inout instead, yet this is nothing else than a bad habit. After all, what one wants to model from an electrical point of view is just an output, not a bidirectional. No HDL language restriction should make us obscure our honest intentions.

5.7.6 Present state vs. next state

Most synthesis models of finite state machines include two signals for capturing a state variable, namely a first one that holds its present value and a second one that predicts the value it is going to assume in response to the next active clock edge. It is imperative not only to identify such signals as being closely related, but also to make it clear which is which.

While this is immediately obvious in schematics, identifiers are the sole means of conveying this kind of information in HDL models. We therefore suggest using the same signal name but telling present and next state apart with the aid of suffixes P and N.

5.7.7 Syntactical conventions

We have now compiled a number of attributes for making a signal's nature evident. A practical problem is how to permanently attach those attributes to a signal. About the only way that can be expected to work across different EDA platforms consists in blending them right into the signal's identifier. A natural idea is to separate the name per se from its attributes by inserting some non-alphanumerical character in between, i.e. to write FOO_DB for an active-low data signal, for instance. To make a long story short, there is no special character that would qualify for the job without restrictions.[29] As a workaround, we recommend using an x instead, possibly combined with the popular mixed lower/upper-case style. We must admit, however, that parsing such identifiers becomes difficult if they happen to get processed by some program that has the side effect of mapping all lower-case letters to upper-case ones or vice versa.

```
signal_identifier    ::= signal_name "x" signal_attributes
signal_name          ::= signal_name_part { signal_name_part }
signal_name_part     ::= upper_case_letter { letter | digit }
signal_attributes    ::= class_char [ signal_waveform ] state_char
                         trist_mode active_low_char io_mode
class_char           ::= "R" | "A" | "C" | "S" | "D" | "T"
signal_waveform      ::= "Q" | "M" | "G"
state_char           ::= "N" | "P" | ""
trist_mode           ::= "Z" | ""
active_low_char      ::= "B" | ""
io_mode              ::= "I" | "O" | "IO" | ""
```

---

[29] The underscore _ is about the only non-alphanumerical character tolerated in node names across many EDA tools. The IEEE 1076.4 VITAL standard unfortunately reserves that very character as a delimiter in generics, e.g. in an identifier such as tpd_FOO_BAR for the propagation delay from an input FOO to some output BAR. VITAL models must not contain any underscore in an entity's port declaration. This choice is all the more unfortunate as the underscore is the one and only non-alphanumerical allowed within VHDL identifiers.

The situation is not too bad for the majority of people who are just simulating with VITAL models rather than writing their own ones. This is because any underscore in a user-defined port name always appears in the actual part of the port maps of VHDL component instantiation statements, and never in the formal part subject to VITAL ruling. The aforementioned reservations relating to the underscore character do not apply in this case.

Yet, beware of the one tool or netlist language that interprets the underscore as a hierarchy delimiter. For the sake of unlimited cross-platform portability, we thus recommend spelling node names with no special characters.

Note: In order to facilitate the decoding of signal names by human beings, no attribute character has been assigned more than one interpretation. We nevertheless recommend ordering attribute characters according to the syntax given above and, more particularly, always making the mandatory `class_char` come first in the `signal_attributes` substring.

Examples

`FOOxC` identifies a clock signal.

`BARxD` indicates this is a regular data signal within some clock domain. As opposed to this, the name `BARxA` refers to the same signal before its being synchronized to the local clock and, hence, subject to toggle at any time.

`SNAFUxRB` refers to some asynchronous active-low reset signal.

`ADDRCNTxSN` and `AddrCntxSN` are legal names for the next state of an address counter as long as the signal remains within the clock domain where it originates.

`IRQxAMI` identifies an interrupt request input that emanates from a foreign clock domain and that is meant to be stored in a positive-edge-triggered flip-flop until it is serviced.

`CarryxDB` denotes an active-low carry signal that remains confined to within some VHDL architecture. As explained earlier,[28] a duplicate signal named `CarryxDBO` may be necessary for the sole purpose of driving an output port of the entity.

`GeigerCntxDZO` reflects an output-only data signal with high-impedance capability.

`ScanModexT` stands for an active-high scan enable signal within some clock domain.

□

For any programmer, it comes most naturally that every variable in a software module must be given its own unique identifier, and the same holds true for HDL code. It is important that the same principle also be enforced in schematic drawings and netlists because two nodes that carry identical names in the same circuit block are considered by EDA tools as **connected by name**, i.e. as one and the same circuit node. Figure 8.18 nicely illustrates that connection by name often helps to render schematic diagrams easier to read.

Hint: While you are encouraged to take advantage of connections by name to avoid cluttering your schematic diagrams with clocks, resets, and other signals that connect to a multitude of subcircuits, beware of shorts that will arise whenever a signal name gets reused for two electrically distinct nodes. Inadvertently or consciously doing so ("... but logically they are the same ...") is a typical beginner's mistake.

5.7.8 A note on upper- and lower-case letters in VHDL

The case of letters is insignificant in VHDL, so identifiers `FOOBARxDB`, `FooBarxDB`, `FOOBARXDB`, and `foobarxdb` are all the same. For the sake of legibility, we nevertheless encourage VHDL programmers to write their source code in the following way:

- Reserved words: lower case throughout, e.g. `signal`, `elsif`.
- Subprograms, i.e. functions and procedures: mixed casing, e.g. `FooBar`.
- Constants and VHDL variables: lower case throughout, e.g. `foobar`.
- VHDL Signals: mixed casing, e.g. `FooBarxDB`, or pure upper casing, e.g. `FOOBARxDB`.

5.7.9 A note on the portability of names across EDA platforms

Porting netlists and schematics from one EDA platform to another is often painful as each vendor has his own syntax rules for the naming of circuit nodes. Translation steps are inevitable unless one finds a common subset that is accepted by all tools involved. Here are a few items to watch out for.

First and last characters. Most tools have specific rules as to what non-alphanumerical characters they accept as part of a node name and in what position. Luckily, VHDL names tend to be widely accepted as the syntax for signal identifiers is rather restrictive.

Busses. One finds brackets [ ], parentheses ( ), and curly brackets { } in combination with a colon :, a to, or a downto for indicating the index range of a vector.

Hierarchy delimiters. Slash /, period mark ., and dollar sign $ are commonly used.

Aliases. Although typically legal, multiple names for the same node should be avoided because many software tools are not able to handle them correctly and because of the confusion this practice tends to create for humans.

Chapter 6

Clocking of Synchronous Circuits

6.1 | What is the difficulty in clock distribution?

Up to this point, we have have ignored the difficulties of distributing a clock signal over a chip or a major portion thereof. We were in good company as systems engineering, automata theory, and other theoretical underpinnings of digital design assume simultaneous updating of state throughout a circuit. Physical reality is different from such abstractions, though.

Consider a population of flip-flops or other clocked subcircuits that make part of one clock domain in a synchronous design as shown in fig.6.1. A common clock tells them when to transit into the next state. Ideally, all such bistables are supposed to react to the clock instantly and all at exactly the same moment of time.

In practice, however, switching will be retarded due to many small delays inflicted by drivers and wires in the **clock distribution network**. As most clock signals connect to a multitude of storage elements spread out over an entire clock domain, individual switching times will differ because delays along the various clock propagation paths are not quite the same. This scattering over time is loosely referred to as **clock skew**. To make things worse, those delays will slightly vary from one clock cycle to the next, thereby giving rise to **clock jitter**.

Many causes contribute to the timewise scattering of clocks:

- Unevenly distributed fanouts and load capacitances.
- Unequal numbers of buffers and/or inverters along different branches.
- Unlike drive strengths and timing characteristics of the clock buffers instantiated.
- Unbalanced interconnect delays due to dissimilar layout parasitics
  (R: wire length and thickness, via count; C: plate, fringe, and lateral capacitance).
- Unequal switching thresholds of bistables (translate clock ramps into staggered switching).
- Process, temperature, voltage (PTV), and — more so — on-chip variations (OCVs).
- Supply noise as caused by ground bounce and supply droop.
- Crosstalk from switching activities in the surrounding circuitry.

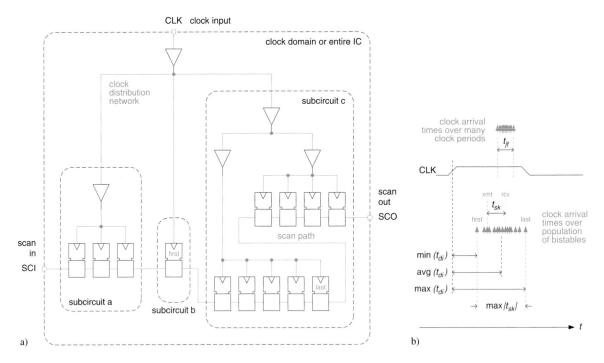

Fig. 6.1 Clock distribution. Clock domain with flip-flops, clock distribution network, and scan path shown (a), relevant timing quantities (b).

It is evident that excessive scattering of switching events compromises the correct functioning of a digital circuit. Design engineers thus try hard to eliminate any systematic disparities among clock arrival times.[1] However, depending on fabrication depth and design level (board, field-programmable logic, semi-custom IC, full-custom IC, hand layout), they are never able to control all of the underlying phenomena, so that a certain amount of unevenness remains.

6.1.1 Agenda

Designing dependable circuits in spite of clock skew and jitter involves two issues, namely
(a) knowing and lowering the vulnerability of a design, and
(b) minimizing scattering by distributing clock signals over a domain in an adequate way.

The two issues will be discussed in sections 6.2 and 6.3 respectively. For simplicity, we will focus on signals that circulate within one clock domain, thereby dropping any input and output signals from our analysis of section 6.2. Synchronous I/O is addressed in section 6.4, whereas the problems

---

[1] Experienced designers sometimes introduce clock skew on purpose either to accommodate RAMs and other subcircuits with larger-than-normal setup/hold times or to allow for faster clocking by adapting to uneven path delays. A better tolerance with respect to delay variations may also be sought in this way, see problem 2. The process of tuning a clock distribution network to local timing requirements is termed **clock skew scheduling**. Please refer to problem 8 and to the specialized literature [148] [149] [150] for more details on this optimization technique.

associated with assimilating data that arrive asynchronously are postponed to chapter 7. How to safely implement clock gating is the subject of section 6.5.

6.1.2 Timing quantities related to clock distribution

In order to study the clock-related characteristics of synchronous circuits, we need to introduce three timing parameters that specifically relate to clock distribution networks, see fig.6.1 for an illustration.

> t_{di} **Clock distribution delay**. The time lag measured from when a clock edge appears at the clock source until a state transition actually takes place in response to that edge. When referring to an IC, some package pin is normally meant to act as clock source.
>
> t_{jt} **Clock jitter**. The variability of consecutive clock edges arriving at the same location.
>
> t_{sk} **Clock skew**. The inaccuracy of the same clock edge arriving at different locations within a given clock domain. There is a local and a more global view.
>
> In a narrow sense, skew refers to the clock terminals of two subcircuits connected by some signal propagation path. Skew is considered positive if the receiver is clocked after the transmitter, and negative if it is clocked before, i.e. $t_{sk} = t_{di\,rcv} - t_{di\,xmt}$.[2]
>
> From a wider perspective, one is interested in knowing the largest difference in clock arrival times between any two clock terminals within a clock domain. The term clock skew then takes on the meaning of overall skew and is defined as $\max|t_{sk}| = \max(t_{di}) - \min(t_{di})$.

6.2 | How much skew and jitter does a circuit tolerate?

6.2.1 Basics

Numerous schemes for driving synchronous digital circuits have been devised over the years. Some of them are more vulnerable to clock skew and jitter than others, some ask for more hardware resources, some have a different impact on performance, and this section aims at comparing them. Note that while selecting a clocking discipline typically amounts to finding an optimum choice between conflicting goals, it is not concerned with functionality as any decent functionality can be combined with any decent clocking scheme.

Before entering the discussion of individual clocking schemes, let us recall from observation 5.1 that any digital signal must be allowed to settle to a valid state before it is accepted into a memorizing subcircuit. Correct and dependable circuit operation is otherwise not possible.

The time interval during which data at the input of a flip-flop, latch, RAM or other sequential subcircuit must remain valid is referred to as the **data-call window**, aka aperture time, and is defined by the setup and hold times there. The term **data-valid window** designates the time

---

[2] The sign of clock skew is controversial, some authors have elected to define clock skew the other way round as $t_{sk} = t_{di\,xmt} - t_{di\,rcv}$. We prefer to have data and clock delays share the same sense of counting.

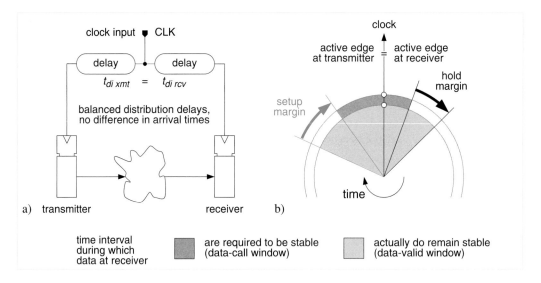

Fig. 6.2 Clocking in the absence of skew and jitter. Circuit (a) and Anceau diagram (b).

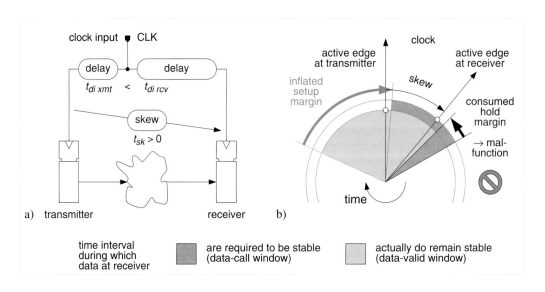

Fig. 6.3 Impact of excessive positive clock skew. Circuit (a) and Anceau diagram (b).

interval during which data at the receiver actually do remain valid and is dependent on a circuit's propagation and contamination delays. As becomes immediately clear when comparing figs.6.2 and 6.3, the fundamental requirement simply is

Observation 6.1. *Transferring data between two subcircuits requires that the data-call window of the receiving subcircuit be fully encompassed by the transmitter's data-valid window.*

In subsections 6.2.2 through 6.2.7, each clocking discipline will be introduced by outlining its operation and elementary characteristics. Then follows a more detailed analysis which establishes conditions for how much clock skew can be tolerated within a circuit without causing malfunctioning or undeterministic operation. The resulting inequalities will be termed **skew margins** and are visualized in the Anceau diagram of fig.6.2. While noise margins delimit the safe operating range of a digital circuit with respect to uncertainties in amplitude, skew margins do the same with respect to timewise uncertainties.

Observation 6.2. *A certain amount of clock scattering is unavoidable. What really counts is that clocking discipline and clock distribution network are chosen such that the combined effects of skew and jitter exceed the actual setup and hold margins*
- *nowhere within a clock domain, and*
- *under no operating condition.*[3]

Although there are more clocking disciplines than those discussed here, it should be noted that any other scheme can be analyzed along the same lines. Also, the study is easily extended to include interconnect delays by adding appropriate terms to $t_{cd\,c}$ and $t_{pd\,c}$.

6.2.2 Single-edge-triggered one-phase clocking

HARDWARE RESOURCES AND OPERATION PRINCIPLE

Single-edge-triggered one-phase clocking is the most natural approach from a background in automata theory or abstract systems design. Registers are implemented from flip-flops and all of them get triggered by the same clock edge, henceforth termed the **active edge**. No bistables other than ordinary **single-edge-triggered flip-flops** (SETFFs) are used.

Each computation cycle starts immediately after an active clock edge and ends with the subsequent one so that $T_{cp} = T_{clk}$. All transient phenomena must die out before the active clock edge, that is within one clock period. The exact moment of occurrence of the passive clock edge is immaterial as long as the clock waveform meets the minimum pulse width requirements $t_{clk\,hi\,min}$ and $t_{clk\,lo\,min}$ imposed by the flip-flops.

DETAILED ANALYSIS

Starting from setup and hold requirements of flip-flops, let us now find out how much clock skew and jitter can be tolerated without exposing a circuit to timing problems. The corresponding Anceau diagram is shown in fig.6.5.

Setup condition

Consider a pair of flip-flops with a unidirectional data propagation path in between as shown in fig.6.2. The setup condition of the receiving flip-flop is expressed as

$$t_{di\,xmt} + t_{pd\,ff\,xmt} + t_{pd\,c} \leq T_{clk} + t_{di\,rcv} - t_{su\,ff\,rcv} \tag{6.1}$$

---

[3] Operating conditions are meant to include data patterns, loads (typically capacitive in CMOS), clock frequency, PTV and on-chip variations (OCVs), ground bounce, supply droop, and crosstalk.

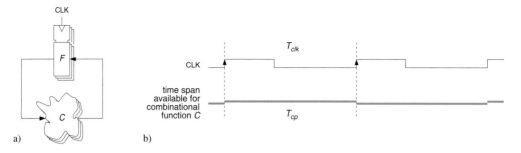

Fig. 6.4 Edge-triggered one-phase clocking (with the rising edge being the active one). Basic hardware organization (a) and simplified timing diagram (b).

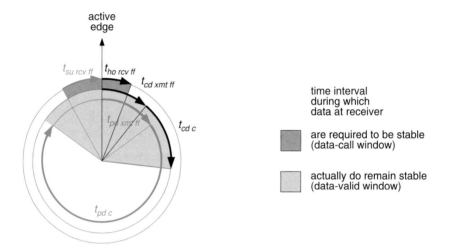

Fig. 6.5 Anceau diagram of edge-triggered one-phase system. Note that arcs of different colors are being used to distinguish between timing quantities that relate to the setup condition and others that matter for the hold condition.

which makes reference to timewise uncertainty of the clock after being recast into

$$t_{pd\,ff\,xmt} + t_{pd\,c} + t_{su\,ff\,rcv} \leq T_{clk} + (t_{di\,rcv} - t_{di\,xmt}) = T_{clk} + t_{sk} \tag{6.2}$$

In the absence of skew and jitter, the inequality stipulates a minimum clock period. It also appears — in this limited context — that positive skew has a beneficial effect and negative skew a detrimental one. This is because a lag of the receiver clock with respect to that of the transmitter facilitates meeting the setup condition, while the opposite is true for a lead. Relation (6.2) thus effectively imposes a lower — and typically negative — bound on clock skew, which becomes even more obvious when the inequality is transformed into

$$t_{sk} \geq (t_{pd\,ff\,xmt} + t_{pd\,c} + t_{su\,ff\,rcv}) - T_{clk} \underset{typ.}{<} 0 \tag{6.3}$$

For the continuation of our analysis, we will work with the equivalent form

$$-t_{sk} \leq T_{clk} - (t_{pd\ ff\ xmt} + t_{pd\ c} + t_{su\ ff\ rcv}) \tag{6.4}$$

From a broader perspective, one has to consider the ensemble of all flip-flops in a circuit which calls for a separate inequality for any two communicating bistables. A static **timing analyzer** essentially is a software tool that takes a gate-level netlist, that checks whether inequalities (6.4) and (6.8) are met, and that flags violations, if any. To that end, the software begins by calculating the numeric delay figures along each signal propagation path in a circuit.[4]

Rather than checking thousands of mathematical relations, however, humans prefer to come up with one simple worst-case condition which, when satisfied, guarantees that setup times are respected throughout an entire clock domain. Skew plus jitter here appear as the maximum over all signal propagation paths and as a magnitude $\max|t_{sk}|$. After all, a practical clocking scheme must accommodate data transfers between any two flip-flops, explicitly allowing for reciprocal data exchange, which renders the distinction between positive and negative skew meaningless.

$$\max|t_{sk}| \leq T_{clk} - \max(t_{pd\ ff} + t_{pd\ c} + t_{su\ ff}) = T_{clk} - t_{lp} \tag{6.5}$$

The expression $\max(t_{pd\ ff} + t_{pd\ c} + t_{su\ ff})$ reflects the delay along the longest signal propagation path between any two adjacent flip-flops in the circuit. For sake of brevity, we will refer to this path together with its delay t_{lp} as the **longest path**.[5]

Inequality (6.5) indicates that longest path, clock skew, and jitter together bound the minimum admissible clock period from below and so define how fast the circuit can be safely clocked. This finding puts us in a position to estimate the performance of a given circuit more accurately than in chapter 2 and gives rise to a first observation:

$$T_{clk} \geq \max(t_{pd\ ff} + t_{pd\ c} + t_{su\ ff}) + \max|t_{sk}| = t_{lp} + \max|t_{sk}|$$
$$\approx t_{id\ ff} + \max(t_c) + \max|t_{sk}| \tag{6.6}$$

Observation 6.3. *Clock skew is at the expense of maximum performance in circuits that operate with edge-triggered one-phase clocking.*

In a pipelined datapath, for instance, $t_{ff} + \max|t_{sk}|$ is nothing else but time unavailable for payload computations, which is why this quantity is often referred to as **timing overhead**. Any positive difference between the left-hand and the right-hand side of (6.6) implies that the clock period is not fully utilized by computations and timing overhead. This surplus amount of time is referred to as **slack** and routinely computed during static timing analysis. While slack must always remain positive, designers strive to minimize it when in search of maximum performance.

---

[4] Please refer to section 12.1.2 for details.

[5] The term **critical path** is often used as a synonym for longest path. We prefer to understand critical path as a generic term for the longest <u>and</u> the shortest path, however, because meeting the timing conditions along the shortest path is as important for the correct functioning of a circuit as meeting those along the longest path.

Example

A Sun UltraSPARC-III CPU implemented in 250 nm 6M1P CMOS runs from a 600 MHz clock, which amounts to $T_{clk} = 1.67$ ns. Some 70% of this time is available for combinational data processing and suffices to accomodate approximately eight consecutive levels of logic, assuming that a 3-input NAND with a fanout of three is representative for a typical gate delay. The rest is taken up by the registers and the necessary allowance for clock skew and jitter [151].

□

Hold condition

Starting from the hold requirement for data travel between two flip-flops,

$$t_{di\,xmt} + t_{cd\,ff\,xmt} + t_{cd\,c} \geq t_{di\,rcv} + t_{ho\,ff\,rcv} \tag{6.7}$$

a second condition is obtained that bounds the acceptable skew and jitter from above, this time because it is positive clock delay that puts the hold condition at risk.

$$t_{sk} = (t_{di\,rcv} - t_{di\,xmt}) \leq t_{cd\,ff\,xmt} + t_{cd\,c} - t_{ho\,ff\,rcv} \tag{6.8}$$

Again from a circuit perspective one finds

$$\max|t_{sk}| \leq \min(t_{cd\,ff} + t_{cd\,c} - t_{ho\,ff}) = t_{sp} \tag{6.9}$$

Here the right-hand side $\min(t_{cd\,ff} + t_{cd\,c} - t_{ho\,ff})$ stands for the **shortest path** t_{sp} through a circuit, which is sometimes referred to as the race path. As opposed to the situation found for the setup margin (6.5), the hold margin (6.9) either holds or doesn't, no matter how much the clock period is being stretched. As both conditions must be met, the latitude to clock skew does not depend on the speed at which the circuit is operated. Another interesting observation is that any combinational logic placed between two adjacent flip-flops facilitates meeting (6.9) because of its inherent contamination delay. The same also applies for slow interconnect lines.

IMPLICATIONS

The most precarious circumstances occur when no combinational logic — and thus also no contamination delay — is present between two consecutive flip-flops or other edge-triggered storage elements. Consider a shift register, for instance. The skew margin then collapses to

$$\max|t_{sk}| \leq \min(t_{cd\,ff} - t_{ho\,ff}) \tag{6.10}$$

which also indicates that there is no way to improve the situation by adjusting clock waveforms. Whenever an IC suffers from insufficient hold margins, a painful and costly redesign is due.

In practice, there is a deplorable difficulty with giving numerical figures for the skew margins defined by (6.9) and (6.10) because most datasheets lack indications on contamination delay.[6] Yet, as the inequality $0 \leq t_{cd} < t_{pd}$ closely bounds it from both sides, it is safe to state that $t_{cd\,ff} - t_{ho\,ff}$ always is a very small quantity. Since any differences in clock arrival times are at the expense of this tiny

---

[6] The reasons for this are given in appendix A.6.

Table 6.1 | Timing problems and their remedies in edge-triggered one-phase circuits.

Remedies for timing problem(s)	if identified during the design process	if found once prototypes have been manufactured
on long path(s)	redesign circuit such as to reduce its max. prop. delay (may be very hard)	extend clock period or renegotiate PTV conditions (most likely unacceptable)
on short path(s)	insert delay buffers between consecutive flip-flops (comparatively easy)	adjusting clock waveform or PTV does not help (fatal)

safety margin, edge-triggered one-phase systems must be considered inherently sensitive to clock skew. Note that interconnect delay has a beneficial effect, however.

Example

Table 6.2 is an excerpt from the datasheet of a CMOS flip-flop.[7] The maximum admissible clock skew between any two such flip-flops where the Q output of one cell directly connects to the D input of the next is 124 ps $-$ (-14 ps) $= 138$ ps. Please keep in mind this is just an estimate that assumes identical MOSFETs and PTV conditions throughout. The benefical impact of interconnect delay is also ignored, on the other hand.

Table 6.2 | Timing characteristics of a standard cell flip-flop in a 130 nm CMOS technology.

D Flip-flop with reset 1x drive propagation delay for fanout 2 @ typ. process, 25 °C, 2.5V / 2.5V	Timing parameter				Max. toggle frequency
	$t_{pd\,ff}$ [ps]	$t_{cd\,ff}$ [ps]	$t_{su\,ff}$ [ps]	$t_{ho\,ff}$ [ps]	[GHz]
from a high-density library	160	124	70	-14	4.35

☐

Observation 6.4. *Matching of clock distribution delays and careful timing analysis are critical when designing circuits and systems with edge-triggered one-phase clocking. Shift registers and scan paths are especially vulnerable to (positive) clock skew.*

The finding is all the more significant as most digital ICs indeed do include shift registers, be it only implicitly as scan paths to ensure testability.

---

[7] As the official datasheets give no indications for contamination delay, the number included in the table is an educated guess obtained from the computer models prepared for simulation and timing analysis. The so-called **timing library format** (TLF) describes a cell's overall delay as a sum of an initial inertia followed by output ramping. Both inertial delay and output ramp time are modelled as functions of switching direction (rise, fall), capacitive load, and input ramp time. The contamination delay figure of table 6.2 is the minimum inertial delay for a cell characterized with no external load attached and for clock ramp time zero, or almost so.

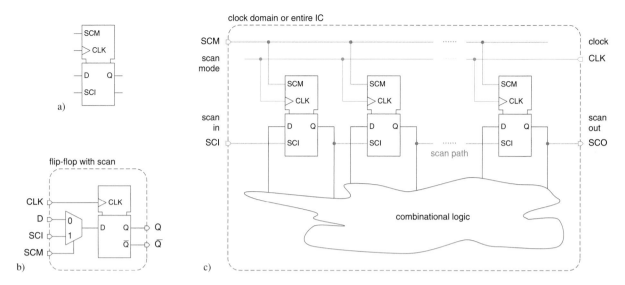

Fig. 6.6 Scan test structures. Scan flip-flop (b), icon (a), and overall circuit structure (c).

SCAN-TYPE TEST STRUCTURES

The goal of all **design for test** (DFT) techniques is to facilitate the search for fabrication defects in manufactured circuits. Scan testing is particularly popular with VLSI designers due to its efficiency and elegance. Every flip-flop gets replaced by an equivalent library cell that includes a multiplexer in front of it, see fig.6.6b. While data are ingested from the regular data input D during normal operation, the scan input SCI gets read instead when a global control signal SCM is active. As depicted in fig.6.6c, these scan-type flip-flops are then chained together so as to form a shift register when the circuit operates in scan test mode.[8]

A built-in **scan path** supports

- Checking the proper functioning of all flip-flops that make part of it,
- Serially reading out the circuit's state at any time for inspection,
- Serially putting the circuit into a user-defined state during a test run, and
- Testing the combinational logic that sits between the flip-flops by combining the application of test vectors prepared beforehand with serial write-in and read-out operations.

That a scan path cannot cross over from one clock domain to a second domain goes without saying, so each clock domain needs its own scan path(s).[9] Scan paths typically traverse many subcircuits

---

[8] One speaks of **full scan** if all flip-flops in a circuit are part of the scan path and of **partial scan** otherwise. Obtaining adequate test vectors for scan testing is a part of automatic test pattern generation (ATPG), a topic beyond the scope of this text. Please refer to the specialized literature such as [5], for instance. Just note that generating test vectors for full-scan circuits is essentially the same as for combinational logic because all inputs and outputs can be observed and controlled via the scan facility, a boon not shared by partial-scan circuits.

[9] There are exceptions, however, as clock domains that run asynchronously during normal operation are often operated from the same clock during volume testing.

and are, therefore, particularly exposed to clock skew at the boundaries. In the example of fig.6.1a, hold-time violations are most likely to occur in the first flip-flop of subcircuit c.[10]

MIXING CELLS FROM LOGIC FAMILIES

Our previous estimates were based on numerical data from table 6.2. Yet, real circuits are assembled from flip-flops of many different types, which further puts hold margins at risk.

A key idea behind the concept of a **logic family** — whether available as a cell library or as a set of physical SSI/MSI components — is the ability to universally combine components with others from the same family in spite of fabrication tolerances. Timing-wise, this implies that the shortest contamination delay must be greater than or equal to the longest hold time over all edge-triggered storage devices.

$$\max(t_{ho\,ff}) \leq \min(t_{cd\,ff}) \tag{6.11}$$

Note that no margin whatsoever is left when this relation holds with equality. Only flip-flops with zero hold time warrant free interchange, **negative hold times** are even better. High-quality standard cell libraries are usually designed along this line, yet beware of exceptions that do not meet the requirement of (6.11).

Observation 6.5. *Timing is particularly at risk when different logic families are mixed so that signals travel from fast storage elements to slower ones.*

Mixing logic families, which has long been standard practice at the board level, is gradually finding its way into IC design when high-speed and low-power cells are being combined on a single chip in search of the optimum performance–energy tradeoff. Also note that the timing parameters of macrocells, such as on-chip RAMs, tend to differ significantly from those of simple and, hence, much faster flip-flops and latches.

As edge-triggered one-phase-clocking is so vulnerable to clock skew and jitter, modern EDA tools support automatic **hold-time fixing**, a technique whereby buffers get inserted into all signal propagation paths found to provide insufficient hold margins until the extra contamination delay suffices to compensate for the deficit. Extensive buffer chains not only inflate circuit size, however, but also waste energy without contributing to computation.

> Hint: Delay hold-time fixing until load capacitances and interconnect delays can be estimated with good precision during the routing phase. Have the P&R tool or the timing verifier list the shortest paths as a function of their respective hold margins. You are now in a good position to find out where delay buffers must be inserted to ensure a reasonable minimum hold margin in your design.

---

[10] Scan flip-flops sometimes feature a separate scan-out terminal SCO that differs from the ordinary Q output by virtue of passing through two extra inverters. We now understand why the dilated contamination delay $t_{cd\,ff}$ so obtained is very welcome. Similarly, automatic scan insertion tools can splice in so-called **lock-up latches** at the boundaries between major subcircuits to bolster up hold margin. We consider this option as a workaround, however.

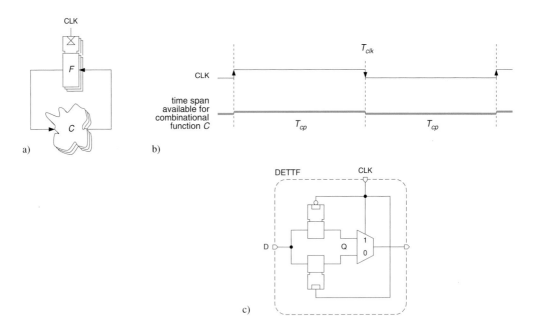

Fig. 6.7 Dual-edge-triggered one-phase clocking. Basic hardware organization (a) and simplified timing diagram (b). Logically equivalent circuit for a dual-edge-triggered flip-flop (c).

For a concluding remark of more fundamental nature, reinspect inequalities (6.1) through (6.11). Notice that there is a law that will be confirmed throughout our timing analyses.

Observation 6.6. *Within timing conditions, quantities always combine as follows:*

inequality	*critical path*	*relevant timing parameters*
setup condition	*longest*	*propagation delays and setup time*
hold condition	*shortest*	*contamination delays and hold time*

6.2.3 Dual-edge-triggered one-phase clocking

HARDWARE RESOURCES AND OPERATION PRINCIPLE

Dual-edge-triggered clocking was for a long time left aside before gaining acceptance along with the buzzword **double data rate** (DDR) in the context of computer memory and mainboard design, where it serves to increase memory bandwith while avoiding excessive clock frequencies. Conceptually, this technique is very similar to single-edge-triggered one-phase clocking but rests on special storage elements that operate on both clock edges. As suggested in fig.6.7c, it is possible to construct a **dual-edge-triggered flip-flop** (DETFF) from a pair of latches and a multiplexer; more detailed circuit diagrams are to be presented in section 8.2.5.

Exactly as in single-edge clocking, each computation period gets bounded by two consecutive active clock edges. Yet, the fact that either edge causes the state to get updated cuts the clock frequency in half for the same computation or memory access rate. The fundamental timing requirement thus becomes $T_{cp} = \frac{1}{2}T_{clk} \geq t_{lp}$, see fig.6.7b.

IMPLICATIONS

As a DETFF is described by the same set of timing parameters as a SETFF, analysis yields much the same findings, only the numerical values are bound to differ. The differences relate to clock frequency and energy efficiency. In a single-edge-triggered circuit, the clock net toggles twice per computation cycle. A DDR circuit carries out the same computation with a single clock edge, which cuts the energy spent for charging and discharging that net in half, at least in theory. This explains the name **half-frequency clocking** sometimes being used.

Unfortunately, the higher capacitive loads of a DETFF on clock and data inputs partially offset this economy. The slightly more complex circuit does not help either. The economy may even turn negative if the data input is subject to intense glitching [152]. Still, reductions on the order of 10% to 20% of overall power over an equivalent single-edge-triggered design seem more typical.

While dual-edge-triggered clocking has gained acceptance with DDR RAM interfaces, it has yet to make it into cell libraries and EDA tools, notably for synthesis and timing verification.[11] As HDL synthesizers do not currently accept dual-edge-triggered circuit models, one must model for single-edge one-phase-triggering and replace every SETFF by an analogous DETFF following synthesis.

Another particularity is that dual-edge-triggered clocking assumes a strictly symmetrical clock waveform that satisfies $t_{clk\,lo} = t_{clk\,hi} = \frac{1}{2}T_{clk}$. Put differently, the clock must be made to maintain a duty cycle δ_{clk} of exactly 0.50 under all operating conditions.

6.2.4 Symmetric level-sensitive two-phase clocking

HARDWARE RESOURCES AND OPERATION PRINCIPLE

Circuits are **latch-based**, which is to say that no bistables other than level-sensitive latches are being used. A subset of them is controlled by clock signal CLK1, the complementary set by CLK2. Data flow occurs exclusively between the two subsets, there is no exchange within a subset. Put differently, the latches together with the data propagation paths in between always form a bipartite graph.[12] Violating this rule might lead to unpredictable behavior as a consequence from zero-latency loops that would inevitably form when the latches at either end of a combinational logic become transparent at the same time.

Ignoring the setup times and propagation delays of the latches, for a moment, one finds the situation shown in fig.6.8. The two clock signals subdivide the clock period T_{clk} into four intervals labeled T_1 through T_4. During each computation period, data complete a full circle from latch set L_1 through

---

[11] One may consider using soft macros where qualified DETFF cells are unavailable. Internal wires shall then be given so much weight that the cells involved always get placed next to each other. As opposed to SETFFs, the output of a DETFF is not connected to a bistable's output directly but passes through a multiplexer, that is through a combinational network. Careless design or routing might thus engender hazards. It is imperative that a highly consistent timing be guaranteed in spite of minor variations in the soft macro's interconnect routing. Also, any decent collection of DETFFs must be made to satisfy (6.11) to ensure interoperability.

[12] A graph is said to be **bipartite** iff it is possible to decompose the set of its vertices into two disjoint subsets such that every edge connects a vertex from one subset with a vertex from the other.

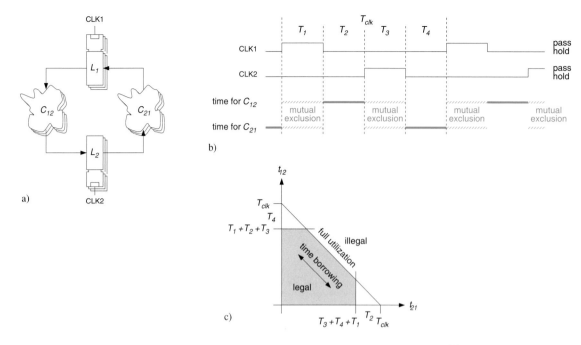

Fig. 6.8 Symmetric level-sensitive two-phase clocking. Basic hardware organization (a), simplified timing diagram (b), and range of operation (c).

logic C_{12} to latch set L_2 and from there through logic C_{21} back to latch set L_1. The computation period and clock period are the same, $T_{cp} = T_{clk}$.

The earliest moment combinational logic C_{21} can take up evaluating a new input is when latch set L_2 becomes transparent, provided logic C_{12} has completed its evaluation at that time. Otherwise this point in time is delayed into clock interval T_3 until C_{12} completes.

The last possible moment for C_{21} to complete its evaluation is right before latch set L_1 stores the result when switching to hold mode, i.e. at the end of T_1. One thus obtains

$$\max(t_{21}) \leq T_3 + T_4 + T_1 \tag{6.12}$$

and analogously for combinational logic C_{12}

$$\max(t_{12}) \leq T_1 + T_2 + T_3 \tag{6.13}$$

The two inequalities seem to suggest that more than one clock period is available to accommodate the cumulated evaluation times of both combinational blocks. Of course, this is not possible as C_{21} and C_{12} must work strictly one after the other to guarantee orderly circuit operation. This condition of mutual exclusion is captured in a third constraint, which requires that

$$\max(t_{12} + t_{21}) = \max(t_{pd\ full-circle}) \leq T_{clk} = T_1 + T_2 + T_3 + T_4 \tag{6.14}$$

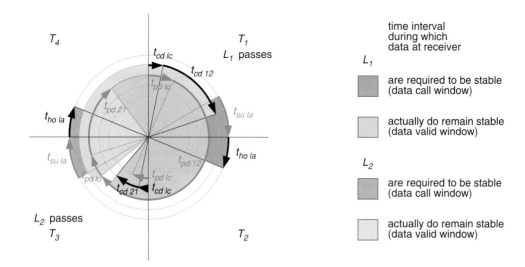

Fig. 6.9 Anceau diagram of symmetric level-sensitive two-phase system shown for a time-borrowing situation where C_{12} occupies approximately $\frac{5}{8}$ and C_{21} $\frac{2}{8}$ of a clock period.

Note that there exists no dead clock interval during which neither C_{21} nor C_{12} is allowed to evaluate new input data. Depending on the actual timing figures, any of the intervals T_1 through T_4 can actually become productive. However, the absence of concurrency between C_{21} and C_{12} implies that each stage in a pipeline must include both latch sets L_1 and L_2. As a rule, two latches are needed in a level-sensitive two-phase system where an edge-triggered system has one flip-flop.

Together (6.12), (6.13), and (6.14) confine the legal operating range with respect to timing as depicted in fig.6.8. Full utilization of the clock cycle is achieved when (6.14) holds with equality, i.e. when $\max(t_{pd\ full-circle}) = T_{clk}$. Further observe from fig.6.8 that symmetric level-sensitive two-phase clocking offers the potential for trading evaluation time left unused by C_{12} against time for C_{21}, and vice versa, without any modification to the clock phases. This somewhat surprising characteristic is referred to as **time borrowing**.

The three characteristics found so far, namely mutual exclusion, alternation with no dead time, and time borrowing, make a symmetric level-sensitive two-phase system resemble a relay race, during which two runners alternate but can decide themselves — within certain bounds — where exactly to hand over the baton.

DETAILED ANALYSIS

Incorrect timing is likely to lead to corrupt circuit states when input data are not properly stored by the target latches. This phenomenon, from which there is no recovery other than resetting the entire circuit, is termed **latch fall-through** or latch race-through. The skew margins necessary to stay clear of this problem are obtained from the setup and hold requirements of the two latch sets L_1 and L_2, also see figs.6.8 and 6.9.

Setup condition

More precise formulations of equations (6.12) through (6.14) are obtained when the various delays introduced by the latches themselves are also taken into consideration.

$$\max|t_{sk}| \leq T_{clk} - T_2 - \max(t_{pd\,lc\,2} + t_{pd\,21} + t_{su\,la\,1}) \tag{6.15}$$

$$\max|t_{sk}| \leq T_{clk} - T_4 - \max(t_{pd\,lc\,1} + t_{pd\,12} + t_{su\,la\,2}) \tag{6.16}$$

$$0 \leq T_{clk} - \max(t_{pd\,ld\,1} + t_{pd\,12} + t_{pd\,ld\,2} + t_{pd\,21}) \tag{6.17}$$

All three of these inequalities must hold in order to ensure correct operation. Whatever values the various timing parameters may have, it is always possible to find a minimum for the clock period T_{clk} that satisfies them all.

Hold condition

Correct timing at the inputs of L_1 and L_2 requires that

$$\max|t_{sk}| \leq T_2 + \min(t_{cd\,lc\,2} + t_{cd\,21} - t_{ho\,la\,1}) \tag{6.18}$$

$$\max|t_{sk}| \leq T_4 + \min(t_{cd\,lc\,1} + t_{cd\,12} - t_{ho\,la\,2}) \tag{6.19}$$

respectively. Like in edge-triggered circuits, the most critical situation occurs when no logic is present. In contrast to the situation there, however, the margin against hold violations is largely determined by the duration of the two non-overlap intervals T_4 and T_2. This quality holds for skew and jitter across a single clock distribution net as well as between the two nets.

IMPLICATIONS

The latitude for skew and jitter is largely determined by the waveforms of the clock signals. Room to accommodate unknown or unpredictable timing variations can be designed into level-sensitive two-phase systems just by sizing the non-overlap intervals accordingly. In principle, the most excessive skew can be coped with even after circuits have been fabricated, provided it is possible to define the two clock waveforms from outside the chip, e.g. by driving CLK1 and CLK2 from two separate pins. As both non-overlap intervals are productive, this entails no loss of speed, only the time-borrowing capability gets somewhat restricted.

This picture sharply contrasts with edge-triggered one-phase clocking, where the hold margin is imposed by flip-flop parameters that cell-based designers are unable to control.

The price paid for this advantage lies in the routing overhead and in the extra design effort for distributing two clock signals. In order to keep variations between the two clocks within reasonable limits, the two nets should be made similar both geometrically and electrically, i.e. they should run close together with similar loads attached at corresponding points. Area is another cost factor because two latches occupy more die area than one flip-flop. Probably the most important handicap, however, is the fact that it is not currently possible to obtain level-sensitive two-phase circuits from HDL synthesis without rewriting the RTL source code.

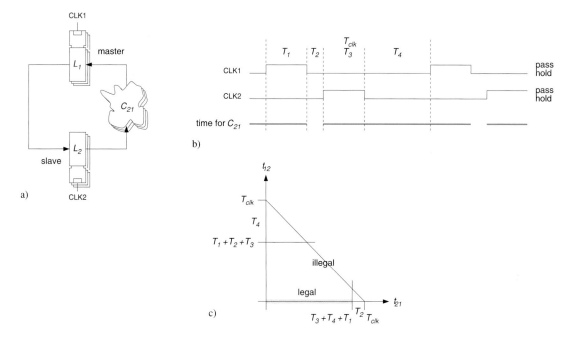

Fig. 6.10 Unsymmetric level-sensitive two-phase clocking. Basic hardware organization (a), simplified timing diagram (b), and range of operation (c).

6.2.5 Unsymmetric level-sensitive two-phase clocking

HARDWARE RESOURCES AND OPERATION PRINCIPLE

Unsymmetric level-sensitive two-phase clocking is obtained from the symmetric scheme by transferring combinational logic C_{12} to the opposite side of latch set L_2 and merging it into C_{21}, thereby making intervals T_3, T_4, and T_1 available for evaluating the combined function. Clock interval T_2 becomes unproductive due to the absence of combinational logic between L_1 and L_2. Most often, T_2 is shortened at the benefit of the other intervals to utilize the clock period as much as possible.

DETAILED ANALYSIS

Setup condition

For latch set L_1 and L_2 respectively is

$$\max |t_{sk}| \leq T_{clk} - T_2 - \max(t_{pd\,lc\,2} + t_{pd\,21} + t_{su\,la\,1}) \tag{6.20}$$

$$\max |t_{sk}| \leq T_2 + T_3 - \max(t_{pd\,ld\,1} + t_{su\,la\,2}) \tag{6.21}$$

Observe that setup time for latch set L_2 is minimal when L_1 accepts fresh data at the very end of T_1.[13] Inequality relation (6.21) is uncritical for any case of practical interest and (6.20) can always be met by selecting an adequate clock period.

---

[13] The finding that (6.21) does not follow from (6.16) whereas (6.20) is identical to (6.15) is irritating at first sight. The underlying reason is as follows.

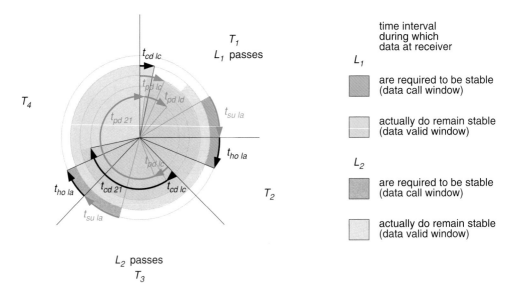

Fig. 6.11 Anceau diagram of unsymmetric level-sensitive two-phase system shown for near-maximum utilization of clock period.

Hold condition

For latch set L_1 and L_2 respectively we find

$$\max |t_{sk}| \leq T_2 + \min(t_{cd\,lc\,2} + t_{cd\,21} - t_{ho\,la\,1}) \tag{6.22}$$

$$\max |t_{sk}| \leq T_4 + \min(t_{cd\,lc\,1} - t_{ho\,la\,2}) \tag{6.23}$$

Either condition provides a non-overlap interval that offers protection in case of excessive skew and jitter. Shortening the unproductive non-overlap interval T_2 in search of maximum performance is limited only by the quest for an adequate skew margin.

IMPLICATIONS

In summary, benefits and costs of unsymmetric level-sensitive two-phase clocking are almost identical to those of its symmetric counterpart. A difference is that wide skew margins are bought at the expense of operating speed because non-overlap interval T_2 is unproductive.

The starting point for formulating setup conditions in the symmetric case was that both C_{12} and C_{21} would take up as much time as possible, i.e. they would start as soon as their input latches become transparent and end just before their output latches switch to hold. As a consequence, (6.17) had to be added as a third condition to make sure the clock period suffices to accommodate the cumulated delays of C_{12} and C_{21}.

As there is no C_{12} in the unsymmetric case, the above way of looking at the setup condition for L_2 is inappropriate. Instead, (6.21) is obtained from the assumption that C_{21} eats as much time as it can, thereby leaving the bare minimum for the propagation path from L_1 to L_2. As a side effect, a third condition is dispensed with.

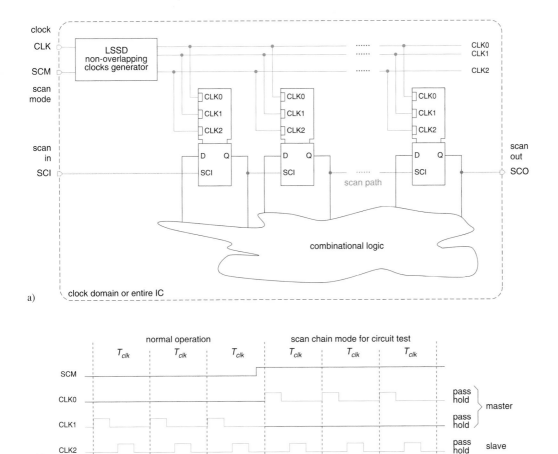

Fig. 6.12 Level-sensitive scan design. Overall circuit structure (a) and clock waveforms (b).

Observation 6.7. *What makes level-sensitive two-phase clocking disciplines more tolerant to clock skew and jitter are their non-overlap intervals. Liberal skew margins can be designed into such circuits by adjusting the waveforms of their clock signals.*

Example

IBM's patented **level-sensitive scan design** (LSSD) technique [154] [155] is a clever combination of unsymmetric level-sensitive two-phase clocking with a scan-type test facility. Figure 6.12 shows the key ideas behind it. The output Q of each LSSD storage element connects to an input labeled SCI on a subsequent LSSD cell much as in an edge-triggered scan path design.

The three clock nets notwithstanding, LSSD follows a two-phase clocking scheme. During normal operation, CLK1 and CLK2 exhibit non-overlapping pulses as in any other unsymmetric level-sensitive two-phase clocking system. Terminal D of each LSSD storage element then acts as data input and

Q as data output. Clock CLK0 is made to remain inactive at logic 0 by the clock generator, thereby disabling all scan inputs SCI.

In scan mode, the waveforms of CLK0 and CLK1 are swapped, which causes all LSSD elements to read data from their respective SCI terminals. Data at the D inputs are being ignored because CLK1 is shut down at 0. The LSSD storage elements together act as a long shift register, which makes it possible to serially write out their state to pin SCO and to read in a new state from pin SCI.

Incidentally, observe that the power dissipated in driving the clock nets is roughly the same as for a two-phase scheme because only two out of the three clock nets are active at any time.
□

Practically speaking, the relative robustness of the two level-sensitive clocking schemes discussed in sections 6.2.4 and 6.2.5 implies that their clock distribution networks need not necessarily be balanced to the same degree of perfection as in the case of edge-triggered circuits. Clock skew even becomes close to harmless when circumstances permit one to drive a circuit from a slow clock that affords ample non-overlap phases. Relaxed timing constraints are very welcome to experienced designers, who take advantage of them for making do with less and lighter clock buffers in order to improve on overall energy efficiency and on switching noise.

Distributing two or three clock signals is not always popular, however, because of the extra wiring resources required, the inferior EDA support, the less intuitive circuit operation, and the more complicated timing. Also note that considerable skew can build up between the various clock nets if they are driven from distant buffers or if they significantly differ in their electrical or geometric characteristics. Thus, one can't help asking

"Is it possible to design latch-based circuits that are driven by a single clock signal, and what characteristics would such circuits have?"

The answer is positive and two very different solutions are going to be examined next.

6.2.6 Single-wire level-sensitive two-phase clocking

HARDWARE RESOURCES AND OPERATION PRINCIPLE

Single-wire level-sensitive two-phase clocking comes in a symmetric and in an unsymmetric variation. Either variation differs from its two-wire counterpart in that all latches are being driven from one common clock signal. The usage of latches that pass and hold on opposite clock polarities does away with the need for a second clock net.

As far as timing is concerned, this approach is equivalent to driving the two subsets of latches in the original configuration of fig.6.8 with complementary signals or, which is the same thing, with clock signals whose non-overlap intervals have been removed.

DETAILED ANALYSIS

The critical hold margins are those of (6.18) and (6.19) with zero substituted for T_2 and T_4. Assuming identical timing figures for either latch bank, these two inequalities reduce to

$$\max|t_{sk}| \leq \min(t_{cd\,lc} + t_{cd\,c} - t_{ho\,la}) \tag{6.24}$$

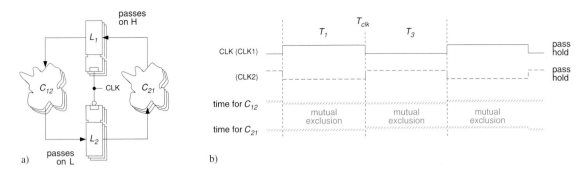

Fig. 6.13 Symmetric single-wire level-sensitive two-phase clocking. Hardware organization (a) and simplified timing diagram (b).

where $\min(t_{cd\,c})$ is a shorthand for $\min(t_{cd\,21}, t_{cd\,12})$.

$$\max|t_{sk}| \leq \min(t_{cd\,lc} - t_{ho\,la}) \tag{6.25}$$

then holds in the absence of combinational logic between the latches. This constraint can be relaxed only if the circuitry is organized in such a way as to never connect two latches directly with no logic in between, inserting dummy buffers for their contamination delay where necessary.

IMPLICATIONS

With all non-overlap intervals that could protect against timewise variability gone, single-wire level-sensitive two-phase clocking is essentially as exposed to skew and jitter as edge-triggered one-phase clocking is. On the positive side, there is an unrestricted capability for time borrowing between C_{12} and C_{21}.

Yet, level-sensitive two-phase clocking offers another and less obvious opportunity for improving circuit performance. Assume the data input of a latch is being driven from a non-inverting logic gate, such as a 3-input AND, for instance. We then have a cascade that consists of a 3-input NAND, a NOT, and the bistable's own inverting input buffer. As shown in fig.8.21, the circuit can be improved by merging the AND operation into the bistable and by collapsing the cascaded inverters. The usage of a "3-input AND function latch", as the combined cell is called, does away with two inverter delays, or almost so.[14] Energy efficiency also benefits from the circuit modification.

What sets symmetric level-sensitive two-phase clocking apart from other schemes is that it becomes possible to play this trick <u>twice</u> per computation period by embedding logic in both of the two latch subsets. Whether one or two nets are being used for clock distribution does not matter in this context, so most of what has been said here applies to standard (two-wire) level-sensitive two-phase clocking as discussed in section 6.2.4 too.

---

[14] You may want to see sections 2.4.3 and 8.2.2 for more information on function latches.

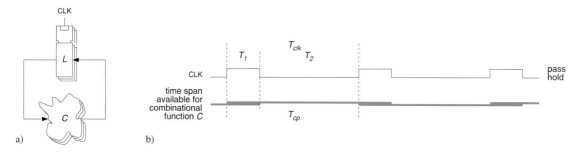

Fig. 6.14 Level-sensitive one-phase clocking. Basic hardware organization (a) and simplified timing diagram (b).

Example

Probably the most prominent VLSI circuits constructed along this line were among those of the Alpha processor family introduced by the now defunct Digital Equipment Corp.[15] Please refer to section 6.3 for details on how the clock is distributed over each of those chips.

□

6.2.7 Level-sensitive one-phase clocking and wave pipelining

We will now turn our attention to a clocking discipline the characteristics of which sharply differ from those discussed so far. In spite of advantages in performance and energy efficiency, level-sensitive one-phase clocking cannot be recommended for ASICs, where limiting the design effort and the design risk are paramount. The subsequent analysis will show why.

HARDWARE RESOURCES AND OPERATION PRINCIPLE

Level-sensitive one-phase clocking stores state in latches exclusively and uses a single clock to drive them. The circuit arrangement corresponds to edge-triggered one-phase design with all flip-flops replaced by latches, see fig.6.14a. The same arrangement is also obtained by bypassing all slave latches in an unsymmetric level-sensitive two-phase design.

The fact that a latch must become transparent before it can accept a new data item for storage, together with the simultaneous clocking of all bistables, implies that the combinational logic is fed with new data (at the beginning of T_1) <u>before</u> the previous results have been latched (which takes place at the end of T_1). As a consequence, the combinational network must assure that data at its output remain unchanged while another wave of data has begun to propagate from the input through that very network. Correct circuit operation rests on inertial effects as transient phenomena are no longer allowed to die out before clocking occurs.

Spinning this idea further, one may want to arrange for two or more data waves to propagate through the combinational logic at any time, a bit like a juggler who keeps several balls at a time

---

[15] Depending on the authors, the clocking discipline used in the 21064 is referred to as "level-sensitive single phase with no dead time" or as "single-wire two-phase clocking".

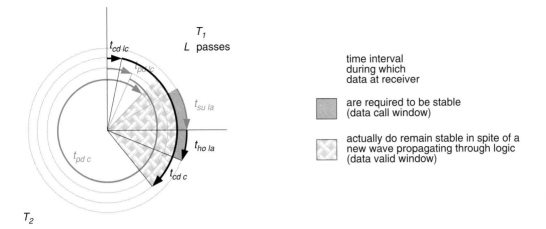

Fig. 6.15 Anceau diagram of level-sensitive one-phase system. The formation of multiple waves becomes apparent by virtue of $t_{pd\,c}$ covering more than a full circle.

in the air. This bold approach is known as **wave pipelining** [156] [157] and attempts to break the upper bound that normally limits throughput by accepting path delays in excess of the clock period.[16]

$$T_{cp} = T_{clk} < t_{lp} = t_{id\,la} + \max(t_c) \qquad \Theta = \frac{1}{T_{cp}} = \frac{1}{T_{clk}} > \frac{1}{t_{lp}} \qquad (6.26)$$

DETAILED ANALYSIS

Setup condition

Combinational logic is fed with new data at the beginning of transparent interval T_1 and must yield a stable result no later than at the end of the subsequent transparent interval.

$$\max|t_{sk}| \le T_{clk} + T_1 - \max(t_{pd\,lc} + t_{pd\,c} + t_{su\,la}) \qquad (6.27)$$

Once more, the setup condition is easily satisfied by acting on T_{clk}. Relation (6.27) exhibits a more intriguing characteristic, however, especially when compared with (6.5) or (6.20). More than a full clock period becomes available for the combinational logic, unless the cumulated latch delay, clock skew, and jitter eat up more time than T_1 provides. In other words, the circuit can be made to operate in a wave pipelined regime as shown in fig.6.15.

---

[16] While it is indeed possible to design feedforward wave pipelines that satisfy (6.26), recall from section 2.7 that first-order feedback loops are not amenable to pipelining (but must be tackled with loop unfolding instead). Since that finding does not depend on how the necessary latency is obtained, it applies to wave pipelining too. Also note that the clock frequencies and throughputs made possible by wave pipelining will not exceed those obtained from a register-based fine-grained pipeline because some minimum separation must always be maintained between any two consecutive waves. Wave pipelining manages with far fewer registers, though.

Hold condition

Data applied to latches at the beginning of transparent interval T_1 must not change before the end of this very transparent interval.

$$\max|t_{sk}| \leq -T_1 + \min(t_{cd\,lc} + t_{cd\,c} - t_{ho\,la}) \tag{6.28}$$

In comparison with edge-triggered one-phase clocking, the meagre skew margin has been reduced further by T_1 and is almost certain to become negative in the absence of combinational logic. It becomes clear now, at what expense the extra leeway in the setup condition has been bought.

IMPLICATIONS

Equation (6.28) imposes an <u>upper</u> bound for clock high phase T_1 which contrasts with all other clocking schemes analyzed so far. As any bistable comes with minimum clock width requirements, the high phase gets bounded from below too. We thus end up with a two-sided constraint for T_1.

$$\max(t_{clk\,hi\,min\,la}) \leq T_1 \leq \min(t_{cd\,lc} + t_{cd\,c} - t_{ho\,la}) - |t_{sk}| \tag{6.29}$$

Whether there is a solution or not depends on the detailed timing figures of the cells being used. Yet, in the absence of combinational logic and relevant interconnect delay, the remaining upper bound $\min(t_{cd\,lc} - t_{ho\,la}) - |t_{sk}|$ leaves very little room indeed for a comfortable clock high time. The problem is further aggravated by the necessity to find a valid timing for a variety of latch types and under all possible operating conditions.

As a consequence, it is impossible to build a shift register — or a scan path — in the normal way, even if skew and jitter are optimistically assumed to be zero. For the circuit to work properly, delay elements must be inserted into each overly short propagation path until (6.28) is safely satisfied. Redundant gates have to be added for their contamination delay alone.

There is another peculiarity that it is worthwhile to note. Slowing down a circuit's operation is often helpful for locating problems. For all clocking disciplines discussed earlier, this could be achieved simply by stretching the clock's waveform. With (6.28) specifying an upper bound, this is no longer possible. Slowing down a level-sensitive one-phase clocked circuit requires keeping one clock phase constant (T_1 in the occurrence of fig.6.14) while extending the other (T_2).

From our analysis of level-sensitive one-phase clocking, we conclude:

+ When compared with two-phase designs, the number of latches and clock nets is cut in half.
+ In theory, and with the exception of computations organized as first-order feedback loops, there is a potential for shortening the clock period to below the logic's propagation delay.
− As a consequence of the feedback loop being closed during the transparent intervals, correct operation critically depends on clock pulse width, gate delays, and interconnect delays.
− All library cells must be bindingly characterized for their contamination delays.
− Interconnect delays, and hence also layout parasitics, must be precisely controlled.
− Signals that arrive too early must be delayed artificially at extra costs in terms of area and energy.
− Sufficient skew margins must also be bought with additional contamination delay.

Delay tuning is a nice word for the process of adjusting circuit delays until a design appears to work for a given set of timing parameters. This practice is not compatible with regular high-productivity EDA design flows. While some HDL synthesizers and automatic place and route (P&R) tools are capable of hold-time fixing, most EDA software is designed to meet long path constraints at minimum hardware and energy costs. Also, contamination delays are neither guaranteed by manufacturers nor normally indicated in library datasheets. Most simulation models do not accurately reproduce contamination delay either.

Observation 6.8. *Level-sensitive one-phase clocking critically depends on fine-tuning propagation and contamination delays through both combinational logic and interconnect. This practice is incompatible with EDA design flows, with existing cell libraries, with reliable circuit operation, and with the drive towards ever higher productivity in VLSI design and test.*

However, note that level-sensitive one-phase clocking was applied successfully to supercomputers such as the Amdahl 580 and the Cray-1 [158] at a time when circuits were assembled at the board level from SSI/MSI parts. A milder form is also being discussed in problem 6.

6.3 | How to keep clock skew within tight bounds

6.3.1 Clock waveforms

The main problem of **clock distribution** is to drive many thousands of clocked subcircuits, such as latches, flip-flops, and RAMs, spread over a die, a board or a system while keeping the unavoidable skew within narrow limits. Slow clock ramps are problematic for several reasons:

- As illustrated in fig.6.16, unavoidable disparities of the switching thresholds across clocked cells translate ramp times into skew.
- The timing figures published in datasheets and stored in simulation models are obtained from stimulating cells with fast clock ramps of 50 ps and less during library characterization. Setup and hold times tend to grow when a cell is being clocked with slow ramps.
- Correct behavior and accurate timing of flip-flops, memories, and — to a somewhat lesser extent — latches are put at risk.

As a consequence, clock signals must literally snap from 0 to 1 and back again.

Observation 6.9. *Among all signals on a chip or in a system, clocks typically feature the largest fanout, travel over the longest distance, and operate at the highest speed. Yet, their waveforms must be kept particularly clean and sharp.*

The most innocent idea for distributing a clock is to treat it like any other signal, i.e. to use a standard minimum-width line to connect all clocked subcircuits to a common source. That such an approach is not viable becomes immediately clear from a numerical example.

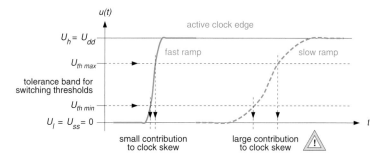

Fig. 6.16 Sluggish clocks translate into extra skew.

Warning example

Consider a clock domain in a CMOS IC that includes a modest 500 flip-flops. Clock distribution is via a metal line 130 nm wide that meanders through the chip's core area much as in fig.6.17a. Sheet resistance is 70 mΩ/□, a typical value for a second- or third-level metal. For simplicity, assume that the flip-flops are connected to that wire at regular intervals, thereby forming an RC network of ladder type with $\#_{sct} = 500$ identical sections of length 100 □ and with a capacitance of 12 fF each. The delay and the ramp time at the end of such a net in response to a step at the input then roughly are [159]

$$t_{pd\,wire} \approx 0.4\,R_{sct}\,C_{sct}\,\#_{sct}^{2} = 0.4\,R_{wire}\,C_{wire} = 0.4 \cdot 3500\ \Omega \cdot 6\ \text{pF} = 8.4\ \text{ns} \qquad (6.30)$$

$$t_{ra\,wire} \approx R_{sct}\,C_{sct}\,\#_{sct}^{2} = R_{wire}\,C_{wire} = 3500\ \Omega \cdot 6\ \text{pF} = 21\ \text{ns} \qquad (6.31)$$

where R_{sct} and C_{sct} refer to the lumped resistance and capacitance respectively of one section as illustrated in fig.6.18a.[17] Obviously, such figures are totally inadmissible for a clock.
□

The above example suggests two starting points. Firstly, the clock net must be reshaped and resized so as to cut down, control, and balance interconnect delays in a better way. Secondly, interconnect lengths, capacitances, and resistances must be cut down by subdividing a chip-wide clock net into many smaller nets, each with a driver of its own. These are indeed the basic ideas behind two alternative approaches that we are going to discuss next.

6.3.2 Collective clock buffers

The collective approach has a single buffer that connects to all clocked cells directly via metal lines. To handle the huge fanout, the buffer must be huge and consist of multiple stages of increasing drive strengths. The idea failed in the above example because the clock net was shaped like a long, narrow, and winding alpine road. The picture improves if length, width, and shape are chosen more carefully. As the ensemble of clocked cells sets a lower bound for the load capacitance, it is the resistance of the distribution network that must be kept low, while, at the same time, attempting to make all interconnect delays about the same.

---

[17] More interconnect models are to be presented in section 12.8.

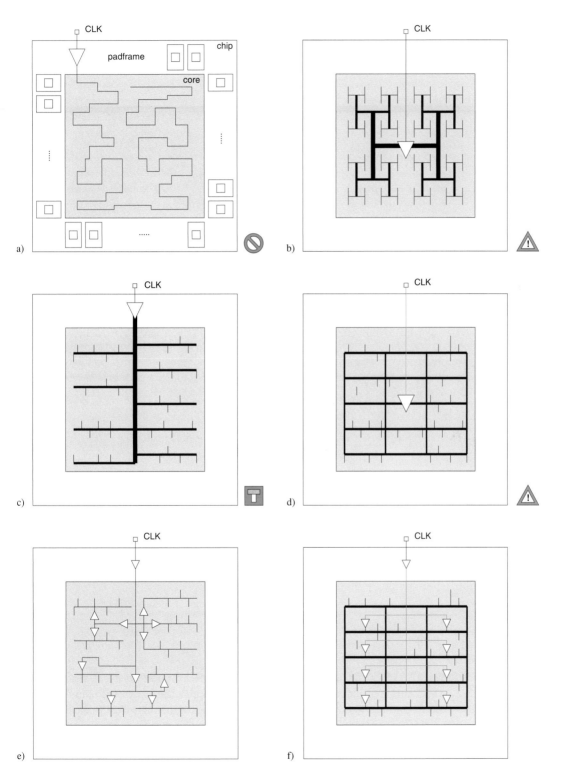

Fig. 6.17 On-chip clock distribution schemes (simplified). Narrow meander (a), central buffer combined with a balanced H-tree layout (b), collective buffer combined with spine-like wiring (c), central buffer combined with grid-like wiring (d), distributed buffer tree (e), distributed buffer tree combined with a grid (f).

The following countermeasures all help to improve the situation:

- Keep clock wires as short as possible, place the driver close to the center of the circuit.
- Make clock distribution wires reasonably wide. Narrow wires yield a high resistance whereas plate capacitance dominates in excessively wide wires.
- Use the upper metal layers as only they combine low resistance with low capacitance.
- Avoid unnecessary layer changes as contacts and vias contribute to resistance.
- Equalize delays by making all clock paths electrically and geometrically similar.

Three layout arrangements developed with these rules in mind are shown in figs.6.17b to d. Although the recursive H-tree immediately strikes one as the optimal solution, it is difficult to implement in practice unless clock loads are uniformly distributed and unless one is prepared to set aside one metal layer for the sole purpose of clock distribution. Spine, comb, and grid topologies similar to those adopted for power distribution purposes are or — rather — were more realistic. Yet, wide clock lines unnecessarily inflate parasitic capacitance and, hence, power dissipation. Another difficulty is that the important switching currents associated with driving large loads concentrate at a few points in a chip's floorplan.

Example

Consider a circuit with 10 000 bistables. Clocking is from a collective buffer with a voltage swing of 1.2 V. Let the aggregate load capacitance per bistable be the same as in the previous example since the clock lines are shorter but also much wider this time. Further assume that the drive current follows a triangular waveform during the signal's ramp time of 100 ps. It then peaks out at

$$\hat{I}_{clk} \approx 2\,\frac{C_{clk}\,U_{dd}}{t_{ra\,clk}} = 2\,\frac{10\,000 \cdot 12\ \text{fF} \cdot 1.2\ \text{V}}{100\ \text{ps}} \approx 2.9\ \text{A} \tag{6.32}$$

At a clock frequency of 500 MHz, the final inverter stage driving the clock net dissipates

$$P_{clk} > \frac{\alpha_{clk}}{2}\,C_{clk}\,U_{dd}^2\,f_{clk} = 1 \cdot 10\,000 \cdot 12\ \text{fF}\,(1.2\ \text{V})^2\,500\ \text{MHz} \approx 86\ \text{mW} \tag{6.33}$$

☐

Current spikes as strong as this necessitate special precautions. Older cell libraries used to provide special **clock drivers** designed to fit into a chip's padframe, where they were fed from dedicated power and ground pads in order to avoid ground bounce problems.

Example

The original Alpha processor, the 21064, followed a collective driver approach in its purest form. Total capacitive load of the clock node was 3.25 nF, requiring final driver transistors with $W_n = 100$ mm and $W_p = 250$ mm [161]. Power dissipation was 30 W from a 3.3 V supply at a clock frequency of 200 MHz. Clock rise and fall times were 500 ps and a peak switching current of 43 A had been measured. Clock distribution roughly followed fig.6.17d with the central driver extending along more than half of the chip's centerline. Transmitting a clock edge from the chip's center to a corner was found to take less than 300 ps, which compared favorably with the 5 ns clock period and paved the way for single-wire level-sensitive two-phase clocking where there are no non-overlap intervals that could provide extra room for skew.

☐

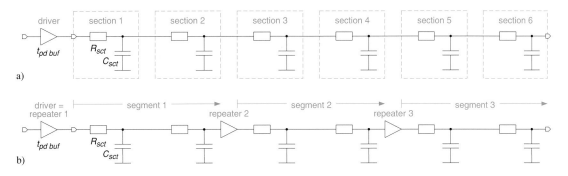

Fig. 6.18 Lumped RC model of a bare wire (a) vs. that of a repeated wire (b). $\#_{sct} = 6$ and $\#_{seg} = 3$ in this example.

6.3.3 Distributed clock buffer trees

The concern for energy efficiency subsequently mandated the introduction of gated clocks and drove designers away from the flat clock distribution schemes popular at the time when the Alpha 21064 was developed.

What's more, observe from (6.30) that interconnect delay grows quadratically with line length l because $R_{wire} \propto l$, $C_{wire} \propto l$, and $\#_{sct} \propto l$. Let us examine what happens when a long line gets turned into a **repeated wire** by subdividing it into $\#_{seg}$ shorter segments separated by $\#_{seg} - 1$ inverting or non-inverting buffers in between as shown in fig.6.18b. To first order, overall propagation delay then becomes

$$t_{pd\,rept} \approx \#_{seg} \left(t_{pd\,buf} + 0.4\,R_{sct}\,C_{sct} \left(\frac{\#_{sct}}{\#_{seg}} \right)^2 \right) = \#_{seg}\,t_{pd\,buf} + \frac{0.4}{\#_{seg}} R_{wire}\,C_{wire} \quad (6.34)$$

which means that the overall delay can be made a <u>linear</u> function of line length just by chosing $\#_{seg} \propto l$. To minimize the overall delay, the propagation delay of each wire segment should be made to match that of one repeater, in which case one speaks of an optimally repeated wire. In practice, repeaters are typically spaced further apart so as not to unnecessarily inflate the overhead in terms of area and power.

A distributed clock tree thus consists of a multitude of repeaters of moderate size physically located close to the loads they drive and inserted into every major branch, see fig.6.17e. As opposed to the collective approach, no large currents need to flow over large distances. Also, the power dissipated for driving the clock net is distributed over the chip's area. Conversely, a tree-like structure requires careful delay matching. The goal is to equalize the delays along all branches in spite of unevenly distributed loads and distances.

Techniques for equalizing branch delays include:

- Hierarchically partition the design in such a way as to balance clock loads to a reasonable degree.
- Plan for local subtrees and size clock buffers as a function of the load they must drive.
- Prefer the upper metal layers for their lower resistance and capacitance.

- Retard overly fast clock distribution paths by inserting extra buffers.
- For fine-tuning, consider adding dummy loads and detours to early branches.

One difficulty in implementing the distributed buffer approach is that accurate data on which to base detailed delay calculations do not become available until late in the design cycle. Another problem is that the clock tree must be readjusted whenever a design modification entails a minor change in the loads to drive or in their geometric locations.

The EDA industry has come to the rescue by developing automatic software tools that are capable of inserting balanced clock trees into circuits of substantial size, largely doing away with the need for manual delay budgeting and iterative tuning. Such specialized **clock tree generators** are run as part of physical design, e.g. between place and route, so as to base their calculation on fairly accurate estimates for parasitic capacitances and interconnect delays. Such tools have significantly contributed towards making clock distribution trees popular.

Hint: Excluding inverters from clock trees in favor of non-inverting repeaters does away with any risk of ending up with branches that differ in the number of negations.

6.3.4 Hybrid clock distribution networks

Hybrid schemes have been devised in an attempt to combine the best of both worlds. Instead of having the leaf cells of a distributed buffer tree drive the clocked cells directly, their outputs are shorted together by a domain-wide metal grid as sketched in fig.6.17f. The grid minimizes local skew by providing low-resistance current paths between nearby points. The fairly stiff and uniform structure so obtained facilitates circuit design as timing does not depend too much on details of cell placement. This approach results in a small distribution delay and acceptable power without eating too much wiring resources [162] [155].

Examples

The utmost degree of sophistication has been achieved with Intel's Itanium 2 9000, an impressive dual-core design clocked at 1.6 GHz with $1.72 \cdot 10^9$ transistors manufactured in a 90 nm bulk CMOS process with seven layers of Cu interconnect. The sub-100 nm technology, the huge die size of 27.7 mm by 21.5 mm, and the maximum power dissipation of 104 W in conjunction with local hot spots all contribute to important and unpredictable on-chip variations (OCVs). While an oversize clock grid would have helped to contain skew, designers were afraid of the large equalizing currents and of the power overhead this would have entailed [163]. Instead of contenting themselves with statically balancing clock distribution delays at design time, their regional active deskew scheme adaptively fine-tunes delays in the different clock tree branches at run time with the aid of multiple phase comparators and adjustable delay lines.

A rather different concept has been pursued by Sun Microsystems in their Niagara processor that includes $279 \cdot 10^6$ transistors manufactured on a die of 378 mm^2 in 90 nm 9LM (Cu) bulk CMOS technology. Maximum power is indicated as 63 W at 1.2 GHz and 1.2 V. By cleverly combining a global H-tree with multiple regional clock-gated grids and with local subtrees, its designers were able to keep clock skew below 50 ps without resorting to complex active deskewing circuitry [164]. While overall power density is comparable to the above example, the smaller die and the better

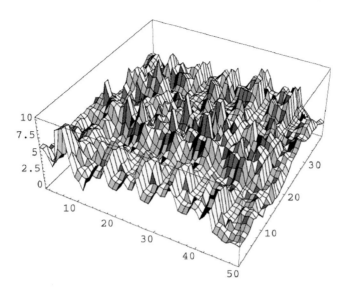

Fig. 6.19 Clock skew map for a Cell processor chip measuring roughly 17.8 mm by 12.3 mm. Horizontal dimensions are given in arbitrary units, skew is in picoseconds (photo copyright IEEE, reprinted from [165] with permission).

spreading of heat afforded by the concurrent operation of eight relatively modest processor cores were certainly helpful in avoiding inordinate delay variabilities. Another particularity is post-layout hold-time fixing with the aid of metal-programmable delay buffers. Following static timing analysis, delays get statically adjusted using a local metal with no impact on the previously frozen detailed signal routing.

□

6.3.5 Clock skew analysis

Functional simulation is inadequate for uncovering clock skew problems. As exhaustive simulation is not practical, it is very likely that not all critical patterns get applied and that some skew-related timing problems pass unnoticed. The good news is that there is no need to simulate a circuit functionally to find out whether timing conditions are met or not in synchronous designs. Timing verification, aka static timing analysis, is much more appropriate as it

- Circumvents all coverage problems,
- Quantifies the timing margins on all signal propagation paths, both short and long,
- Locates slack, if any, and
- Entails a more reasonable computational effort.

Do not forget to include layout parasitics, to use adequate interconnect models, and to account for crosstalk, PTV, and OCV variations during static timing verification.

Observation 6.10. *While both collective clock buffers and distributed clock trees have been made to work in commercial circuits, trees prevail because they support clock gating, help to keep distribution* *delay low, and tend to be less demanding in terms of routing resources, overall power, and current crowding. In high-performance VLSI, there is a trend towards combining the efficiency of a distributed clock tree with the robustness of a domain-wide metal grid. Careful design and timing verification are critical in any case.*

6.4 | How to achieve friendly input/output timing

6.4.1 Friendly as opposed to unfriendly I/O timing

We now extend our discussion to the board-level interface of entire chips that are part of the same clock domain, see fig.6.20a.[18] The basic requirement of observation 6.1 that any data-call window must be fully enclosed by the pertaining data-valid window remains exactly the same as for data transfers between simple flip-flops within a chip.

Interfacing with outputs that provide valid data for most of a clock period is straightforward as there is plenty of time for the receiving circuit to evaluate and store the acquired data. Similarly, most of the clock period is available for the transmitting circuit to settle if the receiver calls for stable data during a brief time slot just prior to the active clock edge. Note the ample setup and hold margins in fig.6.20b.

As opposed to this, interfacing with a circuit that just flashes output data for a brief moment of time is a real nuisance. Data are likely to settle too late or to vanish too early for the receiver to get hold of them. Similarly, inputs that ask for stable data for an extended period leave little time for output logic and interconnect delays. A long hold time proves especially awkward since it obliges the transmitter to maintain its former output long after the active clock edge has sparked off a transition to a new state. The two I/O timings shown in fig.6.20d are in effect incompatible: It is impossible to transfer data if transmitter and receiver are to be driven from a common clock. Engineers are then forced to resort to delay lines, stopover registers, adventurous clocking schemes, and other makeshift improvisations.

The desiderata for friendly input–output timing are as follows.

I/O timing	data-call window	$t_{su\,inp}$	inputs $t_{ho\,inp}$	data-valid window	$t_{pd\,oup}$	outputs $t_{cd\,oup}$
friendly	narrow	small	close to zero or, better, negative	wide	small	large, as close to $t_{pd\,oup}$ as possible
unfriendly	wide	large	large (positive)	narrow	large	close to zero

How to express timing constraints for commercial synthesis and timing verification tools has been the subject of section 4.3.6.

---

[18] Note that much the same argument also applies to virtual components and other major circuit blocks that contain a clock distribution network of their own. How to synchronize data at the boundaries between distinct clock domains is to be addressed in chapter 7.

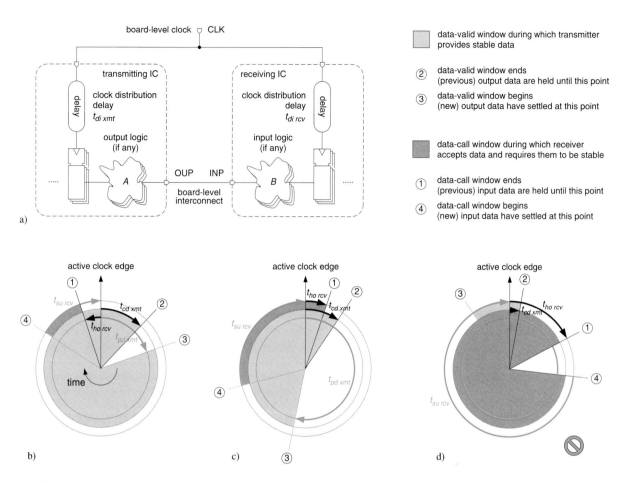

Fig. 6.20 Two VLSI chips with a data transfer path (a). Anceau diagrams for friendly (b), average (c), and incompatible I/O timing characteristics (d). Any counterclockwise arrow implies a negative sign.

6.4.2 Impact of clock distribution delay on I/O timing

Driving the huge capacitances associated with clock nets necessitates large multi-stage drivers which bring about important clock distribution delays. So far, we have not cared much about this effect as only skew matters within a circuit. Yet, the impact of clock distribution delay on I/O timing is twofold, just compare figs.6.21 and 6.22.

On outgoing signals, distribution delay simply adds to both propagation and contamination delays. The resulting data lag remains largely uncritical unless a system is to operate at high clock frequencies, in which case the prolonged propagation delay will be felt.

On incoming signals, distribution delay shortens setup and extends hold time. Even a moderate distribution delay is likely to render setup time negative on those inputs that directly connect to a register. More importantly, however, the same amount of delay will inflate hold time way beyond any realistic value for a transmitter's contamination delay.

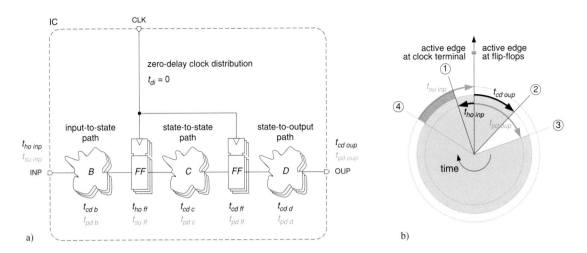

Fig. 6.21 I/O timing of a VLSI chip in the absence of clock distribution delay. Circuit (a) and Anceau diagram (b). Note: Events refer to input and output terminals of a single circuit here, which view contrasts with all Anceau diagrams shown so far in this chapter, where events related to the interface between two circuits.

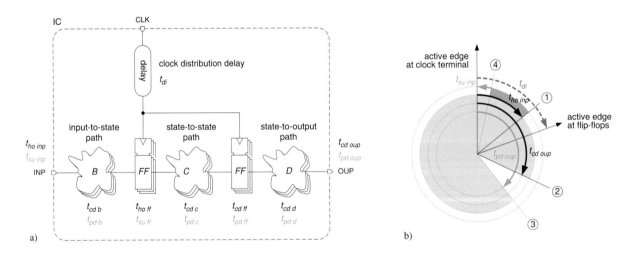

Fig. 6.22 Impact of excessive clock distribution delay. Circuit (a) and Anceau diagram (b).

For a quantitative analysis, assume the circuit operates with edge-triggered one-phase clocking. The circuit's input and output terminals then exhibit these timing parameters

$$t_{su\,inp} = \max(t_{pd\,b} + t_{su\,ff}) - t_{di} \tag{6.35}$$

$$t_{ho\,inp} = \max(-t_{cd\,b} + t_{ho\,ff}) + t_{di} \tag{6.36}$$

$$t_{pd\,oup} = \max(t_{pd\,ff} + t_{pd\,d}) + t_{di} \tag{6.37}$$

$$t_{cd\,oup} = \min(t_{cd\,ff} + t_{cd\,d}) + t_{di} \tag{6.38}$$

In conclusion, distribution delay adds to input hold time and may thus easily turn a circuit with perfect I/O timing characteristics into an unacceptable one. What's more, within a population of digital components, any differences between clock distribution delays translate into clock skew unless they get counterbalanced within the board- or system-level clock distribution network.

Observation 6.11. *While it is imperative to minimize clock skew, it is also important to keep on-chip clock distribution delay within tight bounds to avoid the awkward timing characteristics at the chip's I/O interface that otherwise result.*

Example

FPGAs of the Actel SX-A family include three dedicated low-skew clock networks. One of them is referred to as hardwired, which is to say that there are no more than two programmable links between the clock input and any on-chip flip-flop. This is done to minimize not only clock skew but also distribution delay. The resulting skew is said to be as low as 0.1 ns and the pin-to-pin delay from the clock input to any register output is specified as 5.3 ns (A54SX72A-3 under worst-case commercial conditions).

☐

6.4.3 Impact of PTV variations on I/O timing

Keeping a receiver's data-call window within the transmitter's data-valid window under all circumstances is particularly difficult at the interface between two components because

- I/O timing is subject to variation with the off-chips loads attached,
- I/O timing is affected by board-level interconnect delays,
- What matters is the <u>difference</u> between the distribution delays of receiver and transmitter,
- The PTV conditions of two chips are bound to differ, and
- PTV variations notoriously have a large impact on timing figures.[19]

While clock distribution delay is of little importance as long as all clocked subcircuits are affected alike, it becomes highly critical at the interface between two components and at the interface between two supply domains. It is not uncommon to find that maintaining all components properly synchronized within a clock domain is impossible when those components are subject to moderate but non-uniform PTV variations.

Observation 6.12. *PTV variations put data transfers between ICs at risk. A healthy latitude to timing variations is thus even more important at the I/O interfaces than within a chip itself.*

Before discussing a more radical solution for all the above problems in section 6.4.8, let us suggest a few countermeasures that alleviate the undesirable effects of clock distribution delay.

---

[19] It is not exceptional to see timing quantities vary by a factor of three from the slowest to the fastest operating conditions acceptable for some given CMOS technology. See chapter 12 for actual figures.

6.4.4 Registered inputs and outputs

Output signals should not be made to propagate through deep combinational networks. Including an **output register** right before the pad drivers maximizes the data-valid window and — as a welcome side effect — also yields hazard-free signals. A price to pay is the extra latency. If needed, the circuit's contamination delay can be extended by driving the output register from a somewhat delayed local clock.

Input registers, on the other hand, must not be allowed to suffer from substantial and/or variable clock distribution delay as input hold time otherwise becomes unmanageable. The workarounds to be presented next are, therefore, particularly important on the input side.

6.4.5 Adding artificial contamination delay to data inputs

As data lag compensates for clock distribution delay, any combinational network inserted between the input pads and the first bistable helps because of its contamination delay. After all, it is the timing of the clock relative to the data signals that decides when exactly data voltages get interpreted, also see problem 8.

Examples

Various Xilinx FPGA families include a user-configurable multiplexer in each I/O block that selects or bypasses a delay line in the data input propagation path, see fig.6.23. Note that the data-call window is being shifted and observe the impact on the FPGA's hold-time requirement.

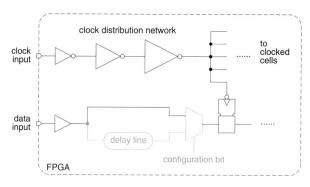

Data input configured	Input timing if			
	straight		delayed	
Xilinx FPGA	$t_{su\,inp}$ [ns]	$t_{ho\,inp}$ [ns]	$t_{su\,inp}$ [ns]	$t_{ho\,inp}$ [ns]
XCS05-4	1.2	1.7	4.3	0.0
XCS40-4	0.4	3.5	5.3	0.0
XC2V40-6	0.84	−0.36	3.24	−2.04

Fig. 6.23 Compensating clock distribution delay with configurable delay lines on FPGA data inputs. Principle (top) and excerpts from Xilinx data sheets (bottom).

6.4.6 Driving input registers from an early clock

The interfacing of ICs or virtual components (VCs) that share a common clock would be easier if their input registers — and possibly their output registers too — could be driven from a separate clock that does not suffer from a distribution delay as important as that of the rest of the circuit. This is indeed possible on the basis of the observation that I/O registers account for only a small fraction of a circuit's total clock load. An early clock is tapped from the main tree close to its root, see fig.6.24. The resulting lead of the input registers must be taken into account when designing the remainder of the circuit, however.

6.4.7 Tapping a domain's clock from the slowest component therein

Complex circuits often feature comparatively long hold times, making it difficult to interface them with smaller and faster components. This is mainly because clock distribution delay tends to grow with circuit size and clock load. The practice of mixing high-density CMOS core functions with fast SSI/MSI components — such as high-speed CMOS, bipolar or BiCMOS — for glue logic functions on a circuit board exacerbates the problem.

A workaround exists provided that the component that exhibits the longest clock distribution delay within the clock domain can be arranged to have a special output tapped from the clock distribution network close to its leaves and brought to an extra package pin, see fig.6.24. The original clock is then first fed into that slower part and redistributed from its clock tap to all faster components within the same clock domain.

☐

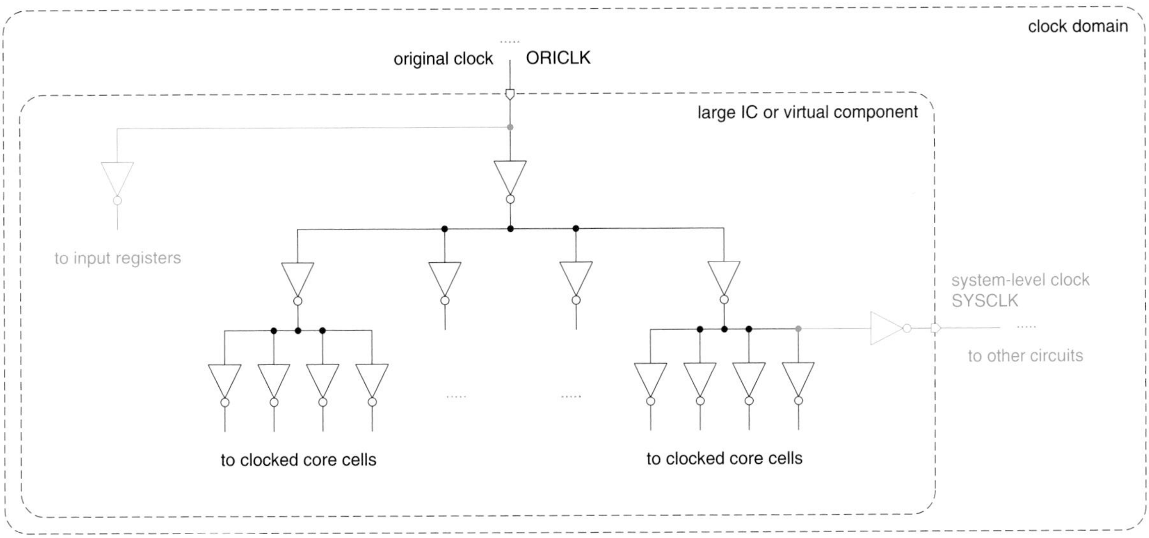

Fig. 6.24 Two more options for compensating clock distribution delay (data I/O not shown.)

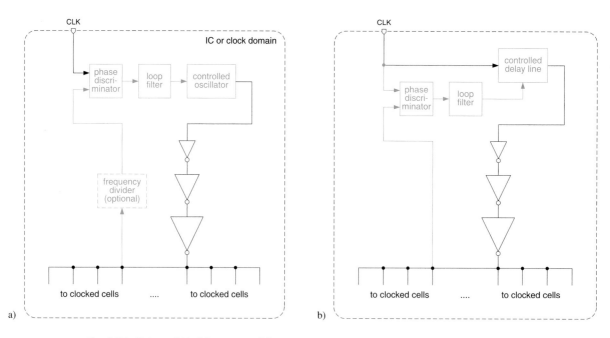

Fig. 6.25 Using a PLL (a) or a DLL (b) to minimize and stabilize clock distribution delay.

6.4.8 "Zero-delay" clock distribution by way of a DLL or PLL

All problems of I/O timing would be gone if clock distribution delay could be made zero as the events that trigger state changes and the external clock would then perfectly align. This is indeed possible with an on-chip **delay-locked loop** (DLL) or **phase-locked loop** (PLL). Once locked, both DLLs and PLLs act as a servo loop that keeps a local signal in phase with a reference signal supplied from externally. The difference essentially relates to the actuator in the servo loop: PLLs use a controlled oscillator whereas DLLs include an adjustable delay line or — which is the same — a controlled phase shifter. In either case, the servo loop is made to compensate for the cumulated propagation delays of clock buffers and interconnects, see fig.6.25. The overall circuit then behaves like a distribution network with zero or close to zero insertion delay. Better still, PTV variations are automatically compensated for.

As every overclocker knows, microprocessors typically operate at rates that are an integer multiple of the clock frequency distributed over the motherboard. This is made possible by using a PLL with a frequency divider inserted into the feedback loop as indicated in fig.6.25a. Both DLLs and PLLs take a number of cycles to lock and have a limited lock range, which makes them intolerant of abrupt frequency changes and vetoes single-cycle operation. Unless frequency multiplication is truly required, DLLs are preferred for their unconditional stability, faster lock times, and low jitter.

DLL and PLL subcircuits are not yet routinely available as part of ASIC design libraries. They can be designed either from voltage-controlled analog delay lines and oscillators or around a numerically adjustable delay line. [481] is a reference on the design of integrated PLLs. Digitally controlled delay lines with a resolution finer than one gate delay will be addressed in section 8.4.6. A particular difficulty with adjustable delay lines in the context of clocking applications is that the respective propagation delays for rising and for falling edges must closely match.

Examples

The PowerPC 603 and 604 by Motorola were among those CPUs that combined an on-chip PLL with an H-tree distribution network driven by a collective buffer [166]. Lock range was reported to span from 6 to 175 MHz with lock times below 15 µs. Instead of driving thousands of leaf cells directly, the H-tree was designed to distribute a low skew clock to 266 local clock regenerators, each of which drives its share of clocked cells. With 80 fF load from each regenerator, the H-tree was measured to cause a delay of 230 ps and a clock skew of 135 ps.

An analog and a mostly digital DLL circuit designed for clock alignment in the context of a memory system have been compared in [167] and the results are summarized below. Both circuits have been fabricated on the same chip with a 400 nm 3.3 V standard CMOS process. Nominal clock frequency is 400 MHz, yet data are transferred on both edges of the clock so that the time interval is a mere 1.25 ns per bit. Also consider that the digital circuit takes less effort to design and is more easily ported to a new process than the analog alternative.

DLL architecture	analog	digital	
area	0.68	0.96	mm^2
power dissipation	175	340	mW @ 3.3 V
maximum clock frequency	435	> 667	MHz
minimum supply voltage	2.1	1.7	V
lock time	2.0	2.9	µs
clock jitter	195	245	ps peak-to-peak

The clock distribution subsystem in Intel's Itanium CPU is by itself very complex and organized into global, regional, and local clock distribution. It makes use of one PLL at the global and of 30 (sic!) DLLs at the regional level. Implemented in 180 nm CMOS technology, the sophisticated distribution scheme limits clock skew to 28 ps for a 1.25 ns (800 MHz) clock [168].

Actel's ProASIC^{PLUS} FPGAs include two clock conditioning blocks, each of which consists of a PLL, four programmable clock dividers, and a few delay lines. The output frequency range is 6 to 240 MHz with a maximum acquisition time of 20 µs. A lock signal indicates that the PLL has locked onto the incoming clock signal. Delay lines are programmable in increments of 250 ps. Supply voltage must be between 2.3 V and 2.7 V, nominal power dissipation is 10 mW.
□

6.5 | How to implement clock gating properly

6.5.1 Traditional feedback-type registers with enable

In most designs, part of the flip-flops operate at a lower rate than others. Just compare a pipeline register against a mode register that is to preserve its state for millions of consecutive clock cycles or even until power-down. Other registers occupy positions in between. This situation suggests the usage of enable signals to control register operation as illustrated in the VHDL code fragment below.

```
     .....
     -- activities triggered by rising edge of clock
     if CLK'event and CLK='1' then
        if ENA='1' then    -- control signal to govern register update
           PREST <= NEXST;
        end if;
     end if;
     .....
```

When presented with such code, synthesizers call upon a multiplexer to close and open a feedback loop around a basic D-type flip-flop under control of the enable signal as shown in the example of fig.6.26b. As the resulting circuit is simple, robust, and compliant with the rules of synchronous design,[20] this is a safe and often also a reasonable choice. Some libraries indeed offer an E-type flip-flop that combines a flip-flop and a MUX in a single cell.

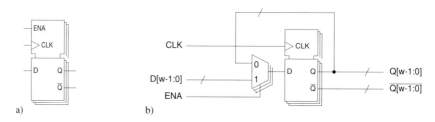

Fig. 6.26 Register with enable. Icon (a) and safe circuit built from D-type flip-flops and multiplexers (b) or, which is the same, from E-type flip-flops.

On the negative side, this approach takes one fairly expensive multiplexer per bit and does not use energy efficiently. This is because any toggling of the clock input of a disabled flip-flop amounts to wasting energy in discharging and recharging the associated node capacitances for nothing. The capacitance of the CLK input is not the only contribution as any clock edge causes further nodes to toggle within the flip-flop itself. Most of that energy could be conserved by selectively turning off the clock within those circuit regions that are currently disabled.[21] Any such conditional clocking or **clock gating** scheme must be implemented very carefully, however, as safe circuit operation is otherwise compromised.

6.5.2 A crude and unsafe approach to clock gating

Using a simple AND gate for the clock seems particulary tempting in positive-edge-triggered circuits, but is also particulary dangerous, see fig.6.27a. This is because any glitch of the enable signal ENA is passed on to the gated clock CKG while the clock input CLK is 1. All downstream bistables are exposed to hazards during the clock's first phase where hazards are particularly likely, jeopardizing

---

[20] Notably with the dissociation principle presented in section 5.4.

[21] Refer to [169] for a conceptually different approach to conditional clocking that does not depend on the presence of a special enable signal but compares the data value at the input of a flip-flop against that stored.

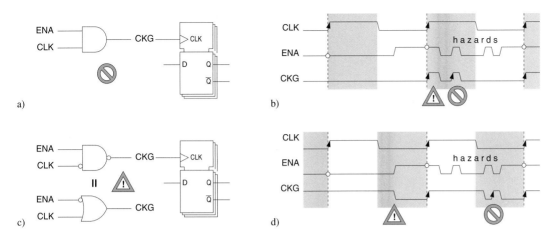

Fig. 6.27 Unsafe clock gating circuits (a,c) along with the waveforms they produce (b,d).

the correct functioning of the entire system. To make the circuit work, designers would have to resort to extensive hazard control with all its undesirable implications.[22]

6.5.3 A simple clock gating scheme that may work under certain conditions

The circuit of fig.6.27c fares better because the NOT-NAND gate in the clock net is transparent for glitches during the second clock phase rather than during the first phase. This approach can be made to operate safely provided

- The timewise position of the passive clock edge on clock CLK is always well defined,
- All transients on the enable signal ENA are guaranteed to die out before the end of the first clock phase under any circumstance,
- The gated clock CKG is rechecked for hazards following layout, the enable signal ENA is checked for hazard conditions during timing analysis, and
- The extra delay introduced by the gate is accounted for in the clock distribution network.

Making sure all of these requirements are consistently met complicates the design process, however. On the positive side, this approach minimizes circuit overhead, clock load, and energy dissipation during those periods when the clock remains disabled.

6.5.4 Safe clock gating schemes

More robust schemes are based on a special **clock gate**, a sequential subcircuit that accepts the enable signal ENA plus the main clock CLK and that outputs an intermittent clock CKG which is to toggle only when the enable asks it to do so. Here come the specifications for such a contraption in the context of edge-triggered clocking:

---

[22] Such as complicated timing analysis, redundant logic for hazard suppression, finicky delay tuning, layout dependencies, difficult design verification, low design productivity, and all the other side effects of ad hoc (i.e. non-self-timed) asynchronous operation.

Table 6.3 Truth tables for safe conditional clocking in single-edge (left) and dual-edge triggered circuits (right).

ENA	CLK	CKG	
1	↑	↑	forward active clock edge
–	0	0	clamp output to zero
↑	1	CKG	maintain output
↓	1	CKG	idem

ENA	CLK	CKG	
1	↑	$\overline{\text{CKG}}$	toggle output
1	↓	$\overline{\text{CKG}}$	idem
0	–	CKG	maintain output
↑	–	CKG	idem
↓	–	CKG	idem

- No latency, that is the enable input must affect the next active clock edge.
- The gated clock output CKG must be free of hazards under all circumstances.
- The enable input ENA must be immune to hazards
  (the absence of glitches from the input clock CLK is taken for granted).
- The only timing constraints imposed on the enable input ENA must be the observation of reasonable setup and hold times (exactly as for an ordinary E-type flip-flop), any signal that emerges from combinational logic must qualify as enable.
- The gated clock must have the same duty cycle as the input clock
  (applies to single-edge-triggering exclusively).
- The propagation delay from CLK to CKG must be small and fixed.
- Overall power dissipation must be low, especially while the clock is disabled.
- The gated clock should come up with a predictable polarity after circuit reset.

The corresponding subcircuits are being shown in fig.6.28. Make sure you understand that all flip-flops in a register — or in a larger storage array — are made to share one such clock gate. The overall power dissipated for clocking a w-bit register will be much lower than if w individual E-type flip-flops were being used as in fig.6.26. The more flip-flops are served by one clock gate, the more important the economy.

It goes without saying that the propagation delay from CLK to CKG must be taken into account during clock tree generation much as with any regular clock buffer. Substantial interconnect delays between the bistable and the combinational gate may also put at risk the correct functioning of the subcircuits of fig.6.28. A better approach than assembling them from standard cells is to add them to a design library as elements in their own right. This is to say that compliance with the original specifications must be checked and that numeric timing figures must be obtained by way of transistor-level simulations exactly as for any other library cell.[23]

Table 6.4 compares all four options. The quest for optimum energy efficiency puts pressure on designers to resort to a multitude of local clocks with a small fanout each and, as a consequence, to prefer NOT-NAND-type gates over latch-based clock gates. Keep in mind this is dangerous unless you know exactly what you are doing. Further note that situations where a hierarchy of conditions is to

---

[23] A soft macro shall be considered only as a workaround when clock gates are unavailable as library elements. The internal wires must then be given so much weight that the cells involved get placed next to each other such as to tightly control interconnect delays and their variations. Even more importantly, soft macros must be prevented from being torn apart by later logic (re)optimization steps, see observation 4.28 for practical advice.

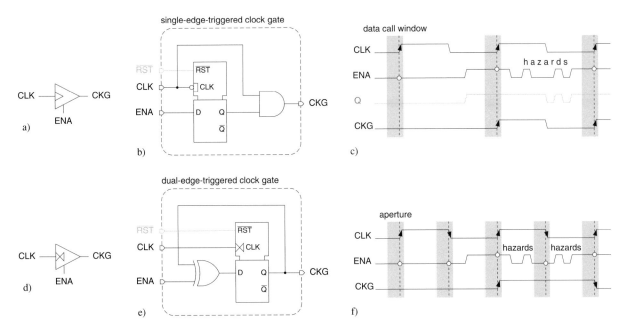

Fig. 6.28 Safe clock gates for single-edge- (b) and dual-edge-triggered one-phase clocking (e). Suggested icons (a,d). Pertaining clock waveforms (c,f).[24]

exert control over local clocks require special attention, see problem 10. As a last word of caution, recall that scan testing mandates that all flip-flops of a scan path be enabled during scan-in and scan-out operations.

Hint: Before opting for a clock gating scheme, check what kind of subcircuits the EDA tool suite available can handle properly during clock tree generation.

6.6 | Summary

- The prime problem in clock engineering is not the lack of solutions but the plethora of divergent alternatives, see table 6.5. Strictly adhering to a consistent clocking discipline throughout is paramount to successfully designing large and dependable digital circuits.

  - Edge-triggered clocking disciplines are most simple conceptually. However, their small latitude with respect to clock skew and jitter requires careful balancing of the clock distribution network all the way to physical design. Careful timing verification is crucial, hold-time fixing may help.

---

[24] Though unnecessary from a purely functional point of view, a reset facility serves to bring the subcircuit into a predictable start state following power-up.

Table 6.4 | Conditional clocking techniques for single-edge-triggered clocking compared.

Enable/disable mechanism	feedback via MUX	AND gate	NOT-NAND gate	latch-based clock gate
Illustration	fig.6.26b	fig.6.27a	fig.6.27c	fig.6.28b
Robustness	safe	unsafe	vulnerable	safe
Circuit overhead for a w-bit register [GE]	$1w \ldots 3w^a$	1.5	1.5	≈ 4.5
Energy balance	wasteful if disabled most of the time	optimal	optimal	good if many flip-flops get gated together
Impact on clock tree synthesis	none	must compensate for skew	must compensate for skew	must compensate for skew
Design effort	minimal, HDL only	not an option	recurring for each clock net	one-time to establish cell

[a] The lower bound refers to an E-type flip-flop library cell, the upper to a separate MUX.

- Single-edge-triggered one-phase clocking is most popular with both ASIC designers and EDA vendors.
- All latch-based clocking disciplines with non-overlapping clocks boast a large tolerance with respect to skew and jitter. This is at the expense of performance in the unsymmetric case only.
- Symmetric level-sensitive two-phase clocking provides much room for optimizing the clock subsystem either for performance or for energy efficiency.
- Translating an edge-triggered one-phase design into an equivalent unsymmetric level-sensitive two-phase design is straightforward, which greatly helps in designing circuits of the latter category.
- Level-sensitive one-phase clocking is hardly applicable to ASICs due to its fatal dependency on contamination and interconnect delays.

• Macrocells, megacells, and subcircuits synthesized from third-party HDL code tend to exhibit unexpected, undesirable, and often undocumented timing features such as excessive hold times, massive internal clock distribution delays, absence of adequate clock buffers, a mix of rising- and falling-edge-triggered flip-flops, amalgams of level-sensitive and edge-triggered bistables, and more. Watch out for such peculiarities.

• Experience tells us that any departure from a plain one-phase-edge-triggered-flip-flops-only approach complicates the design process and asks for a more substantial engineering effort. Just about everything from interface design down to timing verification and design for test (DFT) tends to become more onerous to handle in the presence of

- Asynchronous subcircuits and constructs,
- Hybrid clocking schemes (e.g. mixes of edge-triggered with level-sensitive bistables),
- Unclocked memories or other macrocells that have timing characteristics incompatible with standard flip-flops,

Table 6.5 Comparison of the most important clocking disciplines along with their compatibility and usage with clock distribution schemes.

Characteristics	Synchronous clocking discipline					
	edge-triggered		level-sensitive			
	single edge	dual edge	two-phase			one-phase
			sym-metric	unsym-metric	single wire	
Bistables per computation stage	SETFF	DETFF	two latches			latch
Clock nets, wiring resources	1	1	2	2	1	1
Relative clock power (approx.)[a]	$1 \cdot 1 \cdot 2$	$1 \cdot 1.3 \cdot 1$	$2 \cdot \frac{1}{2} \cdot 2$	$2 \cdot \frac{1}{2} \cdot 2$	$1 \cdot 1 \cdot 2$	$1 \cdot \frac{1}{2} \cdot 2$
All timing constraints one-sided	yes	yes	yes	yes	yes	no
Relatively tolerant of skew	no	no	yes	yes	no	no
Allows for time-borrowing	n.a.	n.a.	yes	no	no	n.a.
Utilization of clock period	≤ 1	≤ 1	≤ 1	< 1	≤ 1	> 1
Function latches	[0,1]	[0,1]	[0,1,2]	[0,1]	[0,1,2]	[0,1]
EDA support	full	limited	limited	limited	limited	poor
Key quality	slim, straight	slower clock	highly flexible	easily derived	high perf.	risky, onerous
Applied in conjunction with collective buffer	past ASICs				DEC Alpha	not for ASICs
Applied in conjunction with distributed tree	most ASICs	DDR circuits	some ASICs	IBM LSSD[b]		not for ASICs

[a] expressed as $\#_{clock\_nets} \cdot \#_{FF\_load\_caps\_per\_net} \cdot \#_{edges\_per\_computation\_cycle}$.
[b] Three clock nets, two of which are active at any time.

- Multiple clock domains,
- Multi-cycle paths,[25]
- Conditional clocking, and — to a lesser extent — also
- Latch-based design, and
- Dual-edge triggering.

Especially in combination, such departures tend to strain EDA tools beyond what they have been designed for.

- Ideally, one could think of a synthesis tool that accepts some functional model and that lets the designer select among the more popular clocking disciplines before generating a circuit

---

[25] A **multi-cycle path** is a signal propagation path that extends over more than one clock period (computation period in the case of dual-edge triggering). Multi-cycle paths come into existence when VLSI architects decide to have parts of a circuit run at a lower rate than others, or when they accept having to wait two — or more — clock cycles for the results from certain combinational operations to appear at the output of the pertaining logic in exchange for being allowed to clock the entire circuit at a faster rate than the longest path would normally permit.

netlist. Commercial HDL software does not work in this way, though, and designers have to adjust their RTL source code whenever they opt for a different clocking discipline.

- A good clock distribution network exhibits

  - Low clock skew and jitter,
  - Fast clock slopes,
  - Modest distribution delay, and
  - A good tolerance towards PTV and OCV as well as towards diverse circuit arrangements that emanate from automatic placement and routing (P&R).

  While distributed trees and hybrid clock distribution networks prevail, careful design and post-layout timing verification are critical in any case.

- Designing clock distribution networks must be considered part of back-end design. In the case of distributed clock buffering, use a trustworthy clock tree generator to synthesize, place, and route the clock distribution networks. Never entrust clock tree insertion to standard logic synthesis tools as such products have not normally been designed to handle clock distribution.

- Circuit simulation does not suffice to detect skew-related problems. Rather, the circuit together with its clock distribution network must be subject to static timing verification. In synchronous designs, this basically implies the checking of setup and hold conditions along all signal propagation paths between any two bistables. Layout parasitics must be accounted for by working from post-layout netlists as obtained from layout extraction. Software tools for doing so are commonplace, see section 12.1.2 for details.

- Conditional clocking requires particular attention as it interferes with scan test, and as the simplest and most efficient clock gates are highly exposed to malfunction. The presence of logic gates other than buffers and inverters in clock distribution networks should always alert the expert during a design review.

- Most cell libraries are written with no particular clocking discipline or clock distribution scheme in mind. Thus, when evaluating a clocking strategy, do not let yourself be misguided by some arbitrary feature of your design environment. Put your system requirements first. That is, take into account

  - The compatibility with other subsystems to be integrated on the same chip, if any,
  - The library elements available such as flip-flops, latches, LSSD elements, memories, clock drivers, clock gates, PLLs, etc. along with the timing data that go with them,
  - The tools available for load estimation, layout extraction, circuit simulation, timing verification, clock tree generation, and circuit synthesis in general,
  - The accuracy of the timing model(s) being used,
  - The available software tools for inserting test structures,
  - The available EDA tools for generating and grading of test vectors,
  - The hardware test equipment at your disposal, and, last but not least,
  - Personal experience.

- Clock skew and jitter have become more acute with the ever shrinking feature sizes and clock periods. Luckily, today's semiconductor manufacturing processes provide many low-resistance metal layers of which designers can take advantage for better clock distribution.

6.7 | Problems

1. For all clocking disciplines discussed in this chapter visualize the skew margins in their Anceau diagrams. Use different colors for margins that (a) are a function of the operating speed of the circuit, (b) vanish unless the components involved have non-zero delays, and (c) do not depend on either one.

2. Old hands among IC designers have long known that edge-triggered shift registers are particularly in danger of malfunctioning. As a remedy, they used to oppose the flows of clock and data during physical design, e.g. they had the clock wire run from right to left in a shift register where data travel from left to right. Explain why this helped.

3. Reconsider the flip-flop timing data given in table 6.2. What do you think, is this a well-behaved flip-flop or does it impose timing conditions that are awkward to meet?

4. Some designers like to combine rising-edge-triggered flip-flops with others that trigger on the falling edge in the same clock domain. Analyze the timing characteristics of this scheme and discuss its merits (wrt skew tolerance, performance, scan path insertion, energy efficiency, design verification, engineering effort, etc.). Does it offer any advantage over the disciplines discussed in this chapter?

5. Yet another approach to edge-triggered clocking uses flip-flops that are built from three latches instead of two, see fig.6.29b. Consider a circuit where all standard flip-flops have been replaced by such bistables and carry out the same analysis as before.

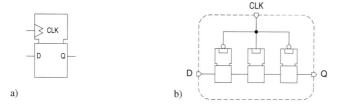

a)   b)

Fig. 6.29 Master–slave–slave flip-flop. Proposed icon (a) and logically equivalent circuit (b).

6. The circuit of fig.6.30 occupies a position somewhere between single-edge-triggered and level-sensitive one-phase clocking. The motivation behind replacing part of the flip-flops by (pulse-clocked) latches is the search for improved energy efficiency. Savings are expected from a lighter clock load and from the reduced node count and switching activities in a latch when compared with those in the more complex master–slave flip-flop. Substitution takes place only where the combinational logic between two consecutive bistables can be demonstrated to meet the setup- and hold-time conditions of the receiver due to the circuitry's inherent contamination delay alone, i.e. with no extra gates or wiring detours.

Draw the Anceau diagram assuming that clock frequency remains the same as in the original circuit. Establish all relevant setup and hold margins. Formulate adequate replacement rules in a more formal way. Comment on the viability and effectiveness of this hybrid clocking scheme.

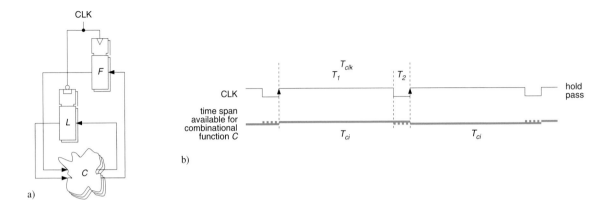

Fig. 6.30 A hybrid clocking scheme that combines latches with flip-flops. Basic hardware organization (a) and simplified timing diagram (b).

7. Reconsider the timing data in table 6.23 and calculate the length of the FPGA data-call windows for the various situations. What do you observe? Do you have an explanation?

8. Assume you want to include an on-chip RAM in a clock domain to be designed from a standard cell library. The edge-triggered synchronous RAM is to interface with a 14 bit address register and two 24 bit data registers, one for reading and one for writing. For simplicity, let's assume all flip-flops are identical. Both RAM and flip-flops trigger on the rising edge; their timing data are given next.

| Cell | Timing parameters | | | |
	t_{pd} [ns]	t_{cd} [ns]	t_{su} [ns]	t_{ho} [ns]
Flip-flop	0.5	0.2	0.3	0.1
RAM macrocell	2.5	1.5	1.0	0.8

a) Assuming zero skew, estimate the setup and hold margins if the circuit is to run at a clock frequency of 250 MHz. Where's the difficulty?

b) Make several proposals for solving the problem and evaluate them!

c) Study the options of inserting some delay τ in the various input and output lines of the RAM macrocell. Compile a table that lists the impact on the RAM's apparent timing figures if delay is inserted (α) in address and data inputs, (β) in the data outputs, and (γ) in the clock input.

d) What is the result of deliberately designing the clock tree so as to slightly retard or advance the clock signal to the RAM? Does one of these help?

e) Can you imagine situations where it makes sense to apply such tricks? Is it possible to improve performance by designing a carefully imbalanced clock distribution network?

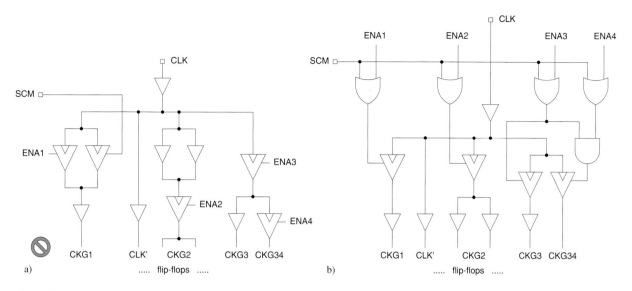

Fig. 6.31 Hierarchical clock gating and scan test.

9. A commercial synthesis tool from the 1990s claimed to construct FSMs with hazard-free outputs by adding latches to all input and output lines of a Mealy-type automaton. These extra latches were controlled by the same clock signal that drove the edge-triggered state register so as to always have either the input latches transparent and the output latches on hold, or vice versa. Find out when outputs are supposed to switch! As both latch sets are not allowed to be transparent simultaneously, outputs were believed to be isolated from asynchronous input changes. Show that this is not always true!

   Note: If you find it hard to analyze this state machine, imagine the difficulty of designing a larger circuit that mixes level-sensitive with edge-triggered clocking.

10. Figure 6.31 shows two clock distribution trees that make use of clock gating. Whether local clocks CKG1, CKG2, CKG3, and CKG34 are to be active or not obviously depends on enable signals ENA1 through ENA4 generated from within the circuit itself. Circuit testing is supposed to happen via a full scan path that gets turned on from the SCM scan mode terminal. Unfortunately, though, fig.6.31a includes four oversights. Locate them and explain what's wrong!

Chapter 7

Acquisition of Asynchronous Data

7.1 | Motivation

Most digital systems that interact with the external world must handle asynchronous inputs because events outside that system appear at random points in time with respect to the system's internal operation. As an example, a crankshaft angle sensor generates a pulse train regardless of the state of operation of the electronic engine management unit that processes those pulses.

Synchronization problems are not confined to electromechanical interfaces. Much the same situation occurs when electronic systems interact that are mutually independent, in spite — or precisely because — of the fact that each of them is operating in a strictly synchronous way. Just think of a workstation that exchanges data with a file server over a local area network (LAN), of audio data that are being brought to a D/A converter for output, or of a data bus that traverses the borderline from one clock domain to another. Obviously, the processing of asynchronous inputs is more frequent in digital systems than one would like.

The following episode can teach us a lot about the difficulty of accommodating asynchronously changing input signals. The account is due to the late Charles E. Molnar, who was honest enough to tell us about all misconceptions he and his colleagues went through until the problems of synchronization were fully understood.

Historical example

Back in 1963, a team of electronics engineers was designing a computer for biological researchers that was to be used for collecting data from laboratory equipment. In order to influence program execution from that apparatus, a mechanism was included to conditionally skip one instruction depending on the binary status of some external signal. The basic idea behind that design, sketched in fig.7.1a, was to increment the computer's program counter one extra time via a common enable input of its flip-flops iff the external signal was at logic **1**.

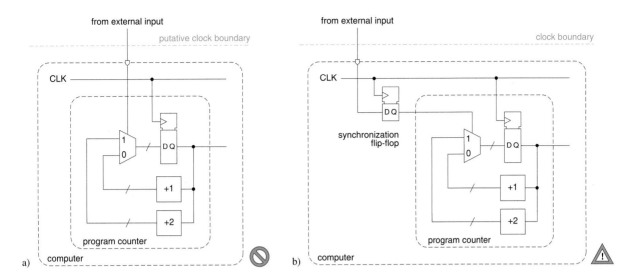

Fig. 7.1 Control unit with conditional instruction skip mechanism (simplified). Original design without synchronization (a), improved design with external signal synchronized (b).

When monitoring the computer's operation, the engineering team observed that the next instruction was occasionally being fetched from a bogus address after the external signal had been asserted, thereby causing the computer to lose control over program execution. As an example, either of $0\,1111 = 15$ or $1\,0000 = 16$ is expected to follow after address $0\,1110 = 14$, depending on the external input. Instead, the program counter could be observed to enter a state such as $1\,1101 = 29$ or $0\,0010 = 2$. The team soon found out that such failures developed only if the external signal happened to change just as the active clock edge was about to occur, causing the program counter to end up with a wild mix of old and new bits.

The obvious solution was to insert a synchronization flip-flop so as to make a single decision as to whether the external level was 0 or 1 at the time of clock. Although the improved design, sketched in fig.7.1b, performed much better than the initial one, the team continued to observe sporadic jumps to unexpected memory locations, a failure pattern for which it had no satisfactory explanation at the time. It took almost a decade before Molnar and others who worked on high-speed interfaces[1] dared publicly report on the anomalous behavior of synchronizers and before journals would accept such reports that contrasted with general belief.

□

Two failure mechanisms are exposed by this case, namely data inconsistency and synchronizer metastability, both of which will be discussed in this chapter together with advice on how to get them under control.

---

[1] What was considered high speed yesterday, no longer is high speed today. In the context of synchronization, "high-speed" always refers to circuits that operate at clock frequencies and data rates approaching the limits of the underlying technology.

7.2 | The data consistency problem of vectored acquisition

We speak of vectored acquisition when a clock boundary — the hypothetical line that separates two clock domains — is being traversed by two or more electrical lines that together form a data, control, or status word, or some other piece of information.

7.2.1 Plain bit-parallel synchronization

Consider the situation of fig.7.2 where data words traverse a clock boundary on w parallel lines before being synchronized to the receiver clock **CLKQ** in a register of w flip-flops. Due to various imperfections of practical nature,[2] some of the bits will switch before others do whenever the bus assumes a new value. If this happens near the active clock edge, the register is bound to store a **crossover pattern** that mixes old and new bits in an arbitrary way.

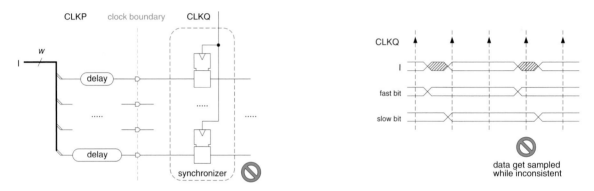

Fig. 7.2 Non-simultaneous switching in a parallel bus may result in inconsistent data.

Depending on context, the impact of occasional crossovers on a circuit's operation ranges between benign malfunction and fatal disaster:

○ Data error for one cycle before being overwritten with correct value.
○ Episodic disturbance lasting for several clock cycles (e.g. in a filter).
○ Value outside legal range (e.g. if item of data is a complementary pair, follows one-hot encoding, or is coded in some other redundant format).
○ Finite state machine (FSM) deflected to a state unplanned for.
○ Finite state machine (FSM) trapped in a lock-up situation.
○ An address counter pointing to mistaken memory locations.

Warning example

A stream of digital audio data available in a 16 bit parallel format is resynchronized to a 44.1 kHz output clock before being fed to a D/A converter. As audio samples are coded in 2's complement

---

[2] Such as unmatched gate delays, unlike loads, distinct wire routes, unequal layout parasitics, clock skew, PTV variations, OCV, and crosstalk.

format (2'C), their range covers the interval $[-32'768, +32'767]$. Only one sample will be affected in the occurrence of a synchronization failure. Yet, to apprehend the impact of crossover, imagine two consecutive samples of low amplitude end up intermingled as follows.

sample	decimal	relative		2's complement code
$s(t)$ (correct)	+47	+0.0014	→	0000 0000 0010 1111
$s(t+1)$ (correct)	−116	−0.0035	→	1111 1111 1000 1100
maximum mix	+32 687	+0.9976	←	0111 1111 1010 1111
random crossover	+5 388	+0.1644	←	0001 0101 0000 1100
minimum mix	−32 756	−0.9997	←	1000 0000 0000 1100

☐

Observation 7.1. *Data that cross a clock boundary on parallel lines cannot be synchronized with the aid of parallel registers alone. Inconsistent patterns may otherwise develop from crossover between individual data bits and upset the downstream data processing logic.*

7.2.2 Unit-distance coding

We first observe that any crossover pattern from two data words necessarily matches either of the two words iff their Hamming distance is one or less. The consistency problem can thus be solved by adopting a unit-distance code[3] provided data are known never to change by more than one step at a time in either direction, a requirement that confines unit-distance coding to applications such as the acquisition of position and angle encoder data.

	unsigned binary		Gray coding	
clock count	decimal	code	decimal	code
$c(t)$	+47 →	0010 1111	+47 →	0011 1000
$c(t+1)$	+48 →	0011 0000	+48 →	0010 1000
maximum mix	+63 ←	0011 1111	+48 →	0010 1000
minimum mix	+32 ←	0010 0000	+47 →	0011 1000

Example

A sampling rate converter IC for digital audio applications included a subfunction for tracking the ratio of the two sampling frequencies in real time. The circuit was built on the basis of an all-digital phase-locked loop (PLL) and asked for a counter clocked at frequency f_p, the state of which had to be read out periodically with frequency f_q. The frequency ratio of the two clocks was known to be contained in the interval $[\frac{1}{2}...2]$ and to vary slowly. Experience had shown that any crossover in

---

[3] **Unit-distance codes** have the particularity that any two adjacent numbers are assigned code words that differ in a single bit. They include the well-known Gray code (2^w) and the Glixon, O'Brien, Tompkins, and reflected excess-3 codes (binary coded decimal (BCD)). 4 bit Gray coding, for instance, goes as follows:

decimal	Gray code	decimal	Gray code	decimal	Gray code	decimal	Gray code
15	1000	11	1110	7	0100	3	0010
14	1001	10	1111	6	0101	2	0011
13	1011	9	1101	5	0111	1	0001
12	1010	8	1100	4	0110	0	0000

the clock count data would impair the final audio signal in a critical way, even if increment and read-out operations happened to coincide only sporadically. An error-safe solution had thus to be sought. Figure 7.3 shows the relevant hardware portion, which eliminated crossovers by having the clock count data traverse the clock boundary as Gray code numbers.

☐

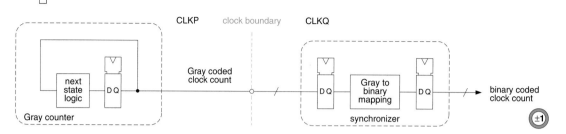

Fig. 7.3   Frequency ratio estimator with vectored synchronizer based on a unit-distance code (partial view).

7.2.3   Suppression of crossover patterns

The idea here is to detect and ignore transients by comparing subsequent data words at the receiver end. If any mismatch is detected, data are declared corrupted and are discarded by having the synchronizer output the same value as in the cycle before. A comparison over just two words as in the circuit of fig.7.4 suffices, provided input data are guaranteed to settle within one clock period. To avoid loss of data, any data item must get sampled correctly at least two times in a row, which implies that up to three clock intervals may be necessary until a new item becomes visible at the synchronizer output. A similar but more onerous proposition is to use error detection coding to find out when a freshly acquired data word should be ignored.

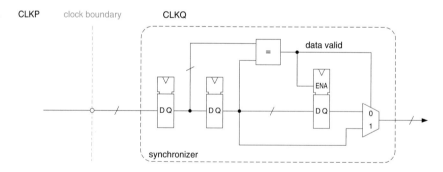

Fig. 7.4   A vectored synchronizer that detects and ignores inconsistent data.

Note that specialized hardware is not the only possible approach to reject crossover patterns. The same idea can also be implemented in software provided data rate, data transition time, and sampling rate can be arranged to be consistent with each other.[4]

---

[4]   Observe the resemblance to the debouncing of mechanical contacts to be discussed in section 8.5.5.

7.2.4 Handshaking

This approach contrasts with the previous one in that it prevents inconsistent data from being admitted into the receiving circuit. The key idea is to avoid sampling data vectors while they might be changing. Instead, the updating and the sampling of the data get coordinated by a **handshake protocol** that involves both the producing and the consuming subsystems. As will become clear later in this section, handshake protocols also have other applications than just avoiding the emergence of crossover patterns at clock boundaries. **Full handshaking** is essentially symmetrical and requires two control lines, termed **request REQ** and **acknowledge ACK** respectively, that run in opposite directions, see fig.7.5a.

Observation 7.2. *The rules of full handshaking demand that any data transfer gets initiated by some specific event on the request line and that it gets concluded by an analogous event on the acknowledge line.*

Handshake protocols come in many flavors, let us focus on a few important ones.[5]

Transition or two-phase handshake protocol

The waveforms and event sequences of figs.7.5b and c show a possible scenario. Let both control lines be at logic 0 before the first data transfer begins. When the producer has finished preparing a new data word, he stores it in his output register. By toggling the REQ line, he then informs the consumer that the DATA vector has settled to a new valid state and requests it to ingest that data item. He thereby becomes liable of maintaining this state until data reception is confirmed by the consumer as the latter is free to accommodate and process the pending data item whenever he wants to do so.

When, some time later, the consumer has safely got hold of that data item and when he is ready to accept another one, he toggles the ACK line running back to the producer to make this manifest.[6] Upon arrival of this confirmation, the producer no longer needs to hold the present data item but

---

[5] What follow sare comments on further variations of the basic handshake protocol. Notice from fig.7.5 that the request line has the same orientation as the data bus. This assumption, which we have made throughout our discussion, is referred to as a **push protocol** because any data transfer gets initiated by the producer. The data- valid message is transmitted over the request line, and the data-secured message over the acknowledge line.

In a **pull protocol** everything is reversed. The request line runs in the opposite direction to the data bus, transfers get initiated by the consumer, the request line carries the data-secured message (that prompts the producer to deliver another data item), and the data-valid message is encoded on the acknowledge line. Yet, it is important to understand that these designations relate to naming conventions only; by no means do they imply that an active producer is driving a passive consumer, or vice versa.

Further observe in fig.7.5d that the data bus holds valid data for only half of the time. The precise name of this scenario is "prolonged early push protocol". The freedom in deciding when to withdraw or overwrite a data vector, represented in fig.7.5d by the event labeled VAL-, suggests this is not the only option. You may want to refer to [170] for an exhaustive discussion of handshake protocols and their terminology.

[6] Note that this does not necessarily imply that the processing of the last data item is actually completed. A pipeline, for instance, can accommodate new data as soon as the result from the processing of the previous item by its first stage has been properly stored in the subsequent pipeline register. There is no need for this data item to have finished traversing the second or any later stage. It should be obvious, however, that data must have propagated through any combinational logic placed between the producer's output register and the first data register in the consumer before the acknowledge is asserted.

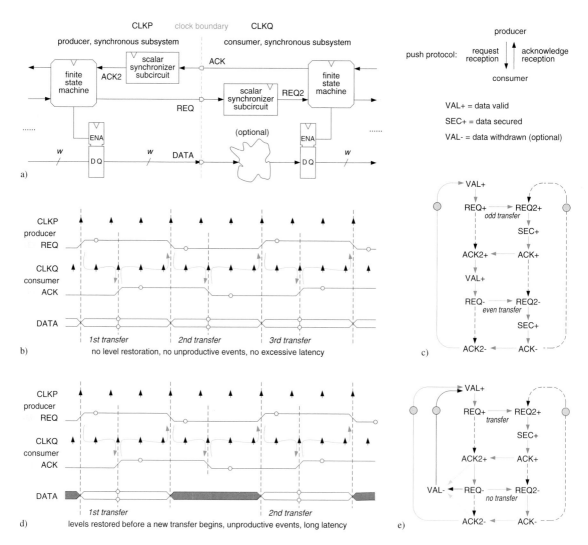

Fig. 7.5 Full handshaking put to service for vectored synchronization. Circuit (a), waveforms for two-phase protocol (b), signal transition graph (STG) (c), same for four-phase protocol (d,e).

becomes free to withdraw or to overwrite it at any time. When the first data transfer comes to an end, both control lines are at 1, ready for a second transfer.

Notification occurs with transition signaling and each data transfer operation involves one event on each of the two control lines. This explains why the two-phase protocol is also known as the two-stroke, transition, or non-return-to-zero (NRZ) protocol. An implication is that it always takes two consecutive data transfers before the control lines return to their initial logic values.

Level or four-phase handshake protocol

The alternative scenario depicted in fig.7.5d and e contrasts with the above in that the control lines get restored to their initial values between any two consecutive transfer operations. This variation is based on level signaling and involves two extra events, hence the synonyms four-stroke, level, and return-to-zero (RZ) protocol. Although the data rate is cut to half of what is achievable with transition signaling at the same clock frequency, this approach is more common because the hardware tends to be somewhat simpler and easier to understand.

With either signaling scheme, the handshake protocol makes sure that data vectors get sampled only while the electrical lines are kept in a stable and consistent state. The need for synchronization — together with the risk of metastability — is confined to two scalar control signals no matter how wide the data word is. Two-stage synchronizers are typically being used.

Full handshaking works independently of the relative operating speeds of the subsystems involved.[7] If, for instance, the producer were replaced by a much faster implementation, data transfers would continue correctly with no change to the consumer or to the interface circuits. Although the producer would spend much less time in preparing a new data item, the mutual exclusion principle inherent in the full handshake protocol would take care that data items do not get produced at a rate that is incompatible with the consumer.

Observation 7.3. *The strict sequence of events imposed by a full handshake protocol precludes any loss of data, confines synchronization problems to two control lines, and makes it possible to design communicating subsystems largely independently from each other.*

The last property greatly facilitates the modular design of complex systems as nothing prevents VLSI designers from taking advantage of full handshaking to govern data transfer operations even when all subsystems involved are part of the same clock domain. On the negative side, the symmetric protocol requires that both subsystems involved be **stallable**. That is, they must be capable of withholding their data production and consumption activities for an indeterminate amount of time if necessary. This is not always possible, though.

7.2.5   Partial handshaking

Consider a digital data source (sink) operating with a fixed sampling rate. An example is given in fig.7.6a (f) with an A/D (D/A) converter acting as producer (consumer) of audio samples. Such a data source (sink) works spontaneously and neither needs nor cares about protocols. Input terminal **ACK** (**REQ**) thus becomes meaningless and must be dropped, thereby rendering the circuit and the protocol unsymmetrical.

More importantly, a fixed-rate producer (consumer) is not stallable. This is to say that it does not wait, thereby forcing its counterpart to always complete the data processing in time before the next data item becomes available (is needed). Data items may otherwise get lost (get read multiple times). By making assumptions about the response time of the partners involved, partial handshaking becomes exposed to failure when these assumptions change. In the example of fig.7.6c, **SEC+** must occur before **VAL+** under <u>any</u> circumstance, which imposes restrictions on the mutual relationship of producer and consumer clock rates and latencies.

It is important to understand that porting an existing design to a new technology, reusing part of it in a different context, or applying dynamic voltage and/or frequency scaling is likely to challenge

---

[7] A questionable implementation where this is not the case is discussed in problem 4 of section 7.6.

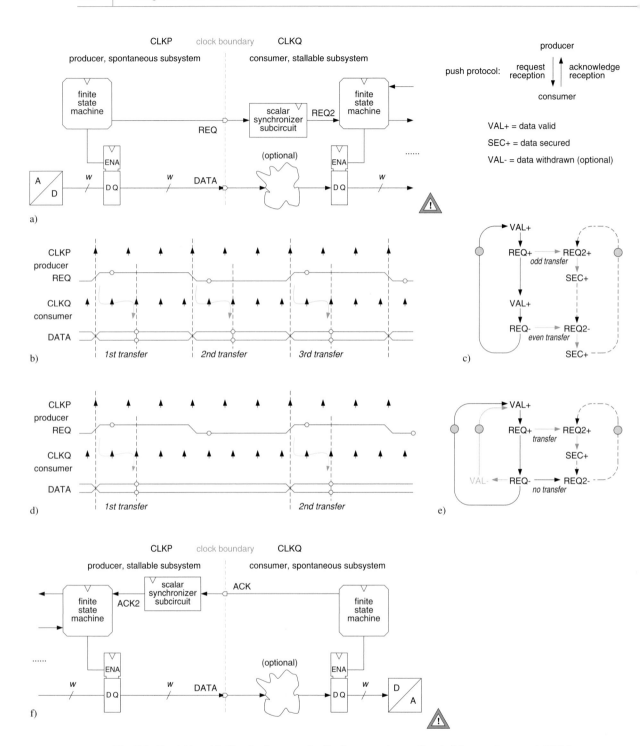

Fig. 7.6 Partial handshaking in the case of a fixed-rate producer. Circuit (a), waveforms and STG for a two-phase protocol (b,c), and the same for a four-phase protocol (d,e). Circuit in the case of a fixed-rate consumer (f).

such assumptions. As opposed to this, full handshaking scales with frequency and operating speed because no assumptions are made.

Observation 7.4. *Partial handshake protocols cannot function properly unless the reaction times of the communicating subsystems are known a priori and can firmly be guaranteed to respect specific relationships under all operating conditions.*

7.3 | The data consistency problem of scalar acquisition

As the name suggests, scalar acquisition implies that a clock boundary is being traversed by just one line. It may appear surprising that the acquisition of a single bit should give rise to any problem that is worth mentioning, yet, there are a few subtle pitfalls. In order to better understand the peculiarities, let us first examine how unsophisticated schemes fail in the presence of asynchronously changing inputs before proceeding to more adequate approaches.

7.3.1 No synchronization whatsoever

In the circuit of fig.7.7a, a scalar input signal I is being fed into two combinational subcircuits g and h that are part of a synchronous consumer circuit without any prior synchronization to the local clock CLKQ. Two deficiencies are likely to lead to system failure.

Firstly, the outputs C_g and C_h emanating from g and h respectively will occasionally get sampled during the time span between contamination and propagation delays when their values correspond neither to the settled values from the past interval t nor to those for the upcoming interval $t + 1$.[8] In the timing diagram of fig.7.7a such unfortunate circumstances apply to the central clock event.

Secondly, even though C_g and C_h may happen to be stable at sampling time, they may relate to distinct time intervals if $t_{cd\,g} > t_{pd\,h}$. If so, an inconsistent set of data gets stored in the two registers before it is passed on to the downstream circuitry for further processing. This undesirable situation typically occurs when one of the paths includes combinational logic whereas the other does not. For an example, check the rightmost clock event in fig.7.7a.

7.3.2 Synchronization at multiple places

Adding synchronization flip-flops in front of all combinational subcircuits as in fig.7.7b improves the situation but does not suffice. This is because the flip-flops involved would sample the input slightly offset in time as a consequence of unbalanced delays along the paths $I \rightarrow I_g$ and $I \rightarrow I_h$, clock skew, unlike switching thresholds, noise, etc. Every once in a while, input data would get interpreted in contradicting ways as depicted in the timing diagram.

7.3.3 Synchronization at a single place

To stay clear of inconsistencies across multiple flip-flops, synchronization must be concentrated at a single place before any data are distributed. This is the only way to make sure that all downstream circuitry operates on consistent data sampled at a single point in time. It is, therefore, standard practice to use synchronizers similar to those depicted in figs.7.7c and d.

---

[8] Remember from section A.5 that combinational outputs do not necessarily transit from one stable value to the next in a monotonic fashion, rather, they may temporarily assume arbitrary values due to hazards.

Observation 7.5. *Any scalar signal that travels from one clock domain to another must be synchronized at one place by a single synchronization subcircuit driven from the receiving clock. Similarly, when two complementary signals are required, it is best to transmit only one of them and to (re)obtain the complement after synchronization.*

7.3.4 Synchronization from a slow clock

A synchronizer clocked at rate f_{clkq} is bound to miss part of the input data unless all data pulses are guaranteed to last for at least $T_{clkq} = 1/f_{clkq}$. Note the incongruity between data rate and sampling rate in fig.7.7c and the ensuing data loss. Actually, the sampling period of the synchronizer must provide sufficient leeway to accommodate setup and hold times as well as data transients. This can be a problem when no clock of adequate frequency is available on the consumer side. Various workarounds have been developed.

- Convert the data stream from its bit-serial into a bit-parallel format with the aid of a shift register, thereby making it possible to transmit data at a much lower rate. Then use one of the vectored acquisition schemes found to be safe.
- Use an analog phase-locked loop (PLL) to clock the synchronizer at a faster rate.
- Resort to dual-edge-triggered one-phase clocking if clocking the synchronizer at twice its original frequency suffices.

Even with the best synchronization scheme, an active clock edge and an input transition will occasionally coincide when signals get exchanged between two independent clock domains, see fig.7.7d. For the system to function correctly, the synchronizer must then decide for either one of the two valid outcomes as the downstream circuitry cannot handle ambiguous data. This is precisely the subject of the next section.

7.4 | Metastable synchronizer behavior

7.4.1 Marginal triggering and how it becomes manifest

Reconsider the various synchronization schemes discussed. What they all have in common is a flip-flop — or a latch — the data input of which is connected to the incoming signal. As no fixed timing relationship between data and clock can be guaranteed, the data signal will occasionally switch in the immediate vicinity of a clock event, thereby ignoring the requirement that input data must remain stable throughout a bistable's data-call window.

"What happens when an input change violates a bistable's setup or hold condition?"

Technically, this kind of incident is referred to as **timing violation** or **marginal triggering**.[9]

---

[9] Please recall that clocked (sub)circuits impose not only
- minimum setup and hold times t_{su} and t_{ho} but also
- minimum clock pulse widths $t_{clk\,lo\,min}$ and $t_{clk\,hi\,min}$,
- maximum clock rise and fall times $t_{clk\,ri\,max}$ and $t_{clk\,fa\,max}$, and
- absence of runt pulses and glitches on clock and asynchronous (re)set inputs.

Disregarding any such timing condition is likely to cause marginal triggering. The customary abstraction of flip-flops, latches, and RAMs into bistable memory devices the behavior of which is entirely captured by way of truth tables, logic equations, and the like then no longer holds.

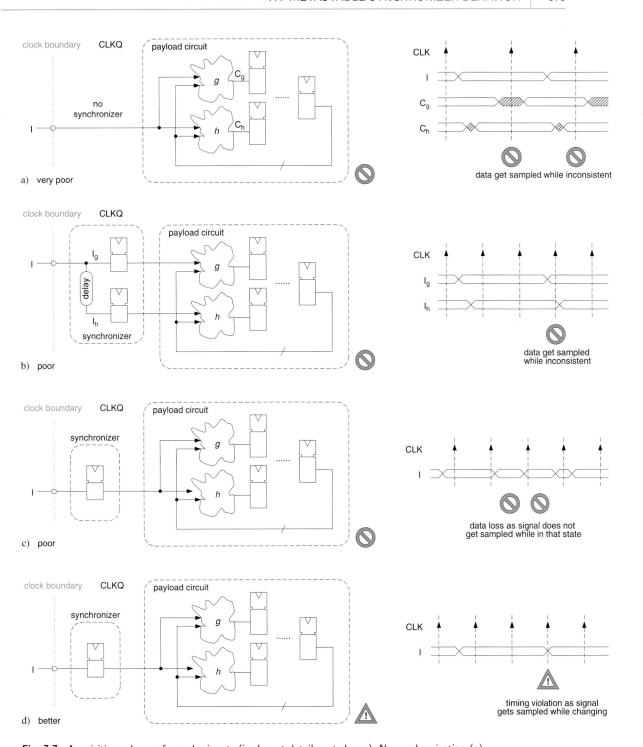

Fig. 7.7 Acquisition schemes for scalar inputs (irrelevant details not shown). No synchronization (a), synchronization at multiple places (b), synchronization at a single place with inadequate (c) and with adequate sampling rate (d).

Until the mid 1970s, it was generally believed that the bistable would then decide for either of the possible two outcomes right away and output either a 0 or a 1. As stated in the introduction to this chapter, it took a long time until the design community began to understand that this was not necessarily true.

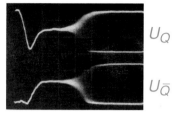

Fig. 7.8 Measured behavior of a CMOS latch in response to marginal triggering (photo reprinted from [171] with permission).

What had been observed were intermediate voltages and excessive delays before the two complementary outputs of the bistable eventually settled to a normal steady-state condition in response to marginal triggering. It was found that delays occasionally exceeded officially specified propagation delay figures by orders of magnitude, which made it clear that a better understanding of the process was imperative. We will, in the remainder of this chapter, summarize important results from empirical measurements and theoretical analysis of this phenomenon.

Today's digital subcircuits are inherently analog networks, and bistables are no exceptions. The data-retention capability of latches, flip-flops, and SRAM cells is essentially obtained from closed feedback loops built from two inverting amplifiers.[10] Figure 7.9b shows the transfer chararacteristics of two CMOS inverters where the output of either inverter drives the input of the other. Two stable points of equilibrium reflect the binary states 0 and 1 respectively. A third and unstable point of equilibrium exists in between. More precisely, the state space features two "valleys of attraction" separated by a line of unstable equilibrium. Marginal triggering implies bringing a bistable very close to that separation line before leaving it to recover. The bistable is then said to hang in an evanescent **metastable condition.**[11]

Mathematical analyses of the metastable behavior of cross-coupled inverters can be found in the literature such as in [172] [171] [47]. While correctly modelling the behavior of the feedback loop, they do not necessarily describe the waveforms observable at the output terminals of actual latches and flip-flops. Similarly, node voltages half-way between zero and U_{dd}, such as those depicted in fig.7.8, are not normally visible at the outputs of a library cell or physical component. This is because of the various auxiliary subcircuits inserted between the metastable memory loop itself and the I/O terminals. Output buffers, for instance, tend to restore intermediate voltages to regular logic 0s and 1s. Device characteristics and physical layout also matter. Figure 7.10 shows how metastable conditions normally become manifest in various bistables.

---

[10] Transistor-level circuit diagrams are to be presented in sections 8.2 and 8.3.

[11] A philosophical concept known as Buridan's Principle states that a discrete decision based upon input having a continuous range of values cannot be made within a bounded length of time. It is named after the fourteenth-century philosopher Jean Buridan, who claimed that a donkey placed at the same distance from two bales of hay would starve to death because it had no reasons to choose one bale over the other.

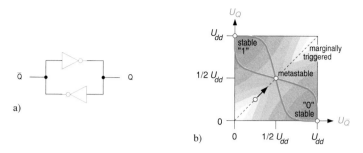

Fig. 7.9 Points of equilibrium and marginal triggering. Cross-coupled inverting amplifiers (a) and state space with superimposed transfer curves (b).

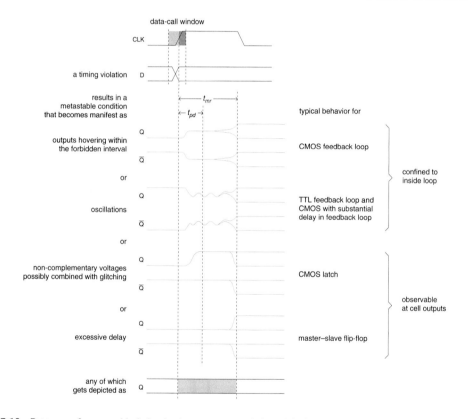

Fig. 7.10 Patterns of metastable behavior in response to timing violations.

Eventually the circuit returns to either of the two states of stable equilibrium. Yet, not only is the outcome of the decision process unpredictable, but the time it takes to recover from a metastable equilibrium condition necessarily exceeds the bistable's customary propagation delay, see fig.7.11. In fact, **metastability resolution time** has been reported to outrun propagation delay by orders of magnitude on occasion, so that we must write $t_{mr} > t_{pd}$ and sometimes even $t_{mr} \gg t_{pd}$. Most

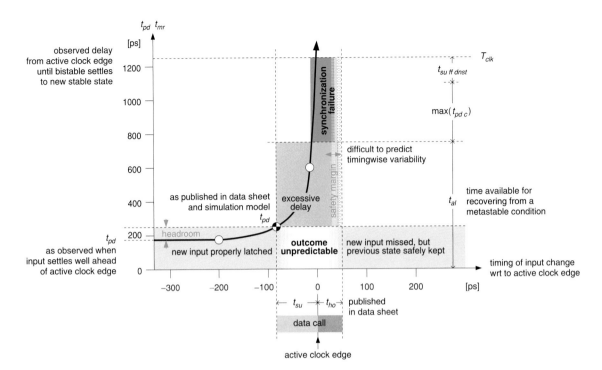

Fig. 7.11 Measured timing characteristics of a bistable when input data change in the vicinity of the data call window (arbitrary numerical data).

alarming are the facts that it is not possible to guarantee any upper bound for the time a bistable takes to recover and that the behavior of a physical part is no longer consistent with the one published in the datasheet and captured in the simulation model.

Observation 7.6. *Metastability is a problem because of unpredictable delay, not because of unpredictable logic outcome.*

While it is true that actual behavior greatly varies from one type of bistable to the next, it should be understood that the phenomenon of metastability is a fundamental one observed independently from circuit and fabrication technology. There are currently no flip-flops or latches that are free of metastable behavior.

7.4.2 Repercussions on circuit functioning

As a consequence of a synchronizer hanging in the metastable condition,

- o The downstream circuitry may process wrong data, or
- o May be presented with voltages within the forbidden interval, or
- o May find itself short of time for settling to a steady-state value before the next clock event arrives, see fig.7.12.

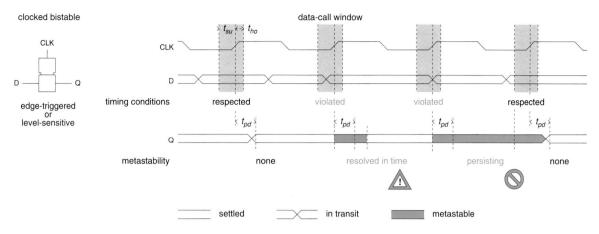

Fig. 7.12 Recovery from marginal triggering and the impact on downstream circuitry.

In the last two cases, some downstream storage elements will be subject to marginal triggering themselves, thereby permitting metastability to spread further into the clock domain. In any of the three cases, part of the system is likely to malfunction.

Warning example

The erratic behavior that Molnar and his colleagues had observed with their laboratory computer even after they had added a flip-flop for synchronizing the external input was in fact the result of metastability in the program counter. The problem was exacerbated by the slow germanium transistors then available and the poorly designed flip-flop circuits. Yet, it is unfair to blame the circuit designers for this as nobody was aware of metastability and of the importance of quick recovery in synchronizers at the time.
☐

Observation 7.7. *Metastability together with its fatal consequences is completely banned from within a clock domain iff care is taken to meet all timing requirements under all operating conditions. As opposed to this, there is no way to truly exclude metastability from occurring at the boundaries of independently clocked domains.*

7.4.3 A statistical model for estimating synchronizer reliability

As metastability has been found to be unavoidable in synchronizers, it is essential to ask

"How serious is the metastability problem really?"

As illustrated by the two quotes below,[12] even experts do not always agree.

---

[12] By Peter Alfke and Bruce Nepple respectively.

"Having spent untold hours at analyzing and measuring metastable behavior, I can assure you that it is (today) a highly overrated problem. You can almost ignore it."

"Having spent untold hours debugging digital designs, I can assure you that metastable behavior is a real problem, and every digital designer had better understand it."

Within their respective contexts, both statements are correct, which indicates that we must develop a more precise understanding of the phenomenon. Theoretical and experimental research has come up with statistical models that can serve as a basis for estimating the probability of system failures due to metastability. Consider a flip-flop driven by a clock of frequency f_{clk} that is connected to an asynchronous data signal with an average edge rate of f_d. We will speak of a **synchronization failure** whenever a metastable condition persists some time t_{al} after the synchronizer has been clocked, i.e. when $t_{mr} > t_{al}$. The **mean time between errors** (MTBE) for such an arrangement has been found to obey the general law

$$t_{MTBE} = \frac{e^{K_2 t_{al}}}{K_1 \, f_{clk} \, f_d} \tag{7.1}$$

where parameters K_1 and K_2 together quantify the characteristics of the synchronizer flip-flop with respect to marginal triggering and metastability resolution [172] [173]. K_2 is an indication of the gain–bandwidth product around the memory loop of its master latch. K_1 represents a timewise window of susceptibility for going metastable.

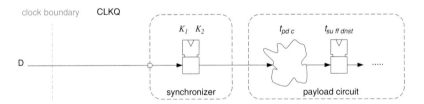

Fig. 7.13 Circuit diagram of single-stage synchronizer.

Now consider a single-stage synchronizer like the one of fig.7.13 and ask yourself for how much time the first flip-flop is allowed to rave in an out-of-the-normal condition for the downstream logic to stay clear of trouble. The timewise allowance for resolving metastability t_{al} is simply the clock period T_{clk} diminished by the setup time of the downstream flip-flop(s) $t_{su\,ff\,dnst}$ and the propagation delay $\max(t_{pd\,c})$ along the longest path through any combinational circuitry inserted between the flip-flops. You may want to refer back to fig.7.11 for an illustration.

$$t_{al} = T_{clk} - \max(t_{pd\,c}) - t_{su\,ff\,dnst} \quad \text{with} \quad T_{clk} = \frac{1}{f_{clk}} \tag{7.2}$$

Example

The metastability characteristics of a flip-flop in an XC2VPro4 FPGA from the Xilinx Virtex II Pro family are listed in table 7.1 for typical operating conditions. For $f_{clk} = 100$ MHz, $f_d = 10$ MHz, $t_{su\,ff\,dnst} = 0.5$ ns and $t_{pd\,c} = 4$ ns, the calculated MTBE exceeds the age of the universe. Just doubling the clock frequency reduces the MTBE to a mere 4 s, which demonstrates that the impact of metastability is extremely dependent on a circuit's operating speed. Whether the single-stage synchronizer of fig.7.13 is acceptable thus turns out to be a quantitative question.

□

7.4.4 Plesiochronous interfaces

Being a statistical model, (7.1) assumes that there is no predictable relationship between the frequencies and phases of the data and clock signals. This is does not always apply, though. A notable exception is **plesiochronous** systems where data producer and consumer are clocked from separate oscillators that operate at the same nominal frequency. Think of a local area network (LAN), for instance. Once clock and data have aligned in an unfortunate way at the receiver end, successive data updates are subject to occur during the the critical data-call window many times in a row, thereby rendering the notion of MTBE meaningless.

Observation 7.8. *Plesiochronous interfaces are exposed to burst-like error patterns and are not amenable to analysis by simple statistical models that assume uncorrelated clocks.*

Plesiochronous interfaces require some sort of self-regulating mechanism that avoids consecutive timing violations and repeated misinterpretation of data. A tapped delay line may be used, from which an adaptive circuit selects a tap in such a way as not to sample a data signal in the immediate vicinity of a transient [174]. Two related ideas are adaptively shifting the data or the clock signal via a digitally adjustable delay line [175], and oversampling the input data before discarding unsettled and duplicate data samples. None of these approaches is free of occasional coincidences of clock events and data changes, however. The problem is just transferred from the data-acquisition flip-flop to the subcircuit that adaptively selects a tap or controls sampling time.

7.4.5 Containment of metastable behavior

Limiting the harmful effects of metastability is based on insight that directly follows from (7.1) and (7.2). Keep the number of synchronization operations as small as possible and allow as much time as practical for any metastable condition to resolve. More specific suggestions follow.

Estimate reliability at the system level

Having a fairly accurate idea of the expected system reliability is always a good starting point. It makes no sense to try to improve synchronization reliability further when the MTBE already exceeds the expected product lifetime. However, always remember that (7.1) and all further indications in this text refer to one scalar synchronizer and that a system may include many of them. Also keep in mind that statistical models do not apply to plesiochronous operation.

In practice, a frequent problem is that only few bistables come with datasheets that specify their metastability resolution characteristics. Luckily, there exists a workaround.

Observation 7.9. *As a rule of thumb, synchronization failure is highly infrequent if a flip-flop is allowed three times its propagation delay or more to recover from a metastable condition.*

Table 7.1 Metastability resolution characteristics of various CMOS flip-flops. Note: The above figures have been collected from distinct sources and do not necessarily relate to the same operating conditions. Still, a massive improvement over the years is evident.

| reference or vendor | D-type flip-flop | | | | Metastability | |
	cell type or name	technology	F [nm]	K_1 [ps]	K_2 [GHz]	
Horstmann *et al.* [172]	n.a.	std cell	1500	47 600	3.23	
VLSI Technology	DFNTNS	std cell	800	140 000	12.3	
Ginosar [175]	n.a. "conservative"	std cell	180	50	100	
Xilinx [173]	XC2VPro4 CLB	FPGA	130	≈100	27.2	

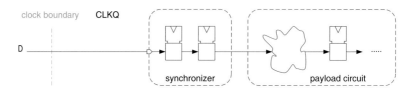

Fig. 7.14   Two-stage synchronizers obtained from adding an extra flip-flop.

In the absence of numerical K_1 and K_2 values, refraining from detailed analysis is probably safe if the application is not overly critical and if $t_{al} \geq 3\, t_{pd}$ can be guaranteed throughout.

Select flip-flops with good metastability resolution

Flip-flops optimized for synchronizer applications feature a small K_1 and, above all, a large K_2. They shall be preferred over general-purpose flip-flops with inferior or unknown metastability recovery characteristics.[13] Unfortunately, component manufacturers and library vendors continue to be extremely reticent when it comes to disclosing quantitative metastability data.[14]

Remove combinational delays from synchronizers

Recall from (7.2) that any extra delay $t_{pd\,c}$ between two cascaded flip-flops is at the expense of recovery time t_{al} for the first bistable. As a consequence, the one-stage synchronizer circuit of fig.7.13 is far from being optimal at high clock rates. When the MTBE proves insufficient, a better solution must be sought.

---

[13] [176] finds that static bistables should be preferred over their dynamic counterparts.

[14] There are two reasons for this. For one thing, the issue is no longer perceived as urgent now that the speed and metastability resolution characteristics of flip-flops have improved so much when compared with older fabrication technologies. For another thing, it takes a considerable effort and degree of sophistication to properly determine the K_1 and K_2 parameters for a cell library. The burden of doing so is thus left to VLSI designers in critical high-speed applications, see [172] [173] for measurement setups. It is important that such characterizations be carried out under operating conditions that are as identical as possible to those actually encountered by the synchronizer when put into service in the target environment. Relevant conditions include capacitive load, layout parasitics, clock slew rate, fabrication process, and, last but not least, PTV conditions.

Two flip-flops cascaded with no combinational logic in between extend the time available for metastability resolution to almost an entire clock period. The low error rates typically obtained in this way have contributed to the popularity of **two-stage synchronizers** as shown in fig.7.14. In case the additional cycle of latency resulting from the extra flip-flop proves unacceptable, try to reshuffle the existing registers or check [177] for a proposal that operates multiple two-stage synchronizers in parallel.

Drive synchronizers with fast-switching clock

Experience has shown that clocking synchronizers with fast slew rates tends to accelerate recovery from a marginal triggering condition [172]. What's more, overly slow clock ramps tend to dilate setup and hold times beyond their nominal values as obtained from library characterization under the assumption of a sharp clock edge with zero or close-to-zero ramp time. Yet, these are the figures stored in simulation models and listed in datasheets on which design engineers necessarily base all their reasoning.

Free synchronizers from unnecessary loads

Not surprisingly, capacitive loading has been found to slow down the metastability resolution process in a bistable. It is therefore recommended to keep the loads on synchronizer outputs as small as possible by using buffers and buffer trees where necessary.

Lower clock frequency at the consumer end

As a minor change of the clock frequency has a large impact on the MTBE, it is always worthwhile to check whether it is not possible to operate the entire consumer from a somewhat slower clock.[15]

Use multi-stage synchronizers

In those — extremely infrequent — situations where a two-stage synchronizer does not leave enough time for metastability to resolve, the available time span can be extended well beyond one clock period by resorting to synchronizer circuits that make use of multiple flip-flops. Please refer to [178] where the merits of cascaded and clock-divided synchronizers are evaluated.

Keep feedback path within synchronizers short

Digital VLSI designers normally work with predeveloped cell libraries. If you must design your own synchronizers at the transistor level, consult the specialized literature on the subject [179] [180] [172] [171] [178]. For the purpose of analysis, the two cross-coupled amplifiers can be replaced by a linear model in the vicinity of the metastable point of equilibrium. As a rule, the internal feedback path should be kept as fast as possible. A fast master latch is desirable because a high gain–bandwidth product tends to improve recovery speed, K_2, and MTBE. This is also why a higher supply voltage has been found to be beneficial.

---

[15] A more exotic proposal is the concept of a **pausable clock**, where metastability is detected by way of analog circuitry designed for that purpose, and where the consumer's clock is frozen until it has been resolved [178].

A variety of misguided approaches to synchronization, such as a "metastability blocker" and a "pulse synchronizer", for instance, are collected in [175].

7.5 | Summary

- Asynchronous interfaces give rise to two problems, namely inconsistent data and metastability. While it is always possible to avoid data inconsistencies altogether by making use of adequate data-acquisition schemes and protocols, the metastability problems that follow from marginal triggering can be tackled in a probabilistic fashion only.

- Metastability is often put forward as a welcome explanation for synchronization failures because of its unavoidable and unpredictable nature. Yet, we tend to believe that most practical cases of malfunctioning circuits actually result from data consistency problems that have been overlooked.

- Metastability becomes a threat to system reliability when synchronizers are being operated close to their maximum clock and data frequencies, and/or when synchronizers are involved in very large quantities. Circuits that can afford a few extra ns of clock period for the synchronizers to recover should be safe, on the other hand. Two-stage synchronizers normally prove more than adequate in such situations.

- To steer clear of the imponderabilities and the extra costs associated with synchronization, implement the rules listed below.

  - Eliminate uncontrolled asynchronous interfaces wherever possible,
    that is partition a system into as few clock domains as technically feasible.
  - If you must cross a clock boundary, do so where data bandwidth is smallest.
  - Estimate error probability or mean time between errors at the system level.
  - Whenever possible, set aside some extra time for synchronizers to settle.
  - Within each clock domain, strictly adhere to a synchronous clocking discipline.
  - Avoid (sub)circuits that tend to fail in a catastrophic manner
    when presented with corrupted data.

7.6 | Problems

1. What is wrong with the two-stage synchronizer circuit of fig.7.15?

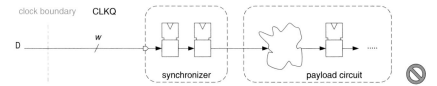

Fig. 7.15 Bad synchronizer circuit.

2. Reconsider Molnar's original circuit of fig.7.1a and recall that the computer failed when its program counter became filled with a bogus address as a consequence from a synchronization failure. At first sight, it appears that only a scalar signal is being acquired from externally, so that there should be no chance for an inconsistent address word to develop. Find out why this is not so.

3. Establish detailed state diagrams for the two finite state machines in fig.7.5. Generate all necessary control signals and try to keep latency small. Depending on how the interfaces with the surrounding circuitry are defined, there may be more than one acceptable solution. Can you design the models such as to minimize the differences in the HDL codes of two- and four-phase protocols?

4. Figure 7.16 shows an arrangement that has a long tradition for carrying data vectors from one clock domain to another. What sets it apart from the handshake circuit of fig.7.5 is a bistable that sits right on the clock boundary. This has earned the circuit names such as "shared flip-flop" or "signaling latch" synchronizer although an unclocked data-edge-triggered seesaw is typically being used (a level-sensitive seesaw is sometimes also found). The shared bistable functions as a flag, set by the producer and reset by the consumer, that instructs one partner to carry out its duty and the other to wait. Much as in fig.7.5, each of the two control signals gets accepted into the local clock domain by a standard two-stage synchronizer. Compare the two circuits and their detailed operation.

The correct functioning of the circuit of fig.7.16 rests on an assumption that may or might not hold in real-world applications, however. Find out what that assumption is. Establishing a signal transition graph (STG) may help. Hint: Consider situations where the two clock frequencies greatly differ from each other.

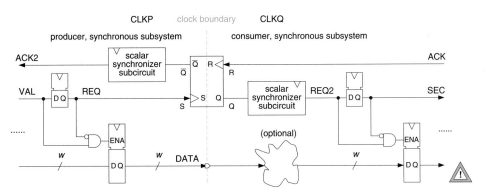

Fig. 7.16 Vectored synchronization on the basis of a shared bistable.

5. In fig.7.11, three points are marked by empty circles or a propeller respectively. Explain what the three points have in common. What sets the propeller apart from the other two marks?

Chapter 8

Gate- and Transistor-Level Design

The focus of attention in digital VLSI has moved away from low-level circuit details since the advent of HDL synthesis and virtual components. Library developers are the only ones who face transistor-level circuits on a daily basis. Yet, understanding how logic gates, bistables, memories, I/O circuits, and other subcircuits are built and how they operate continues to be a valuable asset of any VLSI engineer that helps him make better design decisions and imagine solutions otherwise unthought of.

Sections 8.1 through 8.4 attempt to explain just that for a variety of CMOS subcircuits. It is also hoped that the richness and beauty of this subject become manifest. Any reader who is looking for an in-depth exposure will have to consult more detailed and more comprehensive texts such as [181] [182] [159] [183] [184]. More specifically, we have elected to skip all circuit styles that rely on short-time charge retention.[1] As an exception, section 8.3 not only discusses static memories but also gives a glimpse on dynamic memories. Section 8.5, finally, serves to make digital designers aware of a variety of pitfalls that are associated with certain (sub)circuits.

Before we can begin, we must know what transistors can do for us. A very basic thought model is introduced next while a discussion of calculation and simulation models is available in appendix 8.7.

8.1 | CMOS logic gates

CMOS logic is built from enhancement-type n- and p-channel MOSFETs. While the physics and electrical characteristics of transistor devices are quite complex, simple abstractions generally suffice to understand and draft digital subcircuits such as logic gates, bistables, and memories.

---

[1] While **dynamic CMOS** logic [185] offers an advantage in terms of operating speed, designing and testing circuits is definitely more laborious and error-prone than with **static CMOS** logic. Due to leakage and other undesirable phenomena, electrical charge is bound to quickly disappear unless it gets refreshed on a periodical basis, which imposes a lower bound on the admissible clock frequency and impedes conditional clocking. Sensitivity to glitches, switching noise, and charge-sharing phenomena are three more difficulties of dynamic logic. All that, together with the lack of appropriate cell libraries, puts this logic style off limits for typical ASIC projects and, at the same time, explains its popularity in top-notch microprocessors and switching equipment.

8.1.1 The MOSFET as a switch

A crude approximation for a MOSFET is a relay where a contact operates under control of a voltage applied to the **gate** terminal so as to make or break the current path between the **source** and **drain** terminals. In the case of the predominant enhancement devices, the contacts are open (transistor "off") when the gate is at the same potential as the source, and closed ("on") when the full supply voltage is being applied between those two terminals. To become truly useful, this primitive **switch model** must be refined in two ways.

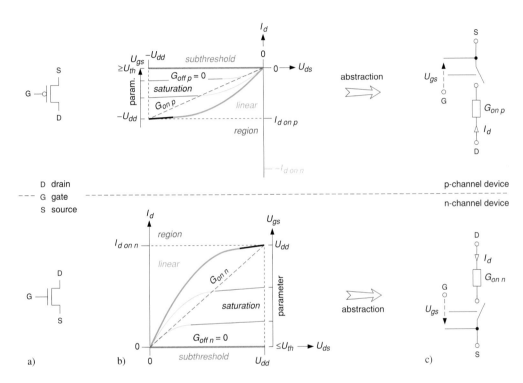

Fig. 8.1 Enhancement-type n- and p-channel MOSFETs. Icons (a), DC transfer characteristics $I_d = f(U_{gs}, U_{ds})$ (b), and switch models (c).

Firstly, while the "off"-state conductance G_{off} of the device is extremely low, conductance is far from infinite in the "on" condition. G_{on} actually is a function of transistor geometry and scales with the width-to-length ratio $\frac{W}{L}$. Also, it is smaller for a p-channel transistor than for a same-sized n-channel device by a factor of two or so. This gets reflected by the drain current plots of fig.8.1b and is a consequence of unlike mobilities of electrons and holes.

Secondly, the input and output side of a MOSFET switch share the source terminal as a common node, see fig.8.1c. This contrasts with an electromagnetic relay — where winding and contacts are insulated from each other — and impairs the MOSFET's use as a switch. As a net result, n-channel devices perform well as current sinks (pull-down, low-side switch) but poorly as current sources (pull-up, high-side switch), and vice versa for p-channel transistors.

For an explanation, assume an n-channel MOSFET connected as high-side switch were to charge a large capacitance from 0 to U_{dd}. Initially, one would measure the full supply voltage between gate and source $U_{gs} = U_{dd}$ which would drive the transistor into full conduction. As the charging progresses, however, the gate–source voltage would gradually diminish, $U_{gs} < U_{dd}$, thereby lessening the current delivered to the load. Charging would become slower and slower and eventually come to a standstill at an output voltage slightly below U_{dd}.[2]

In a nutshell, a MOSFET operating as a switch can be summarized as follows.

	enhancement-type MOSFET	
electrical behavior	n-channel device	p-channel device
turns "on" when	gate voltage is positive relative to source	gate voltage is negative relative to source
then acts	as poor pull-up but good pull-down	as good pull-up but poor pull-down
with an "on"-state conductance	$G_{on\,n} \propto W/L$	$G_{on\,p} \propto \approx \frac{1}{2}W/L$

8.1.2 The inverter

A static CMOS inverter is made up of two complementary MOSFETs. Not surprisingly, the n-channel device acts as low-side switch while its p-channel counterpart operates as high-side switch, see fig.8.2. Their gate terminals are tied together to form the logic gate's input terminal. Most often the body terminals are omitted from schematic drawings of digital circuits as they are almost always tied to VSS and VDD respectively.[3]

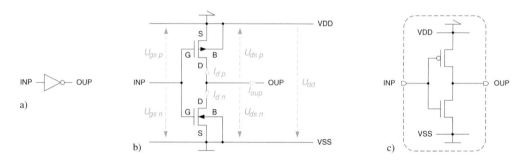

Fig. 8.2 Static CMOS inverter. Icon (a), detailed schematic with MOSFET body terminals and counting conventions (b), and simplified diagram as often used in digital design (c).

From the switch model, the circuit's logic operation as a Boolean inverter is immediately obvious. Before putting that model to service to study more complex subcircuits, however, we want to develop a more detailed picture of the inverter's electrical and timing characteristics.

---

[2] In theory at $U_{dd} - U_{th\,n}$, but subthreshold conduction tends to make the result less predictable. Also note that another approach to explaining the above limitation is pursued in problem 1.

[3] See section 11.6.3 for an explanation.

With node VSS acting as reference potential, we have the circuit equations

$$U_{gs\,n} = U_{inp} \qquad\qquad U_{gs\,p} = U_{inp} - U_{dd} \qquad\qquad (8.1)$$
$$U_{ds\,n} = U_{oup} \qquad\qquad U_{ds\,p} = U_{oup} - U_{dd} \qquad\qquad (8.2)$$

Counting currents in the conventional way one also obtains

$$I_{oup} = -I_{d\,n} - I_{d\,p} \qquad\qquad (8.3)$$

Static inverter behavior

Our analysis will be confined to situations with no resistive load connected to the inverter's output terminal. Static behavior implies the absence of transient currents. Even with a substantial capacitive load attached, static analysis applies as long as voltage changes occur so slowly that [dis]charge currents remain negligible, a regime which is termed **pseudostatic**. One then has

$$I_{d\,n} = -I_{d\,p} \qquad\qquad (8.4)$$

Incidentally, note that any current that flows from supply node VDD to ground node VSS without ever seeing the load will be referred to as **crossover current** I_{cr}.

To understand how the circuit behaves, let us find out what happens when input voltage U_{inp} slowly ramps from 0 to U_{dd}. Most of our analysis will be based on the simple **Sah model**, to be more thouroughly discussed in appendix 8.7.2. The Sah model distinguishes between the three operating regions introduced in fig.8.1b and essentially uses a gain factor β and a threshold voltage U_{th} to characterize any given MOSFET. The combined operation of two complementary devices in the inverter then makes it necessary to discern five ranges labeled A to E. You may want to skip to fig.8.4 for the overall results.

Operating range A $\qquad\qquad\qquad\qquad 0 \leq U_{inp} \leq U_{th\,n}$

The n-channel transistor is "off", hence there can be no current flowing from VDD to VSS. The negative gate–source voltage being applied to the p-channel turns it "on", thereby pulling the inverter's output terminal up towards the power rail. More precisely, the p-channel device operates in its linear region because of the strongly negative gate voltage $U_{gs\,p} \approx -U_{dd}$ and the zero drain–source voltage $U_{ds\,p} = 0$. The transistor's equivalent conductance is

$$G_{lin\,p} = \frac{dI_{d\,p}}{dU_{ds\,p}} \approx \beta_p(U_{dd} + U_{th\,p}) \qquad\qquad (8.5)$$

and the inverter's output voltage firmly rests at the maximum as defined by the power supply $U_{oup} = U_{dd}$.

Operating range B $\qquad\qquad\qquad\qquad U_{th\,n} < U_{inp} < U_{inv}$

The n-channel transistor enters the saturation region and begins to conduct, thereby causing a crossover current to flow from VDD to VSS. The current through the n-channel device grows quadratically with U_{inp}.

$$I_{d\,n} = \frac{\beta_n}{2}(U_{inp} - U_{th\,n})^2 \qquad\qquad (8.6)$$

The p-channel transistor continues to operate in its linear region.

$$I_{d\,p} = -\frac{\beta_p}{2}[2(U_{inp} - U_{dd} - U_{th\,p})(U_{oup} - U_{dd}) - (U_{oup} - U_{dd})^2] \tag{8.7}$$

Equating the two drain currents according to (8.4) and solving for the output voltage one obtains

$$U_{oup} = (U_{inp} - U_{th\,p}) + \sqrt{(U_{inp} - U_{dd} - U_{th\,p})^2 - \frac{\beta_n}{\beta_p}(U_{inp} - U_{th\,n})^2} \tag{8.8}$$

The net result is that U_{oup} drops below U_{dd} and that the diminution accelerates as U_{inp} grows.

Operating range C (basic model) $\qquad\qquad U_{inp} = U_{inv} \approx \frac{1}{2}U_{dd}$

Both devices operate in saturation. Their respective drain currents are given as

$$I_{d\,n} = \frac{\beta_n}{2}(U_{inp} - U_{th\,n})^2 \qquad\qquad I_{d\,p} = -\frac{\beta_p}{2}(U_{inp} - U_{dd} - U_{th\,p})^2 \tag{8.9}$$

which, when substituted, yield

$$U_{inp} = \frac{U_{dd} + U_{th\,p} + U_{th\,n}\sqrt{\frac{\beta_n}{\beta_p}}}{1 + \sqrt{\frac{\beta_n}{\beta_p}}} = U_{inv} \tag{8.10}$$

This result indicates that the circuit functions in this regime for only one value of U_{inp}, which implies that the operating range degenerates to a single point. That singular voltage is termed **inverter threshold** U_{inv}. Assuming electrically symmetric transistors where $\beta_p = \beta_n = \beta$ and $-U_{th\,p} = U_{th\,n} = U_{th}$, one obtains $U_{inv} = \frac{1}{2}U_{dd}$. Another finding is that the crossover current reaches its maximum at this point. Back to the general case where devices are not necessarily symmetric, current peaks at a value of

$$I_{d\,n} = \frac{\beta_n}{2}\left(\frac{U_{dd} + U_{th\,p} - U_{th\,n}}{1 + \sqrt{\frac{\beta_n}{\beta_p}}}\right)^2 \tag{8.11}$$

Equation (8.10) says nothing about U_{oup}, which means that the output voltage no longer depends on the input voltage. In fact, differentiating (8.10) yields $\frac{dU_{inp}}{dU_{oup}} = 0$, which suggests that the inverter's voltage amplification v is infinite at this point. For obvious reasons, the output voltage cannot grow indefinitely, however. Bounds are imposed by the condition that both transistors stay saturated

$$U_{th\,n} \leq U_{gs\,n} < U_{ds\,n} + U_{th\,n} \qquad\qquad U_{th\,p} \geq U_{gs\,p} > U_{ds\,p} + U_{th\,p} \tag{8.12}$$

Substituting (8.1) and (8.3) and combining the relevant terms one obtains

$$U_{inp} - U_{th\,n} < U_{oup} < U_{inp} - U_{th\,p} \tag{8.13}$$

which condition is shown in fig.8.4a by a pair of parallel dashed lines. In conclusion, our calculations tell us the output voltage is free to vary within this range as long as $U_{inp} = U_{inv}$.

The reason for this intriguing finding lies in the oversimplification of the Sah model that describes a saturated MOSFET as an ideal current source. In the inverter circuit, there are two such ideal sources connected in series which results in an unstable behavior. Yet, what the simple model describes correctly is that the transfer characteristic is very steep at this point.

Operating range C (improved model) $\qquad U_{inp} = U_{inv} \approx \frac{1}{2} U_{dd}$

For a more accurate analysis, we have to resort to the **Shichman–Hodges model** which is to be introduced in appendix 8.7.3 along with its heuristic channel-length modulation factor λ. This refinement leads to the equivalent circuit of fig.8.3.

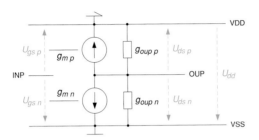

Fig. 8.3 Equivalent circuit for a CMOS inverter operating in range C.

The **voltage amplification** for small signals is given as

$$v = \frac{dU_{oup}}{dU_{inp}} = -\frac{g_{m\,n} + g_{m\,p}}{g_{oup\,n} + g_{oup\,p}} \tag{8.14}$$

Transconductances g_m are obtained from differentiating transistor equations wrt U_{gs},

$$g_{m\,n} = \beta_n(U_{gs\,n} - U_{th\,n})(1 + \lambda_n U_{ds}) \quad g_{m\,p} = -\beta_p(U_{gs\,p} - U_{th\,p})(1 + \lambda_p U_{ds}) \tag{8.15}$$

which, for a rough hand calculation, may be approximated by

$$g_{m\,n} \approx \beta_n(U_{gs\,n} - U_{th\,n}) \qquad g_{m\,p} \approx -\beta_p(U_{gs\,p} - U_{th\,p}) \tag{8.16}$$

Output conductances g_{oup} follow from differentiating transistor equations wrt U_{ds} this time.

$$g_{oup\,n} = \frac{\beta_n}{2}(U_{gs\,n} - U_{th\,n})^2 \lambda_n \qquad g_{oup\,p} = -\frac{\beta_p}{2}(U_{gs\,p} - U_{th\,p})^2 \lambda_p \tag{8.17}$$

As opposed to what was found for the Sah model, the voltage amplification for small signals obtained with the Shichman–Hodges model is in fact finite. Assuming electrically symmetric transistors once again, the small-signal amplification in range C can be estimated as

$$v \approx -\frac{4}{\lambda(U_{dd} - 2U_{th})} \tag{8.18}$$

Observe that none of the above equations includes transistor width W or length L. An IC designer can thus influence voltage amplification only indirectly by lessening or by emphasizing channel-length modulation because λ varies as a function of L. Typical values of v range between -10 for short and -1000 for long transistors.

Operating range D $\qquad\qquad U_{inv} < U_{inp} < U_{dd} + U_{th\,p}$

When compared with range B, the two transistors have swapped their modes of operation. The n-channel device enters the linear region whereas the p-type transistor operates in saturation.

$$I_{d\,n} = \frac{\beta_n}{2}[2(U_{inp} - U_{th\,n})U_{oup} - U_{oup}^2] \qquad (8.19)$$

$$I_{d\,p} = -\frac{\beta_p}{2}(U_{inp} - U_{dd} - U_{th\,p})^2 \qquad (8.20)$$

which together yield an output voltage of

$$U_{oup} = (U_{inp} - U_{th\,n}) - \sqrt{(U_{inp} - U_{th\,n})^2 - \frac{\beta_p}{\beta_n}(U_{inp} - U_{dd} - U_{th\,p})^2} \qquad (8.21)$$

As U_{inp} grows, U_{oup} continues to diminish, albeit at a slower and slower rate, eventually converging to 0 as U_{inp} approaches $U_{dd} + U_{th\,p}$. The crossover current follows a similar pattern.

Operating range E $\qquad\qquad U_{dd} + U_{th\,p} \le U_{inp} \le U_{dd}$

Analogously to range A, there is no crossover current as the p-channel device is turned off this time. The n-channel transistor behaves much like a linear conductance of

$$G_{lin\,n} = \frac{dI_{d\,n}}{dU_{ds\,n}} \approx \beta_n(U_{dd} - U_{th\,n}). \qquad (8.22)$$

that pulls the inverter's output down to the ground potential $U_{oup} = 0$.

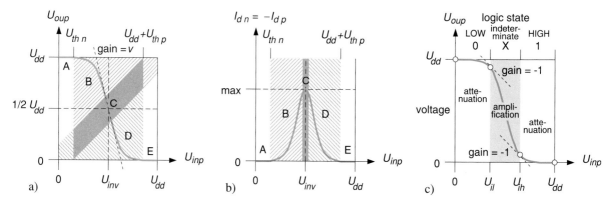

Fig. 8.4 CMOS Inverter. Operating ranges with static transfer characteristic (a), crossover current (b), and the abstraction of voltages to logic states (c).

THE INVERTER AS A BINARY DISCRIMINATOR

What we have found is that an inverter essentially acts as a voltage amplifier.[4] Why, then, is it possible to use it as a digital device that discriminates between logic 0 and 1? How are continuous voltages mapped to discrete logic states?

Look at the inverter's static **transfer characteristic** $U_{oup} = f(U_{inp})$ in fig.8.4c and note the presence of two points, one each in ranges B and D, where voltage amplification is unity. While any voltage deflection between the two unity gain points gets amplified, those below U_{il} and above U_{ih} get attenuated. That is, any input voltage close to U_{ss} results in an output even closer to U_{dd}, and vice versa. This property is called **level restoration** and supports defining a logic 0 as $0 \leq U_{inp} \leq U_{il}$ and a logic 1 as $U_{ih} \leq U_{inp} \leq U_{dd}$.[5]

The interval in between where $U_{il} < U_{inp} < U_{ih}$ acts as a "no man's land" that separates logic 0 from 1. The logic value associated with any such voltage is said to be unknown and denoted with an X in the IEEE 1164 standard. Albeit predictable electrically, the inverter's output is considered undefined from a digital or logic point of view.

Observation 8.1. *A transfer characteristic that exhibits two flat tails separated by a steep crossover band where $|v| > 1$ is a prerequisite for binary signal discrimination and restoration of logic levels to proper voltages. This trait is shared by all static CMOS gates.*[6]

THE IMPACT OF TRANSISTOR SIZES

While the transfer characteristic of a CMOS inverter always looks the same qualitatively, the exact run of the curve depends on the relative gain factor $\frac{\beta_n}{\beta_p}$ of the two MOSFETs involved.

$$\frac{\beta_n}{\beta_p} = \frac{\beta_{\square n} \frac{W_n}{L_n}}{\beta_{\square p} \frac{W_p}{L_p}} \tag{8.23}$$

Equation (8.10) indicates the inverter threshold voltage as a function of $\frac{\beta_n}{\beta_p}$. Numerical figures are given below under the assumption that $-U_{th\,p} = U_{th\,n} = \frac{1}{5}U_{dd}$, that is for any inverter where $U_{dd} = 3.3$ V and $U_{th\,n} = 0.66$ V or where $U_{dd} = 2.5$ V and $U_{th\,n} = 0.5$ V, for instance.

$\frac{\beta_n}{\beta_p}$	100	10	5	3	2	1.5	1	0.67	0.5	0.33	0.2	0.1	0.01
$\frac{U_{inv}}{U_{dd}}$	0.25	0.34	0.39	0.42	0.45	0.47	0.50	0.53	0.55	0.58	0.61	0.66	0.75

Sizing MOSFETs such that the n- and the p-channel transistor exhibit the same drive strength[7] puts the inverter threshold midway between power and ground potentials $U_{inv} = \frac{1}{2}U_{dd}$ and maximizes the overall noise margin. Another argument in favor of doing so is that rise and fall times are made

---

[4] Inverters are actually used as amplifiers in applications such as crystal oscillators, for instance, by biasing them to the point of maximum gain with the aid of a high-ohmic resistor that runs between input and output.

[5] It goes without saying that non-inverting gates restore inputs close to U_{ss} to even closer to U_{ss}, and those close to U_{dd} to even closer to U_{dd}.

[6] The transmission gate to be introduced in section 8.1.5 is an exception and not considered a logic gate.

[7] That is, saturation currents of the same magnitude but opposed signs.

equal in this way. All this is obtained from selecting $\frac{\beta_n}{\beta_p} = 1$, which mandates a shape factor ratio of

$$\frac{\frac{W_p}{L_p}}{\frac{W_n}{L_n}} = \frac{\beta_{\square\,n}}{\beta_{\square\,p}} \tag{8.24}$$

With ϵ_{ox}, t_{ox}, and L being identical for n- and p-channel transistors, the rule simplifies to

$$\frac{W_p}{W_n} = \frac{\mu_n}{\mu_p} \tag{8.25}$$

which means that the p-channel transistor ought to be roughly twice as wide as its n-channel counterpart to fully compensate for the lower mobility of holes.

However, insisting on perfect electrical symmetry by beefing up the p-channel MOSFET in accordance with (8.25) would unnecessarily inflate area occupation and parasitic capacitances, which is also detrimental to both switching speed and energy efficiency. Gate performance has been found [186] to be optimal when transistor widths comply with

$$\frac{W_p}{W_n} = \sqrt{\frac{\mu_n}{\mu_p}} \tag{8.26}$$

$\frac{W_p}{W_n}$ ratios that range between 1.2 and 1.7 resulting in $\frac{\beta_n}{\beta_p} \approx \sqrt{2}$ or so are thus more typical. As suggested by the above figures, noise margins do not suffer much from a minor imbalance.

DYNAMIC INVERTER BEHAVIOR

The dynamic behavior refers to the time-dependent processes associated with (dis)charging the two capacitances that are present in any inverter circuit, see fig.8.5a. Taking into consideration transient currents, (8.3) becomes

$$0 = I_l + I_m + I_{d\,n} + I_{d\,p} = (C_l + C_m)\frac{dU_{oup}(t)}{dt} - C_m\frac{dU_{inp}(t)}{dt} + I_{d\,n} + I_{d\,p} \tag{8.27}$$

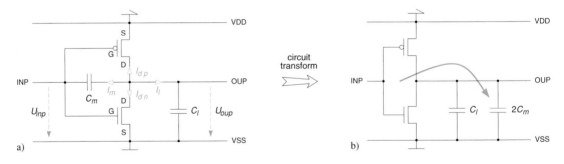

Fig. 8.5 Static CMOS inverter circuit with load and Miller capacitances shown (a) and with Miller capacitance transformed to the output node (b).

Load capacitance C_l has one end connected to ground and is meant to include parasitics from the inverter itself.[8] Further note the presence of a coupling capacitance between the input and the

---

[8] A comprehensive list of contributions is given in section 9.1.1.

output node that is commonly referred to as **Miller capacitance** C_m. For the purpose of delay estimation, one would want to transform it to the inverter's output by introducing a delay-equivalent load $C_k > C_l$ such as to restate the problem in a simplified form as

$$0 \approx C_k \frac{dU_{oup}(t)}{dt} + I_{dn} + I_{dp} \tag{8.28}$$

There is an important difference between load capacitance and Miller capacitance, however. While C_l is being charged to U_{dd} and discharged back to zero, the voltage across C_m changes from $+U_{dd}$ to $-U_{dd}$ and back again, which means that the amplitude across C_m is twice that across C_l. As this also doubles charge and discharge currents, we must account for the Miller capacitance with two times its nominal value for the purpose of delay calculation.

$$C_k = C_l + 2C_m \tag{8.29}$$

Given behavioral models for the two transistors and some input waveform $U_{inp}(t)$, differential equation (8.27) or (8.28) must be solved for the output voltage $U_{oup}(t)$ as a function of time. Even using the simple Sah MOSFET model, doing so analytically is impractical, however, because I_{dn} and I_{dp} are themselves nonlinear functions of $U_{oup}(t)$.

A precise output waveform is not always sought, a couple of numeric quantities often suffice to characterize the timing of an inverter (or some other logic gate). The most interesting question is how circuit performance depends on overall load capacitance, transistor sizes, supply voltage, and MOSFET thresholds. As a first-order approximation, the rise and fall times of a CMOS inverter can be obtained under the following set of simplifying assumptions [186].

- The input voltage does switch abruptly, that is $t_{ri\,inp} = t_{fa\,inp} = 0$.
- Output rise and fall times are measured from 0 to $\frac{1}{2}U_{dd}$ and from U_{dd} to $\frac{1}{2}U_{dd}$ respectively.
- The current-carrying MOSFET is operating in its saturation region throughout.
- A Sah model is a good enough approximation for the transistors.
- The p-channel (n-channel) device sources (sinks) a fixed current throughout.

The derivation then immediately follows as

$$t_{ri\,oup} \approx -\frac{C_k\,U_{dd}}{2\,I_{oup}} \tag{8.30}$$

$$t_{fa\,oup} \approx \frac{C_k\,U_{dd}}{2\,I_{oup}} \tag{8.31}$$

$$-I_{oup} \approx I_{d\,on\,p} = \frac{\beta_p}{2}(U_{dd} + U_{th\,p})^2 \tag{8.32}$$

$$I_{oup} \approx I_{d\,on\,n} = \frac{\beta_n}{2}(U_{dd} - U_{th\,n})^2 \tag{8.33}$$

$$t_{ri\,oup} \approx \frac{C_k\,U_{dd}}{\beta_p(U_{dd} + U_{th\,p})^2} \tag{8.34}$$

$$t_{fa\,oup} \approx \frac{C_k\,U_{dd}}{\beta_n(U_{dd} - U_{th\,n})^2} \tag{8.35}$$

Though rather rudimentary, this approximation is often sufficient for manual calculations. Remarkably, the more general relation

$$t_{pd} \propto \frac{C_k \, U_{dd}}{(U_{dd} - U_{th})^2} \tag{8.36}$$

has been found to hold for numerous CMOS subcircuits fairly independently of the logic function and circuit style (static vs. dynamic) involved. Yet, keep in mind that these approximations do not cover subthreshold operation, that is they are invalid unless $U_{dd} > U_{th}$.

Also, all of the above calculations apply to long transistors because they have been derived on the basis of the Sah model that does not account for velocity saturation. For today's short transistors, (8.36) takes on the form of

$$t_{pd} \propto \frac{C_k \, U_{dd}}{(U_{dd} - U_{th})^\alpha} \tag{8.37}$$

with α a function of L and U_{dd}. Rough indications are as follows:

L [nm]	2000	1000	500	250	130
$\alpha \approx$	$1.6 - 1.65$	$1.45 - 1.6$	$1.3 - 1.5$	$1.1 - 1.4$	$1.0 - 1.25$

You may want to refer to sections 8.7.4 to 8.7.6 for further explanations on velocity saturation, α, and on short-channel effects (SCE) in general. More sophisticated inverter models are being proposed in [186] [187] [188].

8.1.3 Simple CMOS gates

In a CMOS inverter, we have two transistors that act as voltage-controlled switches. When the input voltage is at a logic high level, the n-transistor is "on" and the p-transistor "off", thereby establishing a current path from the output terminal to ground. For a low input, one has the complementary situation with the output high. A variety of logic gates is obtained from generalizing this concept to circuits of four, six, and more transistors.

NAND GATES

One obtains an i-input NAND function from the inverter by replacing the single n-channel MOSFET by a series network of i transistors and the single p-channel MOSFET by i transistors connected in parallel. Consider a 2-input NAND gate, for instance, that implements the logic equation $\mathtt{OUP} = \overline{\mathtt{IN1} \wedge \mathtt{IN2}}$.

2-input NAND			switch model			
IN1	IN2	OUP	N1	N2	P1	P2
0	0	1	off	off	on	on
0	1	1	off	on	on	off
1	0	1	on	off	off	on
1	1	0	on	on	off	off

3-input NOR			
IN1	IN2	IN3	OUP
0	0	0	1
0	0	1	0
0	1	0	0
0	1	1	0
1	0	0	0
1	0	1	0
1	1	0	0
1	1	1	0

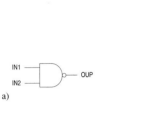

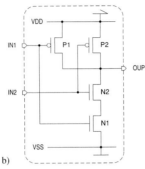

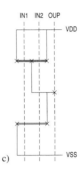

Fig. 8.6 CMOS 2-input NAND gate. Icon (a), circuit diagram (b), symbolic layout (c).[9]

NOR GATES

NOR gates result when the n-type pull-down network has its transistors connected in parallel and vice versa for the p-type pull-up network. A 3-input NOR gate implements the operation OUP = $\overline{\text{IN1} \lor \text{IN2} \lor \text{IN3}}$.

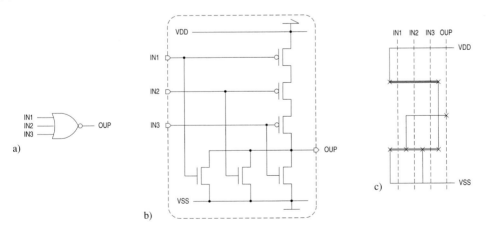

Fig. 8.7 CMOS 3-input NOR gate. Icon (a), circuit diagram (b), symbolic layout (c).

---

[9] The stick-diagram notation shown in (c) is to be formally introduced in section 11.5.2.

UNDERLYING PRINCIPLES

A closer analysis of inverter, NAND gate, and NOR gate reveals the following:

- There is one n- and one p-channel MOSFET for every argument in the Boolean equation and they both have their gates connected to the logic gate's corresponding input terminal.
- For any combination of logic values at the input, there exists a conducting path from the output terminal to either VSS or VDD.
- The output is never left in a floating condition.
- The output voltage can swing from one supply rail to the opposite one.
- No input vector can establish a direct conducting path from VDD to VSS.[10]
- Except for leakage currents, there is no power dissipation under steady-state conditions.

Definition 8.1. *Fully complementary static CMOS means that a logic gate is composed from a pair of n-channel pull-down and p-channel pull-up networks that turn on and off in a mutually exclusive fashion for any binary vector that can possibly be applied to their input terminals.*

Two such pull-down and pull-up networks are best characterized as an **antagonistic** pair. This property also explains why fully complementary static CMOS logic is **ratioless**.[11]

Definition 8.2. *A (sub)circuit is ratioless, if the geometric sizes and current drive capabilities of its transistors do not affect its logic functionality as specified in a truth table, for instance.*

Put differently, the impact of geometric modifications and fabrication tolerances is limited to electrical and timing parameters such as switching thresholds and delay figures.

The antonym **ratioed** refers to (sub)circuits the logic behavior of which is dependent on the relative transistor sizes. We will not get to know ratioed CMOS circuits until section 8.2. Because

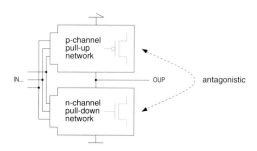

Fig. 8.8 General circuit structure of fully complementary static CMOS gates.

---

[10] This property forms the basis of I_{ddq} **testing**, a technique to tell apart defective circuits from intact ones by comparing their supply currents in steady-state condition [189] [190]. A stimulus vector is applied to the **circuit under test** (CUT) which is then given sufficient time to settle before the supply current gets measured. The quiescent currents of a defect-free static CMOS circuit would in fact be zero were it not for leakage. The operation is repeated many times with a series of vectors carefully selected so as to check for the most likely faults. Any out-of-the-normal reading — non-zero or well above that of a leaky but intact reference circuit — indicates a fabrication defect.

[11] This is not so in NMOS, PMOS, and dynamic (precharge and evaluate) CMOS logic, where one of the two networks is replaced by a simple pull device that does not operate under control from the gate's inputs. CVSL and CPL are other exceptions. None of these design styles is being addressed in this text, however.

of their nonpermanent drive, logic gates with three-state outputs are another notable exception to definition 8.1; they will be discussed shortly in section 8.1.5.

8.1.4 Composite or complex gates

AOI GATES

In the aforementioned NAND and NOR gates, all MOSFETs of identical polarity did form a series chain while their counterparts were connected in parallel. This property is overly restrictive, however, as any pair of **dual networks** will do. Examples are given in fig.8.9.

Definition 8.3. *Two transistor-level networks are dual iff there is a one-to-one correspondence between their edges such that any terminal-to-terminal chain in one network cuts all terminal-to-terminal chains in the other network.*[12]

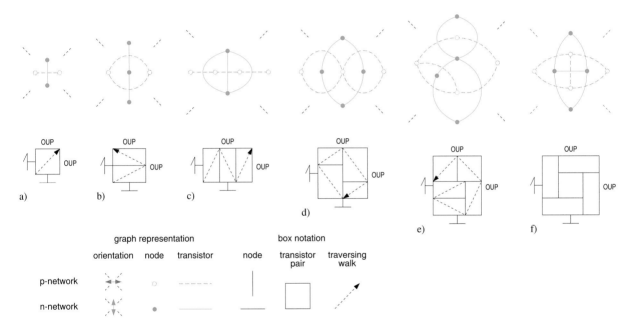

Fig. 8.9 A collection of dual graph pairs (top row) and their notation as stacked boxes (bottom row).[13] The associated CMOS gates are: Inverter (a), 2-input NAND (b), 3-input NOR (c), a 4-input AOI gate (d), a 5-input AOI gate (e), and a 3-input minority gate (f).

---

[12] The definition of duality in mathematical graph theory is more general: Two graphs are said to be dual if there is a one-to-one correspondence between the edges of the two graphs such that any circuit in one graph corresponds to a cut-set in the other graph [191]. This definition accomodates pendant vertices and self-loops that do not exist in well-formed electrical circuits. Also, before the definition can be applied to circuit analysis, one fictive edge must be added to each graph to interconnect its two terminal nodes.

[13] The interpretation of the box notation and of the traversing walks therein is left to the reader as an exercise.

This insight allows one to implement functions more general than NAND and NOR in a single gate. The circuit of fig.8.10, for instance, conforms to fig.8.9d and implements

$$\mathtt{OUP} = \overline{(\mathtt{IN1} \wedge \mathtt{IN2}) \vee (\mathtt{IN3} \wedge \mathtt{IN4})} \tag{8.38}$$

Any monotone decreasing logic function can be carried out in this way after having been recast into a nested expression of AND and OR operations followed by one final negation at the outermost level. This explains why composite gates are also known as **and-or-invert** (AOI) gates. That name is understood to include OAI functions and similar functions with deeper nestings such as the AOAI gate (8.39) of fig.8.11. Given some AOI-type logic equation, constructing the gate's inner circuitry is straightforward and follows a process of hierarchical composition.

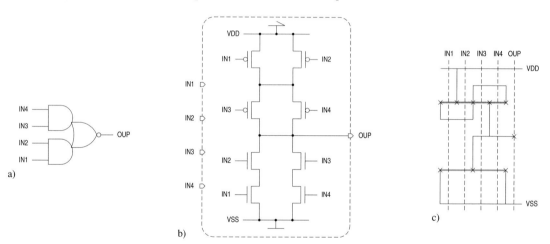

Fig. 8.10 A CMOS 4-input AOI gate. Icon (a), circuit diagram (b), symbolic layout (c).

Example

$$\mathtt{OUP} = \overline{((\mathtt{IN1} \wedge \mathtt{IN2}) \vee \mathtt{IN3}) \wedge (\mathtt{IN4} \vee \mathtt{IN5})} \tag{8.39}$$

□

Composite gates are typical CMOS subcircuits. They contribute to layout density by combining two or more levels of logic operations in a single cell. The quest for high-speed operation limits the size of composite gates, however, as more series transistors augment the resistance through which the output node is being (dis)charged. In addition, node capacitances also tend to grow with cell size. The following table indicates the number of logic gates that can be built without exceeding a given number of transistors connected in series [192].

maximum number of transistors in series	total number of AOI functions
1	1
2	7
3	87
4	3521

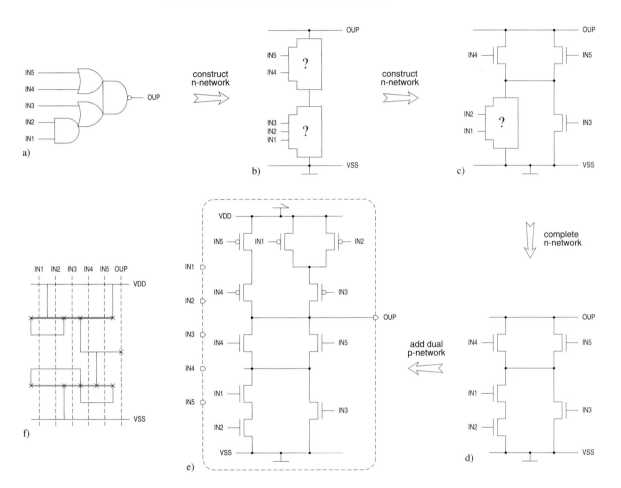

Fig. 8.11 Stepwise construction of a CMOS 5-input AOI gate. Icon (a), successive refinement of circuit diagram (b,c,d,e), symbolic layout (f).

Observation 8.2. *Cumulated resistance and the back gate effect tend to make MOSFET stacks slow. CMOS subcircuits such as logic gates and bistables are thus not normally designed with more than three MOSFETs connected in series.*

Complex combinational operations that would ask for more series transistors are broken down into smaller subfunctions and implemented as cascades of simpler gates. This also applies to NAND and

NOR gates with four inputs or more. Design rules for very-high-speed circuits even prescribe that no more than two p-channel MOSFETs be connected in series.

The truth tables of 3-input **majority** and **minority function** are as follows.

<table>
<tr><th colspan="4">3-input MAJ</th><th colspan="4">3-input MIN</th></tr>
<tr><th>IN1</th><th>IN2</th><th>IN3</th><th>OUP</th><th>IN1</th><th>IN2</th><th>IN3</th><th>OUP</th></tr>
<tr><td>0</td><td>0</td><td>0</td><td>0</td><td>0</td><td>0</td><td>0</td><td>1</td></tr>
<tr><td>0</td><td>0</td><td>1</td><td>0</td><td>0</td><td>0</td><td>1</td><td>1</td></tr>
<tr><td>0</td><td>1</td><td>0</td><td>0</td><td>0</td><td>1</td><td>0</td><td>1</td></tr>
<tr><td>0</td><td>1</td><td>1</td><td>1</td><td>0</td><td>1</td><td>1</td><td>0</td></tr>
<tr><td>1</td><td>0</td><td>0</td><td>0</td><td>1</td><td>0</td><td>0</td><td>1</td></tr>
<tr><td>1</td><td>0</td><td>1</td><td>1</td><td>1</td><td>0</td><td>1</td><td>0</td></tr>
<tr><td>1</td><td>1</td><td>0</td><td>1</td><td>1</td><td>1</td><td>0</td><td>0</td></tr>
<tr><td>1</td><td>1</td><td>1</td><td>1</td><td>1</td><td>1</td><td>1</td><td>0</td></tr>
</table>

The fact that the majority function is monotone increasing precludes it from being implemented as an AOI gate and, instead, suggests a cascade of two subcircuits where an inverter follows the monotone decreasing minority function

$$OUP = \overline{(IN1 \wedge IN2) \vee (IN1 \wedge IN3) \vee (IN2 \wedge IN3)} \tag{8.40}$$

which indeed has nice and surprising CMOS circuit realizations. For a first and somewhat uninspired solution, consider the 6-input function

$$OUP = \overline{(IN1 \wedge IN2) \vee (IN3 \wedge IN4) \vee (IN5 \wedge IN6)} \tag{8.41}$$

and note that the desired minority function follows from tying together $IN1 = IN4$, $IN2 = IN5$, and $IN3 = IN6$. This proposal results in an AOI gate that takes twelve transistors out of which up to three connect in series, yet we can do better.

A GATE WITH SELF-DUAL TRANSISTOR NETWORKS

A remarkable circuit is shown in fig.8.12. The n- and p-channel networks both look like a bridge circuit, making the ensemble match fig.8.9f. The two networks differ from the other examples presented so far in that they are identical, self-dual, and not amenable to series/parallel decomposition. Taken as it is, the circuit computes the four-level-deep AOI function

$$OUP = \overline{(IN1 \vee IN4) \wedge (IN2 \vee IN5) \wedge (IN3 \vee ((IN1 \vee IN5) \wedge (IN2 \vee IN4)))} \tag{8.42}$$

which seems of no practical interest. Only when one sets $IN1 = IN5$ and $IN2 = IN4$ can the circuit be recognized as an elegant implementation of the 3-input minority function.

GATES WITH NON-DUAL TRANSISTOR NETWORKS

A third alternative shown in fig.8.13c follows from the AOI gate for (8.41) after a couple of equivalent circuit transforms. The new circuit is an improvement over the original one because it makes do

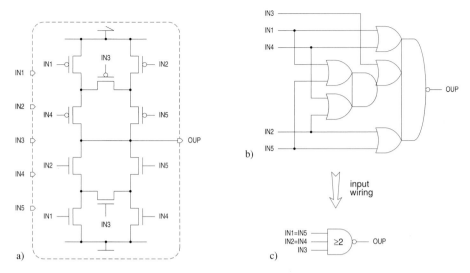

Fig. 8.12 3-input minority function. H-topology circuit with 5 inputs (a) and logic equivalent drawn from NAND and NOR operators (b). Same circuit rewired as MIN function (c).

with a total of ten MOSFETs, no more than two of which ever connect in series. Observe that the resulting pull-down and -up networks are antagonistic but no longer dual to each other!

Observation 8.3. *Electrical antagonism between pull-down and pull-up networks is entirely sufficient for a fully complementary static CMOS gate. Structural duality is a stronger criterion that implies electrical antagonism.*

Another variety of static CMOS gates that abandon duality of pull-down and pull-up networks while maintaining electrical antagonism are known as **branch-based logic**. Both the n- and the p-networks are made up of one or more branches that are connected in parallel where each branch consists of one or more transistors connected in series. The converse of connecting clusters of parallel transistors in series is disallowed.

Transistor count, cell area, capacitance, and dissipated energy are not normally minimal because of the necessary duplications of controlled switches and the extra interconnections. These drawbacks notwithstanding, advantages with respect to layout regularity, cell characterization, and other criteria have sometimes sufficed to tip the balance in favor of this approach [193].

8.1.5 Gates with high-impedance capabilities

TRANSMISSION GATES

T-gate and pass gate are synonyms for transmission gate.[14] A transmission gate has two data terminals interconnected by a pair of complementary MOSFETs, see fig.8.15a and b. Two signals

---

[14] A single MOSFET, typically of n-type, being used for the same purpose is referred to as a **pass transistor**.

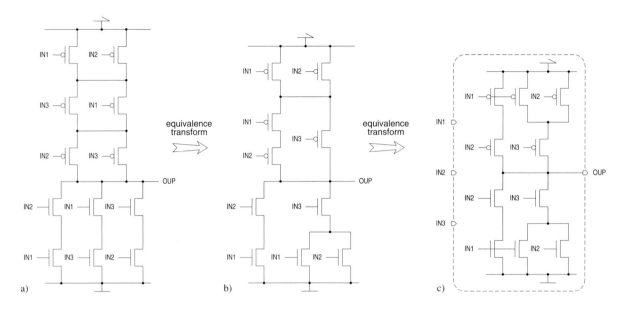

Fig. 8.13 3-input minority function. Simplified circuit (c) obtained from the AOI gate for (8.41) with the aid of equivalent circuit transforms (a,b).

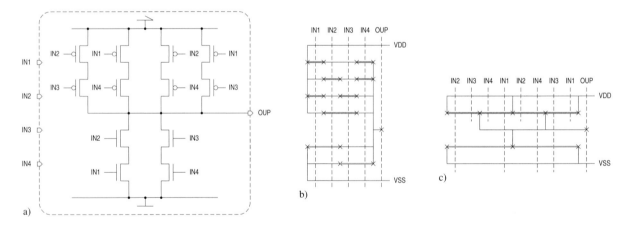

Fig. 8.14 The AOI gate of fig.8.10 reimplemented in branch-based logic. Circuit diagram (a), symbolic layout alternatives (b,c).

of opposite voltages $0 \leq U_{ena} \leq U_{dd}$ and $U_{dd} - U_{ena}$ are applied to the gate electrodes of those MOSFETs and so steer the conductance of the electrical path between the data terminals. Note that a T-gate has no ground and supply terminals and thus can provide neither voltage amplification nor level restoration, which contrasts with all other logic gates presented so far.

As the two data terminals are perfectly interchangeable, there is no input and no output either. Put differently, it is not possible to establish a directed dependency, which also implies that there is no way to capture the functionality of a pass gate in a logic function or a truth table.

Strictly speaking, a transmission gate must not be called a logic gate but rather resembles a contact that makes and breaks a conducting path under the direction of a control voltage. As such, it is very convenient and finds many applications in multiplexers, XOR gates, adders, latches, and flip-flops, for instance. The undirectedness also brings about numerous peculiarities and perils, however, that will be discussed in section 8.5.2.

T-gate	path between	switch model	
ENA	TRM1 and TRM2	N1	P1
0	break	off	off
1	make	on	on

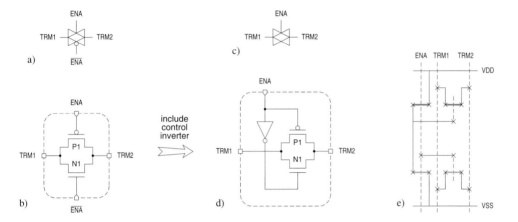

Fig. 8.15 CMOS transmission gate. Icons (a,c), circuit diagrams (b,d), symbolic layout (e).

The circuit of fig.8.15c and d is just a minor extension that includes an ancillary inverter for obtaining the drive voltage for the second longitudinal transistor when complementary signals are unavailable from externally.

THE CONTROLLED INVERTER AND OTHER GATES WITH THREE-STATE OUTPUTS

As opposed to transmission gates, three-state gates do have a logic function and a truth table associated with them. What sets them apart from regular gates is the extra control input that enables or disables the output(s) as a function of the logic value applied. Three-state buffers and three-state inverters are often used as bus drivers. Section 8.5.1 explains the difficulties that arise from this practice and gives recommendations on how to avoid them.

An arbitrary logic gate is easily extended to include a three-state capability by inserting a T-gate into its output as shown in fig.8.16b for an inverter. As a variation, it is always possible to omit the connection between the original gate's n- and p-channel transistors as shown in fig.8.16c. Circuit operation remains almost the same except for a minor difference in speed. This new circuit arrangement is referred to as a **controlled inverter** or multiplexer arm. The term **clocked inverter** applies to situations where two MOSFETs get controlled by a clock signal; examples will be shown in section 8.2.

three-state inverter		
ENA	IN	OUP
0	0	Z
0	1	Z
1	0	1
1	1	0

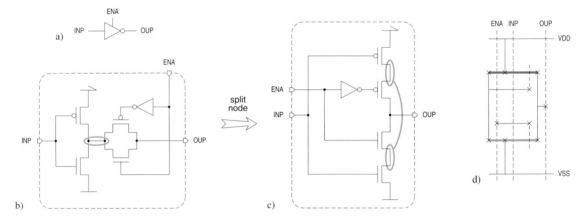

Fig. 8.16 CMOS three-state inverter. Icon (a), alternative circuits differing in the highlighted nodes (b,c), and symbolic layout for the controlled inverter of circuit c (d).

8.1.6 Parity gates

XOR AND EQV GATES

The 2-input exclusive-or function $OUP = IN1 \oplus IN2$ is also known as XOR, EOR, antivalence function, or even parity.[15] Exclusive-nor, XNOR, EQV, and odd parity are synonyms for its counterpart, the 2-input equivalence function $OUP = \overline{IN1 \oplus IN2}$. Both functions are non-monotone and more costly to implement than the monotone NAND, NOR, and AOI functions presented earlier in this chapter. A couple of circuit alternatives are shown in fig.8.17.

Figure 8.17b makes use of 2-input NAND gates exclusively and has the merit of being perfectly symmetrical but is unattractive in terms of size, delay, and energy. The more economic solution of

---

[15] The term parity is actually more general and extends to functions with more than two arguments. The attributes "even" and "odd" refer to the total number of 1s in a data word composed from the original bits and the parity bit. As an example, consider the data word 10110. The even parity is 1 so that 10110 1 includes a total of four bits of logic value 1. Conversely, 10110 0 is said to be of odd parity.

fig.8.17c is essentially built from two controlled inverters and acts as a multiplexer that selects the inverted or the original input under control of the second input.

The circuits of fig.8.17d and e include one AOI gate each. The difference is that fig.8.17e takes advantage of bubble pushing to save one more pair of transistors. This, together with the fact that an inversion operation has been spared, makes it also more energy-efficient.

Figure 8.17f conditionally negates input `IN2` by combining an inverter with a transmission gate in an intricate way. While the schematic appears to do with six transistors, correcting the various problems that are associated with T-gates will raise its overall transistor count to 10 or more. Please refer to section 8.5.2 for details.

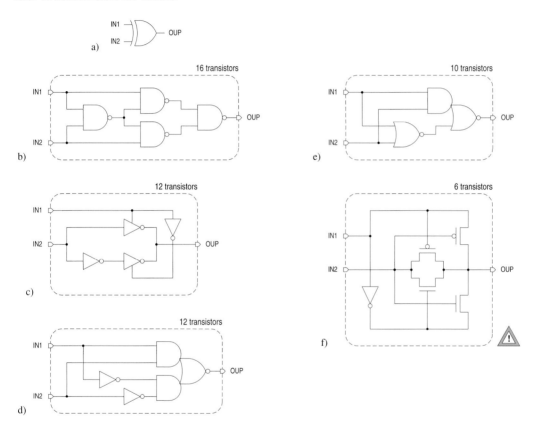

Fig. 8.17 CMOS 2-input XOR gate. Icon (a) and alternative circuit diagrams (b,...,f).

8.1.7 Adder slices

MIRROR ADDER

Computing more than one output in a single cell often gives designers the opportunity to replace individual subcircuits by one joint transistor network, thereby improving on overall transistor count, parasitic capacitances, propagation delay, and/or energy efficiency. As a popular example, consider the full-adder function that accepts three input bits and produces a sum and a carry bit.

full adder					
INA	INB	INC	OUPC	OUPS	carry
0	0	0	0	0	absorb
0	0	1	0	1	*idem*
0	1	0	0	1	propagate
0	1	1	1	0	*idem*
1	0	0	0	1	*idem*
1	0	1	1	0	*idem*
1	1	0	1	0	generate
1	1	1	1	1	*idem*

The static CMOS circuit of fig.8.18 is very elegant indeed. It includes the minority gate of fig.8.13c as a subcircuit to compute the more time-critical carry bit $\overline{\text{OUPC}}$ from the inputs directly. The sum bit $\overline{\text{OUPS}}$ then gets derived from this intermediate result by a transistor network that combines a controlled inverter with elements from branch-based logic in a clever way. Note that the carry path has no more than two transistors in series. The perfect structural symmetry of this elegant 24-transistor circuit earned it the name "mirror adder". Two inverting buffers drive the outputs and compensate for logic inversions by the preceding circuitry.

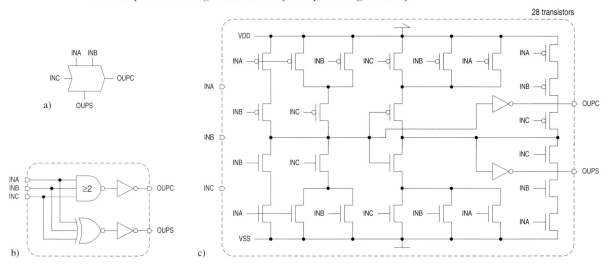

Fig. 8.18 Mirror adder. Icon (a), logic equivalent (b), and circuit diagram (c).

$$\overline{\text{OUPC}} = \overline{(\text{INA} \wedge \text{INB}) \vee (\text{INA} \wedge \text{INC}) \vee (\text{INB} \wedge \text{INC})} \tag{8.43}$$

$$\overline{\text{OUPS}} = \text{INPA} \oplus \text{INPB} \oplus \text{INPC} = \begin{cases} 0 & \text{if } (\text{INPA} \wedge \text{INPB} \wedge \text{INPC}) = 1 \\ 1 & \text{if } (\text{INPA} \vee \text{INPB} \vee \text{INPC}) = 0 \\ \text{OUPC} & \text{otherwise} \end{cases} \tag{8.44}$$

Observation 8.4. *Inverting all three inputs of a full adder amounts to both outputs being inverted.*

This follows immediately from the truth table or from the logic equations and suggests that there is room for further improvements if one is willing to design an adder slice that processes more than one binary digit at a time. A full adder built from transmission gates is shown in fig.8.45, while references such as [181] [186] [182] [194] [195] discuss many more transistor-level circuits.

Two-bit adder slices

Many commercial cell libraries include two-bit adder slices and fig.8.19 explains why. By cascading $\frac{w}{2}$ such slices rather than w full adders, the time-critical carry chain in a ripple-carry adder can be relieved of unnecessary inverters.

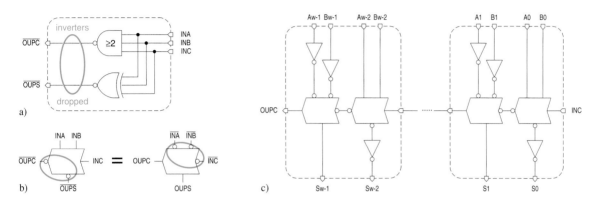

Fig. 8.19 Ripple-carry adder. Adder slice with inverting output buffers dropped (a), two equivalent views (b), ripple-carry adder assembled from two-bit adder slices.

Cascading transmission gate adder slices with no level restoring results in energy-efficient albeit very slow circuits [181] [196]. Hybrid solutions where a mirror adder is inserted after every two or three transmission gate adders exhibit attractive delay and energy figures and make it possible to find interesting tradeoffs [197]. Other full-adder circuits involve significant departures from the static and fully complementary CMOS design style put forward in this text.[16]

Ripple-carry adders are the simplest and most area-efficient, but also the slowest adder structures. Today's automatic synthesis tools select and adapt the circuit structures of adders and other arithmetic units as a function of the timing constraints imposed before resorting to gate-level optimizations. As they typically find reasonable compromises among longest path delay, circuit area, and energy efficiency, we do not want to go into computer arithmetics but refer the reader to the specialized literature [199] [200] [68] [201] [66] [202] [85] [67] [145] [69].

8.2 | CMOS bistables

A storage element is said to be static if it preserves its current state indefinitely provided the supply voltage remains applied. With current electronic devices, any **static memory** circuit requires the

---

[16] Either by resorting to dynamic logic (fast but energy-hungry) or by accepting threshold voltage losses [198].

presence of a local feedback loop.[17] This contrasts with a **dynamic memory**, where it is the presence or absence of an electrical charge on a capacitance that reflects the current state. This section is concerned with static latches and flip-flops exclusively; larger memory circuits (SRAMs and DRAMs) are to be briefly touched upon in section 8.3.

8.2.1 Latches

Bistable circuit behavior is obtained from connecting two inverting gates so as to form a positive feedback loop. The two stable states of equilibrium then naturally correspond to two memory states.[18] In order to admit a new data item, there must be some mechanism that temporarily suspends or overrides feedback so that the present input can determine the states of all circuit nodes along the loop. Let us study five approaches to doing so.

Switched memory loop

The idea behind the circuit of fig.8.20d is to interrupt feedback under control of the clock. Two complementary transmission gates open the loop and admit the voltage at data terminal D into the loop while CLK = 1. Conversely, the feedback loop is closed and the input locked out while CLK = 0. The resulting behavior is indeed that of a (level-sensitive) latch. Minor variations of this highly popular circuit are obtained from substituting clocked inverters for one or both of the inverter-plus-T-gate pairs as demonstrated in fig.8.16. Another possible modification consists of duplicating the forward inverter so as to have one instance drive the output and the other drive the backward inverter [155].

Overruled memory loop

A more important departure is to dispense with the steering of the feedback path by dropping the backward transmission gate, see fig.8.20e. Instead, transistors are resized in such a way as to make the backward inverter so weak that it is easily overriden by the controlled buffer in the forward path whenever a new data item is to be stored. This principle of operation strongly contrasts with that of fig.8.20d and makes this circuit our first example of ratioed logic, also see observation 8.6. The weak backward inverter is sometimes referred to as a **trickle inverter**. In comparison with to fig.8.20d, overruling reduces transistor count and clock load. As will be made clear in observation 8.6, there are limitations to this approach, however.

Jamb latch

The circuit of fig.8.20f also works by overruling a memory loop of two cross-connected inverters to bring in new data. This is achieved here by selectively pulling down one end or the other of the loop while CLK = 1 with the aid of four n-channel MOSFETs. The circuit is clearly ratioed, yet the fact that there are n-channel transistor networks with no p-channel counterparts is another striking departure from the customary static CMOS design style.

---

[17] A feedback loop is considered local if it remains confined to within the memory subcircuit itself. Non-local feedback has been discussed in section 2.7 in the context of architecture design.

[18] Two stable states of equilibrium imply the existence of a point of metastable equilibrium in between; please refer to section 7.4 for more details.

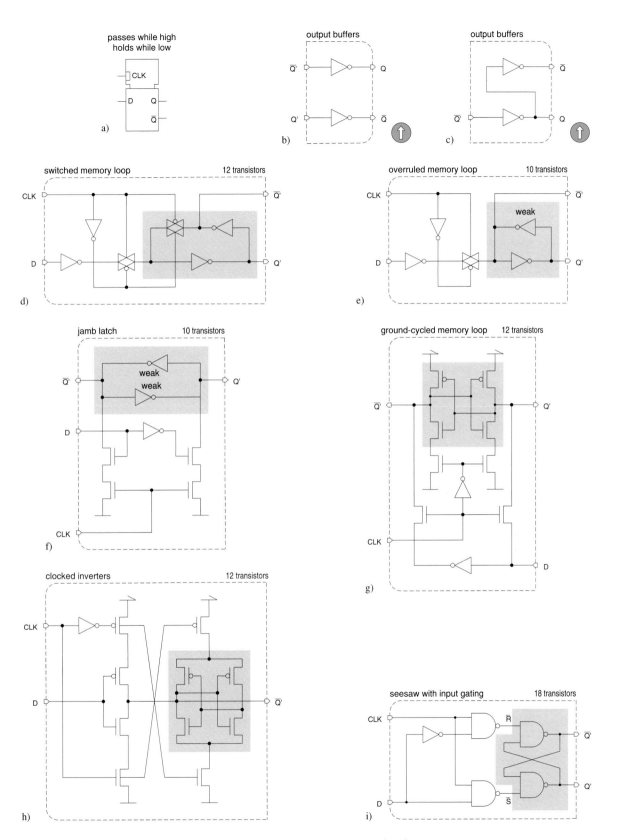

Fig. 8.20 CMOS latch. Icon (a), alternative circuit diagrams of static latches (d...i), and compulsory output buffer stage (b,c).

Power-cycled memory loops

The idea behind the next two circuits is to power down the memory loop to admit a new data item and to power it up again when the latch is to maintain its state. Figure 8.20g resembles a jamb latch where the cross-coupled inverters are cut from the ground potential under control of clock input CLK while two pass transistors bring in the data being applied to input D.

Figure 8.20h essentially consists of three clocked inverters. Two of them are connected back to back and share a pair of enable transistors. The output node $\overline{Q'}$ is driven either from the input inverter or from the memory loop at any time. Please note that no complementary node Q' with the same property exists in this circuit.

Seesaw with input gating

Instead of inverters, the memory loop of fig.8.20i uses cross-connected NAND gates in a seesaw configuration. There is no overruling as three more logic gates switch the seesaw back and forth between the storage state and a "set" or "reset" condition under control of the clock input CLK. In addition, they make sure that the seesaw is never presented with $\overline{S} = \overline{R} = 0$ as this would force it to produce an illegal output where $Q' = \overline{Q'} = 1$. When compared with the designs presented before, this one gets penalized by its higher transistor count, larger parasitic capacitances, and inferior loop gain, and thus also by its lower speed. It used to be popular in TTL circuits where T-gates and clocked inverters were unavailable.

It is important to note that none of the above circuit structures can be used "as is" because all of them are exposed to loss of memory under certain circumstances. The problem is that a brief voltage impulse applied at either output terminal might cause the memory loop to flip from one equilibrium state to the opposite one.[19] Even long after the disturbance has gone away, the original state will not get re-established. To protect the data stored from whatever might happen at the output, the memory loop needs to be decoupled from the output terminals Q and $\overline{Q}$ by way of an extra buffer stage such as the ones shown in fig.8.20b and c.

Observation 8.5. *Memory loops are exposed to data losses from backward signal propagation. Buffers must provide isolation from temporary disturbances that might occur at the output.*

As a welcome side effect, delays at one output become independent of the load connected to the complementary output. With some further precautions and refinements, the subcircuits shown are being used as part of most latches and flip-flops found in ASIC cell libraries.[20]

8.2.2 Function latches

Consider fig.8.21a and observe that three gate delays cumulate as the AND gate is perforce composed from a NAND and an inverter, which are then followed by the input inverter of the latch. Embedding the NAND operation into the latch does away with two inverters along with their node

---

[19] Note that SRAMs — which are to be discussed in section 8.3.1 — take advantage of this property because read and write operations occur over the same bit lines with no distinction between input and output.

[20] More sophisticated circuit designs, both of static and dynamic nature, are being presented in [159] along with a catalogue of possible failure modes. Not surprisingly, dynamic bistables are found to be much more exposed to malfunction than their static counterparts.

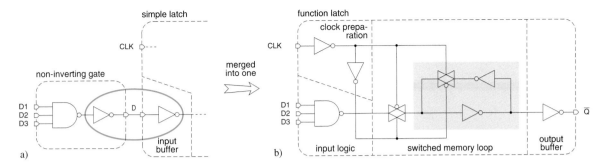

Fig. 8.21 Embedding a 3-input AND into a latch in search of maximum performance or energy efficiency.

capacitances and crossover currents, which explains why the trick is occasionally being played in the context of low-power design. The idea has a longer tradition in high-speed, circuits, where it proves beneficial whenever a few gate delays can be collapsed into one which, while larger than the individual contributions, is smaller than their sum.[21]

Taking advantage of these benefits in cell-based designs requires an extended cell library that includes not only the regular bistables but also combined functions such as 2- and 3-input OR-latches, AND-latches, an XOR- and an EQV-latch, plus various AOI-latches.[22]

8.2.3 Single-edge-triggered flip-flops

Most flip-flops operate as single-edge-triggered bistables, a behavior that can be obtained in several ways. The most popular approach cascades two (level-sensitive) latches and drives them from a common clock such that one of them is in hold mode while the other is in pass mode and vice versa, see fig.8.22. The up-stream latch reads from the input terminal and is referred to as **master** while the down-stream latch termed **slave** drives the output, hence the name **master–slave flip-flop** for this circuit arrangement. The two latches cooperate as follows.

	flip-flop triggered on			
	rising edge		falling edge	
CLK	master	slave	master	slave
0	pass	hold	hold	pass
1	hold	pass	pass	hold

---

[21] You may want to refer back to section 6.2.6 for information on how function latches relate to clocking disciplines. Incidentally, note that embedding logic into a bistable seems particularly attractive in conjunction with dual-rail high-speed logic because a total of four distinct operations can be obtained from a single NAND-NOR-latch simply by crossing over the input and/or output lines.

[22] It is obviously possible to play the same trick once again at a bistable's output Q or $\overline{Q}$ by substituting a NAND or a NOR gate for the inverter there [203]. As an ultimate consequence, this would ask for a cell library that encompasses the Cartesian product of all input and output functions of up to three variables including, as an example, the "3-input OR latch NANDed with a 2-input AND latch". It thus seems preferable to write a software tool that assembles function latches on the fly from logic gates and basic memory loops with no input and output buffers as the overall number of library cells might otherwise get out of hand.

At the active clock edge, the master stores the data present at its input while the slave becomes transparent. At the passive clock edge, it is the master that becomes transparent while the slave holds the previous output. The transfer of data from the master to the slave is internal to the circuit and cannot be observed from outside. Thus, the overall behavior is indeed that of a single-edge-triggered flip-flop (SETFF). In the occurrence of fig.8.22b, the active edge is the rising or — which is the same — the positive one. The opposite applies to fig.8.22d.

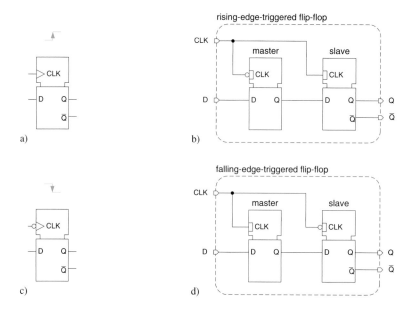

Fig. 8.22 Single-edge-triggered D-type master–slave flip-flops. Icons (a,c) and general circuit arrangements (b,d).

Detailed schematics are readily obtained from assembling the necessary building blocks from fig.8.20. This has been done in figs.8.23 and 8.47 for a D-type flip-flop equipped with an asynchronous reset mechanism. The table below summarizes the resulting transistor counts.

transistor count for a static CMOS master–slave flip-flop	circuit design style					
	switched memory loop	over-ruled loop	jamb latch	ground-cycled loop	clocked inverters	seesaw w. input gating
input buffer/inverter	2	2	2	2	0	2
master latch	8	6	8	8	6	16
slave latch	8	6	8	8	6	16
output buffers	4	4	4	4	4	4
clock preparation	4	4	4	4	8	2
async. reset facility	4	4	2	2	2	4
total	30	26	28	26	26	44
in gate equiv. [GE]	7.5	6.5	7	6.5	6.5	11
toggling with clock	12	8	8	12	8	10

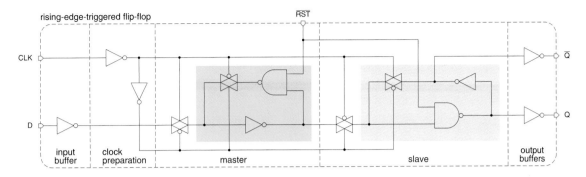

Fig. 8.23 A static CMOS master–slave flip-flop (SETFF) built from transmission gates (with asynchronous reset and complementary outputs).

The data transfers from master to slave need particular attention because of the absence of any logic and, hence, contamination delay in between.[23] In order to improve on the hold margin there, the clock is often distributed in such a way as to introduce a small negative skew. That is, the slave is clocked slightly before the master, e.g. by driving the master's clocks from the slave's via (resistive) poly lines.

The oscillating clock and the toggling data input are responsible for almost all dynamic energy that gets dissipated by a flip-flop; proportions vary as a function of data statistics and circuit details. Techniques for trimming unnecessary activities of internal nodes are discussed in [204]. [205] compares nine flip-flop circuits in terms of energy efficiency and performance. One finding is that differential or dual-rail circuits dissipate more energy than single-ended alternatives. Still, differential latches are being used in commercial high-performance low-power microprocessors such as the StrongARM. Also, semi-dynamic flip-flops built from a dynamic master and a static slave have been found to combine better performance with an energy efficiency comparable to that of fully static implementations. [206] evaluates D-type flip-flops in terms of circuit complexity, delay, energy dissipation, and metastability resolution. A not-so-surprising finding is this:

Observation 8.6. *The temporary contention that occurs while a memory loop is being forced into a new state tends to inflate ramp times and dissipated energy. Overruling has also been found to be detrimental to low-voltage operation (below 2.5 V or so) and to render bistables much more susceptible to marginal triggering and metastability.*

8.2.4 The mother of all flip-flops

The D-type flip-flop with asynchronous reset forms a fundamental building block from which any other type of edge-triggered storage element can be derived with the aid of a few extra gates. Examples of enable or E-type flip-flops, toggle or T-type flip-flops, scan flip-flops, and counter slices are shown in fig.8.24.

For the sake of performance, layout density, and energy efficiency, library elements are always designed and optimized at the transistor level, however. As an example, scan flip-flops have their

---

[23] The situation can be viewed as a case of unsymmetric two-phase clocking, explained in section 6.2.5, where the non-overlap time has been reduced to zero.

input multiplexer blended into the master latch much as explained before in section 8.2.2. This is done in an attempt to minimize the impact of scan testing on the computation rate of the final circuit as a D to Q or $\overline{Q}$ delay is almost certain to be part of the longest path.

As opposed to this, E- and T-type flip-flops and counters are not normally found in cell libraries but assembled from D-type flip-flops. With the advent of synthesis technology this is no longer done by hand. Yet, keep in mind that well-formed HDL code must produce circuit structures equivalent to those presented in fig.8.24.

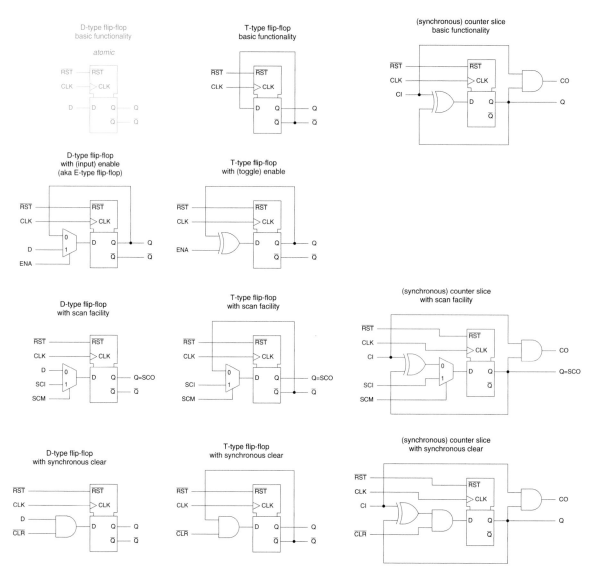

Fig. 8.24 Various bistable subfunctions assembled from a D-type flip-flop and logic gates.

8.2.5 Dual-edge-triggered flip-flops

The DETFF has been introduced as a more energy-efficient alternative to the SETFF.[24] As shown in fig.8.25, there is no master and no slave. Instead, two latches connect to the input and alternate in accepting and holding data; a clock-controlled multiplex function selects the datum currently being held for output.

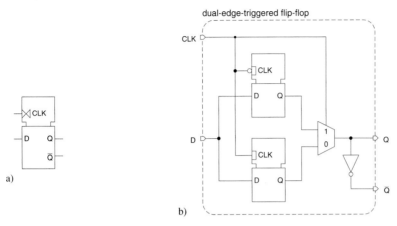

Fig. 8.25 Dual-edge-triggered D-type flip-flop. Icon (a) and circuit concept (b).

A remarkably slim circuit can be built from clocked inverters, see fig.8.26. However, note that the clever economy of an actual multiplexer subcircuit renders this particular design more delicate than others because of the undirectness of the two four-transistor memory loops. While the clock is ramping up or down, the voltages in the cross-coupled inverter pair being powered up are in fact biased from both sides at the same time. Conflicts are bound to occur. Careful transistor sizing must ensure that the on-going memory loop indeed settles to the state determined by the input data rather than by its off-going counterpart, making this circuit yet another example of ratioed logic.

transistor count for a static CMOS dual-edge-triggered flip-flop	circuit design style		
	switched memory loop	overruled memory loop	clocked inverters
inverting input buffer	2	2	0
low-transparent latch	8	6	6
high-transparent latch	8	6	6
output multiplexer	4	4	0
output buffers	4	4	4
clock preparation	4	4	8
async. reset facility	4	4	1
total	34	30	25
in gate equivalents [GE]	8.5	7.5	6.25
toggling with clock	16	12	8

---

[24] Section 6.2.3 discusses its usage as a part of synchronous systems and its impact on energy efficiency. More DETFF implementations are suggested in [207] [208] [152] [209] and other references.

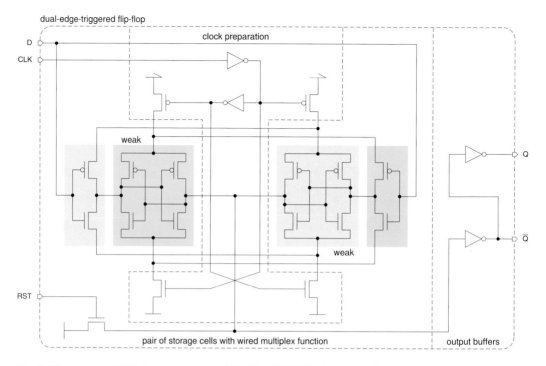

Fig. 8.26 A static CMOS dual-edge-triggered flip-flop (DETFF) circuit built from clocked inverters, with asynchronous reset.

8.2.6 Digest

Observation 8.7. *There is no such a thing as the optimum bistable design. Any choice is a compromise between conflicting requirements such as circuit size, speed, energy efficiency, supply voltage range, cell inventory, overall design effort, testability, friendly timing, skew tolerance, compatibility with clocking disciplines, and low susceptibility to marginal triggering.*

Probably the most comprehensive evaluation of latch and flip-flop circuits can be found in [210] on a background of advanced clocking disciplines while [211] extends the discussion to DETFFs. Before you set out to design your own bistables (latches, flip-flops, counter slices, and the like), recall the important desiderata related to timing, clocking, and metastability resolution made explicit elsewhere in this text.[25] Also, a word of caution on the usage of transmission gates will be given in section 8.5.2.

8.3 | CMOS on-chip memories

8.3.1 Static RAM

For economic reasons, minimizing the die area occupied per bit is imperative in memory design as the basic storage cell gets repeated over and over again. Figure 8.27 compares the

---

[25] In sections 6.2.2 and 7.4.

transistor-level circuits of a latch and a flip-flop with those of RAM cells. In a static RAM (SRAM), each bit is stored as one of the two stable states that develop in a positive feedback loop built from cross-coupled inverters, see fig. 8.27a.

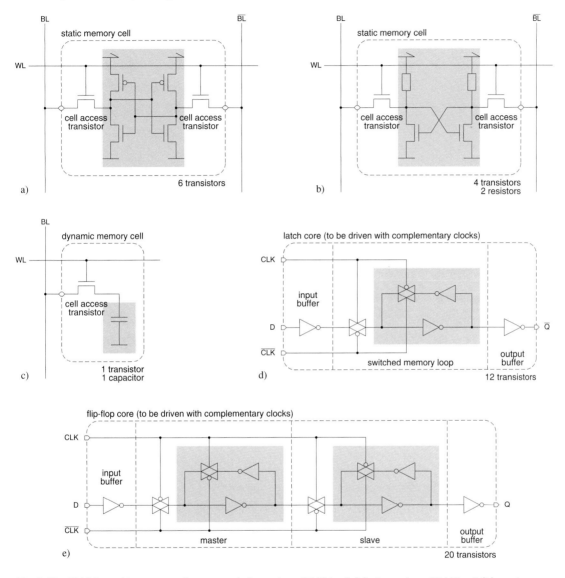

Fig. 8.27 CMOS one-bit storage cells compared. 6-transistor SRAM cell (a), 4-transistor SRAM cell (b), and DRAM cell (c); latch (d) and master–slave flip-flop (e) shown for reference.

Figure 8.27b shows a variation where the p-channel MOSFETs have been replaced by passive pull-ups. Dispensing with the complementary transistors along with the necessary separation between their wells saves a substantial proportion of area. Drawbacks are static power dissipation, slower readout due to an inferior current drive, and the extra fabrication steps required to manufacture resistors of extremely high resistance, on the order of many GΩ, in a small area.

As suggested in fig.8.28, many one-bit storage cells are arranged in a two-dimensional array so that each memory location is uniquely defined by a row and a column address. Bundles of columns often operate in parallel such as to store and retrieve an entire data word at a time. With w_{data} denoting the data word width and w_{addr} the address word width, the capacity of a standard[26] RAM immediately follows as

$$\#_{bits} = \#_{words} \cdot w_{data} = 2^{w_{addr}} w_{data} = \#_{rows} \cdot \#_{columns} = 2^{w_{rowaddr}} \cdot 2^{w_{coladdr}} w_{data} \ (8.45)$$

where $w_{addr} = w_{rowaddr} + w_{coladdr}$. A preferred choice is $\#_{rows} \approx \#_{columns} \approx \sqrt{\#_{bits}}$ because close-to-quadratic shapes of the cell array tend to minimize layout parasitics.

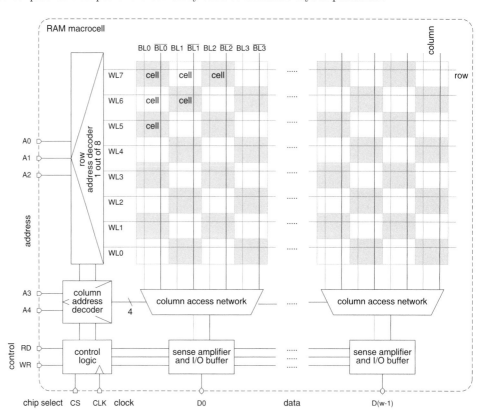

Fig. 8.28 Organization of a RAM macrocell (32 words by 2 bits drawn, simplified).

Accessing a specific memory location

Provided the chip select line is active CS=1, part of the address bits are fed into the row address decoder, where they get translated into a one-hot code. One out of the $2^{w_{rowaddr}}$ **word lines** WL is

---

[26] Some macrocell generators are also capable of constructing memories with **fractional cores** where the word count is not an integer power of two, $\#_{words} < 2^{w_{addr}}$. To better adapt to specific storage requirements, such a generator might be used to construct a 1280 × 22 bit RAM, for instance, whereas a less sophisticated tool would not be able to generate any configuration between 1024 × 22 bit and 2048 × 22 bit.

so driven to logic **1**. In the toy example of fig.8.28, where $w_{rowaddr} = 3$, this would involve address bits **A2...A0**. As becomes clear in the more detailed view of fig.8.29, the active word line brings all n-channel access transistors attached into a low-impedance ("on") condition. One memory cell in each column is so coupled to the **bit lines** BL and $\overline{\text{BL}}$ via a pair of access transistors while all other memory locations remain electrically isolated.

The remaining $w_{coladdr}$ address bits **A4** and **A3** are decoded into column select signals **SEL#** that turn on one pair of column switches while deactivating all others that belong to the same bundle of columns. By the combined effects of row and column addressing, two electrically conducting paths, involving BL and $\overline{\text{BL}}$ respectively, are established between the cross-connected inverters within the cell currently being addressed and the associated read–write circuitry.

Data write operation

The write-in buffers are enabled in such a way as to have them control the bit lines. More precisely, by setting **WE=1**, **OE=0**, **PHIP=0**, and **PHIS=0**, the data terminal **D** is made to drive one BL and the conjugate $\overline{\text{BL}}$ line in direct and in complemented form respectively. Activating a word line causes the cross-connected inverters in the selected cell to assume the condition imposed from externally. After the cell has been deselected, the positive feedback mechanism maintains that state indefinitely until the cell's content is overwritten or the supply voltage gets turned off.

Data read operation

As opposed to the write operation, setting **WE=0** disables the write-in buffers, thereby permitting the memory cell selected to exert control over bit lines BL and $\overline{\text{BL}}$. Further setting **OE=1** propagates the state stored in that cell to the data terminal **D** via the enabled output buffer.

While such a simple scheme can be made to work, two effects make it awfully slow. Firstly, the bit lines are long and connect to hundreds of bit cells so that they represent heavy capacitive loads. The driving transistors in each bit cell cannot be sized accordingly, however, but must be kept as small as possible for reasons of economy. Secondly, the n-channel access transistors are not good at pulling a bit line up. More sophisticated circuits have thus been devised to do away with these limitations, see fig.8.30 for the waveforms.

1. After having selected one column, the two bit lines are first **precharged** to U_{dd} by setting the auxiliary signal **PHIP=1**.[27] A third MOSFET that directly connects BL to $\overline{\text{BL}}$ helps establish a perfect voltage equilibrium more rapidly.

2. The three precharge transistors are then turned off by switching to **PHIP=0** just before one of the word lines gets activated. As the one memory cell selected begins to act on the two bit lines, the voltage of either BL or $\overline{\text{BL}}$ starts to decline so that an imbalance develops.

3. At that time, the **sense amplifier** is powered up by setting **PHIS=1**. Its inner four transistors form a positive feedback loop that exhibits two stable states of equilibrium as shown in fig.7.9. Working on the two conjugate bit lines, this differential contraption will intensify and accelerate any existing voltage disparities until BL and $\overline{\text{BL}}$ reach fully opposite voltage levels. The short wait time before the sense amplifier gets turned on serves to avoid marginal triggering

---

[27] Some RAMs have their bit lines precharged to a voltage closer to $\frac{1}{2}U_{dd}$, yet the concept remains the same.

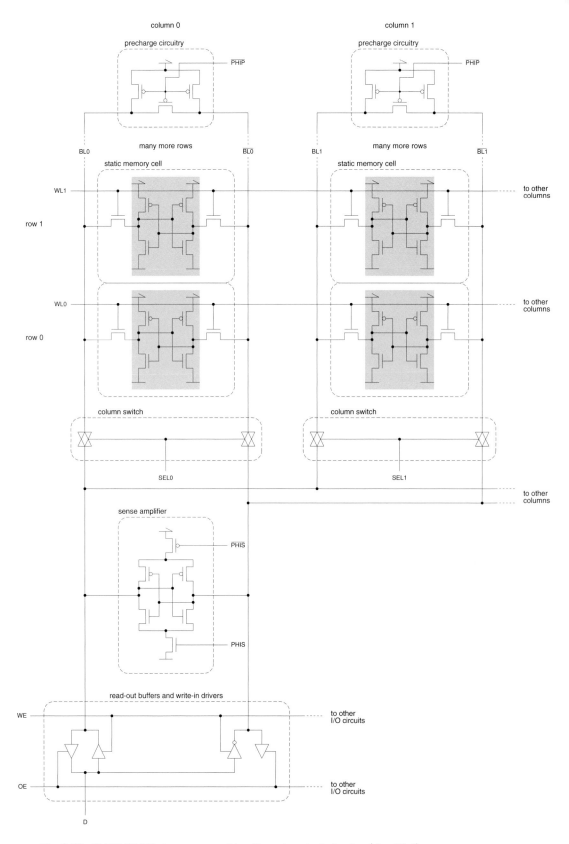

Fig. 8.29 CMOS SRAM storage array with write and readout circuitry (simplified).

conditions. An amplifier with minimal offset, and perfect electrical and layout symmetry is critical, however, as much as protecting the entire readout circuitry from ground bounce and crosstalk.

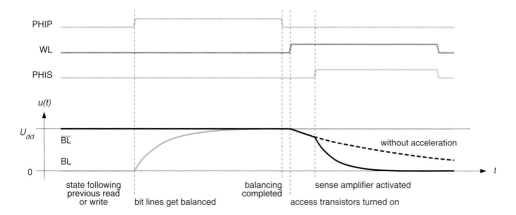

PHIP

WL

PHIS

$u(t)$

U_{dd}

$\overline{BL}$

BL

0

state following
previous read
or write

bit lines get balanced

balancing
completed

sense amplifier activated

access transistors turned on

without acceleration

t

Fig. 8.30 Read signal waveforms with and without sense amplifier (simplified).

8.3.2 Dynamic RAM

The overall organization of a dynamic RAM as a two-dimensional array of memory cells is very much the same as for an SRAM. The main difference is that each bit of information gets represented by the presence or absence of an electric charge on a tiny capacitor in one of the many memory cells, see fig.8.27c. While most economic, the concept of charge storage has a few peculiarities. Most importantly, the tiny charge is exposed to all sorts of **leakage** effects and needs to be periodically refreshed to prevent it from vanishing over time.

Data readout requires one such storage capacitor to be electrically connected to a bit line by its associated access transistor. Yet, as the line's capacitance is much larger, the original charge gets lost in the ensuing charge-sharing process. One thus speaks of **destructive readout**, and the information must be written back immediately.

Another consequence of the unfavorable capacitance ratio is that the voltage deflections observable on the bit line are rather small. Also, the readout amplifier must essentially weigh the bit line signal against the precharge voltage because DRAM storage cells are **single-ended**. Comparing against a fixed reference, rather than against the counter-acting signal of a conjugate bit line, gives away part of the speed and robustness of the differential readout scheme implemented in SRAMs.[28]

---

[28] In search of maximum electrical and layout symmetry, DRAM designers have actually come up with sophisticated schemes that combine single-ended bit cells with conjugate bit lines and differential amplifiers. Bit cells are hooked up to BL and $\overline{BL}$ in an alternating pattern. While one line, say BL, is impinged upon by the access transistors selected during some read operation and, hence, participates in the charge-sharing process, no access transistor is allowed to turn on on its counterpart during the same read operation. The precharged $\overline{BL}$ line can thus act as a reference for readout via BL, and vice versa for readout operations via $\overline{BL}$. Another benefit of this scheme is that the capacitive load attached to each bit line is cut in half over a fully one-sided circuit arrangement.

The fact that a sense amplifier has common terminals for input and output is particularly welcome in DRAMs because the amplifier restores the small signal received to its full logic level before reflecting it to the storage cell over the same bit line once it has snapped to one state or to the other. Data **write back** so becomes an implicit part of every read operation, which has the extra benefit of providing a nice hook for **memory refresh**. An array of DRAM cells essentially gets refreshed by accessing each data word on a regular basis for a — silent — readout operation; a counter that periodically sweeps the address range is sufficient to do so.[29]

The refresh operations are typically hidden from the outside world by having them take place during idle time slots or within sections of the memory array that are not currently being accessed. Still, depending on the application and on the refresh scheme, refresh operations can get in the way of regular memory accesses and necessitate the introduction of wait cycles. Another major disadvantage is the permanent drain of power associated with DRAM refresh.

More facts and figures about DRAMs:

- Key design goals are large capacitance in a small area and low leakage.
- Minimum storage capacitance is 25 to 35 fF per bit cell,
  with 40 to 50 fF being more typical.
- Planar capacitor arrangements had been used until the 1 Mibit generation before special process steps were introduced to fabricate trench or stack capacitors that can take advantage of the third dimension to economize on silicon area.
- Charge retention times are specified to be on the order of 64 to 256 ms.
- A 10% loss of charge is considered a data failure.
- An entire refresh cycle typically takes 64 ms for a 64 Mibit DRAM.
- To minimize leakage, DRAM processes feature higher threshold voltages than logic processes and make use of negative back-biasing.
- The ratio defined by the bit line capacitance to the storage capacitance is no more than 15:1 and typically ranges between 5:1 and 10:1.

8.3.3 Other differences and commonalities

Storage density. To put things into perspective, fig.8.27 includes not only SRAM and DRAM cells but also the core circuits of a latch and of a D-type flip-flop (both with a single output, no reset, and no clock preparation circuitry). As everyone knows, it is in the DRAM that the complexity of a 1-bit memory circuit is reduced to a minimum, with the SRAM far behind. The fact that RAM storage arrays get assembled from abutting layout tiles[30] further contributes to their superior layout densities when compared with individual latches and flip-flops. These findings substantiate the numbers that have been given earlier in the context of architecture design in section 2.5 and notably in table 2.9 there.

Circuit overhead. The radical simplification of the storage cell in semiconductor memories is bought at the price of substantial auxiliary peripheral circuitry for address decoding, precharging, readout acceleration, internal sequencing, I/O interfaces, etc. invisible from

---

[29] Most DRAMs have a one sense amplifier per memory column thereby making it possible to address one row after the other and to refresh all cells of a row at a time.

[30] The concept of tiled layout is to be explained in section 11.5.5 but also obvious from fig.11.26c.

fig.8.27 but indicated in fig.8.28. DRAMs further call for on-chip provisions for memory refresh and back-biasing. While these circuit overheads cause large RAM macrocells to occupy roughly 1.5 times as much area as the storage cell array alone, this factor may exceed 2 for smaller on-chip RAMs.

Memory clocking. RAMs come in clocked and unclocked varieties. **Synchronous RAMs** have a clock input terminal that must be driven from a periodic signal. Write operations are to be timed with reference to edges on the clock. An extra write input acts as enable. Timing diagrams resemble those of a flip-flop. An analogous scheme often applies to read operations, but there are exceptions where read transfers occur independently from the clock.

Asynchronous RAMs feature no clock input. Instead, address and data signals must be applied with reference to the $\mathtt{WR/\overline{RD}}$ input or to some address select inputs. Any such input is sensitive to hazards as it combines the roles of a <u>what</u> signal and of a <u>when</u> signal. Still, it cannot be driven from a system clock directly because read and write transfers typically alternate in an irregular fashion. [212] suggests a number of safe driver circuits for asynchronous RAMs.

ASIC developers do not normally design embedded memories from scratch, they either a use macrocell generator made available by a library provider or commission the silicon vendor to generate the macrocell views for them. Our discussion of RAMs has, therefore, been limited to the very basics. Please refer to the specialized literature for a more in-depth exposure to memory circuit and layout design [213] [214] [215] [216] [46] [45] [217] [218] [219] [47] [220] [221]. A few examples of detailed layouts are to follow in section 11.5.5.

8.4 | Electrical CMOS contraptions

The building blocks collected in this section have one thing in common, they are being used in digital circuits for their particular electrical characteristics rather than for some logic function.[31]

8.4.1 Snapper

A snapper essentially consists of a positive feedback loop implemented from two cross-coupled inverters, see fig.8.31. The bistable nature of this mechanism attempts to maintain the sole terminal $\mathtt{TRM}$ in its most recent state, that is either at logic 0 or at 1. As will be explained in section 8.5.1, this opportunistic contraption may serve to prevent the voltage on a three-state node from drifting away when the node is not being driven.

The drive strength of the snapper is chosen low enough that its output is easily overridden by any other gate wanting to impose a new logic state. This is obtained by making the two MOSFETs that are driving the terminal node significantly weaker — that is longer — than those found in the output stage of any other cell, another occurrence of a ratioed circuit.

---

[31] The transmission gate is not reiterated here because it has been introduced earlier to explain the construction of three-state and parity gates, bistables, and the like. Please refer to section 8.1.5 for details.

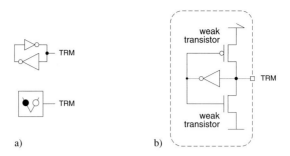

Fig. 8.31 Snapper. Icons (a) and circuit (b).

As it is impossible to indicate any dependency between an input and an output, no truth table can be given. A snapper is not a logic gate but the simplest form of an unclocked bistable. Snappers are also known as bus holders, holding amplifiers, and level retention devices.

Observation 8.8. *Its bistable nature notwithstanding, a snapper is not intended for data storage. Spurious signals such as glitches may cause a snapper to toggle at any time. Once driving of its terminal node has ceased, any state previously assumed must be regarded as corrupted.*

Much like any other bistable subcircuit, a snapper exhibits metastability when subjected to marginal triggering. For the memory loop to reliably snap into a new state, any data transition must reach a valid logic state and rest there for some minimum time. Timing conditions are, therefore, expressed in terms of minimum low and high times for the snapper's terminal.

8.4.2 Schmitt trigger

What sets Schmitt triggers apart from buffers and inverters is their hysteresis, see fig.8.32. As opposed to ordinary gates, the output is logically defined for any input voltage with no indeterminate or forbidden interval. Instead, distinct thresholds apply to rising and to falling input voltages, which means that a Schmitt trigger's output partly depends on past input.[32]

$$U_{hy} = U_{ih} - U_{il} > 0 \tag{8.46}$$

Hysteresis is desirable in certain situations to obtain ampler noise margins and/or better level restoration than ordinary buffers or inverters would provide. External inputs, for instance, are often passed through Schmitt triggers to filter out noise and to reshape waveforms. Three alternative circuit diagrams are shown in fig.8.33. Observe that positive feedback is essential in obtaining hysteresis and that its exact amount U_{hy} is a function of the transistor sizes involved. Figure 8.33b makes use of weaker-than-normal MOSFETs to obtain a limited degree of feedback whereas c and d can be designed to make do with minimum-length transistors.

---

[32] The assignment of the mathematical symbols U_{ih} and U_{il} to a Schmitt trigger's pair of threshold voltages is debatable. Some sources of information assign them the other way round so that $U_{ih} < U_{il}$. With the definition $U_{hy} = U_{il} - U_{ih}$, a hysteresis of $U_{hy} < 0$ naturally follows for all regular gates and $U_{hy} > 0$ for Schmitt triggers. While this is meaningful, we prefer the assignment of fig.8.32 because U_{il} thus consistently defines the upper bound below which all input voltages get interpreted as logic 0, and vice versa for U_{ih}.

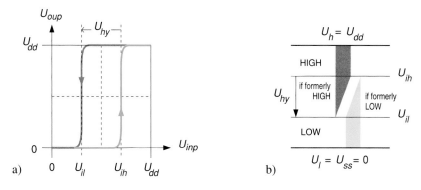

Fig. 8.32 Non-inverting Schmitt trigger. Static transfer characteristic (a) and switching levels (b), compare with fig.8.4.

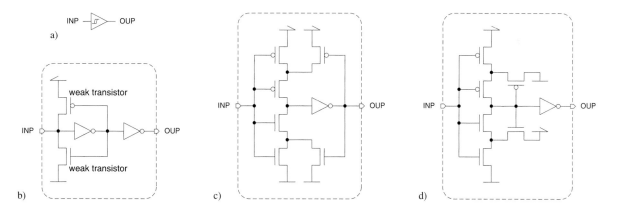

Fig. 8.33 Non-inverting Schmitt trigger. Icon (a), circuit built around a snapper (b), circuit built from series transistors (c), and alternative circuit with two transistors operating from reversed voltages so as to weaken feedback further (d).

8.4.3 Tie-off cells

Even with today's exuberant cell libraries, it is sometimes necessary to permanently tie a cell's input to logic 0 or 1. Imagine you specify a 3-to-1 multiplexer function in your HDL code. Very likely, the synthesizer will then instantiate a 4-to-1 multiplexer from the library and connect the unused input to ground. Yet, gate dielectrics have become so thin in deep submicron processes that even a minor overvoltage might damage them. To prevent this from happening, manufacturers prohibit low-ohmic connections from MOSFET gates to vss or vdd.

Instead, a tie-off cell such as those shown in fig.8.34 must be used. The RC-section formed by the MOSFET's "on" resistance together with the gate capacitance(s) attached suppresses fast transients and so protects the vulnerable thin oxides from electrostatic discharge (ESD) and other transient overvoltages. Please figure out yourself what the other transistors serve for.

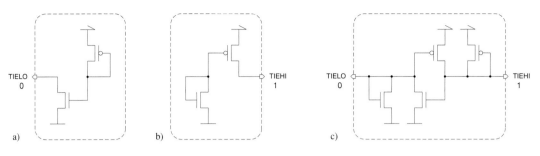

Fig. 8.34 Tie-off cells.

8.4.4 Filler cell or fillcap

To allow for signal routing in congested areas, it is sometimes necessary to provide extra room between standard cells. From a signal integrity point of view, on the other hand, it is highly desirable to intersperse bypass capacitors over the entire die area to momentarily supply the necessary energy for local switching activities. Filler cells laid out such as to fit into standard cell rows and to abut with other cells there serve both purposes.

In designing fillcaps, as filler cells are also called, the goal is is to obtain a useful amount of capacitance with a low (equivalent) series resistance (ESR) from a very limited area. The most effective choice is to use the thin gate oxide layer as dielectric.[33] The gate material constitutes the upper capacitor plate while the inversion layer that forms underneath when properly biased acts as lower plate. Nearby diffusion areas serve as bottom terminals. As shown in fig.8.35b, the arrangement may be viewed as a MOSFET that is permanently turned on and that has its drain and source terminals shorted together. The name **MOSCAP** has been coined for such structures. The nonlinearity of its $C(U)$ characteristic does no harm for decoupling purposes.

To prevent damage from voltage spikes on the supply rails, the fragile gate dielectrics are protected with the aid of small series resistors on the order of 100 Ω. An alternative arrangement depicted in fig.8.35c avoids area-consuming resistors in much the same way as the tie-off cell discussed before.

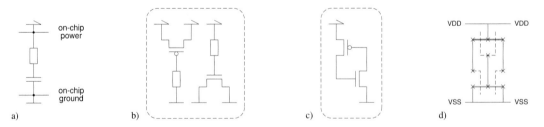

Fig. 8.35 A pair of MOSCAPs acting as on-chip bypass capacitor or fillcap. Equivalent circuit (a), two implementations (b,c), and symbolic layout (d).

---

[33] A single layer of poly (1P) is typical for purely digital CMOS fabrication processes. More sophisticated processes targetted towards analog and mixed-signal circuits provide designers with an extra dielectric layer for obtaining on-chip capacitors of better quality. Sandwiched either between two poly layers (2P) or between two metal layers (MIM), a thin layer of dielectric material makes it possible to construct capacitors that are linear and free-floating, that is, connected neither to the VSS nor to the VDD node.

8.4.5 Level shifters and input/output buffers

Level shifters translate logic signals at the interface of two subcircuits that operate at distinct voltage levels. They have a long tradition in input and output buffers, e.g. to translate between TTL and CMOS switching levels or when a chip's core logic operates from a 1.2 V supply while the surrounding board-level circuitry insists on 2.5 V swing. More and more, level shifters have also appeared within cores to accommodate multiple supply voltages. They are further being used as part of reduced-swing clock distribution networks [222].

The circuit of fig.8.36 functions much like a jamb latch that is permanently kept in pass mode. The circuit has the nice property that the same general arrangement can be used for voltage step-up and for voltage step-down. Better still, the circuit can handle variable supply voltages as transistors need not be resized as a function of voltage ratio. Gate dielectric thicknesses and drain–source separations must of course always be chosen such as to be commensurate with the maximum voltage applied.

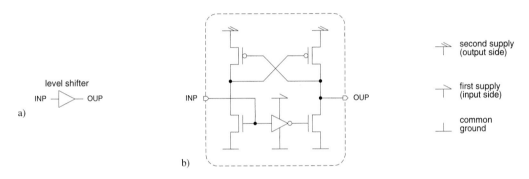

Fig. 8.36 Non-inverting level shifter. Icon (a) and circuit (b).

I/O buffers not only translate between distinct voltage levels, but also must provide good protection against electrical overstress, acceptable crossover currents, and low leakage. They must further be safe from latch-up in spite of the higher voltages typically used at the board level. While all these issues are being addressed separately in this text, the reader is referred to the specialized literature for an evaluation of mixed-voltage I/O buffer circuits [223] [224].

8.4.6 Digitally adjustable delay lines

There are situations where one needs to adjust the delay along some signal or clock propagation path to adapt to conditions unknown at design time, e.g. to compensate for PTV variations or to track a time-varying phase relationship. Just consider PLLs and DLLs, for instance.

Three digitally adjustable delay lines are shown in fig.8.37 and each of them functions according to a different idea. The first one must be controlled with one-hot encoding and supports delay variations in fixed increments of two inverter delays. The second one uses transmission gates to select among two different load conditions for the inverter in each delay element. This approach allows for a finer resolution as the delay increments get defined by MOSCAPs of graded sizes. A **current-starved**

inverter is at the heart of a third adjustable delay element. Turning on and off transistors from a bank of graded MOSFETs modulates a (static) control current I_c. That current gets mirrored into a pair of n- and p-channel MOSFETs that limit the inverter's maximum sink and source current respectively and, hence, also its output slew rates.

Before preferring one circuit design over another, watch out for

- Adequate delay resolution,
- A good match between the respective propagation delays for rising and for falling edges,
- Monotonic speed-up and slow-down in response to a new digital delay setting, and
- The usual area and energy requirements.

Detailed accounts on how to design digitally adjustable delay lines with timing resolutions finer than one gate delay can be found in [225] [226] [227].

8.5 | Pitfalls

Certain subcircuits, such as three-state busses, transmission gates, bus interfaces, and mechanical contacts, exhibit treacherous peculiarities. Being aware of the pitfalls continues to be important because a piece of HDL code may cause a potentially unsafe circuit to be produced during synthesis without the code's author being aware of this.

8.5.1 Busses and three-state nodes

Designers frequently face the necessity to access data from multiple locations distributed over a chip or system. They then essentially have three options:[34]

- Route a wire from data source to destination for each transfer path that might be required.
- Use multiplexers to grant one source at a time access to one net that connects all locations.
- Have the data sources drive a common net as a function of need via three-state outputs.

The simplest example of a **multi-driver node** is a bidirectional line with a driver/receiver at each end; a **bus** typically comprises multiple lines and multiple points of access, see fig.8.38. Having multiple drivers alternate in driving a common circuit node avoids the wiring overhead that can grow to become prohibitive with the other two options, but brings about a number of complications that need special attention.

A **drive conflict**, aka bus contention, occurs when two or more drivers attempt to impose incompatible logic states — such as 0 and 1 — on a common multi-driver node. Drive conflicts are highly

---

[34] A fourth technique can no longer be considered an option. An open-drain output consists of an n-channel MOSFET with no p-channel counterpart. Two or more such outputs connect to a common pull-up node, thereby resulting in an implicit wired-AND operation. The idea was known as open-collector output and very common in TTL technology, but the static currents render it unpopular in the context of CMOS circuits.

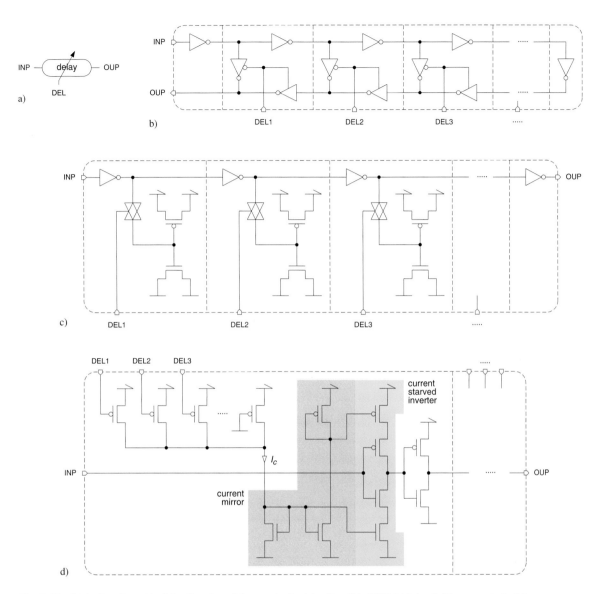

Fig. 8.37 Digitally adjustable delay line. Icon (a), round-trip delay line (b), MOSCAP-loaded inverter chain (c), and current-starved inverter (d) (simplified).

undesirable because the logic outcome is unpredictable, because a short-circuit current flows from VDD to VSS directly, and because unnecessary ground bounce is produced.[35]

---

[35] More precisely, one must distinguish between stationary and transient conflicts. A drive conflict is said to be stationary if the drivers continue to fight for control over a node even after the circuit has settled to its steady-state condition. Stationary contentions always point to deficiencies in the logic circuits and/or protocols in charge of controlling access to the bus. They often continue for a prolonged period of time and cause needless power

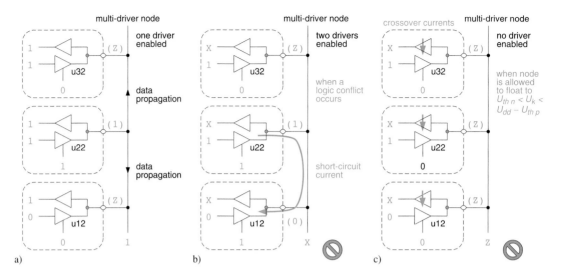

Fig. 8.38 Three-state bus. Regular operation (a), drive conflict (b), and floating bus (c).

The converse situation where a node is left **floating** because none of the drivers attached gets enabled is also undesirable. Any charge initially present on the node's capacitance will fade over time due to leakage, alpha particles, cosmic rays, and other unpredictable phenomena. A node left undriven after charging may discharge, and vice versa. If its potential drifts towards the logic family's switching threshold, the receivers become susceptible to crosstalk from the nearby circuitry and may start to develop undesirable crossover currents and/or oscillations.

Observation 8.9. *Under no circumstance must a multi-driver node be driven by more than one output at a time. Also, some mechanism must be provided to prevent a node from floating when it is left undriven for a prolonged period of time.*

Floating nodes also compromise the accuracy and dependability of simulation as there is no way for a logic simulator to predict physical effects such as leakage, noise coupling, radiation, etc. with the resulting charge decay times. Simulated waveforms may suggest a node continues to be charged to its former logic value, while, in reality, its voltage has long since drifted away.[36] Similarly, gate-level simulation typically fails to correctly predict circuit behavior during drive conflicts.[37]

---

dissipation and local overheating. Permanent damage may result when maximum current density or temperature ratings are exceeded.

As opposed to this, a <u>transient</u> conflict may develop in any bus circuit when one driver is turning on while another is going off-line. While stationary contentions must absolutely be avoided, brief transient conflicts can be tolerated because no out-of-the-normal currents flow, provided timewise overlaps do not exceed regular signal transition times. After all, the crossover current in an inverter or any other static CMOS gate may be understood as the result of a transient drive conflict between two antagonistic MOSFET networks. With today's CMOS technology, this means that transient conflicts must come to an end within less than a nanosecond.

[36] A common stopgap is to set up simulation in such a way that any undriven node immediately assumes an unknown logic state. Actually, this is the case in the IEEE 1164 standard where charged high, charged low, and charged unknown are collapsed into a single logic value Z. While this is certainly prudent, it is an overly conservative approximation for dynamic CMOS circuits.

[37] In a fight between logic 0 and 1, logic simulation typically yields a logic X whereas the 0 would win in many practical cases because most CMOS outputs can sink more current than they can source.

Observation 8.10. *In the presence of drive conflicts, floating nodes, or snappers, signal values and waveforms predicted by logic simulation may significantly differ from physical reality.*

As a consequence, and because of other issues related to timing, energy efficiency, and testing, VLSI engineers generally prefer to stay away from multi-driver nodes whenever possible. Many companies in fact veto the usage of on-chip three-state busses. Let us nevertheless examine circuit techniques for staying clear of floating nodes and drive conflicts.

Passive pull-up or pull-down

The usage of pull-up or pull-down resistors to prevent an undriven node from floating around has a long tradition in digital electronics. TTL inputs have a built-in pull-up characteristic, in MOS technology weak transistors can be made to act as passive pull devices. Pulling towards some fixed voltage is unpopular in VLSI, however, because of serious drawbacks:

- Pull devices dissipate static power when being overridden. The stronger the pull device, the more power gets dissipated.
- Ramp times are relatively slow and difficult to control. The weaker the pull device, the longer it takes for the node to transit across the forbidden voltage interval.
- The multi-driver node toggles more frequently than necessary, which translates into extra dynamic energy dissipation and switching noise.
- Static currents compromise I_{ddq} testing.

Active snapper

A snapper avoids both static power dissipation and unnecessary charge/discharge activities. Yet, its bistable nature also brings about two minor difficulties of its own:

- Spurious signals such as glitches may cause a snapper to toggle at any time.
- Snappers are exposed to marginal triggering. A critical situation occurs when a three-state driver is disabled while transiting the forbidden interval between 0 and 1.[38]

Centralized as opposed to distributed bus access control

Adding a snapper or a passive pull-up/down is not much more than treating the symptoms that result from leaving a node undriven. More importantly, neither a snapper nor a pull-up/down helps to avoid drive conflicts. A better approach is to organize bus access around a controller designed so as to guarantee that exactly one driver gets enabled at any time.

Example

Consider a microcomputer I/O bus connected to a number of external interface circuits or other hardware units. The existence of idle cycles during which no data transfer is scheduled to occur is typical for such applications.

In fig.8.39a, each driver gets its enable signal from a local controller, all of which coordinate their activities via some <u>distributed</u> bus access protocol. There is no hardware mechanism that could

---

[38] Some bus systems have indeed been reported to suffer from this kind of complication [228].

prevent drive conflicts should the protocol fail due to transmission errors, initialization failure, defective parts, software bugs, hot plug-in/out operations, and the like. Similarly, the bus is left undriven when none of the units has meaningful data to send, which calls for a pulling or snapping device. Local control is typical for large systems with numerous units distributed over several boards.

In the <u>centralized</u> scheme, in contrast, one controller is in charge of producing the enable signals for all drivers, see fig.8.39b. Clock cycle by clock cycle, some form of finite state machine (FSM) generates the address of the one and only driver that is to be enabled. The address is then fed into a decoder, thereby excluding the risk of stationary conflicts and minimizing overlap times. Still, during idle cycles the address may assume a value not assigned to any hardware unit, e.g. address code "00" points to an inexistent instance u02. A simple OR function prevents the node from floating and does away with the need for a snapper or pull-up/down. The data values put on the bus during idle cycles are immaterial as they are ignored anyway.

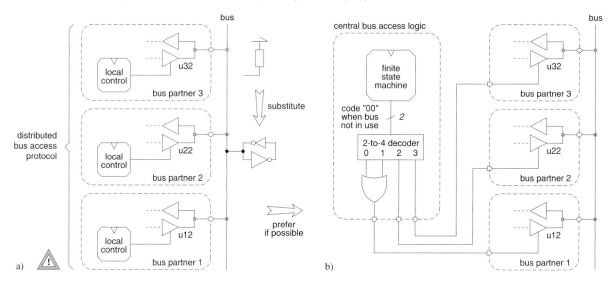

Fig. 8.39 Three-state bus access control schemes. Distributed (a) and centralized (b).

□

Observation 8.11. *On-chip multi-driver nodes complicate IC design and test, they should not be adopted light-heartedly. If opting for a multi-driver bus, however, prefer central over distributed access control whenever possible because it is simpler to implement and inherently safe.*

Centralized bus control does, unfortunately, not scale very well to systems where the numbers of hardware units are large and subject to variation. It is therefore, typically limited to smaller and hardwired systems such as those confined to within one chip.

8.5.2 Transmission gates and other bidirectional components

Much like three-state drivers, poorly designed transmission gate circuits develop drive conflicts and floating nodes. The 2-to-1 multiplexer of fig.8.40a, for instance, leads to a conflict when SEL1 = SEL0 = 1 and INP1 ≠ INP0. Conversely, when SEL1 = SEL0 = 0, node OUP remains undriven.

The design may, therefore, behave in an unpredictable way as long as control terminals SEL1 and SEL0 are allowed to evolve independently from each other. Deciding whether it is actually safe or not requires precise information about those combinations of input values that will occur and those that won't. A better circuit will be shown shortly in fig.8.41a.

The same network with inputs and outputs swapped is inherently unsafe, see fig.8.40b. As one or more nodes always remain undriven except for SEL1 = SEL0 = 1, this design is no good as a 1-to-2 demultiplexer, or for any other purpose.

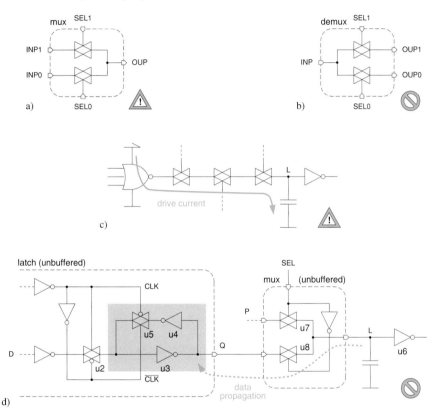

Fig. 8.40 Unsafe circuits with transmission gates. Drive conflict and floating node problems (a,b). Low drive capability because of accumulated "on"-state impedances (c). Vulnerability to backward signal propagation (d).

Transmission gates have no **drive capability** of their own. The current necessary for driving a load attached to one data terminal has to come from the circuit that is driving the other data terminal. When t-gates are cascaded, their "on"-state impedances accumulate, thereby lowering the drive currents and slowing down the charging and discharging of load capacitances. The capacitor of fig.8.40c, for instance, is charged through a total of six p-channel transistors connected in series. A voltage drop would result in steady state if a resistive load were attached.

Transmission gates are also lacking the **level restoration** capability of other CMOS subcircuits. Regenerating a logic signal with depressed voltage levels to its normal rail-to-rail swing requires a steep and saturating voltage transfer characteristic that only an amplifier can provide.

A particularly perfidious phenomenon is **backward signal propagation**. Consider fig.8.40d, where a latch drives a substantial capacitive load over a t-gate multiplexer, and assume the following initial situation. A logic 0 is latched, u8 is turned off, and the capacitance is charged high, i.e. CLK = 0, D = −, Q = 0, SEL = 1, and L = 1. What happens when SEL goes low? First, u8 gets turned on. Due to the important charge stored on node L the voltage there will not change immediately. Instead, the transistors of u4 and u8 will form a voltage divider between power and ground. Depending on whether the voltage at node Q exceeds the switching threshold of u4 or not, the latch may flip. Put in other words, the output terminal of the latch might unexpectedly act as an input! A more subtle form of backlash may be caused by the momentary low-resistance path that forms between the multiplexer data inputs when P ≠ Q while SEL is slowly transiting from one state to the opposite one.[39]

Undirectness also complicates **logic simulation**. Both RTL and gate-level simulation essentially work by re-evaluating the inputs to a subcircuit whenever one or more of them have changed and by updating the subcircuit's output values if necessary. Logic simulation, therefore, accommodates directed subcircuits only. Undirected subcircuits must be simulated at the transistor level or at the switch level. In either case, the computational burden is much higher than with logic simulation, and no timing conditions are being checked.

Last but not least, it has been found that t-gate circuits give rise to more severe **testability** problems than others. This is due to the t-gate's undirectedness and to the innate redundancy of having two transistors connected in parallel [229].[40]

Observation 8.12. *A transmission gate is not a logic gate but the CMOS equivalent of a relay. Designing with transmission gates requires particular attention in order to avoid unpleasant surprises. In addition, its built-in redundancy makes it next to impossible to test.*

Warning example

In search of the reasons behind the malfunctioning of fabricated circuits, engineers of a major semiconductor company found that part of their on-chip shift registers were actually running backwards. Closer investigation revealed that workload and information had been split among different parties in an unfortunate way.

Under pressure to help minimize manufacturing costs and power dissipation, library developers had devised a series of new flip-flops with (a) no inputs buffers, (b) no decoupling buffers at their outputs, and (c) close-to-minimum-width transistors in both the feedforward and feedback gates. Each such cell had been individually subject to SPICE-type simulations to verify its functioning and to quantify its timing characteristics before eventually being included in the latest release of a standard cell library. No time was left to confirm the cooperation of several such cells as parts of larger subcircuits, however.

Consistent with its general objective of meeting given timing constraints at the lowest possible circuit complexity, the VHDL synthesis tool had regularly preferred the "low-cost low-energy" flip-flops over the safer but more onerous alternatives during the library mapping step.

The ASIC developers were concentrating on getting the functionality of their RTL code right. To them, one flip-flop was as good as any other flip-flop. They were simply not aware of the existence of

---

[39] Note the analogy with a make-before-break type switch.
[40] See problems 11, 9, and 10 for this and further shortcomings of transmission gates.

special trimmed-down bistables in the target library, let alone of their peculiarities and vulnerability as, nothing in the documentation and simulation models indicated them.

□

It should be clear from the above accounts that great care must be taken when designing with bidirectional primitives. A number of vendors assume a very conservative attitude by rejecting transmission gates and other bidirectional constructs altogether.

A more liberal position disallows them as standard cells of their own, but, at the same time, tolerates their usage when embedded into larger subcircuits in such a way that

- Bidirectionality remains invisible when the subcircuit is viewed from outside,
- No two drivers can ever enter into a conflict situation,
- There is no way for any node to float,
- Output levels are always properly restored,
- No unwanted repercussions on the subcircuit's state are possible,
- It is possible to characterize the subcircuit by way of a truth table or a logic function,
- No more than three MOSFETs are connected in series.

Most standard cell libraries are in fact designed along these lines in order to take advantage of transmission gates while protecting designers and automatic synthesis tools from the pitfalls associated with free bidirectional primitives.

Examples

Figure 8.41a shows the 2-to-1 multiplexer of fig.8.40a redesigned to meet the above requirements. The single control input in conjunction with inverter u3 guarantees $\texttt{SEL0} = \overline{\texttt{SEL1}}$, thereby doing away with all unsafe combinations that might lead to a stationary drive conflict or a floating node condition. Output inverter u2 provides the same low impedance drive capability that is found in every other CMOS gate. Inverters u1 and u0 bring about several benefits. Firstly, they compensate for the data inversion inflicted by u2. Secondly, the two buffers prevent the transmission gate impedance from adding to that of the driving gate, which would make the multiplexer's switching speed and propagation delay depend on the context in which it is being used. Thirdly, they isolate the inputs from any kind of backward signal propagation. Lastly, u1 and u0, together with u2, yield a steep transfer characteristic.

For the sake of completeness, alternative solutions that make use of unidirectional gates exclusively have also been included in fig.8.41b through d.

A flip-flop that makes extensive use of transmission gates has been shown earlier in fig.8.23. Please check yourself how this design has been made to comply with the various safety rules.

□

8.5.3 What do we mean by safe design?

Our discussion of bus access protocols and the differences between poor and improved cell designs boils down to one thing. A (sub)circuit is to be considered **unsafe** if its functional behavior may depend on <u>external</u> <u>circumstances</u>. Deciding whether such a circuit will actually develop a problem or not requires making assumptions about the surrounding circuitry, the signals being applied, ambient conditions, and the like.

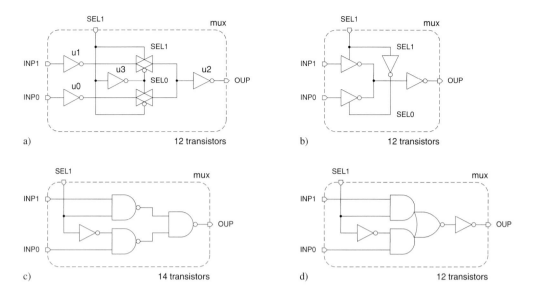

Fig. 8.41 Safe 2-to-1 multiplexer circuit on the basis of transmission gates (a). Alternative solutions built from three-state inverters (b), simple logic gates (c), and an AOI gate (d).

A **safe** (sub)circuit, in contrast, behaves as expected by virtue of its <u>innate construction</u> alone and, therefore, stays clear of potential risks and side effects as much as possible.

This distinction is by no means restricted to busses and transmission gate circuits, but also applies to clocking disciplines, PTV and OCV variations, clock gating, interface design, synchronization, state machine design, and even software engineering.

Observation 8.13. *A circuit that can be proven to consistently behave as specified and not to generate unexpected side effects from its schematics alone should always be preferred over an alternative circuit that requires hypothesizing about the circumstances in which it is going to operate. This is especially true when designing library cells.*

Turning unsafe designs into safe ones often augments circuit complexity and energy dissipation, though, which is not what is wanted. Minimizing those overheads or attempting to do without them, on the other hand, typically mandates more substantial engineering and verification efforts. The challenge consists in finding an optimum balance.

8.5.4 Microprocessor interface circuits

Assume you are given some microcomputer that will act as **host computer** in a system. It is your job to design an IC that interfaces with the microcomputer over the customary address bus, data bus, and control lines.[41] Your circuit is thus to become one of the host's peripheral devices. In

---

[41] The reader is assumed to have a basic understanding of microcomputer organization and operation. A summary of the three classic I/O transfer protocols — polling, interrupt-driven, and direct memory access (DMA) — is available in appendix A.7.

such a situation, it always pays to look at the host's prospective I/O circuitry not only from the IC designer's point of view but also from the perspectives of a software programmer and of a test engineer. What follows are some guidelines based on practical experience.

- Always keep configuration-related, event-related, and data-related bits in separate registers, see fig.8.42. Lengthy instruction sequences are otherwise required to read, extract, process, re-assemble, and write back a register's content.

- Never design write-only registers. In an attempt to minimize the address space occupied by the peripheral, certain designs from the pioneer days had the same address serve dual purposes. A write operation would access a command register while a read would access a status register. This kind of address multiplexing must be considered a bad habit, though, as it precludes reading back the most recent command from the I/O port, which forces the programmer to maintain an external copy of the register's content.

- Always render the current state of data transfer operations observable from the outside world. As a bad example, consider an IC that ingests and outputs 32 bit words in sets of four subsequent bytes without making it possible for the host computer to infer whether the current transfer refers to byte 0, 1, 2 or 3. This coerces programmers of software drivers — and test engineers too — into resorting to off-chip shadow counters just to keep track of a circuit's internal operation. Also, a risk of losing synchronization always remains.

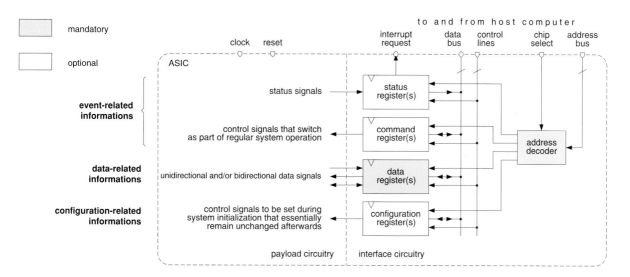

Fig. 8.42 A well-organized processor bus interface.

Example

Consider a controller for some optical disk drive. A variety of information gets exchanged between the controller and its host. This information can be structured as follows.

Type	Information
Status	transfer finished, data ready/needed for transfer, disk at speed, disk write-protected, head over track #, error flags, etc.
Command	spin up!, spin down!, eject disk!, search track #!, read sector!, write sector!, etc.
Data	data word read from disk, data word to be written to disk.
Configuration	number of tracks, number of sectors, recording format, etc.

Table entry "transfer finished" is an event that would typically be handled by an interrupt, in the occurrence, whereas direct memory access (DMA) transfers are more appropriate to handle "data ready/needed for transfer" requests since any reaction to such events must occur within a few microseconds to prevent any data losses.

□

8.5.5 Mechanical contacts

Signals that emanate from mechanical contacts almost always exhibit spurious pulse trains instead of clear-cut edges. This is because mechanical contacts tend to recoil a couple times before closing for good. The process of breaking a contact is not clean either. Contact bouncing typically extends over a period of 4 to 20 ms and calls for special precautions to suppress or neutralize unwanted signal transitions.

The traditional approach to **debouncing** consisted in filtering out spurious signals by way of an SR-seesaw in conjunction with a double-throw switch (with standard break-before-make contacts), see fig.8.43a. The circuit works on the grounds that the memory loop preserves its previous state whenever the contact is broken. This continues until the blade has travelled all the way to the opposite side and has made a first — and most likely ephemeral — contact there, in which case the output flips. Figures 8.43b and c show variations on the theme. Debouncing is indeed one of the rare occasions where a zero-latency loop finds a safe and useful application.

The problem with all these approaches is that they are incompatible with the level shifters that are normally inserted between a chip's pads and its core circuitry. Figure 8.43d shows a totally different solution that uses a Schmitt trigger and a very low sampling rate to suppress consecutive spurious transitions. Note that it does not matter whether a signal in transit gets interpreted as logic 0 or 1. Another benefit over figs.8.43a and b is that this design makes do with a single-throw switch and a simplified wiring. What's more, this kind of debouncing lends itself well to implementation in software, thereby doing away with the necessity for any special circuitry except for a static pull-up and a level shifter, preferably with hysteresis.

8.5.6 Conclusions

What has contributed to the long-lasting popularity of CMOS is not only the scaling property of MOS devices, but also the many benefits offered by a fully complementary static CMOS circuit style.

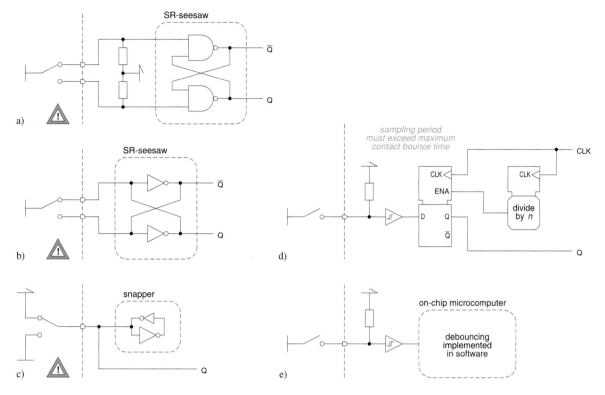

Fig. 8.43 Auxiliary circuits for the debouncing of mechanical contacts.

- Simple and elegant (compare with TTL and ECL circuits),
- Robust (ratioless and insensitive to leakage currents),
- Operational over a wide range of supply voltages,
- Modest in terms of interconnect resources (single-rail signal, composite gates), and
- Energy-efficient (low activity when compared with dynamic CMOS logic).

May the many circuit variations and design tricks that have been presented in this chapter and that are listed below serve as a source of inspiration for young designers.

- Antagonistic pull-down and pull-up networks,
- Composite gates (and-or-invert AOI, mirror adder),
- Transmission gates,
- Controlled, overruled, and power-cycled memory loops,
- Jamb latches,
- Function latches (storage and logic combined),
- One-transistor data storage cells,
- Differential readout combined with amplification,
- MOS capacitors,
- Digitally adjustable delays,
- Hysteresis and level shifting.

8.6 | Problems

1. The (poor) performance of an n-channel MOSFET as a high-side switch has been explained in section 8.1.1 by studying how the gate–source voltage evolves while a capacitive load is being charged. Visualize the process in fig.8.1 and compare it with a situation where the same n-channel MOSFET is being put to service as a low-side switch.

2. Implement the logic function below with the most simple static CMOS circuit you can think of. Also suggest a reasonable layout arrangement in gate-matrix style. Compare your solution with an alternative that makes do with traditional 1- and 2-input gates.

$$\text{OUP} = \overline{\text{IN1} \lor (\text{IN2} \land ((\text{IN3} \land \text{IN4}) \lor (\text{IN5} \land \text{IN6})))} \tag{8.47}$$

3. Consider the truth table below and assume all of x, y, z are available in complemented and non-complemented form. Is it possible to implement this function in a single gate in static CMOS technology?

	f	yz 00	01	11	10
x	0	1	1	0	0
	1	0	1	0	1

4. Have a closer look at the circuit of fig.8.44. Are the n- and p-transistor networks dual? What is the circuit's functionality? Explain the role of each circuit element shown. Do you see any advantage?

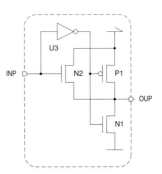

Fig. 8.44 Five-transistor CMOS circuit.

5. Figure 8.45 shows a full adder that greatly differs from the circuit of fig.8.18. Begin by trying to understand its structure and operation. How many transistors does the circuit include? Do you see any weaknesses? Can you find a remedy? How does the redesigned circuit compare in terms of transistor count and energy efficiency?

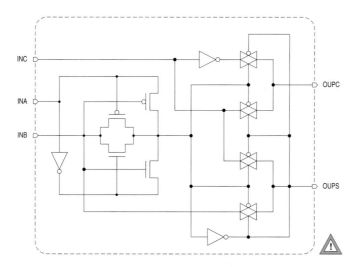

Fig. 8.45 Transmission gate adder circuit.

6. With no more than 12 transistors, the flip-flop proposed in fig.8.46 appears to be a very attractive alternative to those discussed in section 8.2. Yet, the design is not acceptable as a library component because it is exposed to failure. Find all potential problems, specify the preconditions that would allow this circuit to operate as intended, and suggest improvements that make it more robust.

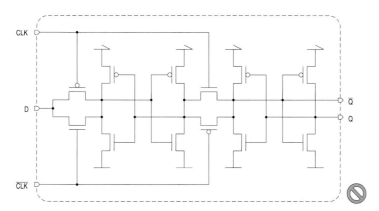

Fig. 8.46 Unsafe flip-flop circuit.

7. Figure 8.47 shows a more complete flip-flop circuit. Analyze its organization and functioning.

8. The general idea behind LSSD has been introduced in section 6.2.5. The circuit of fig.8.48b implements the functionality of an LSSD cell but is not optimal for CMOS.
   (a) Design a static CMOS circuit at the transistor level. Decide yourself whether you prefer to use switched or overruled memory loops.
   (b) What characteristics are important to make a bistable safe, friendly, and fast?

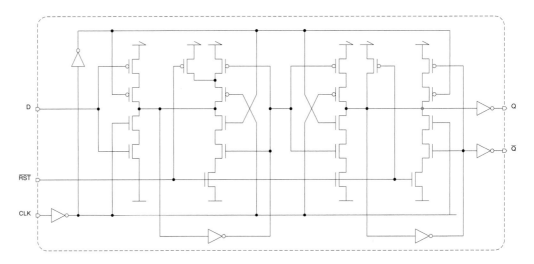

Fig. 8.47 Industrial flip-flop circuit.

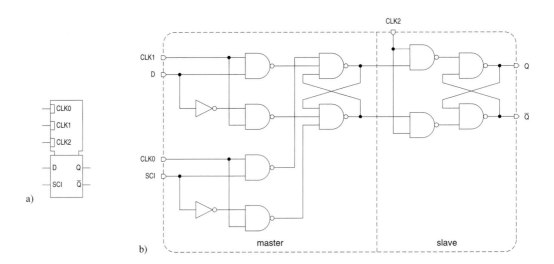

Fig. 8.48 LSSD storage element. Proposed icon (a) and logically equivalent circuit (b).

9. Assume the circuit of fig.8.40c were to drive the D-input of the latch subcircuit shown in fig.8.40d. Do you see any problem with that? Formulate a rule for safe design.

10. Reconsider the circuit of fig.8.40d. Let CLK = 0, Q = 1, and D = 0. Further assume the output of u1 undershoots heavily as a consequence of strong ground bounce. What might happen in an extreme case?

11. Consider a digital IC that includes transmission gates such as those shown in fig.8.15. Assume the p-channel MOSFET in one of them is stuck in a non-conducting state as a consequence

of some local fabrication defect. Examine the impact on the circuit's functioning and performance. List all causes that may result in the MOSFET being stuck in (a) a non-conducting and (b) a conducting condition irrespective of the t-gate's control input. Which of the defects listed must be assumed to be permanent and which are subject to change?

8.7 | Appendix I: Summary on electrical MOSFET models

with contributions by C. Balmer

The electrical behavior of a MOSFET with its four terminals and various operating regimes is quite intricate in itself, and the move towards ever smaller geometries has brought forth additional phenomena. Technological variety, the desire to have computers simulate and optimize MOSFET circuits, mathematical concerns, but also commercial interests have complicated things further. As a result, a plethora of transistor models has been developed over the years. Emphasis in this overview will be on first-order models amenable to hand calculations.

8.7.1 Naming and counting conventions

Trivial as it may appear, the naming of MOSFET terminals is controversial. Most integrated devices exhibit no physical difference whatsoever between **source** and **drain** since the two are manufactured in the same way and at the same time, see fig.8.49e. As a consequence, the two terminals are electrically interchangeable and device characteristics are symmetrical with respect to drain and source, e.g. $I_d(-U_{ds}) \equiv -I_d(U_{ds})$.[42] There are two alternative ways to cope with this property in a transistor model.

1. The two terminals at either end of a MOSFET channel are labeled as source and as drain in an <u>arbitrary but immutable</u> fashion. The model must then be devised in such a way as to produce correct results both for forward operation where U_{ds} is positive (negative) in the case of an n-channel (p-channel) device, and for backward operation where U_{ds} is of opposite sign. An example is the EKV transistor model that consistently refers all voltages to the local substrate or, which is the same, to the transistor **body**.

2. The labeling of source and drain is not permanent but made <u>dependent on the voltages</u> currently present across the MOSFET channel. In the case of an n-channel (p-channel) device, the more negative (positive) terminal is considered to be the source, and the more positive (negative) one the drain at any time. The requirement of a symmetrical transistor model is so dispensed with because equations need only cover forward operation and may safely reference all voltages to one terminal, normally the source. On the other hand, one needs to find out which of the two possible orientations applies before transistor equations can be evaluated.[43] The counting of voltage and currents then follows the orientations shown in fig.8.49d. A majority of models actually being used

---

[42] This is why we prefer the symmetrical icons of fig.8.49a and b over the unsymmetrical ones of fig.8.49c.

[43] Subcircuits where U_{ds} changes sign during operation exist, just consider the transmission gates in a bistable.

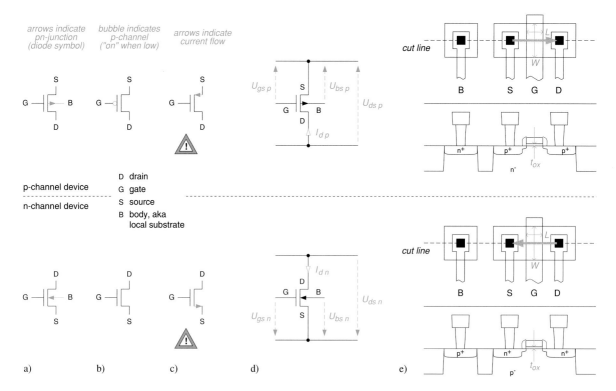

Fig. 8.49 MOSFET conventions. Icons with body terminals drawn (a) and abstracted from (b,c), customary orientations for voltage and current counting (d), layout and cross section (e).

in circuit simulator programs follow this approach, and so will the ones to be presented in sections 8.7.2 through 8.7.4.

Figure 8.49e illustrates how the geometric size of the MOSFET channel is defined. Note that **channel length** L is always measured in the direction of current flow and **channel width** W perpendicular to it. For the sake of performance and layout density, L is typically chosen to be minimal in digital design, so almost all transistors have $W > L$.

8.7.2 The Sah model

Transistor models are essential for circuit analysis, simulation, and optimization. They attempt to approximate measured device characteristics with a set of mathematical equations, which is a matter of finding workable tradeoffs between accuracy and tractability.

A simple approximation is the Sah model, aka Shockley model [230]. Its limited accuracy notwithstanding, simplicity has made it the most common model for hand calculations. The Sah model distinguishes between three operating regions that are termed subthreshold, saturation, and linear region respectively, see fig.8.51. Within each such region, the MOSFET's output characteristic gets

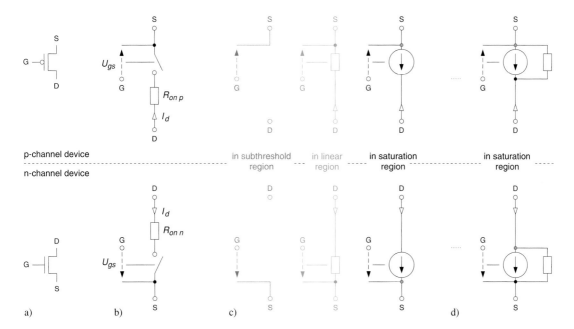

Fig. 8.50 Simple MOSFET models. Equivalent circuits for the DC characteristics of fig.8.1. Icon (a), switch model (b), Sah model (c), alpha-power law model (c), and Shichman–Hodges model (d).

captured by one equation.

$$\text{subthreshold region}: \qquad I_d = 0 \qquad \text{if} \quad U_{gs} - U_{th} \le 0 \qquad (8.48)$$

$$\text{saturation region}: \qquad I_d = \tfrac{1}{2}\beta(U_{gs} - U_{th})^2 \qquad \text{if} \quad 0 < U_{gs} - U_{th} \le U_{ds} \qquad (8.49)$$

$$\text{linear region}: I_d = \tfrac{1}{2}\beta[\,2(U_{gs} - U_{th})U_{ds} - U_{ds}^2\,] \text{ if} \quad 0 < U_{ds} < U_{gs} - U_{th} \qquad (8.50)$$

Linear region. The linear region is also called the resistive, ohmic, triodic, or nonsaturation region. In fact, the attribute "linear" is somewhat idealized because I_d is proportional to U_{ds} only as long as the quadratic term U_{ds}^2 is very small, i.e. when $2(U_{gs} - U_{th}) \gg U_{ds}$. The transistor then effectively acts as a linear resistor the effective resistance of which gets controlled by the voltage being applied between gate and source.

Saturation region. A saturated MOSFET is modelled as a current source that operates under control of the gate–source voltage applied. A MOSFET is at the verge of saturation when $(U_{gs} - U_{th}) = U_{ds}$, or, which is the same, when $U_{gd} = U_{th}$. The borderline between the linear region and the saturation region is given by $I_d = \tfrac{1}{2}\beta U_{ds}^2$, which function is plotted on top of the device's output characteristics in fig.8.51. The saturation region is sometimes referred to as the pinch-off, pentodic, or active region.

Subthreshold region. For any gate–source voltage below a predefined threshold U_{th}, drain current is assumed to be zero in this model, which explains why the subthreshold regime is often called the cut-off region.

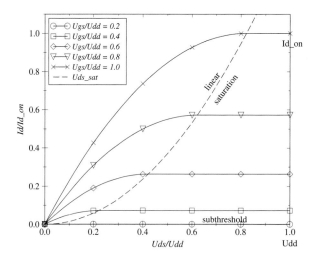

Fig. 8.51 MOSFET transfer characteristic $I_d = f(U_{gs}, U_{ds})$ as obtained with a Sah model.[44]

Equations (8.48) through (8.50) refer to n-channel devices. For p-channel devices, the right-hand sides must be rewritten with a minus sign in front, and all condition operators must be reversed. This is because currents and voltages are consistently oriented the other way round when compared with n-channel transistors, refer to fig.8.49d. After all, power dissipation must never become negative for either type of device. Incidentally, also note that it is always possible to discriminate between the three operating regions by looking at the expression $(U_{gs} - U_{th})$.

Further explanations on the underlying physical mechanisms that cause the semiconductor device to change between various regimes of operation as a function of the electrical conditions will have to wait until section 14.1.4. For the time being, it is sufficient to understand that conduction in a MOSFET is obtained from attracting so many mobile minority carriers to the silicon volume underneath its gate that a thin inversion layer forms between source and drain. The term **channel** refers to this conducting layer and a MOSFET's **threshold voltage** U_{th} is nothing else than the voltage necessary for inversion to occur. Threshold voltage is largely, though not exclusively; determined by the fabrication process; more details are to follow shortly in subsection 8.7.5.

An n-channel MOSFET with a positive threshold voltage $U_{th\,n} > 0$ is said to be of **enhancement type**, and the same applies to a p-channel transistor with a negative threshold $U_{th\,p} < 0$; otherwise one speaks of a **depletion-type** MOSFET. The key difference is that enhancement transistors do not conduct when $U_{gs} = 0$ whereas it takes a negative (positive) gate voltage $U_{gs\,n} < 0$ ($U_{gs\,p} > 0$) to turn an n-channel (p-channel) depletion device off. Digital CMOS circuits consist of enhancement transistors exclusively, depletion devices are not used.[45]

---

[44] The output characteristics of a MOSFET may remind you of those of a bipolar junction transistor (BJT) as they indeed look quite similar, see fig.8.55. The fact that the designations "linear region" and "saturation region" are permuted between the two can be confusing. Please refer to section 8.8 for explanations.

[45] Icons of both enhancement and depletion devices are given below along with typical threshold voltages. In order to give more room for circuit optimization, many CMOS fabrication processes make enhancement transistors available with two or three distinct threshold voltages. We use special symbols to identify low-threshold devices in

Quantity β is called the **MOSFET gain factor** and is a function of both process parameters and layout geometry.

$$\beta = \frac{\mu \, \epsilon_{ox}}{t_{ox}} \frac{W}{L} = \mu \, c_{ox} \frac{W}{L} = \beta_\square \frac{W}{L} \tag{8.51}$$

where t_{ox} is the thickness of the insulating material underneath the gate and where $\epsilon_{ox} = \epsilon_0 \, \epsilon_{r\,ox}$ denotes its permittivity. The two quantities are identical for n- and p-channel MOSFETs. μ is the effective **carrier mobility** in the inversion layer that forms the channel. The mobility of electrons varies between 400 and 700 $\frac{cm^2}{V\,s}$ while it is in the range of 100 and 300 $\frac{cm^2}{V\,s}$ for holes.[46] This explains the superior "on"-state conductance of n- over p-channel devices noted earlier.

μ, ϵ_{ox}, and t_{ox} are often combined into a **process gain factor** $\beta_\square$ that the IC designer must accept as a process-specific constant; $c_{ox} = \frac{\epsilon_{ox}}{t_{ox}}$ indicates the gate capacitance per unit area. Note that channel width W and channel length L are the sole parameters IC designers can act upon, subject to the condition that they are allowed to intervene at the layout level.

Practitioners often express a transistor's **strength** with either one of the quantities below:

$I_{d\,on} = I_d(U_{gs} = U_{ds} = U_{dd})$, aka $I_{d\,sat}$, states the maximum drive current (in the saturation region).

$G_{on} = I_{d\,on}/U_{dd}$ indicates a MOSFET's equivalent conductance when fully driven and saturated.

$I_{d\,on}/W$, sometimes termed **drivability**, relates a transistor's maximum drive current to its channel width and is convenient for comparing devices across fabrication processes.

In summary, the Sah model substitutes simple electrical equivalents for the MOSFET in the following manner, also see fig.8.50c.

region	equivalent circuit
subthreshold	open circuit
saturation	voltage-controlled current source
linear	voltage-controlled resistance

schematics. The gray-shaded area is meant to be suggestive of their off-state leakage and their closer resemblance to depletion-type transistors. Incidentally, note that with typical values on the order of $U_{th\,n} = 1.2\text{--}30$ V the threshold voltages of power MOSFETs are significantly higher than those found in CMOS logic.

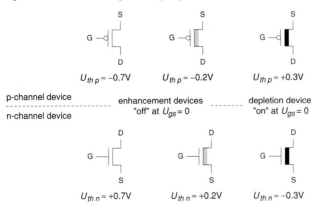

[46] The above data taken from [231] apply to low doping concentrations and low fields. Please refer to the same reference for an in-depth discussion of carrier mobility.

8.7.3 The Shichman–Hodges model

A striking deficiency of the Sah model occurs in the saturation region where the drain current is supposed to remain totally unaffected by the drain–source voltage being applied, which behavior would correspond to an ideal source of constant current. In practice, channel length modulation, a physical effect to be explained later in section 14.1.4, causes the $I_d(U_{ds})$ traces of real devices to become slightly inclined as shown in fig.8.52. The Shichman–Hodges model [232] takes that into account in a heuristic fashion by introducing a corrective term $(1 + \lambda U_{ds})$:

$$\text{subthreshold region} : \qquad\qquad I_d = 0 \qquad\qquad\qquad (8.52)$$

$$\text{saturation region} : \qquad I_d = \tfrac{1}{2}\beta(U_{gs} - U_{th})^2\,(1 + \lambda U_{ds}) \qquad (8.53)$$

$$\text{linear region} : I_d = \tfrac{1}{2}\beta[\,2(U_{gs} - U_{th})U_{ds} - U_{ds}^2\,](1 + \lambda U_{ds}) \qquad (8.54)$$

The **channel length modulation factor** λ accounts for the variation of the effective channel length. As variations have a relatively larger impact on short channels than on long channels, the magnitude of λ decreases with channel length L. Note that λ has a positive value for n-channel transistors and a negative value for their p-channel counterparts.[47] In the equivalent circuit, the corrective term manifests itself as a conductance connected in parallel to the voltage-controlled current source of the Sah model, see fig.8.50d.

8.7.4 The alpha-power-law model

While the Sah and Shichman–Hodges models often suffice to approximate MOSFETs with long channels, of say 2 µm and more, they fail to faithfully reproduce the static characteristics of submicron transistors. The customary square law $I_d \propto (U_{gs} - U_{th})^2$ no longer holds for short channels. Instead, the dependency becomes a subquadratic one for the following reasons.

Sah and Shichman–Hodges models assume fixed carrier mobilities in the inversion layer. While this is fine at low electrical fields, it does not apply for the strong longitudinal fields found in today's MOSFETs. Above 0.7 V/µm or so, the average velocity of electrons that travel through p-type silicon increases more slowly than the driving field and eventually saturates around $7 \cdot 10^6$ cm/s [231] [187]. Holes in n-type silicon exhibit the same effect at fields beyond approximately 1.5 V/µm. This phenomenon is called **velocity saturation**, aka mobility reduction or mobility degradation of carriers. Another effect that contributes to mobility degradation emanates from the vertical field in the channel because carriers attracted to the interface between bulk and gate dielectric material tend to collide more often with lattice defects there.

Ultimately, I_d becomes a linear function of $(U_{gs} - U_{th})$. Figure 8.53 compares static MOSFET characteristics obtained from the Sah model with and without a correction for reduced electron

---

[47] Most authors state the corrective term as $(1 + \lambda|U_{ds}|)$, which makes λ a positive number for both n- and p-channel devices. Taking the magnitude is actually unnecessary for source-referenced transistor models because in forward operation U_{ds} is positive for n-channel and negative for p-channel transistors by definition. Omitting the superfluous magnitude operator has the extra benefit that algebraic manipulations of (8.53) and (8.54) are facilitated. Clearly, model characteristics are not affected by this alteration. Note, however, that taking the magnitude is indeed binding in those models that cover both forward and backward operation such as in the original form of equations published by Shichman and Hodges [232].

Also, (8.54) is often quoted with no corrective term. We do not follow this mistaken omission because it is contradictory to the authentic Shichman–Hodges model and because it gives rise to a discontinuity in the $I_d(U_{gs}, U_{ds})$ function at the borderline between the linear and the saturation region.

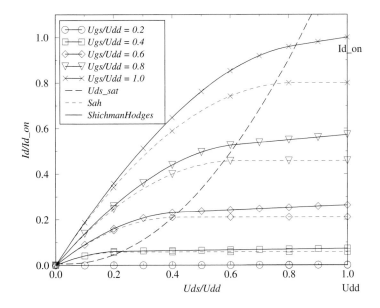

Fig. 8.52 Transfer characteristic as obtained with a Shichman–Hodges model.

mobility. The alpha-power law has been developed as a heuristic hand model that accounts for this [233].

$$\text{subthreshold region:} \qquad I_d = 0 \qquad \text{if} \quad U_{gs} \le U_{th} \qquad (8.55)$$

$$\text{saturation region:} \qquad I_d = I_{d\,on}\left(\frac{U_{gs}-U_{th}}{U_{dd}-U_{th}}\right)^{\alpha} \qquad \text{if} \quad U_{ds\,sat} \le U_{ds} \qquad (8.56)$$

$$\text{linear region:} \; I_d = \left(2 - \frac{U_{ds}}{U_{ds\,sat}}\right)\frac{U_{ds}}{U_{ds\,sat}}\, I_{d\,on}\left(\frac{U_{gs}-U_{th}}{U_{dd}-U_{th}}\right)^{\alpha} \; \text{if} \quad U_{ds} < U_{ds\,sat} \qquad (8.57)$$

where saturation voltage $U_{ds\,sat}$ depends on the gate–source voltage U_{gs} in the following way:

$$U_{ds\,sat} = U_{ds\,sat\,on}\left(\frac{U_{gs}-U_{th}}{U_{dd}-U_{th}}\right)^{\frac{\alpha}{2}} \qquad (8.58)$$

The model has four parameters that must be adjusted in such a way as to obtain a good match between the models' characteristic curves and those of the actual device. U_{th} is the threshold voltage introduced with earlier models. $I_{d\,on}$ is the maximum saturation current in full drive condition ($U_{ds} = U_{gs} = U_{dd}$) while $U_{ds\,sat\,on}$ stands for the saturation voltage when the transistor's gate is fully driven ($U_{gs} = U_{dd}$). α is termed the **velocity saturation index** and covers a range of (full velocity saturation, i.e. short channel) $1 \le \alpha \le 2$ (unimpeded mobility, i.e. long channel). The evolution of α over a couple of process generations has been documented in section 8.1.2.

Please keep in mind that the above equations refer to n-channel MOSFET; the sign and condition operators need to be adapted for p-type transistors.

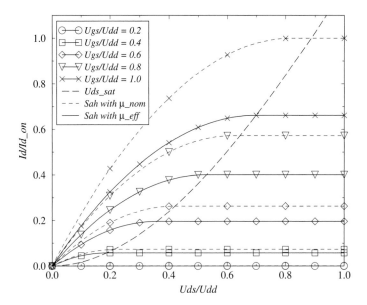

Fig. 8.53 Impact of mobility reduction on transfer characteristic.

In its present form, the alpha-power-law model does not account for channel length modulation, making it inferior to the Shichman–Hodges model in that respect. As an extension, (8.56) and (8.57) are sometimes made to include a corrective factor similar to those in (8.53) and (8.54).

8.7.5 Second-order effects

The simple hand models presented so far ignore many higher-order phenomena that also contribute to MOSFET characteristics. Transistor models intended for computer simulations, aka **compact models**, can be more accurate, of course. We briefly sketch the most important of those effects before giving an overview of prominent simulation models in table 8.2.

Subthreshold conduction

Although very small, drain current is not zero in the subthreshold region but is found to grow exponentially with $(U_{gs} - U_{th})$ in physical devices.

$$I_d = \mu c_{ox} \frac{W}{L} (m-1) U_\theta^2 \, e^{\frac{q_e (U_{gs} - U_{th})}{m \, k \, \theta_j}} \, (1 - e^{-\frac{U_{ds}}{U_\theta}}) \quad \text{if} \quad U_{gs} - U_{th} \leq 0 \qquad (8.59)$$

where $U_\theta = \frac{k\theta_j}{q_e}$ denotes the thermal voltage, k the Boltzmann constant, q_e the electron charge, and θ_j the absolute junction temperature. The so-called **body effect coefficient** m, aka body factor, depends on the capacitances that exist above and below the inversion channel.

$$m = \frac{c_{ox} + c_{dm}}{c_{ox}} = 1 + \frac{c_{dm}}{c_{ox}} = 1 + \frac{\epsilon_{\text{Si}}}{t_{dm}} \frac{t_{ox}}{\epsilon_{ox}} \qquad (8.60)$$

Numerical examples

Table 8.1 Numerical data for enhancement-type MOSFETs in a 130 nm 1P6M triple-well poly-gate CMOS process (for illustrative purpose only).

model	parameter	typical values for n-channel devices	p-channel devices	measurement conditions $\theta_j = 27\,^\circ\mathrm{C}$
	W_{min} [nm]	150	150	
	L_{min} [nm]	130	130	
Sah	U_{th} [V][a]	0.351	−0.300	$W = 10\ \mu\mathrm{m}, L = L_{min}$
	$\beta_\square$ [μAV^{-2}]	222	103	*idem*
	$I_{d\,on}/W$ [μA/μm]	670	−320	*idem*
	R_{on} [Ω]	179	375	*idem*
Shichman–	λ [V^{-1}]	0.145	−0.486	$L = L_{min}$
Hodges	λ [V^{-1}]	0.125	−0.054	$L = 20 L_{min}$
Alpha-	U_{th} [V]	0.34	−0.29	$W = 10\ \mu\mathrm{m}, L = L_{min}$
power	$U_{ds\,sat\,on}$ [V]	0.62	−1.10	*idem*
law	$I_{d\,on}/W$ [μA/μm]	666	−312	*idem*
	α	1.04	1.20	*idem*

[a] The manufacturing process provides MOSFETs with three distinct threshold voltage levels, all numbers given here refer to the high-speed (i.e. low-threshold) variety. Nominal core supply voltage is $U_{dd} = 1.2$ V.

□

The gate dielectric of permittivity ϵ_{ox} is t_{ox} thick and yields a unit capacitance of c_{ox}. Similarly, ϵ_{Si}, t_{dm}, and c_{dm} refer to the capacitance formed by the channel and the bulk material where the depletion layer underneath the inversion channel acts as dielectric. You may want to refer to fig.14.10 for an illustration. Equations (8.59) and (8.60) are derived and explained in full detail in [187], which further indicates that m typically varies between 1.1 and 1.4.

The error caused by dropping the rightmost factor from (8.59) is less than 5% if $3U_\theta < U_{ds}$, which is normally the case for CMOS inverters in steady-state condition. Focussing on those quantities that circuit designers actually can control and changing exponentiation to basis 10, the equation then simplifies to

$$I_d \approx \beta_\square \frac{W}{L} (m-1)\, U_\theta^2\, 10^{\frac{q_e(U_{gs}-U_{th})}{m\,k\,\theta_j}\log_{10}e} \propto \frac{W}{L} 10^{\frac{U_{gs}-U_{th}}{S}} \tag{8.61}$$

The extra MOSFET parameter S is termed **subthreshold slope**, aka subthreshold swing.

$$S = \frac{\partial U_{gs}}{\partial(\log_{10} I_d)} = \frac{1}{\log_{10}e} \frac{\partial U_{gs}}{\partial(\ln I_d)} \approx \ln 10 \frac{m\,k\,\theta_j}{q_e} \approx 2.3\,m\,U_\theta \tag{8.62}$$

A value of $S = 90$ mV, for instance, implies that the subthreshold current swells by one order of magnitude for each 90 mV increase of U_{gs} and, analogously, for each 90 mV reduction of U_{th}. Note that S is proportional to θ_j, which causes leakage to augment exponentially with absolute temperature. Accounting for all sorts of conditions, 70–120 mV/decade can be considered a realistic range in bulk CMOS. Figure 8.54 compares the subthreshold behaviors of different MOSFET models.

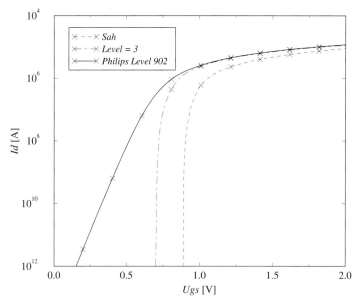

Fig. 8.54 Modelling of subthreshold conduction.

Short- and narrow-channel effects

Below a drain–source separation L of 500 nm and even more so below 250 nm, the observed threshold voltage tends to fall off,[48] causing the off-state leakage current $I_{d\,off} = I_d(U_{gs} = 0)$ to augment exponentially in accordance with (8.61). Several physical effects contribute to this undesirable phenomenon. What they have in common is that the horizontal electrical fields of the source–body and drain–body junctions interact with the vertical field from the gate and impact the flow of electrons in the inversion layer.

Similarly, threshold voltage U_{th} is bound to vary as a function of transistor width as W approaches the lower admissible bound. As a result of various counteracting phenomena, this dependency is not necessarily monotonic, however.

Back gate effect

A MOSFET's threshold voltage is further subject to variation when a voltage is applied between body and source $U_{bs} \neq 0$ which suggests that the field underneath the inversion channel acts like a second gate. In an n-channel (p-channel) device one observes a shift of the threshold voltage U_{th} towards more positive (negative) values when the source–body junction gets <u>reverse-biased</u> more strongly, e.g. by making the body more negative (positive) than the source potential so that $U_{bs\,n} < 0$ $(U_{bs\,p} > 0)$. Reverse back biasing also takes place when the source potential gets lifted (lowered) relative to that of the embedding well or substrate.

---

[48] To be precise, U_{th} becomes less positive for n-channel and less negative for p-channel devices.

This indeed occurs in transmission gates and in the stacked transistors found in multi-input CMOS gates. Depending on how a given transistor connects to the surrounding circuitry, U_{bs} may thus remain fixed to zero or be subject to variation during circuit operation.

Incidentally, note that <u>forward</u> biasing may lead to loss of insulation and latch-up.

Table 8.2 | Popular MOSFET compact models compared.

| Physical effect | modelling based on | | | | | |
| | threshold voltage | | | inversion charge | | surface potential |
	Level1-3[a]	BSIM3v3	MM9	EKV	BSIM5	PSP
Channel length modulation	yes	yes	yes	yes	yes	yes
Mobility reduction	no	yes	yes	yes	yes	yes
Velocity saturation	no	yes	yes	yes	yes	yes
Drain-induced barrier lowering	no	yes	yes	yes	yes	yes
Impact ionization	no	yes	yes	yes	yes	yes
Poly depletion	no	yes	no	yes	yes	yes
Gate current	no	no	no	yes	yes	yes
Quantum-mechanical effects on charges	no	no	no	yes	yes	yes
Bias-dependent overlap capacitances	no	no	no	yes	yes	yes
Self-heating	no	no	no	no	no	yes
Gradual channel approximation	yes	yes	yes	yes	yes	no
Charge sheet model	-	-	-	yes	yes	yes
symmetrical	-	-	-	yes	yes	yes

[a] SPICE Level 1 essentially corresponds to the Shichman–Hodges model explained in section 8.7.3.

8.7.6 Effects not normally captured by transistor models

Even the most sophisticated MOSFET models have limitations as to what physical effects they do account for. In fact, several phenomena do not come to bear as long as the transistor is operated within its legal operating range.

- **Gate dielectric breakdown** refers to the permanent destruction of the thin oxide layer underneath the gate electrode by an excessive vertical field.

- A MOSFET may enter an irregular conducting state when excessive voltages are applied between drain and source. The gate electrode loses control over the current that flows through the channel in this **avalanche breakdown** regime, aka punch-through.

- In the presence of overly strong fields, an electron can gain so much energy that it gets ejected from the channel into the gate dielectric, where it may get trapped. Once the accumulated charge becomes sufficiently important, it causes a shift of the built-in threshold voltage towards higher values. This **hot-electron degradation** curtails currents in all MOSFET operating regions and is experienced as an increased gate delay in digital logic. Some simulation models account for this long-term wearout effect by providing two parameter sets, one for pristine MOSFET devices and one that is supposed to match their electrical characteristics after 10 years of operation.[49]

Manufacturers define **absolute maximum ratings** to make sure that circuits are safe from these and other destructive effects. It is most important to respect them during both device operation and handling. A group of effects totally unaccounted for by compact models is brought about by the various parasitic devices that come along with the regular thin oxide MOSFETs,[50] and that do or do not become manifest depending on a circuit's operating conditions.

- The MOSFETs **drain–body junction** and a **source–body junction** are reverse-biased during normal operation. Their sole contribution then is of capacitive nature and included in the drain and source capacitance respectively. Extra devices might be needed to model forward operation should a device ever enter this regime in the application at hand.

- CMOS ICs further include field oxide MOSFETs and parasitic BJTs that are at the origin of voltage clamping, abnormal current flow, and other effects. Standard MOSFET models do not account for those devices, however, because they do not come to bear unless a circuit is about to enter electrical overstress conditions. Circuit designers do everything possible to prevent this from happening during normal operation.

  Still, those parasitic devices must be modelled if one wants to study exceptional events that approach or transgress a circuit's regular range of operation such as latch-up or electrostatic discharge (ESD). They must then be approximated either as separate networks of lumped circuit elements or — more accurately — as spatially distributed semiconductor devices using technology CAD (TCAD) simulation software.

8.7.7 Conclusions

- The observed behavior of semiconductor devices depends on many factors in a complex way. Physical effects that contribute to electrical device characteristics can be classified into three categories:

  - First-order effects that are satisfyingly approximated by simple hand models,
  - Second- and higher-order effects not easily amenable to hand calculation but honored to various degrees by a multitude of established simulation models,
  - Effects ignored by common models and thus confining their range of validity.

  It is important to realize which effects matter for a given situation or application before opting for some device model, equivalent circuit, or simulation tool.

- Careful calibration of transistor models over all operating conditions of interest may be onerous, but is absolutely essential.

- Highly sophisticated models are a mixed blessing because of their many parameters that must be obtained either from measurements or from device simulations. Physical reality and

---

[49] p-channel MOSFETs are not as susceptible as their n-channel counterparts because hole mobility is inferior.
[50] Please refer to sections 11.5.2, 11.6.3, and 11.6.2 for illustrations and details.

simulation are likely to diverge unless all such parameters are available with good accuracy and statistical relevance.

A more in-depth coverage of MOSFET operation and compact models can be found in numerous textbooks such as [234] [231] [187] [235] [236], while [237] gives a thorough overview on the evolution of compact models over the years.

8.8 | Appendix II: The Bipolar Junction Transistor

BJTs are also known as bipolar transistors or simply as bipolars. While CMOS logic makes no use of them functionally, they are present as parasitic devices in any CMOS circuit and participate in ESD protection and in latch-up, two phenomena that are to be explained in sections 11.6.2 and 11.6.3 respectively. We will limit our discussion here to a brief comparison of BJTs with MOSFETs to put the reader into a position to understand the respective roles played by BJTs in these two phenomena.

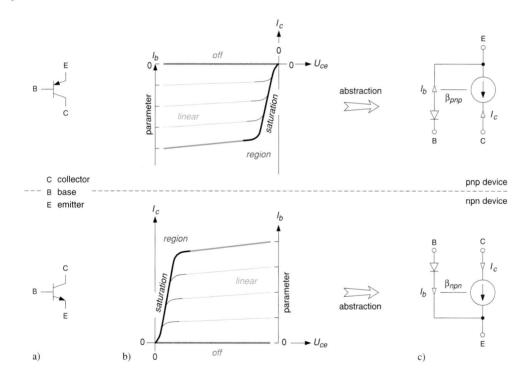

Fig. 8.55 Icons (a), DC transfer characteristics (b), and equivalent models (c).

- While a MOSFET can be approximated as a voltage-controlled current source (VCCS), a BJT — to first order — acts like a current-controlled current source (CCCS). When viewed from a purely digital perspective, the MOSFET abstracts to a voltage-controlled switch and the BJT to a current-controlled switch.

- A comparison of fig.8.55b and fig.8.1b reveals that the output characteristics of a BJT, $I_c = f(I_b, U_{ce})$, resemble those of a MOSFET, $I_d = f(U_{gs}, U_{ds})$. Yet, note that the designations for the linear and saturation regions are permuted! In the case of a BJT, the name "linear" refers to the proportional relationship between base and collector currents $I_c = \beta \cdot I_b$ whereas for a MOSFET it alludes to the behavior as a linear resistor $U_{ds} = R \cdot I_d$ exhibited as long as the drain-to-source voltage U_{ds} remains small.

- As a consequence of the above, a BJT operated as a switch settles in saturation when fully turned on whereas a MOSFET operates in the linear region in the same situation.

- Normally, BJTs are unsymmetric devices by construction with the emitter more strongly doped than the collector.

- A typical current gain $\beta = \frac{I_c}{I_b}$ in the linear region is on the order of 100. The thinner the base layer that separates the collector from the emitter, the higher a BJT's current gain.

Chapter 9

Energy Efficiency and Heat Removal

Energy considerations are no longer confined to battery-operated circuits. Power dissipation of high-performance CPU chips is on the order of 50 to 120 W, see table 9.1, roughly as much as a craftsman's soldering iron. Removing that much thermal power necessitates sophisticated packages, heat sinks, heat pipes, forced ventilation, and other costly options. On the input side, fat supply rails and elaborate multiphase step-down converters built from numerous and bulky power transistors, inductors, and capacitors are required to handle massive supply currents without critical voltage drops. Not only costs but also packing density suffers.

Observation 9.1. *The problem in battery-operated circuits is where to get the energy from whereas getting the heat out is a major problem in high-performance circuits.*

The first section in this chapter analyzes what CMOS circuits spend energy for at a fairly detailed level. Section 9.2 then gives practical guidelines for how to improve energy efficiency before section 9.3 summarizes the very basics of heat flow and heat removal.

9.1 | What does energy get dissipated for in CMOS circuits?

Note, to begin with, that power is not an adequate yardstick when it comes to evaluating alternative schemes for the processing of information because it does not relate to performance in any way. A circuit's efficiency is better defined as dissipated **energy per** computational **operation** or, which is the same, as **energy per** processed **data item**.[1] Our discussion will thus center around the amount of energy E_{cp} expressed in [J] that gets dissipated in a given circuit during one computation cycle rather than around power dissipation P in [W]. Obtaining one quantity from the other is trivial.

$$P = f_{cp} E_{cp} \qquad\qquad E_{cp} = P\,T_{cp} \qquad\qquad (9.1)$$

---

[1] The same quantity has been used for evaluating alternative architectures throughout chapter 2.

Numerical examples

Table 9.1 Power figures of commercial VLSI chips.

part name	year	litho-graph. size [nm]	tran-sistor count [M]	die size [mm^2]	power dissi-pation [W]	at clock freq. [MHz]	core voltage [V]	supply current [A]	power density [W/cm^2]
Intel IA-32 (general purpose CISC; top performance, superscalar, superpipelined, 32/64 bit)									
Pentium	1993	800	3.1	296	15	60	5.0	3	5
PentiumPro	1995	350	5.5	197	28	200	3.3	11	14
Pentium 4	2002	130	42	146	54	2000	1.5	36	37
Pentium 4 560	2004	90	125	112	115	3600	1.385	83	103
Core 2 Extreme	2006	65	291	143	75	2933	1.28	59	52
Sun UltraSPARC (RISC for servers; high performance, eight cores, 32 bit)									
Niagara	2005	90	279	378	63	1200	1.2	52	17
Niagara II	2007	65	503	342	84	1400	1.1	76	25
ARM (general purpose RISC processor; performance-power tradeoff, no FPU, 32 bit)									
XScale	2001	180	6.5	25	0.45	600	1.3	0.35	1.8
dspfactory (dedicated audio processor for digital hearing aids; locally optimized word widths)									
Delta-2	2000	180	0.28	10	0.24m	1.34	1.2	0.2m	2.4m

For comparison: A typical kitchen hotplate has a diameter of 18 cm and dissipates 1800 W, which results in a power density of 7 W/cm^2. The current drawn from the 400 V mains is 4.5 A.

□

As defined earlier, a computation period T_{cp} is the time span that separates two consecutive computation cycles. Computation rate f_{cp} denotes the inverse, that is the number of computation cycles per second.[2] Let us now study the four phenomena that dissipate energy in static CMOS circuits, namely

- Charging/discharging of capacitive loads,
- Crossover currents,
- Driving of resistive loads, and
- Leakage currents.

9.1.1 Charging and discharging of capacitive loads

FOR A SINGLE NODE

We begin by quantifying the amount of energy E_{fa} that gets dissipated while the logic state of some circuit node k changes from 1 to 0. Standard two-valued CMOS logic features **rail-to-rail**

---

[2] In the occurrence of single-edge-triggered one-phase clocking, a computation period begins just after an active clock edge and ends with the next. Computation cycle and clock cycle are the same so that $f_{cp} = f_{clk}$. As opposed to this, dual-edge-triggering fits two consecutive computation cycles into every clock period, which implies that $f_{cp} = 2f_{clk}$. See sections 2.3.7 and 6.2.3 for details.

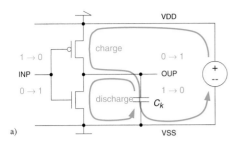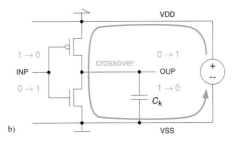

Fig. 9.1 Charge/discharge (a) and crossover currents (b) in a switching CMOS inverter.

outputs, which means that voltages swing back and forth between VDD and VSS, see fig.9.1a. The energy stored in the node's capacitance C_k when fully charged to supply voltage U_{dd} is

$$E_{C_k} = \frac{1}{2}C_k U_{dd}^2 \qquad (9.2)$$

Energy conservation postulates that all this must get converted from electrical into thermal energy in the resistive circuit elements, that is essentially in the n-channel MOSFET(s), when the capacitance is discharged to voltage zero. We thus have $E_{fa\,k} = E_{C_k}$. For reasons of symmetry, the same amount of energy gets turned into heat in the p-channel transistor(s) during the process of charging the node, hence $E_{ri\,k} = E_{C_k}$. The total energy that gets dissipated per charge–discharge cycle immediately follows as

$$E_{ch\,cyc\,k} = E_{ri\,k} + E_{fa\,k} = C_k U_{dd}^2 \qquad (9.3)$$

It is worthwhile to note that this quantity depends neither on the (dis)charging circuitry nor on the signal waveforms involved. Except for their contributions to the overall node capacitance, the exact electrical and geometrical properties of the MOSFETs do not matter.

Except for the minority of multi-driver nodes, an EDA tool typically approximates overall node capacitance C_k as follows.[3]

$$C_k \approx C_{gate\,k} + C_{wire\,k} \qquad (9.4)$$

where

$C_{wire\,k}$ The total interconnect or wiring capacitance of node k made up of
contributions from the wire to ground and from the wire to adjacent wires.

$C_{gate\,k}$ An energy-equivalent capacitance that accounts not only for the output of the
gate that drives node k but also for all its inputs and inner capacitances
after they have been transformed to the output of that very gate.

Numerical values for C_{gate} are tabulated in datasheets and cell models, those for C_{wire} must be obtained from layout extraction.[4] They get consolidated into C_k values for every circuit node and

---

[3] Please refer to appendix 9.4 for more details.

[4] Incidentally, observe that C_{gate} scales with the geometric width — and length — of the MOSFETs whereas C_{wire} does not. This difference notwithstanding, C_{wire} also tends to grow when many transistors are sized in a more

included into a gate-level netlist during back-annotation, which is carried out as part of physical design verification.

FOR A COLLECTION OF NODES IN A CIRCUIT

We now extend our analysis to an entire VLSI circuit. As not all nodes within a circuit change state at the same rate, we introduce individual **node activities**, aka toggle rates.

Definition 9.1. *A node's activity α_k indicates how many times per computation cycle node k switches from one logic state to the opposite one when averaged over many computation cycles.*

On average, some binary signal with activity α_k takes $\frac{2}{\alpha_k}$ computation cycles to complete one full charge–discharge cycle.[5] The average energy dissipated per computation cycle for node k immediately follows as

$$E_{ch\,k} = \frac{\alpha_k}{2} E_{ch\,cyc\,k} = \frac{\alpha_k}{2} C_k U_{dd}^2 \tag{9.9}$$

The energy dissipated for the charging and discharging of nodes across an entire digital circuit is obtained by summing up (9.9) over all K nodes in that circuit.

$$E_{ch} = \sum_{k=1}^{K} E_{ch\,k} = U_{dd}^2 \sum_{k=1}^{K} \frac{\alpha_k}{2} C_k \tag{9.10}$$

---

generous way because inflated cell areas also imply longer interconnect lines. A general oversizing of transistors and gates does, therefore, no good to energy efficiency.

[5] This simply is because the signal must switch forth and then back again. Numerical examples follow.

○ Ungated clock in case of single-edge-triggered clocking: $\alpha_k = 2$ (toggles twice per computation cycle).

○ Ungated clock in case of dual-edge-triggered clocking: $\alpha_k = 1$ (toggles once per computation cycle).

○ Output of a T-type flip-flop that is permanently enabled: $\alpha_k = 1$.

○ Output of a D-type flip-flop fed with random data: $\alpha_k = \frac{1}{2}$.

Next, we give some figures that refer to the ensemble of flip-flop outputs in various counting circuits. All registers are assumed to be permanently enabled, next state logic is not included.

○ Binary counter (w bit wide, 2^w states):

$$\sum_{k=0}^{w-1} \alpha_k = \sum_{k=0}^{w-1} \left(\frac{1}{2}\right)^k \overset{w \to \infty}{=} 2 \tag{9.5}$$

○ Gray counter (w bit wide, 2^w states):

$$\sum_{k=0}^{w-1} \alpha_k = 1 \tag{9.6}$$

○ Shift register that is part of a binary pseudo-random sequence generator (w bit wide, $2^w - 1$ states):

$$\sum_{k=1}^{w} \alpha_k \approx \sum_{k=1}^{w} \frac{1}{2} = \frac{1}{2}w \overset{w \to \infty}{=} \infty \tag{9.7}$$

○ Moebius counter, aka twisted ring counter and Johnson counter (w bit wide, $2w$ states):

$$\sum_{k=1}^{w} \alpha_k = 1 \tag{9.8}$$

Again, this overall figure is meant per computation cycle and averaged over many cycles.

Equation (9.10) has two benefits. Firstly, it quantifies what almost always is the most significant contribution to a VLSI circuit's overall energy dissipation. Secondly, the few arguments on its right-hand side are fairly straightforward to come up with. Remember that it has been obtained under five simple assumptions:

- All node capacitance values C_k are fixed.
- The switching devices (MOSFETs) are of resistive nature when "on".
- Supply voltage U_{dd} is constant and the same throughout the (sub)circuit considered.
- Nodes are always fully charged and discharged (full swing from rail to rail).
- All activity numbers refer to the same computation cycle and clock signal.

Several observations about node activities are due.

1. A circuit node's activity is not the same as the probability of finding that node in the opposite logic state at the end of a computation period as digital circuits are subject to glitching. Investigations on various adder structures fed with random numbers have resulted in activities 10% to 20% above what is anticipated with glitching ignored. While that much extra unrest is typical for many circuits, data activities can even grow beyond the intuitive bound of $\alpha_k \leq 1$. Node activities in excess of 6 have been reported in circuits where signals propagate along paths of markedly different depths before converging in combinational operations. Unbalanced delays occur in multipliers and even more so in cascades of multipliers or other arithmetic units with no registers in between. As an example, consider the isomorphic architecture of a lattice filter in fig.9.2. How to mitigate this effect by way of delay balancing and signal silencing will be a subject of section 9.2.2.

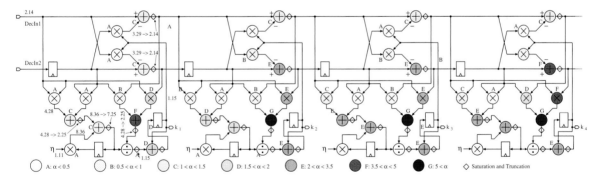

Fig. 9.2 Inflation of node activities in a lattice filter (reprinted from [238]).

2. Node activities are statistical data. They are typically obtained during gate-level simulation runs, see fig.9.3.[6] For every circuit node, the toggle counts get collected over what is believed to be a representative sequence of operation for the circuit being analyzed.

---

[6] The alternative of probabilistic power estimation ignores glitching and is less accurate.

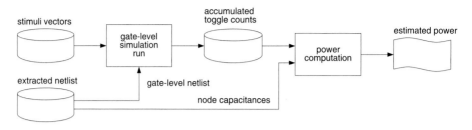

Fig. 9.3 Simulation-based power/energy estimation.

3. Almost all circuits exhibit significant temporal (over time) and spatial (across signals and bits) correlations among the toggling of their nodes. This is because most flip-flops are part of registers and because many registers are updated on a regular basis. Also, activities may greatly differ among the bits of a data word. Consider an accumulator, for instance, and watch how bits evolve from one computation cycle to the next. LSBs will typically behave like random variables whereas MSBs will be strongly correlated.

Numerical examples

Statistical data collected from the entire population of flip-flops in six benchmark circuits that have been fed with representative stimuli.

Benchmark circuit	Average node activities	
	D inputs	Q outputs
ARES	0.178	0.066
Bongo	0.076	0.054
FIR	0.50	0.30
SST	0.40	0.11
Shiva	0.54	0.24
CCDChip	0.38	0.26

Figure 9.4 plots the average bitwise activities of a noisy speech signal quantized with 16 bit resolution, 16 kHz sampling rate, and SNR = 40 dB. Note the huge difference between MSB and LSB. Also observe that 2's complement (2'C) encoding entails higher node activities than sign-and-magnitude (S&M) representation in signals such as speech that fluctuate around zero for much of the time with modest amplitudes.

☐

In more general terms, the finding is

Observation 9.2. *Node activities are distributed very unevenly. Circuits typically include a number of flip-flop outputs that toggle with $\alpha_k \approx \frac{1}{2}$ while most other circuit nodes exhibit significantly lower activities. Nodes with an activity close to zero are also common.*

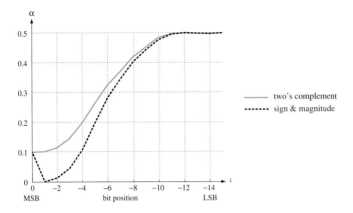

Fig. 9.4 Average activities in noisy speech signals (reprinted from [239]).

This is because most VLSI circuits contain numerous registers the contents of which change very infrequently when compared with the busy datapath and pipeline registers. This includes registers that hold configuration data, calibration parameters, mode or status information, or slowly-varying coefficients. Most nodes that are part of register files or RAMs also exhibit low activities and the same applies to exception-handling circuitry and human interfaces.

9.1.2 Crossover currents

As illustrated in fig.8.4b, both the n- and the p-channel transistor in a CMOS inverter are partially conducting when the voltage at the input satisfies $U_{th\,n} < U_{inp} < U_{dd} + U_{th\,p}$. This means that charge is then allowed to flow from VDD to VSS via the two MOSFETs without ever reaching the load, see fig.9.1b. We will refer to this phenomenon as crossover current, but the reader is cautioned that many synonyms exist.[7]

The energy dissipated by crossover currents per charge–discharge cycle depends on numerous factors. The approximation

$$E_{cr\,cyc\,k} = \frac{\beta}{12} \left(U_{dd} - 2U_{th} \right)^3 t_{ra\,k} \tag{9.11}$$

was derived by the author of [219] under a number of simplifying assumptions:

- Analysis refers to CMOS inverters exclusively.
- Electrical symmetry is assumed, which is to say that gain factors and threshold voltages are the same, i.e. $\beta = \beta_n = \beta_p$ and $U_{th} = U_{th\,n} = -U_{th\,p}$.
- The output load is assumed to be zero.[8]
- The input voltage rises and falls linearly with ramp time $t_{ra\,k} = t_{ri\,k} = t_{fa\,k}$.

---

[7] Such as overlap current, short-circuit current, shootthrough current, rushthrough current, contention current, Class A current, and even dynamic leakage current.

[8] Any capacitance attached to the output slows down the build up of drain–source voltage across the off-going MOSFET and so curbs the current that flows from VDD to VSS directly. This effect effectively makes (9.11) an upper bound for $E_{cr\,k}$, a finding confirmed in [240], where a much more detailed power model is presented.

• A Sah model is a good enough approximation for the transistors.

As one naturally expects, losses augment with input ramp time and with transistor size because of $\beta = \beta_\square \frac{W}{L}$. This gives rise to a dilemma. The slower the input ramps to a gate, the more energy gets wasted by crossover currents. Making transitions faster helps to cut dissipation in the gate being driven, but the generous sizing of the MOSFETs in the driving gate necessary to do so inflates the energy losses there. A compromise must be sought. As a general guideline, circuits should be designed so as to

(a) make signal rise and fall times approximately the same, and
(b) make them comparable to the propagation delay of a typical gate
    from the cell library being used [241].

Only drivers that handle very heavy loads such as off-chip loads, vast clock nets, or long busses require a more sophisticated analysis to better balance area, delay, and dissipated energy.[9] Otherwise, any step taken in pursuit of a lower charge/discharge dissipation E_{ch} — such as cutting down U_{dd}, α_k, C_k, transistor sizes, and total node count K — at the same time also helps to abate the energy losses E_{cr} that are due to crossover currents. The argument is in support of a rough but popular approximation:

$$E_{cr\,k} = \frac{\alpha_k}{2}E_{cr\,cyc\,k} \approx \sigma_k\frac{\alpha_k}{2}E_{ch\,cyc\,k} = \sigma_k E_{ch\,k} \qquad (9.12)$$

$$E_{dyn\,k} = E_{ch\,k} + E_{cr\,k} \approx (1+\sigma_k)E_{ch\,k} = (1+\sigma_k)\frac{\alpha_k}{2}C_k U_{dd}^2 \qquad (9.13)$$

As a rule of thumb, σ_k is generally between 0.05 and 1.5 [241], with the larger numbers applicable to those situations where (almost) no load is attached to the driving gate. The average value for digital VLSI circuits with adequately sized buffers has been found to be $\sigma_k \approx 0.2$ or less [182]. With supply voltages and overdrive factors being lowered from one process generation to the next, crossover losses and hence also σ_k continue to diminish.

Observation 9.3. *Crossover currents are not normally addressed in any specific way during digital CMOS circuit design, apart from*
a) keeping ramp times within reasonable bounds and
b) pad drivers that handle very heavy loads and more substantial voltages.

From a practical perspective, note that library vendors refrain from modelling capacitive and crossover currents separately. Instead, they characterize each library cell with a single energy figure Ψ_{gate} in their datasheets. This quantity is obtained by dividing the cell's power dissipation P_{gate} by the frequency f_{oup} of the signal at the output, which explains why it is expressed in µW/MHz rather than in pJ.

$$\Psi_{gate} = \frac{P_{gate}}{f_{oup}} \approx E_{dyn\,cyc\,gate} = (1+\sigma)C_{gate}U_{dd}^2 \approx 1.2\,C_{gate}U_{dd}^2 \qquad (9.14)$$

Before comparing such numbers or calculating with them, it is extremely important to understand a vendor's tacit assumptions about output loads, ramp times, node activities, contributions from input and clock activities (included or excluded), definition of toggle rate (equal to or two times frequency), and static currents (neglected or included in P_{gate}).

---

[9] It is needless to say that drive conflicts cause significant crossover currents, they should be avoided anyway. Examples of soft-switching pad drivers are given in section 10.4.

Examples

Ψ_{gate} figures disclosed by Alcatel Microelectronics state the overall energy spent on cycling the output of a cell with no external load attached and seem to conform with (9.14).

Older datasheets issued by austriamicrosystems indicated 1.34 µW/MHz for an inverter with 1x drive strength operated at 3.3 V and fabricated in 600 nm CMOS technology. This figure is by no means comparable to $E_{dyn\,cyc\,gate}$, though, as all library cells were characterized for an assumed output load of 100 fF and with input ramps of 1 ns. Also, input capacitances were not lumped to the output. Chargµing and discharging a capacitance of 100 fF at a frequency of 1 MHz calls for 1.09 µW. As an inverter has no dependent inner nodes, the difference of 0.25 µW must get spent in driving the gate's own output and Miller capacitances and in crossover currents.

A product summary of IBM's Cu-11 ASIC library for their 130 nm CMOS 8SF process characterizes the typical power dissipation per logic gate as 0.005 µW/MHz. While not unlikely when averaged over an entire chip, this indication is not directly comparable with $E_{dyn\,cyc\,gate}$ either as it rests on an assumed output activity $\alpha_{oup} = 0.1$.

□

The charge/discharge and crossover currents of fig.9.1 have one thing in common: There is no dissipation without switching. Most of the energy dissipated by past and present CMOS circuits is in fact absorbed by these two dynamic phenomena. Yet, there also exist two dissipation mechanisms of static nature.

9.1.3 Resistive loads

Static currents result from the presence of DC paths from VDD to VSS, see fig.9.5a for an example. In theory, the electrically dual transistor networks that are the trademark of CMOS logic offer no such current paths, yet notorious exceptions exist. These involve

- Pseudo NMOS/PMOS subcircuits (included in many RAMs, ROMs, and PLAs),
- Any kind of amplifier (including read and regeneration amplifiers in memories),
- Current sources and current mirrors,
- Voltage dividers, voltage converters, and voltage regulators,
- Oscillators, clock generation and conditioning subcircuits, transceivers,
- Continuous servo loops (such as PLLs, DLLs, and other regulators),
- Low-swing current logic (sometimes found in RF prescalers and front ends),
- LVDS receivers/transmitters and other current-mode I/O subcircuits,
- Termination resistors (whether on-chip or not),
- On-chip loads of resistive nature (such as passive pull-up/downs),
- Off-chip loads (TTL, bipolars, LEDs, relays, etc.), and possibly also
- ESD protection structures.

Observation 9.4. *Pure CMOS logic circuits provide no direct current paths from VDD to VSS. Departures abound, however.*

When relating resistive dissipation to a single computation cycle, one obtains

$$E_{rr\,k} = P_{rr\,k}\,T_{cp} = \frac{U_{dd}^2}{R_k}\,\delta_k\,T_{cp} \tag{9.15}$$

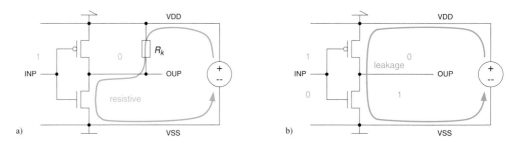

Fig. 9.5 Resistive load current (a) and drain–source leakage (b) in a stationary CMOS inverter.

where R_k indicates the load resistance and δ_k the **duty cycle**, that is the average proportion of time during which it is being energized. A pull-up resistor hooked up to some data signal that is logic 1 for three quarters of the time, for instance, has a duty cycle δ_k of 0.25.

9.1.4 Leakage currents

Though normally minute, four phenomena contribute to leakage in a bulk CMOS process.

- Subthreshold conduction in MOSFET channels that are turned off, $I_{ds\,off}$.
- Leakage currents through reverse-biased drain–bulk and source–bulk junctions, $I_{db\,rev}$.
- Leakage currents through reverse-biased well–well and/or well–substrate junctions, $I_{bb\,rev}$.
- Electron tunneling through the gate dielectric $I_{g\,tun}$, also known as gate leakage.

$$I_{lk} = \sum_{chip} (I_{ds\,off} + I_{db\,rev} + I_{bb\,rev} + I_{g\,tun}) \tag{9.16}$$

Leakage currents have always been critical in DRAMs and other dynamic circuits where charge retention is absolutely essential. From a power perspective, however, leakage traditionally contributed close to nothing to a chip's overall energy dissipation as the other effects discussed in sections 9.1.1 through 9.1.3 used to predominate in almost all applications.

$$E_{lk} = P_{lk}\,T_{cp} = U_{dd}\,I_{lk}\,T_{cp} \ll E_{ch} + E_{cr} + E_{rr} \tag{9.17}$$

Wristwatches have always been a notable exception since these must operate from a tiny battery on the order of 25 or 30 mAh for a very long time, thereby mandating an average total power dissipation below 1 μW. The picture has begun to change with the 250 nm and later technologies, and leakage power has now become a widespread concern for several reasons.

For one thing, even minute leaks add up to a substantial stand-by current when zillions of MOSFETs are involved. Note that off-state currents not only drain batteries and contribute to heat, but also compromise I_{ddq} testing.

For another thing, subthreshold current exponentially depends on the threshold voltage, and $U_{th} = U_{th\,n} = -U_{th\,p}$ has continually been lowered in search of better performance from lower supply

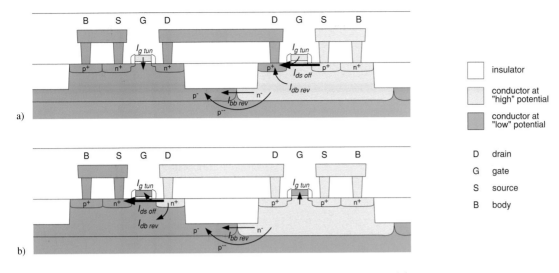

Fig. 9.6 Leakage paths in a bulk CMOS inverter with output 0 (a) and output 1 (b).

voltages as will be explained shortly in section 9.1.6. Starting from (8.59) and assuming that all transistors are of the same length L, we can estimate

$$I_{lk} \approx \sum_{g=1}^{G} I_{ds\,off\,g} \approx \frac{\Delta I_{ds\,off}}{\Delta W} \sum_{g=1}^{G} W_g \propto 10^{\frac{U_{gs\,off} - U_{th}}{S}} \sum_{g=1}^{G} W_g = 10^{\frac{-U_{th}}{S}} \sum_{g=1}^{G} W_g \qquad (9.18)$$

where $I_{ds\,off\,g}$ stands for the leakage current of gates g and G for the total number of gates in a circuit. $\frac{\Delta I_{ds\,off}}{\Delta W}$ essentially expresses a MOSFET's leakage current per unit width, while W_g denotes the width of a representative transistor in each logic gate and scales with the relative drive strength of that gate.[10] S is the subthreshold slope defined in (8.60). The final equality in (9.18) implies $U_{gs\,off} \equiv 0$, which is indeed true for standard CMOS logic.[11]

For a third thing, subthreshold current $I_{ds\,off}$ strongly grows when a chip heats up during operation. This is because S has been found to be proportional to the absolute temperature θ_j in (8.60). As an example, the power wasted due to leakage in a 100 nm top-performance microprocessor operating from a 0.7 V supply has been reported to grow from 6% or so to 127% of the (fixed) dynamic power as junction temperature rises from 30 to 110 °C.

---

[10] This is to say that a 4x drive gate leaks four times as much as a 1x gate, for instance. Interestingly, the leakage current of a logic gate averaged over all possible input states is roughly the same as the subthreshold current of one of its MOSFETs, which makes (9.18) workable as a first-order approximation. A mathematical model that accounts for threshold variations across a die is proposed in [242].

[11] But not necessarily for more sophisticated circuit structures; see fig.9.14 for a counterexample.

Table 9.2 Typical subthreshold currents for two CMOS processes.

process generation	junction temp. θ_j [°C]	off-current $\frac{\Delta I_{ds\,off}}{\Delta W}$ n-channel [pA/μm]	p-channel [pA/μm]	common measurement conditions
250 nm	25	3	2	$W = 10$ μm, $L = 240$ nm
	85	88	35	$U_{ds} = 2.5$ V, $U_{gs} = 0$ V
	125	420	190	
130 nm[a]	25	18 000	12 000	$W = 10$ μm, $L = 130$ nm
	85	89 000	62 000	$U_{ds} = 1.2$ V, $U_{gs} = 0$ V
	125	103 000	137 000	

[a] The manufacturing process from which the data are taken provides MOSFETs with three distinct threshold voltage levels, the numbers given refer to the high-speed (i.e. low-threshold) variety.

9.1.5 Total energy dissipation

We are now in a position to put together all four contributions that make up the total amount of energy dissipated by a CMOS circuit during a computation cycle.

$$E_{cp} = \sum_{k=1}^{K}(E_{ch\,k} + E_{cr\,k} + E_{rr\,k}) + E_{lk} \approx$$

$$U_{dd}^2\,(1+\sigma)\sum_{k=1}^{K}\frac{\alpha_k}{2}C_k + T_{cp}\left(U_{dd}^2\sum_{k=1}^{K}\frac{\delta_k}{R_k} + U_{dd}\frac{\Delta I_{ds\,off}}{\Delta W}\sum_{g=1}^{G}W_g\right) \quad (9.19)$$

This finding tells us that the overall amount of energy that gets dissipated in carrying out some given computation roughly grows with U_{dd} squared, which gives rise to the subsequent observation.

Observation 9.5. *The most important single factor that affects the energy efficiency of full-swing CMOS circuits is supply voltage.*

The other enemies of energy-efficient CMOS circuits are unproductive node activities, oversized transistors or drive strengths, and excessive loads, both resistive and capacitive.

Numerical example

Consider a CMOS IC of modest size, say of 50 000 gate equivalents, which is to be manufactured in a 130 nm technology, operated at 1.2 V, and driven from a 100 MHz single-edge-triggered one-phase clock. Overall node activity is $\alpha = \frac{1}{4}$, which means that the average node charges and discharges within eight clock cycles. Let each of the 100 000 or so internal nodes have a capacitance of 18 fF. The circuit is to drive 16 off-chip loads of 25 pF and 3.3 kΩ each that toggle at the same rate as the core nodes do. These 2.5 V outputs equally share their time between the "on" and "off" conditions. Further assume that the MOSFET off-state current is on the order of 80 nA/μm at 70 °C junction temperature with an average n-channel width of 1 μm.

How much energy is being dissipated per computation cycle and what is it spent on?

$$E_{cp} \approx U_{dd}^2 \left(1 + \sigma\right) \sum_{k=1}^{K} \frac{\alpha_{k \, on\_chip}}{2} C_{k \, on\_chip} + U_{bb}^2 \left(1 + \sigma\right) \sum_{k=1}^{K} \frac{\alpha_{k \, off\_chip}}{2} C_{k \, off\_chip} +$$

$$T_{cp} \left(U_{bb}^2 \sum_{k=1}^{K} \frac{\delta_{k \, off\_chip}}{R_{k \, off\_chip}} + U_{dd} \frac{\Delta I_{ds \, off}}{\Delta W} \sum_{g=1}^{G} W_g \right) =$$

$$(1.2 \text{ V})^2 (1 + 0.2) \sum_{k=1}^{100\,000} \frac{\frac{1}{4}}{2} 18 \text{ fF} + (2.5 \text{ V})^2 (1 + 0.2) \sum_{k=1}^{16} \frac{\frac{1}{4}}{2} 25 \text{ pF} +$$

$$10 \text{ ns} \left((2.5 \text{ V})^2 \sum_{k=1}^{16} \frac{\frac{1}{2}}{3.3 \text{ k}\Omega} + 1.2 \text{ V} \cdot 80 \frac{\text{nA}}{\mu\text{m}} \sum_{g=1}^{50\,000} 1 \, \mu\text{m} \right) =$$

$$389 \text{ pJ} + 375 \text{ pJ} + 152 \text{ pJ} + 48 \text{ pJ} = 964 \text{ pJ} \approx 0.96 \text{ nJ} \qquad (9.20)$$

where the four terms stand for contributions due to on-chip capacitive loads, off-chip capacitive loads, off-chip resistive loads, and leakage respectively. Note that the first two terms relating to dynamic dissipation outweigh the two static ones for the computation rate assumed in this example. Yet, this is not always so, as we will learn shortly.

What is the overall power consumption?

$$P = f_{cp} \, E_{cp} \approx 100 \text{ MHz} \cdot 0.96 \text{ nJ} = 96 \text{ mW} \qquad (9.21)$$

Observe that not all of this power is being dissipated within the chip as the resistive loads are external. They do not, therefore, contribute to heating up the driving circuit directly, and there is no need to account for them when selecting an adequate package and heat sink for the IC.[12]

□

Figure 9.7 shows the energy spent per computation cycle as measured on a microprocessor. The anticipated impact of U_{dd} on E_{cp} is clearly visible. Further observe that the energy per operation — not to be confounded with power dissipation — decreases with clock frequency. While this may appear unexpected at first sight, it is in fact easy to understand from the presence of the term T_{cp} in (9.19) and (9.20). The dynamic energy always remains the same, but the longer a computation cycle, the more static energy gets wasted on leakage currents and other unproductive DC contributions.

Observation 9.6. *Quiescent currents most burden the energy balance of those electronic appliances that operate in short bursts but do not allow for complete shut-down in between.*

9.1.6 CMOS voltage scaling

While observation 9.5 is a compelling argument in favor of the lowermost voltage levels, the matter is not quite as simple as it may appear.

The supply voltage of digital CMOS ICs has long been maintained at $U_{dd} = 5$ V for compatibility reasons and — in later years — also to maximize switching speed by operating the MOSFETs with ever stronger fields. Beginning with the 350 nm process generation, ever thinner gate oxides

---

[12] Heavy off-chip loads have an indirect impact, however, by inflating voltage drops and heat generation in the driver circuits and by increasing the ambient temperature.

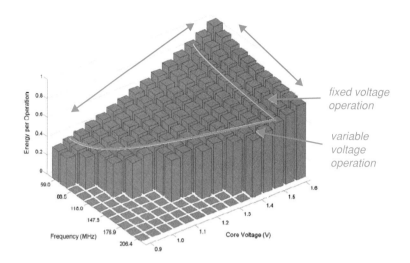

Fig. 9.7 Energy per operation as a function of supply voltage and clock frequency for a StrongARM SA-1100 processor (photo copyright IEEE, reprinted with permission from [243] with permission).

have made it necessary to lower supply voltage accordingly in order to stay away from dielectric breakdown.[13] Constant-field scaling has since then been adopted by necessity.

The velocity saturation that also came to bear in deep submicron circuits was yet another argument for moving in the same direction. Recall from (8.37) that CMOS gate delays are given by

$$t_{pd} \propto \frac{C_k U_{dd}}{(U_{dd} - U_{th})^\alpha} \tag{9.22}$$

because a MOSFET's maximum drain current grows with the difference $U_{dd} - U_{th}$ rather than with the absolute voltage U_{dd}. With carrier velocity saturated, that is with $\alpha \approx 1$ instead of the bygone $\alpha \approx 2$, the accelerating effect of strong lateral fields is largely lost. The predominant effect of a strong **overdrive** $U_{th} \ll U_{dd}$ is to inflate energy losses with just a minor gain in speed. Ideally, one would want to scale supply and threshold voltages proportionally in order to maintain drive currents and hence also switching speed. For the 65 nm CMOS generation, the overall power dissipation (dynamic plus leakage) has been found to be minimal at a supply voltage of 0.5 V [247].

Unfortunate difficulties prevent industry from continuing along this path. As stated in (9.18), sub-threshold current is exponentially dependent on threshold voltage U_{th}. Below a certain point, "off"-currents literally explode. As a rule, applications where off-state currents are very critical mandate threshold voltages of 0.6 V or more, while leakage tends to become generally unacceptable when the thresholds fall below 0.3 V or so.

---

[13] The dielectric strength of amorphous SiO_2 is approximately 500 kV/mm. For comparison, dry air is considered a safe insulator for 1 kV/mm while the Kapton and Teflon materials being used in high-voltage cables support fields on the order of 10 kV/mm.

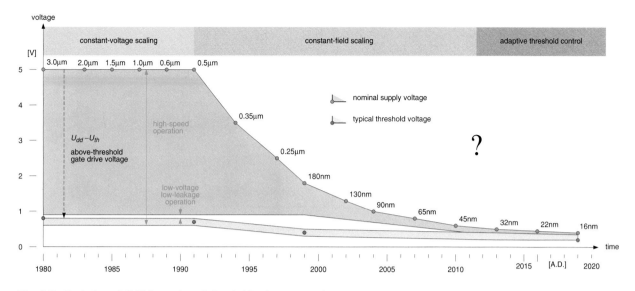

Fig. 9.8 Evolution of CMOS supply and threshold voltages over the years.

With a more ample threshold voltage, on the other hand, switching speed rapidly deteriorates when the supply voltage U_{dd} is scaled to below $3U_{th}$. Table 9.3 illustrates this dilemma from a process developer's view.

Table 9.3 Scaling options when moving from one CMOS process generation to the next.

Parameter	Symbol	Option 1	Option 2
gate dielectric thickness	t_{ox}	scaled down	scaled down
supply voltage	U_{dd}	scaled down	scaled down
threshold voltage	U_{th}	scaled down	constant
leakage current	$I_{ds\,off}/W$	greatly inflated	roughly as before
current drive	I_{d0}/W	scaled down	greatly reduced
voltage swing	$U_{oh} - U_{ol} = U_{dd}$	scaled down	scaled down
operating speed	t_{pd}, t_{su}, etc.	as before[a]	deteriorated

[a] Pessimistically assuming identical load capacitances.

To make things worse, the same exponential dependency renders "off"-currents extremely sensitive to threshold voltage variations. Whether a fabricated part meets the specifications depends on relatively minor and difficult-to-control processing tolerances, which puts yield at risk and ultimately frustrates further voltage scaling.

Observation 9.7. *Subthreshold conduction has become a major concern for CMOS circuits because of the depressed MOSFETs threshold voltages along with their precarious variability.*

As a stopgap, today's fabrication processes provide circuit designers with MOSFETs of two or even three distinct threshold voltages to support optimum adaptation to a variety of needs, see fig.9.9. Circuit techniques that take advantage of multiple threshold voltages to fight leakage will be the topic of section 9.2.3. Voltage scaling is discussed in [496] while [246] is more concerned with the limits of CMOS low-voltage operation.

9.2 | How to improve energy efficiency

9.2.1 General guidelines

A good starting point for improving energy efficiency is (9.19). What follows is a compilation of practical measures with the most effective steps listed first.

POWER BUDGETING

Find out where energy goes in the target application. Take into account the various operating modes along with their relative frequencies of occurrence. Make the individual contributions explicit and decide where lowering dissipation pays most.

Example

Assume you are designing a baseband processor IC for a mobile phone. Before taking any major design decisions, you would want to break up overall energy dissipation as follows.

- Hardware subsystems: radio frequency, intermediate frequency, baseband processing, A/D and D/A conversion, audio section, display, backlighting.
- Baseband computation: speech processing, ciphering, channel coding, antenna combining.
- Operating modes: talk & transmit, receive & listen, standby = contacting base station at regular intervals, off = timers continue to work.
- Subcircuits: on-chip RAMs, datapaths, controllers, clocking generation and distribution, output drivers; clock and supply domains; random logic, bistables, interconnect.

☐

Leakage is most pressing for those battery-operated circuits that sit idle for much of the time, and to which we refer as **low-activity circuits**. This is the case when the clock period extends much beyond the long path delay $T_{cp} \gg t_{lp}$ such as in a watch circuit, for instance. This is also the case for appliances that operate with a low duty cycle because they are busy during occasional bursts but require that most of their VLSI circuits remain powered up for prolonged periods of time in between. While static currents may be acceptable during operation, they are tantamount to draining batteries for nothing whenever the circuits idle in standby mode. With a cellular phone or a pager, this is actually the case for most of the time.

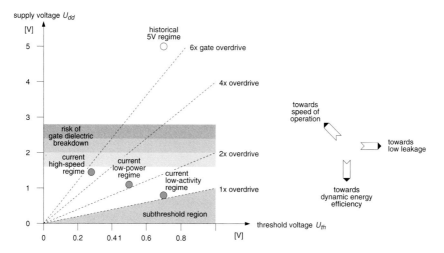

Fig. 9.9 Options and tradeoffs for selecting supply and threshold voltages.

PROCESS AND SUPPLY VOLTAGE SELECTION

- Downscaled fabrication processes not only reduce parasitic capacitances in comparison with older processes but typically also permit one to operate circuits from lower supply voltages. Watch out for static leakage currents, though.

- Deciding on supply and threshold voltages for digital CMOS logic is always a compromise between conflicting goals such as fast circuit operation, low energy losses per node toggling (dynamic dissipation), low leakage current (static dissipation), low sensitivity to process variations, reliability, and electrical compatibility [490]. Before opting for specific supply and threshold voltages, get the activity profile and other energywise boundary conditions of your application straight, see fig. 9.9.

  - High-speed design:
    Ensure vigorous current drive by maintaining a healthy gate overdrive of $4U_{th} \leq U_{dd}$.
  - Low-power design:
    Maximize efficiency by operating the circuit from a voltage no higher than what is necessary to meet speed requirements. The optimum choice is where the increase in leakage energy compensates for the savings in dynamic energy, and vice versa.
  - Low-activity design:
    Minimize the energy dissipated by static currents using high-threshold MOSFETs. Making their channels somewhat longer than minimum reduces leakage further.

- Design for the lowest possible supply voltage compensating unacceptable losses of throughput with a faster architecture and a better arithmetic/logic design. For an overview of the options available, refer to sections 2.9 and 9.5.1.

- If the computational burden is subject to important variations over time, you may want to opt for **dynamic voltage and frequency scaling**, a scheme whereby both supply voltage and clock frequency get adjusted as a function of the speed requirement at that time.[14] See fig.9.10 for a block diagram and note the added complexity.

In an open-loop approach, the controller uses some kind of predefined look-up table $U_{dd} = f(f_{clk})$ to match supply voltage and clock frequency. A safety margin ensures correct operation in spite of process, temperature, and local variations.

Closed-loop operation, in contrast, admits and detects occasional timing violations and uses them as error signals to adjust the supply voltage [246]. As an extra benefit, process and temperature variations are inherently compensated for.

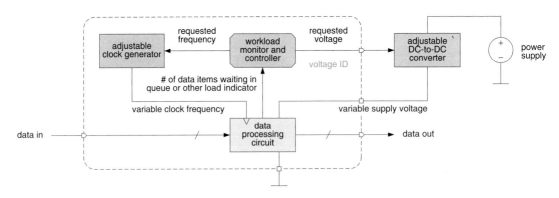

Fig. 9.10 Dynamic voltage and frequency scaling (simplified).

9.2.2 How to reduce dynamic dissipation

Lowering dynamic dissipation remains a top concern in all high-speed and low-power designs. The goal is to lower the sum over all node activities in a circuit weighted with the respective node capacitance.

REDUCE NODE ACTIVITIES AT THE ALGORITHM AND ARCHITECTURE LEVELS

- Minimize the overall computational effort for the data processing required. Check to what extent it is possible to relax the requirements on throughput, numerical precision, perceived audio/picture/video quality, coding gain, response time, or whatever figure of merit is applicable, in exchange for better energy efficiency.

- Check to what extent it is possible to confine flexibility requirements in exchange for better energy efficiency. Program-controlled general purpose processors have been found to waste orders of magnitude more energy than dedicated architectures.[15] Try to find a good compromise between the flexibility of a software-programmed processor and the more modest node activities of a hardwired architecture.

---

[14] Transmeta was first to implement this concept in its Crusoe processors under the name LongRun.

[15] The reasons why have been explained in section 2.4.8, please see there for further details.

- Prune or simplify all activities that do not directly contribute to data processing (e.g. control flow, multiple modes of operation, instruction fetch and decode, address computation, caching, multiplexing, branch prediction, speculative execution, concurrent testing).

- Carefully optimize all word widths and storage capacities involved (data widths, cofficients, filter orders, addressing capabilities, ALUs, register banks, all sorts of busses, etc.). The goal is to minimize overall switching energy as much as possible while keeping overall implementation losses within acceptable bounds. Pay special attention to off-chip busses. Automatic rescaling of data — periodic or demand-driven — helps to to manage with a narrower datapath in certain applications.

- Avoid off-chip communication wherever possible, keep data exchange local instead. Use interface protocols that entail as little activity overhead as possible.

- Put subcircuits that remain inactive for prolonged periods of time into a **sleep mode** by temporarily disabling the clock and/or by selectively turning off the supply voltage. Note that the former approach preserves circuit state while the latter does not. Consider a multi-processor architecture where a low-power processor monitors low-rate protocol and interface requests and selectively activates more specialized high-performance hardware units when bulk data need to be processed.

- Stay away from DRAMs with their memory refresh cycles. Check RAMs and ROMs for their dynamic <u>and</u> static currents.

- Evaluate the impact of splitting large memories into smaller chunks. While overall access rate remains the same, the parasitic capacitances associated with reading and writing are bound to grow with memory size.

- Check whether carrying out arithmetic operations on data encoded with a different number representation scheme might help to lower the overall node activities (consider block-wise floating point, canonic signed digit, sign-and-magnitude, etc.).[16]

Example

Figure 9.11 refers to a subsystem from telecommunications, more precisely to a channel estimator for an IEEE 802.11a OFDM WLAN[17] receiver. A method called "repeated interpolation" can be employed to filter white Gaussian noise from estimates of the channel's transfer function. The main computational burden comes from repeated matrix–vector multiplications that ask for a large number of concurrent multiply/accumulate units in hardware. Figure 9.11 plots the mean square error (MSE) over a relevant range of signal to noise ratios (SNRs) for six channel estimates computed with different levels of accuracy.

---

[16] The ubiquitous 2's complement (2'C) number representation scheme, for instance, has most bits of a data word flip whenever the numerical value changes from positive to negative, or back. As audio and many other real-world signals tend to vary around zero with small amplitudes for much of the time, this is clearly suboptimal from the point of view of energy efficiency. Sign-and-magnitude (S&M) representation can cut overall dissipation by more than 25% in transversal filter circuits [239], yet experience has also shown that too many conversions between 2'C and S&M formats tend to render this approach ineffective in other signal processing applications.

[17] Orthogonal Frequency Division Multiplexing Wireless Local Area Network.

The lowermost curve refers to noise filtering with full precision (using MATLAB), the topmost curve to unfiltered channel estimation. Intermediate curves visualize the impact of fixed-point arithmetics. Whereas degradation is unacceptable when product terms are truncated from 19 to 7 bits, calculating with 10 bits yields almost the same result as the floating-point reference model. This finding has not only permitted one to save 36% of the area for the matrix–vector multiplication circuitry, but has also been shown to reduce dynamic power by one third when compared with an analogous circuit that calculates with product terms of 19 bits [247].

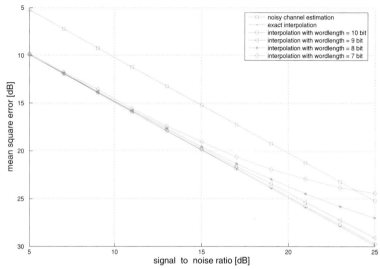

Fig. 9.11 Results from wordlength analysis of a OFDM channel estimator.

☐

Be aware of the fact that much of the power dissipation of a VLSI circuit is committed by the time the RTL design is finished; the beneficial impact of lower-level optimizations is relatively minor. Please refer to section 2.3 *et seqq.* for more material on how high-level design decisions affect energy efficiency.

ENERGY-EFFICIENT CLOCKING

- Clocking a register dissipates energy regardless of the extent of state change because each clock edge causes nodes within each bistable to toggle, even when the register is disabled or when there are no fresh data to act upon. Energy thus gets wasted whenever substantial portions of a design are characterized by low data activities in conjunction with an unnecessarily fast clock. If so, consider the options below.

  - Distribute computations more evenly over time, possibly combined with lowering the clock frequency.
  - Resort to clock gating for registers with enable.
  - Use multiple clock frequencies. Derive all distinct frequencies from a common master clock, however, in order to keep the various clock domains synchronized.

- Asynchronous prescalers or clock dividers. Such departures from synchronous operation are acceptable only if strictly confined to within a local and fairly simple subcircuit.[18]

- Referring to single-edge-triggered designs, note that power dissipation slightly varies with the clock's duty cycle in glitch-intensive circuits where many path delays come close to the clock period at which the circuit is being operated. This is because part of the nodes within each flip-flop are bound to toggle with the D input while the master latch sits in pass mode during the second phase — the one that immediately precedes the active edge — in every clock period. While the impact typically is just 1% or 2% of the overall power, one can easily avoid unsymmetric clock waveforms that cause those nodes to toggle more than necessary as a consequence of intense glitching in the upstream logic.

- As a more radical change, you may want to consider the selective replacement of master–slave flip-flops by pulse-clocked latches where the situation permits.[19]

- Slow ramps are a side effect of operating circuits from very low voltages, yet sharp clock waveforms have been found to be important in edge-triggered circuits because these are highly vulnerable to clock skew.[19] Unfortunately, generous clock buffering — in terms of both buffer count and sizing — translates into extra energy dissipation. Level-sensitive two-phase clocking schemes with non-overlapping phases are more forgiving. They can thus be made to work with lighter clock buffering and still tolerate more skew and jitter. [248], which analyzes a RISC processor design, reports significant energy savings over single-edge triggering. Also, clock currents can be distributed better over a clock period. Do not exaggerate, however, as overly slow clock ramps cause energy to be wasted in crossover currents. What's more, datasheets and simulation models of bistables do not normally account for excessively sluggish clock waveforms.

- Dual-edge-triggered one-phase clocking is an option if much of the overall energy is spent for clocking. How the flip-flops are constructed internally matters.

REDUCE NODE ACTIVITIES AT THE REGISTER-TRANSFER AND LOGIC LEVELS

- Do not feed data that have propagated along paths of largely different delays into a combinational subcircuit of substantial depth as this promotes intense glitching. Chains of multipliers and/or adders are notorious examples, see fig.9.2. Try to rearrange circuits so as to prevent nodes from unnecessarily switching back and forth by better **delay balancing**.[20] The associativity transform often helps.

- Consider a large multiplier or a cascade of combinational subcircuits as shown in fig.9.12a. Any data change at any input causes a wave of toggling activities to propagate through the circuit

---

[18] Such as a clock divider in a wristwatch or a prescaler in a PLL.

[19] See problem 6 in section 6.7.

[20] A related technique consists in reordering (data and control) inputs so as to confine the impact of glitches to small portions of a circuit (largely stable inputs upstream, glitchy inputs downstream). With the same general intention, [249] reports that power dissipation may be lowered by an average of 11% over normally optimized logic networks by selectively adding redundant gates or redundant inputs to existing gates in logic circuits following technology mapping. [250] proposes filtering out glitches before they cause energy to be lost on signal ramping by deliberately downsizing the transistors of every logic gate as a function of the timewise offset present at the input of that gate. Energy savings come at the cost of a longer path delay, however.

before nodes settle to a new state. Unless the result is being stored in a register for further processing in each and every clock cycle, unnecessary switching occurs and energy gets wasted. This can be avoided by way of **signal silencing**, a technique whereby a circuit is isolated from glitches and irrelevant transitions at the input until the downstream logic indeed accepts a new data item. Silencing can be obtained with a bank of simple gates, of three-state gates, or of latches, or by disabling existing upstream registers. The latter approach is combined with delay balancing and clock gating in fig.9.12b. Similarly, prevent heavily loaded busses from switching while they sit idle.

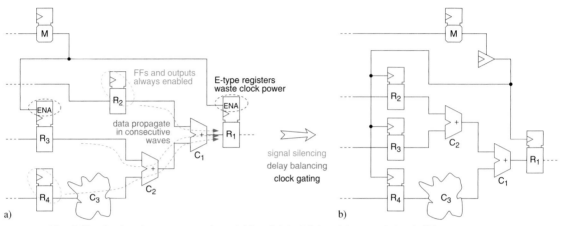

Fig. 9.12 Cutting down excess node activities. Original (a) and improved circuit (b).

- Adopt suitable **bus encoding** schemes on busses that are wide, heavily loaded, and highly active; I/O busses prove especially rewarding. **Bus invert coding** uses $w + 1$ electrical lines to represent w bits of information. The extra bit always tells the receiver whether the original bits are being transmitted in direct or in inverted format. For each clock cycle, the transmitter calculates the Hamming distance between the past and the present data word and selects either format such as to minimize the number of toggles on the bus. More sophisticated bus invert coding schemes require no extra line but use the presence or absence of an extra data toggle in the middle of the clock period to signal the present transmission format [251].

  In the case of an address bus, consider adding an increment bit to the bus. For consecutive addresses, just assert the increment line and have the receiving end compute the new address locally; use the parallel load facility only for non-consecutive addresses. This technique is more energy efficient than even unit-distance coding, provided memory addressing predominantly occurs in ascending order.[21]

- Select appropriate data and state encoding schemes. A general idea is to assign similar codes to those subsets of states or numbers that alternate most frequently.

---

[21] More sophisticated encoding schemes that further take advantage of correlations between consecutive patterns are discussed in [252] [253]. Also note that low-activity bus encoding also benefits switching noise reduction.

- If a wide datapath is indispensable but not being fully exploited throughout a computation run, try to temporarily disable the operation of those bit slices that do not contribute significantly to certain intermediate results.

CUT DOWN PARASITIC EFFECTS AT THE ELECTRICAL AND PHYSICAL LEVELS

- Again, avoid excessive capacitive loads by not going off-chip with highly active signals.

- Avoid resistive loads in bus systems, static pull devices, I/O pads, off-chip loads, RAMs, ROMs, PLAs, PLLs, amplifiers, voltage converters, and the like.

- Avoid the general adoption of cells with overly strong outputs. Profile ramp times and use the smallest acceptable drive strengths except where delay is critical.

- Where heavy buffers are unavoidable, that is primarily on output pads and clock drivers, use break-before-make drivers to keep crossover currents small.

- Basically, the smaller the MOSFETs, the lower the gate and node capacitances and, hence, the dynamic energy dissipated. So, if given the freedom to design subcircuits at the transistor level, try to downsize MOSFETs wherever speed requirements permit. Size p-channels barely wider than their n-channel counterparts and accept the sacrifice of electrical symmetry. Yet, beware of short- and narrow-channel effects that tend to inflate leakage currents in minimum-sized MOSFETs. Use simulation and calibrated transistor models to find a good compromise.

- Most ASIC designers work with cell libraries and are not given the freedom to opt for a circuit style other than static CMOS. [254] has evaluated a total of 23 circuit styles and concluded that static CMOS is a good choice for most applications — and so has the rest of the world. Pure transmission gate logic makes sense only where delay is uncritical, but circuit styles that combine t-gates with level-restoring gates may sometimes offer better tradeoffs between performance and energy efficiency [255] [256].

- Prefer energy-efficient library cells where available, paying particular attention to latches or flip-flops (depending on clocking discipline). The relative activities of data signals and (gated) clock may influence your choice. Combine logic gates and bistables into function latches or function flip-flops where possible.

- While half-voltage swing may be an option for clock nets, see section 9.5, stay clear of steady-state crossover currents by fully driving normal CMOS inputs to VDD or VSS.

- Try to trim node capacitances by better circuit and layout design. Clock nets deserve particular attention because of their high node activities.

- Make adjacent metal layers run perpendicularly to each other so as to maximize the vertical separation between any two parallel signal lines. To reduce lateral capacitances, you may want to extend horizontal separations beyond the minima stipulated by layout rules. Allowing wires to run at angles of $45°$ (Boston geometry) has been reported to reduce total wire length by about 20%.

- Shuffling the wires in a long bus helps lower the maximum coupling capacitance between bits. If bits are known to exhibit patterns of correlation, route strongly correlated bits next to each other and do the exact opposite for strongly anticorrelated bits.

- Consider tiled layout as an alternative to cell-based synthesis for datapath circuits.

9.2.3 How to counteract leakage

As found earlier, the growth of leakage currents from one process generation to the next is imposed on us as by device physics as a consequence of constant-field scaling, so VLSI designers have begun to look for alternatives at the architecture, arithmetic/logic, and circuit levels.[22]

Fast architectures built from low-leakage cells

As stated in (9.18), leakage current is exponentially dependent on threshold voltage U_{th}. Most fabrication processes thus provide MOSFETs with two or even three distinct threshold voltages to accommodate different needs. A secondary effect that allows one to trade current drive for low leakage during library design is the fact that leakage diminishes with channel length L. Standard cells typically come in two or more varieties. High-speed cells feature superior current drives but their leakage suffers from the low threshold voltages and minimum length channels of their MOSFETs while the opposite is true for low-leakage cells.

The simplest way to avoid excessive leakage is to use as few low-threshold MOSFETs as possible. As an example, inserting pipeline registers into long signal progation paths can make it possible to obtain the same throughput from slower cells, thereby lowering overall leakage in spite of a higher transistor count. Similarly, leakage is likely to benefit from replacing a ripple-carry adder built from high-speed cells by a faster structure built from low-leakage cells.

The countermeasures presented below are more demanding and should be considered only when ambitious performance goals mandate a widespread adoption of low-threshold logic.

Variable-threshold CMOS (VTCMOS)

Reconsider fig.9.9 and observe that it would be highly desirable to alternate between two distinct points, namely between a low-threshold regime while busy and near-perfect cutoff while at rest. Variable-threshold CMOS does so by taking advantage of the body effect to adjust the effective threshold voltage $U_{th}(U_{bs})$ on the fly as a function of the current mode of operation, see fig.9.13. This explains why the technique is also known as dynamic back-biasing (DBB).

As an additional benefit, a self-adjusting servo loop can be designed to compensate for fabrication tolerances of the thresholds that become particularly critical with today's low voltage levels [257]. What's more, VTCMOS can be combined with dynamic voltage and frequency scaling.

Example

VTCMOS was pioneered by Kuroda and his colleagues on a low-voltage low-power high-performance Discrete Cosine Transform (DCT) circuit of 120 000 or so transistors [258]. Fabricated in 900 nm 2M CMOS technology, operated with a 150 MHz clock, and dissipating no more than 10 mW at

---

[22] Fabricating ultra-thin-body or dual-gate MOSFETs in silicon-on-insulator (SOI) substrates are alternatives on the grounds of fabrication technology and discussed elsewhere in this text.

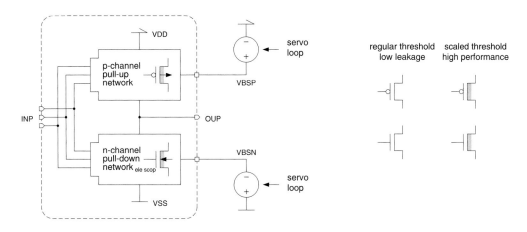

Fig. 9.13 Variable-threshold logic, a technique that adjusts MOSFET thresholds by dynamic back-biasing (simplified).

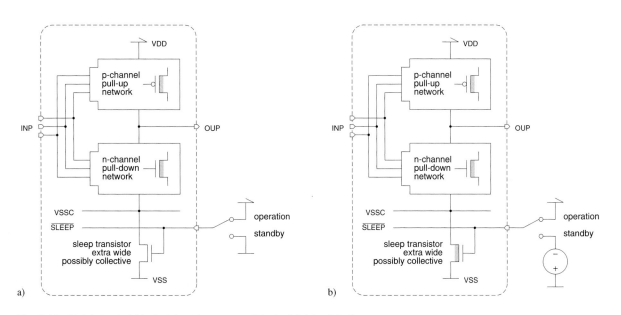

Fig. 9.14 Multi-threshold logic (a) and super cutoff logic (b) (simplified).

0.9 V, their design nicely demonstrated the effectiveness of DBB in bulk CMOS VLSI. The on-chip back-biasing circuitry, which includes the charge pumps and current sources required to charge and discharge the p- and n-well capacitors under control of leakage current monitors, occupied 0.19 mm^2 of silicon, less than 5% of the data processing logic. The voltages and leakage currents in this design are collected below.

condition	U_{dd} [V]	U_{th} [V]	U_{bs} [V]	I_{lk}
nominal of process	3.3	0.15 ± 0.1	0	n.a.
effective while busy	0.9	0.27 ± 0.02	-0.5 ± 0.2	0.1 mA
effective in standby	0.9	> 0.5	-3.3	10 nA

☐

A more typical application is in SRAMs to reduce leakage power while in sleep mode [259]. While the body effect in bulk CMOS technology unfortunately becomes less effective with the on-going down-scaling of device geometries [260], VTCMOS has been applied to circuits manufactured in silicon-on-insulator (SOI) technology [261]. This rather unexpected achievement has been obtained with the aid of an ultra-thin buried oxide (BOX) layer that separates the devices from more or less conventional wells underneath that accept the bias voltage. According to the authors, a supply voltage of 0.8 V should be safe with SRAMs in 65 nm technology.

Multi-threshold CMOS (MTCMOS)

A more popular alternative is to use MOSFETs with distinct threshold voltages $U_{th\,logic} < U_{th\,sleep}$. Low-threshold transistors are used in all speed-critical subcircuits. Extra-high-threshold **sleep transistors** connected in series serve to cut the leakage paths while a (sub)circuit is in standby, see fig.9.14a. Multiple thresholds can be obtained in various ways, namely
○ from a true dual-threshold fabrication process where $U_{th\,logic} \neq U_{th\,sleep}$ comes naturally,
○ from back-biasing sleep transistors with a fixed voltage $U_{bs\,sleep} < 0$, or
○ by resorting to a special fabrication process and precharged floating gates.

You may want to refer to [262] for further details on dual-threshold logic. [263] devised an enhancement to make leakage currents more predictable. Each cell comes in two variations that differ in terms of where the sleep transistor is placed: one variation has an n-channel footer, the other a p-channel header. Technology mapping picks one or the other depending on whether the cell's output is mainly at logic 1 or 0 while the circuit waits in suspended state. The goal is to have two or more series-connected MOSFETs in "off"-condition at any time with, in addition, one of them being of low-leakage high-threshold type.

A limitation of MTCMOS is that the sleep transistors severely limit current drive with supplies below 0.7 V or so because of the high threshold voltage necessary to obtain near-perfect cutoff.

Super cutoff CMOS (SCCMOS)

The circuit structure is the same as with MTCMOS but low-threshold MOSFETs are being used throughout, see fig.9.14b. The leaking of the sleep transistors in standby mode is reduced to below $I_{lk} = I_d(U_{gs\,off} = 0)$ by driving them into a super cutoff regime with the aid of a negative gate voltage $U_{gs\,sleep} < 0$. This technique currently appears to have the most promising long-term perspectives because it does not suffer from excessive performance degradation at very low voltages and because it is compatible with silicon-on-insulator (SOI) technology [264] [265].

Example

Intel uses n-channel sleep transistors in SRAM cache memories to shut off subsections of the storage array while not in use. A leakage reduction by a factor of three has been reported.
□

A general problem remains, however, as all bistables lose their state when cut from the supply voltage during sleep mode. Five workarounds are known today:
○ implement important registers from high-threshold MOSFETs and accept their poor speed,
○ refrain from putting important registers into sleep mode and accept their leakage currents,
○ temporarily store their state elsewhere (e.g. flash memory, disk) while in standby,
○ augment each bistable with an auxiliary low-leakage bistable, similar to an SRAM cell and often referred to as a **balloon**, that maintains its state while the principal bistable is put to sleep [266] [267], or
○ use specially designed low-leakage latches and flip-flops.

Triple-S logic

The astute circuit of fig.9.15 works much like any other super cutoff gate. What sets it apart is a pair of additional high-threshold MOSFETs connected in parallel to the sleep transistors. These bypass transistors ensure that the inverter remains operational even while in standby, albeit with a far lower current drive. The inventors of this patented technique have coined the name **triple-S logic** which stands for "smart series switch" [268].

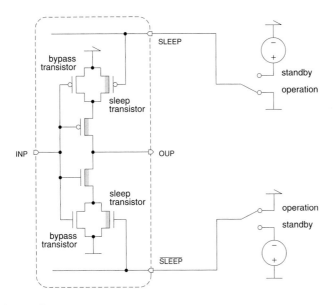

Fig. 9.15 Triple-S (inverter).

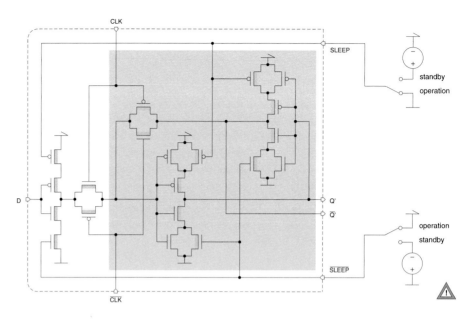

Fig. 9.16 Low-leakage latch built from triple-S inverters (simplified).

A state-preserving latch with hardly any leakage is shown in fig.9.16.[23] Both the feedforward and the feedback inverters are implemented as triple-S circuits. The input inverter is of symmetric super cutoff design to avoid any leakage current across the subsequent transmission gate. Note the absence of VDD to VSS leakage paths while in sleep mode and observe that only low-threshold MOSFETs participate in regular circuit operation. Variations where just the more time-critical forward inverter is implemented in triple-S logic also exist in the literature [269].

[270] has analyzed various MTCMOS and SCCMOS latches and has compared them with conventional CMOS. The authors concluded that (a) the threshold voltages of the low-threshold transistors have to be very low to compensate for the speed that is lost due to the significant circuit overhead, and (b) MTCMOS and SCCMOS are beneficial only when node activities are fairly low. Yet, remember these statements refer to latches, not to combinational logic.

Virtual power/ground rails clamp

The proposal of fig.9.17a combines dynamic back-biasing with cutoff transistors in an elegant way [271]. The circuit remains powered while in sleep mode, albeit via two diodes connected in parallel to the sleep transistors. While the voltage drops across the diodes cause a negative back bias in the logic transistors, so quenching leakage, the remnant supply voltage is sufficient to let the bistables maintain their states. For a 250 nm 2M CMOS circuit operated at 1 V, the authors report a leakage reduction of 94% and a speed degradation of a mere 2%.

---

[23] It goes without saying that observation 8.5 that postulates the insertion of output buffers applies.

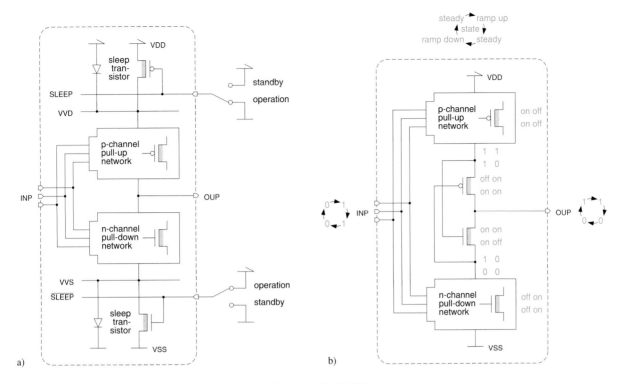

Fig. 9.17 Virtual power/ground rails clamp technique (a) and LECTOR (b).

Self-reverse biasing

Two turned-off MOSFETs connected in series exhibit much less leakage than one transistor alone. Consider a stack of two n-channel MOSFETs, for instance, but note that the same applies for p-channel transistors. Any current flow will cause a voltage drop across the lower of the two and will so inflict a negative bias on the upper transistor $U_{gs\,n} < 0$. As subthreshold current is exponentially dependent on the gate–source voltage, a small bias suffices to quench leakage considerably. Self-reverse biasing works even better with stacks of three or more "off"-state transistors but returns diminish; [272] includes a nice illustration.

Figure 9.17b depicts yet another idea based on the same observation. The two extra leakage control transistors (LECTOR) are in "on"-condition during transients only. Once a steady state is reached, one or the other of them gets turned off, thereby reducing the direct current from supply to ground. Leakage is not exactly zero as the two extra MOSFETs must be of low-threshold type, yet the authors report reductions of 80% with an area overhead of 14% and no speed penalty [273].

For a more comprehensive account of low-power VLSI design in general you may want to consult [241] [274] [275] [276] [277] while [249] [278] [279] discuss specific optimization techniques. The focus in [280] is on reducing leakage with [272] specifically addressing SRAM cache memories. [281] [282] [283] [242] [284] are of interest to those engineers who are given the liberty to redesign circuits at the transistor level and to adjust MOSFET sizes and threshold voltages.

9.3 | Heat flow and heat removal

Heat is generated where electrical power gets converted into thermal power and flows from a zone of higher temperature to a lower-temperature region. The mechanisms that participate in heat transfer are conduction, convection, and radiation, although the latter plays a minor role in the context of VLSI. A convenient way to model their combined effect is to map the thermal flow to an equivalent electrical circuit, see fig.9.18.

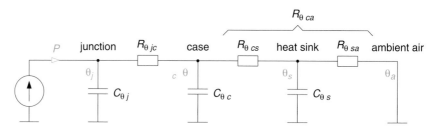

Fig. 9.18 Electrical equivalent circuit for thermal conditions (simplified).

P stands for the heat flow that emanates from a silicon die and is referred to as **thermal design power** (TDP). Because of inertial effects associated with die, package, and heat sink, what is most relevant for the thermal design engineer is the maximum sustained power dissipated by the chip, and this may be less than the maximum short-time peak of electrical power absorbed. In fig.9.18, the inertial effects are modelled with the aid of two thermal capacitances C_θ.

The temperatures of the junction,[24] the case and the ambient medium — typically air — are denoted θ_j, θ_c, and θ_a respectively. The quantities that impede the flow of heat are the **thermal resistances** between junction and case $R_{\theta\,jc}$ and between case and ambient $R_{\theta\,ca}$. Under steady-state conditions, those quantities relate to each other as follows:

$$\theta_j = \theta_a + (R_{\theta\,jc} + R_{\theta\,ca})P = \theta_a + R_{\theta\,ja}P \qquad (9.23)$$

For reasons of circuit lifetime and reliability, **junction temperature** must not exceed some maximum admissible value which typically lies in the range between 85 and 125 °C for VLSI chips for information processing applications.[25] The fact that the operating speed of CMOS circuits degrades with temperature while leakage currents multiply further tends to limit the range of acceptable junction temperatures.

With θ_j fixed, an upper bound on the admissible thermal design power P gets imposed by the overall thermal resistance $R_{\theta\,ja}$ along with the highest **ambient temperature** θ_a at which the semiconductor device is required to operate. Conversely, if P is given, designers must provide a

[24] For historical reasons, the semiconductor die is commonly referred to as junction in the context of thermal engineering. A tacit assumption is that temperature is the same across an entire die.

[25] Power ICs are typically specified with upper bounds between 125 and 150 °C while Si-based discretes can withstand up to 150 to 170 °C junction temperature.

thermally conducting path that satisfies

$$R_{\theta\,jc} + R_{\theta\,ca} = R_{\theta\,ja} \leq \frac{\theta_j - \theta_a}{P} \qquad (9.24)$$

$R_{\theta\,jc}$ is mainly a function of die size and of a variety of package-related factors such as bonding techniques (solder or epoxy resin compound; wire bonding or ball grid array), package geometry, presence or absence of a heat spreader, encapsulating materials, and of number, geometry, and material of package leads. $R_{\theta\,ca}$ is dependent on external circumstances such as air circulation, package orientation, mounting technique (socketed or soldered), lead count, and board layout. After all, heat is not only convected and radiated from the package surface but also flows to the board via the package pins.

In still air, that is on the order of forced ventilation, the overall thermal resistance $R_{\theta\,ja}$ ranges between 20 K/W for a 208-pin MQUAD package and 180 K/W for an 8-pin SSOP package, with 80 K/W a typical figure for a PQFP64 package.

By slashing thermal resistance between case and ambient air, a **heat sink** reduces the temperature difference necessary to transfer a given thermal power. This is basically obtained by extending the overall surface in contact with the air. $R_{\theta\,ca}$ reaches down to approximately 0.55 K/W for generously-sized aluminum heat sinks and further down to 0.25 K/W when combined with intense **forced air cooling** such as in top-performance desktop processors. As the best packages available offer junction-to-case resistances $R_{\theta\,jc}$ of 0.3 K/W or so, the maximum practical power dissipation currently is on the order of 125 to 150 W with forced air cooling.

> Hint: When using a heat sink, make sure the fins are aligned with the direction of air flow, whether natural or imposed.

Air flow is otherwise disturbed, which cuts back the active surface and makes the effective thermal resistance significantly differ from published K/W figures. Heat sinks where fins are too closely packed or that exhibit a cross-cut fin pattern also tend to obstruct (laminar) air flow.

A **heat pipe** essentially is an evacuated copper tube the inner walls of which are lined with a wick. The tube is loaded with a small quantity of water or some other working fluid before being sealed. Heating one end stimulates a circular process whereby liquid evaporates and travels to the cooler end, where it condenses before being returned by the capillary forces developed in the wick. Thermal conductivity is many times that of an equivalent massive piece of copper. Heat pipes work best in upright position when heat is fed in at the bottom end because gravity then helps to return the condensed fluid to the evaporator.

In any case, watch out for the thermal contact resistance $R_{\theta\,cs}$ that is being introduced between the IC package and the heat sink's surface accepting the thermal flow.

$$R_{\theta\,ca} = R_{\theta\,cs} + R_{\theta\,sa} \qquad (9.25)$$

$R_{\theta\,cs}$ and hence also $R_{\theta\,ca}$ depend on whether the heat sink is mounted with or without electrical insulation (such as a sheet of mica, Mylar or Kapton), on the mechanical pressure, and on the presence or absence of a thermal interface material. The surfaces of dies, cases, and heat sinks are never perfectly flat, so microscopic gaps form in between. With air being a poor thermal conductor, these gaps inflate $R_{\theta\,cs}$. Either a **heat-conducting compound** — e.g. a paste of silicone grease with silver oxide powder — or a graphite pad can serve to displace air and to fill those gaps with a material of better thermal conductivity as expressed in W/K m.

Another common mistake is to underestimate the ambient temperature. θ_a refers to the air next to the component and its heat sink under worst-case conditions. All too often, this reading is significantly above the room temperature measured outside the equipment's casing.

As a final remark, any lowering in overall thermal resistance is generally paid for with higher costs.

9.4 | Appendix I: Contributions to node capacitance

Section 9.1.1 has introduced the concept of node capacitance. In this appendix, we examine in more detail what capacitances are associated with some circuit node k for the purpose of energy computations. Consider a typical digital CMOS subcircuit where I_k logic gates drive a total of J_k logic gates. Figure 9.19a illustrates a case where $I_k = 1$ and $J_k = 3$. We then find:

$C_{mil\,i}$ The Miller capacitance of gate i (drains or sources to gate electrodes) transformed into an equivalent capacitance at the gate's output node k.

$C_{oup\,i}$ The output-to-ground capacitance of the ith logic gate that drives node k, that is the junction (drains to ground) and overlap capacitances there.

$C_{wire\,k}$ The total interconnect or wiring capacitance of node k made up of contributions from the wire to ground and from the wire to adjacent wires.

$C_{inp\,j}$ The input-to-ground capacitance of the jth gate that is driven from node k which is dominated by the thin-oxide areas (gate electrodes to ground) there.

$C_{dep\,j}$ Extra capacitances internal to logic gate j that are due to logically dependent inner nodes (drains or sources to ground plus overlaps).

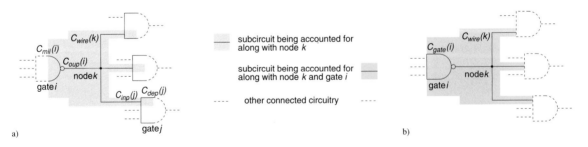

Fig. 9.19 A small excerpt from a larger logic network with the various contributions to node capacitance (a) and with all contributions from a gate transformed to the output (b).

$C_{mil\,i}$ and $C_{dep\,j}$ are not part of the capacitance of node k itself but have been introduced to account for (dis)charge processes that take place within gates i and j respectively as a consequence of the switching of node k. Note that a Miller capacitance must get mapped to the output with almost four times its value to account for the fact that the voltage amplitude across is twice that of a regular node-to-ground capacitance.[26]

Energywise, the total equivalent capacitance of circuit node k is obtained as

$$C_k = \sum_{i=1}^{I_k} (C_{mil\,i} + C_{oup\,i}) + C_{wire\,k} + \sum_{j=1}^{J_k} (C_{inp\,j} + C_{dep\,j}) \tag{9.26}$$

where $I_k > 1$ and $J_k > 1$ reflect multiple drivers and fanout respectively.

Many EDA tools use a simplified model, shown in fig.9.19b, whereby all capacitive contributions from library cells are concentrated into the respective gate models instead of being attributed to the nodes in between. The effects of C_{mil}, C_{oup}, C_{inp}, and C_{dep} are hence being lumped into a single quantity.

$C_{gate\,i}$ An energy-equivalent capacitance that accounts not only for the output of the ith gate that drives node k but also for all its inputs and inner capacitances after they have been transformed to the output of that very gate.

Under this approximate model, (9.26) is replaced by

$$C_k \approx \sum_{i=1}^{I_k} C_{gate\,i} + C_{wire\,k} \tag{9.27}$$

which simplifies to (9.4) in the case of a single gate driving node k.

9.5 | Appendix II: Unorthodox approaches

9.5.1 Subthreshold logic

Section 9.1.5 has identified voltage swing as a major cause of energy dissipation in CMOS circuits. This immediately raises the question of the lowest possible supply voltage at which logic can be made to operate. Eric Vittoz, one of the pioneers of subthreshold logic, has determined that CMOS should work with supply voltages as low as 100 mV thanks to the MOSFET's exponential current–voltage characteristics in the weak inversion regime, thereby slashing dynamic energy by a factor of 100 when compared with a 1 V supply [244].

Going that far is not easy, however. The same exponential dependency makes circuit operation extremely sensitive to process variations and requires that the threshold voltages be adjusted by a control loop similar to dynamic back-biasing (DBB). Also, the effective threshold voltages must be chosen so low as to make leakage of the same order of magnitude as switching currents.[27] The important loss of switching speed is welcome in some sense as it helps to keep ground bounce and

---

[26] A factor of 3.6 has been found to be more realistic than 4.0 [285]. The reason is that a fraction of the charge initially stored on the Miller capacitance flows back into the power net during a brief overshoot phase at the very beginning of each output transition. Another subtlety is that part of the energy from the Miller capacitance of gate i, k is being dissipated in the logic gate itself and part in the gate that is driving it (not shown in fig.9.19).

[27] Jokingly, one might speak of "Leakage Current Modulation Logic".

crosstalk in accordance with the depressed noise margins. Yet, compensating for the inferior speed with many concurrent hardware units will not always prove economically feasible, and would inflate leakage further. Less radical compromises are thus probably more practical.

Example

A microchip capable of carrying out 1024-point FFTs with a word width of 16 bit has been designed using a standard 180 nm six-layer metal CMOS process [286]. Thresholds are fixed at 450 mV with no biasing. Standard cells have been redesigned such as to exhibit dependable operation in spite of process variations, parallel leakage paths, and transistor stacking. The circuit consists of 670 000 transistors placed on a die of 2.6 mm by 2.1 mm and is fully functional over a supply voltage range from 180 to 900 mV. The minimum energy point occurs at 350 mV and is 155 nJ/FFT, 350 times less than a software implementation running on a low-power microprocessor. Maximum clock frequencies are 164 Hz, 5.6 MHz, and 10 kHz respectively.

□

9.5.2 Voltage-swing-reduction techniques

Another idea is to reduce the voltage swing per bit. **Multi-valued logic** uses intermediate voltage grades to encode multiple bits of data with a single signal. As an example, think of a four-valued logic where two bits of information get represented by nominal voltages of 0, 1, 2, and 3 V respectively. A one-bit increment would then materialize as a voltage step of $\frac{1}{3}U_{dd}$.

While multi-valued circuits find applications in high-density flash memories [51], the concept is not being adopted in random logic because the important circuit overhead required to discriminate among multiple voltage grades tends to annihilate any improvements in energy efficiency. Repeated reductions of supply voltage over the last process generations have not only rendered the handling of intermediate voltages even more difficult, but also dramatically improved the energy efficiency of two-valued CMOS logic.

Reduced voltage swing is more likely to pay off in nets where no more than two logic states must be told apart and where node capacitance is huge. This holds true for clock nets [279] [222] and for bit lines in large memory arrays [287]. The same applies to low-voltage differential signaling (LVDS), a popular technique for input/output lines that run between chips. The fact that LVDS uses two lines to encode one bit of information simplifies circuit design.

9.5.3 Adiabatic logic

Throughout our analysis, we have accepted as a fact that a fixed amount of energy defined by $E_{ri\,k} = E_{fa\,k} = \frac{1}{2}C_k U_{dd}^2$ gets dissipated in a logic gate's MOSFETs whenever its output toggles, and this irrespective of the waveforms and ramp times involved. That assumption, reflected by (9.3), is indeed a valid one for level-restoring static CMOS logic that works from rail to rail.

In adiabatic logic, in contrast, switching is supposed to follow a different paradigm for which (9.3) no longer holds. For simplicity, assume node capacitance C_k is being charged via some fixed resistance

R so as to maintain a constant current throughout the process. It is easy to show that the energy dissipated in the resistor then amounts to

$$E_{rik} = \frac{RC_k^2 U_{dd}^2}{t_{ri}} \tag{9.28}$$

What is most intriguing about this equation is the denominator. It implies that the electrical energy lost to heat can be made arbitrarily small by allowing more time for the charging, and — by the same argument — for the discharging as well. This is because a low current through the resistance also minimizes the voltage drop across it and, hence, the overall energy dissipated.

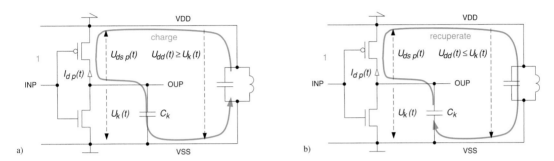

Fig. 9.20 Adiabatic switching. Operation during computation phase (a) and energy recuperation phase (b) (greatly simplified).

The idea behind adiabatic logic is to take advantage of such energy-conserving charge and discharge processes. In order to come close to constant current transients, the supply voltage must be allowed to vary over time, a radical departure from standard CMOS operation.

The operation of all gates is carried out under control of a special periodic signal $U_{dd}(t)$ that combines traits of supply and clock. Phases where energy is being fed into the logic to charge the node capacitances there, and where it gets taken back into a suitably designed resonant tank circuit alternate, see fig.9.20. This explains why adiabatic logic is also referred to as charge recovery and energy-recycling logic. MOSFET switches are being used in lieu of fixed resistors but the small drain–source voltages justify their approximation as linear resistances.

Observation 9.8. *Adiabatic logic attempts to save energy over standard CMOS by recuperating the electric charge made available for logic evaluation into some tank circuit after evaluation has come to an end. The exchange of electric charge between supply-clock and logic circuitry is made exceedingly slow to minimize the resistive losses in doing so.*

The nice thing about adiabatic logic is that it permits one, at least in theory, to trade energy for switching speed by reducing the supply-clock frequency, and this at any time during circuit operation. Asymptotically, one could obtain a zero-energy infinite-time computation. This contrasts with regular CMOS logic that essentially operates in a fixed-energy bounded-time regime as a result of design decisions about supply voltage, circuit design style, transistor sizes, and node capacitances. Note, however, that adiabatic logic cannot be expected to do better than regular CMOS logic if it is to operate at the full speed of the latter. If speed requirements are modest, on the other hand, operating static CMOS from a lower voltage seems a more natural proposition.

It thus remains questionable whether the theoretical benefit of this unorthodox concept will materialize in practice because of all the overhead associated with more complicated circuitry, state retention, multiphase operation, supply-clock generation, and others. Dynamic voltage and frequency scaling, see fig.9.10, provides energy–speed tradeoffs at a much lower cost. Results from actual circuit evaluations are rather pessimistic [288] [289] or have found no continuation [290]. Still, you may want to refer to [241] [291] for more comprehensive assessments of the promises and limitations of adiabatic logic which, by the way, comes in many varieties such as split-level charge recovery logic (SCRL), quasi-static energy recovery logic (QSERL) [292], efficient charge recovery logic (ECRL) [293], and others [294] [295].

More on the positive side, significant power savings have been reported with **resonant clocking** where a sinusoidal waveform obtained with a tank circuit is used to drive the clock net(s) in otherwise static CMOS circuits [293] [296].

Chapter 10

Signal Integrity

10.1 | Introduction

Noise generally refers to unpredictable short-term deviations of a signal from its nominal value. Although noise is not nearly random in digital circuits, the same word is nevertheless used. To comprehend the impact, noise generation as well as a circuit's tolerance to noise must be studied. This chapter aims at understanding potential failure mechanisms, at quantifying their repercussions, and at learning how to keep switching noise below critical levels.

10.1.1 How does noise enter electronic circuits?

One can distinguish four mechanisms that convey noise from a source to a receptor, see fig.10.1.

Conductive coupling develops when a wire collects noise outside an electronic circuit and brings it to sensitive nodes there. A power rail corrupted by spikes or ripple from a poorly filtered power supply is a classic example. A data cable that picks up electromagnetic radiation from a nearby motor, chopper circuit, or RF transmitter — effectively acting like an antenna — also falls into this category.

Electromagnetic coupling occurs when noise sources impinge upon a circuit by the immediate effects of the electromagnetic field. No external line acts as conveyor in this case, rather, the receiving antenna is within the victim itself. The tiny dimensions of microelectronic circuits tend to render them relatively immune to externally generated fields.

Crosstalk is a particular form of electromagnetic coupling. Polluter and victim sit close to each other on the same die, package, or printed circuit board (PCB). Crosstalk effects are typically modelled in terms of lumped elements such as coupling capacitances and mutual inductances.

Common impedance coupling requires that polluter and victim share common power and/or ground (return) lines. The parasitic series impedances of those lines turn rapid supply current variations into noise voltages that then propagate across a chip (or board).

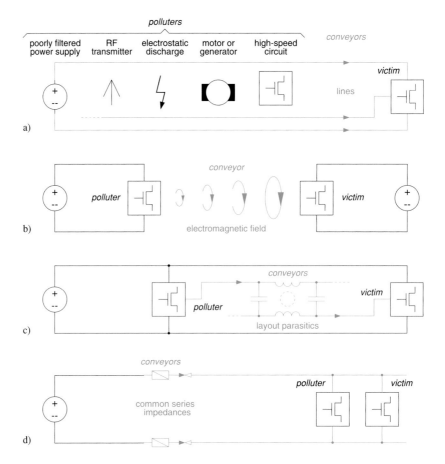

Fig. 10.1 Noise coupling via conductors (a), via electromagnetic fields (b), via crosstalk (c), and via common supply lines (d).

As crosstalk and common impedance coupling have their roots in nearby switching activities, we will occasionally use the term **switching noise** as a generic term for both.

10.1.2 How does noise affect digital circuits?

Noise impacts digital circuits both when settled in a steady state and while transiting from one stable state to the next. Let us analyze those two situations separately.

IMPACT OF NOISE ON CIRCUITS UNDER STEADY-STATE CONDITIONS

A two-valued circuit cannot work properly unless <u>all</u> of its logic gates

1. flawlessly keep apart the logic states 0 and 1 at the input, and
2. restore both logic states to proper electrical levels at the output.

Excessive noise compromises these requirements. Consider the predominant single-rail logic where data get transmitted over a single line, see fig.10.2.[1]

Let U_{oh} denote the lowest output voltage produced by a subcircuit when driving a logic 1 and, analogously, U_{ol} the uppermost voltage when at logic 0. Further let U_{ih} indicate the lowest voltage that gets safely interpreted as a 1 by the subcircuit driven, and U_{il} the highest voltage that gets recognized as 0.[2] Voltages between U_{il} and U_{ih} could be interpreted as 0 by some subcircuits and as 1 by others. They are, therefore, said to form a **forbidden interval**.

$U_{ol} \leq U_{il} < U_{ih} \leq U_{oh}$ holds by necessity. Any non-zero difference between the respective output and input thresholds provides welcome latitude for uncertainties and variations of signal voltage, see (10.1) and (10.2). The lesser of the two differences, defined in (10.3), determines the maximum noise that can safely be admitted without compromising the correct functioning of the circuit and is, therefore, known as the static **noise margin**. As documented by table 10.1, noise margins have shrunk a lot over the years because supply voltages have repeatedly been lowered.

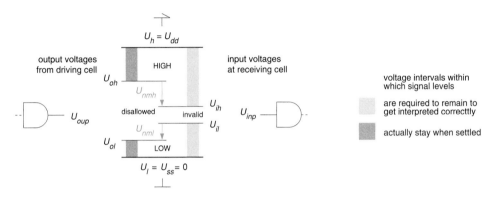

Fig. 10.2 Noise margins in two-valued digital circuits.

$$U_{nml} = U_{il} - U_{ol} \tag{10.1}$$

$$U_{nmh} = U_{oh} - U_{ih} \tag{10.2}$$

$$U_{nm} = \min(U_{nml}, U_{nmh}) \tag{10.3}$$

IMPACT OF NOISE ON SIGNALS IN TRANSIT

Noise also affects the waveforms of transiting signals as any spurious voltage coupled into a circuit node may either speed up or slow down an on-going signal transition, see fig.10.3. As a result, the victim signals are subject to jitter, that is to random timing variations. The impact is most severe when the coupling from a polluting line occurs while the victim signal is traversing the high-amplification

---

[1] The picture is different for dual-rail circuits that rely on complementary signaling and differential input stages, such as the LVDS scheme to be discussed later in this chapter.

[2] As explained in section 8.1.2, U_{ih} and U_{il} are typically defined by the unity gain points in the subcircuit's transfer characteristic.

Table 10.1 Standard switching thresholds for digital interfaces. Note that the labels CMOS and TTL in this context refer to agreed-on voltage levels, not to fabrication technology. In fact, many CMOS circuits have been designed to accept signals with TTL levels for reasons of compatibility. For similar reasons, off- and on-chip voltage levels often differ.

Voltage levels	U_{dd}	U_{ol}	U_{il}	U_{ih}	U_{oh}	U_{nm}
CMOS	5 V ±10%	0.1 V	$0.3U_{dd}$	$0.7U_{dd}$	$U_{dd}-0.1$ V	1.40 V
TTL	5 V ±10%	0.4 V	0.8 V	2.0 V	2.4 V	0.40 V
CMOS	3.3 V ±10%	0.1 V	$0.2U_{dd}$	$0.7U_{dd}$	$U_{dd}-0.1$ V	0.56 V
TTL	3.3 V ±10%	0.4 V	0.8 V	2.0 V	2.4 V	0.40 V
CMOS	2.5 V ±0.2 V	0.2 V	$0.35U_{dd}$	$0.65U_{dd}$	$U_{dd}-0.2$ V	0.68 V
CMOS	1.8 V ±0.15 V	0.2 V	$0.35U_{dd}$	$0.65U_{dd}$	$U_{dd}-0.2$ V	0.43 V
HSTL Class I[a]	1.5 V ±0.1 V	0.4 V	0.65 V	0.85 V	1.1 V	0.25 V
GTL[b]	1.2 V nominal	0.5 V	0.75 V	0.85 V	1.2 V	0.25 V
CMOS[c]	0.9 V nominal	0.1 V	$0.35U_{dd}$	$0.65U_{dd}$	$U_{dd}-0.1$ V	0.22 V

[a] High-Speed Transceiver Logic, a technology-independent interface standard.
[b] Gunning Transceiver Logic, a low-swing interface standard for backplane busses on the basis of CMOS.
[c] No established standard yet, table entries based on extrapolation.

region of the receiving gate(s). Also note that noise and jitter not only affect the behavior of physical circuits, but also add to the uncertainty of timing verification.

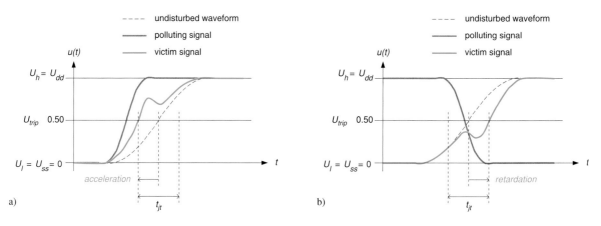

Fig. 10.3 The effect of crosstalk when polluter and victim signals ramp in the same direction (a) and when they ramp in opposite directions (b).

The signals most vulnerable to noise-induced jitter are

- Clock signals,
- Signals with a small setup margin (long paths, vulnerable to retardation), and
- Signals with a small hold margin (short paths, vulnerable to acceleration).

Observation 10.1. *Switching noise affects digital circuits in two ways:*

- *Settled nodes are immune to disturbances only as long as their noise margins are respected. If not, circuit behavior may become entirely unpredictable.*
- *The data-dependent jitter inflicted on signals in transit renders timing data uncertain, which is detrimental to performance as it eats away from timing budgets.*

10.1.3   Agenda

A brief discussion of crosstalk follows next. Being predominant among the four perturbations, common impedance coupling will be explained in more detail in section 10.3 before various countermeasures against switching noise are presented in section 10.4.

10.2 | Crosstalk

As a consequence from the evolution of the interconnect stack depicted in fig.12.7, unpredictable timing variations due to crosstalk between adjacent signal lines have become a concern for digital design with VLSI technologies of 180 nm and less.

While noise of sufficient amplitude on a clock or reset signal can cause a circuit to change state erratically, their large node capacitances and strong drivers tend to make settled clock and reset nets fairly robust against crosstalk from regular signal lines. Clocks and busses are important polluters, however, because their fast ramps and long lines have them generously couple into many of the other signals.

Countermeasures:

- Make lines on adjacent metal layers run perpendicularly to each other
  in order to reduce coupling capacitances and mutual inductances.
- Keep more than minimum lateral spacing between critical victims and aggressive polluters.
- Minimize the separations between current paths and the pertaining return paths.
- Intersperse shield lines and, if need be, also shield layers around critical signals.
- Provide more ample setup and hold margins (often a costly proposition).
- Prefer processes that provide low-permittivity interlevel dielectrics.
- Use specialized EDA tools to carry out crosstalk analyses.

Please refer to [297] [298] [299] for a comprehensive discussion of crosstalk. [300] gives an analytical formula for inductive coupling and compares different shield-line-insertion schemes.

10.3 | Ground bounce and supply droop

10.3.1   Coupling mechanisms due to common series impedances

In any VLSI chip, thousands of logic gates are hooked to VSS and VDD over the same interconnect lines. Figure 10.4a depicts a situation where gates u1 and u3 share a piece of ground line with

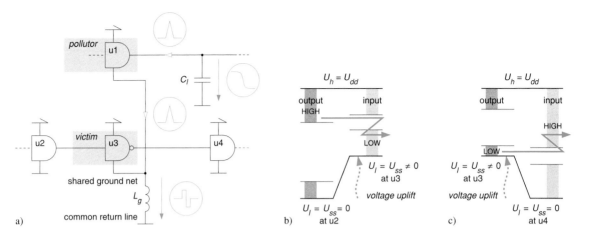

Fig. 10.4 Ground bounce. Basic coupling mechanism (a) and distorted switching levels (b,c).

parasitic self-inductance L_g. Depending on the number of gates driven (fanout), the transistor sizes there, and the interconnect geometry, the capacitive load C_l gate u1 must handle may be rather important. Assume gate u2 outputs a legal but not entirely perfect logic 1 so that gate u3 is presented with $U_{oh} < U_{inp\,u3} < U_{dd}$. Gate u3 then correctly recognizes the signal as logic 1 and produces a legal 0 with $0 < U_{oup\,u3} < U_{ol}$ because of its inverting nature.

"What happens when gate u1 switches from logic 1 to logic 0?"

Look at u3's <u>input</u> behavior first. C_l, previously charged to logic 1, is rapidly being discharged by u1, resulting in a transient current to ground. This current causes a voltage drop over the ground wire's inductance L_g, which affects u3 as much as u1 due to their interconnected ground terminals. The uplift of u3's ground potential will raise its input thresholds to unknown values, see fig.10.4b. If sufficiently strong, the unplanned rise of U_{ih} and U_{il} may cause the input to get misinterpreted as a logic 0 and, therefore, the output of u3 to temporarily change to 1. That unforeseen spike may in turn incite u4 to glitch and so propagate into the downstream logic.

To make things worse, a second mechanism is at work at u3's <u>output</u> that intensifies this phenomenon or that may by itself provoke false switching of u4. This is because the uplift of u3's ground terminal directly impinges upon its output by displacing the voltage from below U_{ol} to some higher value. If the uplift consumes the noise margin, as shown in fig.10.4c, then gate u4 is likely to flip, thereby temporarily producing an unexpected logic value at its output.

In either case, a transitory deviation of the reference potential jeopardizes the correct functioning of logic gates by electrically coupling individual data signals in an undesirable way. The effect is nicely termed **ground bounce**. Owing to the common series impedances present in the VDD network, circuit operation is further affected by an analogous mechanism referred to as **supply droop** or as power bounce. For the sake of simplicity, we usually make no distinction and subsume both effects under the term ground bounce.

Under extreme conditions, the coupling of a subcircuit onto itself may grow so strong that the circuit begins to oscillate in an uncontrollable way. Another potential threat from strong voltage over/undershoots is the loss of stored data. Excessive ground bounce thus causes a digital circuit's behavior to become unpredictable and so renders it useless as part of a system. Milder forms have the circuit function correctly under optimum conditions but make it susceptible to external circumstances. It is, therefore, absolutely essential to keep ground bounce within the bounds circumscribed by the noise margins of the technology.

10.3.2 Where do large switching currents originate?

In static CMOS circuits, the most important switching currents are typically drawn by **output pad drivers** due to the important off-chip load capacitances they must handle and the extremely wide transistors they include to do so. This explains why ground bounce is also known as simultaneous switching output noise (SSO noise or SSN). $di(t)/dt$ noise is yet another synonym, and we will shortly see why.

Huge switching currents are by no means limited to pad drivers, since a thousand or so internal nodes that switch at the same time can cause spikes of similar magnitude. A major contribution to switching currents arises from **clocking**. Keep in mind that a clock event causes many circuit nodes other than the clock net nodes to toggle too. In the case of the CPU 21064 by Digital Equipment Corp., for instance, it was found that ground bounce was equivalent to a cumulated capacitance of 12.5 nF while the capacitance of the clock net alone was 3.2 nF.

Another important source of transient currents that is often overlooked stems from transitory **buffer contentions** where the off-going three-state driver has not yet released a net before the on-going driver starts to pull in the opposite direction. Although very short-lived in logically correct designs, transient currents matter when strong buffers are involved.

10.3.3 How severe is the impact of ground bounce?

A ROUGH FIRST-ORDER APPROXIMATION

The subsequent table taken from [301] quantifies the parasitic elements for a few packages of different construction including contributions from bond wires.

package type (64–68 pins)	ceramic DIP	plastic DIP	worst pin of pin grid array	chip carrier
series resistance [Ω]	1.1	0.1	0.2	0.2
parallel capacitance [pF]	7	4	2	2
series inductance [nH]	22	36	7	7

Numerical data of on-chip wiring parasitics are given next for two digital CMOS processes. A generous power/ground line of fixed width and signal lines of minimum widths are compared across all metal layers. Where two numbers are given, the first one refers to the bottom and the second one to the top metal layer. Line lengths and line spacings are the same throughout.

line length 10 mm, spacing 1 μm	process A 250 nm 5M1P Al		process B 130 nm 8M1P Cu	
	supply	signal	supply	signal
line width [nm]	50 000	320 to 420	50 000	160 to 400
series resistance [Ω]	11 to 7.2	1700 to 860	14 to 5.4	4400 to 680
capacitance to substrate [pF]	15 to 2.4	0.46 to 0.14	28 to 2.7	0.52 to 0.12
series inductance[3] [nH]	3.8	8.2 to 8.0	3.8	8.8 to 8.0

Now consider an output pad driver as shown in fig.10.5. The series impedance Z_g in the current path to system ground is formed by on-chip interconnect lines, bond wire, and package lead. It includes a resistive part R_g and an inductive part L_g. The ground bounce voltage, i.e. the difference between on-chip and system ground potentials, is obtained as

$$u_g(t) = R_g i_g(t) + L_g \frac{di_g(t)}{dt} \tag{10.5}$$

where $i_g(t)$ is the superposition of the driver's output current and the crossover current that flows through the complementary MOSFETs while they are in the process of switching.

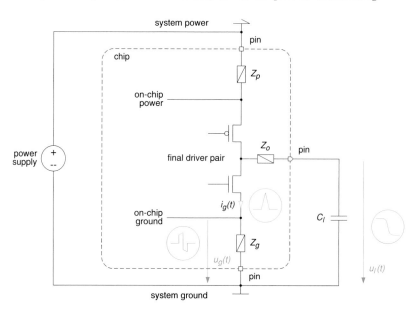

Fig. 10.5 Equivalent circuit for a CMOS output that drives a capacitive load.

---

[3] According to [302], the **line inductance** was estimated as

$$L \approx 200 \cdot 10^{-9} \, l \log \left(8\frac{h}{w} + 4\frac{w}{h} \right) \tag{10.4}$$

where l denotes the length and w the width of the line. h indicates the distance that vertically separates the metal line from the back metal surface carrying the chip. Quantities are expressed in [m] and [H] respectively; the numeric figures given above were obtained for $h = 500$ μm.

Example

Let an IC discharge a capacitance C_l of 20 pF from 2.5 V to zero within 5 ns. For simplicity, the crossover current is neglected and the driver's output current is assumed to follow a symmetric triangular waveform. The switching current that flows to ground then linearly rises to 20 mA — twice the value for a uniform discharge — within 2.5 ns before ramping back to zero in another 2.5 ns. Further assume $R_g = 1\ \Omega$ and $L_g = 10$ nH. The maximum uplift is reached after 2.5 ns when the ohmic loss $R_g i_g(t)$ peaks out at 20 mV while the inductive loss $L_g\, di_g(t)/dt$ reaches 80 mV. A negative deflection of similar amplitude follows during the second half of the discharge process. The switching of one output pin thus causes a 100 mV difference in reference potentials between chip and system. Noise amplitude corresponds to 4% of U_{dd} or to approximately 15% of U_{nm}. Imagine the impact when many outputs switch at a time.

□

A MORE ACCURATE SECOND-ORDER APPROXIMATION

Experience has shown that first-order approximations on the basis of (10.5) tend to overestimate noise voltages in CMOS ICs because they do not account for the moderating influence of a series of second-order phenomena. The most important single effect is that any ground uplift eats away from the gate–source voltage of the discharging MOSFET. This negative feedback quenches the drain and source currents and, hence, the excursions of the critical term $di_g(t)/dt$. Note that the same effect also degrades the driver's switching speed, though.

Other beneficial circumstances include input waveforms with finite slopes and the combined effects of various parasitic circuit elements. Quiet outputs with their load capacitances also exert a significant stabilizing influence on the on-chip ground and power potentials unless all outputs switch in the same direction simultaneously.

Ground bounce voltage $u_g(t)$ typically consists of an initial peak followed by a damped oscillation. To estimate those waveforms, the authors of [303] have obtained an improved model under a number of simplifying assumptions detailed in appendix 10.7. The model essentially boils down to

$$\hat{u}_g \approx \frac{U_{dd}}{\frac{t_r}{t_r - t_0}\left(K_{1s} + \frac{1}{L_g}K_{2s}\right)} \tag{10.6}$$

with

$$K_{1s} = 1 + \frac{m\, C_l}{(t_r - t_0)\, n}R_{eq} \qquad \text{and} \qquad K_{2s} = \frac{(t_r - t_0)}{n}R_{eq} \tag{10.7}$$

where

$\hat{u}_g$ is the peak voltage uplift due to ground bounce,
t_r is the rise time from 0 to U_{dd} of the driver's input voltage $u_{gs}(t)$,
t_0 is the time span until $u_{gs}(t)$ traverses $U_{th\,n}$, the on-going n-channel transistor's threshold voltage,
n is the number of simultaneously switching output drivers, each loaded with C_l,
m is the number of drivers connected to the same internal ground node that keep output quiet low,
$R_{eq} = \frac{U_{dd} - U_{th}}{I_{d\,on}}$.

Mutatis mutandis (10.6) and (10.7) also hold for supply droop, of course.

Example

Let us reanalyze essentially the same situation as before, using the new noise model this time. The values for t_r and R_{eq} follow from the original postulate that the n-channel MOSFET is fully turned on after 2.5 ns and capable of sinking a maximum current of 20 mA.

$$R_{eq} = \frac{U_{dd} - U_{th}}{I_{d\,on}} = \frac{2.5 \text{ V} - 0.5 \text{ V}}{20 \text{ mA}} = 100 \text{ }\Omega \tag{10.8}$$

We begin by considering one isolated output, $n = 1$ and $m = 0$, and so get

$$\hat{u}_g \approx \frac{U_{dd}}{\frac{5}{4} + \frac{t_r}{L_g} R_{eq}} = \frac{2.5 \text{ V}}{\frac{5}{4} + \frac{2.5 \text{ ns}}{10 \text{ nH}} 100 \text{ }\Omega} = 95 \text{ mV} \tag{10.9}$$

which comes close to the result obtained from (10.5). The situation is quite different, however, when 16 out of 16 drivers switch simultaneously because for $n = 16$ and $m = 0$ one has

$$\hat{u}_g \approx \frac{U_{dd}}{\frac{5}{4} + \frac{t_r}{L_g} \frac{R_{eq}}{n}} = \frac{2.5 \text{ V}}{\frac{5}{4} + \frac{2.5 \text{ ns}}{10 \text{ nH}} \frac{100 \text{ }\Omega}{16}} = 0.89 \text{ V} \tag{10.10}$$

While still exceeding the noise margin, this figure contrasts favorably with the 1.6 V estimate obtained from the first-order model. The fact that growth is degressive with n is very welcome indeed. As a last exercise, let us plug in the figures for $n = 8$ and $m = 8$, that is for a case where half of the 16 outputs remain quiet low.

$$\hat{u}_g \approx \frac{U_{dd}}{\frac{5}{4} + \left(\frac{25}{16} \frac{m \, C_l}{t_r} + \frac{t_r}{L_g}\right) \frac{R_{eq}}{n}} = \frac{2.5 \text{ V}}{\frac{5}{4} + \left(\frac{25}{16} \frac{8 \cdot 20 \text{ pF}}{2.5 \text{ ns}} + \frac{2.5 \text{ ns}}{10 \text{ nH}}\right) \frac{100 \text{ }\Omega}{8}} = 0.44 \text{ V} \tag{10.11}$$

□

In summary, a number of second-order effects render ground bounce not quite as immense as a crude first-order approximation had suggested. Still, ground bounce is a real threat and the challenge is becoming more demanding with each process generation because

- Supply voltages and noise margins continue to erode,
- Switching times accelerate,
- Designs tend to grow more complex, so that
- More output pins and other nodes switch at a time.

10.4 | How to mitigate ground bounce

A first remedy is to keep the impedance of the power distribution networks low. What matters is the source impedance seen by the load circuit over the entire frequency range where current transients occur. Particular attention must be paid to inductive components and to impedances shared between polluters and potential victims. Package selection, supply routing, and capacitive decoupling are essential. Other countermeasures consist of limiting transient currents to a minimum, e.g. by paying attention to ramp times. Practical guidelines derived from this general insight are going to be presented in the remainder of this chapter, some of them also benefit immunity from crosstalk.[4]

10.4.1 Reduce effective series impedances

Ground and power pads

Include a sufficient number of pads and pins for current return in your designs as parallel conducting paths lower overall impedance. The higher the switching currents, the more supply connectors will be required. Return paths are often allocated more generously to VSS than to VDD. Reasons for this are the fact that ground acts as system reference, the usage of ground as a return for more than one power supply, and — in many older designs — the narrower noise margin on the low side when TTL logic levels were adopted.

Examples

Pin budgets of high-performance CPUs and their evolution over the years.

Microprocessor	total pins	power and ground pins	pro- portion
Motorola MC 68000	64	4	6%
Motorola MC 68020	101	22	22%
Motorola MC 68030	128	24	19%
Motorola MC 68040	179	67	37%
Motorola MC 68060	223	86	39%
Intel Pentium	273	99	36%
Intel Pentium Pro	387	177	46%
Intel Pentium 4	478	265	54%
Intel Core 2 Duo	775	523	67%

□

Observation 10.2. *Before powering up a circuit, make sure you connect <u>all</u> VDD and VSS pins to power and ground respectively, be it on the printed circuit board or on a circuit tester. Always ensure that you obtain solid low-impedance current paths when doing so.*

Package choice

Select a package with low lead impedance. Prefer solder ball connections over bond wires for their lower inductances (0.1 to 0.5 nH vs. 4 to 10 nH) [304]. Lower parasitics were a major driving force behind the move from bulky dual-in-line packages (DIP) to ball grid arrays (BGA) and other compact packages.[5] Do not expect the reduction in ground bounce to be proportional to the lowering

---

[4] Let us briefly review the impact of synchronous design. On the one hand, synchronous operation leads to an undesirable concentration of switching activities in a short time interval that immediately follows the active clock edge, thereby inflating peak current. On the other hand, synchronous design allows all nodes to settle before clocking takes place and is, therefore, less vulnerable to transient phenomena than asynchronous circuits, where data may be accepted by bistables and/or memories at any time.

[5] More material on packaging is given in section 11.4.

of package inductance, however. This is because — as observed earlier — ground bounce augments with series inductance in a degressive way and because package leads are not the only contributions to parasitic impedance. An experimental study found that ground bounce was lessened by a mere 35% when a package with approximately 20 nH was replaced by another one with 2 nH [305].

Examples

Pin parasitics of a few common packages for standard logic ICs.

package type	pin pitch [mm]	self-inductance corners [nH]	self-inductance centers [nH]	cap. to ground corners [pF]	cap. to ground centers [pF]
20 pin packages					
dual-in-line (DIP)	2.54	13.7	3.4	1.49	0.53
small-outline (SOP)	1.27	5.8	3.0	0.85	0.45
shrink-small-outline (SSOP)	0.65	5.0	2.6	0.47	0.30
thin-shrink-small-outline (TSSOP)	0.65	3.3	1.8	0.40	0.21
48 pin packages					
shrink-small-outline (SSOP)	0.635	7.2	3.3	0.74	0.28
thin-shrink-small-outline (TSSOP)	0.5	5.0	2.3	0.56	0.21
100 pin package					
thin-quad-flat (TQFP)	0.5	6.1	4.8	0.33	0.23

☐

Optimum pinout

For a given package, parasitic values vary considerably among pins because different leads take distinct routes within the package. Reserve the low-inductance pins for ground and power, but do not forget to account for routing on the printed circuit board (PCB) that will eventually accommodate the chip. Note that many pin grid array (PGA) packages include a number of low-impedance leads that are specifically optimized for ground and power distribution.

Further observe that the traditional diagonal configuration for dual-in-line (DIP) and small-outline (SOP) packages, termed **corner pinning**, is the worst choice possible, which explains why it is being displaced by the superior **center pinning** pattern in many products such as SRAMs and high-speed logic components. See fig.10.6a and b for illustrations (the benefits of pinout patterns c to f are to be discussed shortly).

Bypass capacitors

Bypass capacitors, aka decoupling capacitors, are connected between power and ground rails to momentarily supply energy for switching activities. Transient currents hence do not have to travel all the way from a distant power supply. Minimizing the impedance of the wiring between ICs and nearby bypass capacitors is absolutely essential to contain noise.

Bypass capacitors placed in the immediate vicinity or underneath IC packages, see fig.10.8a, have a long tradition in digital electronics. Each package must have its own bypass capacitor(s). It is

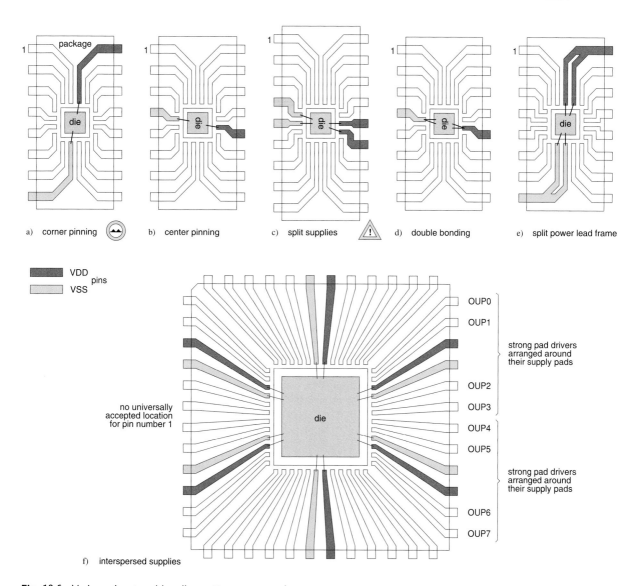

Fig. 10.6 Various pinout and bonding patterns compared.

essential to select a mounting form with a high resonance frequency, that is with low equivalent series inductance (ESL) and low equivalent series resistance (ESR). This is why leadless surface mount devices (SMD) with specially constructed low-inductance plate connections long ago displaced cylindrical forms with axial leads.

Choosing the capacitor's value is a balancing act. On one hand, the value must be large enough to provide the necessary transient energies without much impact on the voltage across the IC. An overly large capacitance, on the other hand, merely lowers the resonance frequency (unless the ESL and ESR values are improved accordingly). Beyond its resonance, a capacitor behaves like

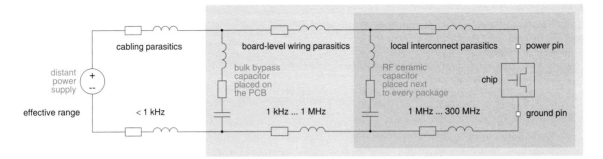

Fig. 10.7 Power distribution network with hierarchically arranged bypass capacitors.

an inductance and is no longer able to supply energy for fast transients. The task is thus divided between small and low ESL capacitors placed as close to each package as possible, and larger but also slower capacitors mounted nearby, see fig.10.7. Depending on the clock frequencies, slew rates, and current levels involved, typical values for the former range between 100 pF and 10 nF while the latter are sized to be many times larger.

A more innovative idea is to distribute bypass capacitance across the PCB itself by having power and ground planes arranged next to each other with just a thin dielectric layer of high permittivity in between [306].

Placing bypass capacitors within the package next to the die pushes resonance to still higher frequencies. This is because self-inductance augments with the area enclosed by the current flowing in the loop. Interconnection is either via a short bonding wire as illustrated in fig.10.8b or via short metal lines if the die is flip-chip mounted on a laminate substrate as in fig.11.14.

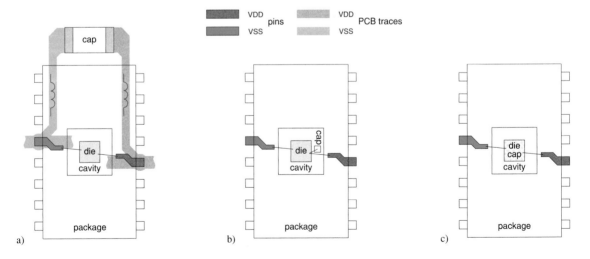

Fig. 10.8 Recommended locations for bypass capacitors. Placed on the printed circuit board in the immediate vicinity of the IC package (a), included within the package next to the die (b), and distributed across the die itself (c).

Observation 10.3. *When selecting bypass capacitors, do not strive for maximum capacitance. Instead, pay much attention to the resonance frequency of the loop formed by the IC, the capacitors, and the wiring. Use low-ESL parts and keep all related interconnects as short as possible.*

On-chip bypass capacitors

With ever higher switching speeds and bus widths, the usage of discrete capacitors alone proved insufficient in high-performance designs, so additional decoupling capacitance had to be brought even closer to the active circuitry. In fig.10.8c, thousands of filler cells, each containing a tiny capacitor, are spread out over the entire core area to obtain an adequate overall capacitance and low-impedance paths for switching currents.[6] [307] have shown that on-chip decoupling helps a lot to mitigate the effects of inductance in the power and ground nets and that the local distribution of fillcaps should best follow the distribution of loads.

Example

The 21364 CPU of the Alpha processor family, implemented in a 180 nm 7M1P Cu CMOS technology, has a total bypass capacitance of 450 nF distributed over the chip. An extra 4000 μF is included within the land grid array (LGA) package. For a low-inductance interconnect, the die is flip-chip-connected to the package base board with 5904 bumps, of which 4962 are allocated to power and ground [308].
☐

It should be noted that switching pad drivers do not benefit from on-chip capacitors as much as core activities do. This is because I/O charge and discharge currents continue to flow through package-related series impedances. Also watch out for critical resonances that might develop.

Low-impedance supply routing

Minimize inductance and resistance of on-chip ground and power networks as follows:

- Run them on low-resistance metal layers throughout.
- Keep lines short while generously sizing their overall widths.
- Avoid unnecessary layer changes because of via and contact resistance.
- Connect multiple vias or contacts in parallel to lower resistance.
- Avoid convoluted inductor-like routing shapes.
- Minimize the area enclosed between power and ground current paths.

Due to their relative thickness and wide pitch, the uppermost metal layers are preferred for supply routing.[7] Several layout arrangements for power and ground distribution are shown in fig.10.9. Good results are obtained from routing power and ground on adjacent parallel lines on a first layer combined with an analogous grid on the next layer below that runs perpendicularly to the first, and where the two grids are generously cross-connected with vias. Superimposing wide VDD and VSS lines minimizes inductance and contributes to on-chip decoupling capacitance but may stand in the way of long signal wires assigned to the same metal layers for routing.

---

[6] How to build CMOS fillcaps is explained in section 8.4.4.
[7] See section 11.2 for a discussion of interconnect layers.

Solid planes are not used in VLSI as they would create manufacturing problems and induce too much mechanical stress due to diverging thermal expansion coefficients of metal, silicon, and dielectric materials. Interdigitated combs that feed cell areas from just one side as shown in fig.10.9b and c used to be popular at a time when very few metal layers were available.

Observe from fig.10.9e how pads generously placed on top of the active circuitry, rather than along the die's edges, help distribute supply currents and reduce the distances these have to flow on the die. The larger the chip, the higher its speed and current drain, the more important the benefit. Flip-chip packaging not only takes advantage of this but also does away with bond wires and their inductance.

Further advice can be found in [304] [309], two textbooks entirely devoted to power distribution.

Example

Intel's Itanium CPU contains 25.4 million transistors including L1 and L2 caches. Global power distribution is with a mesh built from metal5 and metal6 lines with signal lines finely interspersed for shielding. On-chip bypass capacitors are placed in the proximity of high-$di(t)/dt$ subcircuits and in routing channels. Total on-chip bypass capacitance is 800 nF in 180 nm CMOS technology [310]. In addition, extra decoupling capacitance has been included in the package.

□

Solid PCB ground and power connections

Series impedance is minimized by shaping ground and power distribution nets into solid planes or — to a lesser degree — into meshes; the usage of comb-type layouts is discouraged in the context of PCB design.[8] Package or socket pins must always be soldered to the ground or power plane directly; do not use wire-wrapping for supply connections!

Also, never forget that any (forward) current that flows out of a package pin gives rise to a return current, and that **loop inductance** is proportional to the area enclosed by the current flow. The absence of a low-impedance return path will force the return current to find its way through nearby board traces and so generate unnecessary crosstalk. Digital board layouters benefit from looking inside UHF radio or TV equipment. In fact, current spikes with rise and fall times on the order of 1 ns must be treated like GHz signals.

10.4.2   Separate polluters from potential victims

Physical segregation

Not all cells within an IC contribute equally to supply noise, nor are all signals equally susceptible to it. As a rule, separate polluters from vulnerable subcircuits. Important sources of current spikes include electrostatic discharge (ESD) protection circuitry, pad drivers, collective clock buffers, and other drivers of heavy loads. These are typically placed in the padframe and fed via dedicated supply lines, see fig.10.10 for a sketch of a typical CMOS I/O circuit.

---

[8] [311] [312] [313] offer practical hints on grounding, shielding, termination, layout, etc. at the PCB level.

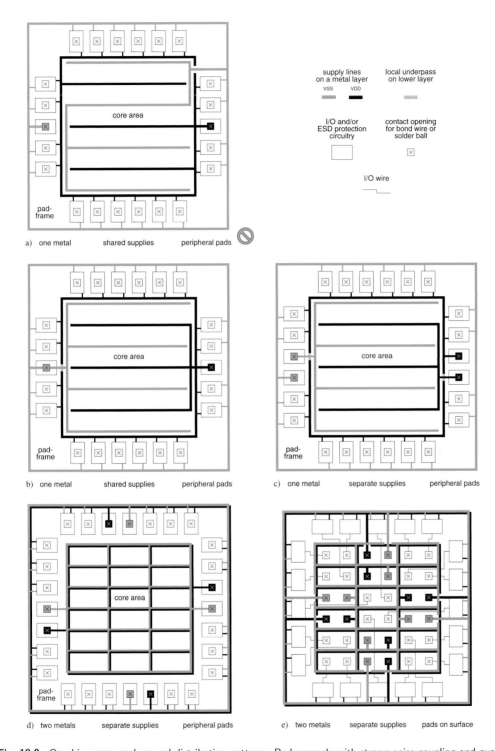

Fig. 10.9 On-chip power and ground distribution patterns. Bad example with strong noise coupling and overly long supply wires (a), better examples that differ in their respective power and ground impedance and in their appetite for pads and metal layers (b to e). Note that lines have been slightly offset in (d,e) to show the usage of two superimposed interconnect layers.

Input buffers, on the other hand, are especially exposed to deviations between external and internal reference potentials. They are, therefore, placed in the core and fed from the relatively clean supply lines there.

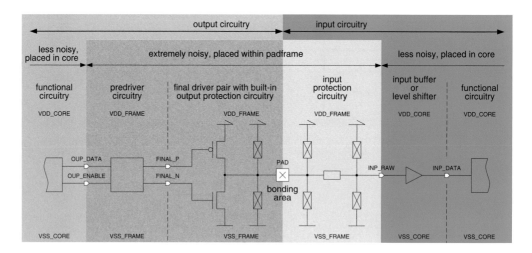

Fig. 10.10 Typical IC input/output circuit (simplified). Note that the figure combines an input on the right-hand side with an output on the left in a single schematic drawing; only bidirectional pads include all items shown.

Electrical decoupling of switching pads from non-switching ones

While the circuit of fig.10.10 nicely isolates the core logic from ground bounce and supply droop in the padframe, stationary outputs continue to suffer from contamination with noise. The situation is improved in fig.10.11. The idea is to provide separate current paths for switching and non-switching outputs by introducing <u>two</u> final transistor pairs per output pad that get activated in an alternating fashion. The first pair drives an output from logic 1 to 0 and vice versa whereas the second pair merely serves to keep an output in its state once it has settled. The hefty transient currents from the first pair are dumped on noisy supply rails while the stationary drivers connect to a second set of rails without causing much noise there. This requires four electrically separate supply rings for powering the padframe (in addition to those feeding the core):

- VDD_FRAME_ST to provide a path to power for pads while stationary,
- VDD_FRAME_TR to source charge currents of pads during transients,
- VSS_FRAME_TR to sink discharge currents of pads during transients, and
- VSS_FRAME_ST to provide a path to ground for pads while stationary.

Electrical decoupling of core and padframe

Minimizing the common series impedances of padframe supplies and core supplies greatly reduces noise coupling. The routing of fig.10.9a is extremely unfortunate in this respect. Figure 10.9b performs much better as currents from the padframe and from the core follow different paths except

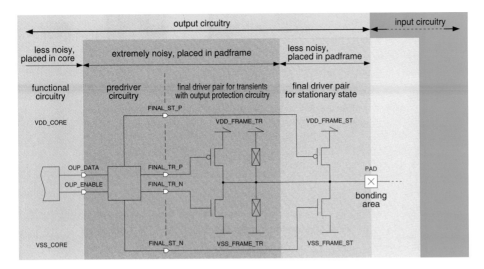

Fig. 10.11 An output circuit that provides separate current paths for transient and for stationary currents (simplified).

for very short segments next to the supply pads. On-chip coupling between core and padframe is virtually eliminated by feeding them via separate pads as shown in fig.10.9c to e.

Note that applying the concept of **split supplies** at the package level implies a major risk. Different parts of a circuit that are being fed from separate pins with no on-chip interconnection, see fig.10.6c, may lead to destruction of the device if some of the pins remain unconnected when the device is powered up in a test fixture or on the final board. A safer way is to **double-bond** supply pins by providing two pad sites for one package pin, as shown in fig.10.6d, thereby eliminating noise contributions caused by common bonding wires.

Specially designed lead frames are sometimes used — mainly in conjunction with corner pinning — to minimize noise contributions from common package leads. Such **split power lead frames** accommodate separate bonding wires for core and padframe but connect them to a single pin near the edge of the package, see fig.10.6e.

Even with a moderate number of simultaneously switching outputs, feeding the padframe from a single pair of supply pads and pins often proves insufficient. A better approach is to provide extra ground and power pads at regular intervals and to group pad drivers around those pads in the layout. This **interspersed supplies** approach illustrated in fig.10.6f further contributes to abating cross-coupling effects between core and pad drivers and among outputs.

10.4.3 Avoid excessive switching currents

Driver sizing

Select pad and clock drivers carefully. Do not use stronger drivers than necessary since this would further increase switching and crossover currents. Begin by estimating the loads the different outputs will have to handle in the target system, including wiring parasitics. Also, do not forget that the specialized hardware test equipment that your IC will have to drive before being put to service may load outputs more heavily than the final application does.

With respect to PTV variations, it is interesting to note that what are the best conditions speedwise are the worst ones noisewise. Slow circuit samples exhibit long delays but produce little switching noise, while the opposite is true for fast samples. This contention together with the importance of PTV variations makes the sizing of large drivers a difficult compromise.

Slew-rate control

Oversizing buffers and transistors in an attempt to account for worst-case timing often leads to unacceptable ground bounce, and vice versa. It is, therefore, necessary to render rise and fall times less dependent on PTV conditions.

A simple solution is to subdivide driver transistors and to selectively disable sections thereof as a function of the process conditions found when testing a circuit copy [314]. A second approach compensates for process variations by controlling the switching currents from voltage reference circuits the outputs of which depend on the process outcome, so as to maintain approximately constant charge and discharge rates [316]. Yet another idea is to deliberately reduce the slew rate by a negative feedback mechanism so as to make it less sensitive to PTV variations. Most modern cell libraries include pad drivers that implement some kind of slew-rate control. Prefer such drivers whenever performance specifications allow.

Soft switching drivers

Standard CMOS inverters and buffers would cause excessive crossover currents if they were sized up to handle off-chip loads. Output pad drivers are thus typically designed so as to avoid crossover currents by minimizing the simultaneous turn-on of n- and p-channel MOSFETs, see fig.10.12 for an example. Similarly, the final transistors are sometimes made to turn off rapidly to minimize crossover and to turn on gradually to contain $di(t)/dt$. Check your cell library for soft switching drivers and see problem 2 for further suggestions in case you are given the opportunity to design your own buffer circuits.

Staggered switching

When there are too many heavily loaded primary outputs, try to spread their switching over a short lapse of time rather than let all of them change simultaneously. There are several ways to do so, see fig.10.13.

- The most radical solution is to multiplex the output data in two — or more — groups over the same pins and pad drivers (b). Besides alleviating the noise problem, this solution reduces pin count and packaging costs at the expense of bandwidth.

- Stay with the full number of pins but withhold half of the output data for one clock cycle, thereby effectively **staggering** the switching of the drivers (c). As with the former solution, throughput is significantly reduced. The subsequent proposals eliminate that bottleneck by reducing the staggering to less than one period of the system clock CLK.

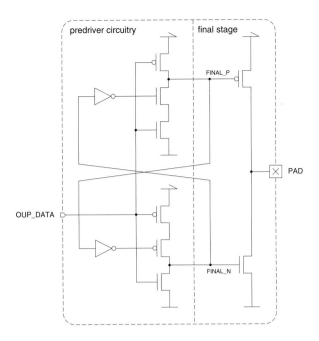

Fig. 10.12 Low-noise pad driver circuit, after [317].

o Postpone the switching of the second half of the output register until the transient currents from the first half have largely died out by driving it with the opposite edge of clock **CLK**.

o If this proves unacceptable, check whether a higher-frequency clock **CLF** is available, from which an auxiliary clock **CLD** retarded by some smaller integer fraction of the system's clock period can be obtained (d, e).

o If no fast clock is available, consider using a carefully tuned delay line to derive the auxiliary clock (f).[9] Be aware that the usage of delay elements violates the rules of synchronous design and is, therefore, more delicate.

The last approach can be understood as introducing clock skew on purpose. Note that it is by no means limited to outputs, but can also serve to better distribute switching activities of the core logic. The authors of [318] report typical peak current reductions of some 20% to 30% with this technique.

Low-Voltage differential signaling

Low-voltage differential signaling (LVDS) was originally developed for interboard communication at distances of up to 10 or 15 m for video and multimedia applications. LVDS, which was accepted as IEEE standard 1596.3 in 1996, combines data rates up to 800 Mbit/s with low power, low noise,

---

[9] Delays can be obtained from an inverter chain, from transmission gates, from extra load capacitances, and/or from resistive interconnect lines. You may also want to consult section 8.4.6.

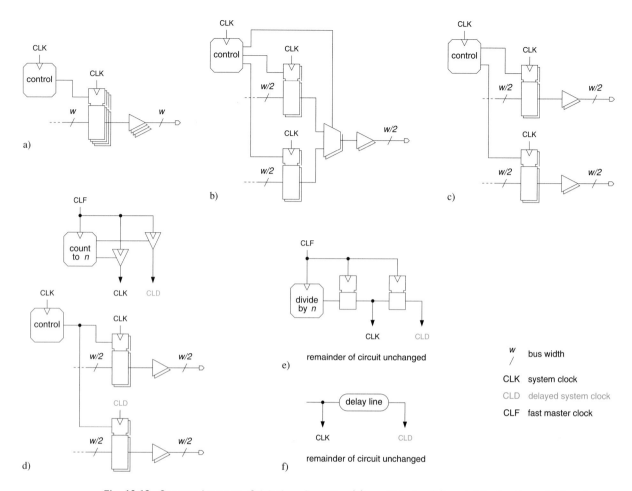

Fig. 10.13 Staggered outputs. Original configuration (a), multiplexing (b), and deferring the switching of a subset of outputs by a full cycle of the system clock (c), by some fraction of it (d,e), and by some absolute delay time (f).

and low cost. Communication is unidirectional over controlled differential impedance media such as twisted pairs, twinax cables, ribbon cables, or matched PCB lines.

Auxiliary subcircuits not drawn in fig.10.14 serve to detect out-of-the-normal conditions such as the absence of a driving signal and shorted or broken lines. The reduced voltage swing of some 350 mV limits static power dissipation to approximately 1.2 mW per link. For comparison, note that traditional NRZ[10] full-swing voltage signaling with 3.3 V absorbs just about the same amount of power while driving a capacitive load of a mere 10 pF at a data rate of 400 Mbit/s.

---

[10] NRZ is an acronym for "non return to zero", a collective term for those waveforms that are obtained from concatenating one bit to the next in the simplest possible way, that is with nothing in between.

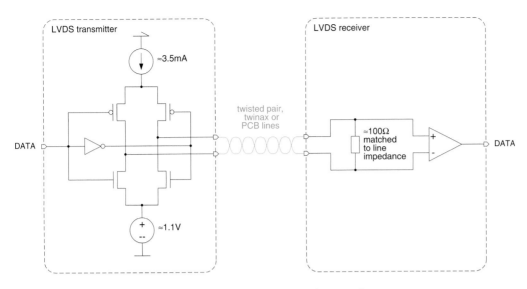

Fig. 10.14 Low-voltage differential signaling communication link (simplified).

The suppression of common-mode noise, the low power dissipation, the absence of current spikes from the transmitter circuit, and the tolerance with respect to parameter variations make it possible to achieve a data rate with one pair of lines that is more than double that of conventional single-rail signals. Please refer to [159] [320] [321] for further discussions.

Examples

The PlayStation 3 board connections between CPU, memory, and graphics processor are implemented with low-voltage differential signaling [318]. Similar techniques have been adopted in the FireWire and Serial ATA (SATA) standards. The digital DVI and HDMI video links use LVDS in conjunction with a standardized 8-to-10 bit encoding scheme to transmit the RGB components with signal waveforms that minimize the number of transitions and that balance the average DC level, hence the name transition-minimized differential signaling (TMDS).

☐

10.4.4 Safeguard noise margins

Switching thresholds and input level shifters

Level shifters must be used on all primary inputs and wherever subcircuits that operate from two different supply voltages exchange data, that is where U_{ol}, U_{il}, U_{ih}, and U_{oh} of transmitter and receiver are incompatible. A level-shifter circuit is depicted in fig.8.36. Maximize noise margins by preferring switching thresholds centered around $\frac{1}{2}U_{dd}$ (CMOS levels) over unsymmetric (TTL-style) levels. Feed input level shifters from the core rather than from the padframe. The reason for this becomes clear from fig.10.4 when g3 stands for a level shifter. Consider using level shifters with Schmitt-trigger characteristics to restore poor input signals.

Warning example

A digital IC was found to malfunction when driven from its on-chip oscillator whereas it worked correctly from an external clock. This was because the slow ramps of the 32 kHz sine waveform from the oscillator — a classical feedback loop built from a CMOS inverter and an off-chip crystal — were corrupted by switching noise. The problem could have been avoided if a Schmitt trigger had been used instead of a normal buffer to shape the signal from the oscillator.
□

Asynchronous resets and gated clocks deserve special attention

Noise is particularly critical on asynchronous reset signals because a spurious pulse could trigger reset of a circuit — or of parts thereof — at any time. To prevent this from happening, the voltage levels of reset signals are often made unsymmetric so as to maximize the noise margin for the reset's inactive state. This is why active-low resets in conjunction with unsymmetric TTL levels were generally preferred when supply voltage was 5 V. A similar reasoning applies to critical inputs subject to impulse or transition signaling such as edge-triggered interrupt request lines and gated clocks.

Unused inputs

Do not leave any unused logic inputs open. Especially in MOS circuits, inputs may otherwise float near the threshold voltage and unnecessarily draw DC. In addition, an open input may pick up AC signals from a nearby source, switch in an unwanted way, draw even more supply current, and so contribute to overall switching noise.

Mixed-signal design

Analog signals are exposed to interference from the switching activities in the digital circuit blocks that coexist on the same die. While many of the countermeasures discussed so far — such as geometric separation, distinct supply nets, and soft switching — help to fight noise injection, let us see what else can be done on the digital side to minimize the impact.

Clocks have been found to be particularly pervasive polluters. The extremely fast clock ramps found necessary to minimize skew in edge-triggered designs in section 6.3.1 result in strong harmonics. Two-phase level-sensitive clocking offers an opportunity to relax slew rates because substantial skew can be accommodated by generously sizing the non-overlap phases. This helps one not only to manage with less aggressive clock waveforms, but also to get by with lighter clock buffers, and to better distribute switching currents over the clock period.

Noise pollution from digital signals can further be reduced by resorting to CMOS **current-mode logic** (CML) families that rely on current switching in conjunction with reduced voltage swings [322][481]. Directing a constant current through either of two branches in a Y-topology network of n-channel MOSFETs reduces current spikes by two orders of magnitude over conventional CMOS [323], but brings about static power dissipation and extra routing overhead.

Common series impedance coupling and crosstalk are not the only conveyors of noise when digital and analog subcircuits operate simultaneously on a common chip. Coupling also occurs via the substrate as rapid voltage fluctuations from the digital part tend to modulate MOSFET threshold voltages in the analog part via the body effect. The impact is highly dependent on the substrate being used (lightly vs. heavily doped; presence of an epitaxial layer; single, twin or triple wells). Please refer to [324] [325] [326] for advice on how to minimize substrate coupling.

Adopting differential signaling on the analog side also helps because differential circuits see most forms of interference as largely uncritical common mode noise signals.

Noise analysis

To compare ground bounce against noise margins, one can come up with a simplified noise-equivalent circuit and carry out electrical simulations using a SPICE-type software tool there. The equivalent circuit must include all significant polluters, the most vulnerable victims, plus the actual parasitics as extracted from the circuit's final layout. Estimating the combined effects of ground bounce and crosstalk on signals in transit is much more demanding, though, as too many signals, layout parasitics, and data patterns are involved. In practice, specialized software tools are being used for the purposes of power grid analysis and crosstalk analysis, see sections 12.4.7 and 12.4.8.

10.5 | Conclusions

- Many of the practical difficulties with ground bounce are due to
  - Poorly sized or carelessly routed power supply current supply and/or return paths,
  - Missing, inadequate, or too distant decoupling capacitors, and
  - Oversized, and hence also overspeed, output pad drivers.[11]

- While it is true that ground bounce is critical in any digital VLSI chip or printed circuit board of more than modest speed or size, a vast collection of countermeasures is available to keep this undesirable phenomenon within safe limits.

- Working out the details of supply networks calls for schematic and layout drawings that highlight power supplies, current return paths, decoupling capacitors, and connectors, along with cross sections and layout parasitics. The customary gate-level diagrams do not suffice as they do not indicate current flow.

- Data-dependent jitter caused by crosstalk is a concern for technologies below 250 nm.

- Detailed noise analysis is a must for all circuits of substantial size, speed, or output loading. Take advantage of specialized signal integrity analysis tools.

10.6 | Problems

1. Numbers for ground bounce in 2.5 V CMOS outputs have been given in section 10.3.3.
   (a) Leaving the other quantities unchanged, repeat the calculation for a traditional process

---

[11] Incidentally, observe that replacing one IC with another of significantly higher speed grade on the same circuit board may also give rise to troubles.

and packaging technology where $U_{dd} = 5$ V and $L_g = 20$ nH, with rise and fall times of 5 ns each. Compare the numbers and draw conclusions about the general trend.

(b) Using a graphing tool, plot (10.6) as a function of n when n out of n (i.e. all), when n out of 16, when n out of 32, and when n out of 64 outputs switch at the same time.

2. Consider the tristate buffer circuit of fig.10.15.

(a) Replace the predriver circuitry by a gate-level equivalent.

(b) Compare the crossover currents in the final pair of MOSFETs when the output switches. What is the benefit of the original circuit over its logic equivalent?

(c) Focus on the final stage now and devise a solution that combines gradual turn-on with rapid turn-off. Hint: Subdividing very wide MOSFETs into a bunch of smaller transistors connected in parallel comes very naturally in VLSI layout. Take advantage of this!

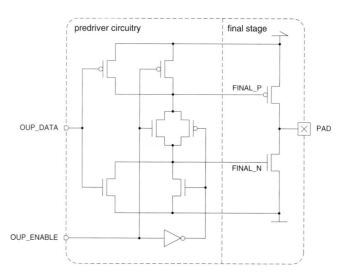

Fig. 10.15 Low-noise tristate pad driver circuit.

3. Think of a capacitor and qualitatively draw its equivalent circuit diagram. Which of the circuit elements matter for the application as a bypass capacitor? Sketch the capacitor's impedance as a function of frequency.

4. Figure 10.16 shows three possible arrangements for metal lines on a board or in a VLSI chip. Discuss their relative merits if those lines are to carry power and ground.

5. Assume you had found that the interconnect delay of some long signal line in your circuit is on a critical path and exceeds the acceptable timing budget no matter how you size the

Fig. 10.16 Cross sections of power and ground lines.

driving buffer. Do you see a way to speed up signal propagation with the help of crosstalk? Hint: Reconsult fig.10.3. Do you see more options?

10.7 | Appendix: Derivation of second-order approximation

This appendix serves to justify equations (10.6) and (10.7). The more onerous part of the derivation has been done in [303], where a few mathematical models for estimating the maximum excursion of the ground potential during the damped oscillations that follow simultaneous switching in a bank of output drivers are proposed and evaluated.

Among those approximations, the one most appropriate for today's short-channel transistors is based on the alpha-power-law model[12] and stated in equation (10.6). The approximation has been obtained under a number of simplifying assumptions.

- Currents get limited by transistors (rather than by interconnect resistance).
- The on-going MOSFET operates in its saturation region throughout the input ramp time.
- Velocity saturation is complete ($\alpha = 1$).

Still, the noise model has been found to be in good agreement with SPICE-type simulations by its authors [303].

The original approximation makes use of two coefficients,

$$K_{1s} = 1 + \frac{m\,C_l}{(t_r - t_0)\,n}\frac{L}{P_c\,W} \qquad \text{and} \qquad K_{2s} = \frac{(t_r - t_0)}{n}\frac{L}{P_c\,W} \tag{10.12}$$

where

P_c is an an empirical MOSFET parameter of the alpha-power-law model,
W is the effective channel width, and
L is the effective channel length.[13]

The problem with applying these two definitions in the context of cell-based design is that P_c, W, and L are not normally known, so a more practically useful alternative is sought. We begin by observing that the alpha-power-law model comes in many variations and that the authors of [303] use a formulation where drain current is modelled as

$$I_d = P_c\,\frac{W}{L}\,(U_{gs} - U_{th})^\alpha \tag{10.13}$$

instead of

$$I_d = I_{d\,on}\left(\frac{U_{gs} - U_{th}}{U_{dd} - U_{th}}\right)^\alpha \tag{10.14}$$

as in (8.56). By equating these two alternative formulations one finds

$$\frac{L}{P_c\,W} = \frac{(U_{dd} - U_{th})^\alpha}{I_{d\,on}} \tag{10.15}$$

---

[12] This and other MOSFET models are introduced in section 8.7.4.
[13] The remaining model parameters have been explained in section 10.3.3.

which simplifies to

$$\frac{L}{P_c\,W} = \frac{U_{dd} - U_{th}}{I_{d\,on}} = R_{eq} \tag{10.16}$$

when one sticks to the assumption of complete velocity saturation. Quantities U_{dd}, U_{th}, and $I_{d\,on}$ are indeed much easier to look up in library and technology manuals. Note that R_{eq} is not quite the same as the MOSFET's "on"-resistance R_{on}.

Chapter 11

Physical Design

11.1 | Agenda

Physical design is concerned with turning circuit netlists into layout drawings that

- Are amenable to fabrication with some given target process,
- Logically function as expected in spite of numerous parasitic effects,
- Meet ambitious performance goals in spite of layout parasitics, and
- Keep fabrication costs down by minimizing die size and by maximizing yield.

The degree to which physical issues are placed under control of IC designers is highly dependent upon fabrication depth and design level. While global interconnect must be planned for in every design project — even when opting for field-programmable logic — few digital designers continue to work with layout at the detail level today. This chapter is organized accordingly. Sections 11.2 through 11.4 cover issues that are relevant in any IC design, such as floorplanning and packaging, while the material on detailed layout is postponed to section 11.5. Section 11.6, finally, collects discussions of various destructive phenomena that must be contained.

11.2 | Conducting layers and their characteristics

The layers made available by VLSI processes greatly differ in their geometric and electrical characteristics. Let us begin by studying those properties and differences.

11.2.1 Geometric properties and layout rules

The transfer of layout patterns to the various layers of material on a semiconductor die is obtained from photolithographic methods followed by selective removal of unwanted material. Numerous effects concur to limit the achievable resolution:

- Tolerances and misalignments of photomasks,
- Wave diffraction and proximity effects,
- Uneven profile together with shallow depth of focus,
- Reflections from underlying layers,
- Tolerances in photoresist exposure,
- Etching along undesired dimensions,
- Lateral diffusion of dopants, and
- Spiking of aluminum.

To ensure reasonable fabrication yields in spite of such imprecisions, process engineers define a set of **layout rules**[1] that must be observed if a design is to be manufactured using their fabrication process. Layout rules must be viewed as a compromise among layout density, yield, and desirable electrical characteristics. A rule deck for a modern CMOS process includes hundreds of rules, and increasing process complexity is driving increasingly complex rule decks.

Layout rules also provide a workable interface between design and manufacturing. While layout designers must know and respect all geometric rules imposed by the target process, they are freed from knowing the target process and its limitations and peculiarities in too much detail. To protect themselves against incongruous designs and improper claims, silicon foundries invariably ask for a proof of compliance with layout rules, termed **design rule check** (DRC), before accepting a design for fabrication.[2]

Observation 11.1. *A set of layout rules provides a clean separation between the responsibilities of VLSI designers and manufacturers. A foundry typically rejects any design that cannot be demonstrated to fully comply with all layout rules of the target process.*

As current rule decks include so many rules, as those differ from process to process, and as numbers change from one process generation to the next, it makes no sense to attempt to enumerate them here. Instead, we will just explain the motivations behind key geometric restrictions. Most rules fall into one of the categories illustrated in fig.11.1.[3]

Minimum width

A minimum size is stipulated for each physical layer essentially to prevent a structure from falling into electrically disjoint pieces as shown in the example of fig.11.1a. Long and narrow lines with no other structure nearby tend to be more susceptible to overetching than densely packed items of the same width and are, therefore, sometimes required to be of larger width.

Minimum intralayer spacing

Minimum spacing constraints between layout structures on a single conducting layer serve to prevent short circuits between adjacent items, see fig.11.1b. Minimum intralayer spacing is sometimes made dependent on whether two structures are electrically connected or not. As an example, one 130 nm

---

[1] Though dated and imprecise, the term "design rule" continues to be in common use.

[2] Rule-based layout checking has its limitations, and these are to be discussed in section 12.4.2.

[3] There is no universally agreed definition of the terms being introduced here. NXP Semiconductors, formerly part of Philips, for instance, uses "space" for intralayer spacing, "separation" for interlayer spacing and "overlap" for enclosure, whereas other sources interpret "extension" and "overlap" the other way round.

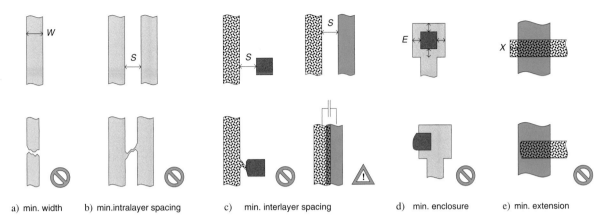

a) min. width   b) min. intralayer spacing   c)   min. interlayer spacing   d)   min. enclosure   e) min. extension

Fig. 11.1   Minimum size rules (top row) and the likely consequences of violating them (bottom row).

process asks for a minimum well-to-well separation of 630 nm if wells are at the same potential, and of 1000 nm otherwise.[4]

Minimum interlayer spacing

Minimum spacing constraints between layout structures on different layers are specified to exclude undesired interactions. Consider a MOSFET, for instance. In the absence of sufficient lateral separation between the poly gate and a nearby source/drain contact, a leakage path might form. A more subtle example also shown in fig.11.1c is a poly wire placed next to an unrelated diffusion area. Zero spacing between the two would result in their edges lying flush on top of each other with only thin oxide in between, thereby inflating mutual capacitance.

Minimum enclosure

Minimum enclosure constraints relate to layout structures on different layers and are uniform in all directions, see fig.11.1d. In the case of a contact or a via, they make sure that the layers involved connect properly in spite of minor misaligments of masks and tolerances of the etching process. More such constraints are concerned with the embedding of diffusion areas within wells. Yet another enclosure rule requires the top-level metal to extend underneath the overglass passivation layer around all pad openings such as to obtain a hermetic seal.

Minimum extension

Much as for enclosures, minimum extension rules apply to overlaps between different layers, yet they have an orientation instead of being uniform. Consider the MOSFET depicted in fig.11.1e,

---

[4] The problem with voltage-dependent rules is that DRC essentially starts from geometric data. Some DRC tools cannot anticipate electrical parameters of a circuit in operation. Others could, but the way the DRC has actually been implemented in the foundry kit prevents them from doing so. Practical DRC runs thus often work from the assumption that all wells of one type are at the same potential, which applies to standard digital circuitry, but not necessarily to analog subcircuits, voltage converters, pad drivers, and the like.

for example, where the polysilicon gate acts as a mask during implantation of source and drain. If the poly were allowed to end flush with the diffusion areas, a narrow conducting diffusion channel might form underneath the gate's edge, thereby preventing the transistor from ever fully turning off. A positive extension of the gate area perpendicular to the source–drain channel provides some safety margin both against lateral diffusion and against misalignment of masks.

Maximum width

Much as in the examples presented so far, most design rules specify <u>minimum</u> dimensions. <u>Maximum</u> width requirements also exist and typically refer to contact and via openings, see fig.11.2a. Contacts and vias are fabricated by etching cuts into a dielectric layer and by filling them with tungsten plugs before depositing the next metal layer on top, see fig.11.2b. A maximum opening is always prescribed to warrant plugs of uniform quality and to favor a planar surface.[5] Often the same number is also specified as a minimum dimension, which implies that a contact/via can be manufactured to one legal size only. Larger contacts and vias, as required on wide interconnect lines, must be broken up into an array of smaller features, a technique which is known as **stipple contact/via** and shown in fig.11.2a.

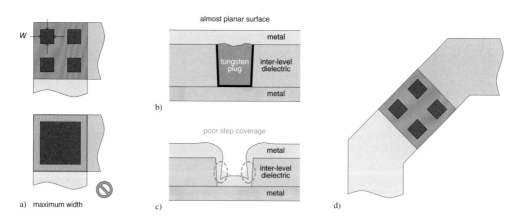

Fig. 11.2 Maximum size rule for contacts and vias. Stipple contact/via versus an oversize single contact/via (a). Cross section of a plugged via (b) and of a historical sink-in via (c). Electromigration-aware stipple contact/via (d), to be explained in section 11.6.1.

Density rules

Planar density is defined as the area occupied by all structures on a layer within a given layout region divided by the overall area of that region. Density rules specify lower and upper bounds for occupancy. As an example, the metal layers could be asked to occupy, on the average, no less than 20% and no more than 80% of each 1 mm by 1 mm region. A multitude of small unconnected

---

[5] Traditional fabrication processes did without plugs. Instead, a contact/via was obtained by having the upper metal sink into a cut previously etched into the intermetal dielectric underneath, see fig.11.2c. As the resulting coating was thinner on sidewalls than on level surfaces, the resistance and maximum current density of such a contact/via were largely determined by the circumference of the opening rather than by its area. Large contacts/vias thus had to be stippled much as with modern plugs, although the reason was quite different.

dummy patterns must be filled in where the payload structures do not occupy a region to a sufficient degree. Density rules have been introduced along with chemical mechanical polishing (CMP) to prevent unacceptable dishing of the softer material when two materials of different hardnesses, such as copper and silicon dioxide, are subject to planarization.

Antenna rules

During manufacturing steps such as reactive ion etching, poly and metal structures collect the charge carriers they get bombarded with, resulting in a build up of voltage. Due to their extremely thin gate dielectrics, any MOSFETs connected to such an "antenna" become exposed to strong vertical fields. While the accumulated charge gets reconducted to bulk via Fowler–Nordheim tunneling, the gate oxide material suffers in the process. To prevent the dielectric from deteriorating to a point where gate leakage, reliability, and threshold voltage shifts might become a problem, manufacturers constrain the ratio of exposed conductor surface to the connected transistor gate area for each layout structure.

11.2.2 Electrical properties

As the conducting layers are made of different materials and with different thicknesses, they necessarily differ in their electrically characteristics, see tables 11.1 and 11.2 for actual numbers. An overriding concern of interconnect design is to obtain low resistances and low parasitic capacitances so as to minimize switching times and interconnect delays.

Conductivity

Observe the huge difference in sheet resistance between metal and silicon layers. Were it not for silicidation, the contrast would be even more significant. In most fabrication processes, the conductivity of source, drain, and gate areas gets improved by depositing a metal **silicide** such as NiSi or $CoSi_2$ on top of the silicon material. The term **polycide** refers to a silicide layer placed over polysilicon whereas silicide deposited on top of both poly and diffusion areas is known as **salicide**, an acronym for self-aligned silicide.

Sustainable current density

Reliability concerns impose a limit on the amount of current a line can handle in practice. This is because metal conductors tend to disintegrate to the point of rupture when subjected to excessive current densities over prolonged periods of time. Section 11.6.1 is devoted to this undesirable phenomenon known as **electromigration**. As a consequence, VLSI designers must plan interconnect networks, particularly VDD and VSS nets, not only as a function of resistance but also as a function of current load.

11.2.3 Connecting between layers

The various layers and fabrication processes further differ in how layout structures on one layer can be made to connect to layout structures on other conducting layers. Figure 11.3 shows a nice view of an IC with five layers of metal.

Examples

Table 11.1 Conducting layers of a 250 nm 2.5 V 5M1P twin-well poly-gate CMOS process. The number of mask layers is 23.

layer	metals (aluminum)			poly	diffusions		wells	
	M5	M4...2	M1	P1	n^+	p^+	n^-	p^-
thickness or depth [nm]	900	600	600	250	180	180	1800	n.a.
min. width [nm]	420	400	320	240	300	300	1200	1200
min. spacing [nm]	540	400	320	320	400	400	2000	2000
min. pitch [nm]	960	800	640	560	700	700	3200	3200
sheet resistance [$\Omega/\square$]	36m	53m	53m	2.5	2.5	2.0	400	n.a.
max. current density [mA/μm]	1.5	0.8	0.8	n.a.	n.a.	n.a.	n.a.	n.a.

Table 11.2 Conducting layers of a 130 nm 1.2 V 8M1P twin-well poly-gate CMOS process. Current densities refer to junction temperature $\theta_j = 100\ ^\circ$C.

layer	metals (copper)			poly	diffusions		wells	
	M8/7	M6...2	M1	P1	n^+	p^+	n^-	p^-
thickness or depth [nm]	800	320	320	180	120	190	1500	n.a.
min. width [nm]	400	200	160	120	160	160	630	630
min. spacing [nm]	400	200	160	300	200	200	1000	1000
min. pitch [nm]	800	400	320	320	360	360	1630	1630
sheet resistance [$\Omega/\square$]	27m	70m	70m	8.0	7.0	7.0	380	n.a.
max. current density [mA/μm]	8.0	2.56	2.56	n.a.	n.a.	n.a.	n.a.	n.a.

□

The conducting paths available in some given target process are best summarized in a graph where a node stands for a conducting layer and where an edge expresses the fact that two overlapping polygons connect.[6] The absence of an edge between two nodes indicates that there is no way to connect the two layers directly. Figure 11.5 shows an example for a baseline CMOS process. Observe from the drawing that there is no way to connect polysilicon to diffusion by way of a simple contact; two contacts and a piece of first-level metal must be used instead.[7]

While **contact** is the name for a connection between a metal and one of the silicon layers, a connection between two superimposed layers of metal is commonly termed a **via**. Placing a via on top of a contact, or two vias on top of each other, always mandates a small piece of regular metal in between. Such constructions are termed **stacked contacts/vias** or **staggered contacts/vias** depending on whether the contacts/vias are aligned or offset laterally, see fig.11.4. Staggered contacts/vias tend to be less area-efficient than stacks.

---

[6] It goes without saying that two overlapping polygons on the same conducting layer are electrically connected and thus form a single electrical node.

[7] There are exceptions to this rule. Fabrication processes optimized for memory design support silicide straps that connect poly to diffusion directly without involving metal1 or a contact plug, see fig. 11.23 for an example. Also, many analog processes offer a second poly and special intermetal dielectrics as extensions of the baseline process of fig.11.5 for constructing linear on-chip capacitors without occupying too much area.

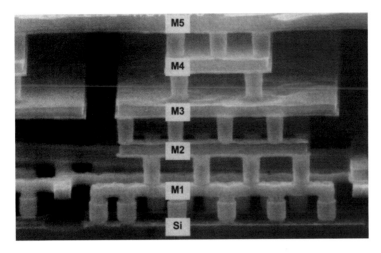

Fig. 11.3 Cross section of a 350 nm CMOS process with five layers of metal (from Jessi News April 1996, cooperation of France Telecom CNET, SGS-Thomson, and Philips Semiconductors).

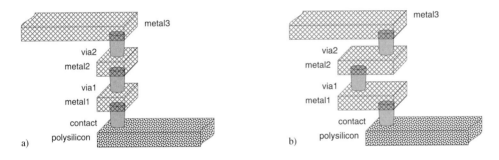

Fig. 11.4 Stacked (a) versus staggered (b) contacts and vias.

Observation 11.2. *Each layer of metal typically connects to the next metal layer above and the next layer underneath; only the first-level metal can connect to the various silicon layers. Connecting two non-adjacent wiring layers requires going through all layers in between.*

11.2.4 Typical roles of conducting layers

We have, so far, discussed the various conducting layers provided by VLSI technology and have learned how to capture their most essential characteristics in a few tables and graphs for any given MOS process. Putting everything together, it is now possible to determine the preferred utilizations for the various layers.

Almost all interconnect is done in metal. Poor conductivities discourage the usage of polysilicon and diffusion layers for nets where low delay and/or low voltage drop are vital. Polysilicon and diffusion are, therefore, confined to strictly local wiring. Metal layers are used as follows.

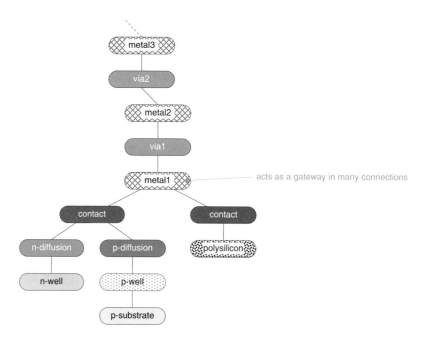

Fig. 11.5 Conducting paths between conducting layers in some bulk CMOS twin-well p-substrate baseline process (example reduced to 3M1P, higher-level metal layers not shown).

- The first-level metal is instrumental for intracell routing because it can freely traverse layout items on any other conducting layer, can connect to all silicon layers with no detour, and acts as gateway to all metal layers above.
- Intermediate-level metals are mostly used for general extracell signal interconnect.
- High-level metals are thicker, which lowers their sheet resistance and has them withstand higher currents. This along with their coarser pitches makes them more suitable for distributing power, clock, and other critical nets over long distances.

With up to nine layers of metal, today's CMOS processes provide generous interconnect resources, and very little area is lost to wiring.[8]

---

[8] This was not so with the single-metal processes of the pioneer days. All but the shortest wires had to be implemented with metal and poly exclusively as no other layer of acceptable conductance was available. Rows of standard cells had to alternate with **routing channels** that were to accomodate the interconnect lines. Standard cells used to have their input and output connectors at their top and bottom edges, as shown in fig.11.20b, so as to facilitate channel routing and cell traversal. With routing channels twice or even three times as high as the cell rows, layout density was very poor, to say nothing of RC delays.

Processes with three layers of metal offered a major improvement as they were the first to allow for **over-the-cell routing**, i.e. placing wires on top of the active cells. With polysilicon and metal1 handling the intracell wiring, two metal layers became available for non-local routing with no obstacles in their way. Yet, three metal layers were not enough to do away with routing channels completely, so more and more have been added. Even with five or six metals, extra space occasionally needs to be set aside between adjacent cells or cell rows when there are just too many nets to make all the wiring fit on top of the cells.

11.3 | Cell-based back-end design

11.3.1 Floorplanning

In analogy with the floorplan of a house, the floorplan of a chip indicates how the silicon die (the construction volume) is going to be partitioned in order to accommodate the various building blocks (the rooms), the busses in between (the hallways), the location of input and output pads (the doors), as well as the distribution of power, clock, and other important nets (the plumbing). The floorplan has a significant impact on performance and costs of the final product.

Floorplanning is the activity of organizing the physical aspects of a VLSI circuit and is a continuous process that accompanies all steps of its conception.[9] At any point in a design, the floorplan documents the actual state of the planning process and keeps track of items such as

- Partitioning into major building blocks (datapaths, controllers, memories, megacells, etc.),
- Number and anticipated sizes, shapes, and placement of all such blocks,
- Package selection and pin/pad utilization,
- Wide busses and electrically critical signals,
- Clock domains (frequency, conditional vs. unconditional clocking),
- Voltage domains (power dissipation, power density, local current needs), and
- On-chip power and clock distribution schemes.

By contributing estimations about the physical characteristics of hypothetical circuit implementations, floorplanning helps VLSI designers make good decisions and so effectively guides their search towards sound and economical solutions. Floorplanning eventually culminates in a fairly detailed and precise final floorplan that then serves as target specifications during place and route (P&R) and chip assembly.

Example step 0: Overall architecture

Floorplanning is primarily a matter of common sense and of sound engineering practices. In order to illustrate the process, let us follow a circuit project, e.g. an add–drop switch for a proprietary packet-based data exchange network. The circuit shall connect to up to four identical network branches. Each port is bidirectional and comprises two pairs of differential signals.[10] When the switch receives a data packet, it analyzes the address header, determines the appropriate destination, does the necessary processing of the header, and sends the packet off over the appropriate network branch.

The circuit shall further connect to a local host via a microcomputer interface. The switch accepts data from that interface and translates them into the standard packet format before transmitting them through the network (add). Conversely, whenever the circuit receives a network packet addressed to the local host, it strips the header off and delivers the payload data (drop). The evolution of a floorplan for an ASIC for this hypothetical example is shown in a series of snapshots, see figs.11.6 through 11.10.

---

[9] The various steps that make up back-end design are put into context in fig.13.15. You may want to consult that figure during the subsequent discussion.

[10] Much like a FireWire interface.

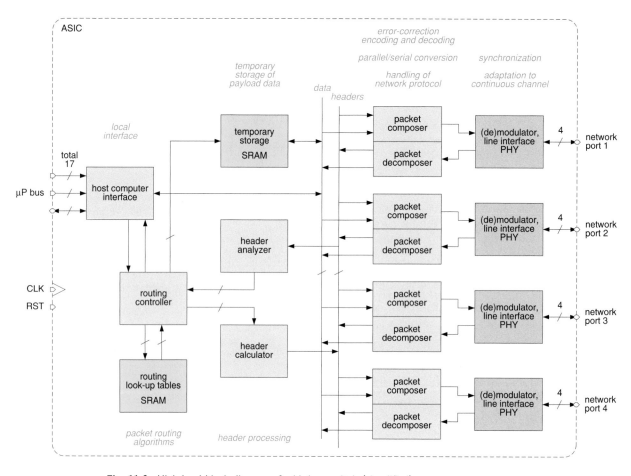

Fig. 11.6 High-level block diagram of add-drop switch (simplified).

□

11.3.2 Identify major building blocks and clock domains

Partitioning means subdividing a system into blocks and subblocks and, if necessary, into several chips. Most often this is a critical issue because total pin count dominates packaging, board, and assembly costs. The large node capacitances found on a printed circuit board (PCB) also limit performance and inflate power dissipation. While these arguments are in favor of integrating a system on as few chips as possible, there are also reasons for not going that far.

Large standard RAMs and ROMs are typical examples as it makes little sense to incorporate functions that are cheaply available as catalog parts, but that would take up a lot of die area when implemented within an ASIC. Also, generic parts that mix and match with others in a modular way tend to find wider markets than more complex and more specialized ones.[11]

---

[11] Please refer to sections 11.4.7 and 13.8 for more thoughts on this.

Example step 1: Circuit partitioning

In the case of the add–drop switch, a physical link interface (PHY) which combines modulator/demodulator with line drivers/receivers is required for each of the four network ports. We plan to reuse a proven design which is available as a megacell (hard layout) from a previous project. Temporary data buffering requires storage space as do look-up tables for holding routing information. We opt for two identical on-chip SRAM macrocells in view of pin count and data bandwidth, and because the required storage capacities are rather modest. The remaining portions of the packet switch are to be captured as HDL code and shall be synthesized into standard cells. Because of its relative simplicity, the project makes do with a single clock domain. It has been decided to use a tree for distributing a one-phase clock.

☐

11.3.3 Establish a pin budget

Deciding on the number of pins is a highly critical issue because the total pin count obviously affects package selection and packaging costs. It also impacts the effective die size and — as a consequence — fabrication costs per unit. In the case of wire bonding where pads are arranged at the perimeter of the chip, two outcomes are possible when a **padframe** is being prepared for a given core, see fig.11.7.

Corelimited design.
> A large core is surrounded by relatively few pads. Overall die size is primarily determined by core size and, hence, by circuit complexity. While porting the design to a denser target process is bound to reduce core size, cost calculations are needed to find out whether this makes sense economically.

Padlimited design.
> A large number of pads encloses a comparatively small core. Their number defines the necessary circumference and so imposes a minimum die size. Whether a denser or a not so dense fabrication process is chosen has little impact on die size but is likely to change the picture in terms of costs.

As an important proportion of die area is lost in highly padlimited situations, floorplanning is again bound to impact architectural decisions at this point. Just consider system partitioning (e.g. on- vs. off-chip memories), input/output communication scheme (e.g. parallel vs. serial, unidirectional vs. bidirectional), and test strategy (e.g. block isolation vs. built-in self-test (BIST)).

As a secondary measure, pad geometry can be chosen in such a way as to minimize die size. Narrow tall pads that get arranged perpendicularly to the chip's perimeter are optimal for padlimited designs. Corelimited situations, in contrast, are better served with wide flat pads placed along the chip's perimeter.[12] Special corner pads, depicted in fig.11.7b, and staggered pads arranged in two concentric circles as shown in fig.11.13 further help to reduce the chip's circumference for a given

---

[12] Most ASIC libraries include pads for padlimited situations exclusively, because situations where the application asks for as many pads as possible tend to prevail with today's extremely dense technologies.

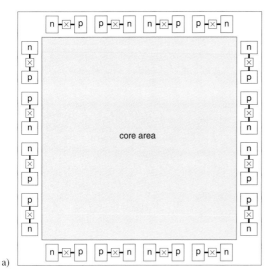 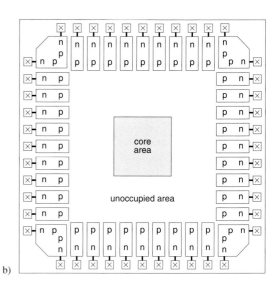

Fig. 11.7 A core- (a) versus a padlimited floorplan (b).

pad count. The more up-to-date option of covering the die with bumps instead of attaching bond wires along its periphery will be discussed in section 11.4.

A good pinout accounts not only for on-chip interconnect but also for board-level wiring. Though most details depend on the package type chosen, it always pays to

- Consciously plan the chip's footprint and signal routing on the PCB,
- Provide power and ground in the package's outermost circle of pins or balls, where they can be made to connect to solid planes or wide lines,
- Arrange signals from the various power domains as contiguous blocks,
- Keep bus signals together and have them follow some logical ordering,
- Provide ample return paths for switching currents, and
- Make differential signals occupy adjacent pins.

11.3.4 Find a relative arrangement of all major building blocks

Local connections are always preferable for reasons of performance and of area efficiency. Consequently, building blocks that exchange data at high bandwidths must be placed next to each other. Ideally a VLSI architecture is designed already with floorplanning in mind.

What is sought at this point is a planar arrangement of building blocks that

- Respects clock and voltage domains (if there are several of them),
- Makes optimum use of die area by minimizing routing overhead, and that
- Minimizes overall delay on all performance-critical paths.

Many designs combine standard cells, macrocells, and megacells on a single die. A major difference lies in a property that might be termed geometric plasticity. A building block assembled from

Example step 2: Pin budget

Table 11.3 | Directory of pins.

Function	Pins	Subtotal
clock and reset	2	
boundary scan test	5	
Basic equipment		7
network ports	4 · 4	
µP port data bits	8	
µP port address bits	4	
µP port control lines	5	
Functionality		33
core ground	4	
core supply	3	
padframe ground	7	
padframe supply	7	
Power		21
Spare		3
Total		64

☐

standard cells can be made to assume almost any shape as long as there is sufficient space to accommodate the cells along with their interconnect. Also, a standard cell block can always be broken into more than one standard cell area. Conversely, several standard cell blocks can be collected into a single area.

As opposed to this, megacells come with fixed dimensions and fixed connector locations. Flipping and rotating are the only options for the floorplan. Much the same also applies to macrocells, although many generators allow for two or three different aspect ratios for the same functionality.

11.3.5 Plan power, clock, and signal distribution

Interconnect planning essentially implies holding back five adverse effects listed in the table below along with their relative importance for various types of nets.

adverse effect	power and ground	clock distribution	signal wiring
resistive voltage losses	high	low	low
ground bounce	high	low	low
electromigration	high	low	low
crosstalk	low	high	high
interconnect delays	low	high	high

Example step 3: Arrangement of building blocks

Figure 11.8 sketches a first tentative floorplan. It has become clear at this point that the design will be corelimited. Also, a decision has been made about the two macrocells' aspect ratios.

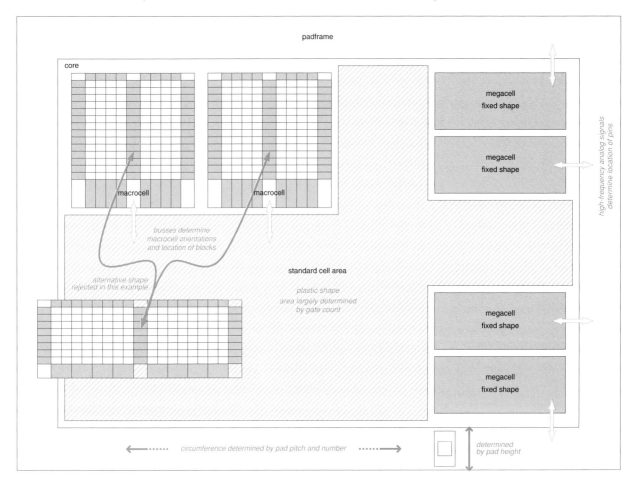

Fig. 11.8 Tentative arrangement of building blocks.

☐

Designers and EDA tools are more or less free to choose

- The routes and shapes of the individual nets,
- The interconnect layers and the layer changes,
- The conductor widths.

The characteristics of the various interconnect layers have been discussed in section 11.2. Performance, cost, and reliability concerns incite designers to make the most efficient possible usage of the available resources. Allocation best follows a priority list that starts with long and critical nets, such as supply and global clocks, and ends with uncritical local nets.

Also, the locations of the connectors for power, clock, and critical signals are now chosen so as to minimize congestion and layout parasitics. These prospective locations will guide the subsequent place and route phase.

Example step 4: Target floorplan
A more detailed floorplan is depicted in fig.11.9. Specific positions have been assigned to all power pads and to key I/O signals. Target regions have been earmarked for all major blocks.

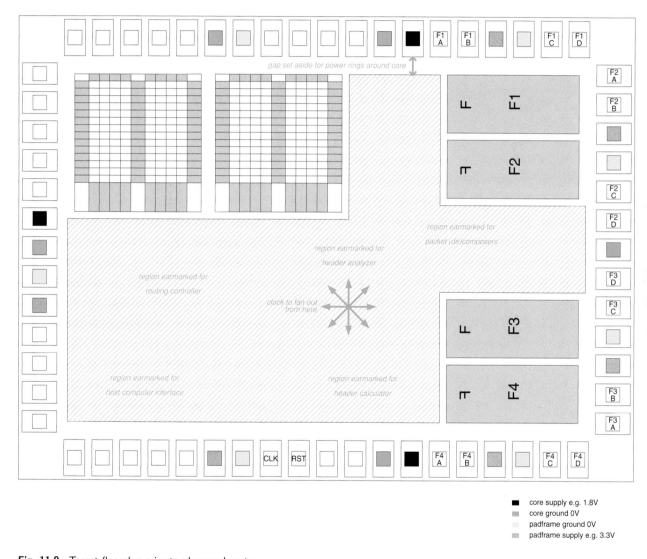

Fig. 11.9 Target floorplan prior to place and route.

11.3.6 Place and route (P&R)

The floorplan serves as a target specification for the placement and routing of cells within each standard cell area. In order to account for unforeseen changes of layout parasitics, modern EDA suites support the reoptimization and rebuffering of synthesized gate-level netlists after initial placement. The dependency is most critical in the case of clock net(s), which is why clock tree generation is postponed until this point in the design flow.

Back-end design remains a tedious exercise, especially when area and/or performance constraints are very tight. Several iterations may be needed to obtain a satisfactory result. The fact that the full set of physical design data must be handled during chip assembly does not help either.[13] For very large designs, good floorplans are structured rather than flat and tend to reflect functional organization. This facilitates decomposition into pieces of manageable size that can be designed and verified on their own by separate persons or teams.

Old hands among physical designers typically begin with those blocks that they expect to remain stable. They often include a number of extra gates and flip-flops in blocks that they suspect will need modifications or bug fixes at a later date. Such spare cells, jokingly referred to as a **sewing kit**, make it possible to make last-minute changes without upsetting the floorplan too much. And should a minor bug surface after prototypes have been manufactured, then chances are that it can be fixed by redoing just a few metal and via masks.

Whenever a major building block has been worked out in detail and its actual characteristics can be stated more precisely, the overall floorplan must be updated and refined accordingly. A major revision of the initial floorplan may prove necessary if area, long path delay, power dissipation, or other critical figures diverge too much from earlier estimates.

Good routing makes a difference in fabrication yield. At the interconnect level, yield enhancement implies:

- Connecting multiple vias in parallel.
- Making the preferred orientations of wires on adjacent metal layers perpendicular to each other to minimize the impact of crosstalk on path delays, both long and short.
- Spacing long lines further apart than required by the minimum separation rule to minimize the chance of shorts and the severity of crosstalk.

---

[13] There was a time when it was very painful to refine a finished design because automatic P&R tools tended to exhibit erratic behavior on flat floorplans. A minor modification made to a netlist, such as resizing a buffer in accordance with its actual load found during layout extraction, for instance, might have led to a totally different arrangement of cells and wires. This, in turn, modified the capacitive load on most nodes, thereby causing radical alterations of path delays and ramp times. What had been a totally unobtrusive path prior to modification all of a sudden became a crucial one, while the path for which the modification had been undertaken proved largely uncritical afterwards. All too often, timing closure remained an elusive goal.

The situation has improved with the advent of EDA tools that do not re-place all cells and re-route all wires after each and every design modification. Instead, the software saves and preserves the overall layout arrangement and limits the impact to locally exchanging a few cells and/or re-routing a number of interconnect lines. **Engineering change order** (ECO) is a name for such belated but relatively minor design changes.

A layout drawing that includes angles of $90°$ exclusively is said to conform to **Manhattan geometry**, while **Boston geometry** implies that any angle is an integer multiple of $45°$. In spite of its potential of shortening interconnect lengths significantly, Boston geometry is not (yet) regularly applied to on-chip wiring.

Scan paths offer some potential for improvement too. It is typically possible to reorder bistables and entire blocks in a scan chain in such a way as to make most scan connections local with no cutback in observability or controllability. A similar reasoning applies to block isolation circuitry and other test structures.

11.3.7 Chip assembly

Chip assembly is the final phase of physical design during which a design gets assembled from building blocks the inner design of which has been completed beforehand. Chip assembly includes the following steps.

- Place all top-level building blocks.
- Interconnect those blocks to obtain the chip's core.
- Prepare the padframe required to electrically connect to the external world.
- Connect core and padframe to complete the chip's design.

Input and output subcircuits are intimately related to critical issues such as ESD protection, latch-up avoidance, ground bounce, and I/O timing. Rather than creating their own pads, drivers, and level shifters, designers are well advised to use proven designs from a commercial library, the layouts of which have been developed by experienced specialists with detailed process knowledge. Also, do not assemble a padframe in such a way that MOSFETs of opposite polarities happen to lie next to each other because this is detrimental to latch-up immunity, as will be explained in section 11.6.3.

Example

The final result is summarized by fig.11.10. Observe the wide metal stripes that run perpendicularly to the cell rows in order to attenutate the current densities and voltage drops on the narrow horizontal supply lines there. Placing VDD and VSS lines on top of each other contributes to on-chip bypass capacitance.

Example step 5: Chip assembly

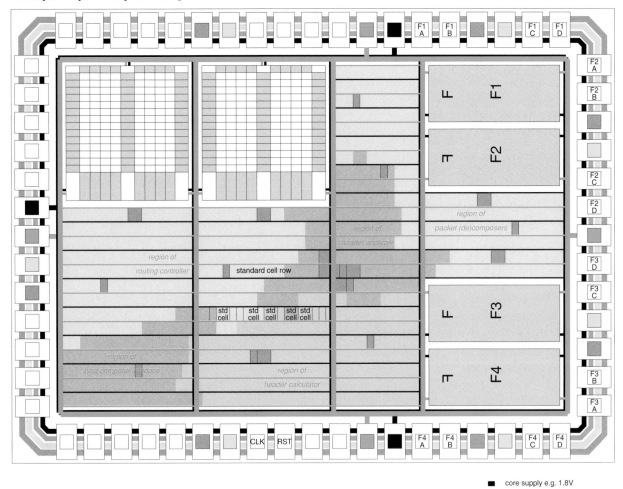

part of clock
circuitry

■ core supply e.g. 1.8V
■ core ground 0V
■ padframe ground 0V
■ padframe supply e.g. 3.3V

Fig. 11.10 Final floorplan. Block boundaries, power distribution, and clock circuitry are highlighted, all other details are abstracted from.

11.4 | Packaging

In the early days of microelectronics, packaging was not much more than encapsulation. Under the influence of ever augmenting pin counts, operating frequencies, power dissipations, and the like, it has evolved into a key element of electrical and thermal design. IC packages serve the following roles.

- Protect semiconductor dies against mechanical stress and other environmental attacks.
- Expand connector geometry so that contacting becomes possible
  at the next higher assembly level.
- Provide electrical connection with the surrounding circuitry,
  with a particular emphasis on low impedance for power and ground nets.
- Facilitate the handling of parts during shipping and board assembly.
- Carry away the thermal power while keeping die at an acceptable temperature.

In order to meet divergent needs, a great variety of package types has been developed over the years and table 11.4 attempts to summarize a few of their most essential characteristics. The reader is referred to the specialized literature such as [483] [484] [485] [327] [328] [329] for more profound information on packaging technology. Datasheets, selection guides, availability, and pricing data must be obtained from industrial package suppliers.[14]

Table 11.4 | An alphabet soup of popular package types, past and present (subset).

		Package terminals		
Package type	mount[a]	location	typ. pitch [mm]	typ. count
single in-line package (SIP)	TH	1 edge	2.54	2–12
dual in-line package (DIP, DIL)	TH	2 edges	2.54	6–40
pin grid array (PGA)	TH	surface	2.54, 1.27	72–478
small outline package (SOP, SOIC)	SM	2 edges	1.27	8–44
thin SOP (TSOP)	SM	2 edges	1.27, 0.8, 0.5	24–86
shrink SOP (SSOP)	SM	2 edges	0.8, 0.65, 0.5	8–70
leadless chip carrier (LLCC, LCC)	SM	4 edges	1.27	16–84
quad flat package (QFP, QFN)	SM	4 edges	1.27, 0.8, 0.65	32 upwards
fine-pitch QFP (FQFP)	SM	4 edges	0.5, 0.4, 0.3	up to 376
small outline J-leaded pack. (SOJ)	SM	2 edges	1.27	24–44
leaded chip carrier (LDCC, JLCC)	SM	4 edges	1.27	28–84
land grid array (LGA)	SM	surface	1.27, 1.0	48 upwards
fine-pitch land grid array (FLGA)	SM	surface	0.8, 0.5	up to 1933
ball grid array (BGA)	SM	surface	1.5, 1.27, 1.0	36 upwards
fine-pitch BGA (FBGA)	SM	surface	0.8, 0.65, 0.5, 0.4	up to 2577

[a] TH stands for through-hole and SM for surface mount.

Instead of going into more detail on individual packages, we will outline the packaging process, advise prospective circuit designers on how to establish the necessary instructions for wire bonding, and give decision criteria for selecting among competing packaging techniques.

Most VLSI chips continue to be packaged individually, that is encapsulation takes place before interconnection with other electronic components does. Small outline packages (TSOP, SOJ) and

---

[14] Such as Amkor, Fujitsu, Kyocera, NEC, NTK, Signetics Korea, and Tessera, to name just a few.

ball grid arrays (BGA, FBGA) that are widely used in memory modules are prevalent today. To gain an appreciation of the standard packaging process, let us follow a processed wafer until chips are encapsulated in a quad flat package (QFP), see fig.11.11.

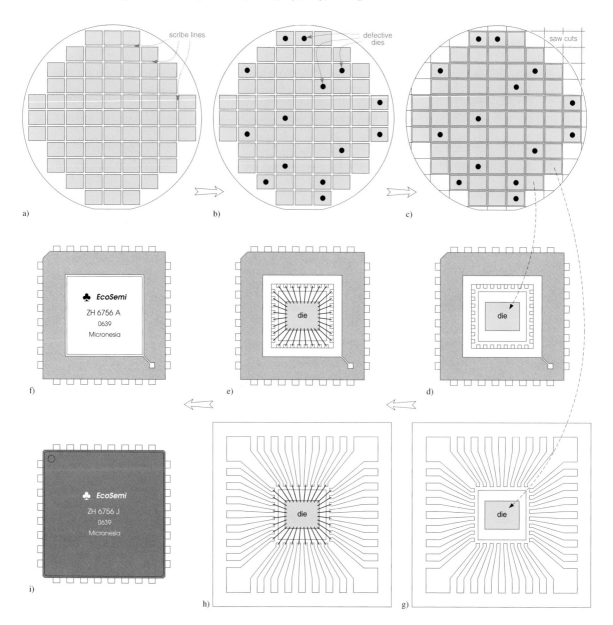

Fig. 11.11 Testing and encapsulation steps (simplified, not drawn to scale). Processed wafer (a), probed wafer with defective circuits inked (b), wafer after sawing (c), a good die attached to package cavity (d), wires bonded to lead frame (e), final IC after sealing, testing, and stamping (f). (g,h,i) correspond to (d,e,f) for a plastic package.

11.4.1 Wafer sorting

The operation of any wafer processing line is constantly being monitored with the aid of generic test structures that get added as part of mask preparation. These **process control monitors** (PCMs) are evaluated after major fabrication steps and one more time before the finished wafers are cleared for packaging. Any wafer found to suffer from fatal defects or from excessive parameter variations is scrapped. PCMs typically get "hidden" in the narrow scribe lines, aka saw lanes, that separate adjacent circuits from each other so as not to sacrifice more of the precious wafer area than absolutely necessary.

11.4.2 Wafer testing

To avoid the costs of packaging defective parts, the prospective ICs are then subjected to a first series of functional tests while still being part of the processed wafer, see fig.11.12. A set of ultrafine needles, firmly held in place by a **probe card**, is lowered onto a wafer until all needles establish electrical contact to the bonding pads of what eventually is to become an IC.[15] Stimuli are applied to the inputs and the actual responses from the outputs are checked against the expected ones. The operation must not take more than a few seconds for reasons of cost and is repeated for all circuits on a wafer. The outcome is kept on record electronically. Traditionally, defective circuits were marked by a droplet of ink and this is reflected in fig.11.11 as wafer testing would otherwise leave no perceptible trace in the drawing.

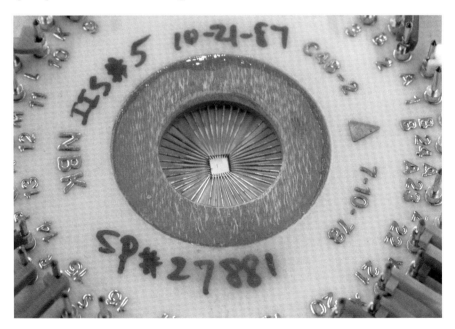

Fig. 11.12 A circuit under test contacted by a probe card (photo courtesy of Dr. Norbert Felber).

---

[15] Observe that each design requires a probe card of its own unless the VLSI designers of a company are willing to agree on a few standard padframes with predefined pad locations.

11.4.3 Backgrinding and singulation

Standard wafers are between 720 μm and 770 μm thick. Not only ultraportable applications such as smart cards and flash memory products but also advanced packaging techniques such as chip stacking mandate that die be thinned. In an optional step, the wafer's back surface is subjected to grinding by disks with embedded diamond abrasives. Final thickness can be as low as 75 μm, although most thin-wafer production averages 250 μm to stay clear of yield losses during grinding, handling, and later packaging steps.

In preparation of the next move, the wafer is placed on a sheet of elastic and sticky material referred to as **blue film** because of its original color.[16] The wafer then gets sawn apart by two orthogonal series of parallel cuts, each of which traverses the wafer from one rim to the opposite one. The depth of the circular saw is adjusted so as to separate the dies without severing the film so that it continues to keep the dies in place. Note that saw lanes continue to be designated by their historical name **scribe line**.

11.4.4 Encapsulation

The blue film is then stretched to increase the separation between the individual dies. A vacuum wand picks one good (non-inked) circuit after the other for encapsulation. In a step termed **die bonding**, aka die attach, a good die is placed in the cavity of the package, where it is mechanically fastened by means of solder or epoxy resin compound. The operation takes approximately one or two seconds per chip on an automatic assembly line.

Next follows **wire bonding**, whereby electrical connections are established between bond pads on the silicon die and their counterparts on the package leads, see fig.11.13. Thin gold or aluminum wires with a diameter of 17, 20, 25 or 33 μm are typically employed. Bonding occurs by way of pressure, ultrasonic energy, and/or heat.[17] Automatic equipment operates at a rate in excess of six wires per second. The minimum pitch imposed ranges between 40 μm and 110 μm. The narrower the pitch, the thinner the bond wires that must be used, and the shorter the admissible wire length. Narrow pitches also are detrimental to packaging yield and manufacturability. Finally, the package cavity is sealed with a lid made of metal or ceramic material.

For plastic packages, the overall procedure is similar except that the die is bonded to a chip support paddle in the center of the metal lead frame. Another major difference consists in the **postmolding process** during which epoxy resins are molded around the lead-frame-and-chip assembly after wire bonding. The leads are then trimmed and bent to their final shapes.

11.4.5 Final testing and binning

The packaged chips are subject to yet another series of functional "go/no go" tests. In addition, electrical and timing parameters are measured to protect against process variations. Only parts that fully qualify are stamped or marked with a laser. Processors, memories, and other catalog parts are typically **binned**, that is classed and labeled with a speed and/or power grade. Finished chips can hence be marketed at different prices as a function of the maximum clock rate they can

---

[16] Blue film is also known as wafer tape and not necessarily of blue color.
[17] [330] discusses the process and its impact on yield and reliability in great detail.

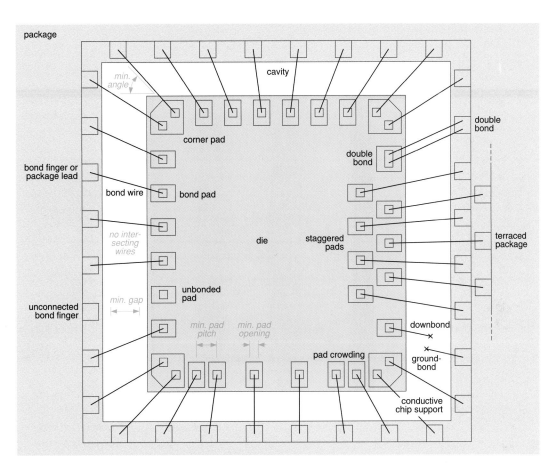

Fig. 11.13 Bonding diagram with related terms and rules.

sustain or of the leakage current they drain, for instance. An individual supply voltage may further be determined to allow running of each circuit at close to its optimum operating point.

11.4.6 Bonding diagram and bonding rules

Instructions on how to connect the pads on the die to the available package leads during wire bonding are conveyed in a **bonding diagram**, see fig.11.13. Practical considerations severely limit the choice of admissible bonding patterns. Intersecting wires, for instance, are disallowed. Concerns regarding package selection, package pinout, floorplanning, and bonding rules thus combine in establishing a valid bonding diagram. As actual bonding limitations differ from one manufacturer to the next, one must always check for instructions.

> Hint: The bonding rules given here do not relate to any specific manufacturer but represent conservative choices that are widely accepted. More aggressive bonding patterns should be adopted only if explicitly approved by the manufacturer.

- Do not design dies with aspect ratios outside the interval $[\frac{1}{2}...2]$.
- Allow for a minimum gap of 0.6 mm between cavity and die on all four sides.
- Make 25 μm bond wires no shorter than 1.0 mm and no longer than 3.5 mm.
- Avoid downbonds, groundbonds, and double bonds.
- Respect a minimum angle of 45° between bond wires and chip edge.
- Make all bond areas square with a minimum overglass opening of 75 μm by 75 μm.
- Respect a minimum pad pitch of 90 μm.

From an electrical perspective, recall that some package types provide special low-impedance leads for power and ground. Also, some packages have the chip support internally connected to one of their pins but most do not. If so, that pin must be clearly identified because it galvanically connects to the substrate from the back side. Depending on whether an n- or a p-well process is being used, it will typically have to get hooked to VSS or to VDD respectively.

> Warning: Be cautioned that the routing of package leads (from bond fingers to pins) need not necessarily be the same for any two packages of the same type and pin count! The lack of a standard for pin grid array (PGA) packages is notorious.

Unless this is checked beforehand, chips bonded in exactly the same way but mounted in packages from two different vendors — and sometimes even from the same vendor — may end up with distinct pinout patterns.

> Hint: Components mounted out of position, i.e. mistakenly rotated by 90°, 180° or 270°, are a frequent mishap, especially during prototype manufacturing. Make the desired orientation immediately visible from the hardware itself, do not rely on accompanying documents that are likely to get lost. This applies both to dies within cavities and to packages on printed circuit boards.

11.4.7 Advanced packaging techniques

Diverse application requirements have led industry to develop a great variety of packaging techniques that go well beyond the basic flow explained before.

HIGH-PERFORMANCE PACKAGES

High-performance packages are driven by the need to accomodate ULSI chips that operate at GHz frequencies and dissipate up to 150 W of power. Wire bonding is impractical in such situations. On-chip power distribution would be very demanding indeed as external connections are confined to the padframe around a chip's periphery and severely limited in number. In addition, bond wires and package lead fingers would excessively add to parasitic impedances.

Table 11.5 compares the key characteristics of on- and off-chip interconnects. Note the superior conductivities offered by the comparatively thick copper sheets embedded in a **laminate substrate**. This observation suggests a more sophisticated solution.

Instead of being surrounded by a padframe, the core area is covered with an array of bumps deposited on small islands fabricated with the IC's uppermost metal layer. This makes it possible to connect to anywhere on the chip's surface, even above active circuitry, and so greatly augments the number of external connections. The bumps can be made of solder, gold, or even a conductive

Table 11.5 On- and off-chip interconnect resources compared (approximate values).

Location	conduct. material	min. pitch [μm]	thick- ness [μm]	sheet resistance [mΩ/□]	insulating material	relative dielectric constant []
on-chip upper metal	Al, Cu	0.8	0.9–1.5	30–50	SiO_2	3.9
ceramic substrate	Ag	250	8–10	0.2	Al_2O_3	7–8
epoxy resin laminate	Cu	70	17	0.01	FR4	≈4

polymer to satisfy different requirements of the application with repect to pitch, costs, temperature profile, mechanical stress, and reliability.

The die is then mounted on an epoxy resin laminate using **flip chip** attachement along with a few local bypass capacitors. The reverse side of that high-density substrate is either equipped with another set of larger bumps for soldering to a printed circuit board (PCB) as in the case of a ball grid array (BGA) package, or designed to fit into a land grid array (LGA) socket with spring-loaded contacts. Figure 11.14 shows a third and more onerous option whereby a mechanical interposer carries pins so as to form a pin grid array (PGA) package.

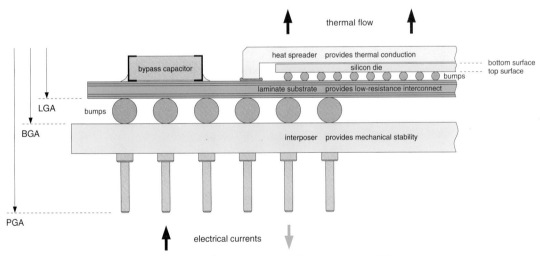

Fig. 11.14 Cross section through a high-performance package (not drawn to scale).

In any case, the six or so metal layers in the laminate serve to collect and redistribute ground, supply, clock, and I/O nets. The laminate substrate is thus understood as an extension of the silicon chip that makes available more metal layers of even lower sheet resistance. A heat spreader placed on the back surface of the die increases the area in thermal contact with a copper or aluminum heat sink that is to be pressed down onto the entire package assembly.

HIGH-DENSITY PACKAGING

High-density packaging uses similar techniques to mount and interconnect bare dies — and often tiny SMD components as well — on a small substrate before encapsulating everything in a common package. Reduction in overall size and lower parasitics over individually packaged components are obvious benefits, but others are even more important.

Many applications ask for a combination of diverse technologies in a single product. A consumer good as common as a cellular phone, for instance, includes CMOS logic, multiple RAMs, flash memory, RF-subcircuits, surface acoustic wave (SAW) filters, and most often also optoelectronics for a camera. Even where this is technically feasible, combining many such features on a single monolithic circuit implies a more complex fabrication process, more lithographic masks, yield issues, difficult engineering tradeoffs, a mixed-everything design flow, costly test equipment, exposure to undesirable interactions, and substantial overall risks.

The alternative is to have each subsystem manufactured with its respective optimal technology and to have them tested separately before mounting them in a common package to obtain a multi-chip module (MCM). The term system-in-package (SiP) has been coined for MCMs that integrate passive and other discrete components too [327] [331]. A more ambitious goal is to combine bypass capacitors, RF thin-film components such as inductors, resistors, filters, and antennas, sensor chips, and possibly even optical waveguides, lenses, and mirrors, in novel composite substrates [332].

Package-on-package (PoP) is a more recent idea whereby individually encapsulated chips with contacts on the bottom <u>and</u> top surface of the package get stacked to assemble larger subsystems. Emerging JEDEC and other industry standards are expected to lead to a wider adoption of chip stacking [333].

In comparison with a system-on-a-chip (SoC) approach, MCM, SiP, and PoP do not inflate up-front costs too much and offer shorter turnaround times as integration happens at the package rather than at the wafer level. Another advantage is the option to assemble multiple product configurations from a limited inventory of monolithic ICs, thereby avoiding the fragmentation of fabrication volume that otherwise results from overcustomized chips.

Example

Hirschmann Electronics had designed an ASIC capable of switching frequencies in the range from 900 MHz to 2.5 GHz between five inputs and four outputs. A new product was to extend these capabilities to nine inputs and to include four digital controllers referred to as Digital Satellite Equipment Control. Coming up with a second ASIC would have meant mixed-signal IC design, expenses for another mask set, and many months of turnaround time. Figure 11.15 shows the system-in-package developed instead. The laminate substrate measures 21 mm by 24 mm, includes four layers, and also carries coupling capacitors, termination resistors, and a tiny glue logic chip [329]. The discretes are soldered whereas wire bonding is used to connect to the seven ICs. Connections to a motherboard are with bumps placed on the reverse side.

□

Observation 11.3. *While high-density packaging competes with monolithic integration to some degree, it is best understood as a technology that complements and extends the benefits of VLSI to products that sell in moderate quantities and/or multiple varieties.*

FOLDED FLEXIPRINTS

Folded flexiprints are particularly popular when a multi-chip circuit must fit into a small or irregular volume. Chips are fabricated, packaged, and tested in the normal way before being surface-mounted

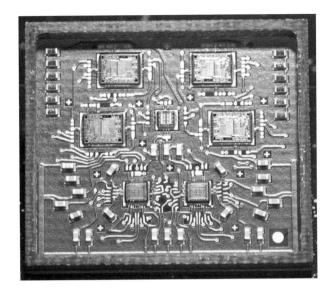

Fig. 11.15 Seven dies mounted on a laminate substrate along with numerous discretes (source *Elektronik* no.12 2004 p.24, reprinted with permission).

on a flexible film substrate. Discrete components, sensors, and the like can also be accommodated. The flexiprint is then cut and folded before being fit into a medical device, a personal telecommunication appliance, or some other space-constrained product.[18]

CHIP STACKING AND CUBING

Hearing aids, mobile phones, and flash memories such as USB memory sticks take advantage of the third dimension by stacking two, three, or more dies on top of each other, see fig.11.16. A digital

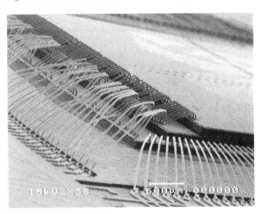

Fig. 11.16 Three chips stacked and interconnected by wire bonding (photo copyright IEEE, reprinted from [259] with permission).

---

[18] As an example, folded flexiprints manufactured by Valtronic find applications in hearing aids.

ASIC can thus be combined with an analog front end and a commodity RAM or a flash memory, for instance. Resorting to wire bonding for the interconnections is cost effective but requires that the individual dies be graded in footprint such as to form a pyramid where all bond pads lie open.

Cubing allows for still higher densities. Bare dies are stacked on top of each other before being interconnected on their outer rims to obtain a cube-like assembly. Burying the vertical interconnections within the chip stack itself is an extra sophistication [334]. To that end, extra deep vias are formed by way of plasma etching before being filled with copper. Cross sections can be as small as 1.2 μm by 1.2 μm. The wafer, initially some 750 μm thick, is then ground down to a mere 13 μm so that these vertical poles become exposed on its back. Multiple wafers are then aligned and bonded together with an unthinned wafer at the bottom providing mechanical support, see fig.11.17. Cubing appears most promising for connecting repetitive functions, such as memory chips, to form larger entities.

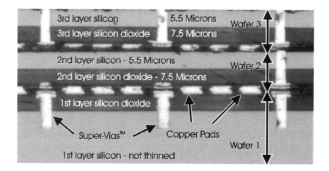

Fig. 11.17 Cross section through chip stack with in-silicon vertical interconnections (photo courtesy of Tezzaron Semiconductor Corp.).

An alternative better suited to accommodate heterogeneous assemblies where dies differ in type and shape first embeds each die in an epoxy resin compound, which is cut into tablets of identical size after curing. Next, the tablets, each with a chip enclosed, are stacked on top of each other. Conductive lines printed on the surface or embedded within the epoxy resin material bring the die's connectors to the desired locations on the cube's outer faces for vertical interconnects.

A similar approach mounts each die on a flexiprint before stacking and molding with epoxy resin takes place. The cube is then sawn so as to expose the flexiprint lines on its outer faces. Vertical interconnection is by way of metal plating and laser cutting. The bottom of the cube is attached to a leadframe or a pin grid array carrier, or equipped with bumps to establish electrical contacts with the printed circuit board underneath.[19]

In any case, signal integrity requires that decoupling capacitors be incorporated into the stack [335]. Heat evacuation, another notorious difficulty in three-dimensional VLSI assemblies, can be improved by embedding copper plates that carry the heat from the chips to the sides of the stack. Overviews on three-dimensional IC assembly techniques are given in [336] [337].

---

[19] Manufacturers of chip stacks include companies such as Irvine Sensor and 3-D Plus Electronics.

11.4.8 Selecting a packaging technique

Opting for an adequate package involves criteria of both technical and economical nature:

- Space available for encapsulated chip(s) on printed circuit board.
- Board mounting technique (surface mount, through-hole, chip-on-board, stacked, etc.).
- Number of pins required and closest number actually available.
- Cavity size along with the range of die sizes that a package can accommodate.
- Electrical characteristics, i.e. package parasitics.[20]
- Maximum power dissipation, ambient temperature, thermal resistance.[21]
- Resistance against mechanical, thermal, and environmental stress.
- Expected lifetime, aging, reliability.
- Graded product range from a limited inventory of parts.
- Required equipment for automatic packaging, testing, and mounting.
- Suitability for replacing a component and/or for repairing a board.
- Yield losses due to packaging and mounting operations.
- Packaging and mounting costs.

Observation 11.4. *Chip, package, board-level wiring, signal integrity, and heat evacuation must be understood together as critical design elements in any advanced electronic system.*

[332] [259] [333] discuss the respective merits of monolithic integration and stacking techniques from a systems perspective. [338] is primarily concerned with yield analysis of high-density packages.

11.5 | Layout at the detail level

One might argue that layout is of negligible importance today and that any discussion of physical design should be confined to higher-level issues such as floorplanning, place and route (P&R), and chip assembly. Manual layout has, after all, become a highly specialized and secluded activity now that cell libraries and P&R software are routinely available. Yet, there are good reasons for gaining a basic understanding of layout design.

Firstly, most VLSI designers will want to know how all those operations that were initially captured as abstract HDL statements eventually materialize in the semiconductor material.

Secondly, it is much easier to understand what floorplanning, P&R, and chip assembly are all about when one has a general appreciation of logic gates along with their layout views.

Thirdly, actual VLSI circuits are exposed to undesirable parasitic effects such as interconnect delay, crosstalk, electrical overstress, latch-up, on-chip variations (OCVs), and the like. Knowing about physical construction helps in keeping those within acceptable bounds.

Finally, somebody has to come up with the layouts of library cells and memories.

---

[20] The impact of package parasitics has been explained in section 10.3.
[21] Thermal design issues are summarized in section 9.3.

11.5.1 Objectives of manual layout design

Detailed layout design is concerned with constructing circuits on the level of fabrication masks. The small polygons that make up transistors, wires, contacts/vias, capacitors, resistors, diodes, etc. are drawn to scale with the aid of a **layout editor**, a specialized color graphics drawing tool. Layout designers must know and respect all layout rules imposed by the target process. Of course, such an activity — sometimes ironically dubbed "rectangle pushing" — is time consuming, cumbersome, and prone to errors.

Observation 11.5. *For economic reasons, manual layout design is not normally justified except in a few particular situations such as*
- *very high production volumes,*
- *library development where cells are going to be reused in many designs,*
- *specific requirements that cannot be met by using library cells,[22]*
- *analog circuit parts,*
- *integrated sensors and actuators, or*
- *test structures for process monitoring and/or device characterization.*

Recall that full-custom fabrication allows one to combine cell-based blocks with manually optimized layout on the same chip. Situations exist where it makes sense to design critical parts by hand in order to meet stringent density and/or performance goals that cannot be met otherwise. This explains why high-end microprocessors, where performance is the ultimate differentiator and for which fabrication quantities are large, continue to include a substantial proportion of hand-crafted layout. Technically, the goals of layout design at the detail level include

- Minimum area or, which is the same thing, maximum layout density,
- Maximum performance, i.e. minimum transistor and interconnect delays,
- Maximum fabrication yield,
- Largely inoffensive parasitic effects, and
- All this with minimum design effort.

In search of better productivity and portability, **procedural layout** is often used to automate the drawing of common layout items such as MOSFETs, capacitors, resistors, contacts/vias, and the like. A small piece of software code, invoked from within a layout editor, generates the polygons on the various mask layers for a given device. The designer's involvement is hence limited to defining device type, location, width, and length.

11.5.2 Layout design is no WYSIWYG business

The fact that wafer processing takes a flat layout drawing and translates that into a complex multi-layer structure of multiple materials makes physical design a rather grotesque business.

- What the designer actually seeks to build is an electrical network.
- What he has to deliver to the foundry is a two-dimensional layout drawing.
- What he eventually gets from fabrication is a three-dimensional structure
  that includes many unwanted parasitic elements.

---

[22] Such as extremely high speed, very low power, non-standard voltages, radiation hardness, uncommon subcircuits for self-timed operation, and the like.

Taking an inverter as example, fig.11.18 opposes all three views. However, before we can develop an understanding of the two latter views, we need to introduce an essential concept underlying solid-state electronics. In any semiconductor device, a few volumes of different electrical characteristics are made to interact locally. The various types of transistors and diodes differ in the relative arrangement and electrical properties of those tiny volumes. **Doping** is the process whereby the characteristics of silicon, or some other base semiconductor material, get adjusted by selectively incorporating atoms of a different element. In drawings, doping is often identified by writing things like n^+ or p^-, where n and p stand for an excess or a deficit of electrons respectively, while $+$, $-$, and $--$ refer to strong, weak, and very weak doping concentrations. Doping will be a major subject of chapter 14.

Observation 11.6. *Understanding layouts and cross sections of semiconductor devices does not require all of the physics and manufacturing details. To gain a fairly useful — though merely quali-tative — appreciation, it suffices to differentiate among*
- *(conductive) metals,*
- *(conductive) polysilicon,*
- *strongly n- and p-doped semiconductor materials,*
- *weakly n- and p- doped semiconductor materials, and*
- *(nonconductive) dielectrics.*

Let us now identify the active semiconductor devices in fig.11.18.

MOSFET devices

An n-channel (p-channel) MOSFET forms wherever p^- (n^-) material is covered by a polysilicon layer with a thin gate oxide layer acting as insulation between the two. The adjoining n^+ (p^+) diffusion areas serve as source and drain respectively, the polysilicon structure acts as gate, and the bulk material forms the body. Complementary MOSFETs are the active devices in inverters and all other digital CMOS subcircuits. They are marked with (a) and (b) in fig.11.19.

A notable detail from CMOS fabrication is that diffusion regions get implanted <u>after</u> polysilicon deposition with poly acting as a shield that prevents dopants from penetrating into the material underneath.[23] As a consequence, there can be no intersection of a polysilicon with a diffusion structure in a layout drawing without a MOSFET emerging at that place. This is something to keep in mind when establishing layout plans.

Observation 11.7. *The fact that a MOSFET forms at <u>any</u> intersection of a diffusion line with a polysilicon line restricts the utility of poly for interconnects and that of diffusion too.*

Now compare figs.11.18b and e. The latter is a simplified view where the dimensions of items are neglected while their topological arrangement is rendered correctly. Wells, body ties, parasitic devices, and other physical details that have no relevance for the operation as a digital circuit are also abstracted from. Such **stick diagrams**, aka symbolic layouts, were promoted in the 1980s by Mead and Conway [339] to facilitate the transition between schematic diagrams and geometric layout drawings, but have been met with little acceptance in industry.

---

[23] How this happens is to be described in section 14.2. Incidentally, observe that the masks shown in fig.14.17 are not quite the same as those on display in fig.11.18. This is because the chapter on CMOS technology refers to the masks that are actually being used during wafer processing, whereas the p^+ and n^+ diffusion areas that matter from a designer's perspective are shown here. Appendix 11.9 indicates how to translate back and forth.

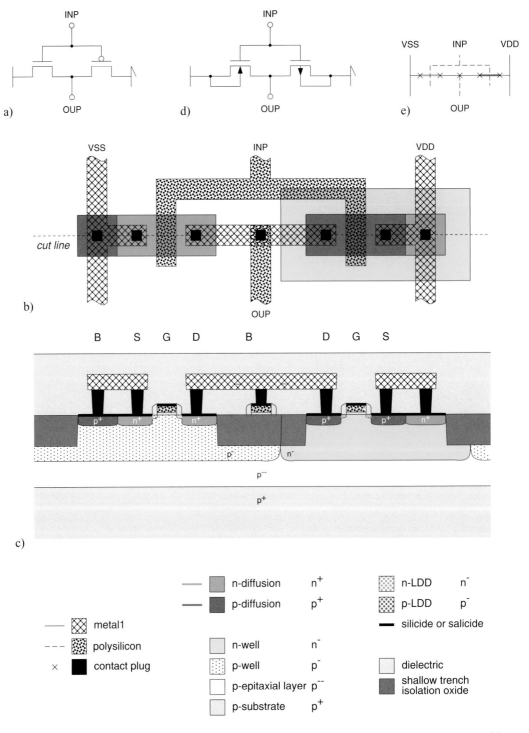

Fig. 11.18 CMOS inverter. Schematics (a,d), geometric layout (b), cross section (c) and stick diagram (e) (approximate, not drawn to scale, only first-level metal shown). The relative placement of the various items has been straightened for demonstration purposes, more practical arrangements are to be discussed in section 11.5.3.

Next, we would like to draw your attention to the presence of **parasitic devices**, that is of active devices other than the gate oxide MOSFETs intentionally sought after to build logic gates and bistables. The active devices that may potentially exist can, again, be visualized in a graph for any given target process. Figure 11.19 provides us with such a graph too.

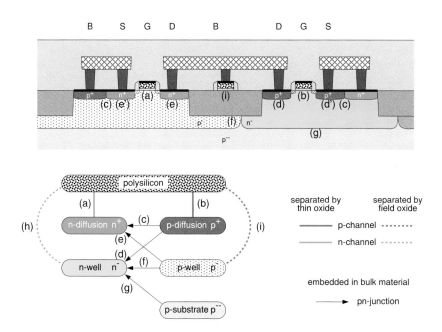

Item	Structure	Bias	Device and typical role
(a)	poly-thox-p$^-$	n.a.	n-channel MOSFET
(b)	poly-thox-n$^-$	n.a.	p-channel MOSFET
(c)	p$^+$n$^+$	zero	diode, not normally used due to low breakdown voltage
(d)	p$^+$n$^-$	reverse	diode, insulates p-MOSFET source or drain from body
(e)	p$^-$n$^+$	reverse	diode, insulates n-MOSFET source or drain from body
(f)	p$^-$n$^-$	reverse	diode, insulates p- from n-channel MOSFETs
(g)	p^{--}n$^-$	reverse	*idem*
(h)	poly-fox-n^{--}	n.a.	field oxide n-channel MOSFET (not shown)
(i)	poly-fox-p^{--}	n.a.	field oxide p-channel MOSFET

Fig. 11.19 MOSFETs and diodes in a bulk CMOS silicon-gate twin-well p-substrate process. Please refer to section 11.6.3 for explanations on parasitic bipolar devices.

pn-junctions

Out of the five pn-junction types available from a bulk CMOS fabrication process, four find applications as diodes, see fig.11.19.

Electrical separation of p- from n-channel MOSFETs is assured by the reverse-biased p^{--}n$^-$ junctions that separate the n-wells from the p-type substrate (g) in conjunction with the reverse biased p$^-$n$^-$ junctions that form where p- and n-wells touch (f).

The unsymmetric p^+n^-- and p^-n^+-type junctions are operated with a reverse bias to provide the necessary insulation between MOSFET drain/source electrodes and the embedding bulk material (d,e); only those that connect to VSS or VDD carry no voltage (d',e'). As will be explained in section 11.6.2, p^+n^-- and p^-n^+-type junctions are also used as clamping devices in electrostatic discharge (ESD) protection networks.

Reverse biased p^+n^+-type junctions (c) are unsuitable for isolation purposes because of the extremely low breakdown voltage that results from the high doping concentrations on either side. In fact, the depletion regions are so thin that strong tunneling may set in well below 1 V. Most process manuals thus disallow p^+n^+-type diodes unless junctions are shorted together to form a pair of butted contacts such as in fig.11.19.

Field oxide MOSFETs

Consider item (i) in fig.11.19 and note that the arrangement of layers is that of a MOSFET much like (a) except that the polysilicon and bulk material are separated by the much thicker field oxide here. As a consequence, this construct may indeed behave as a parasitic transistor. Luckily, its process gain factor $\beta_\square$ is much inferior to that of its gate oxide counterpart. To further prevent such devices from interfering with regular circuit operation in an undesirable way, their threshold voltages U_{th} are adjusted during fabrication so as to disable them as long as the circuit is being operated with voltages approved by the manufacturer. Similarly to p^+n^- and p^-n^+ diodes, field oxide MOSFETs find applications in protecting ICs from overvoltages.

Bipolar and thyristor devices

The graph of fig.11.19 further suggests the presence of $n^+p^-n^-$, $p^+n^-p^-$, and $p^+n^-p^-n^+$ devices where layout structures on three or four conducting layers interact. These parasitic BJTs and thyristors participate in latch-up, a catastrophic and potentially destructive effect. Section 11.6.3 is, therefore, entirely devoted to latch-up prevention.

11.5.3 Standard cell layout

The arrangement of fig.11.18 is not really representative for actual circuits as all layout items have been lined up to make them visible in a single cross section. Actual library cells follow a pattern referred to as **gate-matrix layout** where polysilicon lines traverse a pair of parallel n- and p-diffusion stripes at regular intervals, see fig.11.20a. As a consequence from observation 11.7, a MOSFET necessarily forms at each intersection. Power and ground run from left to right across every cell so that supply rails automatically come into being when such cells are being abutted in a row. Also, those supply rails can be shared between adjacent cell rows.

Figure 11.21b shows a 3-input NOR in gate-matrix style in more detail. Observe that n- and p-channel devices are arranged in two separate horizontal rows. There are several reasons for this. Firstly, MOSFETs must be embedded in wells, and well-separation rules impose relatively large distances. Collecting many transistors of identical polarity in a common well thus allows superior layout density. Secondly, keeping n- and p-channel MOSFETs well apart from each other helps to withhold latch-up. Latch-up prevention is also why a guard bar has been placed underneath each supply rail. Thirdly, cells can thus be designed to differ solely in width depending on the number of

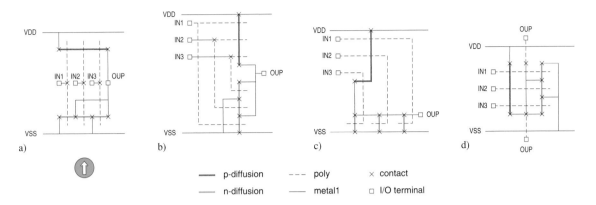

Fig. 11.20 Layout styles compared for a 3-input NOR gate.[24] Gate-matrix topology (a) versus inept patterns (b,c,d).

MOSFETs they must accommodate. Cell height always remains the same, which greatly facilitates placement. The diffusion stripes feature a gap wherever it is necessary to electrically insulate one transistor from its neighbor to the left or to the right in the layout. This is typically the case where two cells adjoin, but also occurs within adders, latches, flip-flops, and other more complex gates.

The desired circuit is formed by connecting the various drain, source, and gate electrodes to VSS, to VDD, to the cell's I/O terminals, or with one another. Metal1 is preferred for this **intracell wiring** because it can freely cross over and connect to any other layer. Being an utterly simple cell, the 3-input NOR gate does not occupy all tracks available for intracell wiring. Wherever possible, contacts are replicated to shunt the important resistance of the diffusion islands and to provide some degree of welcome redundancy.

Inputand output terminals are those nodes of a cell that are accessible for **extracell wiring** on metal2 and higher layers. While they are often arranged near the cell's centerline as shown in fig.11.21, note that modern routers and cell libraries allow connection to a terminal node anywhere on the pertaining metal1 polygon, thereby dispensing with the need to identify specific connector locations.[25]

Full-custom layout gives designers the freedom to specify an individual size for each transistor; note the tapered chain of p-channel MOSFETs in fig.11.21b. In digital design, MOSFETs are almost always drawn with the minimum admissible channel length to maximize layout density, drive strength, and switching speed. Weak transistors, such as those found in snappers and other bistables, are an exception to the rule. A transistor's channel width is chosen so as to adjust its current drive in search of an optimum compromise among switching threshold, operating speed, and energy efficiency. p-channel MOSFETs are generally made somewhat wider than their n-channel counterparts to compensate for their inferior carrier mobility.

---

[24] More examples can be found in section 8.1.

[25] Early standard cells prepared for single metal processes had the vertical polysilicon lines span the entire cell height, making all input/output signals accessible at both the upper and the lower rim, which greatly facilitated extracell wiring with the then-scarce routing resources.

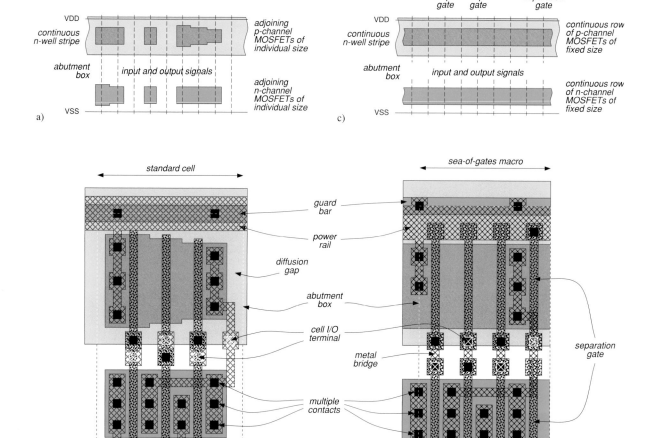

a)

b)

c)

d)

▨ metal1	▨ n-diffusion n⁺
▨ polysilicon	▨ p-diffusion p⁺
■ contact plug	
⊠ via plug	□ n-well n⁻

ground/power rails
and guard bars
are shared between
adjacent cell rows

Fig. 11.21 Gate-matrix layout (approximate). General pattern in full-custom (a) versus semi-custom layout (c). Standard cell implementation of a 3-input NOR function (b) versus sea-of-gates macro (d).

While gate-matrix layout has a long tradition, minor variations have been developed over the years, e.g. to accommodate the multiple metal layers that became available.

The layouts of fig.11.21b and d follow Manhattan geometry, although many library cells and RAM tiles make use of angles of 45° to render them more compact. The limitations of Boston geometry for transistor-level layout are to be discussed shortly in section 11.5.6. The usage of arbitrary angles is generally discouraged because most DRC programs do not properly handle such layouts, because the preparation of photomasks gets more complicated, and because acute angles cause manufacturing problems. Please note that this applies not only to circuit layout but also to texts, company logos, and other on-chip artwork.

11.5.4 Sea-of-gates macro layout

As sea-of-gates masters come with prefabricated silicon layers, only the metal and contact/via layers are available for customization. Sea-of-gates macros thus necessarily differ from standard cells in terms of their layout design. Another difficulty arises from the necessity to electrically insulate adjacent but otherwise unrelated source/drain nodes where two neighboring cells abut, but also within complex cells such as adders, latches, and flip-flops. While it would be possible to prefabricate diffusions with a gap after every fourth transistor or so, this would be inefficient and highly inflexible. Instead, diffusion areas are prefabricated as long and contiguous stripes that periodically intersect with poly lines so as to obtain long chains of MOSFETs. It is nevertheless possible to obtain electrical insulation exactly where required.

The trick is to cut every polysilicon line near the cell's centerline, see fig.11.21c. Where two adjacent nodes need to be separated, i.e. where a diffusion stripe would be discontinued in a full-custom layout, the prefabricated transistor in between is permanently turned off with the aid of a contact. The n-channel poly segment is connected to VSS and the p-channel segment to VDD respectively, thereby insulating the source/drain to the left from that to the right. A pair of MOSFETs sacrificed in this way is said to form a **separation gate**. Another price to pay are the gaps in the poly lines. Most of these gaps are actually unnecessary and are bridged with the aid of a short metal strap to obtain the customary complementary transistor pair from which CMOS gates are built. Fig.11.21d includes three such straps and one separation gate.

Another handicap of semi-custom design is that transistors come with fixed geometries. Strong transistors must thus be obtained from connecting two or more of them in parallel whereas building weak transistors asks for connecting two or more devices in series.

11.5.5 SRAM cell layout

The circuit organization and operation of SRAMs have been discussed in section 8.3. In search of maximum layout density and performance, many different layout topologies have been devised over the years in close collaboration between process developers and layout designers. As their circuit structures are highly regular, RAMs do not make use of as many metal layers as random logic does, but strongly rely on tiling instead. Figure 11.22 shows five symbolic layouts for a 6-transistor cell. Each cell occupies two metals, and is traversed horizontally by the word line and vertically by a pair of complementary bit lines.

Long polysilicon lines, as found in layout topologies b, c, and e, are detrimental to switching speed because of the material's mediocre conductivity. This is why the word line has been doubled with a metal bypass in d. A related approach would be to intersperse a tall narrow cell every eight columns or so just to feed the world line from a copy running on a higher-level metal. Arrangement f has

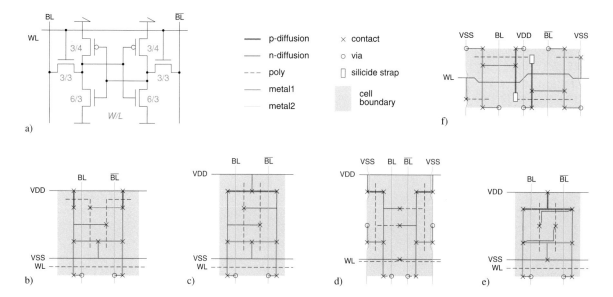

Fig. 11.22 6-transistor SRAM cell. Schematic diagram (a) and stick diagrams of various layout topologies (b to f) (some straight lines drawn cranked to expose layout underneath).

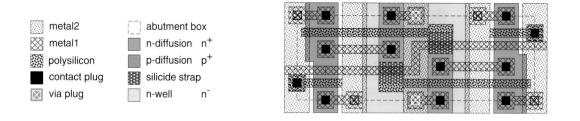

Fig. 11.23 Detailed layout of a 6-transistor SRAM cell (approximate). This design occupies two levels of metal and takes advantage of silicide straps. Note how sharing of metal lines, contacts, and vias between abutting tiles helps to improve layout density.

a metal word line and does not suffer from this problem. Like e, it makes use of silicide straps to boost layout density.

A **silicide strap** electrically connects poly and difusion areas without the customary detour via contact plugs and a metal wire. This is made possible by etching away the side-wall spacer that normally covers the lateral rims of poly gates where indicated by an additional photomask. The thin but conductive silicide film that is to form during salicidation wherever silicon is exposed will thus extend over the side wall of the poly structure and so cover the vertical step that separates it from an adjacent diffusion area. Also known as direct strap or local interconnect technology, this process option is unavailable from baseline CMOS processes due to the extra overhead, but typical for commodity RAMs and also found in microprocessors with large on-chip caches [343]. Figure 11.24 shows a photomicrograph of an SRAM cell similar to topology f that occupies $135F^2$.

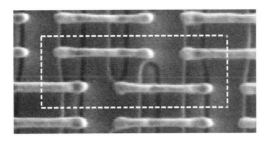

Fig. 11.24 6-transistor SRAM cell after patterning of the polysilicon layer. A third metal (not shown) is used here for the horizontal word line. Cell area is 0.57 μm^2 in 65 nm technology (photo copyright IEEE, reprinted from [340] with permission).

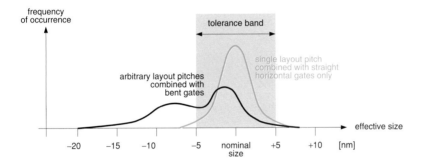

Fig. 11.25 Impact of layout style on geometric and, hence also, on electrical variability.

11.5.6 Lithography-friendly layouts help improve fabrication yield

Design for manufacturing (DFM) is the name for a collection of techniques that aim at enhancing fabrication yield by maximizing process latitude. Layout patterns that are more likely to fail than others or that unnecessarily contribute to geometrical variability are to be avoided. Geometrical variability nurtures electrical and timing-wise variability, and so impacts yield. DFM has become a necessity since the minimum feature size has grown smaller than the wavelength employed in the photolithographic patterning process, see fig.14.22. Note that the measures listed below confine admissible layout patterns further than traditional layout rules do.

- Restrict conductor widths and pitches to a few values known to print well in the lithographic and patterning processes.
- Place narrow features such as MOSFET gates on a periodic grid to better control optical proximity effects. What ultimately results are highly repetitive layouts much as in SRAM/DRAM cores and sea-of-gates masters.
- Stick to a single orientation for MOSFET gate electrodes as distortions are unlikely to affect horizontal and vertical lines in exactly the same way.
- Avoid bends in MOSFET gates as they do not print well and tend to make the final transistor geometries and electrical characteristics difficult to predict.

- Connect multiple contacts or vias in parallel as fabrication defects predominantly manifest themselves as opens.
- Space long lines further apart than required by the minimum separation rule to minimize the chance of shorts and the impact of crosstalk on path delays, both long and short. Take advantage of symmetry to compensate for local process variations. Add dummy patterns next to critical devices to compensate for edge effects.

11.5.7 The mesh, a highly efficient and popular layout arrangement

The floorplans of RAMs are just special cases of a more general concept that can be seen as an extension of gate-matrix layout. The underlying idea is that wires carrying one set of signals cross other wires that carry a set of functionally different signals on a second layout layer, and that the signals interact in one way or another at their points of intersection.

Mesh-type arrangements not only find many applications in software-assembled macrocells, but also prove most useful in manual layout design. Six examples where interactions at the crosspoints range from simple galvanic connections to arithmetic/logic operations have been collected in fig.11.26a to f. Observe that the notion of an orthogonal mesh occasionally needs a liberal interpretation.[26]

Books specializing on layout design include [341] [485] [486], for instance.

11.6 | Preventing electrical overstress

In this section, we will discuss three electrical overstress conditions that lead to component failure unless held back by proper layout design, namely electromigration, electrostatic discharge, and latch-up.

11.6.1 Electromigration

CAUSE AND EFFECTS

Electromigration is a wear-out phenomenon that affects metal conductors subject to excessive current densities. The underlying mechanism is a combination of thermal and electrical effects [342]. In a thin metallic polycrystalline film, a growing proportion of thermally agitated metal ions does exist at temperatures above one-half or so of the melting point of the material. In essence, such agitated ions are free of the lattice and get pushed along by the impact of flowing electrons. Metal is washed away and deposited downstream, thereby diminishing the cross section of the conductor at some point.[27] Those voids lead to a further increase in current density and the vicious cycle accelerates until the wire eventually severs. Disintegration tends to follow lattice imperfections such as dislocations, grain boundaries, and impurities.

The current-carrying capacity of a conductor is a function of material, shape, temperature, and current waveform (continuous DC, pulsed DC, AC) [344]. The higher the operating temperature of

---

[26] The regularity and periodicity of mesh-type layouts are also believed to be enablers for the fabrication of future nanometer-scale circuits of crossbar logic that are to be introduced in section 15.2.2.

[27] Please note that metal atoms are driven upstream wrt the conventional orientation of current flow.

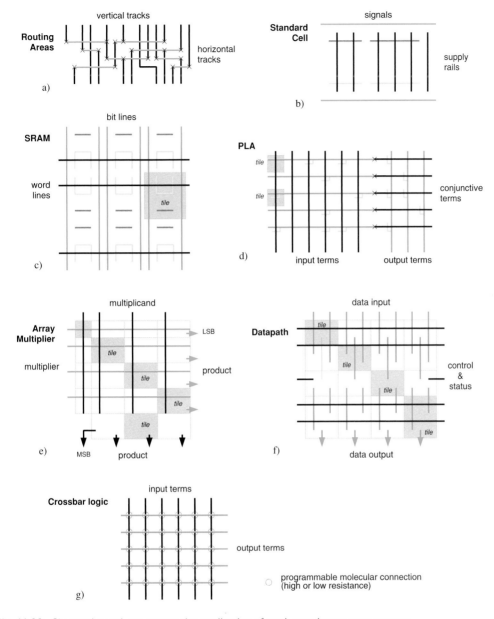

Fig. 11.26 Six popular and one prospective application of mesh-type layout arrangements.

a chip, the more generously must the interconnect lines be sized to handle some given amount of current. To give a rough idea, 5 to 10 mA/μm^2 (= kA/mm^2) is considered a sustainable current density for the traditional aluminum alloys (Al with 0.5% to 4% Cu) being used for interconnects in VLSI. Copper has a higher melting point and can withstand significantly more current, just compare tables 11.1 and 11.2. Yet, keep in mind that safe current densities are also limited by concerns relating to ohmic voltage drop and local overheating.

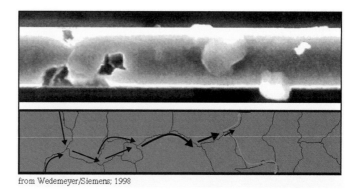

Fig. 11.27 An aluminum line that exhibits severe voiding by electromigration (reprinted from [343] with permission).

Numerical example

Consider a CMOS IC of modest size, say of 50 000 gate equivalents, to be manufactured in a 130 nm technology, operated at 1.2 V, and driven from a 100 MHz single-edge-triggered one-phase clock. Overall node activity is $\alpha = \frac{1}{4}$, which means that the average node charges and discharges within eight clock cycles. Let each of the 100 000 internal nodes have a capacitance of 18 fF. The average current consumption of the circuit's core then amounts to

$$I_{dd} \approx f_{clk}\, U_{dd}\, (1 + \sigma) \sum_{k=1}^{K} \frac{\alpha_k}{2} C_k = 100\text{ MHz}\, 1.2\text{ V}\, (1 + 0.2)\, 100\,000\, \frac{\frac{1}{4}}{2}\, 18\text{ fF} \approx 32\text{ mA} \qquad (11.1)$$

A current density of 3.2 mA/μm flows when the core of such an IC is being fed through on-chip power and ground wires 10 μm wide. Relating that amount of current to the conductor's cross section based on a thickness of 320 nm — a realistic figure for lower-level metals in a 130 nm technology — one obtains $J = 10$ mA/μm² which makes it clear that even modest ICs lead to critical current densities unless supply wires are adequately sized and layers are chosen correctly.

It is very instructive to compare the above figures with overhead power lines. An RMS current of approximately 1500 A flows through each conductor when 1 GW of power is being transmitted over a three-phased 380 kV line. The resulting current density in a round aluminum conductor with a diameter of 50 mm is 0.75 A/mm², i.e. roughly four orders of magnitude less than for IC technology. □

Observation 11.8. *As opposed to overhead lines and household wiring, the current densities found in VLSI interconnects are so high that metal wear-out becomes a major concern.*

Make sure you understand that electromigration is not the same as fusing. In a fuse, metal overheats because of ohmic power losses in a tiny volume $\frac{P}{V} = \rho(\frac{I}{A})^2$ that cannot dissipate, locally reaches its melting point, and breaks up. Depending on the overload factor, this is a matter of seconds or milliseconds. Electromigration, in contrast, is a long-term wear-out process that develops at temperatures way below the material's melting point as a consequence of excessive current densities $J = \frac{I}{A}$. Interestingly, it has been found that the resistance to electromigration improves when a wire is made narrower than the average grain size in the conductor material. This is because grain boundaries then tend to orient themselves perpendicularly to current flow, suggesting a bamboo-like structure [345].

Conclusions

The VLSI designer has no influence on the making of electromigration-resistant alloys nor does he decide on the thickness of the various interconnect layers. To keep current densities within safe limits, he must obtain adequate cross sections by making the lines sufficiently wide. In order to carry a given amount of current with a narrower wire, he is also free to prefer the thicker upper-level metals over the thinner lower-level metals. Specialized CAD tools that calculate the current density in each leg of interconnect help identify those portions that are particularly at risk and that need to be sized more generously. The same tools can also locate any unwanted layout constrictions exposed to meltdown when subjected to surge currents.

Sharp bends and stipple contacts/vias are also exposed to electromigration because of the current crowding that takes place on the path where ohmic resistance is lowest. Taking advantage of Boston geometry makes it possible to distribute currents more evenly, see fig.11.2d.

Metal **slotting** whereby narrow longitudinal openings are carved into heavily loaded power and ground lines may help by allowing for bamboo structures to form in between. In addition, slotting also counteracts dishing.

11.6.2   Electrostatic discharge

Cause and effects

Electrostatic discharge (ESD) is a phenomenon of sudden charge redistribution between a semiconductor component and automated equipment or a person handling that part. To give a rough idea, walking across a synthetic carpet can generate voltages from 100 V up to 30 kV under worst-case conditions. When discharged into an IC package, these voltages give rise to current peaks of several amperes. An ESD event such as this is likely to have two destructive effects:

- Dielectric breakdown as a consequence of excessive fields.
- Local overheating often followed by melting as a consequence of excessive amounts of energy being dissipated in a small volume over a short lapse of time.

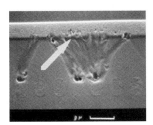 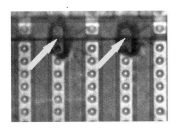 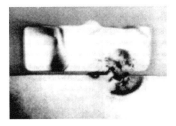

Fig. 11.28   Damage from ESD. Melt filaments shorting a junction (a), evaporated metal lines (b), and broken gate oxide (c) ((a) reprinted from [346] with permission, (b) courtesy of Dr. Andreas Stricker, (c) reprinted from [347] with permission).

Observation 11.9. *The voltages that occur during ESD events are sufficient to force very high currents through any semiconductor circuit. It is better that engineers decide on the paths these currents shall take while they are designing the part.*

As ESD events greatly vary depending on the exact circumstances, industry has come up with standard models for studying their consequences and for assessing the effectiveness of protective circuits [348]. The charged device model (CDM) and the very fast transmission line pulse model (VF-TLP) put more emphasis on very short events that stress gate dielectrics while the human body model (HBM) and the machine model (MM) primarily address slower failure mechanisms that are due to excessive energy dissipation. A part's tolerance with respect to ESD is typically stated as the voltage endured without a change in the electrical characteristics in positive and negative discharge events with a given ESD model, e.g. 500 V CDM, 2 kV HBM, and/or 200 V MM.

Component failures due to ESD must be addressed by a two-pronged approach.

Take handling precautions to reduce the extent of ESD events

Electrostatic charges that build up during the fabrication, handling, shipping, and assembly of semiconductor components must be kept within uncritical limits. This is obtained by an array of preventive measures at the organizational level such as

- Controlled air humidity (30% to 70%) and ionization,
- Floors, carpets, shoes, and clothing made from antistatic, dissipative or conductive materials,
- Dissipative and grounded workbenches, equipment, wriststraps, etc,
- Dissipative bags, boxes, foams, tapes, reels, and other component carriers, and
- Trained workforce, process audits, etc.

Include on-chip ESD protection to make parts more robust

Handling precautions alone do not suffice. MOS circuits must further be protected by auxiliary circuitry that clamps overvoltages and that dissipates undesirable energy without local overheating. The most successful approach is to provide well-defined current paths between any two pins through which ESD currents can safely discharge. Clearly, the impedance must be lower than along any alternative path. A typical ESD protection network is outlined in fig.11.29a and described next.

Input protection. Inputs are kept safe by a Π network that brings down overvoltages in two steps. The primary protection devices are in charge of absorbing — or of diverting — most of the electrostatic energy while the secondary devices limit the voltage excursions across the MOSFET's fragile gate oxide to inoffensive values. A decoupling resistor placed in between safeguards the secondary devices against overstress and, at the same time, provides sufficient leeway for the primary devices to get activated. Typical values range between 200 Ω and 1 kΩ but may be as low as 50 Ω on clock inputs to keep distribution delay small.

Output protection. Much as for an input, the protective network includes energy-absorbing or -diverting devices along with a decoupling resistor. Yet, the range of acceptable values ends at

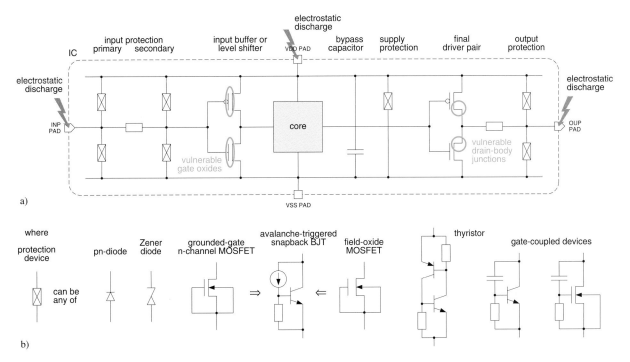

Fig. 11.29 Generic ESD protection network with points of attack shown and with parts particularly exposed to damage highlighted (a). Typical protection devices (b).

50 Ω or so because any more substantial series resistance would seriously impair the output's current drive and ramp times.

Supply protection. The protection circuit clamps any potentially harmful voltages that might develop between the two — or more — supply nodes when a power or a ground pin is involved in an ESD event. It also comes into play when an input or output protection device or a parasitic diode dumps excess charge from an ESD event into the supply net. Incidentally, observe that the parasitic capacitance between the VDD and VSS nodes assists in containing fast voltage spikes and that on-chip bypass capacitors, if any, also help.

While the overall arrangement of ESD protection networks has more or less remained the same over the years, the choice, doping, sizing, and shaping of the protective devices is constantly being re-evaluated to account for the relentless evolution of semiconductor technology. The desiderata for an ideal protection device are [349]:

- Zero on-state resistance.
- Clamp voltage just above the operating supply voltage of the part being protected.
- Instantaneous turn-on.
- Capable of absorbing infinite amounts of energy.
- Triggers only during ESD events, not during regular operation.
- No parasitics that could impair regular circuit operation or performance.
- Zero area requirement.

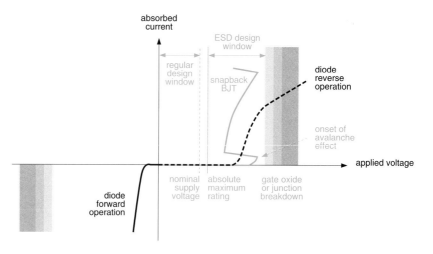

Fig. 11.30 ESD design window along with characteristics of protection devices.

ESD protection must typically make do with devices that come at little or no extra cost. Circuit elements that enter into consideration as protection devices in CMOS ICs thus include

- pn-type diodes operated in forward mode,
- pn-type diodes operated in reverse mode,
- Regular (thin oxide) MOSFETs,
- Thick or field oxide MOSFETs, and
- Thyristors aka silicon-controlled rectifiers (SCR).

See fig.11.29b for illustrations.

Figure 11.31 shows an inverter fabricated in bulk CMOS technology.[28] Note the presence of various **parasitic devices**. The most remarkable item is the lateral npn transistor that hides in every n-channel MOSFET. When the drain–source voltage is being increased beyond the MOSFET's regular operating range, an avalanche current begins to flow through the drain junction. Upon entering the body volume, this current becomes the base current of a parasitic BJT and drives that bipolar into conduction. What makes this tandem device so attractive for ESD protection is its snapback $I(U)$ characteristic along with the resulting on-state resistance of just 2 Ω or so. This unusual trait, sketched in fig.11.28, occurs because the avalanche current gets amplified by the gain of the bipolar and has earned it the name **avalanche-triggered snapback BJT**.

This protection device is often used with its gate terminal grounded, in which case it is also referred to as **grounded-gate NMOS** (ggNMOS). More sophisticated protection schemes have the gate voltage driven by some active triggering circuit to better control the onset of the avalanche effect, an

---

[28] **Bulk technology** makes up the vast majority of CMOS ICs fabricated. Both n- and p-channel MOSFETs are implanted in a single piece of semiconductor material and are electrically insulated by way of reverse-biased well junctions as depicted in fig.11.32a. This is not so in silicon-on-insulator (SOI) technologies, which are to be explained in section 14.3.6.

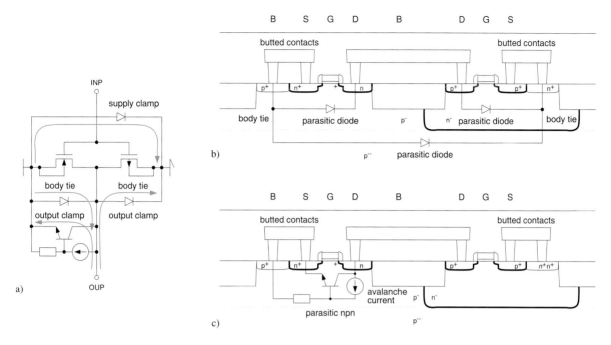

Fig. 11.31 Parasitic devices that can act as clamping devices in bulk CMOS circuits. Cross section of an inverter with parasitic diodes shown (b) and with the parasitic lateral npn BJT that comes with each n-channel MOSFET (c). Overall circuit with forward current paths highlighted (a).

approach known as coupled-gate NMOS (cgNMOS). Make sure you understand that it is the same device that changes from ordinary MOSFET operation to a bipolar regime under ESD conditions.[29]

Observation 11.10. *ESD events tax devices far beyond their customary operating regimes. Behavior under ESD conditions must therefore be simulated either at the level of three-dimensional device physics or with compact models specially designed and calibrated for that purpose. Simulation models intended for regular circuit simulation do not suffice.*

Another parasitic element that comes with each MOSFET manufactured in a bulk CMOS technology is a pn-diode. Note that a diode can serve a dual role. Forward-biased, it can be used to divert or to distribute ESD currents. Reverse-biased, it may act like a Zener diode that clamps the voltage across its terminals. ESD performance in reverse mode is inferior to that of a snapback BJT, though, see fig.11.30.

In a CMOS circuit, two parasitic diodes come with each gate and provide some degree of self-protection to the circuit's output. There was a time when this was sufficient and only inputs required separate protection circuits to be added. Subsequent technological improvements to CMOS such as lithography shrinks, ever thinner oxides, source/drain extensions (LDD), and silicides all were

---

[29] Whether the gate dielectric is made of thin oxide or of field oxide is of minor importance in this context. Its counterpart, the tandem device that consists of a p-channel MOSFET along with a lateral pnp BJT, is not being used as an ESD protection device due to its strongly inferior electrical characteristics.

detrimental to self-protection, however. Table 11.6 is an attempt to summarize the evolution of ESD protection over time.

Table 11.6 | Evolution of ESD protection schemes (simplified, partly after [350]).

timeframe	input protection		supply protection	output protection	
	primary	secondary			
1970s	pn-diode(s)	pn-diodes	none	none	wide geometries
early 1980s	fox MOSFET	pn-diodes	none	none	outputs self-protecting
late 1980s	fox MOSFET	snap BJT	none	snap BJT	absorbs locally
1990s	pn-diodes	snap BJT	snap BJT	pn-diodes	absorbs globally
early 2000s	pn-diodes	snap BJT	active net	pn-diodes	absorbs globally

Placing strong energy-absorption devices locally at every single input pad and output pad is a costly proposition, especially for today's ICs with their high pin counts. A more area-efficient alternative is to include two pn-diodes in every pad. These diodes divert ESD currents onto the supply rails, where the energy gets absorbed by an active power clamp such as a thyristor or a dynamically triggered snapback BJT. More recently, elaborate networks of actively triggered BJTs have been being used as power clamps. A difficulty with all such approaches are the IR drops across long supply lines. To ward off excessive voltage excursions, multiple supply protection devices are to be distributed within a chip's padframe [349].

Last but not least, particular attention must be paid to ESD robustness at the layout level. This is because parasitic resistances with values in the range of ohms that hardly matter during normal operation determine current and heat distribution in an ESD event. Current crowding in obtuse angles or near contacts and vias must be avoided. Similarly, current filamentation in protection devices is to be averted by distributing adequate series resistances over their entire layout width. [183] gives much practical advice on input/output design.

Observation 11.11. *When designing ESD protection circuitry, pay attention to ensuring that you*
- *keep voltages across critical structures to well below damage levels, and*
- *spread out ESD currents and hence also ohmic heating as much as possible.*

CONCLUSIONS

Their exquisitely thin gate oxides and fine layout structures make modern MOS circuits particularly vulnerable to ESD. Handling precautions and built-in energy-absorption circuitry are thus complementary to each other.

Designing effective ESD protection structures is a tricky task that must not be undertaken without good knowledge of device physics, pertinent experience, access to detailed process data, device simulation software, and specialized test equipment. Expert advice must be sought as none of these is routinely available to digital VLSI designers. For those eager to learn more, recommended literature includes [351] [183] [352] [349] [350] [353].

11.6.3 Latch-up

CAUSES AND EFFECTS

A particularity of bulk CMOS technologies is the presence of unwanted BJTs in supposedly pure MOSFET circuits. Note the lateral npn- and the vertical pnp-transistor in fig.11.32a. What the designer wants is the inverter circuit shown in fig.11.32b; what he gets on top of it is the parasitic circuit depicted in fig.11.32c.

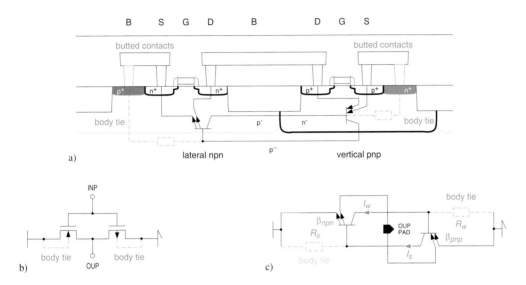

Fig. 11.32 The threat of latch-up in a bulk CMOS circuit. CMOS inverter sought (a), cross section with parasitic BJTs shown (b), resulting circuit with key parameters (c).

The problem with those parasitic BJTs lies in the positive feedback nature of a loop where the collector of one transistor provides the base current to the other. Imagine what happens when one bipolar begins to conduct in a circuit with no protective measures. Currents through either transistor increase until some innate resistance prevents them from growing any further, or until the circuit or part of its bonding wires are destroyed. Even if the triggering disturbance was of ephemeral nature and has long since gone away, the circuit will not recover by itself. In fact, the pnpn-structure is a **thyristor**, aka silicon-controlled rectifier (SCR), and behaves as such.

Provided it has not been damaged, the IC must be temporarily disconnected from its power supply to re-establish normal operating conditions. This situation where a short current path forms between power and ground is termed **latch-up** and must absolutely be avoided.

Observation 11.12. *Two preconditions must be met for a CMOS IC to enter latch-up:*
• *Parasitic bipolar transistors of opposite polarity must form a positive feedback loop and*
• *a disturbance must briefly bias them in such a way as to make the incremental current gain of the loop exceed unity.*

Erratic currents in the well and/or in the substrate provide the mechanism by which a large enough disturbance can bias a parasitic BJT to a critical point. Any of the disturbances below can cause an undesirable displacement of electrons or holes.

- Voltage overshoot and/or undershoot of a drain node,
  e.g. when an inductive load or a ringing line is attached to a chip's output pin.
- ICs or boards being plugged into sockets under power (hot plug-in).
- Overly fast transients of supply voltage at power-up time.
- Excessive ground bounce or supply droop.
- Operating voltages or currents beyond absolute maximum ratings.
- Electrostatic discharge (ESD) phenomena.
- Ionizing radiation such as X-rays, alpha particles, and cosmic rays.
- Alternating magnetic fields from currents in on- or off-chip inductors.

CMOS circuits that operate from supply voltages of no more than 1.5 V are largely immune to latch-up [219]. While this applies to the core logic in many modern ICs, beware of the higher voltages that are likely to occur in their I/O circuitry.

Keeping out disturbances by refraining from hot plug-in maneuvers, by making use of clamping devices, by adequate ESD protection circuitry, etc. is very important. A serious limitation of on-chip protection devices is that the diodes and field oxide MOSFETs put to service to clamp the overvoltages tend to inject charge carriers themselves when taxed.

Of course, process engineers do their best to safeguard CMOS structures against latch-up. Yet, their main objectives of layout density and MOSFET performance tend to conflict with latch-up avoidance. As a consequence, IC layouters are challenged to contribute towards latch-up protection by observing a few simple precautions. Figure 11.33c shows the five starting points, namely R_s, R_w, I_s, I_w, and $\beta_{npn}\beta_{pnp}$. The discussion below follows the same order.

Provide guard structures to divert carriers from where they might harm

A highly effective way of preventing substrate or well currents from forward-biasing a parasitic BJT is to provide a low-resistance bypass to the emitter or to some other node of suitable potential. In the case of an inverter, this implies shortening the MOSFET bodies to their respective source terminals as illustrated in fig.11.32b. Diffusion islands that share the polarity of the embedding substrate or well are used to galvanically connect to VSS or VDD respectively.[30] Such features are referred to as **body ties** with substrate|well|body plugs|contacts|taps and diffusion pickups being used as synonyms.

Body ties are generously distributed over the chip, not a single well must be left floating![31] Body ties are placed close to the parasitic bipolars. The smaller the bypass resistances R_s and R_w are made, the more current can safely be absorbed without having the base–emitter drop exceed the critical threshold of 0.6 V or so at which BJTs begin to conduct.

---

[30] As remains to be seen in section 14.1.3, the highly doped p$^+$ and n$^+$ islands are necessary to avoid the Schottky junction that would form if metal were allowed to connect to lightly doped p$^-$ and n$^-$ regions directly.

[31] Floating wells and/or a floating substrate not only render a chip vulnerable to latch-up, but also open the door to undesirable current leaks, capacitive coupling phenomena, and MOSFET back gate effects.

A popular layout arrangement termed **butted contacts** is shown in fig.11.32a. A p^+ (n^+) island is made to abut with the n^+ (p^+) source region of an n-channel (p-channel) MOSFET. The two are then connected with a short metal strap and tied to VSS (VDD). Larger body ties shaped into elongated stripes help to lower base-to-emitter resistances further by sidestepping the poor conductivity of the lightly doped well and substrate materials. This type of protective structure is referred to as a **guard bar**; an example has been given in fig.11.21. Even more effective are **guard rings**, where a p^+ diffusion stripe fully encloses n-type MOSFETs, and vice versa. Low resistance requires guard structures to be contacted from metal at regular intervals; connecting via diffusion or poly lines must be avoided because of their mediocre conductance.

Body ties aim at protecting parasitic BJTs from **majority carriers**. That is, the p^+ island next to the n-channel MOSFET collects holes that reach the p^- base region of the parasitic npn-bipolar and provides a safe current path to ground. The opposite is true in the n^- well.

PROVIDE GUARD RINGS TO ABSORB CARRIERS WHERE THEY MIGHT GO ASTRAY

A reciprocal approach consists in collecting unwanted carriers close to their place of origin before they could possibly find their way to some vulnerable BJT. The goal is to absorb stray currents so as to keep I_s and I_w small.

To attract unwanted electrons, an n^+ implant is placed around those devices from which dangerous carriers might emanate and connected to VDD, see fig.11.33. The converse applies to holes. Another way of looking at such structures is to consider them as extra collectors added to the parasitic bipolars in order to divert most of their undesirable collector currents to VDD and VSS respectively. The necessary layout structures look very much like guard rings and are in fact subsumed under this more general term, yet they do attract **minority carriers**.

In order to collect carriers that have managed to penetrate deeper into the substrate, minority guards are often prolonged vertically by implanting a well of identical polarity underneath. This is shown as an option on the left-hand side of fig.11.33b. An extra diffusion ring running along the circumference of a chip and contacted from metal at regular intervals further helps to lower overall substrate resistance and to provide maximum dispersal of substrate currents.[32] Such a ring is particularly important when no backside die contact is being used.

FRUSTRATE AMPLIFICATION IN THE FEEDBACK LOOP

Keeping **current amplification** $\beta_{npn}\beta_{pnp}$ in the feedback loop as low as possible also helps. Recall that the current gain β of a BJT is inversely related to the width of its base region. While the current gain of the vertical bipolars is essentially fixed by the fabrication process, it is possible to spoil the lateral bipolars by keeping n- and p-type MOSFETs well apart from each other in the layout. Any arrangement that clusters transistors of the same polarity in a common well thus fares much better than a layout that intertwines n- and p-channel transistors in a random or checkerboard fashion.

A related idea is to more firmly bias the BJTs into "off" condition by applying a negative (positive) back bias voltage to the p-type (n-type) bodies. It then takes larger stray currents through the

---

[32] From the process engineering side, epi substrates, high-dose buried layers (HDBL), and retrograde wells are latch-up prevention measures that follow the same general idea of attracting unwanted currents away from active devices.

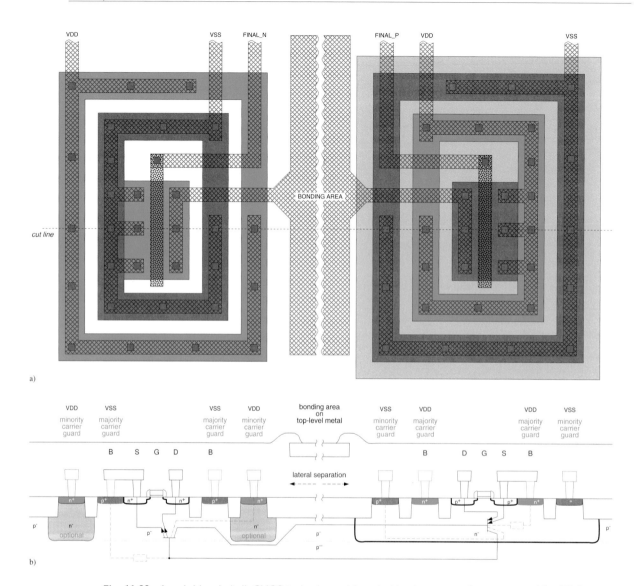

Fig. 11.33 A pad driver in bulk CMOS technology with typical latch-up protection structures (simplified, not drawn to scale). Geometric layout (a) and approximate cross section with parasitic bipolars shown (b). Note that (a) is drawn using a single metal layer.

p-substrate (n-well) before parasitic bipolars can get to a point where their joint current amplification $\beta_{npn}\beta_{pnp}$ exceeds unity.

CONCLUSIONS

Though few digital VLSI designers will be concerned with latch-up prevention at the layout level, they may find the subsequent information helpful for assessing the quality of library cells.

Observation 11.13. *This is what any layout designer should contribute to preventing latch-up:*
- *Provide body ties in generous numbers and place them next to MOSFET sources.*
- *Shape body ties as closed rings or as peripheral bars where possible.*
- *Place n- and p-channel MOSFETs clearly apart from each other.*
- *Provide extra minority guards in areas particulary exposed to external disturbances.*

Roughly speaking, current industrial practice is as follows.

Core cells. The term is meant to include standard cells designed to be placed in the core area of a chip and that do not connect to input or output pads directly, i.e. all bistables and regular logic gates. These are usually protected by pairs of majority carrier guards only.

Most cell libraries include body ties in every cell, even the most humble inverter is designed to feature one pair, larger cells comprise several pairs. The layout of those body ties often takes on the shape of guard bars that run along a cell's perimeter in close proximity to source areas and supply rails as shown in figs.11.21b and d. Input buffers and level-shifters that do connect to I/O pads warrant special attention.

Other library vendors leave out body ties from all standard cells and commit their customers to add them during back-end design instead. A typical rule of thumb would require that each diffusion area electrically connects to a body tie no further away than 20 µm. Modern place and route (P&R) tools can indeed be instructed to intersperse special **body tie cells** into standard cell areas at regular intervals. Yet, this approach unnecessarily burdens back-end designers with another chore and opens a door for mistakes.

Output pads. Note, to begin with, that both input and output cells are particularly exposed to voltage over/undershoot and ESD. Also, these cells continue to operate on supply voltages of 3.3 V or 2.5 V while core voltages have dropped to the vicinity of 1 V. Pad drivers are typically protected by a total of four guard rings, see fig.11.33. The n-channel MOSFETs are enclosed in a p^+ majority carrier guard which is itself surrounded by a-n^+-type minority carrier guard. Two more guard rings of opposite polarities are placed around the p-type MOSFETs. In addition, n- and p-type transistors are set 100 µm or so apart from each other with the gap in between put to service to accommodate the bonding area. Last but not least, adjacent I/O pads are arranged so as to avoid placing transistors of opposite types next to each other during chip assembly, see fig.11.7.

Input pads. Although not susceptible to latch-up themselves, the ESD clamping devices of input pads typically feature the same four-ring protective structures as pad drivers do. This helps catch injected carriers and saves design time by reusing part of the layout.

The focus here has been on what circuit designers can do about latch-up; for a more global discussion the reader is referred to the specialized literature such as [354] [355].

11.7 | Problems

1. Reconsider table 11.1 and note that a via manufactured with the same fabrication process has a resistance of 3.5 Ω. From a resistance point of view, what is the minimum distance for

which it makes sense to bring an interconnect line up from metal4 to metal5? What about metal3 or metal2? Do stippled vias change the picture?

2. Develop a standard cell for the logic function $\texttt{OUP} = \overline{(\texttt{IN1} \wedge \texttt{IN2}) \vee (\texttt{IN3} \wedge \texttt{IN4} \wedge \texttt{IN5})}$, where p-channnel MOSFETs are 1.5 times as wide as their n-channel counterparts. Deliverables include a schematic, a sticks diagram, and a gate-matrix layout with all details.

11.8 | Appendix I: Geometric quantities advertized in VLSI

The **gate length** L of a MOSFET is quantified in either of two ways. L_{drawn} refers to the size of the gate electrode as found in layout drawings whereas L_{eff} is the channel length that becomes electrically effective during circuit operation. Two effects contribute towards making $L_{eff} < L_{drawn}$. Firstly, doping atoms laterally diffuse underneath the gate during the thermal processing steps that wafers undergo as part of fabrication. Secondly, the charge depletion regions that form around the source and junctions eat away from the channel's length.

The **minimum feature size** M indicates the smallest object or interstice that can reliably be manufactured with a given fabrication process and appears in the set of layout rules as the smallest value specified for any width or spacing. All MOS fabrication processes allow for the drawing of gate length to minimum dimensions $\min(L_{drawn}) = M$.[33] As illustrated in fig.11.34, where $\min(W_{\mathrm{poly}})$, $\min(W_{\mathrm{M1}})$, $\min(W_{\mathrm{contact}})$, and $\min(S_{\mathrm{contact}\leftrightarrow\mathrm{poly}})$ are all equal, most fabrication processes support the patterning of more layout features down to the same size M, but there are exceptions to this rule.

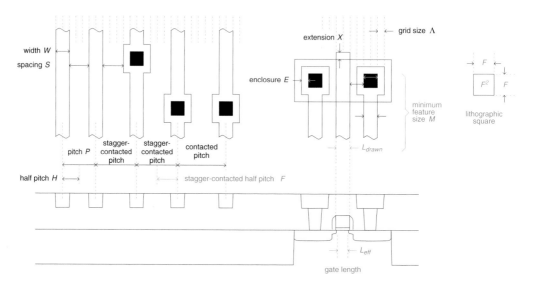

Fig. 11.34 Geometric quantities used in layout design.

---

[33] This is because shorter gates increase drivability and accelerate switching.

Pitch often is a more meaningful quantity than width or spacing because pitch accounts both for layout objects and for gaps. Stating the **half pitch** $H = \frac{1}{2}(W + S)$ makes the numerical value immediately comparable to feature size. In an attempt to give a number that is representative for the layout capabilities of a technology generation, the ITRS roadmap states the minimum half pitch of two lines on the lowest layer of metal (metal1, M1) that carry staggered contacts. Let us use the symbol $F = \frac{1}{2}(\frac{1}{2}\min(W_{M1}) + \min(S_{M1}) + \min(E_{M1 \leftrightarrow contact}) + \frac{1}{2}\min(W_{contact}))$ for that quantity.

F^2 is called a **lithographic square** and is typically used as a reference for comparing the area occupancy of a transistor, a memory cell, a logic gate, or some other layout item across different fabrication processes.

Lambda Λ is a fictive length unit not directly present in the layout. Its numerical value is chosen such that all layout rules are expressed as integer multiples of that basic unit. Any layout can so be drawn on a virtual grid of mesh size Λ.[34] In fig.11.34, $\min(E_{M1) \leftrightarrow contact} = 1\Lambda$, $\min(W_{M1}) = 2\Lambda$, $\min(S_{M1}) = 3\Lambda$, $\min(H_{M1}) = 2.5\Lambda$ (sic!), $M = 2\Lambda$, and $F = 3\Lambda$.

"Now, what does a vendor mean when he advertizes an x nm yMzP CMOS process?"

yMzP is just a shorthand notation that indicates the number of interconnect layers and stands for "y layers of metal plus z layers of polysilicon". Two layers of polysilicon are desirable in analog and memory design.

There is no rule as to what dimension x must refer to. Many foundries indicate the minimum half pitch ($x = \min(H)$, without contacts) of either poly or metal1 while others imply minimum feature size ($x = M$). Mainly for promotional reasons, some vendors prefer to put forward effective gate length so that a 180 nm process ($x = \min(L_{drawn})$) passes for 150 nm ($x = \min(L_{eff})$), for instance. In conclusion, note that any of the above quantities just gives an approximate indication for the achievable layout density because other factors such as the total number of interconnect layers $y + z$ also matter.

11.9 | Appendix II: On coding diffusion areas in layout drawings

Not all authors, foundries, and EDA vendors follow the same notation and terminology with respect to diffusion areas. At least three coding styles coexist.

Physical doping (p^+ & n^+). Diffusion areas are named p^+ and n^+ after their doping type and concentration, and so are the pertaining areas in the layout, see fig.11.35a. This is the coding style used throughout this text. Before wafer processing can begin, the thin-oxide definition and doping selector masks "thox" and "sel" must be obtained from the p^+ and n^+ layout areas in a separate postprocessing step.

---

[34] There was a time when layout rules typically shrank in a proportional fashion from one process generation to the next. Using relative dimensions expressed in Λ, termed **lambda rules**, then allowed rescaling of layouts by a simple change of the basic unit, say from $\Lambda = 175$ nm to 125 nm. Yet, the concept is no longer in use because consistent linear scaling has become an exception rather than the rule.

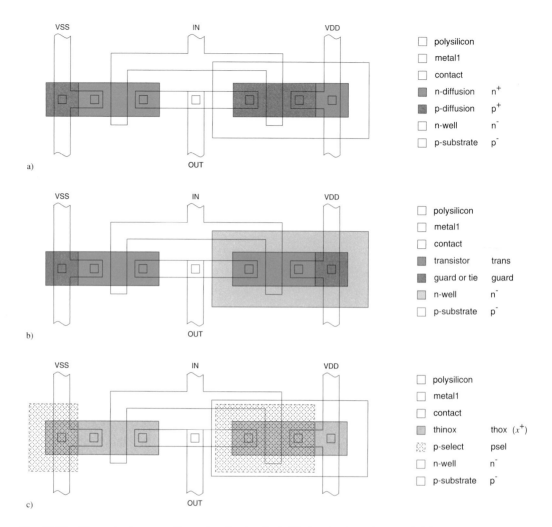

Fig. 11.35 Alternative layout coding styles. Stating final diffusion dopings p^+ and n^+ (a), electrical functions "trans" and "guard" (b), or fabrication masks "thox" and "psel" (c).

Electrical function (trans&guard). Diffusion areas are named "trans" and "guard" after their roles as MOSFET sources/drains, or as guard/tie structures in the semiconductor device, see fig.11.35b. The doping type of these areas follows from the type of the embedding well or substrate material: opposite for "trans", identical for "guard". Again, postprocessing of layout data is necessary prior to manufacturing.

Fabrication masks (thox&psel/nsel). Diffusion areas are described in terms of a thin-oxide definition mask "thox" and of a doping selector mask, which choice most closely reflects the mask set actually used during wafer processing.[35] The selector mask may be defined either

---

[35] Refer to section 14.2 for more information.

to indicate p- or n-type doping and is, therefore, labeled "psel" or "nsel" respectively. Figure 11.35c depicts a situation where the selector mask indicates p-type doping (thox&psel). Further observe from the drawing that the layout is unnecessarily cluttered with edges not relevant to the final result because the selector mask becomes effective solely where placed on top of the thin-oxide layer. This implies that the exact locations of its edges are uncritical except where they separate n^+ from p^+ doping.

It is always possible to translate layout data between the above coding styles on the basis of standard Boolean and resizing operations on polygons. Conversion between p^+ & n^+ and trans&guard styles is given by

$$p^+ := (\text{trans} \cap \overline{\text{pwell}}) \cup (\text{guard} \cap \text{pwell}) = (\text{trans} \cap \text{nwell}) \cup (\text{guard} \cap \overline{\text{nwell}}) \tag{11.3}$$

$$n^+ := (\text{trans} \cap \text{pwell}) \cup (\text{guard} \cap \overline{\text{pwell}}) = (\text{trans} \cap \overline{\text{nwell}}) \cup (\text{guard} \cap \text{nwell}) \tag{11.4}$$

$$\text{trans} := (p^+ \cap \overline{\text{pwell}}) \cup (n^+ \cap \text{pwell}) = (p^+ \cap \text{nwell}) \cup (n^+ \cap \overline{\text{nwell}}) \tag{11.5}$$

$$\text{guard} := (p^+ \cap \text{pwell}) \cup (n^+ \cap \overline{\text{pwell}}) = (p^+ \cap \overline{\text{nwell}}) \cup (n^+ \cap \text{nwell}) \tag{11.6}$$

while translation between p^+ & n^+ and thox&psel/nsel styles follows the rules

$$p^+ := \text{thox} \cap \text{psel} = \text{thox} \cap \overline{\text{nsel}} \tag{11.7}$$

$$n^+ := \text{thox} \cap \overline{\text{psel}} = \text{thox} \cap \text{nsel} \tag{11.8}$$

$$\text{thox} := p^+ \cup n^+ \tag{11.9}$$

$$\text{psel} := \text{bloat}(p^+ - \text{bloat}(n^+)) \tag{11.10}$$

$$\text{nsel} := \overline{\text{psel}} \tag{11.11}$$

where "bloat" stands for a geometric operation that expands a polygon by a small quantity along all dimensions. As mentioned before, the exact amount of oversizing is not critical in this application because it affects the position of uncritical edges only. Note, however, that layout features smaller than the oversizing quantity are lost in the process.

Things are further complicated by the fact that many different terms are being used in the industry for what we have denoted p^+, n^+, "trans", "guard", "thox", "psel", and "nsel" in this text. Thus, before drawing or interpreting detailed layout, make sure you understand what a given layer really stands for and how geometric layout data are being postprocessed as part of **mask preparation** in the foundry's mask shop. Also beware of ambiguous or meaningless names such as "diffusion", "implant", and "well".

11.10 | Appendix III: Sheet resistance

The resistance of a conductor with rectangular cross section $A = wh$ and length l is given by

$$R = \rho \frac{l}{A} = \frac{\rho}{h} \frac{l}{w} \tag{11.12}$$

where ρ indicates the specific resistance of the conducting material used and has dimension Ω m.

As the conducting material and its thickness h are determined by the fabrication process, l and w are the only parameters placed under control of the layout person in VLSI design. A conductive layer is, therefore, best characterized by its **sheet resistance** $R_\square = \frac{\rho}{h}$, from which the actual resistance of any line of rectangular layout is obtained by multiplication by the conductor's shape factor $\frac{l}{w}$:

$$R = R_\square \frac{l}{w} \tag{11.13}$$

Further observe from fig.11.36b that the shape factor and, hence, the resistance of a line of constant width are easily obtained from the layout by counting the number of squares that fit into the line's geometric shape. This is why sheet resistance is commonly expressed in terms of $\Omega/\square$ rather than just Ω.

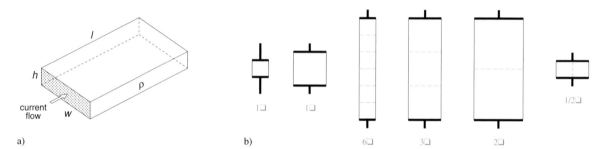

Fig. 11.36 Sheet resistance. Conductor of rectangular cross section (a) and top view of lines of various shapes along with their respective resistance values (b).

Practical sheet resistance measurements typically involve a symmetric layout known as a Van der Pauw structure and a probe with four contact tips.

Chapter 12

Design Verification

While much engineering effort in a VLSI design project goes into checking whether the HDL models developed do indeed capture the desired functionality, this does not suffice to make sure the design data submitted to fabrication are correct. Design flaws may also creep in during netlist synthesis and layout preparation. The extra checks required to uncover such problems and the EDA tools available to do so are the subject of this chapter. Section 12.1 is concerned with locating timing problems while section 12.2 analyzes the accuracy of the timing models being used. Electrical rule check (ERC) and other static design verification techniques are the subject of section 12.3. Section 12.4, finally, addresses post-layout design verification.

12.1 | Uncovering timing problems

As underlined in earlier chapters of this text, getting the timing right is vital for making VLSI circuits work as intended; catching potential timing problems is thus very important indeed. One is easily tempted to accept an error-free simulation run as a proof for a workable design, yet this is not so. We will first demonstrate why before introducing a more effective approach in section 12.1.2.

12.1.1 What does simulation tell us about timing problems?

Abstracting from the fine points that make up the differences between the various clocking disciplines, almost all timing-related difficulties that may occur in a synchronous (sub)system can be attributed to one or more of the causes listed below.

- Inadequate clock waveform (e.g. glitches, sluggish ramps, overly short phases).
- Insufficient setup margin because longest path delay is incommensurate with clock period.
- Insufficient hold margin because shortest path delay does not compensate for hold times of flip-flops, latches, RAMs, etc.
- Excessive clock skew or clock distribution delay.
- Poor synchronization of asynchronous signals from externally.

While uncritical from a functional point of view, overly slow ramps on information signals are undesirable too because they inflate energy losses due to crossover currents.

How timing violations are detected and reported

As explained in section 4.2.4, the search for timing problems during event-driven simulation essentially works by inspecting the simulator's event queue, also see fig.12.1. A message in textual format gets produced whenever a switching event is found to infringe upon setup time t_{su}, hold time t_{ho}, or minimum clock pulse widths $t_{clk\,hi\,min}$ and $t_{clk\,lo\,min}$. As each flip-flop, latch, RAM, etc. imposes its own set of timing requirements, it follows naturally that

Observation 12.1. *The checking of timing conditions is not performed by the simulator itself but is delegated to the circuit models being invoked by its built-in event queue mechanism.*

As a consequence, there can be no reporting of timing violations unless the circuit models are designed to do so. Almost all simulation models found in commercial cell libraries do indeed carry out the necessary timing checks, yet there are two caveats.[1]

Observation 12.2. *Whatever your preferred HDL is, do not forget to include all necessary checks (setup, hold, pulse width, etc.) when writing behavioral models of your own. Please note that this applies to sequential (sub)circuits of any size, not just to flip-flops and latches.*

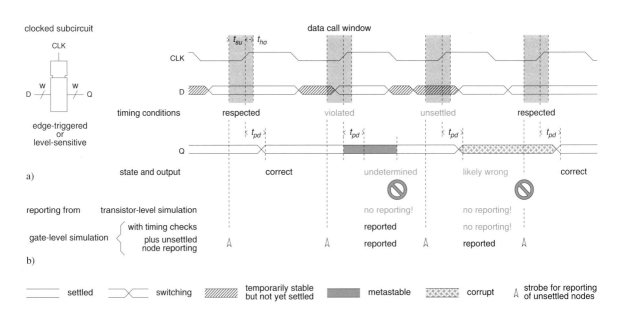

Fig. 12.1 The responses of a clocked subcircuit to correct and incorrect timing (a) and the pertaining reports as obtained from simulating at different levels of abstraction (b).

---

[1] Actually, there is a third caveat as there can be no reporting of timing problems when memorizing behavior is obtained from a zero-latency feedback loop, which is yet another argument for keeping away from this practice.

Transistor-level circuit simulators[2] use analytical equations to model the electrical characteristics of MOSFETs and other circuit elements. It is, therefore, generally accepted that the continuous waveforms they produce are more accurate than the discrete results obtained from gate-level simulation. Yet, neither timing checks nor an event queue constitute part of such simulations because these are digital rather than electrical concepts. If timing problems become manifest at all, they do so by producing waveforms that deviate from the expected ones to a more or less evident degree. Transistor-level simulation is inadequate — and also much too slow — when digital circuits of more than trivial size are to be verified.

Observation 12.3. *Transistor-level — and switch-level — simulations are not in a position to uncover timing violations as such and to report them by way of meaningful error messages.*

UNSETTLED NODES AND THEIR REPORTING

Even if timing checks are rigorously enforced in all simulation models, timing problems may pass simulation undetected. Such a situation is depicted in fig.12.1 for the third data-call window. Suppose the inputs to a combinational network of substantial size have changed in response to the last active clock edge. Further assume that the next active clock event arrives before all circuit nodes have settled to a new steady state. Depending on the exact delay parameters, the logic value at the D input of a subsequent flip-flop may then well remain constant throughout the brief data-call window defined by t_{su} and t_{ho} only to switch shortly thereafter. The flip-flop then stores a bogus value, but no timing violation will occur and no report will tell of it.

As a remedy, simulations are often set up such that the event queue gets inspected once per computation period immediately before the setup interval begins. Any forthcoming transaction that relates to a signal other than the clock then indicates that that signal has not reached its steady state, and points to an overly long signal propagation path. There is no consensus in the industry on what to call this instrument, **unsettled node reporting** and design stability checking are just two of the more common terms.[3]

WHAT IF THE CRITICAL PATHS ARE NEVER EXERCISED?

As shown by the next two examples, dynamic verification continues to suffer from dangerous loopholes even when automatic response checking, timing checks, and unsettled node reporting are combined.

Warning example

An electronics engineer was given a 16 bit arithmetic comparator, i.e. a circuit that determines whether A<B, A=B or A>B, along with the assignment to develop test vectors suitable for finding its maximum operating speed. Consistently with the standard practice of black box probing, he was given a functional description but no details about its inner organization.

---

[2] Such as SPICE, Spectre, ASX, and the like.

[3] The pertaining ModelSim command is **check stable on**. Incidentally, note that setting up simulation for the reporting of unsettled nodes is much more complicated in the presence of multiple clocks and clock frequencies, to say nothing about asynchronous designs.

He knew that the three output bits from arithmetic comparison were straightforward to derive from the "carry" and "zero" flags of subtraction, and figured out that the comparator would work that way. By toggling the two least significant input bits, he made sure that carries would propagate all the way from the LSB to the MSB, so exercising the longest path through a subtractor. He thus found he could make do with just four stimuli–response vectors:

stimuli		expected responses		
A	B	A<B	A=B	A>B
11...1110	11...1101	1	0	0
11...1110	11...1110	0	1	0
11...1110	11...1111	0	0	1
11...1110	11...1101	1	0	0

Measurements confirmed that actual and expected responses matched up to 50 MHz (20 ns), the maximum clock frequency supported by the hardware test equipment available. As specifications had asked for a data rate in excess of 37 MHz (27 ns), the engineer concluded that the circuit performed as requested. Only after circuits had been put into operation on real-world data did it become apparent that the comparator failed for rates beyond some 29 MHz (34 ns).

What had gone wrong? The entire reasoning rested on the assumption that the circuit was patterned after a subtractor. In reality, however, the circuit was organized as an iterative comparator where the ith bit slice evaluates its inputs and delegates the decision to the next lower slice whenever it finds that $A_i = B_i$. Rippling is from the MSB to the LSB so that stimulating the longest path would have asked for vectors entirely different from the ones employed.

□

In the above example, no simulator could possibly report any timing problem even if a register were added right after the comparator because no timing violation and no unsettled node condition would ever occur. The fact that circuit delay and clock rate do not fit together would pass unnoticed as the longest path is never brought to bear with the skimpy set of test vectors.

Observation 12.4. *The absence of timing violations, of unsettled nodes, and of departures from the expected responses during simulation is no guarantee that a circuit is free of timing problems. Similarly to what has been found for functional flaws, there is a coverage problem.*

Warning example

A gate-level netlist was obtained from synthesis after the RTL model had undergone extensive functional simulations. A scan-path and the clock distribution tree were then added by automatic means before physical design was undertaken. Eventually, the extracted netlist was verified reusing the logic gauge previously established during RTL simulations. When prototypes were tested, it was quickly found that they functioned as expected in normal operation mode, but failed to work in scan mode more often than not.

What had happened? The fact that the clock distribution network had been synthesized before physical layout and not readjusted afterwards had led to moderate clock skew. In normal operation, the contamination delays of the combinational logic present in between the flip-flops proved sufficient to compensate for that amount of skew. In scan mode, in contrast, the first of two adjoining flip-flops somewhere in the scan chain got clocked with enough positive skew (i.e. before the subsequent one) to cause repeated hold violations in the second flip-flop. Regrettably, the scan mechanism had never been exercised during post-layout simulations as a consequence of reusing the test vectors from RTL simulations where test structures were not yet present.

☐

Very much as in the example before, dynamic verification missed a timing problem because a critical case was not covered in the test suite. A major difference is that simulation failed to activate the shortest path this time. We can also learn from this example that test structures bring about extra hardware with new failure modes.

In conclusion, whether a timing problem gets detected during circuit simulation or not depends on several preconditions. For a synchronous design, these include:

1. Simulations must be carried out at the gate level (rather than transistor or switch level).
2. The models of all subcircuits used must be coded to report all conceivable timing violations.
3. Zero-latency circular paths through combinational subcircuits are disallowed.
4. The simulator must be properly set up to report unsettled nodes, if any.
5. The longest and the shortest delay must be exercised along every signal propagation path.

Simulation further provides no mechanism that would point to excessively slow clock and signal transitions explicitly. Rather, slow nodes have to be located indirectly from manifestations such as overly long paths, excessive skew, or inadequate waveforms. Attempting to address on-chip variations (OCVs) and crosstalk by way of exhaustive simulations is also impractical.

12.1.2 How does timing verification help?

Our discussion has revealed that simulation alone is not normally sufficient to identify timing problems in a digital design. A much better instrument is static timing analysis (STA) or **timing verification** for short. This analytical technique essentially works by mapping the gate-level circuit onto a constraint graph followed by comparing the maximum and the minimum delays along all signal propagation paths in the graph, see figs.12.2 and 12.3.

For each setup or hold condition, the timewise margin is being obtained as the difference of the respective delays along two distinct signal propagation paths. To that end each component must be characterized with both propagation delay t_{pd} and contamination delay t_{cd}. A negative result typically indicates a timing violation whereas any positive result is a sign of slack.

Interpretation is actually not quite as simple as not every case of negative slack flagged during static analysis necessarily implies that timing violations will indeed develop when the circuit is put to service. This is because certain gate-level circuits include signal propagation paths that are impossible to activate from the inputs [356]. That such **false paths** do not affect the maximum admissible clock rate should be obvious. Similarly, data ranges and formats are sometimes restricted such that not each and every signal propagation path present in a circuit can get exercised in real operation. Designers collect such situations in lists which they feed back into their STA tool in order to exempt false paths from further analysis.

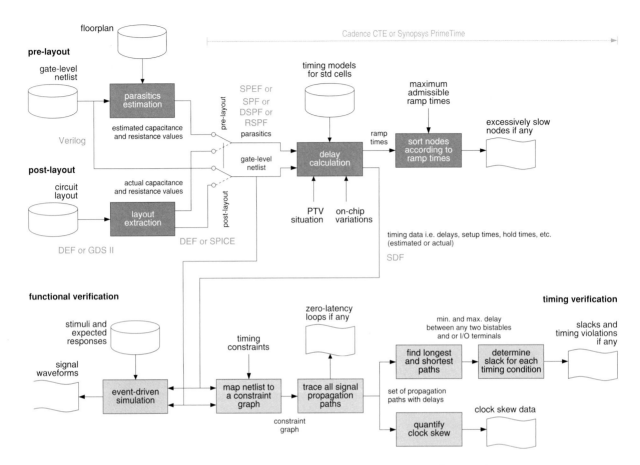

Fig. 12.2 Major steps of static timing analysis along with their context (simplified).

Much like formal verification, timing verification is a purely <u>static</u> technique that works without stimuli and expected responses, so there is no coverage problem. The slacks and other figures obtained do not depend on whether some critical case is being exercised or not, they simply reflect the timings along all paths through a digital circuit.

The results depend on how accurately wiring parasitics are known at the moment of analysis, however. One can distinguish three situations where the parasitics are being

- Eestimated from a gate-level netlist in conjunction with statistical wire load models,
- Estimated from a gate-level netlist complemented by floorplan information, or
- Extracted from actual geometry data (post-layout).

With the data from post-layout timing analysis, it becomes fairly straightforward to check for most of the problems mentioned in section 12.1.1. Better still, error reporting is by directly pointing to those nodes, paths or cells that need to be improved in one way or another. There is no need to dig through interminable vector lists or to analyze lengthy waveform plots.

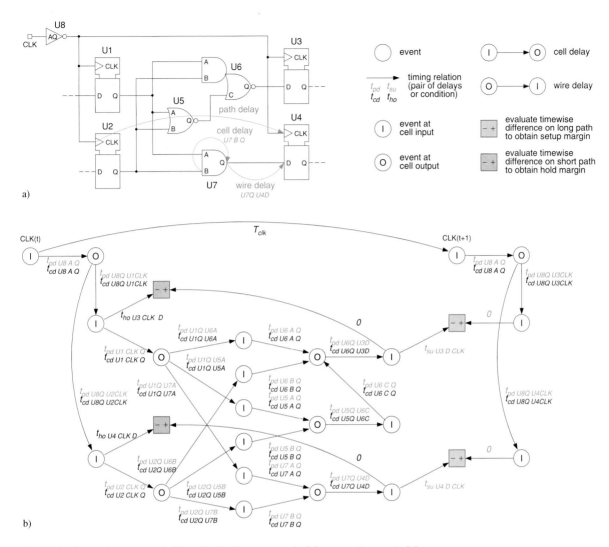

Fig. 12.3 Static timing analysis (simplified). Circuit example (a), constraint graph (b).

12.2 | How accurate are timing data?

The necessity to calculate or estimate delays arises at numerous points in the VLSI design flow, e.g. during gate-level simulation, timing analysis, floorplanning, logic optimization, technology mapping, and place and route (P&R). Entrusting a single piece of software with all delay calculations on the basis of a common timing model saves one from duplicate efforts and ensures consistency across all tools. Most EDA vendors endeavor to implement the idea of central delay calculation. The difficulty in coming up with precise timing figures is that they depend on numerous conditions under which the hardware operates. The most significant effects are discussed below along with explanations on how they are normally taken into account.

12.2.1 Cell delays

ACCOUNTING FOR DIFFERENT INPUT-TO-OUTPUT PATHS AND DRIVE CAPABILITIES

It is standard practice to use a **pin-to-pin delay** model whereby each input-to-output path is characterized individually and with separate figures for rising and falling edges to account for unsymmetrical current source and sink capabilities. Recall that CMOS loads are purely capacitive and that (dis)charging a capacitance with some given current takes non-zero time.

MODELLING OF TRANSIENTS

With voltage discretized to just a few logic states by the IEEE 1164 standard, there is no way to render signal ramps during VHDL simulation. The best approximation is to schedule <u>two</u> transactions per signal assignment. A first transaction turns the signal to "unknown" X when the end of the contamination interval t_{cd} indicates that the previous logic value no longer holds. A subsequent second transaction makes the signal assume its final value, typically one out of 0, 1, and Z, when the propagation delay t_{pd} expires. For an illustration refer to fig.A.25.

VHDL code example   `OUP <= ’X’ after contdelay, INA + INB after propdelay;`

Simulation models do not normally support such two-step schemes, though. Instead, they have the cell output switch from its previous logic value to the new one in zero time at the end of the propagation delay, which implies that t_{cd} is crudely replaced by t_{pd} and that output ramping is not being modelled.

LOAD DEPENDENCIES

The traditional **prop/ramp model** describes propagation delay as a sum of two terms.

$$t_{pd} = t_{it} + r_{cap}\, C_{ext} \tag{12.1}$$
$$t_{cd} = t_{it} \tag{12.2}$$

The product term reflects the time span necessary to charge or discharge the external load capacitance C_{ext} with a finite current. The name for r_{cap} is **load factor** and its measurement unit is $\frac{\mathrm{ps}}{\mathrm{pF}} = \Omega$. The delay contribution from the cell's own output capacitance C_{oup} is included in the cell's **intrinsic delay** t_{it}, a fixed quantity that does not depend on the surrounding circuitry. The prop/ramp model used to be adequate before the 500 nm process generation.

WAVEFORM DEPENDENCIES

A cell's timing also varies with the input waveforms as a consequence of finite voltage amplification and crossover currents in logic gates. The effect is evident in a zero-load condition where input and output voltage "ride" the gate's static transfer function. It has become more pronounced with the advent of deep submicron technologies because channel length modulation and low supply voltages are detrimental to voltage amplification.

The prop/ramp model has thus been supplanted by a more sophisticated **input slope model** that estimates a cell's delay(s) and output ramp time(s) as a function not only of drive strength and load capacitance but also of the ramp times at the input.

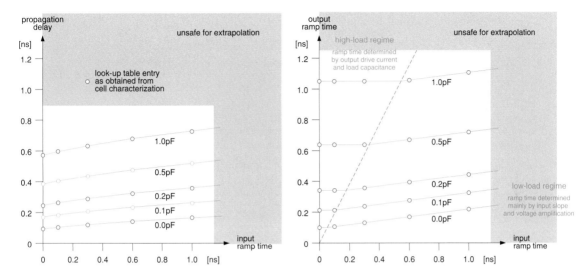

Fig. 12.4 Delay and slope modelling via table lookup.

There is no generally agreed upon analytical model for capturing waveform dependency, but table lookup from 5-by-5 tables with subsequent spline interpolation is a popular procedure, see fig.12.4. Multiple tables are used to account for distinct signal propagation paths, for rising vs. falling ramps, and for the various timing quantities to be modelled. A limitation of this approach is that calculations tend to become inaccurate when excessive load capacitances or overly slow ramps force the delay calculator to extrapolate data far outside the look-up table. Another approach is to extend (12.1) into

$$t_{pd} = t_{it} + s_{ra\,inp}\, t_{ra\,inp} + r_{cap}\, C_{ext} \qquad (12.3)$$

with a dimensionless **slope sensivity factor** $s_{ra\,inp}$ as an extra parameter. Values for all three model parameters are then to be given for each input-to-output path and output ramp orientation. The waveform dependencies of setup and hold times are captured as linear functions of clock and data ramp times in much the same way.

State dependencies

The delay on some given path through a cell further is a function of the voltages or logic states currently applied to the other inputs of that cell, which is why timing is said to be state-dependent. The effect is indeed significant [182] and many modern input slope models use even more look-up tables or parameter sets to account for that.

STATISTICAL VARIATIONS

Statistical process variations (P), junction temperature (T), and supply voltage (V) all impact a MOSFET's drain current and, hence also, the timing characteristics of CMOS circuits. In digital design, it is standard practice to model such **PTV variations** by applying three corrective factors that account for deviations between the actual and the nominal situations.

$$t_{derated} = t_{nominal} \cdot K_P K_\theta K_V \tag{12.4}$$

Functions K_P(fast, typical, slow), $K_\theta(\theta_j)$, and $K_V(U_{dd})$ are referred to as **derating curves**.

Numerical example

The figures below belong to a 130 nm CMOS process. Observe that PTV contributes to an overall variation by a factor of almost three between the fastest and the slowest case!

Process outcome	Derating factor K_P
fast	0.792
typical	1.000
slow	1.341

Junction temp. [°C]	Derating factor K_θ
125	1.139
75	1.070
25	1.000
0	0.962
−25	0.924

Supply voltage [V/V]	Derating factor for K_V
1.08/1.20	1.201
1.20/1.20	1.000
1.32/1.20	0.867

Overall situation		Derating factor
name	PTV condition	$K_P K_\theta K_V$
best case	fast, −25 °C, 1.32 V	0.63
nominal case	typical, 25 °C, 1.20 V	1.00
worst case	slow, 125 °C, 1.08 V	1.83

☐

Derating implicitly assumes that all devices on a die are equally affected by PTV conditions. Unfortunately, this is no longer valid for technologies smaller than 130 nm or so. All sorts of disparities between devices on the same die are subsumed as **on-chip variations** (OCVs).

- Mistracking threshold voltages (especially when multiple thresholds are involved)
- Local tolerances of channel width and channel length due to imperfect lithography and etching steps
- Individual gate oxide thickness variations
- Hot-electron degradation and other long-term wearout effects
- Local mechanical stress (affects carrier mobilities)
- Atomistic variability

Characterization of on-chip variability with statistical models currently is a hot research topic in view of yield enhancement and design for manufacturing (DFM).

TRIP POINTS AND OTHER ASSUMPTIONS UNDERLYING LIBRARY CHARACTERIZATION

Once analog circuit simulations — or actual measurements — have produced signal waveforms such as those shown in fig.12.5, a much more mundane question pops up:

"What is the exact lapse of time that shall be promulgated as propagation delay?"

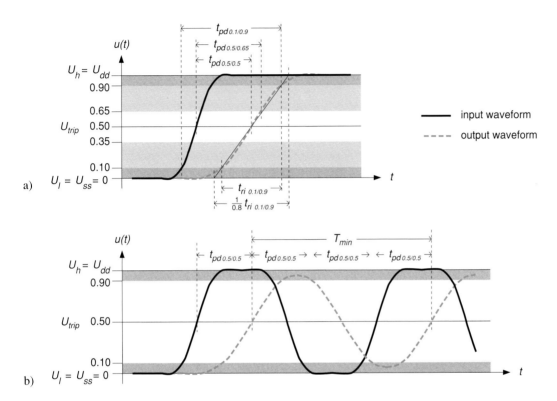

Fig. 12.5 The impact of trip voltages on the timing figures obtained. Different definitions of propagation delay and ramp time (a), low pass effect due to heavy load (b). The drawing refers to a non-inverting signal propagation path through a combinational circuit.

The common practice is to identify t_{pd} with the timewise difference between two crossings of predetermined voltages referred to as **trip points**, see fig.12.5a. Quantities t_{cd}, $t_{ri\,oup}$, and $t_{fa\,oup}$ are determined in a similar manner. As opposed to the load conditions, input waveforms, and other operating conditions discussed before, those trip points do not affect the actual behavior of the (sub)circuit being modelled in any way. Their choice is nevertheless important because they affect the timing figures disclosed in circuit models and datasheets, and so impact the results of timing verification, gate-level simulation, and synthesis.

An extremely conservative attitude is to start counting when the input signal traverses 10% of the voltage swing on a rising edge (90% on a falling edge) and to stop when the output signal settles to 90% (10% respectively). The time span so obtained is referred to as $t_{pd\,0.1/0.9}$ ($t_{pd\,0.9/0.1}$). The opposite attitude has all trip points sit halfway along the voltage swing and states $t_{pd\,0.5/0.5}$. This practice must be considered very optimistic for two reasons. Firstly, there exists no logic family that features identical and perfectly symmetric switching thresholds across all of its members. Secondly, designers are misled into believing that their circuits should work at higher frequencies than those at which they actually will.[4] A more realistic compromise is to indicate $t_{pd\,0.5/0.65}$ ($t_{pd\,0.5/0.35}$).

Observation 12.5. *Vendors of physical components and of design libraries are free to define trip voltages at their own liking, and they often tend to adopt highly aggressive values when under competitive pressure.*

There is no consensus on how exactly to quantify rise and fall times either. Some companies indicate $\frac{1}{0.8}t_{ri\,0.1/0.9}$, whereas others consider $t_{ri\,0.2/0.8}$ to be realistic enough. A majority specifies $t_{ri\,0.1/0.9}$. As stated in footnote 12, the timing-wise characterization of flip-flops, registers, and other sequential subcircuits introduces even more degrees of freedom.

Example

Table 12.1 shows how the conditions for characterizing cell libraries and for establishing datasheets have evolved over the years.[5] While downscaling has led to very impressive speed improvements indeed, it should be understood that nominal, i.e. underated, timing figures in datasheets and simulation models suggest extra speed-ups not justified on the grounds of actual technological progress alone, and unlikely to apply in practice.

Table 12.1 How conditions for library characterization have evolved over the years.

Year of initial release	Process generation	Process outcome	Junction temperature [°C]	Supply actual/nominal [V/V]	Trip points input $\leadsto$ output (rise/fall) relative to U_{dd}	Clock and input ramp time [ps]
1988	2000 nm	slow	85	4.5/5.0	$0.5 \leadsto 0.7/0.3$	1000
1991	1200 nm	slow	70	4.75/5.0	$0.5 \leadsto 0.65/0.35$	500
1994	600 nm	typical	25	5.0/5.0	$0.5 \leadsto 0.65/0.35$	300
1998	250 nm	typical	25	2.5/2.5	$0.5 \leadsto 0.5/0.5$	50
2000	180 nm	typical	25	1.8/1.8	$0.5 \leadsto 0.5/0.5$	24
2004	130 nm	typical	25	1.2/1.2	$0.5 \leadsto 0.5/0.5$	28.4
2005	90 nm	typical	25	1.0/1.0	$0.4/0.6 \leadsto 0.4/0.6$	modelled

Yet another variability comes from the extra parasitic capacitances due to over-the-cell routing. To stay on the safe side, library cells should be characterized with a fully populated routing grid on

---

[4] This is because one naturally expects a buffer with a propagation delay t_{pd} to transmit waveforms with a period as short as $T_{min} = 2t_{pd}$. While this is true on the basis of $t_{pd\,0.1/0.9}$, it is not necessarily so for $t_{pd\,0.5/0.5}$ where the ramp time is not accounted for, see fig.12.5b. Note that pad drivers and other heavily loaded nets are particularly exposed to this kind of misguided conclusion.

[5] All data are from commercial standard cell libraries. As a consequence of mergers and acquisitions, it has not been possible to compile the table from datasheets of any single vendor; a horizontal line thus separates data from distinct companies. Yet, the trend from extremely conservative to more accurate but also overly optimistic characterization is a universal one.

the next two metal layers above. The crosstalk effects due to over-the-cell routing are more difficult to anticipate.

Observation 12.6. *Timing data always refer to those specific circumstances under which they have been obtained. As there is no universally agreed-on standard for measurements and characterizations, make sure you account for diverging conditions when interpreting datasheets.*

Timing data with no indications about the conditions under which they apply are useless.

12.2.2 Interconnect delays and layout parasitics

There was a time when node capacitance was dominated by MOSFET gate capacitances and metal resistance had a negligible impact on interconnect delay, see fig.12.6. As a result, it was perfectly acceptable to take the cumulated cell delays along a signal propagation path as overall delay.

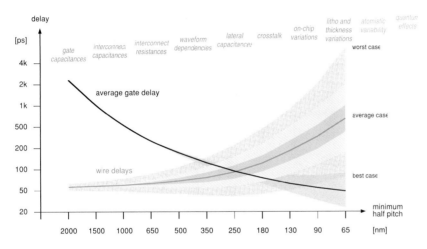

Fig. 12.6 The spreading out of delay figures and the underlying causes of variability (numbers strongly dependent on assumptions made).

With today's deep submicron technologies, however, interconnect lines have become narrower, relatively thicker, and also more tightly packed, see fig.12.7. Yet, die sizes and wire lengths have not shrunk much as a consequence of today's lavish levels of integration. Conversely, gate capacitances have diminished as a consequence from geometric scaling. As a net effect, wire delays prevail over gate delays and predictions on the basis of gate delay models alone have long become unacceptably inaccurate. Interconnect modelling is mandatory.

WIRE MODELS

As for any approximation, the results obtained are only as close to reality as the model is. You may want to see appendix 12.8 for an overview on popular wire models. A reasonable compromise must be found among precision, the effort for finding the numerical values for a model's parameters, and the computational burden of evaluating that model during simulation and timing analysis.

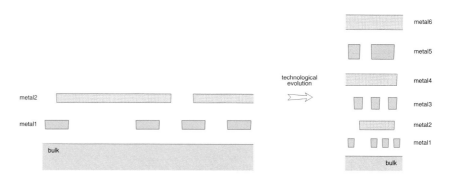

Fig. 12.7 Evolution of the interconnect stack over the years (simplified).

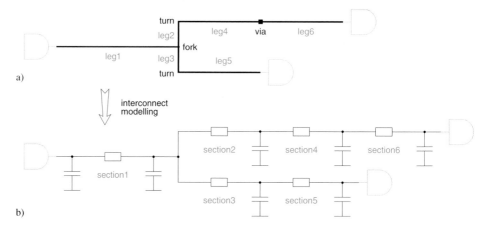

Fig. 12.8 A net (a) and its lumped RC network model (b).

In the **lumped RC network model**, each net is decomposed into multiple sections of uniform electrical characteristics, or very nearly so. Each section is then replaced by a resistance–capacitance pair as shown in fig.12.8. A leg in the layout typically becomes a section in the wire model as the line's electrical characteristics are likely to change at each turn, fork, via, contact, or change of width. Lumped RC models have a long tradition in the context of VLSI timing analysis, and many commercial software tools operate on the basis of such models.[6]

---

[6] The Pearl timing verifier by Cadence calculates interconnect delays as follows. To begin with, the driving gate is being modelled as a controlled voltage source with a series resistance (Thévenin model). The interconnect is then globally replaced by a Π-model and the waveform at the gate's output connector is determined for that load, see fig.12.15g. Next, the delays along all signal propagation paths are calculated from a second interconnect model, of lumped RC type this time, with the aid of Elmore's approximation formula [53] before one RC pair of identical delay is substituted for each such path. Lastly, each RC pair is being driven with a copy of the previously determined gate output waveform to estimate the waveform at the gate input being driven.

As opposed to this, lumping the total capacitance to the driving cell's output is inadequate unless the net being modelled is so short that its series impedance can safely be neglected when compared with that of the driver. Also, obtaining good estimates for the capacitances and resistances in each section is essential.

PARASITIC RESISTANCES

To calculate the parasitic resistance of a line, the line is geometrically decomposed into a series of legs of rectangular shape. The resistance of leg k, for instance, is then defined by the leg's geometry $l(k)$ and $w(k)$ and by $R_\square(k)$, the sheet resistance of the material.

$$R_k = R_\square \frac{l_k}{w_k} + \frac{1}{m(k)} R_{plug\,i,j} \tag{12.5}$$

As each contact or via plug between layers i and j has its own resistance $R_{plug\,i,j}$, we must add an extra contribution wherever two adjacent legs meet on distinct layers so that $i \neq j$. $m(k)$ indicates the number of parallel plugs that together make up the contact or via placed at the end of leg k or, which is the same, between legs k and $k+1$.

A new source of variability has become manifest at the 90 nm technology node. Sheet resistance and plate capacitance data are subject to important variations as chemical mechanical polishing (CMP) no longer produces layers of reasonably uniform thickness across a die.

PARASITIC CAPACITANCES

The external load capacitance attached to an output is

$$C_{ext} = \sum_{j=1}^{J} C_{inp}(j) + C_{line} \tag{12.6}$$

where J indicates the number of cells being driven (fanout) and $C_{inp}(j)$ the input capacitance of the jth such cell. As becomes clear from fig.12.9, a line's overall capacitance consists of contributions from plate, fringe, and lateral fields.

$$C_{line} = \sum C_{plate} + \sum C_{fringe} + \sum C_{lateral} \approx$$
$$\sum_{n,m} \left(A_{overlap\,n,m}\, c_{plate\,n,m} + l_{rim\,n,m}\, c_{fringe\,n,m} + l_{sidewall\,n,n}\, c_{lateral\,n,n} \right) \tag{12.7}$$

$A_{overlap\,n,m}$ denotes the area where layout patterns n and m overlap with no other conductive layer in between and $c_{plate\,n,m}$ the plate capacitance per unit area between the two layers. Similarly, $l_{rim\,n,m}$ stands for the circumference of pattern n that runs immediately on top or below m while $c_{fringe\,n,m}$ indicates the capacitive contribution per unit length of the fringe field between the two layers. Finally, $l_{sidewall\,n,n}$ stands for the length where the edges of two unconnected polygons on layer n face each other at minimum distance as specified by the pertaining layout rule; the capacitance per unit length between two such polygons is $c_{lateral\,n,n}$.

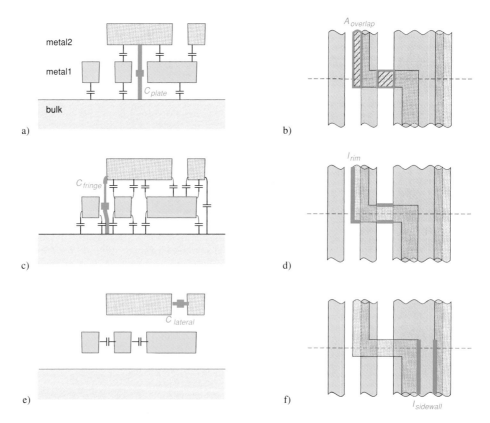

Fig. 12.9 Contributions to parasitic capacitance. Plate (a), fringe (c), and lateral capacitances (e) each with one instance highlighted versus the most relevant dimension in the layout (b, d, and f respectively).

Given a solitary line of width w and length $l \gg w$ that runs on metal1, the overall parasitic capacitance to bulk is then obtained as

$$C_{line} \approx l \left(w \, c_{plate \, M1,bulk} + 2 \, c_{fringe \, M1,bulk} \right) \tag{12.8}$$

PATTERN DEPENDENCIES

So far, each parasitic capacitance has been viewed as just another contribution to a node's total capacitance that must be (dis)charged whenever that node switches. The line under consideration was in fact considered the only switching node. This is obviously not so in a real circuit where multiple nodes toggle at the same time. As explained in chapter 10, crosstalk, ground bounce, and supply droop may significantly distort signal waveforms.

Delay figures are thus dependent not only on layout geometry but also on data patterns, which compromises the accuracy of static timing analysis further. To make things worse, coupling between adjacent signal lines becomes more important when dimensions shrink. Starting approximately with the 130 nm process generation, lateral capacitances between adjacent signal lines have in fact begun to supersede plate capacitances.

12.2.3 Making realistic assumptions is the point

From the above, we must admit that there are practical limits to the accuracy of static timing analysis and simulation because important electrical parameters are subject to all sorts of variations and conditions.

Observation 12.7. *In spite of sophisticated software tools, timing data warrant a fair portion of prudence and disbelief.*

As a rough indication, interconnect delay and noise coupling are estimated to contribute an average of 36% or so to overall stage delay in 250 nm CMOS designs [357], and even more so in denser technologies. This figure does not provide any latitude for sloppiness on the part of the designers but presumes correctly calibrated technology files, transistor models, library cells, and interconnect models as well as correct usage of CAD software.

What's more, a simulation run or a static timing analysis covers a single operating point whereas physical circuits are bound to operate under various operating conditions. As it is next to impossible to carry out analyses for all situations found in real-world operation, one must find a different solution.

Min/max timing analysis is a technique whereby the occurrence of an event is not described as a single moment of time but as a time span. The boundaries of that interval are obtained as the earliest and as the latest possible event times respectively that result from combining assumptions at either extreme of speed. This approach tends to give overly pessimistic results, however. After all, two signal propagation paths may exhibit a fair degree of on-chip variation (OCV) and be subject to severe crosstalk from other signals but will nevertheless operate under similar process, temperature, and voltage (PTV) conditions.

A more pragmatic approach is to predict all variations for which an accurate and efficient model exists and to add an approximate allowance on top for those that are impossible to anticipate with reasonable computational effort. Consider OCV, for instance. Any two signal propagation paths are compared under the assumption that their respective delay figures may deviate in either way but by no more than by some fixed factor of, say, 1.25 from their respective calculated values. In essence, this amounts to asking for more ample margins during timing analysis without overdoing it to handle situations that are extremely unlikely to occur in practice.

The burden of deciding on what is realistic and what not rests with the IC designer. Many default PTV settings from data sheets and simulation models are likely to prove overly optimistic because industry has gradually moved away from conservative characterization conditions to more aggressive ones under the pressure of competition in the marketplace, see table 12.1. Note, in particular, that junction temperatures of 25 °C are unrealistic even if one calls for massive forced cooling. At room temperature, the air contained in any kind of enclosure must be expected to warm up well above 25 °C and the heat-generating chips even more. Similarly, supply voltage must be expected to fall below its nominal value as a consequence of series impedances in power and ground nets.

As a more general conclusion, a broad tolerance with respect to parameter variations must be designed into electronic hardware.[7]

Observation 12.8. *While a design's maximum clock rate necessarily depends on numerous data-dependent, process-induced, and environmental uncertainties, its qualitative functioning must not.*

12.3 | More static verification techniques

12.3.1 Electrical rule check

Experience has led designers to recognize certain circuit structures as safe and others as prone to failure. Over the years, engineers have come up with a number of rules as to what good circuits must look like and as to what structures must be avoided. What exactly is to be considered perilous to safe operation depends — to some extent — on the technology, the circuit design style, the clocking discipline, the surrounding circuitry, and other circumstances. A few truly unacceptable anomalies are given below, a more complete list is available from table 12.2.

- Power and ground fragmented into unconnected segments.
- Shorted power and ground nodes.
- Missing drivers on output pads.
- Permanent drive conflicts.
- Cell inputs left open.
- MOSFET terminals left unconnected or shorted together.

Any such oddity points to a problem that is very likely to prevent a design from operating as intended. Worse than this, fabricated circuits might even behave differently from simulations.

Most structural flaws can be found by careful inspection of the circuit with no need for a simulator and stimuli waveforms. Software tools have thus been developed that accept a gate- or transistor-level netlist and that scrutinize it for violations of **integrity rules**. Such tools are commonly known as **electrical rule checkers** (ERCs) or netlist screeners. What makes them so valuable is that they indicate the existence of potential design problems irrespective of functionality and test patterns, much like static timing analysis (STA).

Example

The problem of using simulation to locate design flaws has been discussed in an earlier chapter of this text; an example where inputs to a multiplexer are permuted by mistake has been given in fig.3.1. As swapping two signals does not violate any integrity rule, no ERC could possibly uncover a problem of this kind. The mistake simply leads to a functionality that deviates from the intended one, uncovering it thus entirely depends on functional verification.

---

[7] It is interesting to note that while fully self-timed operation maximizes the latitude towards uncertain or changing timing parameters, the opposite holds true for clock-as-clock-can asynchronous design styles, with strictly synchronous clocking disciplines occupying reasonable positions in between.

Now consider a case where two wires have inadvertently been shorted together rather than permuted. A node controlled by more than one permanent-drive output is against established principles of logic design unless the two driving gates are connected in parallel and of identical type. Any decent ERC will report this oddity and indicate the shorted drivers whereas tracing the problem back from a few bits of simulation output that do not match their expected values may prove to be an exacting exercise.

□

Observation 12.9. *Most problems uncovered by ERC can also be found by way of simulations, yet ERC has two important benefits. Firstly, ERC being a static technique, there is no coverage problem. An ERC run necessarily uncovers all violations of the integrity rules examined. Secondly, ERC reports are much easier to interpret than simulation outputs, which is why netlists should always be screened prior to simulation.*

In comparison with the early days of VLSI design, ERC has lost part of its former significance as automatic synthesis dispenses with the hand-editing of gate- and transistor-level schematics and so prevents a variety of problems from coming into existence. Also, industrial cell libraries are normally free of structural problems. With the shift to more abstract levels of design entry, static verification techniques had to move upwards as well.

12.3.2 Code inspection

Most design failures today have their roots in mistaken RTL models. While it is true that all simulation and synthesis tools do a thorough syntax checking on the HDL input code, this is by no means sufficient to obtain circuits that are safe, efficient, and functionally correct. Examples of problems that typically slip through syntax checking are given below.

- Naming conflicts in the HDL code that may cause nodes to become short-circuited.
- Reversed, empty, excess or otherwise inconsistent index and/or address ranges.
- Mismatches between endian types (especially when subwords are involved).
- Stationary driver conflicts due to flawed bus access protocols.
- A clocking scheme that does not comply with any of the established disciplines.
- Zero-latency loops through datapaths and their controlling state machines.
- Chip-wide busses and other high-fanout high-load circuit structures.
- Large central multiplexers and other high-fan-in circuit structures.
- Large centralized control circuitry difficult to optimize, maintain, and modify.

Thorough examination of HDL source code prior to synthesis is thus highly desirable to avoid unnecessary and time-consuming design iterations. It is also highly beneficial because decisions at the architecture and RTL level have a much more significant impact on circuit complexity, timing closure, and energy efficiency than gate-level optimizations have.

Code inspection largely relies on human expertise but software aids have also begun to appear. Note that no single code checker searches for all conceivable problems. Few commercial tools consistently check for compliance with synchronous design guidelines, for instance. Always find out what tests are included and which are not when evaluating a new verification product.

Table 12.2 Various design problems and the verification tools supposed to uncover them.

Technique Design flaw	dynamic gate-level simula- tion	code inspec- tion	static electr. rule check (ERC)	timing verifi- cation (STA)
No or inadequate reset mechanism	if covered	yes		
Inconsistent index or address range	if covered	yes		
Mismatches between endian types	if covered	yes		
Unwanted latch in supposedly combinat. logic[a]	if covered	yes		
Driver conflict due to flawed bus access protocol	if covered	yes		
Node left undriven for a prolonged time	maybe[b]	yes		
Signal nodes shorted together[c]	if covered	yes	yes	
Misuse of async. reset for functional purposes		yes		
One-shot, clock chopper, hazard suppressor, etc.		yes		
Cross-coupled gates or other zero-latency loop		yes		yes
Careless clock gating (comb. gate in clock net)		yes		maybe
Lack of synchronization between clock domains		yes		maybe
Excess ramp times on clock or other signal				yes
Excess clock skew	maybe[d]			yes
Long- or short- path problem[e]	maybe[d]			yes
Unfriendly external timing	maybe[d]			yes
Missing level shifter on input pad			yes	
Missing driver on output pad			yes	yes
Lack of driver on test pad[f]			maybe	yes[g]
Cell input left open	if covered		yes	
Cell output shorted to power or ground			yes	
Redundant cell (no output connected)			yes	
MOSFET terminal left unconnected			yes	
MOSFET terminals shorted together			yes	
Floating well due to missing body tie			yes	
Short between power and ground			yes	
Broken supply line			yes	
Missing power or ground pad			yes	

[a] Gets reported also during synthesis.
[b] Provided the simulator is set up to model charge decay correctly.
[c] Due to a naming conflict in the HDL code (or to badly drawn schematics).
[d] Provided netlist is properly back-annotated and problem gets covered by test vector set.
[e] Setup violations/unsettled nodes or hold-time violations respectively.
[f] Capable of handling the load of a probe.
[g] Provided all tests pad get adequately loaded for STA.

12.4 | Post-layout design verification

The final layout obtained with the aid of automatic place and route (P&R) tools ideally

- Matches the original netlist,
- Conforms with all layout rules of the target process,
- Remains within the specified delay and energy budgets, and
- Does not overtax any interconnect lines, contacts or vias electrically.

Incorrect or inadequate layouts occasionally result as a consequence of software bugs, flawed cell libraries, inconsistent design kits, misinterpretations, routing congestions, suboptimal algorithms, misguided manual interventions, and the like. In addition, many designs include portions layouted by hand. Delaying the uncovering of problems until first silicon becomes available would cause enormous cost and time overruns, hence the need to double check whether the above four conditions are indeed satisfied before proceeding to mask preparation.

As shown in fig.12.10, physical verification can be carried out either before or after the detailed layout patterns have been filled in for the cell abstracts. Checking the layout at the **detail level** offers better protection and is, therefore, a mandatory part of any sign-off procedure. However, enabling ASIC designers to do so requires the library vendor to disclose his own layout drawings in full detail to his customer, which proposition is unpopular in a competitive context. Also, the amount of data to be processed is greatly inflated.

Checking at the **abstract level**, in contrast, implies that each cell is epitomized to a box of true size and shape with the connectors correctly located, but with hardly any inner details.[8] Such a simplified view is normally sufficient for place and route (P&R) but perforce confines physical verification to the interconnect layers. Uncovering flaws within library cells and extracting layout parasitics with good precision are not possible in this way.

In order to maximize protection against software bugs, the entire verification process relies on software code other than that included in the EDA tools that serve to establish layout data. Experience has shown that commercially available design kits are not always free of inconsistencies between the various technology files (for synthesis, P&R, DRC, layout extraction, timing verification, etc.) and the documentation for the user. Sorting out such issues is very time-consuming or altogether impossible without detail-level layouts and generates a lot of back and forth between circuit designers, library vendor, and silicon vendor.

> Hint: Always complete a full design cycle on a benchmark design
> before adopting a hitherto unproven design kit for a commercial project.

Next, we will examine what the various software tools contribute to physical design verification and where they fit into the process before summarizing our findings in table 12.3. For illustration, numerous layout flaws have been collected in figs.12.11 and 12.12.

---

[8] The exceptions are called **blockages** and designate areas unavailable to over-the-cell routing on some specified low-level metal layer because of obstructions within the cell itself.

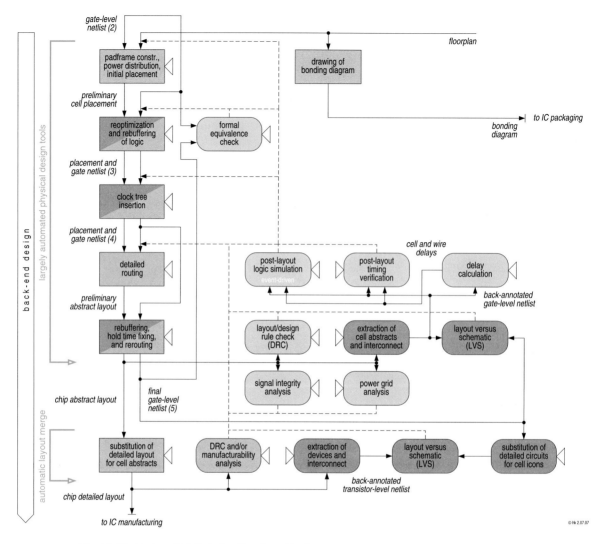

Fig. 12.10 Back-end VLSI design flow (simplified).

12.4.1 Design rule check

A design rule checker (DRC) is a piece of EDA software that verifies whether or not a given design complies with the geometric rules imposed by a target process. DRC software accepts layout data plus a set of layout rules, aka rule deck, and returns a listing of all violations found, if any. Locating DRC violations is greatly facilitated if the software also generates an error file that can be read into a graphical layout editor, where it is overlaid over the flawed layout for inspection. Examples of layout problems include items (d), (h), and (i) in fig.12.11c.

DRC is always carried out before the other physical verification steps because it makes no sense to subject incorrect layouts to further analysis as they are not going to serve for fabrication anyway.

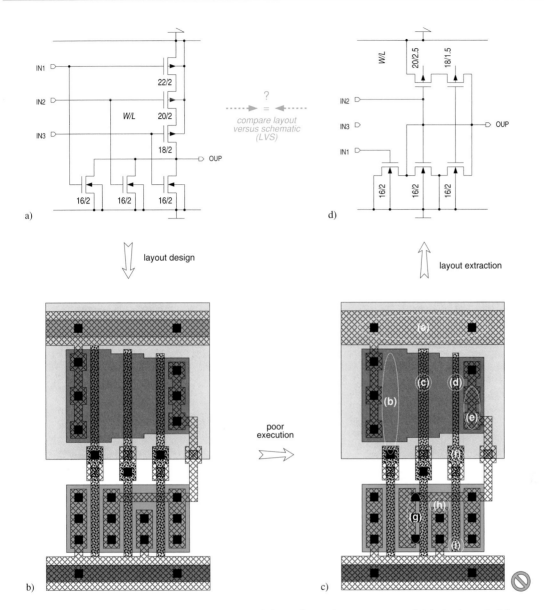

Fig. 12.11 Layout verification at the transistor level shown for a 3-input NOR gate. Original schematic (a), correct layout (b), flawed layout (c), and extracted schematic (d).

Also, invalid layout data are likely to fool circuit netlist extraction and other downstream verification programs.

Most DRC rule decks have been established for detail-level DRC, which is to say that they not only cover, but also require, the full set of layout data as input. While it is true that physical design of cell-based circuits is largely confined to intercell wiring using metal2 and higher layers, the layers below such as wells, diffusions, polysilicon, and metal1 are also affected, albeit indirectly. One can

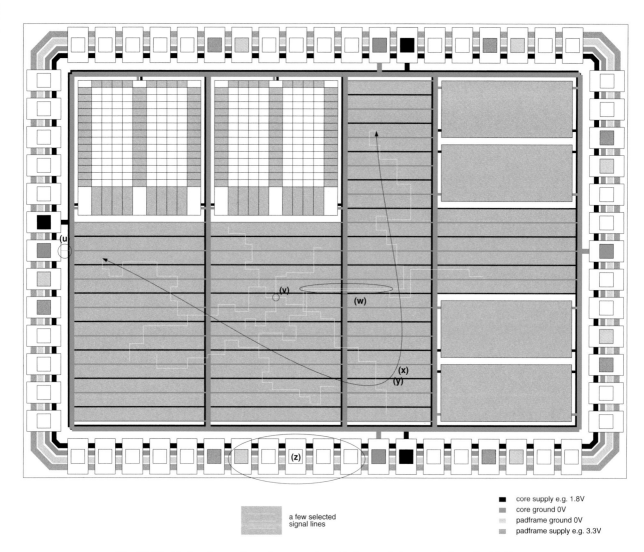

Fig. 12.12 Various layout flaws shown on the add–drop switch example from fig.11.10.

never be 100% sure about geometric correctness unless a layout has indeed been verified on all layers. The process is thus complicated when a third-party library vendor refuses to make detailed cell layouts available to VLSI designers. Having the silicon vendor carry out detail-level DRC runs on behalf of his customer instead significantly inflates turnaround times.

12.4.2 Manufacturability analysis

As stated in observation 11.1, geometric layout rules and traditional DRC have long supported an utterly simple and effective model of collaboration and responsibilities between circuit designers and manufacturers. This proven procedure is unfortunately running out of steam due to the complexities of subwavelength lithography and of sub-100 nm fabrication processes. On a background of yield enhancement and design for manufacturing (DFM), the sharp distinctions among layout design, mask preparation, and wafer processing are blurring [358]. One option, introduced in section 11.5.6, is to restrict the admissible layouts to compositions of a few patterns known to be lithography-friendly.

A complementary idea is to replace the traditional DRC procedure along with its simplistic rules and hard "go/no go" decisions by more elaborate computational models able to predict a design's manufacturability. Lithography compliance checking requires a software tool that accepts layout data plus a sophisticated mathematical model as inputs for simulating the subwavelength lithography process and for approximating the geometric distortions to be expected [359]. The geometry so obtained is then compared against the designer's intentions as reflected by the undistorted layout drawings and target netlists. Problem areas, such as likely shorts and breaks, but also critical dimensions subject to significant variability, are flagged.

Lithography compliance checking can further help to automatically make minor adjustments to electrically uncritical layout patterns so as to improve a design's latitude towards uncontrollable variations, e.g. by slightly relocating existing edges in the layout and/or by adding and optimizing serifs, hammerheads, and other geometric features of purely lithographic purpose illustrated in fig.14.24.

12.4.3 Layout extraction

Layout extraction accepts layout data and returns the corresponding netlist. In addition to the geometrical layout, the extractor software asks for technology files that define all layers and that specify the active and passive devices to look for (e.g. MOSFETs, capacitors, resistors, etc.) along with their electrical characteristics (e.g. sheet resistances, plate and fringe capacitances, process gain factors, MOSFET threshold voltages, etc.).

Depending on whether the input was a detailed or an abstract-level layout, the **extracted netlist** is either a transistor-level or a gate-level netlist. What sets the extracted netlist apart from the original one are the more accurate numerical figures for layout parasitics.

Also note that layout extraction per se does not normally uncover or report any design problems. Extraction just undoes physical design and so serves as a preparatory step for the subsequent scrutiny of a design's true netlist.

12.4.4 Layout versus schematic

Layout versus schematic searches for disparities between two circuit netlists and is typically being used to compare the extracted netlist against the original netlist, hence the name LVS. This powerful instrument is capable of locating a multitude of problems such as:

1. The two netlists do not match in terms of number and/or naming of terminals.
2. There exist circuit entities in one netlist for which no equivalent counterpart can be identified in the other netlist.[9]

---

[9] The word circuit entity here refers to any kind of subcircuit or electronic device such as a building block, a logic gate, or a MOSFET.

3. Entities match but their connectivities have been found to differ between the two netlists.
4. Entities match but their geometric size or electrical parameters differ more
   than some predefined tolerance margin permits.

Examples are shown in fig.12.11c. Items (b), (a), and (f) all cause a mismatch in the circuit's structure while (c) preserves connectivity but causes the electrical characteristics to deviate in an unforseen way.

LVS correctly detects inadvertent shorts and opens in a layout, but cannot normally point to their exact locations. In the occurrence of problem (v) in fig.12.12, for instance, LVS would just flag the names of the two nets being shorted together and would highlight all their polygons in the layout. Identifying the exact place and cause of the short is left to humans, which may take some investigative efforts when the nets concerned are spread out over long distances and when no layout rules are violated.

12.4.5 Equivalence checking

LVS has a strong tradition in VLSI, but its application has become increasingly difficult. This is because today's designs undergo a series of optimization steps such as rebuffering, input reordering, logic reoptimization, scan reordering, and clock tree generation as part of physical design. LVS cannot work properly unless all such changes are propagated back into the original netlist. This is why the extracted netlist is more and more compared against the synthesis model rather than against some intermediate netlist.

Formal verification techniques essentially identify subcircuits, reconstruct their behavior at the RTL level capturing all combinational subcircuits as logic equations, and check for equivalence with the original HDL description. As an extra benefit, any functional flaw that might have crept in during some front-end design step, such as pipelining, retiming, conditional clocking, or the insertion of test structures, is very likely to be found as well.

12.4.6 Post-layout timing verification

As both interconnect and cell delays are functions of layout parasitics, the placement of cells and the routing of wires established during physical design affect a circuit's timing to an extent that is significant, yet difficult to predict. Unexpected deviations from the expected may give rise to all sorts of timing problems and even cause the finished layout to behave quite differently from the original netlist. Every design is, therefore, subject to timing verification on the basis of delay data calculated from the geometries in the finished layout as explained in sections 12.1.2 and 12.2.2. The timing data so obtained are **back-annotated** into the gate-level netlist, that is, they are to overwrite the pre-layout data there.

> Hint: Make it a habit to double check the numerical data that participate in back-annotation and that emanate from post-layout simulation and timing verification.

Warning example

In a particular project, it was discovered that timing analyses were carried out with a fixed propagation delay of 1 ns for all full-adder cells instantiated. Closer inspection revealed that the back-annotation process had failed to become fully effective because of inconsistent namings in the VITAL cell models. While all parameters got assigned correct data from the standard delay format (SDF) files, some assignments actually remained ineffective as the names used in the cell models and those constructed from the cell's terminal names — and hence referenced in the SDF — did not match. As a consequence, a number of timing quantities simply kept their default value of 1 ns. No warning or error message was produced that could have hinted at the problem. To make things worse, with 1 ns being roughly four times the actual delay of a full adder in the target technology, the overall delay reported was well beyond any acceptable tolerance, but not totally implausible. It took the attention of an experienced supervisor and the comparison with a Verilog model to become aware of the problem.

□

12.4.7 Power grid analysis

To locate inadequately sized supply lines, contacts, and vias, specialized EDA tools accept layout geometries, extracted layout parasitics, and node activity data as obtained from simulation runs or from statistical power estimation tools. For each power and ground line they accumulate the DC values before calculating current density and ohmic voltage drop. Output typically is via color maps superimposed on the chip's floorplan.

12.4.8 Signal integrity analysis

Signal integrity analysis begins by calculating ramp times from the current sink and source capabilities of all cell outputs and from the respective loads attached. Starting from the parasitic capacitance, resistance, and inductance values obtained from layout extraction, a software tool then estimates the impact of signal transitions on one line (the polluter) on the waveforms on other nearby lines (the victims) that result from ground bounce and/or crosstalk. Nets found to be at risk get flagged.[10]

12.4.9 Post-layout simulations

Yet another verification technique popular with VLSI designers is post-layout simulations. Such simulations are mainly sought for personal reassurance, however, as they are largely redundant provided meticulous pre-layout simulations have been carried out and provided LVS and equivalence checking have confirmed that pre- and post-layout models are indeed equivalent. Be assured that it is much easier to uncover and locate layout problems from an LVS report than from the output of a post-layout simulation run.

12.4.10 The overall picture

Of all design flaws listed in table 12.3, (e) is very likely to pass unnoticed because it does not violate any layout rule, does not alter the circuit's netlist or functionality, has a minor impact on timing, and does not result in an out-of-the-normal current density or waveform. Contact replication is indeed mostly a matter of yield enhancement.

---

[10] Commercial EDA tools include PrimeTime-SI by Synopsys and CeltIC by Cadence.

Table 12.3 Various layout problems and the verification tools supposed to uncover them.

Design flaw	design rule check (DRC)	layout ex-trac-tion	layout versus schematic (LVS)	equi-valence check	post-layout timing verif.	grid/ signal integr. analysis
(d) insuff. poly width (gate length)	yes		tolerance[a]			
(h) insufficient metal spacing	yes					
(i) insufficient poly extension	yes					
(b) missing gate (p-channel device)			counterp.	yes		
(a) absence of body tie (n-well)	yes[b]		connectiv.			
(f) missing poly contact			connectiv.	yes		
(g) misaligned diffusion contacts	yes[c]		connectiv.	yes		
(v) lines intersecting on same layer			connectiv.	yes		
(c) mismatch of gate length			tolerance[a]			
(x) excessive interconn. resistance					yes	
(y) excessive interconn. capacitance					yes	
(u) constriction in supply line						power
(z) excessive supply line inductance						signal
(w) excessive crosstalk						signal
(e) lack of contact replication						

[a] Provided the original netlist correctly specifies a target value.
[b] Provided the DRC cares about dangling contacts.
[c] Provided a layout rule gets violated as in fig.12.11b.

12.5 | Conclusions

- Code inspection, functional verification — which essentially relies on simulation —, static timing analysis, and electrical rule checking complement one another. As becomes clear from table 12.2, all methods must be combined to maximize the likelihood of finding flaws in HDL models and gate-level netlists.

- No timing figures must be taken for granted before it is made sure that actual operating conditions and those assumed for the purpose of timing analysis are in good agreement.

- No single EDA tool is capable of detecting all potential layout problems, so physical verification is not complete before the subsequent steps have been carried out:

  1. Design rule check and/or manufacturability analysis.
  2. Layout extraction.
  3. Layout versus schematic and/or equivalence checking.
  4. Post-layout static timing analysis.
  5. Power grid analysis.
  6. Signal integrity analysis.

12.6 | Problems

1. Section 12.1.1 has reported a misguided attempt to measure the propagation delay of a 16 bit arithmetic comparator that failed because the test vectors turned out to be inadequate. Can you provide a better set?

2. Figure 12.13 shows a Braun-type array multiplier for unsigned numbers.

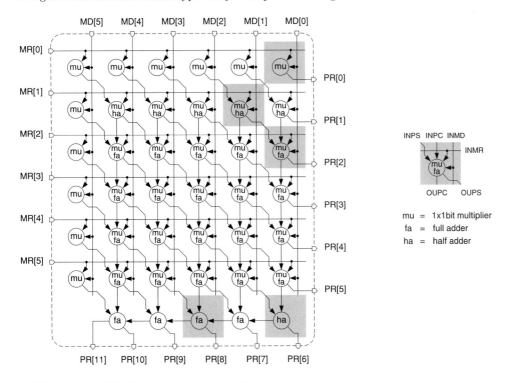

Fig. 12.13 Bit-level DDG of a 6-by-6 bit Braun multiplier.

For the sake of simplicity, let us make a couple of simplifying assumptions:

- Interconnect delays are negligible.
- Routing capacitances are the same for all local wires.
- Load capacitances are the same for all array outputs.
- Delays from local multiplication and local addition are additive.
- Delay of addition is the same for both sum and carry.
- Identical operations have identical delays as given below.

	$t_{pd\,mu}$ [ps]	$t_{cd\,mu}$ [ps]		$t_{pd\,ha}$ [ps]	$t_{cd\,ha}$ [ps]		$t_{pd\,fa}$ [ps]	$t_{cd\,fa}$ [ps]
1×1 bit multiplier	60	20	half adder	80	40	full adder	100	50

a) Find the critical paths, long and short.

b) What are the propagation and the contamination delays of the multiplier?

c) Allow for arbitrary word widths $w_{md} \geq 2$ and $w_{mr} \geq 2$ of multiplicand MD and multiplier MR respectively. Express the overall delays for the MSB and the LSB of the product PR as a function of w_{md} and w_{mr}.

d) Now suppose the multiplier is sandwiched between registers the flip-flops of which feature the timing data tabulated below. Indicate the highest clock frequency at which the circuit can operate.

e) Add a scan path to the circuit. Recalculate the maxmimum clock frequency when the timing of a 2-to-1 multiplexer is as follows on all input-to-output paths.

	$t_{pd\,ff}$ [ps]	$t_{cd\,ff}$ [ps]	$t_{su\,ff}$ [ps]	$t_{ho\,ff}$ [ps]		$t_{pd\,mx}$ [ns]	$t_{cd\,mx}$ [ns]
flip-flop	250	200	110	60	2-to-1 multiplexer	40	10

3. Reconsider the array multiplier of fig.12.13 but assume circuits have been fabricated with $w_{md} = 8$ and $w_{mr} = 8$ this time. Find a suitable set of stimuli and expected responses for measuring the overall propagation delay.

4. (True story) You are employed by an ASIC vendor and in charge of the sign-off process for an LSI design submitted by one of your company's customers. The anticipated sales volume is substantial. Circuits are to be fabricated with a mature and inexpensive single-metal CMOS technology. The major findings of your design review are as follows.

- The design makes use of multiple clocks. At several places, two or more signals combine in a combinational gate to act upon a subordinate clock.

- Logic gates are found in asynchronous reset nets too.

- Extensive pre- and post-layout simulations have been carried out to protect against all sorts of design flaws. No DRC or LVS errors got reported.

- The standard cell library employed had originally been developed and characterized for a similar process by one of your competitors. As management did not grant them the time necessary to redo library characterization, your customer's design engineers have decided to carry out all simulations using a switch-level simulator.

- They did their best to calibrate the switch models to the new target process and to the various MOSFET geometries found in the detailed layout.

Do you accept or reject the design for fabrication? Why?

5. There follows a list of design problems:
   (a) Feedback loop within a network of combinational gates.
   (b) Stationary bus driver conflict.
   (c) Transitory bus driver conflict.
   (d) Logic gate with input left unconnected.
   (e) MOSFET with drain and source electrodes shorted.
   (f) Overly slow signal ramps.
   (g) Excessive clock skew.

(h) Erratic hold-time violation.

(i) Poor performance as a consequence of excessive path delay.

(j) Insufficient extension of metal1 layer around contact openings.

(k) Two interconnect lines shorted together by mistake.

(l) Well with no body tie.

(m) Severe constriction in on-chip power line.

For each of the above flaws, indicate those design verification tools that are most appropriate to locate the problem.

12.7 | Appendix I: Cell and library characterization

Digital designers take it for granted that they can get accurate simulation models, timing models, and datasheets for the many standard cells and macrocells that make up a VLSI design environment. This section briefly describes how the various parameters that capture a (sub)circuit's timewise behavior are calibrated.

Characterization relies on analog simulations once a (sub)circuit's physical layout has been completed.[11] The process is carried out with the aid of SPICE, Spectre, ASX or some other transistor-level circuit simulator and starts from an extracted netlist that includes actual layout parasitics. Time-continuous signals are used to stimulate all input-to-output paths.

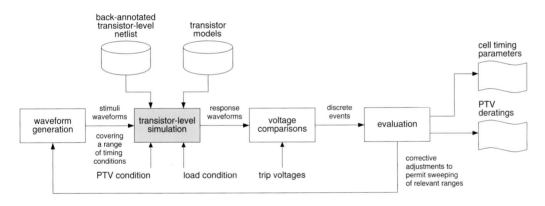

Fig. 12.14 Basic setup for cell characterization (simplified).

Propagation and contamination delays are essentially obtained from applying a pulse at some input and from interpreting the waveform(s) at the output(s). Characterizing for an input slope model makes it necessary to systematically vary the input's ramp times over a predetermined sweep range. In addition, each input to the cell is subject to waveforms with rising and with falling edges, and

---

[11] The alternative of measuring the actual waveforms on a fabricated part with the aid of an oscilloscope is neither very practical nor efficient in the context of cell and library characterization for VLSI.

each output to a variety of load conditions selected to cover the cell's legal range of operating conditions. Figure 12.4 illustrates a typical result.

Setup time, hold time, minimum pulse widths, and the like are somewhat more complicated to determine as two input waveforms are involved.[12] This makes it necessary to sweep relative timewise offsets, further adding to the number of simulation runs.

All or some of such simulation runs are then repeated for different process, temperature, and voltage conditions to obtain the PTV derating curves. Due to combinatorial explosion, the overall computation necessary to characterize an entire cell library may take a couple of days.

12.8 | Appendix II: Equivalent circuits for interconnect modelling

This appendix shows a number of abstract models used for describing, analyzing, and simulating on-chip interconnect lines, see fig.12.15b through h.

Series resistance R, series inductance L, parallel capacitance C, and parallel conductance G together capture a wire's electrical characteristics. While the significance of the former three quantities is pretty obvious, the latter essentially accounts for dielectric losses. In a physical wire, these parasitic elements are not lumped into a single place but distributed along the wire's length and are, therefore, indicated per unit length, that is in Ω/m, nH/m, pF/m, and S/m respectively. This explains why a transmission line model, depicted in fig.12.15f, most closely reproduces reality. Yet, the relevant question is

"Do we need transmission line models for on-chip interconnects?"

As a rule of thumb, a piece of interconnect should be studied on the basis of a transmission line model when a signal ramp extends over a distance that approaches the length of a leg of that net, or that is even shorter than that [360]. A leg refers to a piece of interconnect of uniform electrical characteristics that gets mapped into one section in a lumped network model. With l_{ra} denoting the spatial and t_{ra} the timewise extension of a signal ramp, we are safe as long as

$$l_{leg} \ll l_{ra} = c\,t_{ra} \qquad \text{where} \qquad c = \frac{1}{\sqrt{\epsilon\mu}} = \frac{c_0}{\sqrt{\epsilon_r\mu_r}} = \frac{c_0}{n} \tag{12.9}$$

$\epsilon = \epsilon_0\epsilon_r$ is the permittivity and $\mu = \mu_0\mu_r$ the permeability of the medium. c_0 stands for the speed of light in vacuum of approximately 300 Mm/s. When related to the time spans of interest in electronic circuits, this impressive figure collapses to a bare 30 cm/ns = 0.3 mm/ps. As any medium other than vacuum has an index of refraction $n = \sqrt{\epsilon_r\mu_r} > 1$, propagation is even slower there. The velocity c on actual interconnect lines found in VLSI chips and circuit boards is more like 15 cm/ns.

---

[12] A flip-flop's setup time, for instance, gets determined by repeatedly applying a ramp to the data input under control of the evaluation module in the characterization setup of fig.12.14. The exact switching time, initially well ahead of the active clock edge, is gradually adjusted so as to bring it closer and closer to the clock edge while monitoring whether the circuit continues to behave as expected. Any excessive settling time or out-of-the-normal waveform indicates that the lead time of the ramp on the data input has grown too small, and that the previous reading was in fact the setup time sought. You may want to see fig.7.11 for an illustration. Please note there is a considerable degree of freedom concerning what to regard as out-of-the-normal behavior and in trading setup time for propagation delay when interpreting the output waveform.

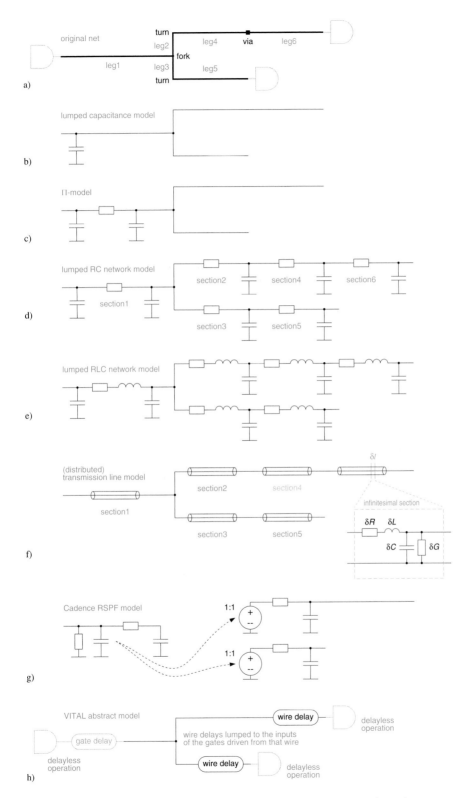

Fig. 12.15 A net (a) and various models used for capturing its delay characteristics (b,…,h).

Numerical example

A signal ramp of 20 ps that propagates on a wire embedded in silicon dioxide extends over

$$l_{ra} = \frac{c_0}{\sqrt{\epsilon_r \mu_r}} \, t_{ra} \approx \frac{300 \text{ Mm/s}}{\sqrt{3.9 \cdot 1}} \, 20 \text{ ps} \approx 3 \text{ mm} \tag{12.10}$$

☐

The above figure is only a rough estimate for a signal travelling on a metal wire within an IC because silicon, air, and further materials with other permittivities also surround the wire. Still, the result supports the subsequent conclusion.

Observation 12.10. *Transmission line models are not at present required to model on-chip nets provided no section in a lumped wire model is allowed to extend further than a fraction of a millimeter. They become a necessity at the board level when high frequencies are involved.*

Various wire models and their respective limitations are discussed in [159] [53] while [361] [362] study the increasing importance of parasitic inductance on interconnect delay, delay uncertainty, dissipated energy, and crosstalk. An analytical model for signal propagation on lossy transmission lines can be found in [363].

Chapter 13

VLSI Economics and Project Management

13.1 | Agenda

Over the last few decades, we have witnessed dramatic evolution and changes in electronics. Figure 13.1 outlines these changes by illustrating

- The displacement of the key driving markets,
- The rise and demise of electronic devices and implementation technologies, and
- The shift of focus from device-level circuit design and physical construction to defining and verifying the functionality that sells best.[1]

As a result, engineers and project managers have never before been presented with so many alternative choices for implementing their circuits and systems. This holds true in spite of the fact that fabrication technology has narrowed down to CMOS in almost all digital applications. Abstracting from lower-level options and commercial products, table 13.1 shows the fundamental options in a highly condensed form.

While there was a time when components used to be very basic and available with a limited choice, today's integration densities have led to a diversification into an almost astronomical number of powerful and highly specialized components. ICs have grown very complex and many of them implement entire systems. It is not exceptional to find that the "datasheet" of a key component such as a CPU, an FPGA, or an ASSP comprises a thousand pages, or almost so.

The emergence of new business models such as virtual components (VCs) has further complicated the process of finding the best approach to implementing an electronic product or system. Making decisions is not just a matter of the technical characteristics evident from the final product, but

---

[1] One reason for this last shift are the design techniques and engineering aids that have evolved from paper and pencil to sophisticated, yet also very complex, electronic design automation (EDA) suites. More from a historical point of view, you may want to refer to [364] [365] for accounts of the genesis of microelectronics and for biographies of the scientists involved. [366] describes the Japanese contributions.

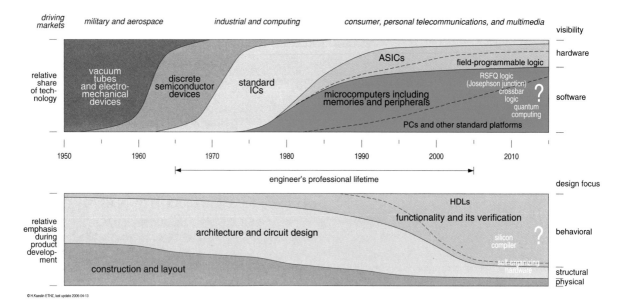

Fig. 13.1 Electronic information processing technology over time (proportions not drawn to scale).

Table 13.1 Options available for implementing a digital circuit (greatly simplified).

	Nonprogrammable standard parts exclusively	Micro- or digital signal processor	Field-programmable logic (FPL)	Semi-custom ASIC	Full-custom ASIC cell-based	Full-custom ASIC hand layout
when introduced	early 1960s	early 1970s	late 1980s	early 1980s	early 1980s	early 1980s
significance today future prospects	low none	high steady	medium rising	medium steady	medium steady	low declining

also of industrial cooperation, EDA infrastructure, design flow, responsiveness to changing requirements, return on investment, risk, logistics, and other issues of financial and organizational nature. Emphasis in this chapter is on the latter issues.

Section 13.2 reviews the models of cooperation between industrial partners for all six options of table 13.1 before section 13.3 elaborates on ASIC projects. Section 13.4 is concerned with virtual components. Sections 13.5 through 13.7 address economic and market issues of ASICs with section 13.6 specifically devoted to parts being sold in smaller quantities. Criteria that should help to make reasonable management decisions are found in sections 13.8 and 13.9.

13.2 | Models of industrial cooperation

The implementation alternatives of table 13.1 greatly differ in terms of design depth, engineering effort, and necessary know-how. At one extreme of the spectrum are the microprocessors and signal processors that present themselves to the programmer as virtual machines. What it takes to tailor such a generic platform to a specific application is essentially an algorithm formulated in some high-level programming language and a compiler for the target instruction set.

At the other end, we find full-custom ASICs where everything from the overall architecture down to layout details is specifically designed for one application. Of course, the better a solution needs to be optimized to the situation at hand, the more design effort will have to be put in at all levels of detail. Semi-custom ICs and library-based designs occupy positions somewhere in between.

To begin with, we would like to ask

"What business partners must be involved and what do they contribute to success?"

Answers are sketched in subsections 13.2.1 through 13.2.5. Each section is devoted to one implementation avenue and a diagram shows the essential activities and requisites for a design implemented with the technique being considered. The drawings also suggest a typical repartition of tasks, although other meaningful models of cooperation are likely to exist.[2]

Watch how control over the final product augments and how unit costs tend to diminish when going from subsection 13.2.1 to 13.2.5. Also observe that this is often bought at the expense of a more onerous development process, a more complex scheme of collaboration, and more varied risk factors.

13.2.1 Systems assembled from standard parts exclusively

The term standard part refers to catalog SSI, MSI, LSI, and ASSP components. What we have in mind here are systems that get designed from such parts with no customization: neither is much software being written specifically for the final product nor are any field-programmable parts being configured.

This time-honored approach gives the system designer much freedom to pick, mix, and match any commercially available components he likes, but also makes him largely dependent on IC vendors for innovation. His opportunities are essentially limited to assembling from standard parts whatever functionality he wants to incorporate into his products. Conversely, he is not required to have a serious understanding of microelectronics, VLSI design, or HDLs.

By opting for ASSP components, a system designer is even freed to a large extent from developing algorithms and hardware architectures. This is because he obtains almost all relevant

---

[2] More particularly, note that the dashed columns in figs.13.2 through 13.6 do not imply that the respective tasks must necessarily be carried out in distinct companies. The "design house" functions, for instance, are often carried out by a team within the "system house" company. Similarly, many "ASIC manufacturers" also carry out "design house" functions. Further keep in mind that the situation grows more complicated when elements from two or more implementation avenues get combined in one design, which is often the case in practice. Just consider mixed-signal circuits and combinations of microprocessors and custom logic.

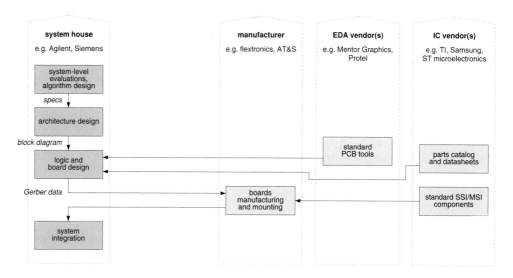

Fig. 13.2 Contributions and responsibilities when electronic circuits are just assembled from standard parts (simplified).

application-specific know-how from the IC vendor shut into the physical parts he buys. His role then essentially becomes that of a system integrator.

Hardware products assembled from standard parts with no significant customization — in terms of software content, FPL configuration, or both — have sunk almost into insignificance over the years because

- It is difficult to make one product stand out from others as
- All design know-how is exposed to competitors,
- System houses largely depend on IC vendors for innovation,
- Integration density is typically not as good as it could be,
- which tends to inflate manufacturing costs per unit,
- Hard-wired circuits are too rigid when requirements change, and
- board-level parasitics stand in the way of maximum performance and energy efficiency.

13.2.2 Systems built around program-controlled processors

This class includes all circuits and systems designed around a microprocessor, a digital signal processor (DSP), or some other central system component that sequentially executes program instructions.

Much of the design scope and of the added value is in software development. The software is what sets one product apart from other products competing in the marketplace. A prime advantage of this approach is the virtually unlimited agility.

Hardware architecture is largely defined by the central processing unit (CPU) being used. Hardware decisions are thus typically limited to the selection of peripheral devices and to minor details of

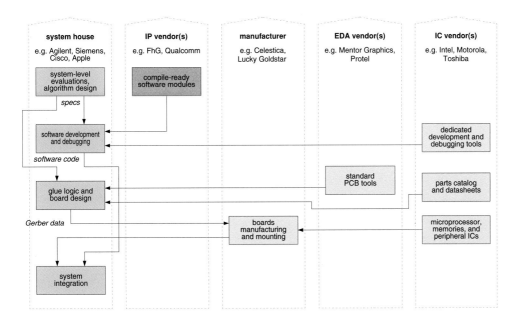

Fig. 13.3 Contributions and responsibilities in electronic design with program-controlled processors (simplified).

memory organization and glue logic. Again, the system house does not need much know-how relating to IC design.

Example

The PC business provides us with a telling example. Much of a PC vendor's activities revolves around marketing, branding, industrial design, supply chain management, sales, and services. Not only does he bank on third parties for the operating system and the application software, he also is at the mercy of the VLSI industry and of the manufacturers of peripherals, such as displays and storage media, for hardware innovation. Cost pressure has become so acute that more and more PC vendors have outsourced the assembly process to specialized electronic manufacturing service (EMS) providers in order to take advantage of competition and economies of scale. Others ended up selling their PC businesses.

□

13.2.3 Systems designed on the basis of field-programmable logic

The term field-programmable logic (FPL) encompasses any electrically configurable logic hardware such as field-programmable gate arrays (FPGAs), complex programmable logic devices (CPLDs), and the much more limited simple programmable logic devices (SPLDs).

The absence of a time-consuming manufacturing cycle is what sets FPL apart from mask-programmed custom ICs. Also welcome is the fact that some FPL devices can be ordered with standard subfunctions hardwired and placed on the same die next to the programmable logic.

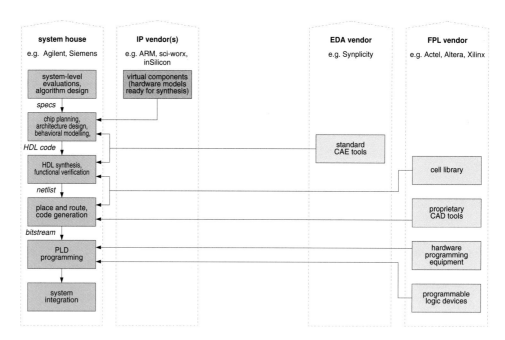

Fig. 13.4 Contributions and responsibilities in electronic design with field-programmable logic (simplified).[3]

Another major benefit is that the cooperation essentially involves just two partners. Not only do FPGA and CPLD vendors make EDA tools available for back-end design with their devices, they also make firm recommendations as to what front-end tools they support. Some of them even integrate standard synthesis and/or simulation tools with their own software so that customers can expect to get a fairly coherent EDA package. Also, in an effort to promote their devices, FPL vendors typically keep up-front charges for such packages at very reasonable levels.

On the negative side, most FPGA and CPLD families provide specific features and impose peculiar architectural constraints. Just consider the limited routing resources, the comparatively slow interconnect, the predetermined logic-to-storage ratio, and the limited selection of packages and device sizes. The costs of large on-chip memories and the lack of fast carry chains also pose problems with some devices. As virtual components (VCs) prepared by independent IP vendors and intended for general consumption do not normally take into account such idiosyncrasies, results are likely to be suboptimal. Addressing low-level design issues thus becomes inevitable whenever performance and/or density are critical.

13.2.4 Systems designed on the basis of semi-custom ASICs

This class includes those ASICs where only a small subset of all layers is custom-made, i.e. mask-programmed structured ASICs, sea-of-gates circuits, and gate-arrays.

---

[3] The box that stands for board manufacturing and mounting in figs.13.2 and 13.3 has been omitted in the drawing as a single package now aggregates the functionality previously distributed over several components. Note that a PCB, albeit much simpler, continues to be required in order to accommodate that package in a system.

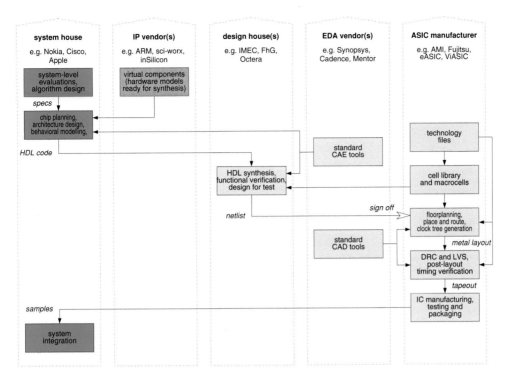

Fig. 13.5 Contributions and responsibilities in electronic design with semi-custom ASICs (simplified).

In comparison with the implementation avenues presented so far, the development process is more challenging because many more aspects must be addressed and because turnaround time is on the order of weeks or months rather than hours or days.[4] As a design cycle encompasses many steps from functional specification down to physical design verification, it is typically subdivided between two or three entities specializing in their fields such as a system house, a design house, and an ASIC manufacturer. Sign-off is essentially confined to gate-level issues.

The necessary macros — such as logic gates, bistables, and memories — are developed by the IC manufacturer and organized into a cell library. Depending on the circumstances, part of the logic will be synthesized from virtual components, i.e. from synthesis-ready HDL models purchased from an IP vendor or placed in the public domain by their authors.[5]

Something that is often underestimated is the effort required to get in place all the necessary contracts and agreements with the various business partners. Also expect a good deal of imponderables in aligning the priorities and delivery schedules.

---

[4] Turnaround is meant here in a wider sense that encompasses everything from locating a design flaw, fixing it, obtaining new circuit samples, testing them, qualifying the revised part, to the updating of software drivers, documentation, and other related material.

[5] A more detailed discussion of the VC business is to follow in section 13.4.

13.2.5 Systems designed on the basis of full-custom ASICs

The term full-custom ASIC is understood to include ASICs where all layers are custom-made.

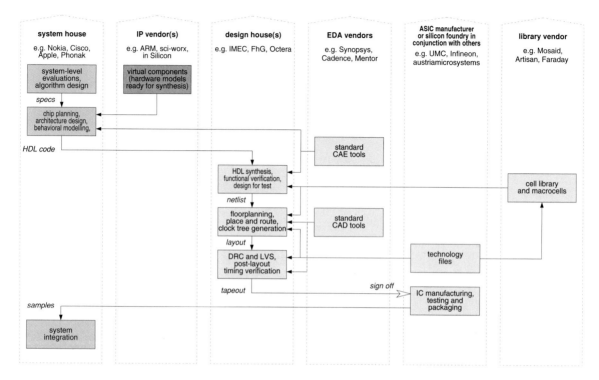

Fig. 13.6 Contributions and responsibilities in electronic design with cell-based full-custom ASICs (simplified).

Much as for semi-custom ICs, a chip design results from a close cooperation of highly specialized partners. Front-end design and back-end design need not necessarily be carried out within the same business unit. What eventually gets transmitted to the IC manufacturer for fabrication are layout data (more details on this are to follow shortly).

Cell libraries were originally made available by the silicon vendors as part of their foundry kits. Yet, the general trend towards focussing on the core business brought about independent **library vendors** that specialize in offering cell libraries for a variety of fabrication processes and silicon foundries. While the same largely applies to macrocell generators too, some ASIC manufacturers generate RAMs, ROMs, and other macrocells on demand and charge that service to their customers. This procedure tends to slow down the design process in an undesirable way, however.

Another difficulty is the quality of the design kits and technology files. The complexity of the interfaces and the rapid progression of technology often prevent them from reaching maturity before becoming obsolete.[6]

---

[6] There is a saying that every wedding is a victory of hope over experience. Seasoned VLSI designers will agree that all too often the same also applies to the adoption of a new design flow.

13.3 | Interfacing within the ASIC industry

When going for a mask-programmed ASIC, important choices relating to cooperation remain.

13.3.1 Handoff points for IC design data

A system house must decide what parts of the design process to commission to its business partner(s) along with the pertaining responsibilities. Alternative points for transferring design data to a specialized partner are evident from fig.13.7.

1. **Full layout handoff**

   The system house is in charge of organizing both front-end and back-end design. Sign-off takes place at the very end of the design cycle. The transfer of layout data, which continues to be called **tapeout** although tapes are no longer used, is via an industry-standard format such as GDS II.

   While predominant in the early days of microelectronics, this approach has become unpopular with companies that need USICs on an irregular basis. Back-end IC design calls for onerous tools, highly specialized expertise, and frequent training, three items that are not only expensive but also subject to rapid obsoletion in a time of relentless technological progress. Full layout handoff today remains confined to larger companies that design many VLSI circuits a year and that have them produced in large quantities.

2. **Netlist handoff**

   The sign-off procedure revolves around a certified netlist, i.e. the customer focusses on front-end design while the ASIC manufacturer is in charge of back-end design. Along with a gate-level netlist, the latter also accepts complementary specifications that relate to timing, electrical characteristics, package, pinout, and the like. As a variation, back-end design is often commissioned to an independent design house that specializes in this kind of activities and that acts as a go-between.

   Netlist handoff used to be the approach of choice for semi-custom ICs and was also standard practice with cell-based full-custom ICs. It became increasingly difficult, however, when interconnect delays started to dominate over gate delays. Still, the model should remain viable for not-so-critical designs and FPL.

3. **Floorplan handoff**

   This is a compromise between full layout and netlist handoff. The customer does a limited degree of floorplanning before handing over an initial cell placement, directions for critical nets, clock domains, power domains, and the like along with a gate-level netlist. These data then serve to guide clock tree insertion, detailed place and route (P&R), and post-layout timing verification by back-end specialists. The fact that feedback on wire lengths, congested areas, and realistic interconnect delays becomes available to front-end designers greatly helps to achieve timing closure.

4. **Architecture or register-transfer-level handoff**

The customer provides HDL models at the architecture or RTL level, which obliges him to decide not only on a circuit's functionality but also on the hardware resources necessary to implement that functionality. The ASIC manufacturer or an independent design house does the synthesis and all subsequent front- and back-end design steps.

This paradigm has quickly become popular in the USIC industry because

- It allows companies to keep focussed on their respective core competences,
- Know-how is better invested and preserved in HDL models than in netlists or layouts,
- There is no need for excessive tool-related expenditures on the customer's side,
- Deep submicron issues are addressed by experts with specialized tools while
- The customer keeps some approximate control over hardware organization and costs.

5. **Behavioral model handoff**

The customer essentially delivers a software model along with performance targets before commissioning one or more contractors to do all the rest for him. While he remains in charge of verifying model behavior, he does not care much about architecture design and even less so about HDL synthesis. The collaboration of an independent design house is almost always sought to bridge the gap between abstract behavioral modelling and physical design. Finding a workable compromise between hardware and software, another key issue in system design, is also left to the design house.

Although a company could — at least in theory — confine its contribution to establishing the specs for the desired USIC and outsource all the rest, this is not common practice. Most firms that successfully take advantage of USICs in their products do address high-level architecture design, HDL modelling, and the verification of such behavioral models themselves. There are good reasons for this:

- Establishing the functionality at the detail level and making sure it conforms to market needs is vital for the success of the final product.
- Deciding on overall system organization calls for profound knowledge in the pertaining domain of application. This represents a core competence of the system house that is not normally available in VLSI design houses.
- Processing algorithms and VLSI architectures should always be developed together.[7]
- HDL models provide an unambiguous interface to the subsequent design steps.
- CAE tools for HDL-level design are not that onerous to buy and operate.

SystemC has recently gained significant interest as an entry point to hardware design simply because the C++ programming language is much more popular with systems people and programmers than hardware description languages such as VHDL and Verilog. Maybe this is the wave of the future.

13.3.2 Scopes of IC manufacturing services

ASIC customers must also decide how to have their design manufactured. Various options exist because VLSI fabrication not only consists of wafer processing but also includes activities such as

---

[7] The reasons for this are given in sections 2.2.1 and 2.3.

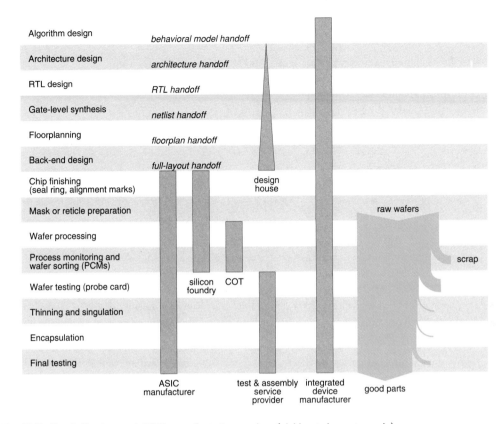

Fig. 13.7 Handoff points and ASIC manufacturing services (yield not drawn to scale).

chip finishing, mask preparation, volume testing, and packaging. How these are being handled bears upon business and liability issues.

1. **ASIC manufacturing service**

In addition to manufacturing, the vendor does the testing of the ASICs on behalf of the customer and typically delivers packaged components ready to go. He sorts out the defective parts and makes the customer pay only for those ICs that conform with functional and electrical specifications. The functional specs are embodied in a set of test vectors to be supplied by the circuit designer. The risk of poor yield and, hence, also the incentive for yield enhancement rest entirely with the manufacturer. This practice, depicted in fig.13.6, represents the standard business model for many commercial full- and semi-custom ASICs.

For obvious reasons, the ASIC manufacturer wants to make sure a design is safe before committing himself to such a venture. Design data and test vectors are, therefore, subject to a series of thorough checks before **sign-off**, i.e. before the design is accepted for manufacturing. As an extra benefit, design flaws and manufacturability problems are more likely to get uncovered when designs get scrutinized by two independent parties. Similarly, a manufacturer encourages his customers to rely on design kits and cell libraries that meet his own quality standards and that have won his approval.

The ASIC manufacturer is basically free to recover his costs by charging a one-time payment or by adding a small fee to the sales price of each chip he delivers. His attitude is likely to depend on the anticipated IC manufacturing business, but design kits are quite often delivered free — or almost so — in order to attract fabrication business.

2. **Foundry service**

According to the alternative foundry service model, testing at the manufacturing site is essentially limited to process-control monitors (PCMs). Provided PCM measurements indicate that process outcome is within normal tolerances, wafers or dies are delivered with nothing but visual inspection. The customer is then in charge of organizing the testing and the packaging of his circuits.

More importantly, the customer not only assumes the risks of design flaws, but also of substandard manufacturability,[8] of mask defects, of inadequate die handling, and of an unsatisfactory packaging process, all of which are detrimental to overall yield. By the same token, the foundry declines any liability for imperfect library cells. In return, foundry service is cheaper and review of incoming design data by the manufacturer is mild when compared with the ASIC manufacturer model. As a consequence, this division of workload and accountability is most appropriate for prototype fabrication.

3. **Customer-owned tooling (COT)**

COT goes one step further in that the manufacturer focusses on pushing wafer lots through his fabrication line on order, followed by evaluating PCM data. Everything else, including mask preparation and supply chain management, falls within the responsibility of the customer. This modus operandi is gaining acceptance with customers that order ASICs in very large quantities because they are thus put in a position to negotiate optimum conditions with separate contractors for licensing of cell libraries, mask preparation, wafer processing, testing, packaging, and the like. Also, the indispensable involvement with design for manufacturing (DFM) gives them more control over yield and other key cost factors.

However, COT is also the most demanding business model in terms of expertise required from customers. Luckily, a number of companies have begun to offer all test and assembly services that come after wafer processing, including supply chain management.[9]

Observation 13.1. *The standard industrial practice is to order fully qualified parts from an ASIC manufacturer. Foundry services appeal to research-type activities that combine it with "home testing" and to large-volume manufacturing. Experienced customers can take advantage of COT in conjunction with extensive subcontracting and supply chain management.*

---

[8] Such as a poor compliance with the lithographic process, an excessive sensitivity to all sorts of parameter variations, and other issues related to design for manufacturing (DFM).

[9] Amkor, ASE, and STATSChipPac are just three examples.

13.4 | Virtual components

Like megacells, virtual components (VC)[10] are higher-level design objects meant to be sold to companies that use them as building blocks for creating their VLSI designs. The difference is that VCs are made available as HDL models for synthesis by customers (soft module) rather than as gate-level netlists (firm module) or as polygon layouts (hard module). This is to warrant portability across fabrication processes and to avoid rapid obsoletion.

A VC must be substantial enough to boost productivity but also general enough to work in many environments. Virtual components work best for standard subfunctions of medium complexity, such as filters, modems, video en/decoders, cipher engines, 3D-graphic accelerators, processor cores, and all sorts of interfaces. Market forecasts had anticipated a rapidly developing business, yet putting virtual components into practical service requires particular attention.

13.4.1 Copyright protection vs. customer information

Assembling systems from purchased components has a long and highly successful tradition in electronics, and designing with VCs seems to be simple extension of this proven approach. Yet, this is unfortunately not so as VCs bring about their own peculiarities. A board designer who buys a physical part does not need to know how that component works internally as long as it behaves as specified. The same holds true when a programmer calls a procedure or takes advantage of a software module from a commercial library.

A VC licensee, in contrast, must complete a design process that others have begun because synthesis code is not a self-sufficient end product but just the entry point for synthesis, place and route (P&R), verification, manufacturing, and production testing. He takes on the responsibility for making a VC work exactly as intended by its original authors, even if he would never have it implemented in the way they have. The result of his efforts is going to be profoundly affected by all sorts of choices about transfer protocols, interface timing, target process, target library, macrocells (if any), synthesis procedure, buffer sizing, clock distribution, layout parasitics, test strategy, and the like. Not only the circuit's performance and timing, but possibly also its functioning are at stake. With the associated risks and liabilities on his side, the licensee acts more like a subcontractor of the VC vendor than as a paying customer.

What if the VC makes use of unsafe design styles that render the circuit susceptible to hazards? What if it imposes awkward timing constraints that are difficult to meet with the target library? What if vital preconditions for correct functioning remain tacitly hidden somewhere in the source code without being mentioned by the vendor? What if the final circuit turns out to be impossible to test? What if fabrication yield is unexpectedly low? Who is to blame if something goes wrong, who pays for redesigns, delays, and lost market opportunities? Most such issues are likely to remain invisible until the process of integrating a purchased VC is well under way.

---

[10] As stated earlier in this text, we prefer the term "virtual component" over the more popular names "intellectual property module" and "IP module".

Making a qualified decision on whether it is worthwhile to purchase some VC for incorporation into an ASIC requires in-depth information that goes well beyond the functional features advertized by VC vendors. All too often, a licensee is forced to extract what he needs to know from the source code in a painful exercise of reverse engineering. In addition, once he has paid for it, a licensee will find it difficult to dismiss a VC even if he finds it inadequate from a purely technical point of view. The question of

"How to render an HDL model amenable to inspection, simulation, and synthesis while, at the same time, protecting it from being copied illegally"

is indeed a major problem of commercial VC licensing. **Code encryption** fails to address inspection by the licensee for the purposes of evaluation, understanding, debugging, adaptation, and further development. In order to embed a VC into an IC and to make it fly with a reasonable chance of success, the circuit's designers occasionally need access to the source code and other detailed information. A workable balance between the legitimate interests of the various parties involved must thus be sought with the aid of **non-disclosure** and other legal **agreements**.

13.4.2 Design reuse demands better quality and more thorough verification

The safe design practices that are being taught throughout this text are even more important when a design is supposed to be reusable. More specifically, things to avoid in VCs include

- Asynchronous, careless, complicated or impure clocking schemes,
- Zero-latency loops (almost always unwanted),
- Incomplete sensitivity lists (always unwanted),
- Multi-driver signals (complicate design and test),
- Non-local timing constraints and multi-cycle paths (complicate design),
- Unportable or inconsistent FSM constructs,
- Components (gates, bistables, macrocells) instantiated from a specific cell library,
- Inconsistent naming of signals, instances, types, parameters, etc.,
- HDL constructs and other idiosyncrasies linked to some specific EDA environment,
- Technology-dependent synthesis constraints,
- Inadequate test structures,
- Functional gauges with insufficient coverage, and
- Poor documentation.

Advice for making HDL code reusable can be found in [367], for instance, but most of it is part of good design practice anyway. A more difficult issue stems from the fact that many VCs are strongly parametrized in order to generate revenues from as many applications as possible. Yet, writing HDL code for multiple configurations is not sufficient as every meaningful configuration also needs to be verified! With exhaustive verification being impractical, checking and ensuring a VC's correct functioning in all of its possible configurations is a major undertaking, and often neglected. Many VCs are thoroughly verified for one or two base configurations only, which practice is an invitation for troubles.

> Hint: As a licensee, make sure you obtain a set of simulation vectors that cover all configurations that are of interest to you. If no such vectors are available, prepare to work them out yourself and prepare to find bugs.

A prudent attitude is to plan for rigorous bottom-up validation of all submodels from the very beginning of a project, which calls for established quality assurance procedures on a solid industrial background. VLSI companies where first-time-right design is part of the corporate culture and is endorsed by management are better prepared than software startups.

From all the above, it should be clear that developing a trustworthy VC requires much more resources than designing the same functionality for one-time usage. Salvaging a piece of RTL code written in the context of a particular VLSI design project under pressure of meeting the deadlines there does not qualify as VC design. A rule of thumb is this:

"If it takes a certain effort to develop a design,
allocate three times that effort to make it reusable in another design,
and nine times that effort to make it generally available as a virtual component."

Management is very unlikely to accept that much overhead as part of regular ASIC design cycles. Moing business with VCs requires a specific organization with engineering teams that focus on VC library development, quality assurance, and pre- and post-sales user support.

13.4.3 Many existing virtual components need to be reworked

Most VCs require moderate to extensive changes before they can possibly fit into a given context. One reason is that functional specifications, timing, and power requirements tend to differ from one application to the next. Another reason that makes it difficult for vendors to provide turnkey VCs is the lack of universally accepted interface standards between building blocks.[11] Only truly universal items such as standard cells, memories, and largely self-contained processor cores tend to escape from this rule.

Making modifications to a VC requires permission by its vendor, access to the source code, and a reverse engineering effort. Also, even minor changes commit one to full verification. The current situation is best summed up by a quote from [368]. Talking about the customers of his company, a major ASIC design house and manufacturer, the author observes

"They usually bring in a shoe box full of intellectual property, dump it on a desk, and say:
'I want to connect this to that — you know how to do that, right?'"

VC customers typically face the choice between wasting much of their time to come up with ugly glue logic between incompatible components and reworking significant parts of the purchased HDL code, test data, synthesis scripts, and software drivers. As neither option is particularly attractive, VLSI designers often prefer to (re)implement the desired functionality from scratch.

13.4.4 Virtual components require follow-up services

As with any purchased piece of hardware or software, there is a possibility of an error in the VC. If the licensee finds a problem, he will face the following difficulties (partly after [369]). He will need to

---

[11] Please refer to section 15.5 for suggestions.

1. Isolate the problem to the VC without being intimately familiar with its design.
2. Convince the vendor that the problem is in the VC rather than in the way it is being used.
3. Find a workaround for the problem until the vendor releases a fix.

Handling such a situation successfully requires excellent on-going arrangements among licensee, vendor, and authors. Direct communication and efficient exchange of both design and simulation data between the engineers of the licensee and those of the VC vendor are absolutely essential.

Warning example

As our first venture into the field, we purchased a VC for the Universal Serial Bus (USB) in 1997 from a vendor that was promoting the USB standard and that was commercializing the pertaining serial interface engine as a VC. The module was generously offered as VHDL source code. However, closer investigation of the code delivered revealed that

- No testbench for making sure that a circuit would indeed work as expected was included,
- The code showed its age by ignoring more recent extensions of the VHDL standard,
- The code included two fairly obvious semantic errors, and, last but not least,
- There was a hidden functional fault that surfaced under specific circumstances.

Upon request, the VC vendor bluntly informed his licensee that

"There is no later revision of the code or any other support including no test code of any kind. None that could review the changes mentioned is around any longer."
□

This is an attitude that one would expect from a cottage industry, not from a serious VC vendor. The necessity for detailed information and for follow-up services together with the difficulty of providing turnkey VCs mentioned earlier suggests that a VC vendor should best be made to operate as an extension of the licensee's ASIC design team.

> Hint: Do not place any purchase order for virtual components before the terms under which the vendor is going to support you and to cooperate with you have been agreed on.

13.4.5 Indemnification provisions

In spite of their shared responsibility for technical success, no licensor wants to get involved in legal litigation with his licensees. A typical VC license agreement thus includes indemnification clauses that oblige the licensee to hold the licensor blameless regarding product liability disputes, patent infringements, and other legal claims.

13.4.6 Deliverables of a comprehensive VC package

- A circuit model in one or more formats amenable to synthesis (VHDL, Verilog).
- A computationally efficient behavioral model for system simulation purposes.

- An evaluation platform for studying the VC's behavior in greater detail.
- An architecture overview that demarcates the various clock domains and that identifies asynchronous subcircuits, if any.
- Information on the clocking discipline adopted and on acceptable clock frequencies.
- Details on parametrization, operation modes, numerical precision, I/O protocols, data formats, endian type, timings, naming conventions, etc.
- A commitment to safe and thus largely delay-tolerant circuit techniques.
- Reports from automatic coding guidelines checking (linting).
- Prerequisites on target technology and cell libraries (macrocells, worst-case timing, etc.).
- Information on the synthesis platforms supported.
- Synthesis scripts and synthesis constraints (critical paths, I/O timing, clock(s), arithmetic options, macrocells, etc.).
- A list of false paths and multicycle paths for post-synthesis timing analysis.
- A test strategy, test or built-in self-test (BIST) circuits, and scripts for inserting them.
- Testbenches and test vectors along with code coverage and expected fault coverage figures.
- Drivers and other software items that are necessary to operate and check the circuit.
- Limitations of any kind (parametrization, functional verification, testability, etc.).
- A well-defined contractual commitment to provide support and accept liabilities.
- An acceptable business model and a successful track record.
- A precise description of each of the above items and a schedule for delivery.

Not all VC vendors are in a position to provide all of the above items, some are even unaware of their significance in the ASIC design flow. Luckily for customers, tools are available on the market that help to assess the quality and completeness of VC packages and synthesis code.[12]

13.4.7 Business models

Multiple sources of intellectual property are common at the hardware level where physical components are being purchased in exchange for money, but they are new for subsystem-type library elements. This gives rise to specific questions such as:

"Who (author, vendor, or licensee) owns what rights in a virtual component?"
"How can one make sure expenditures, proceeds, and risks are shared correctly?"

Special business models are required for ownership and licensing of VCs. Two common but antithetical approaches go as follows (partly after [369]).

1. **For unlimited usage**
   The virtual component gets licensed with the rights for unlimited usage. This approach is easy to implement as there is a single exchange of money against a technology transfer between licensee and vendor. It will result in a high **buyout fee**, however, because the vendor must recover all VC development costs from a very limited number of sales. Also, the licensee

---

[12] HDL Designer by Mentor Graphics, for instance.

assumes the entire risk of never recovering his up-front investment if the overall sales volume falls short of his initial expectations.

Companies selling many products in large quantities are in a better position to amortize the purchase of a module until its impact on final product cost becomes almost negligible. Small companies with few and lower-volume products, on the other hand, are penalized because buyout fees are more likely to inflate product costs in a significant way.

2. **Per unit delivered**

The VC gets licensed per unit sold, that is the licensee is bound by contract to pay a small **royalty fee** to the vendor for each unit delivered. The marketing risk is thus effectively shared between licensee and VC vendor, and all products contribute to development proportionally to their respective sales volume. Many companies shy away from this approach, however, because a fixed per-unit fee seriously restricts the licensee's liberty to price his products. As a remedy, partners may agree on a degressive royalty fee or renegotiate the royalty once a predefined amount has been paid.

These characteristics resemble the traditional trade of ICs and other hardware components, but there is an important difference. All accounting must be accomplished by the licensee who — like any other commercial operation — may be unwilling to disclose details relating to his sales volumes and customers. An audit to convince the VC vendor that he is not being cheated is difficult to implement since hard evidence is not readily available.

In practice, paying an up-front fee is more popular than paying royalties. As an intermediate position, it is also possible to license a VC with the rights for unlimited quantities of one design. Licensee and VC vendor are hence obliged to negotiate the transfer of rights for every new IC.

Observation 13.2. *The virtual component business does not function like the traditional vendor $\rightleftarrows$ buyer model for physical semiconductor parts. A working relationship requires a mutual understanding more like a joint venture where products are being co-developed.*

13.5 | The costs of integrated circuits

Let us begin by asking

"What makes up the overall costs of a microelectronic subsystem or component?"

As for any industrial product, expenses fall into either of two categories.

Non-recurring costs, aka fixed costs, cover everything that must be paid for before volume production can begin. With reference to microelectronic parts, this includes

- Project management, including negotiations with business partners,
- Circuit specification,

- Purchase and assimilation of virtual components, if any,[13]
- Circuit design,
- Design verification at all levels (functional, timing, electrical, and physical),
- preparation of test vectors and testbenches for simulation,
- CAE/CAD related expenses (in-house equipment, renting, or usage of walk-in facilities),
- Sign-off procedure,[14]
- Preparation of fabrication masks,[14]
- Preparation of probe cards,[14]
- Setting up of fabrication and testing facilities,[14]
- Prototype fabrication and testing,
- Redesigns, if any,
- Preparation of test vectors and test equipment for volume production,
- Product qualification (life-cycle tests, burn-in, reliability tests, compliance with JEDEC and other industry standards; requires multiple fabrication lots, very time-consuming).

Recurring costs, in contrast, do depend on the quantity produced, which explains why the term variable costs is used as a synonym. For our application, recurring costs include

- Supply chain management,
- raw semiconductor wafers,
- wafer processing,
- volume testing (wafer testing, final testing, and binning),
- process monitoring and yield enhancement measures,
- packaging (including package, lead inspection, stamping, etc.)
- royalties for virtual components, if any,[13]
- board or other substrate space,
- external catalog parts, if any,
- component mounting.

Let c_0 stand for the sum over all non-recurring expenses. Further assume that recurring costs depend linearly on the quantity and that the cost increment per unit produced is c_1. The total costs per unit from a production run that yields n working circuits then are

$$c = \frac{c_0}{n} + c_1 \qquad (13.1)$$

13.5.1 The impact of circuit size

Putting aside testing and packaging, the expenses for manufacturing one functional integrated circuit c_f can be expressed as the costs for purchasing a raw wafer c_{wr} and for processing that wafer c_{wp} divided by the number n_f of defect-free dies obtained from the wafer.

$$c_f = \frac{c_{wr} + c_{wp}}{n_f} \qquad (13.2)$$

---

[13] It is a matter of negotiations how a licensee pays for the third-party virtual components he incorporates into his designs, see section 13.4 for alternative compensation schemes.

[14] These expenses are billed by IC manufacturers and test houses to customers under the conventional term **non-recurring engineering (NRE) costs**. Make sure you understand that these production-related NRE costs are just a subset of the non-recurring costs associated with putting an ASIC solution into service.

With A_d indicating the total area occupied by one die on a wafer, including the allowance necessary to cut the wafer and to dissociate the individual dies, one may be tempted to think that processing costs would grow linearly with die size $c_f \propto A_d$. Two effects make this an overly optimistic interpretation.

Packing square cookies into a round box

Wafers are round whereas the dies manufactured thereon are of rectangular shape. As there is no way to exactly cover a circular area with rectangles, some silicon is necessarily wasted near the circumference of the wafer. A simplified formula for estimating the number of manufactured dies available from a wafer of diameter d_w is

$$n_m \approx \frac{\pi}{A_d} \left(\frac{d_w}{2} - \sqrt{A_d} \right)^2 \tag{13.3}$$

This approximation diminishes the diameter of the wafer to account for the area near the edge that is unavailable for full reproduction of square tiles. The aspect ratio of the layout and the exact location of the individual circuits are not accounted for. Yet, it becomes clear that relatively more and more wafer area is lost due to the **edge effect** as die size increases.

Functional defects and fabrication yield

Fabrication yield y_f denotes the proportion of functional circuits obtained from a batch of manufactured circuits. Wafers typically have a number of **defects** more or less randomly distributed over their surface. As a design gets larger, the probability of finding a defect within a given circuit increases, thereby lowering fabrication yield. Yield is also a function of the complexity of the fabrication process because the chance of finding a defect on a given circuit increases with each additional processing step. **Yield models** must further account for those wafers that get scrapped as part of wafer sorting. A practical but purely empirical yield model is widely known as the **negative binomial model** and describes fabrication yield as

$$y_f = \frac{n_f}{n_m} \approx \left(1 + \frac{D A_c}{\alpha} \right)^{-\alpha} \tag{13.4}$$

where D is called the **defect density** while the **clustering factor** α reflects the fact that most defects do not distribute evenly over the wafer surface.[15] Smaller values of α imply a higher degree of clustering. Both parameters are best viewed as fitting parameters of a statistical model with no immediate physical interpretation, however. D improves with the maturity of the manufacturing process; a value of 0.004 mm^{-2} was typical for the 90 nm generation in the year 2006 [24]. α tends to grow with manufacturing complexity and the number of lithographic patterning steps. For CMOS processes with multiple layers of metal, α used to be less than 3.0 ten years ago, but 4.0 is now more adequate [24].

Observe the presence of critical area A_c which stands for the die area that is occupied by actual layout structures, discounting unused spaces such as scribe lines and any other empty areas. Yield

---

[15] As a limitation, the simple model of (13.4) cannot easily be made to account for extensive contact replication whereby multiple contacts or vias are systematically connected in parallel to provide redundancy against defects.

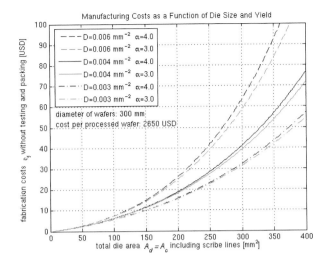

Fig. 13.8 Manufacturing costs per working die as a function of die size and fabrication yield (testing and packaging not included, approximate numbers).

is a function of $A_c < A_d$ rather than of A_d because defects that remain confined to unused silicon areas have no impact on the functioning of a circuit. The difference is particularly significant for pad-limited designs. Depending on circuit size, fabrication technology, maturity of the product, and market pressure, a fabrication yield of 50% to 95% is deemed acceptable.

Putting edge effect and fabrication yield together, the expenses for manufacturing one working die are obtained as

$$c_f = \frac{c_{wr} + c_{wp}}{n_f} = \frac{c_{wr} + c_{wp}}{n_m \, y_f} \approx (c_{wr} + c_{wp})(1 + \frac{D A_c}{\alpha})^\alpha \frac{A_d}{\pi(\frac{d_w}{2} - \sqrt{A_d})^2} \tag{13.5}$$

This function is plotted in the graph of fig.13.8 for various parameter values.

Observation 13.3. *Only for small circuits where the edge effect is of little importance and where the yield remains high do fabrication costs grow roughly proportionally with die size. For truly complex circuits, in contrast, expenses increase in a highly progressive way.*

Circuit complexity indeed becomes a critical factor in estimating the recurring costs of an integrated circuit once the expenditures for manufacturing exceed those for packaging and for component mounting. Further note from fig.13.8 to what extend defect density D determines the largest die that can be manufactured at some given cost; the clustering factor α is of relatively minor importance.[16]

---

[16] High-capacity RAMs often exhibit disproportionate die sizes, yet defect densities are basically the same as for logic chips. This is made possible by extra subcircuits that get selectively switched-in after fabrication in lieu of malfunctioning addresses. Though no more than a few percent of the overall die area is typically being set aside for these spare memory banks, yields close to zero would otherwise result. It is a paradoxical benefit of **hardware redundancy** that it curbs overall fabrication costs in spite of extra silicon occupancy.

13.5.2   The impact of the fabrication process

Impact on non-recurring costs

The growing sophistication and number of photolithographic masks makes their costs explode, see table 13.2. Although mask sets tend to benefit from cost reductions once a process generation reaches maturity, costs continue to escalate. Mask preparation becomes particularly onerous when phase shift masks (PSM) and optical proximity correction (OPC) get involved in the search for sub-wavelength resolution.[17] The geometrically most critical layers such as polysilicon, low-level metals, and contacts are the first to be concerned, but the proliferation of metal layers and extra process options, such as flash memory and silicide straps also take their toll.

Table 13.2 Mask costs for baseline CMOS fabrication processes (approximate numbers).

year of intro- duction	process generation [nm]	metal levels	ranse of costs for a mask set [USD]
1995	350	4	50k to 60k
1997	250	5	90k to 110k
1999	180	6	250k to 300k
2001	130	7	500k to 700k
2004	90	8	800k to 1.1M
2007	65	9	1.2M to 1.5M
2010	45	10	1.6M to 2M

Impact on recurring costs

The variable costs for obtaining one defect-free but untested and unpackaged die are

$$c_f = \frac{c_{wr} + c_{wp}}{n_f} \approx \frac{c_{wr} + \sum_{p=1}^{P} c_{ls}(p)}{n_m \, y_f} \tag{13.6}$$

where P indicates the number of lithographic patterning steps and $c_{ls}(p)$ reflects the cost for that part of wafer processing that comes with the pth such step.

The numbers for P and $c_{ls}(p)$ differ from process to process as they depend on wafer diameter, minimum feature size, the number of interconnect layers, the materials used, and — more generally speaking — the degree of sophistication of a fabrication process. Not only technical differences matter, though. Economic factors such as the invested capital, interest rates, the payback period planned for along with the current degree of amortization, operating costs, plant utilization, market conditions, and competition are even more significant.

---

[17] The reasons are as follows. As opposed to wafer processing, mask manufacturing cannot tolerate a single fatal defect. Yield for a mask is either one or zero. Mask preparation thus involves painstaking inspection and repair procedures. Also, long write times on expensive equipment make costs a function of shot count, that is of the number of geometric features to be exposed. PSM and OPC introduce extra features that make shot counts soar. You may want to refer to section 14.2 for accounts on lithography and on CMOS wafer processing.

Observation 13.4. *Industry typically moves over to the next process generation when the savings from fabricating a given circuit in a more advanced and, hence, denser process compensate for the more expensive masks and wafer processing.*

To be sure, the fact that yield is lower initially is taken into account.

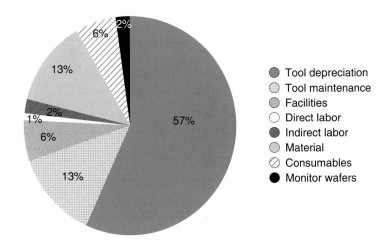

Fig. 13.9 Cost breakdown for 300 mm wafers in 65 nm 1P9M Cu CMOS technology processed by a Taiwan-based fab for 2007 (numbers courtesy of IC Knowledge LLC, a company that offers and regularly updates cost models for various IC product groups).

Figure 13.9 gives a typical cost breakdown for processed wafers. Note the relative importance of

- Capital expenses (tool depreciation + facilities),
- Operating costs (tool maintenance + monitor wafers),
- Cost of materials (raw wafers) and consumables, and
- Labor costs (direct labor + indirect labor).

Observation 13.5. *The high proportion of capital expenses renders processing costs highly dependent on a wafer fab's utilization, and the same applies to profitability.*

Parametric variations and fabrication yield

Beginning approximately with the 90 nm generation, yield and manufacturability are limited not only by fabrication defects, but also by unpredictable parameter variations. As an example, chemical mechanical polishing (CMP) may cause significant thickness variations of metals that in turn affect interconnect delays. Similarly, random variations of oxide thickness and the number of dopant atoms impact key MOSFET characteristics such as threshold voltage, current drive, and leakage currents. Although they essentially perform the intended function, fabricated parts may have to be rejected as **parametric failures** because they do not meet specifications in terms of operating speed, I/O timing, standby power, or the like. None of these effects is accounted for in (13.4) and

(13.5) because the author knows of no uncomplicated mathematical model for them. [490] discusses yield optimization in the presence of leakage.

13.5.3 The impact of volume

Recall (13.1) and let production quantity n be very important indeed. Unit cost c and, hence, the profitable sales price is going to be dominated by the recurring expenses c_1. Lowering them will typically pay off, even if this inflates the initial costs c_0 quite a bit. The higher expenses for labor and more elaborate software tools notwithstanding, extensive optimization at all levels of detail (architecture, logic, circuit, layout) tends to be worthwhile in such a situation. Other steps in this direction are adopting more economical packaging techniques, moving to larger wafers or to a denser and hence more cost-effective fabrication process, trimming test structures, speeding up volume testing, and the like. Die size is a prime concern that often determines the overall profitability of the entire operation. It comes as no surprise that complex general-purpose VLSI parts such as microprocessors and memories are designed along these lines.

Conversely, non-recurring costs prevail in products where manufacturing is limited to relatively few wafers per design. While commodity RAMs exhibit mask usage figures in excess of 5000, 500 wafers per mask set are more typical in the USIC business. It then makes no economic sense to put in extra effort to make the IC smaller or otherwise cheaper to manufacture by getting involved with low-level details during the design process. Instead, the goal is to minimize the total of all non-recurring expenses by relying on well-established, safe, and highly automated design methods and by taking advantage of design reuse as much as possible. Fabrication avenues that involve lower NRE costs are also very welcome in this situation, see section 13.6. Another idea is to avoid using individual mask sets for similar products.

Observation 13.6. *While die size, along with package costs, indeed remains a critical parameter for large chips that are produced in huge quantities, cutting non-recurring expenses is much more important for small to moderate fabrication volumes.*

Numerical example

Consider a full-custom IC measuring 4 mm by 4 mm that is being fabricated on wafers with a diameter of 300 mm. Cutting wafers into dies asks for an allowance of approximately 60 μm on all four sides, so that cost calculations must be based on a square of 4.12 mm by 4.12 mm. The number of dies manufactured on each wafer then is $n_m \approx 3938$. Assuming a yield of 82%, an average of $n_f \approx 3229$ functioning dies are obtained per wafer.

In the year 2007, total expenditures for one 300 mm wafer processed in a 90 nm CMOS technology with 28 mask layers were on the order of 2650 USD, namely 350 USD for the wafer plus 2300 USD for its processing. Package, encapsulation, and volume testing were estimated to add another 1.48 USD so that manufacturing cost is about 2.30 USD per functioning unit.

Further assume that non-recurring expenses amount to $7 \cdot 10^6$ USD. Half of this is to pay for 17 or so man years of management, marketing, design, and verification efforts combined. One production mask set and the EDA infrastructure cost on the order of $1 \cdot 10^6$ USD each. The rest is to pay for virtual components, cell libraries, probe cards, training, sign-off, and setting up fabrication and test facilities for a full-custom IC.

The resulting overall costs per unit are given below as a function of production volume. Note that it takes more than three million units to balance recurring and non-recurring expenditures in this example!

costs per unit [USD]	production volume n					
	1k	10k	100k	1M	10M	100M
non-recurring c_0	7000	700	70	7.00	0.70	0.07
recurring c_1	2.30	2.30	2.30	2.30	2.30	2.30
overall c	7002	702	72.30	9.30	3.00	2.37
mission-critical	design effort		manufacturing costs			

☐

Observation 13.7. *Massive non-recurring and low recurring costs are typical for full-custom ICs. This fact favors large-volume fabrication but, at the same time, also puts at a disadvantage products that sell in minor quantities.*

Yet another effect that works in favor of mass products is the volume discounts that may be obtained from vendors and foundries against placing large orders.

13.5.4 The impact of configurability

Many applications that mandated a custom ASIC just a few years ago fit into a single field-programmable logic (FPL) device today. This trend is to carry on as a consequence of technological progress and continued price reductions. As a consequence of soaring mask and development costs, the sales volume necessary to justify custom parts is gradually moving up. Also, built-in memories, processor cores, interfaces, and other standard functions greatly help to accelerate the FPL design cycle when compared with a custom design where the same functionality must be obtained from macrocells and virtual components.

Yet, unless there is an unforeseen technological breakthrough, field-programmable logic is unlikely to rival hardwired logic on the grounds of integration density and recurring costs. This is because — using essentially the same semiconductor technology — FPL must accommodate extra transistors, antifuses, interconnect lines, vias, lithographic masks, and wafer processing steps to provide for configurability. Also, the required and the prefabricated hardware resources never quite match, leaving part of the manufactured gates unused. This is particularly so when the application asks for large and varied on-chip memories as not all FPL families support the construction of RAMs and ROMs efficiently. Storage must then be pieced together from multiple logic cells or from small configuration memories distributed over the FPGA's core, further tying up hardware resources.

In fact, FPGAs are a factor of 10 to 35 times inferior in terms of density to mask-programmed equivalents manufactured with a similar technology. This explains why large FPGAs continue to be rather expensive, even when bought in substantial quantities. Comparing SRAM-based FPGAs against cell-based ASICs in 90 nm CMOS technology, [370] reports an area overhead of 35, a timing slowdown factor between 3.4 and 4.6, and a dynamic power ratio of 14. Opting for a reconfigurable FPGA, rather than for a mask-programmed ASIC, is thus likely to inflate the AT-product by more than two orders of magnitude. While hardwired multipliers and antifuse technology improve the situation, a significant penalty remains.

Observation 13.8. *In products where recurring costs, energy efficiency, or operating speed are critical, it is a good idea to confine configurable logic to those circuit parts that are truly specific for the application or that are subject to frequent changes.*

13.5.5 Digest

Figure 13.10 compares the recurring and non-recurring costs of various ASIC implementation techniques. As numbers greatly vary from one situation to another, this is no more than a general picture. Also, the drawing is not meant to imply that photomasks alone make up the differences in non-recurring costs among FPL, semi-custom, and full-custom ICs. The extent of reuse and the level of detail at which a design is conducted are as important.

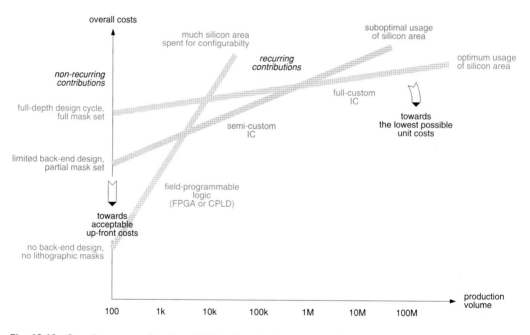

Fig. 13.10 Overall costs as a function of fabrication depth and volume (simplified).

Sophisticated IC size and **cost calculators** have been developed by industry and are continually being kept up-to-date. Yet, most of them are company confidential, only a few are available commercially. A few tools open to the general public are listed in section 13.10.4.

In practice, it is often difficult to draw a clear line between those expenditures that are directly related to a specific project and those that are not. VLSI design expertise, CAE/CAD training, computing and laboratory equipment, software license fees, etc. must all be acquired and paid for by a company, but can hardly be attributed to a single circuit.[18]

More importantly, selecting an implementation technique and a fabrication avenue is not merely a matter of accountancy. The choices made at this point also affect time to market, agility, and risks, all of which strongly impact on the overall profitability of a product. Thus, when it comes to making decisions for some given product, other than purely financial arguments are likely to prevail. In fact, hitting a window of opportunity in the marketplace always is among the highest ranking concerns. A more complete list of criteria will be given in section 13.8.

Figure 13.11 gives an overall picture of the ASIC techniques that summarizes our discussion of technical and pricing issues so far. We would like to emphasize that this is a simplified overview and that numbers are quoted for illustrative purpose only. It should nevertheless become clear that each ASIC implementation technique has its particular niche where it is a good compromise.

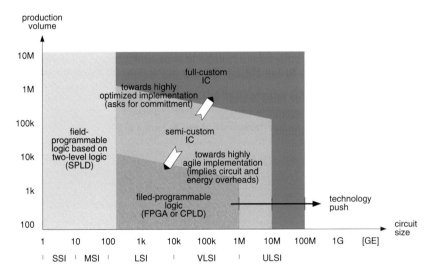

Fig. 13.11 Typical scopes of various ASIC implementation techniques (simplified).

Observation 13.9. *Performance, energy efficiency, and costs of an ASIC are largely determined by functional specifications, architecture design, and fabrication avenue. The impact of the subsequent logic and physical design phases is relatively minor.*

---

[18] Note that leasing of hardware equipment and software tools is an interesting alternative to buying, especially for smaller companies that could not use them to full capacity over a prolonged period of time. As a second benefit, it thereby becomes possible to ascribe expenses to individual projects in a more objective way.

13.6 | Fabrication avenues for small quantities

The initial costs associated with IC manufacturing discourage the design of USICs for products with low to moderate sales volumes. Also, most silicon foundries would not accept orders below 1000 wafers per year. While field-programmable logic (FPL) nicely fills the gap, there are limitations that sometimes justify the search for other options.

- Most FPL devices provide no support for analog circuit blocks.
- FPL devices typically come with a very limited selection of packages.
- FPL is often inferior in terms of energy efficiency.

13.6.1 Multi-project wafers

A **multi-project wafer** (MPW) accommodates several designs, thereby making it possible to have them processed together in a single fabrication run. This contrasts with the standard practice from volume production where a single design is repeated many times over the entire wafer surface, see fig.13.12a and b. The intention behind MPWs is to share the massive costs for an entire set of fabrication masks among different projects. MPWs primarily serve for **prototype fabrication**.[19]

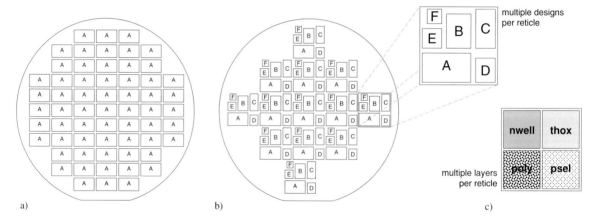

Fig. 13.12 Sharing of mask costs among several designs. Regular single-project wafer (a) versus multi-project wafer (b), arrangement of multiple layout layers on a multilevel mask (c).

Foundries and fabless vendors specializing in this kind of business provide MPW runs on a regular basis, say between 3 and 8 times a year for a given process. Some 6 to 12 designs are combined on a wafer, which explains why some companies refer to MPWs as "shuttle" service. Customers typically receive an engineering lot of 10 or 20 packaged but untested circuits (blind assembly), with an option to order another 20 to 50 dies if testing proves the design correct. Billing normally

---

[19] There have been attempts to put MPWs to service for low-volume production to promote the application of microelectronics in small and medium-size enterprises (SME), but these were met with limited success.

is per area with a minimum charge that corresponds to a few square millimeters. NRE charges are very low or zero, which implies that they are included in the area charges.

13.6.2 Multi-layer reticles

Here, multiple fabrication layers are made to share one photomask or reticle. The available optical field — currently up to 25 mm by 25 mm or so — is typically subdivided into four quadrants, each of which is made to accommodate a different layer, see fig.13.12c. Ideally, overall mask count is thereby reduced by a factor of four. By adapting the lithographic procedure, each mask is reused four times, bringing into play a different quadrant each time. As opposed to MPWs, fabrication runs are not tied to a fixed schedule and the intellectual property of each customer is kept on a separate mask set.

A related idea is to minimize the optical field and, hence, reticle costs by placing a lower number of dies on a reticle than technically feasible. While this is detrimental to wafer through put because of the more frequent step-and-repeat operations and the longer wafer exposure times, it contributes to a more favorable cost structure for low-volume ASIC manufacturing [482]. The same reference also suggests selectively simplifying optical proximity correction (OPC) patterns for cutting overall lithography costs.

13.6.3 Electron beam lithography

The general idea is to avoid costly masks by using an electron beam for drawing layout patterns directly into the photoresist. The fact that layout patterns have to be redrawn on each single die limits throughput, however, and would confine **electron beam direct-write lithography** to engineering lots and very small production quantities. The technology is thus typically used to pattern the metal layers of otherwise prefabricated sea-of-gates masters.

Example

IMS Chips in Stuttgart offers direct-write lithography for the personalization of mixed-signal gate-arrays with up to 100 000 GEs manufactured in 0.8 μm CMOS technology (0.5 μm was in preparation at the time of writing (early 2006)). Electron beam lithography is being used for volumes from 1 to 100 wafers per year. To accommodate products that sell in larger quantities, the company also provides a smooth transition path to photolithographic customization with just two or four masks.
□

13.6.4 Laser programming

Laser structuring also does without custom masks. The additive **makelink** approach promoted by LaserLink Technology starts from specially designed and fully processed wafers that share common traits with those of antifuse FPGAs. The difference is that a laser beam gets focussed on annular layout structures to fuse the two superposed metal layers where required. Connections are formed at a rate of roughly 20 000 links per second. Incidentally, note that the same technology is also being used to switch in redundant columns to improve the fabrication yield of RAMs.[20]

13.6.5 Hardwired FPGAs and structured ASICs

In an attempt to attract business away from traditional ASIC manufacturers, FPL vendors offer mask-programmable semi-custom parts as a pin-compatible but more economic replacement for their electrically programmable components to customers that are ready to order more substantial quantities. Actel, Altera, Xilinx, and others accept a netlist established by the customer for one of their popular FPGA families and retarget the design for their own masters. Depending on the order, this service is provided at little or no cost.[21]

From a customization point of view, hardwired FPGAs are much the same as structured ASICs as only the upper two, three or four metal layers are made to order, thereby limiting the number of application-specific masks to four, six or eight. The difference is that the hardware resources available on the prefabricated masters follow their electrically programmable counterparts.

Observation 13.10. *The exploding cost of a full mask set currently observed works in favor of techniques that achieve customization from a small subset of masks or with no masks at all.*

Example

eASIC is a vendor that supplies parts with no up-front charges by minimizing the number of photomasks necessary to customize the prefabricated masters of their Nextreme SL products. Combinational logic and bistables are built from SRAM-configurable logic cells similar to those found in traditional coarse-grained FPGAs. On the masters, those cells are preconnected using segmented lines running on metal layers 1 through 6. Customization occurs through a single via layer (via6) with e-beam lithography before metal7 gets deposited and patterned using a standard mask. Confining e-beam patterning to one layer that, in addition, features a relatively low density, avoids a bottleneck with throughput. Also, there is the option of substituting a custom mask, once sales are known to grow to higher volumes. Similarly, the LUTs can be frozen using a second custom mask for the via1 layer, thereby saving the recurring expenses for an external storage device. eASIC supports this transition with their Nextreme VL fabrics.
□

13.6.6 Cost trading

More on the grounds of their pricing policy, some ASIC manufacturers are willing to waive part of the NRE charges provided the customer commits himself to purchasing a predetermined quantity

---

[20] The earlier subtractive **laser-programmable gate arrays** (LPGA) technique worked exactly the other way round. Starting from specifically designed gate-array-type masters that carried all possible metal connections, a laser micromachine was employed to selectively cut the unwanted ones. Early promoters had abandoned LPGAs both for technology and for business reasons by 2004, however. Selectively ablating metal using a laser beam became increasingly difficult as the track widths grew smaller relative to the laser's wavelength and as the number of metal layers increased. More importantly, the low throughput restricted laser programming to prototyping, a business that did not offer a significant potential for growth.

[21] Marketed as "HardCopy" by Altera and "HardWire" by Xilinx, for instance. "EasyPath" by Xilinx shares the objectives of fast turnaround and low up-front costs but follows a somewhat different approach.

over a certain timeframe. Others prefer to trade off NRE charges in exchange for slightly higher unit costs.

Observation 13.11. *FPL and semi-custom ICs are the natural choices for having ASICs produced in smaller quantities. In addition, a handful of companies specializing in the business offer full-custom fabrication with more favorable NRE charges. Maskless customization techniques further make it possible for some of them to operate with short turnaround times.*

What matters in the end is that the overall cost structure fits the market. As commercial quotes and prices are extremely short-lived items, the reader is referred to small-volume price estimation forms made available by specialized vendors on their webpages, see section 13.10.3.

13.7 | The market side

13.7.1 Ingredients of commercial success

Let the following truism begin our discussion of some marketing aspects of microelectronics.

Observation 13.12. *Price and cost are unrelated. Price is determined by what the market will bear. Cost depends on how smart a company is at manufacturing a product. It is the difference between the two that determines the profit.*

The list below, attributed to Matti Otala, states the desirable properties of an invention from the perspective of an existing industrial company.

Observation 13.13. *A good invention*
- *pops up just in time,*
- *is an improvement rather than revolutionary,*
- *fits into the company's product portfolio,*
- *fits into the company's manufacturing strategy, and — above all —*
- *reduces costs.*

Examples and counterexamples

Electronic calculators were extremely successful from the onset because they not only offered superior performance, precision, and ease of use but, at the same time, also could be manufactured and sold at more affordable prices than their mechanical predecessors, especially once sales volume took off. With their then revolutionary LSI circuits radically departing from precision mechanics, electronic calculators were a sharp mismatch with the competences of established manufacturers, and none of them has made it into the age of microelectronics.

More recently, manufacturers of photographic films and cameras such as Agfa, Fuji, Kodak, Polaroid, and Konica-Minolta have lived through a similar landslide towards digital imaging. Not all of them have made a successful transition, some have turned to other business activities. The same applies to traditional processing laboratories too.

An on-going struggle is that of light emitting diodes (LED) versus incandescent lamps. Their poor luminous output had long confined LEDs to indicators and small alphanumeric displays. Beginning around 1997, authorities started to convert red and orange traffic lights to LEDs; green was to follow a few years later. Yet, the slow adoption in other applications indicates that better efficiency, longevity, graceful degradation, and lower maintenance costs do not suffice to displace incandescent lamps as long as initial costs are not at par, or almost so.

The transition from cathode ray tubes (CRT) to liquid crystal displays (LCD) illustrates a positive feedback effect of market forces. For decades, the CRT was fine as a display device in TV sets, computers, point-of-sale terminals, information panels, and other applications. Large flat-panel color displays remained a distant vision unable to spur significant demand or industrial interest. It was the laptop computer that imposed the need for a reasonably-priced light-weight display and that created a credible market for it. For many other purposes, however, LCDs began to supplant CRTs only years later when their prices had become roughly the same.

Magnetic levitation trains, though highly innovative and addressing an obvious gap between airplanes and track-and-wheel trains on mid-range distances, are unlikely to meet broad success. This is because their deployment requires substantially higher investments while the public is unlikely to accept prices much beyond the level established by competing means of transportation. The electrical vehicle seems locked in a similar situation until energy prices, tax policies, and/or emission control laws change in a dramatic way.

The development of X-ray photolithography had begun in the late 1970s when everyone believed optical lithography would soon come to an end. So far, with optical lithography still being prevalent thanks to a multitude of improvements and with other alternatives being investigated for the time after, the huge investments into X-ray lithography have not been recovered. Inventions that are too much ahead of their time do not pay off either.

☐

Neither technical novelty nor competence alone warrants economic success. Costs, pricing, and timing are even more important. A quote from [371], a book worth reading on surviving in dynamic industries, aptly describes reality.

> As disconcerting as this may be, it is important to realize that there is no safe and permanent success in technological or business matters. Success is only the opportunity to compete in the next round.

Observation 13.13 also helps us to understand why startup companies tend to be much more innovative than well-established large corporations, a subject addressed in more detail in [372].

13.7.2 Commercialization stages and market priorities

Market expectations are liable to evolve as a product or a service approaches maturity. Consider cellular phones, for instance. Early customers such as doctors, field staff, express mail services, and the like, were perfectly willing to accept bulky equipment, high battery drain, and limited speech quality.[22] They were even prepared to pay substantial prices for all that if only they had access to a

---

[22] You might be too young to remember that the first cellular phone to be commercialized in 1983 was the Motorola DynaTAC 8000X. It was 33 cm tall, weighed 800 g for a standby time of 8 h and sold for 3995 USD.

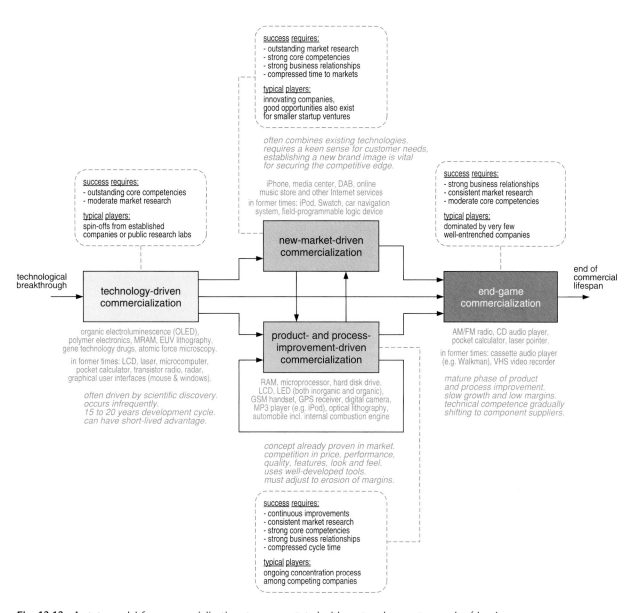

Fig. 13.13 A state model for commercialization stages annotated with past and present examples (drawing adapted from [373] with permission; the four stages of commercialization go back to research by Ralph Gomory, William Howard, and Bruce Guile [371] while the criteria for competitiveness have been defined by Rosabeth Kanter).

telephone wherever they needed to go, simply because this greatly contributed to their professional efficiency.

Attitude and user profile have changed a lot since that. Although much more sophisticated technologically, mobile phones today are just another consumer good, which implies that profit margins are small and that esthetic design, brand name, gizmos for entertainment, and intangible lifestyle attributes are as important as ruggedness, user interface, battery type, and other factual characteristics.

In general terms, fig.13.13 shows four typical commercialization stages that apply to most industrial goods. As a consequence, one must understand what characteristics of a product or a service represent the highest-ranking values in the eyes of the customers before committing oneself to major product and business decisions.

Observation 13.14. *Always ask yourself whether the key priorities of the markets are*
- *technical innovation,*
- *time to market,*
- *total conformity with customer preferences,*
- *low costs, or*
- *overall balance.*

As for the costs, it is important to know whether customers primarily decide on the basis of purchase price, operating costs, total cost of ownership, or cost/benefit ratio.

Example

Albeit eventually successful, the Swiss watch industry had been in great difficulty in the 1970s, struggling to reorient itself towards the development, manufacturing, and marketing of electronic watches. Remember that first-generation electronic wristwatches were equipped with LED displays that offered poor contrast, consumed a lot of batteries and, therefore, required the user to push a button to read the hour. This, together with poorly engineered user interfaces and unesthetic designs, led the traditional watchmakers to underestimate the potential of the new technology, which attitude delayed their reaction in a dangerous way.
□

Example

Only a few of the manufacturers that once dominated the computer business with mainframes and minicomputers continue to be major players in today's networked computing markets. Until 1990 or so, high-performance computers were built on the basis of the fastest-switching semiconductor devices then available, and the comparatively slow MOS microprocessors were believed to remain confined to not-so-demanding low-cost applications such as embedded microcontrollers and home computers. This was when Seymour Cray and his fellow engineers excelled in designing great supercomputers from fast ECL and GaAs components.

With the advent of VLSI, MOSFETs fabricated in large quantities on a monolithic piece of silicon began to offset the handicap of their lower transit frequency by slashing wire lengths, layout parasitics, and interconnect delays. Their abundant availability made it possible to come up with novel and highly parallel architecture designs. Even more importantly, the costs for a given performance level had gradually but irrevocably tipped in favor of CMOS microprocessors and personal computing, radically shaking up the business.
□

Observation 13.15. *Several different stages of commercialization may arise during the life of a product, but they need not all occur within the same company or region of the world.*

Adapted from [373]. In fact, many industrial products transit from technology-driven commercialization to end-game commercialization within just a few years. Things that once seemed miraculous soon become mundane commodities. Just consider wireless LANs, flat panel displays, notebook computers, mobile phones, digital high-definition TV, a pocketable jukebox, the Internet, anti-skid brakes, laser diodes, chess computers, scientific pocket calculators, and ball bearings, for instance.

However, intense price competition can also be understood as a sign of absence of smart innovations in the marketplace and as an incentive to identify hidden or untapped potential for new value creation [493]. In fact, several companies have repeatedly managed to protect their products from falling prey to lower-price competition by combining technical innovation, ease of use, fascinating design, and clever branding. Examples from industry include Apple, Bang & Olufsen, Dyson, Hilti and Sony.

13.7.3 Service versus product

Observation 13.16. *Marketing the same technology as a service or as a product can make the difference. Developing adequate business models is as important as commanding the technology.*

This is because the technical involvement and financial investments that are required from the customer's side to obtain the benefits he seeks from some given technology may totally differ.[23]

Example

ChipX, which was formerly known as Chip Express Corporation, is an ASIC company that emphasizes time to market. Throughout the 1990s, short turnaround times were obtained from a subtractive approach to manufacturing whereby prefabricated wafers got customized by cutting unwanted metal interconnections with a laser.

A similar technology had been developed and put to service by Lasarray in Switzerland as early as 1983. As opposed to ChipX who owned the equipment and offered manufacturing as a service, Lasarray focussed on selling the equipment for laser customization to companies that would develop and/or use USICs as part of their products, a strategy that was met with limited commercial success and ultimately drove the company out of business.

It is interesting to note that the founders of ChipX had initially planned to adopt the same approach until they were advised by venture capitalists to convert their business plan into a service operation instead. In retrospect, the CEO of ChipX attributed much of the company's success to this early strategy change [375]. Even though the technology was ultimately abandoned,[24] it was a solid contributor to the company's overall business for more than a decade.

□

Example

Advanced RISC Machines (ARM) today is a leading provider of embedded RISC processor cores. Their microprocessors are found in products as diverse as mobile phones, car electronics, handheld

---

[23] Albeit exaggerated, an advertisement found on some website hits the mark by asking "Why buy the cow if you can download the milk for free?"

[24] See section 13.6.4 for details.

MP3 players, digital still cameras, and Nintendo games. The company started in 1990 as a joint venture of Apple, VLSI Technology, and Acorn, a moderately successful British manufacturer of early low-cost personal computers. Developing and advancing a CPU along with its own RISC instruction set for those computers, Acorn's engineers had gained expertise in designing efficient hardware. Yet, there was no clear idea about how to exploit that technology commercially at the time. An early ARM product was the ARM 610 that powered Apple's Newton personal digital assistant (PDA).

The company rose to its strength precisely because it did not attempt to compete head-on with established vendors such as Intel, Motorola, Texas Instruments, and the like. Instead of manufacturing CPUs and selling them as packaged chips, ARM developed the idea of marketing them as virtual components. This implied that the company could not afford to focus on engineering alone, but also had to pioneer new business models for licensing, copyright protection, co-development, support and training in close collaboration with their customers. This fresh approach opened the door to emerging markets such as embedded computing and systems-on-a-chip (SoC) and eventually turned out to be as important as, if not more so than, the CPU's architecture and instruction set.

☐

Many manufacturers of telecommunications, multimedia, and building control equipment, to name just a few, do not confine themselves to one-time sales of hardware systems but generate on-going revenues from service agreements with their customers. They essentially contribute to keeping their customers' equipment up to date by delivering updates to support new data formats, transfer protocols, signal processing algorithms, and the like on the existing hardware. Remote installation is also part of their responsibilities in some cases.

This business model is patterned after the software industry. Adapting it to microelectronic circuits and systems asks for early and far-reaching decisions. Hardware must be designed as a platform with a significant proportion of program-controlled processors and/or FPL as it is not otherwise possible to accommodate updates in the form of software downloads and/or revised configuration files.

13.7.4 Product grading

Consumers differ in terms of the number and kind of features they are willing to pay for. While some are prepared to shell out money for a product with more features than they will ever learn to use, others are content with the most basic functionality. Rather than develop a product for each such clientele anew, industry is keen to serve them all with one graded range of products that can be sold at markedly different price tags.

The cost structure of mask-programmed ASICs does not, unfortunately, justify a multitude of versions as a slightly smaller die obtained from dropping a few ancillary features does not pay for an extra mask set unless really large sales volumes are involved. Ideally, one would have a single platform on which one can add or remove features depending on the product. A common practice is to reuse a chip designed to more demanding specifications throughout a product family and to downgrade simpler models via firmware or software. Structured ASICs and one-mask customization are other options.

Example

Consider a family of small-office/home-office (SOHO) document processing devices.

Required features	Supported functionality (incremental)				
	color printer	photo printer	scan & copy	telefax function	network capability
USB interface	yes	yes	yes	yes	yes
color space transform	yes	yes	yes	yes	yes
4-color print engine control	yes				
LED & pushbutton MMI	yes	yes	yes		
6-color print engine control		yes	yes	yes	yes
JPEG decoding		yes	yes	yes	yes
Picture bridge interface		yes	yes	yes	yes
Flash card interface		yes	yes	yes	yes
Scan engine control			yes	yes	yes
JPEG encoding			yes	yes	yes
Telephone modem				yes	yes
LCD & keypad MMI				yes	yes
Ethernet interface					yes

The overall architecture will always revolve around a microprocessor core with on-chip memories as most variations relate to peripheral interfaces. Yet, deciding on how many ASICs to develop and to put into production so as to manufacture the entire range of devices in the most economical way remains a tough problem with many unknowns.

☐

Observation 13.17. *The dilemma of addressing different markets with one family of graded products is to under-perform in the high end or to over-cost in the low end, or both.*

13.8 | Making a choice

13.8.1 ASICs yes or no?

Although there are many commonalities between ASSPs and USICs, this section will primarily be concerned with the latter because many companies in the electronics industry face the decision between assembling a product from standard parts — often in conjunction with field-programmable logic — and designing a user-specific IC instead.

ASICs offer many advantages over circuits assembled from catalog parts:

+ Reduced parts count.
+ Reduced assembly costs.
+ Improved reliability.
+ Reduced space requirements.
+ Full control over the package and all associated issues.

+ Superior performance for applications that warrant a dedicated hardware architecture.
+ The freedom to optimize inter-chip data exchange channels for speed, energy efficiency, and noise immunity by adopting signaling conventions tailored to a specific situation.
+ Tight control over parasitic circuit elements (important in high-speed logic and analog design).
+ Improved energy efficiency.
+ An opportunity to command innovation and to make products stand out from others'.
+ Better protection of proprietary know-how.
+ Excellent protection against unauthorized manipulations (tamper-proof).
+ An opportunity for implementing a consistent test strategy.

There are also handicaps associated with incorporating USICs into a product:

− Little flexibility for accommodating indeterminate and changing specifications.
− The long turnaround times of mask-programmed ICs when compared with FPL.
− A cost structure that favors large-volume fabrication but, at the same time, penalizes products that sell in smaller quantities.
− Mask-programmed ICs offer little support for product grading.
− The technical compromises that typically result from the integration of distinct subsystems or of technologically different devices on a single die.[25]
− The need for highly specialized design engineers that sharply contrasts with programming, where software developers need not know much of the underlying hardware and can work with standard languages and compilers, and with widely available software libraries.
− A need for a stable environment where ICs are being designed on a regular basis.
− The multiple parties being involved in design, fabrication, and testing.
− Usually one single source and high dependency on one business partner.
− Stronger technical challenges and financial risks, hence more demanding at all levels (marketing, specification, engineering, management, business partners).

In industrial practice, it is normally found that these arguments boil down to two typical situations that are in favor of having recourse to ASICs.

• **Advanced products** that are impossible to manufacture without customized microelectronic circuits because of demanding requirements with respect to performance, space, energy efficiency, integrated sensors, reliability, and the like.
   Examples:
   - Hand-held cellular phones (GSM, UMTS).
   - Truly digital hearing-aids (signal processing, remote control).
   - Switching equipment for digital high-speed communication.
   - Spread-spectrum systems on the basis of code-division multiple access (CDMA).
   - Spatial diversity receivers with their adaptive or "smart" antenna arrays.
   - Real-time video (de)compression and (de)coding equipment.

---

[25] Consider mixed-signal circuits, for instance, where digital and analog circuits must be made to coexist. Another example is BiCMOS technology, where BJTs are fabricated alongside MOSFETs. While very valuable in certain applications, these bipolars cannot compete with optimized discrete devices in terms of their electrical characteristics.

The design of this kind of ASICs typically implies an engineering challenge associated with a certain risk of not meeting the initial specifications and deadlines, which makes them inappropriate as entry-level projects. However, as no alternative solutions exist in this situation, the key decision is about finding suitable ASSPs, going for USICs, or forgetting about the product altogether.

- **Cost reduction** where the prime motivation for utilizing ASICs is to lower the recurring costs in comparison with alternative solutions.

Examples:
- Networked room controller as part of air conditioning and building control systems.
- Remote control transmitter/receiver for light dimmers.
- Demodulation, error correction, and audio processing in a compact disk (CD) player.
- Networked fire and smoke detector with programmable alarm criteria.
- Electronic ballasts for fluorescent lamps.
- Scan conversion (100 Hz) and video quality enhancement for home television sets.

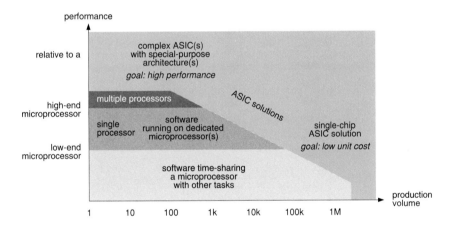

Fig. 13.14 ASIC solutions versus processor-based alternatives as a function of performance requirements and fabrication volume (simplified).

Case study

Adapted from [376]. Pen-size dispensers enable diabetic people to give insulin themselves injections wherever they are. Such dispensers were produced by Disetronic Medical Systems until 2003, when this business was transferred to the newly-founded Ypsomed. Dosing accuracy and fool-proof operation are absolutely essential as the blood-sugar level must be kept within tight limits to avoid long-term damage. Experience with an earlier purely mechanical model had made it clear that operating tiny knobs and reading small scales was asking too much of customers most of whom are in their senior years.

Disetronic thus opted for the following features: simple push-button operation, numerical indication of the selected dose on a comparatively large display, automatic battery monitoring, and self-deactivation at the end of a predetermined lifetime. After a transitory phase with a low-voltage gate-array, the actual product was built around a full-custom IC of 11.6 mm^2 fabricated using a low-voltage CMOS technology. The naked die was mounted directly onto a tiny printed circuit board (chip-on-board). The board also carried a few external components, the most important of which were a two-digit LCD and a small watch crystal for the oscillator. Power was supplied by a 1.5 V silver oxide battery. The current drawn was a mere 5 μA during operation and 1 μA in standby mode.

The decision for a USIC was a fairly logical consequence from the experience with the mechanical device and from the restrictions of the pen format. Chip design was contracted to an external design house and took nine months. Disetronic estimates to have invested 300 000 CHF in non-recurring expenses for the project. Some 200 000 electronic insulin dispensers had been produced in 1993, which represented an eightfold increase over its mechanical predecessor.

☐

Case study

Adapted from [377]. VingCard, a Norwegian company, produces recodable lock systems for hotels and cruise ships that are based on magnetic cards. Compared with purely mechanical locks, an electronic solution provides many benefits in terms of comfort and security. Guests and employees may be selectively granted access to a single room, to some specific set of rooms, to recreational facilities, or to an entire zone. Any key can be made to expire at a predetermined date and the activities of all locks can be logged. Lost keys can be rendered inoperative at any time and at no extra cost.

VingCard's door lock is battery-powered, communicates with a central service terminal at the reception desk, periodically monitors battery condition, and includes a magnetic card reader, motor drives, and a calendar/timer function. Until 1995 the lock control electronics was made of standard components, the most important of which were a microcomputer with an 8 bit A/D converter, amplifiers, and filters for processing the signals from the magnetic head, motor drive electronics, and a UART, plus additional functions for saving power during standby operation.

Preserving the well-proven functionality, a USIC now integrates most electronic circuits and components on a single chip. Only motor drivers have been left out because of the high currents involved. The microprocessor core used is functionally compatible with the old version so that the software code can be reused. An on-chip PLL makes it possible to operate the circuit at 10 MHz from a cheap 32 kHz watch crystal. For the sake of battery conservation, the clock frequency gets reduced to 32 kHz whenever the lock is in standby mode.

The prime motivation for going for a USIC was cost reduction with respect to the traditional version then being produced. As a matter of fact, the total expense for components was cut by more than 50% by adopting the USIC solution, so that development costs were recovered within 8 months (pay-back period). As an extra benefit, the size of the circuit board and that of the lock were cut in half.

☐

Observation 13.18. *ASICs are essentially designed either*
o *because demanding requirements cannot be met otherwise or*
o *for lowering costs with respect to alternative solutions in stable products.*

13.8.2 Which implementation technique should one adopt?

While any approach to designing and fabricating electronic hardware necessarily is a compromise between conflicting goals, some options are more adequate than others in a given situation. There can be no general answer as the "best" choice depends on many factors. The subsequent checklist of criteria may help you not to overlook important issues.

Characteristics of the final product

- Performance as characterized by throughput or operating speed
- Energy efficiency
- Package count
- System partitioning
- Board space requirements
- Electromagnetic compatibility (emission <u>and</u> vulnerability)
- Acceptable operating conditions
- Tolerance with respect to all kinds of parameter variations
- Reliability (related to total package and pin counts)
- Maintainability (e.g. diagnosing, repair, and updating, possibly from remote)
- Compliance with existing or future industry standards and government regulations

Product development issues

- Overall circuit complexity
- The option of including analog subcircuits, on-chip sensors, etc.
- Availability of subfunctions on the market (as physical parts, VCs, or software)
- Reusability of design data from former projects and for future ones
- Aptitude for hardware and software co-design
- Modifiability (in terms of effort, costs, and turnaround time)
- Turnaround time (from design data sign-off to first prototype delivered)
- Testability
- Design effort
- Design risk
- The limited lifetime of fabrication processes and the impact
  of future advances in semiconductor technology[26]

---

[26] The lifetime of IC fabrication processes is typically limited to five to six years. For an ASIC, this means that the design needs to be ported to a more modern process after that time. Most often, however, a redesign will allow one to upgrade a circuit's functionality while, at the same time, taking advantage of recent improvements of fabrication technology with respect to performance, integration density, energy efficiency, etc. In either case the burden of a redesign or shrinking operation is essentially placed onto the customer, possibly with assistance from the manufacturer. In contrast to this, performance and pricing of microprocessors, memories, and other well-established standard parts automatically benefit from advances in semiconductor technology because it is their manufacturer who takes care of porting the device to an up-to-date process.

- Protection of know-how against unauthorized copying
  (from the final product or from subcontracted design data)
- Availability of adequate CAE/CAD tools and support
- Availability of adequate cell libraries and cell generators
- Present in-house expertise and any need to acquire more know-how
- Investments into equipment and personnel
- In-house design vs. subcontracting

Production and supply chain management

- Overall sales volume
- Number of parts per fabrication lot
- Fabrication time for one lot
- Dependability of delivery dates agreed
- Second sourcing
- Availability of components or fabrication processes over the product's lifetime[27]
- Portability of design data to new fabrication processes, FPL devices, or microprocessors
- Potential export restrictions on components manufactured abroad
- Size, handling, and costs of parts inventory
- Potential for reuse of unsold fabricated parts
- Hazards for personnel and environment

Marketing aspects

- Time to market
- Acceptance by market, predictability of sales volume
- Importance of features that depend on a specific way of implementation
- Agility with respect to changing requirements

Investment

- Recurring and non-recurring costs (see section 13.5)
- Overall investment
- Return on investment, payback period, net present value, etc.
- Financial risks of going for a project and also of abandoning it

Similarly, deciding on the most appropriate degree of integration for an electronic system is an act of balancing

- Overall space and power requirements (obviously decreasing with integration degree),
- ASIC manufacturing costs (progressively increasing with integration degree),
- Overall mounting and testing costs (decreasing with integration degree),

---

[27] Swift obsoletion is by no means limited to semiconductor technology. Highly specialized catalog parts, such as ASSPs or microcomputers that exhibit not-so-popular features, for instance, are often replaced by more up-to-date but not truly compatible parts at a rapid pace. Worse than this, a manufacturer is free to discontinue products at any time should their sales volumes and/or revenues fall short of expectations.

- Overall development costs and risks (both tend to increase with integration degree),
- Sales potential for ASICs (more specific functionality tends to imply narrower markets),
- Flexibility (higher integration degree often implies more specific functionality),
- Potential for migrating to future components and/or fabrication processes, and
- Recurring vs. non-recurring costs.

High-density packaging often is a better option than full system-level integration.[28]

13.8.3　What if nothing is known for sure?

As established in subsection 13.5, production volume is crucial for deciding on how to obtain the lowest price per part. However, it is not uncommon in industrial practice that sales prospects of a product are virtually unknown, and its specifications have not been settled in detail, yet the product should hit the market as soon as possible.

A wise approach in such a situation is to begin with a first implementation based on field-programmable logic, general purpose microprocessors or signal processors, or both. Produced in modest quantities, this first version will serve to test the market and to verify the functionality at low risks and at reasonable initial costs. The inherent agility will also permit one to respond to change and enhancement requests almost instantaneously.

Only then, as annual sales volume can be anticipated better and as specifications have been firmly established, can a truly informed decision on the best implementation avenue be made. A semi- or full-custom version marketed somewhat later serves to reduce package count, increase performance, improve battery runtime, etc., while, at the same time, lowering costs per part due to volume production.

Observation 13.19. *A preliminary implementation on the basis of FPL and/or microcomputers often provides a good starting point for making sound design and business decisions.*

In a move to better support this strategy, several vendors have made migrating to mask-programmed semi-custom ASICs easier by offering function-compatible hardwired counterparts for popular FP-GAs, see section 13.6.5. **Drop-in replacement** parts that maintain pin compatibility are of course even more desirable but not always available.

A difficulty in high-performance applications is that it may be impossible or uneconomical to get the necessary operating speed from an FPGA, CPLD, or instruction-set processor.[29] Tight space or power constraints may also preclude non-ASIC prototypes. In such a situation, relying on high-level synthesis and reuse as much as possible may help to reduce design time and design expenses for a first version. Critical parts may later be refined at lower levels of detail. Compared with a field-programmable prototype, much of the flexibility is lost, however.

---

[28]　Refer to section 11.4.7 for details.

[29]　A potential workaround consists in making the surrounding equipment run at a reduced clock frequency or speed. This is often impractical, however, when real-time processes such as disk drives, video equipment, communication channels, control loops, physical processes, and the like are involved.

13.8.4 Can system houses afford to ignore microelectronics?

Many system houses design their products from standard semiconductor components such as program-controlled processors, memories, ASSPs, and FPGAs. They do not feel pressed to design their own VLSI chips because this cannot be justified on short-time economics grounds. Though much of the expertise of these companies resides in system integration,[30] they also generate essential know-how in the areas of proprietary data or signal processing algorithms — or in fine-tuned implementations of published algorithms — hardware architecture, circuit and board design, and software engineering.

As ever larger subsystems move onto chips, the traditional arena of semiconductor companies, the system houses may feel intimidated by the resources needed for VLSI or may dismiss the ASIC option on the basis of negative experience from the past. However, a failure to tap the potential of microelectronics in one way or another is likely to expose a system house to being hollowed out by any semiconductor vendor who is willing to move up the ladder of value [378]. Superior quality alone is unlikely to protect a traditional implementation once an integrated alternative becomes available at a fraction of the cost, space, and power.

Warning examples

- Ciphering: Dedicated chips are not only faster than software but also more tamper-proof.
- Viterbi decoders: Convolutional decoding algorithms once were at the forefront of telecommunication. Today, extremely powerful and energy-efficient Viterbi decoders are available as ready-to-go microchips.
- Audio sampling rate conversion: same as above.
- xDSL equipment: Higher integration degree lowers manufacturing costs.
- Image capture: Traditional film has lost to integrated electronics. Some companies have made the transition, others sticking to photochemicals face difficult times.

A particularly telling example is best captured by a quote from Federico Faggin, one of the key designers of the first microprocessor, the Intel 4004, aka MCS-4.

> I vividly remember a trip to Europe to visit customers with Hank. This trip occurred in the late summer of 1971, a few months before the MCS-4 introduction. I found out that the more computer literate the customer, the more resistant he was to consider using the microprocessor. The worst meeting was at Nixdorf Computer where they nearly ridiculed us for the poor architecture of our machines. Some of their criticism was valid, to be sure, but the level of hostility was only justified by the more or less conscious awareness that a turf war was beginning with the semiconductor guys. Well, we know who is left standing 30 years later!

□

---

[30] System integration is meant to include answering market needs, establishing complex specifications, system partitioning, overall architecture design, finding good hardware vs. software tradeoffs, manufacturing hardware equipment or commissioning that to outside partners, ongoing customer support, offering after-sales services such as running or maintaining installations on behalf of customers, and the like. System houses are neither component manufacturers nor software houses or service companies but combine some elements of each.

Many results of system-level research and development will be met with far more success on the marketplace if made available as VLSI chips or as virtual components (VCs) rather than as slow software executables or as bulky and costly board assemblies. The entire personal communication industry is built upon this experience, for instance.

Observation 13.20. *A system house must not spend resources creating components (hard and soft) that the company could have acquired elsewhere. Conversely, having developed a solution for a problem of fairly broad appeal, a company must capitalize on the investment by marketing that intellectual property, even if this ultimately implies going for silicon.*

A system house that competes with semiconductor or VC vendors on the open market — or that is subject to being exposed to such competition in the near future — basically can choose among seven alternative strategies for what to do with the know-how acquired internally.

1. Design VLSI circuits and market know-how in the form of ASSPs under their own brand, thus effectively becoming a fabless vendor.
2. Team up with a semiconductor vendor for designing, fabricating, and marketing ASSPs.
3. Market know-how in the form of virtual components, preferably in collaboration with an established VC vendor.
4. Protect know-how by patents, be prepared to defend intellectual property by legal means, and license to selected semiconductor and/or VC vendors.
5. Protect know-how by patents and deny the market to semiconductor and VC companies.
6. Take advantage of in-depth system expertise to always stay ahead of ASSPs and VCs in terms of performance, quality, and — most likely impossible — costs.
7. Withdraw from certain business areas in favor of higher-level system design and prepare to rely more heavily on subfunctions purchased from the semiconductor or VC industry.

As stated before, only subfunctions of fairly broad interest for which a market exists or can be created are likely to elicit the attention of the semiconductor industry and/or of VC developers. Much as for ASSPs, merchant markets for VCs remain confined to functions that find fairly wide applications such as microprocessor cores and peripherals, popular interfaces, and all sorts of standardized functions from data and signal processing, telecommunication, networking, audio, video, and multimedia [379].

Algorithmic expertise beyond such "mainstream" applications is not exposed to the same extent. The challenge for system companies operating in niche markets is to take their existing in-house know-how, to turn it into a product by whatever is the most appropriate means of implementation, and to make such an operation profitable [380].

Examples

Qualcomm develops, markets, and promotes new wireless telecommunication algorithms and standards. What makes them special, however, is that they not only license this expertise but also sell it encapsulated into ASSPs of their own design.

Phonak, a highly innovative manufacturer of hearing aids located in Switzerland, holds important expertise in signal processing for noise cancellation, for beam forming, for compensating for aural impairments, and for adjusting hearing aids. Rather than extending its activities into the unfamiliar realms of VLSI, Phonak stayed focussed on improving the signal processing chain under the tight limitations imposed by hearing aid electronics. An external company was commissioned to develop an architecture that combines analog and digital subsystems and to integrate all that into silicon. CSEM was selected because of its experience with low-power microelectronics. It is safe to guess that this approach was much faster and more cost-effective than if Phonak had attempted to solve everything internally.

☐

13.9 | Keys to successful VLSI design

13.9.1 Project definition and marketing

- Develop a good idea about what products and features customers are willing to pay for and not just what they mark as desirable when presented with a marketing questionnaire.

- Try to anticipate sales volume, product lifetime, and future change requirements. This is not only instrumental in selecting an implementation avenue but also helps to prioritize between minimizing one-time investments and per-unit costs of manufacturing.

- Let the people involved in project definition understand <u>all</u> options for doing business in a given application area (marketing systems, equipment, or components; ASSPs, USICs, or VCs; hardware, software, or services). Check how these options match with market needs and with your company's potential.[31] Try to think ahead.

- Evaluate all implementation alternatives available (program-controlled processors, field-programmable logic (FPL), semi- or full-custom IC) with great care. Estimate costs and risks for all scenarios. Before opting for a mask-programmed approach, make sure this is indeed the most appropriate choice in the actual situation and the foreseeable future.[32]

- Carefully select design house, silicon vendor, IP vendor, and any further business partners.[33] Make sure they share the same long-term interests as you do. Decide on the fabrication process and business model together with them.

- Fight for stable specifications. Once actual VLSI design has begun, each modification means shifting plans in midstream, redoing existing HDL code, schematics, layout, test vectors, etc. and jeopardizing integrity of the design. A system the specifications of which cannot be frozen at some point is not suitable for a hardwired circuit implementation.

- Elaborate on specifications until they are unmistakable and as complete as possible. It is always a good idea to start with some sort of behavioral model or functional prototype.[34] Also keep in mind that it is a human deficiency to expand upon known needs in great detail, while, at the same time, glossing over unknown or difficult but all the more critical matters in a few words.

---

[31] See sections 13.2, 13.4, and 13.7.
[32] See section 13.8.
[33] Detailed checklists can be found in section 13.10.1.
[34] See section 3.1.2.

- If you are designing digital circuits for some given DSP application, do not underestimate the effort for finding a good compromise between numerical accuracy and the word widths, number representation formats, calculation schemes, and other computational resources that significantly impact hardware costs and energy efficiency.

- Resist the temptation of featuritus, the symptoms of which are an overloaded wish list, overly many modes of operation, countless interface protocols, oversized memory capacities and — in more general terms — the quest for a universal circuit. If specifications ask for all-embracing hardware there is something wrong with them, most likely because the target application and the circuit's embedding are not sufficiently clear. After all, lean development also implies lean specifications.

13.9.2   Technical management

- Have systems engineers and VLSI designers closely collaborate from the onset of a project. Processing algorithms typically need extensive reworking to find a balance between the theoretically desirable and the economically feasible in terms of hardware.[35]

- Carefully evaluate the respective benefits and limitations of general- and special-purpose architectures.[35] Be prepared to move the borderline between hardware and software from where you initially anticipated it would be.

- Establish bit-true software models and insist on thorough functional testing.

- Contract to external partners those activities that they do better or more efficiently than can be done in-house. Set aside sufficient time for negotiating all the details.[36]

- Always draw high-level block diagrams of the entire chip and the surrounding system. Clearly identify clock domains and all signals traversing the boundaries between them. If you plan to use multiple supply voltages or to temporarily power down part of the circuit, do the same for the voltage domains.

- Follow a design methodology that makes best use of design automation software, cell libraries, virtual components (VC), and available expertise:
  - Work on the highest level of abstraction compatible with technical and economical needs. Please note this is not normally the most abstract level in absolute terms.
  - Reuse proven components as much as possible (e.g. HDL libraries from previous designs, from commercial sources, and from within the current project itself).
  - Know the preconditions for and the limitations of using a given instrument, be it an EDA tool, a library element, or any other subcircuit about to be used.

- Implement a formal version control mechanism to make sure design data are kept consistent by all members of the team at any time. It is a good idea to transfer the final design data to a special repository for tapeout and later use.

- Use a coherent CAE/CAD environment. Do not explore any new design flow (design methodology, tool suite, design kit, cell library) or any new fabrication avenue (process, silicon vendor, foundry) on a critical project. Do so with fairly modest test vehicles first.

---

[35] See sections 2.3 and 2.2 respectively.
[36] See sections 13.2 and 13.4 and refer to figs.13.15 and 13.16.

- Organize design reviews at critical milestones of the project, e.g.
  - to agree on final specifications,
  - to approve the system architecture being proposed,
  - to locate deficiencies in the HDL code established,
  - to assess the functional coverage of the simulations being carried out,
  - once gate-level netlists are available following RTL synthesis,
  - before submitting design data to fabrication, and
  - after prototypes have been tested and measured.

- Make sure a circuit's design team remains available for the project until prototypes have been manufactured, tested, and debugged.

- Do everything possible to avoid redesigns, but prepare for them.

13.9.3 Engineering

- Plan to use powerful yet simple interfaces. Avoid the need for shadow registers and/or lengthy instruction sequences just for handling data transfer protocols.

- It always pays to view hardware from a programmer's perspective too. Do not compromise on the performance and dependability of software drivers just to save a couple of logic gates in an interface. Keep user interfaces simple and manageable.

- Resort to rapid prototyping to check circuit operation within the target context.

- Similarly, make sure interfacing to your design is safe and easy timing-wise.[37]

- Consider floorplanning issues early on in the design process.[38] In particular, estimate chip size, pin count, power dissipation, and package early on in the project and update the figures whenever a major design step is completed.

- For large chips to be fabricated with advanced processes, do not underestimate the domination of interconnect delay over gate delay. Make the planning of signal distribution part of architecture design.

- Be fluent in digital circuit design. System-wide synchronous operation avoids most timing problems and permits one to separate functionality from timing to a limited but helpful extent.

- Adhere to one of the well-known clocking disciplines, refrain from mixing them.[39]

- Include a reset mechanism that unconditionally puts all bistables into a known state.

- Design for test. Economizing on test structures may be found very expensive later.

- Control parasitic effects which could make a circuit behave differently from anticipations, namely

---

[37] See chapter 7.
[38] See section 11.3.1.
[39] See section 6.2.

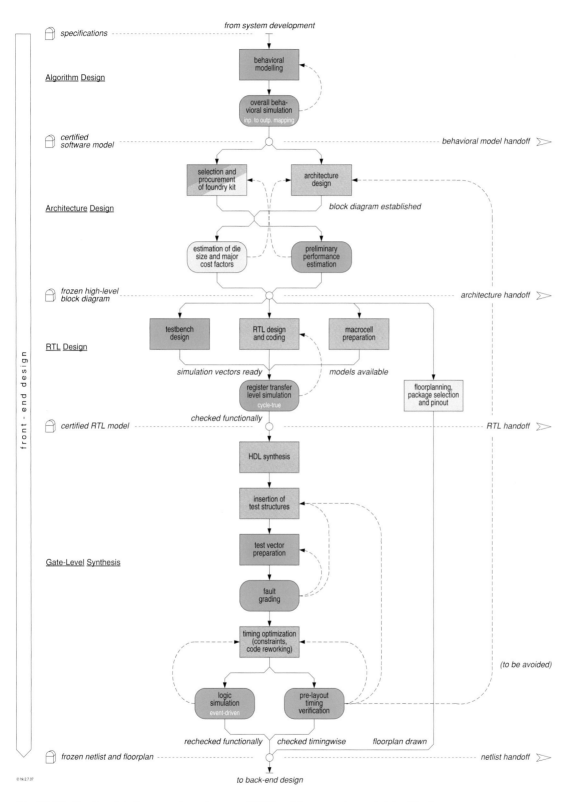

Fig. 13.15 Front-end design procedure shown as PERT activity-on-node graph (simplified).

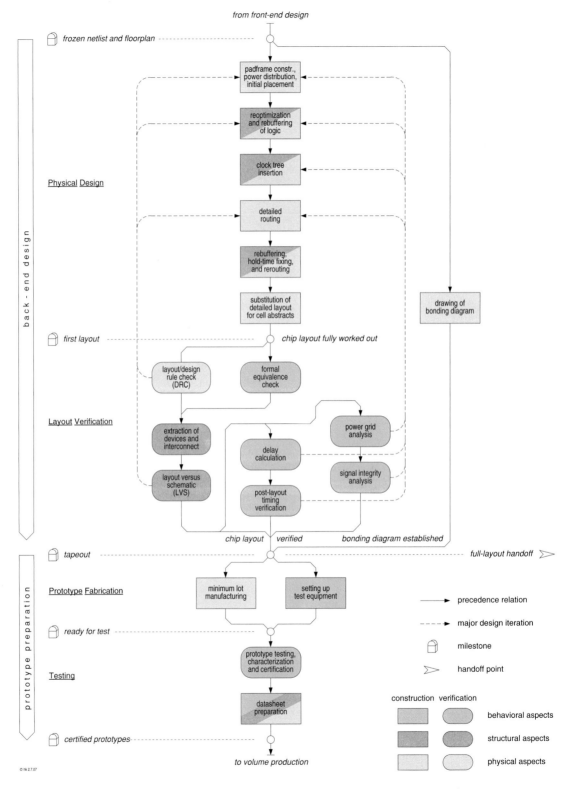

Fig. 13.16 Back-end design procedure shown as PERT activity-on node graph (simplified).

- switching noise (from ground bounce and crosstalk),
- clock skew (on-chip and particularly between chips),
- PTV variations (particularly between chips) and on-chip variations (OCVs),
- layout parasitics (especially in power and clock distribution),
- latch-up of CMOS structures,
- electrostatic discharge (ESD),
- electromigration, and
- radiation effects.

Most parasitic effects are not or not sufficiently taken into consideration by standard CAE/CAD tools. Specific modelling techniques may be necessary, e.g. noise analysis at the electrical network level, skew analysis of clock distribution network, device simulation for latch-up and ESD analysis.

- Keep design iteration loops short and tight, do not put off verification steps.

13.9.4 Verification

- Have all parties involved (customers, marketing, front-end designers, back-end designers, foundry and test engineers) agree on the test cases relevant for the acceptance of the hardware. Use this as a starting point for elaborating a verification plan.

- Beware of having preconceptions from the design and/or programming phases of a system creep into the phases of functional verification and test. Have different persons or teams handle the two in an independent way.

- Be as thorough as possible in your simulations. Try to cover all operating modes the circuit may encounter, do not fail to verify any functions just because they appear to be straightforward. Experience shows that a design is likely to fail in those situations that have not been sufficiently addressed during functional verification.[40]

- Simulation is not good at detecting problems that occur where a design must cooperate with the outside world. Use FPGA-based prototypes as a complement where possible.

- Carry out static timing analysis on top of functional verification.[41]

- Prepare production test vectors as part of the design process, do not postpone this until first silicon arrives.

13.9.5 Myths

HDLs HAVE MADE HARDWARE AND SOFTWARE DESIGN BECOME THE SAME

The idea that today's hardware description languages have eliminated the traditional barriers between hardware and software development is often advanced to imply that automatic synthesis puts any software person in a position to design VLSI circuits. As a matter of fact, the arrival of HDLs has led to significant commonalities between hardware and software design:

---

[40] See chapter 3 for more detailed advice.
[41] See chapter 12.

+ Design entry is by way of formal languages.
+ Typical software development tools are being used
  (e.g. text editors, compilers, code analyzers, debuggers).
+ Established principles from software engineering must be observed (modularity, hierarchy, parametrization, information hiding, version control, documentation, code reviews, etc.).

These apparent similarities notwithstanding, several factors remain that set the two disciplines well apart from each other.

− Hardware design must address behavioral, structural, and physical issues, whereas software design is confined to a purely behavioral view.
− A variety of parasitic effects must be contained to make circuits behave as modelled (i.e. specified and simulated) in abstract and purely logical terms.
− Defects from manufacturing necessitate the incorporation of test structures into hardware.
− Economics and cost metrics are totally different. Fabrication costs and hence also die size are much more critical to VLSI than code size is to software.
− The turnaround times and financial losses associated with fixing errors in VLSI circuits necessitate total commitment to first-time-right design.

In conclusion, the advent of HDLs means that the VLSI designer has to become familiar with software engineering, but it does not do away with the need to master typical hardware design practices at multiple levels of abstraction. Hardware designs starts with a behavioral model; that is exactly where software design ends.

THERE IS NO SUCH THING AS DIGITAL DESIGN

As voltages and currents are continuous quantities in today's electronic circuits, it is sometimes argued that education and experience in analog design would form the optimal background for digital VLSI design as well. There are important technical and cultural differences, however.

Analog IC designers must and do have detailed knowledge about the electrical characteristics of transistors, conducting materials, dielectrics, and parasitics. They also know about limitations of transistor models and about imperfections of fabrication processes. Their main preoccupations are to ensure linearity, to minimize noise, to maximize dynamic ranges, to control fringe effects, to match transistor, resistor, and capacitor pairs, to compensate for PTV and on-chip variations, to maximize layout density, and the like. In order to do so, they need to care about every single device and every single node in great detail.

Digital VLSI designers are not normally allowed to do so. Their prime concerns are first-time-right design, time to market, yield, and overall costs in the face of highly complex circuits. For the sake of productivity, they must take full advantage of design automation tools, work at higher levels of abstraction, and reuse as many subcircuits as possible. Digital designers must refrain from using tricks that would compromise the validity of the assumptions and simplifications that implicitly underlie their models and tools. The key idea is not to get involved with low-level details unless absolutely necessary. This is achieved by following design practices that are dependable, robust, and independent of a specific technology, a given electrical context, or some particular layout arrangement. Relevant techniques such as synchronous operation, safe clocking disciplines, HDL synthesis, separation of functional verification from timing verification, self-checking testbenches, design for test (DFT), tolerance against PTV and on-chip variations, safe design practices, noise control, and more have been discussed throughout this text.

Clearly, being knowledgeable in analog circuit design and device physics always is a great asset, especially with deep submicron technologies where many second-order phenomena begin to be felt in digital circuits. Our point is that tackling digital VLSI with standard approaches from analog design seldom yields satisfactory results.

13.10 | Appendix: Doing business in microelectronics

13.10.1 Checklists for evaluating business partners and design kits

Silicon vendor

- Company name, ownership, geographic location, and webpages.
- Management-level, commercial, and technical contact persons.
- Strategic importance and long-term commitment accorded to the ASIC business.
- Financial health (in view of remaining in business, staying independent, and investing in future technologies).
- Track record (with respect to quality, timely delivery, customer support, etc.).
- Fabrication avenues and design flows supported.
- EDA tool suites supported for use by external customers versus tools used in-house.
- Competence, availability, and manpower of application support.
- Non-disclosure of design data and of business information.
- Assistance with back-end design (place and route, layout verification, etc.).
- Assistance with test vector generation.
- Details on sign-off procedure (data formats, back-annotation, "golden simulator", etc.).
- Availability of engineering samples and/or of a multi-project wafer service.
- Production testing (ASIC manufacturing service vs. foundry service).
- Packaging service with package types supported.
- Export restrictions, terms of delivery and of payment.

Design house

Many of the above criteria also apply to design houses, but there exist some more specific requirements on top of those:

- Critical mass, multiple carriers of know-how.
- Expertise in the ASIC's field of application, general system-level know-how.
- Potential entry levels (informal specifications, software model, HDL model, etc.).
- Expertise in digital, analog, RF, and mixed-signal design.
- Know-how in the specific design niche (high-speed, low-power, full-custom, etc.).
- HDLs being supported.
- Close and preferential links to silicon foundries (privileged access to information, involvement in library and process deployment).

- Adequate cell libraries, macrocell (generators), virtual components, and the like.
- Compatibility between available libraries and the EDA tools in routine use.
- Design steps covered in-house versus tasks being subcontracted.
- Expertise in design for test and in test vector generation.
- Coherence of and experience with the design flow planned for.
- Help with package selection and packaging.
- On-going peer review of design quality and test structures.
- Formal design reviews at critical milestones.
- Quality of functional, parametric, and physical verification.
- Competence in prototype testing and debugging.
- Technical and financial participation in case of malfunctioning, redesign or yield problems.
- Timelines, charges, payement schemes, sharing of commercial risk and benefit.

Fabrication process

- Minimum feature size, number of poly and metal layers.
- Bulk or SOI process, in the former case also n-well, p-well or twin wells.
- Process options (Flash, silicide straps BiCMOS, SiGe, etc.).
- Pitch, material, and electrical characteristics of each interconnect layer.
- Supply voltage range for core and I/O, interfacing with other voltages.
- n- and p-channel drivabilities, subthreshold slopes, and leakage currents.
- PTV variations, admissible temperature range.
- Yield model.
- Maturity and expected lifetime of process.
- Location of fabrication line, export restrictions if any.
- Second source.
- Minimum order and turnaround time.
- Detailed quotation with NRE charges and fabrication costs.

Design kits and cell libraries

- Fabrication depth (semi- vs. full-custom ICs).
- Top priorities (layout density, speed, energy efficiency, safeness, and testability).
- Propagation delays and switching energies of important library elements.
- Conditions assumed for library characterization (PTV, load, trip points, ramp time).
- PTV derating curves.
- Design flows, CAE/CAD tool suites, and tool versions being supported.
- Computing environment assumed (operating system, shell, scripting language, batch queues),
- Available support (hotline, bug fixing, adaptations, etc.).
- Compatibility with your EDA environment and in-house expertise.

In order to operate a design kit you will need to procure:
- Cell models for simulation and synthesis (VHDL, VITAL, Verilog, Synopsys, etc.).
- Technology files (layout rules, transistor models, layout parasitics).
- Digital cell library (logic gates, bistables, adder slices, etc.).
- Pad library (in/out/bidirectional, voltage levels, sink/source currents, ESD protection, etc.).

- Macrocells (SRAM, DRAM, ROM, multiplier, datapath, etc.).
- Megacells (microprocessor cores and peripherals, filters, etc.).
- Analog cell library (opamps, comparators, D/A- and A/D-converters, etc.).

13.10.2 Virtual component providers

Vendors range from large companies with products of wide interest to smaller companies that focus on very specific needs and applications. As frequent business changes would rapidly render a printed list obsolete, we limit ourselves to information hubs that have been around for many years.

- Design and Reuse, http://www.design-reuse.com
- Opencores.org, http://www.opencores.org

A list of VC deliverables has been given in section 13.4.6.

13.10.3 Selected low-volume providers

The companies and institutions listed below accept orders for ICs in small or moderate quantities because they offer structured ASICs or because they run multi-project wafer (MPW) services.

- Altera, http://www.altera.com
- AMI Semiconductor, http://www.amis.com
- ChipX Corp., http://www.chipx.com
- Circuits Multi-Projets (CMP), F-38031 Grenoble, http://cmp.imag.fr
- DELTA Electronics Testing, DK-2970 Hørsholm, http://www.delta.dk
- eASIC Corp., http://www.easic.com
- Europractice IC Manufacturing (ICMS), IMEC, B-3001 Leuven, http://www.europractice-ic.com
- Faraday Technology, http://www.faraday-tech.com
- Fraunhofer-Institut für Festkörpertechnologie (FhG-IFT), D-80686 München.
- Fraunhofer-Institut für Integrierte Schaltungen (FhG-IIS), D-91058 Erlangen.
- Fujitsu Microelectronics, http://www.fma.fujitsu.com
- Institut für Mikroelektronik Stuttgart (IMS), D-70569 Stuttgart, http://www.ims-chips.de
- Leopard Logic, http://www.leopard-logic.com
- Light Speed Semiconductor, http://www.lightspeed.com
- Microdul AG, CH-8045 Zürich, http://www.microdul.com
- MOSIS, http://www.mosis.org
- NEC Electronics, http://www.necel.com
- ViASIC, http://www.viasic.com
- Xilinx Inc, http://www.xilinx.com

13.10.4 Cost estimation helps

- Die size estimation, http://eproject.umc.com/dse
- Wafer cost calculator, http://www.sematech.org/ismi/modeling/wfrcalculator.htm
- Size, power, leakage, yield, and cost calculator, http://www.chipestimate.com
- Manufacturing cost estimator, http://www.icknowledge.com/our_products/cost_model.html

- Project cost estimation spreadsheet, http://dz.ee.ethz.ch/support/ic/asiccostestimator.en.html
- FPGA vs. ASIC project cost calculator, http://www.altera.com/products/devices/cost/cst-cost_step1.jsp
- Small-volume price estimation, http://www.iis.fraunhofer.de/EN/bf/ic/smallrol/preis-jsp
- Price quotes for DRAMs and flash memories, http://www.dramexchange.com
- Standard component price tracker, http://www.isupply.com/catalog/L2_prtr.asp

Chapter 14

A Primer on CMOS Technology

The prime objective of this chapter is to put VLSI design and test engineers in a position to understand how MOSFETs, diodes, and contacts operate as part of digital circuits. Readers shall also be enabled to study, understand, and sketch layout drawings and circuit cross sections. Neither a comprehensive insight into solid-state physics nor overly detailed information on a specific fabrication process is necessary for doing so. It generally proves sufficient to distinguish between a handful of materials and to know in what order these are being manufactured and patterned. This is precisely what the present text attempts to convey.

14.1 | The essence of MOS device physics

14.1.1 Energy bands and electrical conduction

All solids have energy bands which indicate at what levels of energy electrons can exist. Only the **conduction band**, the uppermost, and the **valence band**, the next one beneath, are relevant here, see fig.14.1. Their separation, that is the amount of energy necessary to transfer an electron from the valence band to the conduction band, is known as the **bandgap**. The relative locations of valence and conduction bands largely determine a material's electrical characteristics.

Insulators. The valence band is fully occupied whereas the conduction band is empty; an important bandgap of typically more than 5 eV separates the two. The valence electrons form strong bonds between adjacent atoms. As these bonds are difficult to break, there are no free electrons that could float around and participate in electrical conduction. Resistivity ρ is 10^8 Ω m or more.

Metals. The valence band is only partially filled, so the electrons are free to move through the material. Some metals actually have the valence band and the conduction band overlap with

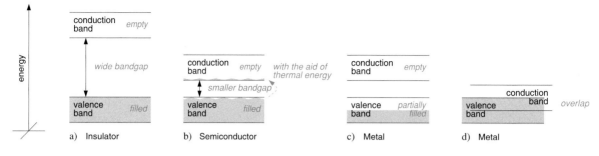

Fig. 14.1 Energy band configurations of insulators (a), semiconductors (b) and metals (c,d) (simplified).

no gap in between, but the effect on conduction is essentially the same. Resistivity ρ is 10^{-6} Ω m or less.

Semiconductors. The conduction and valence bands are separated by a bandgap, yet, 1 eV being a typical value, the difference in energy is not as large as for insulators.[1] Electrons cannot jump the gap easily at low temperatures. At higher temperatures, however, more of the valence electrons do traverse the gap and the semiconductor's conductivity augments accordingly. At room temperature, resistivity ρ is on the order of 10^3 Ω m.

14.1.2   Doping of semiconductor materials

What makes semiconductors technically interesting is that their electrical properties can be profoundly modified by doping them. The term **doping** refers to "impurities" that are selectively incorporated into the crystal lattice of the base material in order to control its local electrical characteristics.

Consider silicon — the predominant base material today — and recall from fig.14.2 that its atoms have <u>four</u> electrons in the outer shell. In a crystal of pure silicon, electrons from adjacent atoms form covalent binding pairs, see fig.14.3b. Though somewhat augmenting with temperature, electrical conductivity remains very limited because of a considerable bandgap.

Extra electrons that come as part of impurity atoms that count <u>five</u> electrons in their outer shells cannot be accommodated in the regular bonding structure of the crystal lattice, see fig.14.3a. The amount of energy required to remove such an **excess electron** from the lattice and to lift it into the conduction band is therefore much lower. As a result, conductivity increases with doping concentration. Impurities are said to act as **donors** of free electrons or as **n-type dopants** in this case.

Conversely, **p-type dopants** have only <u>three</u> electrons in the outer shells, which makes them act as **acceptors** for free electrons. Each dopant atom is then said to create a **hole**, that is a vacant electron position or — which is the same — a positive unit charge, see fig.14.3c. Observe that the presence of holes in the crystal lattice makes it possible for electrons to jump into vacant positions, leaving fresh holes behind. If the process is kept going by an electrical field, the effect can be viewed as a current of (positive) holes that has the same magnitude but flows in the opposite direction to that in which the (negative) electrons do.

---

[1]   Appendix D.3 lists the bandgap energies of selected materials and also discusses carbon allotropes.

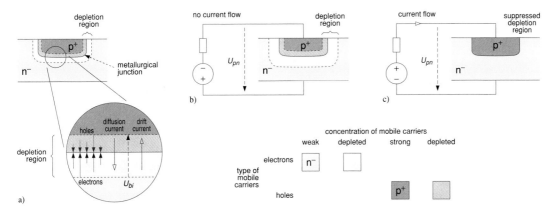

Figure 14.4 The pn-junction in thermal equilibrium with customary orientation of voltage counting (a), in backward-biased (b), and in forward-biased condition (c) (not drawn to scale).

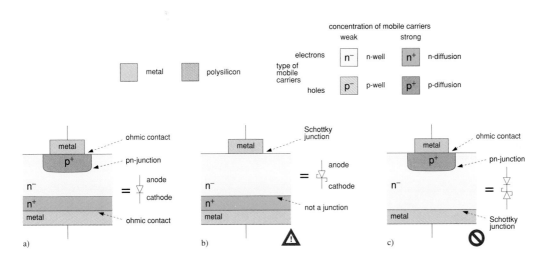

Figure 14.5 Cross sections of diodes. (Rectifier) diode (a), Schottky diode (b), and a malformed device that would result if a high-implant layer were forgotten (c) (not drawn to scale).

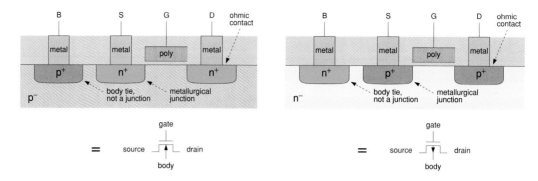

Figure 14.6 Cross sections of bulk MOSFETs. n-channel (a) and p-channel (b) device (not drawn to scale).

Figure 14.7 n-channel device in thermal equilibrium.

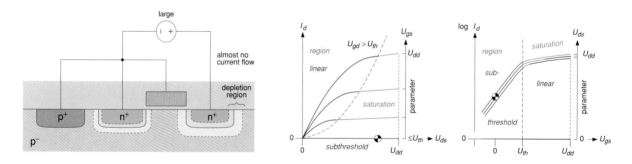

Figure 14.8 n-channel device in cutoff condition (the ⊕ mark identifies an operating point).

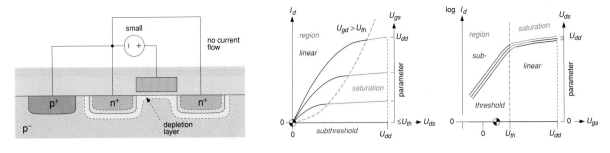

Figure 14.9 n-channel device in weak inversion.

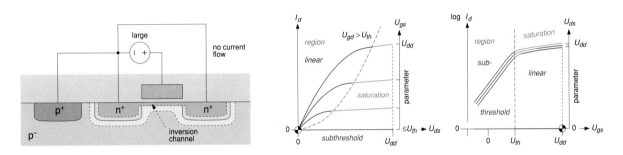

Figure 14.10 n-channel device in strong inversion.

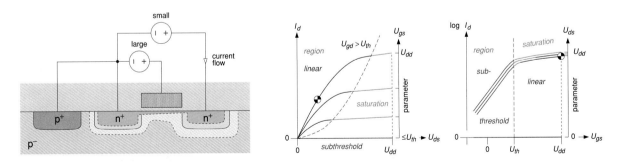

Figure 14.11 n-channel device in linear regime.

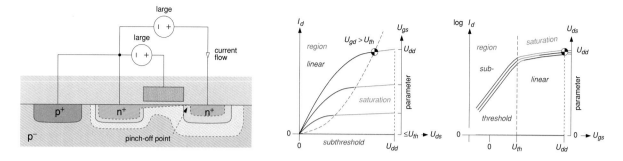

Figure 14.12 n-channel device at pinch off.

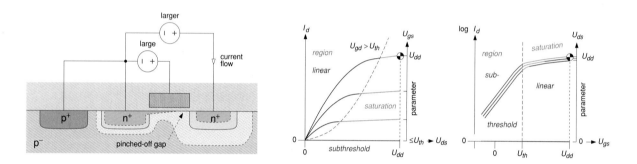

Figure 14.13 n-channel device in saturated condition.

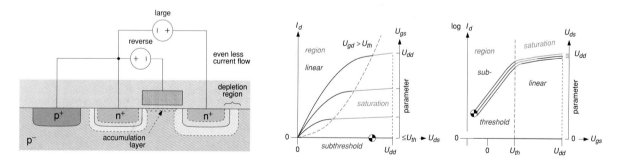

Figure 14.14 n-channel device in super-cutoff condition.

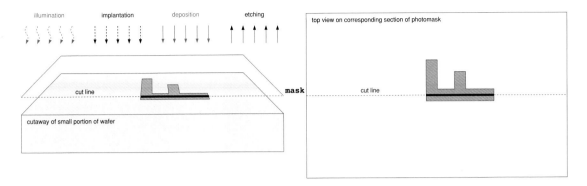

illumination   implantation   deposition   etching

top view on corresponding section of photomask

cut line

mask

cut line

cutaway of small portion of wafer

Figure 14.15 3D view of the situations illustrated in figs.16 through 19.

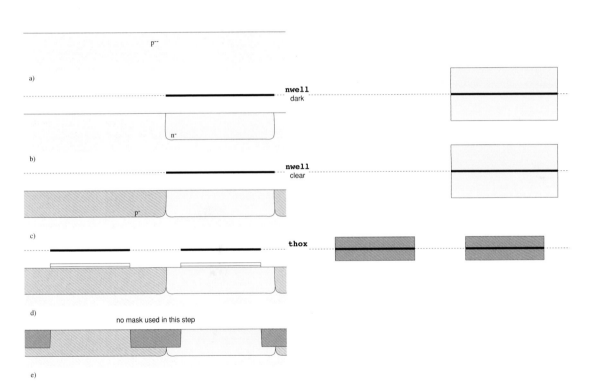

a)

p⁻⁻

nwell
dark

b)

n⁻

nwell
clear

c)

p⁻

thox

d)

no mask used in this step

e)

Figure 14.16 CMOS fabrication steps. Well preparation (greatly simplified, not drawn to scale.).

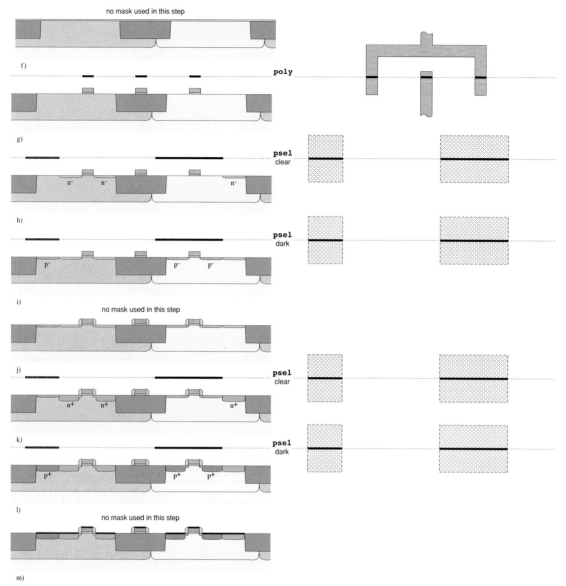

no mask used in this step

f)

poly

g)

psel
clear

h)

psel
dark

i)

no mask used in this step

j)

psel
clear

k)

psel
dark

l)

no mask used in this step

m)

Figure 14.17 CMOS fabrication steps. Transistor formation (greatly simplified, not drawn to scale.).

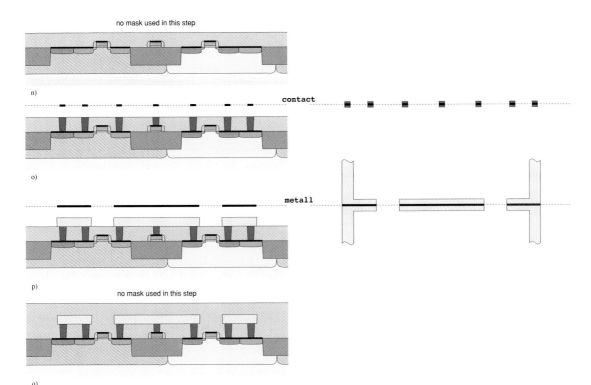

no mask used in this step

n)

contact

o)

metal1

p)

no mask used in this step

q)

Figure 14.18 CMOS fabrication steps. Lower metal interconnects (greatly simplified, not drawn to scale.).

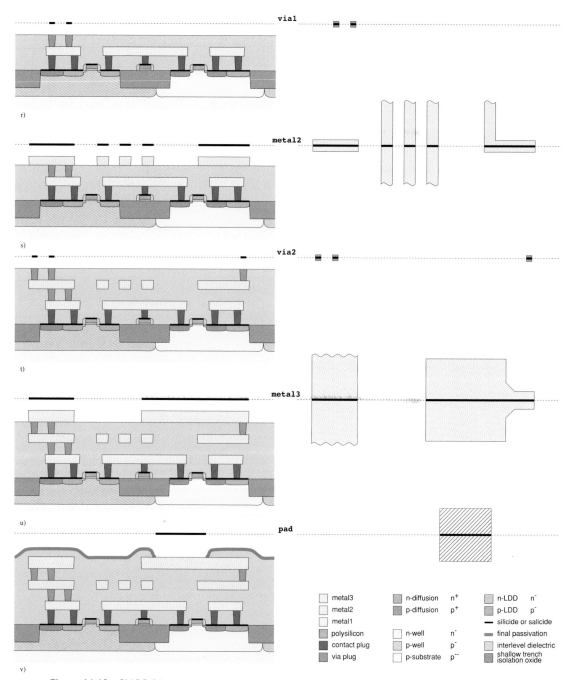

Figure 14.19 CMOS fabrication steps. Upper metal interconnects (greatly simplified, not drawn to scale.).

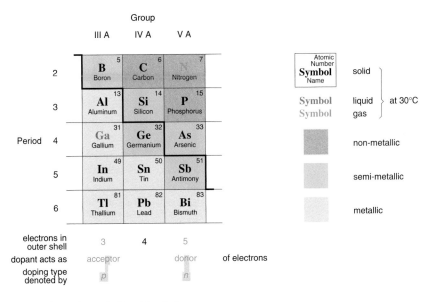

Fig. 14.2 Excerpt from the periodic table of elements.

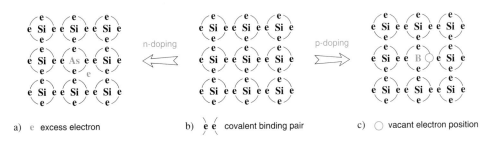

a) e excess electron   b) e e covalent binding pair   c) ○ vacant electron position

Fig. 14.3 n-doped (a), undoped (b), and p-doped (c) silicon (symbolic).

Observation 14.1. *Both holes and excess electrons contribute to electrical conductivity, what really matters for electrical conductivity is their imbalance.*

Doping essentially provides the crystal with mobile charge carriers. A semiconductor where holes and excess electrons exactly compensate has the electrical properties of an undoped material. The stronger the concentration of mobile carriers, the higher the material's conductivity. There is a difference, however, in that it takes less energy to move an electron through the crystal lattice than to move a hole. Electrons are thus said to have a higher **carrier mobility**.[2]

---

[2] A classical analogy is that of a banquet table where a seat stands for a place in the crystal lattice and a guest for an electron. There is little social interaction as long as the numbers of seats and people match, but an excess of seats encourages people to drift around. An excess of guests engenders even more agitation.

The manufacturing of semiconductor devices necessitates a set of doping steps to obtain volumes of well-defined electrical characteristics and shapes. Technologists capture a material's doping in a shorthand notation by writing things like n^+ or p^-, where n and p stand for the doping type while $+$, $-$, and $--$ refer to strong, weak, and very weak doping concentrations respectively.[3]

A particular material that is worth mentioning is **polysilicon**, or **poly** for short, a polycrystalline or amorphous silicon material typically deposited on a wafer's surface from a gas. Depending on the nature and concentration of its doping, a polysilicon film can be made into a conductive layer, a resistive layer, or a semiconductor. CMOS VLSI circuits include one or sometimes two polysilicon layers that are made conductive by generous addition of dopants.

14.1.3 Junctions, contacts, and diodes

All semiconductor devices include both junctions and contacts, so developing a basic understanding of their construction and electrical properties is essential.

pn-Junction. A pn-junction is obtained by implanting p-type dopants into some n-type base material, or vice versa, as shown in figs.14.4 and 14.5. The term **metallurgical junction** refers to the borderline that separates the p-type from the n-type region where the concentrations of donors and acceptors balance.

When the junction is in **thermal equilibrium**, i.e. with no voltage applied and with no light shining on it, a region depleted of mobile carriers forms on either side of the metallurgical junction, see fig.14.4a. This is because charge carriers tend to **diffuse** from regions of high concentration to regions of lower concentration. Free electrons from the n-region thus diffuse

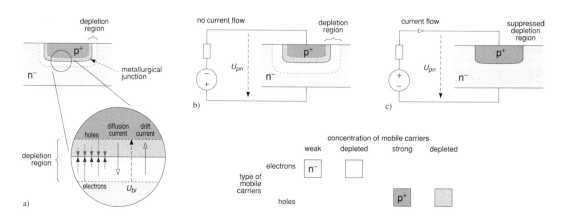

Fig. 14.4 The pn-junction in thermal equilibrium with customary orientation of voltage counting (a), in backward-biased (b), and in forward-biased condition (c) (not drawn to scale). See color plate section.

---

[3] [381] proposes a notation that distinguishes among five levels of doping:
very weak (denoted n^{--} [p^{--}]) if the concentration of dopant atoms is $N_{D[A]} < 10^{14}$ cm^{-3},
weak (n^-[p^-]) if $N_{D[A]}$ is 10^{14}–10^{16} cm^{-3}, moderate (n [p]) if $N_{D[A]}$ is 10^{16}–10^{18} cm^{-3},
strong (n^+ [p^+]) if $N_{D[A]}$ is 10^{18}–10^{20} cm^{-3}, very strong (n^{++} [p^{++}]) if $N_{D[A]} > 10^{20}$ cm^{-3}.

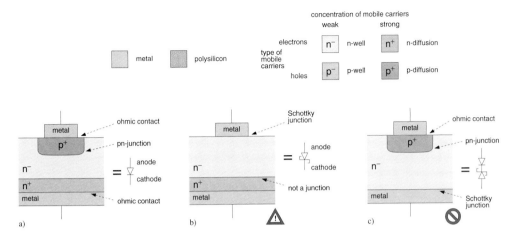

Fig. 14.5 Cross sections of diodes. (Rectifier) diode (a), Schottky diode (b), and a malformed device that would result if a high-implant layer were forgotten (c) (not drawn to scale). See color plate section.

towards the p-region, and vice versa. When they meet, an electron can drop into a hole, causing both to vanish in an event called **recombination**.

What they leave behind is a **depletion region**, aka space-charge region, where the immobile dopant atoms are ionized. The number of ionized donors on the n-side equals the number of ionized acceptors on the p-side. An equilibrium point is reached when the electric field created by those ionized impurities causes carriers to **drift** back at the same rate as they diffuse away. The potential difference across the junction present under thermal equilibrium conditions is called **built-in voltage** U_{bi}, is always negative, and is on the order of -0.7 V for a typical silicon pn-junction.[4]

Observation 14.2. *In a semiconductor, one observes two antagonistic currents that compensate when in thermal equilibrium:*

cause	*primary effect*	*secondary effect*
nonuniform carrier concentration	*diffusion current*	*builds up potential difference*
electric field	*drift current*	*attenuates potential difference*

A pn-junction is said to be **reverse biased** when an external voltage $U_{pn} < 0$ gets applied, see fig.14.4b. The depletion region widens because the voltage draws electrons and holes away from each other. This explains why (almost) no current flows in a reverse-biased condition $I_r \approx 0$. The junction then electrically resembles a small capacitor with the depletion regions acting as a dielectric.[5,6]

---

[4] Diode voltages and currents are counted as positive when the device is in forward-biased condition. It is unfortunately not possible to measure the built-in voltage by connecting a voltmeter to the diode's terminals because the various contact potentials in a closed loop cancel out. See [234] for a more complete rationale.

[5] As the negative bias is increased, the depletion regions extend, thereby widening the gap that separates the two imaginary capacitor plates, and the effective capacitance value diminishes. This is how **varicap** diodes work.

[6] Any junction will fail when a sufficient reverse voltage is applied. Here is why: A few thermally generated electrons and holes always exist in the depletion region. At higher voltages, these will get accelerated to a point

A **forward bias**, on the other hand, pushes electrons in the n-doped material and holes in the p-side towards the metallurgical junction, thereby compressing the depletion region. The layer devoid of mobile carriers vanishes when the forward voltage applied from externally more than compensates for the junction's negative built-in voltage, see fig.14.4c. Electrons and holes then combine at the junction to form a continuous current. For a significant forward current to flow $I_f > 0$; we thus have the condition $U_{pn} + U_{bi} > 0$. In more practical terms, one must accept a voltage drop of $U_f > -U_{bi} \approx 0.7$ V for a typical silicon diode to conduct.

Observation 14.3. *The most prominent property of a junction is that electrical current is allowed to flow from the p-doped anode to the n-doped cathode, but not in the opposite direction.*

Schottky junction. Metal deposited on lightly- or moderately-doped semiconductor material exhibits electrical characteristics that are similar to those of a pn-junction (with the metal acting as anode when adjoined to a n^- material and as cathode when adjoined to p^- material). The most notable difference is the low built-in voltage of a mere -0.3 V or so. The popular name "Schottky contact" for this arrangement is misleading because a diode characteristic is definitely not what one expects from a contact. Schottky junctions find applications in power electronics and fast TTL logic families.

Ohmic contact. Ohmic contacts between metals and semiconductors are obtained by heavily doping the semiconductor material where it adjoins the metal. Connecting to lightly-doped n^- material, for instance, requires that a heavily-doped n^+ zone be sandwiched between the metal and the n^- volume. This is shown at the bottom of figs 14.5a and b that clarify the construction of a rectifier and of a Schottky diode respectively.

14.1.4 MOSFETs

CONSTRUCTION

Field effect transistors (FETs) exploit the fact that the current through a thin layer of semiconductor material can be modulated by the electrostatic field originating from an electrode placed in the immediate proximity. The controlling electrode is called the **gate**. As opposed to junction FETs (JFETs), **MOS field effect transistors** (MOSFETs) have their gate electrode electrically insulated from the conducting channel by a thin dielectric layer known as gate oxide, thin oxide, or — more generally — **gate dielectric**. The acronym **MOS** refers to the materials in the **gate stack**, that is, in the sandwich structure at the heart of these devices:

- Metal for the gate electrode,
- Oxide (traditionally SiO_2) for the gate dielectric, and the
- Semiconductor material (Si) underneath in which the conducting channel forms.

---

where they create new electron–hole pairs when they collide with atoms of the semiconductor lattice. The diode enters **avalanche breakdown** when the carriers brought forth by this **impact ionization** process cause a chain reaction that floods the depletion region. The depletion region then literally collapses and the junction loses its ability to block current. Breakdown per se is not necessarily destructive, yet the heat generated can cause damage unless reverse currents are limited to uncritical levels. Both avalanche and **Zener diodes** take advantage of breakdown effects. The difference is a fine point that we abstract from here, all the more as the two terms are often used interchangeably.

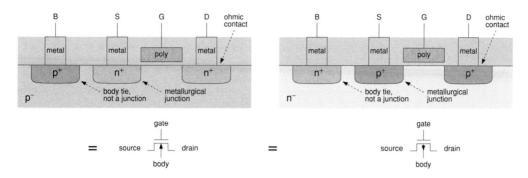

Fig. 14.6 Cross sections of bulk MOSFETs. n-channel (a) and p-channel (b) device (not drawn to scale). See color plate section.

Note that the acronym has become a misnomer because the MOSFETs found in almost all VLSI chips today have their gate electrodes made of polysilicon rather than metal.[7] Also, traditional silicon dioxide gate dielectrics are more and more being displaced by materials with higher relative permittivities.

With some simplifications, fig.14.6 shows the cross sections through complementary MOSFETs fabricated in a bulk process. The **source** and **drain** electrodes at the ends of the channel consist of two identical diffusion islands embedded in a substrate material of opposite doping. A third diffusion island doped like the substrate but with a higher concentration serves to electrically contact the transistor's **body** volume, aka bulk.

OPERATION

To gain insight into the functioning of MOSFETs, let us watch where and why charge carriers concentrate and flow in such devices under a variety of electrical conditions. More specifically, the subsequent discussion refers to an enhancement-type n-channel MOSFET with no substrate bias $U_{bs} = 0$. That is, we will assume source and body terminals to be electrically connected as is the case in inverters, for instance. Analysis of a p-channel device is analogous but with dopings, charge carriers, currents, and voltages reversed.

Thermal equilibrium. As shown in fig.14.7, depletion regions form around the metallurgical junction of the source and of the drain islands. Drain and source are electrically insulated from each other by two zero-biased junctions connected back to back.

---

[7] Aluminum gates were being used in the venerable CD4000/MC14000 logic family and in NMOS parts before being abandonded in favor of poly gates in the 74C00 and other CMOS families. A first reason was the smaller work function of polysilicon, which contributed to lower MOSFET threshold voltages. It thus became possible to reduce the operating voltage (CD4000: 3–18 V, 74C00: 2–6 V) and to improve on energy efficiency without inflating delays. Perhaps even more importantly, the adoption of self-aligned gates simplified the fabrication process but mandated a gate material that can withstand the high-temperature processing associated with the implantation of dopants and with annealing, a requirement not met by aluminum. This is because the manufacturing of self-aligned gates implies that gate electrodes be patterned before source/drain doping occurs, see section 14.2.2. Interestingly, metal gates are to reappear with the 45 nm generation in order to do away with undesirable phenomena that are associated with polysilicon as a gate material, see section 14.3.5.

Fig. 14.7 n-channel device in thermal equilibrium. See color plate section.

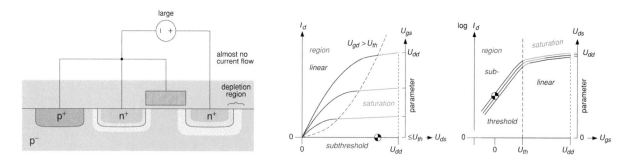

Fig. 14.8 n-channel device in cutoff condition (the ⊕ mark identifies an operating point). See color plate section.

Cutoff. Enhancement-type n-channel MOSFETs exhibit positive threshold voltages $0 < U_{th}$. At gate voltage zero, $0 = U_{gs}$, such a device operates well within the **subthreshold region** as defined by $(U_{gs} - U_{th}) \leq 0$. Upon applying a positive drain voltage, the drain junction gets reverse-biased and essentially no current flow takes place between drain and source, $I_d(0 = U_{gs}, 0 < U_{ds}) \approx 0$. Ideally, the only effect is that the depletion region around the drain electrode grows wider than that around the source as shown in fig.14.8.

In real devices, perfect cutoff is frustrated by **subthreshold conduction**, a phenomenon due to free electrons that always exist in the bulk material. Only a small fraction has enough thermal energy to traverse the barrier of the reverse-biased drain junction. Still, such electrons make up a minute drain-to-body current $I_d(0 = U_{gs}, 0 < U_{ds}) > 0$ referred to as subthreshold or **channel leakage** current that augments with temperature.

Weak inversion. A positive voltage gets applied to the gate. Yet, it is so small that the transistor remains in the subthreshold region because $(U_{gs} - U_{th}) \leq 0$ continues to hold. The main difference is that the holes in the p⁻ body material underneath the gate are being repelled from the surface by the slightly positive gate $0 < U_{gs} \leq U_{th}$. The negatively charged acceptor atoms that are left behind form a depletion layer, see fig.14.9.[8]

Strong inversion. A larger positive gate voltage $U_{th} < U_{gs}$ attracts so many electrons to the silicon surface underneath that a thin layer gets flooded with negative mobile carriers. The material there, originally manufactured with a gentle p⁻ doping, therefore behaves much like

---

[8] Were it not for subthreshold conduction, drain current would remain zero even if a substantial voltage were applied between drain and source such as in fig.14.8.

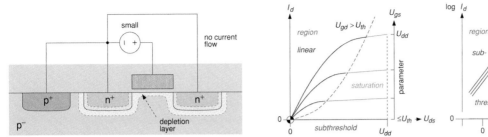

Fig. 14.9 n-channel device in weak inversion. See color plate section.

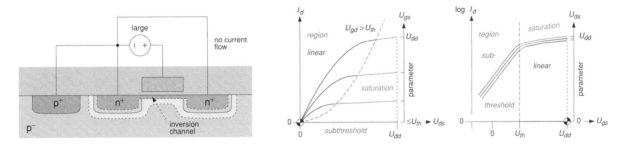

Fig. 14.10 n-channel device in strong inversion. See color plate section.

n$^+$ material, and the name "strong inversion" aptly describes its situation. As net result, a conducting inversion channel bridges the gap between source and drain, see fig.14.10. The MOSFET essentially behaves like a (linear) resistor the conductance of which is controlled by playing on the voltage $(U_{gs} - U_{th})$ to vary the thickness of the inversion layer. Yet, no current flows in the absence of a driving voltage $I_d(0 = U_{ds}) = 0$.

Note that the device can be viewed as a capacitor the upper plate of which is the gate electrode and where the inversion layer forms the lower plate. Not surprisingly, a transistor operated in this way is referred to as a MOSCAP.

Observation 14.4. *The conducting channel that comes into existence in p-doped bulk material is of n-type, and vice versa, as a consequence of field-induced inversion.*

Linear regime. What's new compared with the previous situation is a drain–source voltage that propels free electrons through the resistive inversion layer. Current flows from drain to source $I_d(U_{th} < U_{gs}, 0 < U_{ds}) > 0$. There is a side effect, however, because the voltage between gate and drain is less than that between gate and source. The vertical field thus tapers off towards the drain and so does the thickness of the inversion layer, see fig.14.11. The linear region is defined by $0 < U_{ds} < (U_{gs} - U_{th})$. Also observe that electrical linearity suffers towards the borderline of the region as a consequence of the inversion channel becoming more and more distorted.

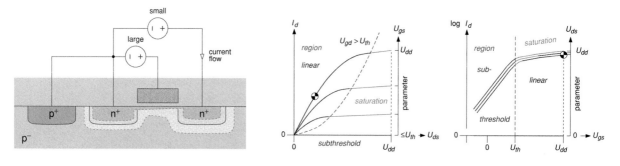

Fig. 14.11 n-channel device in linear regime. See color plate section.

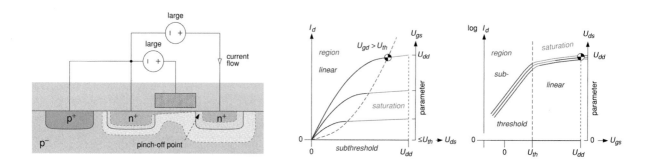

Fig. 14.12 n-channel device at pinch off. See color plate section.

Pinch off. When the voltage drop along the channel is augmented to a point where $U_{th} = U_{gd}$, the channel becomes pinched off next to the drain because the vertical field no longer suffices to obtain strong inversion there, see fig.14.12. Note that $U_{th} = U_{gd}$ is synonymous with $U_{ds} = (U_{gs} - U_{th})$ simply because $U_{gd} = U_{gs} - U_{ds}$. That condition defines the borderline between the linear and the saturation region.

Saturation. Increasing drain-to-source voltage further so that $U_{th} > U_{gd}$, a continuous channel ceases to exist, and one would expect current flow to come to a standstill. Somewhat surprisingly, though, charge transport continues. This is because the lateral field in the narrow pinched-off gap becomes so intense that the few electrons present there get so strongly accelerated towards the drain that they tunnel through the gap. As shown in fig.14.13, a further increase of the drain-to-source voltage justs widens the gap and makes the two effects cancel out. Ideally, current should saturate at a maximum value $I_d = I_{d\,sat}$.

In reality, however, one finds that drain current continues to grow with drain voltage, although only to a minor degree. The reason is that a higher voltage not only expands the pinched off gap but at the same time also widens the depletion region around the drain, thereby eating away from the effective channel length. A slightly shorter channel exhibits a lower resistance which, in turn, allows for more current. The effect is known as **channel length**

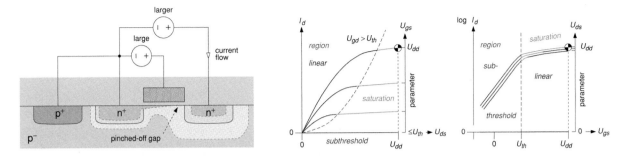

Fig. 14.13 n-channel device in saturated condition. See color plate section.

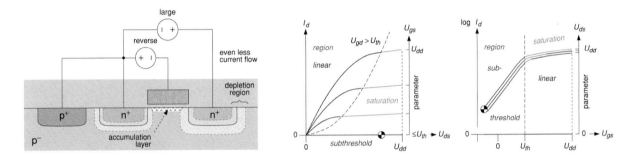

Fig. 14.14 n-channel device in super-cutoff condition. See color plate section.

modulation in the context of MOSFET devices but is sometimes also the referred to as Early effect, a term originally created to describe a similar dependence of collector current on the widening of depletion regions in BJTs. The saturation region as a whole is defined by $0 < (U_{gs} - U_{th}) \leq U_{ds}$.

Super cutoff. As opposed to regular cutoff, a small negative bias is applied to the gate $U_{gs} < 0$, thereby attracting holes to the silicon surface as illustrated in fig.14.14. Because of the reverse-biased drain junction, this **accumulation** effect has little impact. More importantly, any thermal electrons present in the body volume between drain and source get swamped by the reverse field. The fact that subthreshold current exponentially depends on the gate–source voltage makes underdriving the gate a highly effective countermeasure against channel leakage. In fact, subthreshold current grows (lessens) by a factor of 10 to 15 for each 100 mV increase (decrease) of $(U_{gs} - U_{th})$.

Observation 14.5. *The voltage applied to the gate electrode of a MOSFET controls the presence of mobile carriers in the channel area underneath, thereby varying conductivity and current flow in the drain–source channel by several orders of magnitude. Also, the close-to-perfect insulation between gate and channel makes the MOSFET appear as a capacitive load.*

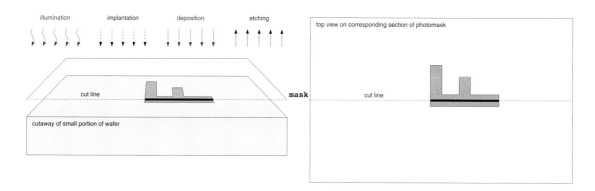

Fig. 14.15 3D view of the situations illustrated in figs.14.16 through 14.19. See color plate section.

Detailed discussions of MOSFET physics are available from textbooks and webpages specializing in this field including [231] [382] [236] [220] [383] [384]. [234], in particular, not only gives a very thorough analysis, but also provides the reader with a vivid analogy from fluid dynamics for a MOSFET's regimes of operation. A real pleasure is [385] because it illustrates the operation of various semiconductor devices with the aid of interactive animations.

14.2 | Basic CMOS fabrication flow

14.2.1 Key characteristics of CMOS technology

The acronym **CMOS** stands for "complementary metal oxide semiconductor". The technology is termed **complementary** because both n- and p-channel MOSFETs are being manufactured side by side in a common silicon substrate and because they closely cooperate in each logic gate. This contrasts with the PMOS and NMOS technologies from the pioneer days of microelectronics that were limited to either p- or n-channel devices.

CMOS has long been and continues to be the technology that dominates VLSI and this section describes the major steps in CMOS wafer processing. Illustrations and explanations refer to a hypothetical fabrication process that exhibits the features listed below and is representative of the 250 nm to 130 nm generations.

- A bulk process (as opposed to a silicon-on-insulator (SOI) technology).
- Twin wells (as opposed to single or triple wells).
- Shallow trench isolation (STI) (as a replacement
  for the obsolete local oxide isolation (LOCOS) technology).
- Planarization by way of chemical mechanical polishing (CMP).
- Poly(silicon) gate material (as opposed to metal gates).
- Lightly-doped source/drain extensions.
- Salicided gate, drain, and source areas.
- Contacts and vias fabricated with tungsten plugs (as opposed to sunk-in aluminum).
- Subtractive metallization (as opposed to (dual) damascene metallization).

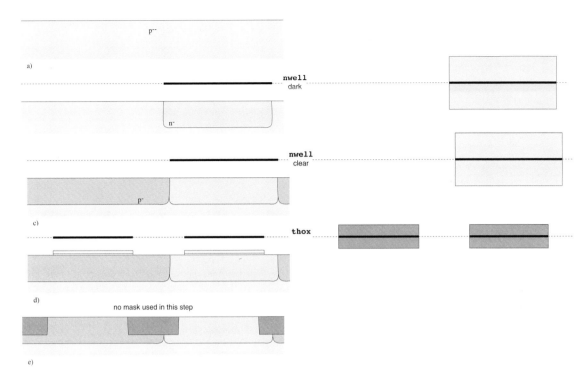

Fig. 14.16 CMOS fabrication steps. Well preparation (greatly simplified, not drawn to scale.). See color plate section.

Many technical details have been abstracted from. In reality, layout geometries are transferred onto silicon in a series of steps such as coating a wafer with a photoresist material, baking, photographic exposure through a photomask or reticle and selective removal of either the exposed or the unexposed portions of the resist.[9] Also, certain doping steps comprise a thermal drive-in of the dopants previously implanted and/or some form of annealing to remove the damage caused by ion implantation. For the sake of conciseness, all such substeps have been collapsed in the illustrations and in most of the accompanying text.[10]

Illustrations show an inverter that includes the mandatory body ties, one power line, a few signal lines, and a small pad opening just wide enough to accommodate a probe. Each processing step is documented with a cross section and the name and geometry of the pertaining fabrication mask. Figures 14.16 and 14.17 are concerned with the so-called front-end-of-line (FEOL) processing where the transistors are being formed in the silicon material, whereas figs.14.18 and 14.19 relate to the back-end-of-line (BEOL) steps during which those devices get interconnected with the aid of metal

---

[9] For **positive photoresists**, those zones that have received a sufficient dose of light become completely soluble in the developer while unexposed areas essentially remain intact. The opposite applies for **negative photoresists**. Further observe that some process steps necessitate auxiliary oxide or nitride layers into which the mask pattern is etched before being copied into the material underneath in a subsequent processing step.

[10] Further details abstracted from include wafer preparation and preprocessing, epitaxial layers, lateral diffusion phenomena, channel stops, liners, and dual-doped poly (n^+ vs. p^+). The omission of multiple threshold voltages (to allow compromising between current drive and gate leakage), of threshold adjustment steps, and of gate dielectrics with multiple thicknesses (to make I/O circuitry withstand higher voltages than the core logic) are further simplifications.

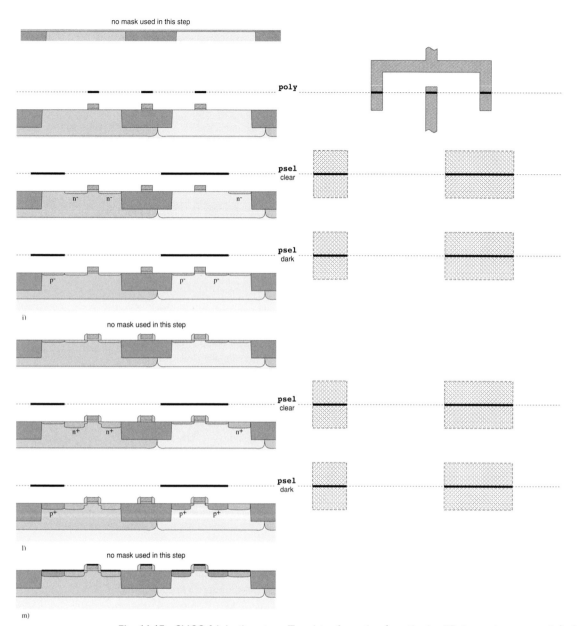

Fig. 14.17 CMOS fabrication steps. Transistor formation (greatly simplified, not drawn to scale.). See color plate section.

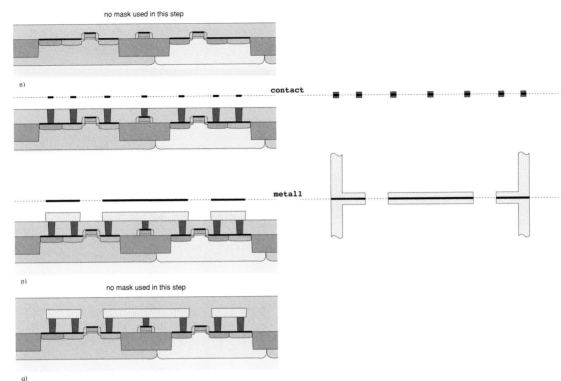

Fig. 14.18 CMOS fabrication steps. Lower metal interconnects (greatly simplified, not drawn to scale.). See color plate section.

lines. For the sake of clarity, the number of metal layers has been confined to three in the drawings although six to eight metals are more standard for 180 nm processes.

14.2.2 Front-end-of-line fabrication steps

a) Initial wafer. Processing is assumed to start from a p-type wafer (roughly 0.72 mm to 0.77 mm thick and with a diameter of either 200 or 300 mm) that carries an epitaxial layer (a few μm thick) of very gentle p-type doping. All subsequent fabrication steps are going to happen within and above this epi layer.

b) n-well formation. The first mask being used is `nwell`. It defines the geometry of the n-wells that will later accommodate the p-channel MOSFETs. Donor atoms are implanted (down to a depth of 1.8 μm or so) where indicated by that mask.

c) p-well formation. The p-wells are obtained as geometric complements of their n-type counterparts. To this end, the `nwell` mask is reused in a second lithographic step in conjunction with a photoresist of opposite polarity. The cross section shows the outcome after dopant implantation and a subsequent thermal drive-in step.

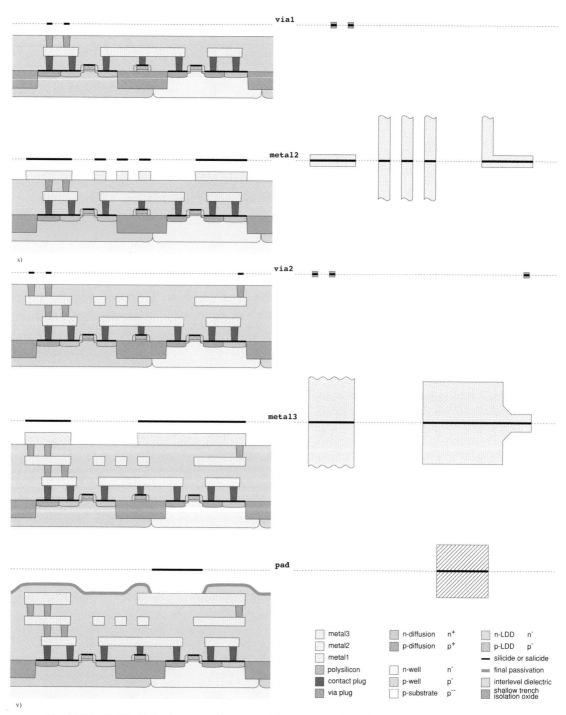

Fig. 14.19 CMOS fabrication steps. Upper metal interconnects (greatly simplified, not drawn to scale.). See color plate section.

d) **Active area definition.** The subsequent mask `thox` essentially defines those areas that are to become MOSFETs, body ties, or other diffusion areas. Silicon nitride is deposited on top of a thin buffer oxide layer. Both layers are initially made to cover the entire surface before being etched away except where indicated by the mask.

e) **Shallow trench isolation (STI).** Next, trenches (approximately 300 to 400 nm deep) are cut into the bulk material with the nitride layer protecting the silicon underneath from etching. In a series of substeps, the open trenches are then lined and filled with an oxide material before the wafer surface is planarized by way of **chemical mechanical polishing**. Obtaining a perfectly flat surface is crucial for the subsequent photolithography steps. CMP combines a chemical reaction with mechanical abrasion to produce the most planar surface of any known technique.[11] The protective sheets of nitride and buffer oxide are also removed in the process, which results in the cross section shown.

f) **Gate dielectric formation.** The wafers are subjected to a delicate oxidation step which produces an extremely thin oxide layer (2 to 5 nm) wherever the bare silicon lies open.

g) **Polysilicon deposition and patterning.** The entire surface is covered with a polysilicon film (of roughly 200 nm) which is then patterned after the `poly` mask. The field-controlled **channel areas** will later form in the bulk material wherever it is separated from a polysilicon structure by a layer of gate dielectric.

h) **n-channel source/drain extensions.** Basically, the wafer is now ready for the n^+ and p^+ implantation steps because all areas that are occupied neither by polysilicon nor by STI at this point are destined to become MOSFET sources, drains, or body ties. Were it not for the source/drain extensions, processing could immediately continue with step j).

These lightly doped protractions serve to avoid excessive lateral fields and to better control short-channel effects (SCE) that would otherwise occur in submicron devices. During implantation, those regions that are to be bombarded with donors are defined by photomask `psel` in clear form. Note that the polysilicon gate acts as a shield that keeps dopant atoms out of the channel area. The edges of the future source and drain areas are thus automatically kept in perfect alignment with the gate electrode above, which benefit has earned poly-gate MOS processes the attribute **self-aligned gate**.

i) **p-channel source/drain extensions.** Essentially the same as before with opposite photoresist and opposite doping. By being used in clear and then in dark form, photomask `psel` effectively defines the boundary between n^+-doped areas and p^+-doped ones.

j) **Side-wall or oxide spacers.** An oxide layer is deposited and etched away so as to leave an insulating wall on either face of the poly gate.

---

[11] Broadly speaking, the chemicals soften up the film to be removed and the slurry particles then carry away the softened material. The advantage of CMP over purely mechanical polishing is that the chemicals can be chosen such as to attack specific materials while sparing others.

k) n-type doping. Now follows a heavier and deeper implant (roughly 180 to 200 nm) that brings forth the drains and sources of the n-channel MOSFETs and the body ties for the n-wells. The side-wall spacers prevent doping atoms from penetrating into the bulk material and so preserve the lightly doped implants next to the channel areas. The n-channel MOSFETs with their source/drain extensions are now complete. Recall that a junction is obtained where an n^+ area is implanted into p^- material whereas an ohmic contact results when an n^+ island is embedded within n^- material.

l) p-type doping. The same as before but with everything reversed.

m) Salicidation. This step serves to lower the electrical resistivity of the source, drain, and gate regions by covering them with a thin but highly conductive silicide film. The so-called salicide approach — an abbreviation for self-aligned silicide — is particularly attractive because no masking step is required and because diffusion and poly get silicided at the same time. The process involves depositing a thin Ni, Co or Ti film and uses a metal reaction with silicon followed by a selective etch that removes all unreacted metal.

14.2.3 Back-end-of-line fabrication steps

n) First interlevel dielectric. A generous layer of silicon dioxide gets deposited with no intervention from any mask. Following another CMP planarization step, this insulating material is to become the first interlevel dielectric (ILD) providing electrical separation between the bulk structures and the first layer of metal.

o) Contact plug formation. Mask `contact` defines those locations where metal1 shall connect to a diffusion or polysilicon region underneath. Cuts are anisotropically etched open where indicated before tungsten is deposited to form plugs. The excess tungsten is then removed and the surface planarized to prepare for the subsequent metallization step.

p) Deposition and patterning of first metal layer. A metal layer (300 nm or thicker) is deposited over the entire surface. Much of this layer is then selectively removed so as to leave behind those parts that are defined by the `metal1` mask, which explains why this step is qualified as **subtractive metallization**. More precisely, metal deposition includes a series of substeps to form a metal stack where aluminum, copper or tungsten is sandwiched between thin liners of metal nitrides or other materials that improve adhesion, abate the formation of hillocks, act as diffusion barriers, or otherwise help suppress undesirable phenomena. Such details are abstracted from in the figure.

q) Second interlevel dielectric. The second interlevel dielectric is deposited and planarized.

r) Via plug formation. Mask `via1` defines those locations where a first-layer-metal structure shall connect to the next metal above. Much as for contact formation, a cut is etched open, then filled with tungsten, and the excess material is removed and the surface planarized.

s) Deposition and patterning of second metal layer. The second layer of metal is obtained by way of subtractive metallization.

t) Third interlevel dielectric followed by via plug formation. From this point on, the steps of dielectric deposition, planarization, plug formation, metal deposition, and metal patterning alternate for each additional metal layer. Note that two masks are required per metal layer. The figure shows the situation after the `via2` plugs have been formed.

u) Deposition and patterning of third metal layer. The third metal layer — which is the topmost one in the case shown — is now complete. As a rule, higher-level metals are fatter (up to 1 μm) and must respect more important minimum widths and spacings than their lower-level counterparts.

v) Overglass and bond pad openings. The entire wafer surface is covered by depositing a final layer of silica. The overglass is complemented by a final passivation layer placed on top to better protect the stack from environmental attacks such as humidity, chemical agents, and scratching. To allow bonding and probing of the chip, generous openings get etched into the protective layers where indicated by the **pad** mask. Note that only the topmost metal can be contacted in this way.

Please keep in mind that there exist many variations to the basic CMOS process described in the above series of drawings. Most likely, the reader will have to make adaptations to account for departures in a specific fabrication process at hand. Comprehensive and authoritative references on VLSI technology include [386] [387] [381] [388] [389]. What sets [219] apart from others is a comparison of CMOS and BiCMOS. Photographs and/or computer animation are available from websites such as [385].

14.2.4 Process monitoring

Between processing steps, wafers are subject to optical inspection and electrical tests. The electrical characteristics of MOSFETs of different sizes are measured and kept on record, and the same applies for an assortment of pn-junctions. Capacitance–voltage characteristics are obtained from MOSCAPs. Van der Pauw structures serve to determine the resistivities of all conductive layers while various shapes of capacitors are included to monitor the dielectric layers. Long interdigitated serpentines and chains of many series-connected contacts or vias help to check for electrical shorts and continuity. Such elementary test structures are collected into **process control monitors** (PCMs) along with inverters and ring oscillators.

Wafers that are found to suffer from fatal defects or from excessive parameter variations are sorted out. Even more importantly, the data gathered are used to constantly monitor fabrication equipment and procedures.

14.2.5 Photolithography

The manufacturing process illustrated in figs.14.17 through 14.19 heavily relies on the ability to transfer layout patterns from a photomask or a reticle[12] onto a semiconductor wafer. Any photolithographic apparatus comprises four major parts, see fig.14.20:

---

[12] A **photomask** is meant to carry the layout patterns for an entire wafer whereas patterns from a **reticle** are to be stepped and repeated many times over the wafer's surface. As this is of little importance to IC designers, we will indiscriminately speak of masks, although reticles prevail today.

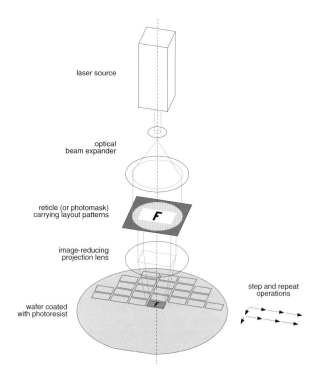

Fig. 14.20 Apparatus for optical lithography (greatly simplified).

- An illumination source.
- A mask either the same size as chip patterns (1×) or magnified by some factor (4× or 5×).
- An exposure subsystem in charge of geometric reproduction.
- Photoresist materials, typically organic polymers.

Traditional optical lithography has had a long history in VLSI. It implies projection printing with image-reducing refractive optics (mostly lenses) from a mercury (Hg) arc lamp. Emission occurs at the following wavelengths:

546 nm E-line.
436 nm G-line, used between 1980 and 1990 or so.
405 nm H-line.
364 nm I-line, used between 1989 and 1998 or so.

Deep UV lithography carries the principles of projection printing with image-reducing refractive optics to shorter wavelengths obtained from an excimer laser [390], namely

248 nm (KrF_6), in production for 250, 180, and 130 nm processes.
193 nm (ArF_6), in production for 90 and 65 nm, probably workable
       for the manufacturing of devices as small as 32 nm or even 22 nm.
157 nm (F_2), laser sources demonstrated, no longer under consideration.

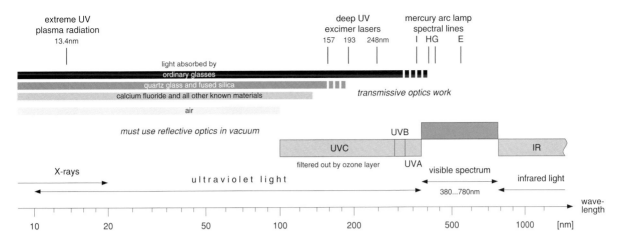

Fig. 14.21 Spectrum of light with wavelengths relevant to VLSI lithography.

Shorter wavelengths get strongly absorbed and poorly reflected by most materials, see fig.14.21. The quartz and fused silica materials from which lenses, beamsplitters, prisms, and masks are traditionally manufactured are opaque below 190 nm or so. For some time, calcium fluoride CaF_2 had been considered for optical components at a wavelength of 157 nm but the material is soft, hygroscopic, prone to chipping, and features a thermal expansion 36 times greater than that of fused silica. Currently, 157 nm lithography is no longer seen as a solution [1].

RESOLUTION ENHANCEMENT TECHNIQUES

Pushing lithographic resolution further ran into severe limitations when layout dimensions had shrunk to become approximately the same as the optical wavelength being used, see fig.14.22. The term resolution enhancement techniques (RETs) refers to a variety of approaches that aim at obtaining a higher resolution from a given wavelength.

Phase shift masks

Light that shines through adjacent apertures separated by a narrow bridge tends to diffract, thereby illuminating the supposedly shaded area from both sides. Diffraction thus compromises the reproduction of narrowly spaced features, see fig.14.23.[13] An extremely shallow depth of focus is another problem. Phase shift masks (PSMs) aim at improving line separation by making light waves cancel out by destructive interference where they shine through nearby mask apertures. Illumination must be from a (partially) coherent source. The necessary selective $180°$ phase shifts are obtained from partial coating with a thin transparent film or from local etching. In either case, this explains the technique's name alternating aperture phase shift mask.

---

[13] Note the analogy between diffractive spillover in photolithography and intersymbol interference in data transmission.

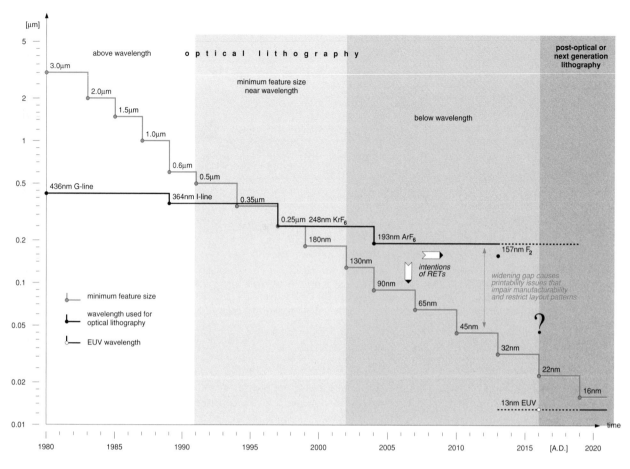

Fig. 14.22 Evolution of VLSI lithography.

Alternating aperture PSM is best suited for the periodic layout patterns that are typical in memory circuits. More sophisticated techniques, known as multi-phase, rim-shifting, attenuated-phase, half-toned, and double-exposure PSM have been developed to improve resolution further and to accommodate the irregular patterns found in logic circuitry.

Off-axis illumination (OAI)

Much like PSM, off-axis illumination (OAI) takes advantage of destructive interference to improve line separation. Yet, there is no extra phase shift coat or etch. The basic idea is to slightly tilt the illumination axis instead so that light shining through adjacent openings reaches the photoresist with a phase shift of 180°. As this cannot work for masks with arbitrary openings, layout patterns are confined to predefined pitches. Also, more sophisticated illumination schemes are adopted to make the idea work along two orthogonal dimensions [391].

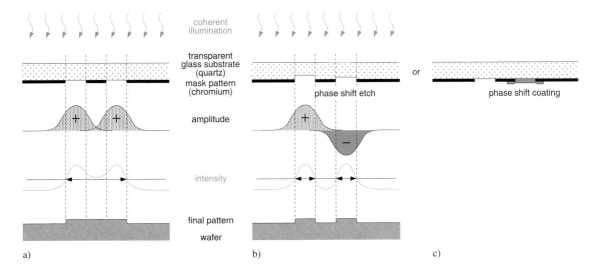

Fig. 14.23 Pattern transfer from mask to photoresist in sub-wavelength lithography with traditional binary mask (a) and with an alternating aperture phase shift mask (b,c) (simplified).

Optical proximity correction

Transferred patterns suffer from severe low-pass-type distortions from the original mask geometries. What appears as a sharp corner in the layout ends up as a washed-out curve; overly small features are absorbed altogether. Optical proximity correction (OPC) applies an inverse distortion to the mask patterns to precompensate for such imperfections of the lithographic steps [392]. Overaccentuated corners, referred to as serifs and hammerheads, are typical for OPC, see fig.14.24. Sub-resolution assist features (SRAFs), narrow lines that diffract light but do not print, are also common.

Computerized resolution enhancement

During the 1990s, PSM and OPC mask patterns essentially had to be optimized by trial and error thereby confining these techniques to high-volume products with highly repetitive layout patterns such as RAMs. In what must be considered a major breakthrough, it has since become possible to combine PSM and OPC in a systematic way.[14]

Immersion lithography

Filling the space between lens and object slide with oil has a long tradition in microscopy. This is because an optical medium with a refraction index larger than that of air makes it possible to push back the total reflectance, thereby augmenting the numerical aperture and improving resolution. A medium suitable for VLSI lithography must be not only transparent at the projection wavelength,

---

[14] A key element of sub-wavelength lithography is an image processing program that computes the necessary corrections to each mask layer prior to mask preparation. The algorithm solves the problem of going from a desired geometry on a wafer to a — rather different looking — pattern on the mask such that the end result conforms to the original layout by compensating for the distorting effects of the optical projection system.

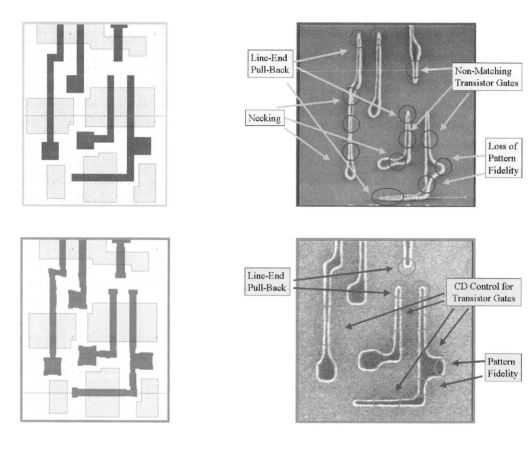

Fig. 14.24 Mask geometry vs. fabricated patterns in sub-wavelength lithography with traditional conformal mask (top row) and with optical proximity correction (bottom row) (reprinted from [393] with permission).

but also uniform, non-contaminating, and compatible with photoresists and the quartz material of the projection lens. The usage of ultrapure water ($n_{H_2O} \approx 1.44$) is currently being put to service to extend optical lithography at 193 nm wavelength down to a linewidth of 45 nm. The search for fluids, lens materials, and photoresists with refraction indices on the order of 1.6 to 1.9 is on.

Submerging the entire lithographic apparatus in a pool turned out to be impractical mostly because the liquid quickly suffers from contamination. Instead, a small puddle of liquid is dispensed between the lens and the wafer, and sucked back before the wafer steps to a new position for exposing the next chip.

Post-optical lithography

Transiting towards post-optical lithography, aka next-generation lithography, will become inevitable. Why industry spares no effort to postpone that day and to buy optical lithography a new lease of life becomes clear when considering the options.

Extreme UV lithography

EUV lithography, aka soft X-ray lithography, may be viewed as a natural extension of optical projection printing to a wavelength of 13.4 nm, yet the differences are dramatic, see fig.14.25. Reflective optics (mirrors) must be used since no transparent materials exist from which refractive optics (lenses) could be built for such short wavelengths. Reflection of EUV waves is obtained from multilayer Bragg interference. Mirrors are to be made of 80 or so alternating layers of silicon and molybdenum separated by a quarter wavelength and smoothed to single-atom tolerances, and the same holds for the reticle or photomask too.

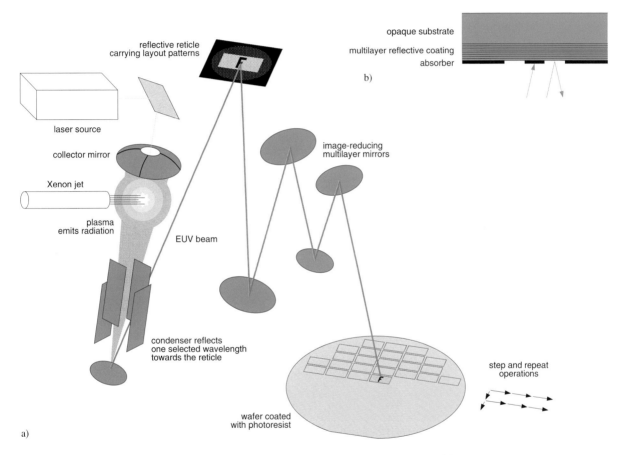

Fig. 14.25 Apparatus for EUV lithography (a) and cross section through a reflective mask (b), mirrors are the same with no absorbing coating (greatly simplifed).

While the wavelength of 13.4 nm has been chosen because this is where Mo–Si multilayer mirrors exhibit peak reflectivity, the reflectance of an individual mirror is just 70% [394]. As the optical path comprises a total of ten mirrors or so, a highly intense source of light must compensate for those losses. One way to obtain a plasma source is from a laser beam aimed at a jet either of xenon gas or of tin droplets while a competing approach uses electrical discharge to produce the plasma.

Holding the 220 000 K plasma together by magnetic fields is a related option. In any case, spectral density is poor and radiation occurs in all directions. Plasma radiation is collected by condenser mirrors and spectrally purified before being projected onto the reflective 4× mask, from where it gets cast onto wafer and photoresist by a cascade of mirrors.

An electrical input of 100 kW is likely to be required for 100 W of optical power [395] and experts currently worry that 100 W might not suffice to meet commercial throughput requirements. The heat that develops in the optical path as a consequence of the enormous power losses makes it extremely difficult to ensure mechanical and optical precision. Further note that everything must be carried out under high vacuum because even air is opaque for EUV radiation. To make things worse, it takes several days to re-establish the vacuum condition following repair or servicing.

Electron beam direct-write lithography (EBDW)

is also known as maskless lithography. Layout patterns get written into the photoresist layer by an electron beam, which technique offers the possibility of higher resolution than UV lithography because of the small wavelength of less than 0.1 nm of electrons at 10 to 50 keV. Resolution is not limited by diffraction but by electron scattering in the target materials and by aberrations in the electron optics. While EBDW lithography finds applications in prototype manufacturing, scanning a single beam over a wafer's surface is much too slow for volume production. Writing with multiple parallel beams is being investigated as an alternative especially for chips that do not sell in huge quantities.

The usage of electron beams in conjunction with masks is no longer considered a potential solution for post-optical lithography [1].

Nano imprint lithography

Instead of using optical imaging to project mask patterns onto a photoresist, nano imprint lithography transfers patterns into a polymer coating previously applied on the wafer's surface by pressing a stencil into the soft polymer. The absence of optical size reduction necessitates a stencil of the same size and resolution as the circuit itself. Throughput being largely uncritical, the nickel, quartz or silicon material of the stencil is patterned by way of electron beam lithography.

There are two options for the polymer. Photochemical nano imprint lithography (P-NIL) uses a flash of UV light to cure the highly viscous polymer film immediately after patterning. Thermoplastic nano imprint lithography (T-NIL) starts from a solid polymer that is made moldable by warming to 150 to 180 °C prior to patterning.

Current belief is that nano imprint lithography should allow resolutions down to 5 nm. Yet, the defect densities of the two patterning steps — wafer and stencil — are unacceptable today and the endurance of a stencil is not yet sufficient for mass production.

At the time of writing (late 2007), EUV appears to be the most promising candidate for next-generation lithography but still is several years away from maturity as industry struggles to produce sources of sufficient optical output and to develop more sensitive photoresists. Please refer to [396] [488] [397] [398] [1] [399] [400] [391] for more detailed accounts on lithography.

14.3 | Variations on the theme

14.3.1 Copper has replaced aluminum as interconnect material

For more than 30 years, aluminum had remained the predominant **interconnect material** for microelectronic circuits while SiO_2 served as dielectric material. The picture began to change with the 180 nm process generation or so. The reason is that lines become more resistive and exhibit larger coupling capacitances as the geometric scaling associated with moving from one process generation to the next makes them narrower and brings them closer together, see fig.12.7. Local wires, such as those running within a cell or between a few adjacent cells, scale down in length, which partly compensates for this.

Global wires, on the other hand, do not scale in length because packing more functionality into a chip prevents die size from shrinking proportionally. Expressed in lithographic squares, global wires are even bound to grow in length. Had industry continued with the traditional materials, their RC delays would have grown dramatically in comparison with gate delay, see fig.14.26.

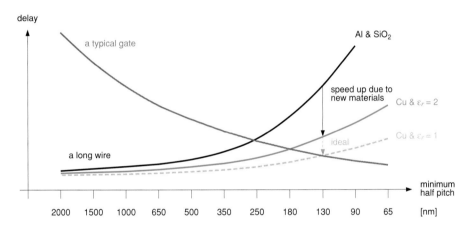

Fig. 14.26 Gate and interconnect delays compared for different conductor and dielectric materials (overall trend).

Observation 14.6. *Line resistance and line capacitances of long wires have become highly critical with the advent of ULSI cross sections and integration densities.*

The search for a lower-resistivity conductor and a lower-permittivity interlevel dielectric has prompted industry to replace Al and SiO_2 with new materials, also see tables D.3 and D.4. Second only to silver in conductivity, copper has displaced aluminum as interconnect material in high-performance logic circuits.

As Cu easily diffuses through SiO_2 and Si, though, the trenches etched into the dielectric must be sealed with an extra **liner** acting as diffusion barrier before they can accommodate the copper conductor. Another major obstacle was vulnerability to corrosion because Cu does not possess a tight self-passivating oxide as Al does, which calls for a special passivation layer. At the time of introduction, measurements indicated that the effective resistance of Cu interconnect was 30% to 45% lower than that of traditional Al alloys. Cu also supports significantly higher current densities. In electromigration tests conducted by IBM, reliability was found to improve by more than two orders of magnitude.

Example

Among the first fabrication processes to abandon aluminum (Al) was CMOS 7S, a 180 nm CMOS process put into volume production by IBM in 1999. The upper six metal layers are made of copper (Cu) while tungsten (W) is used for local interconnects in the bottom layer, see fig.14.27. TaN linings approximately 30 nm thick act as diffusion barriers. The overall process comprised approximately 26 masks.

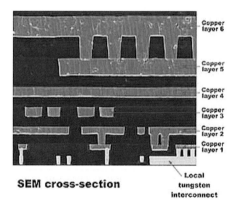

SEM cross-section

Fig. 14.27 Cross section of CMOS 7S interconnect stack (source IBM, reprinted with permission).

☐

14.3.2 Low-permittivity interlevel dielectrics are replacing silicon dioxide

Interlevel dielectric (ILD) materials must combine low permittivity with good mechanical strength to withstand the stresses that occur during CMP and wire bonding. Process compatibility, thermal stability, and low moisture absorption are other desirable qualities. More and more, organosilicate glasses and organic synthetics are replacing silica ($\epsilon_r = 3.9$) as ILD.[15]

Another noteworthy approach is to include nanoscale closed bubbles into a base material. As the permittivity of air is only marginally larger than that of vacuum ($\epsilon_r \approx 1.00055$ versus $\epsilon_r = 1$), the more porous the material, the lower its overall permittivity. Many nanoporous materials being marketed as aerogels, xerogels, nanoglasses, and airgap materials come with adjustable permittivities. A limitation is that mechanical strength and surface smoothness tend to deteriorate with porosity.

---

[15] Although the customary symbol for permittivity is ϵ, low-permittivity (high-permittivity) materials are colloquially referred to as low-k (high-k) dielectrics.

Example

Intel's 65 nm process presented at IEDM 2004 combined up to eight levels of copper interconnect with a low-permittivity carbon-doped oxide (CDO) ILD ($\epsilon_r = 2.9$). Tungsten plugs are used for contacts to poly and diffusion. Metal pitches increase gradually from bottom to top for optimum density vs. performance.

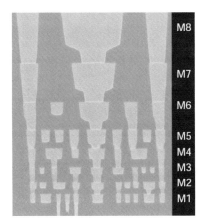

Fig. 14.28 Cross section of an interconnect stack that combines Cu lines with carbon-doped oxide (CDO) interlevel dielectric layers. Note graduated metal pitches and thicknesses (source Intel, reprinted with permission).

□

For all the enthusiasm about superior metals and dielectrics, note that copper combined with vacuum — the lowest-possible-permittivity material — can provide, at best, a signal delay improvement by a factor of 6. A more realistic scenario where $\epsilon_r \approx 2$ leads to a one-time improvement of approximately 3. The overall trend of global interconnect delays outrunning gate delays remains unbroken.[16] Also, the long term perspectives for copper as interconnect material are not entirely clear as any kind of lining is at the expense of the conductor's effective cross section. The relative proportion occupied by liners are unfortunately bound to grow as line width shrinks.

14.3.3   High-permittivity gate dielectrics to replace silicon dioxide

For SiO_2 layers below 3 or 2 nm, electrons begin to tunnel through the gate dielectric.[17] Tunneling current through a 1.5 nm sheet of SiO_2 is on the order of 0.1 $\frac{A}{mm^2}$ at 1 V [401] and increases exponentially when thickness is shrunk further. Tunneling, quantum mechanical effects, and reliability concerns thus preclude silicon dioxide dielectrics thinner than 0.7 to 1.2 nm [402]. A better gate dielectric material must feature a larger permittivity than that of SiO_2. This is because the

---

[16] Please refer to section 12.8 for the fundamental bound on interconnect delay.

[17] Incidentally, notice that two effects overlap. Tunneling-in occurs from the drain to the gate electrode. In addition to this, there is tunneling-out from gate to source. As the two leakage phenomena compensate at some gate potential roughly half-way between source and drain, the net gate current becomes zero at that point.

depressed MOSFET gain factor incurred with a more generous thickness t_{ox} gets compensated by a higher permittivity ϵ_{ox}.

$$\beta = \beta_\square \frac{W}{L} = \frac{\mu \, \epsilon_{ox}}{t_{ox}} \frac{W}{L} = \frac{\mu \, \epsilon_r \, \epsilon_0}{t_{ox}} \frac{W}{L} \tag{14.1}$$

A gate dielectric with an ϵ_{ox} twice that of SiO_2 can be made twice as thick as a silicon dioxide film and still offer the same degree of control over the inversion channel. Any gate oxide thinner than this improves channel control thereby relieving the threshold voltages from the pressure towards proportional downscaling with each new process generation.

Example

Intel reports that 45 nm MOSFETs with a Hafnium-based oxide provide either a 25% increase in driveability at the same subthreshold conduction or more than fivefold reduction in leakage for the same drive current when compared to 65 nm transistors with their traditional gate stacks [492]. At the same time, gate oxide tunneling has been reduced by more than a factor of ten.

☐

Incidentally, note that semiconductor physicists often operate with the **equivalent oxide thickness** (EOT) for an assumed SiO_2 dielectric instead of the physical t_{ox} so as to facilitate comparisons and calculations across various materials. For some gate dielectric material "ox", the equivalent oxide thickness is obtained as $EOT = t_{ox} \frac{\epsilon_{SiO_2}}{\epsilon_{ox}}$.

Table 14.1 | Selected candidates for gate dielectric materials, most numbers from [401] [403] [404] [491]. Mitrovica, I. Z., *et al.* Electrical and structural properties of hafnium silicate thin films. Microelectronics Reliability, **47(4–5)**: 645–648. April–May 2007.

material	relative permittivity ϵ_r	
silicon dioxide SiO_2	3.9	traditional
nitrided silicon oxide aka oxynitride SiO_xN_y	≈ 5.1	
silicon nitride Si_3N_4	7.5	
aluminum oxide Al_2O_3	8–11.5	
hafnium silicon oxynitride HfSiON	≈ 9–11	
hafnium silicate $(HfO_2)_x(SiO_2)_{1-x}$	≈ 12	for $x = 0.6$–0.7
hafnium aluminum oxynitride HfAlON	≈ 18	
hafnium oxide HfO_2	≈ 21	
zirconium oxide ZrO_2	22–28	Intel "TeraHertz"
lanthanum aluminum oxide $LaAlO_3$	25.1	
tantalum pentoxide Ta_2O_5	27	
titanium dioxide (rutile) TiO_2	>25	
barium strontium nitrate $Ba|Sr(NO_3)_2$	>25	
strontium titanate[a] $SrTiO_3$	≈ 200	ceramic capacitors, DRAMs

[a] Strontium titanate belongs to the family of perovskite ceramics and is given here for reference.

While silicon dioxide was the sole dielectric available in the past, other materials optimized for their respective roles are now being introduced.

Observation 14.7. *A material of low permittivity is desirable for interlevel dielectrics where the lowest possible parasitic capacitances are being sought. Exactly the opposite is true for the gate dielectric in order to minimize gate leakage while maximizing MOSFET drivability and, hence, the switching speed of logic circuits.*

14.3.4 Strained silicon and SiGe technology

Electron mobility in Si augments when the crystal lattice is subjected to tensile stress, whereas compressive stress tends to improve hole mobility [405]. Stress can be applied mechanically as part of the packaging process. Stress can also be built right into the semiconductor crystal by incorporating comparatively large Ge atoms into the narrower Si lattice or by combining two layers of materials with distinct lattice spacings. Straining is entirely different from doping and works because of the different lattice constants. The proportion x of Ge in a strained $Si_{1-x}Ge_x$ crystal is on the order of 15% to 90%.

Example

Intel's 65 nm process that went into mass production in early 2006 with the Core Duo processor implements strained MOSFET channels. n-channels are put under tensile stress with the aid of a SiN capping film while p-channels are compressed by using epitaxial SiGe films for the adjacent source and drain areas [340]. The table below lists key transistor characteristics.

parameter	typical values for n-channel devices	typical values for p-channel devices	measurement conditions
effective channel length L_{eff} [nm]	35	35	
drivability $I_{d\,on}/W$ [μA/μm]	1460	−880	$U_{dd} = 1.2$ V
@ leakage $I_{d\,off}/W$ [nA/μm]	100	−100	low thresholds[a]
drivability $I_{d\,on}/W$ [μA/μm]	1040	−630	$U_{dd} = 1.2$ V
@ leakage $I_{d\,off}/W$ [nA/μm]	1	−1	higher thresholds[a]
strain-induced improvement	18%	50%	
subthreshold slope S [mV/decade]	≈100		

[a] Not numerically specified.

□

SiGe technologies further make it possible to combine heterobipolar junction transistors (HBT) for analog and RF front-end circuits with standard CMOS logic on a single die by growing an epitaxial layer of SiGe on a silicon wafer after device isolation has been completed.

Example

austriamicrosystems (AMS) started to offer a SiGe option for its 800 nm BiCMOS process family in the year 2000. The maximum transit frequency f_T went from 12 GHz for a Si BJT (process BYE) to 35 GHz for a SiGe HBT (process BYS). This improvement gives room for handling higher frequencies or for lowering collector currents and hence power dissipation.

□

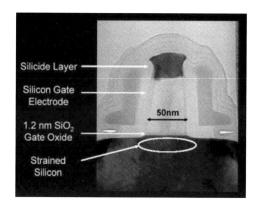

Fig. 14.29 Cross section of a MOSFET with strained silicon (source Intel, reprinted with permission).

14.3.5 Metal gates bound to come back

A number of undesirable phenomena are associated with polysilicon as a gate material [406]. Firstly, thin gate dielectric layers cannot always prevent **boron penetration** from p$^+$ polysilicon into the material underneath, where boron can alter the device's threshold voltage. Secondly, when a voltage is applied to a MOSFET gate electrode, we find not only an inversion channel that forms in the bulk material underneath but also a space-charge region that comes into existence within the gate itself. This **depletion effect** is more acute in polysilicon than in metals because the carrier concentration remains inferior even when the material is heavily doped. The depleted layer of approximately 0.4 nm makes the device behave as if its gate dielectric were thicker than it physically is. Most important, however, are the difficulties of combining poly gates with high-permittivity gate dielectrics. A 40% to 50% reduction in electron mobility has been reported for poly-Si/HfO$_2$/Si gate stacks, for instance.

As metal gates do away with these problems, they are to reappear along with high-permittivity gate dielectrics such as HfO$_2$ [407]. Metal gates bring their own difficulties, however. A MOSFET's threshold voltage gets determined by the work function of the gate material, and that work function is fixed by the choice of the metal.[18] As opposed to the case with polysilicon, there is no way to adjust the threshold voltages of n- and p-channel transistors separately by doping their gate materials. Unless a way is found to adjust a metal's work function, one will have to use two different metals, which complicates fabrication [408].[19] Also, self-alignment of gate electrodes with the pertaining source/drain regions is not possible unless all materials in the gate stack can withstand the temperatures associated with the drive-in of dopants and with subsequent annealing steps.

Example

In January 2007, Intel announced that their forthcoming 45 nm process would combine a hafnium-based gate dielectric with metal gates of two different but undisclosed compositions.

---

[18] The term **work function** denotes the energy needed to remove an electron from the Fermi level in a metal to a point outside at infinite distance. The German term "Austrittsarbeit" nicely reflects that notion.

[19] Candidate materials for the gates of n-channel MOSFETs include TiN, Ti, Ta, Zr, Hf, and IrO$_2$/Hf while TaN, WN, Mo, Pt, Ir, Ni, and IrO$_2$ are being investigated for p-channel gates [406] [409] [407].

 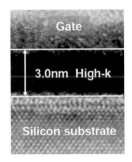

Fig. 14.30 Traditional poly-Si/SiO$_2$/Si stack (90 nm, left) versus prospective metal/HfO$_2$/strained-Si stack (45 nm, right) (source Intel, reprinted with permission).

Incompatibilities between the high-permittivity dielectric material and polysilicon seem to have incited the company to make both changes at the same time.

☐

14.3.6 Silicon-on-insulator (SOI) technology

Rather than implanting source and drain islands into a wafer's base material as in traditional bulk processes, SOI wafer processing occurs within a thin layer of silicon — less than 200 nm thick — that rests on top of an insulating material such as silicon dioxide or sapphire, see fig.14.31.[20] The n- and p-channel MOSFETs then get separated from each other by etching away the silicon film in between before replacing it by a nonconductive oxide.

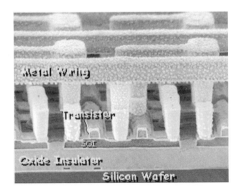

Fig. 14.31 Cross section through an SOI circuit (source IBM, reprinted with permission).

---

[20] A thin, yet perfectly crystalline silicon layer is essential. Various techniques for fabricating SOI wafers compete [410]. The UNIBOND process takes two silicon wafers and grows an oxide layer on each of them before bonding the wafers together without using any intermediate glueing layer so as to obtain a silicon–oxide–silicon sandwich. The one side later to accept the circuit is then ground and polished down to a thickness of one μm or less [411]. The SIMOX process starts from a regular silicon wafer and its full name "separation by implantation of oxygen" largely explains the rest. Ultra-thin silicon (UTSi) is a modern implementation of silicon-on-sapphire (SOS) technology whereby a thin film of silicon is deposited onto a wafer of synthetic sapphire Al$_2$O$_3$. The film is then recrystallized in a solid-phase epitaxy step.

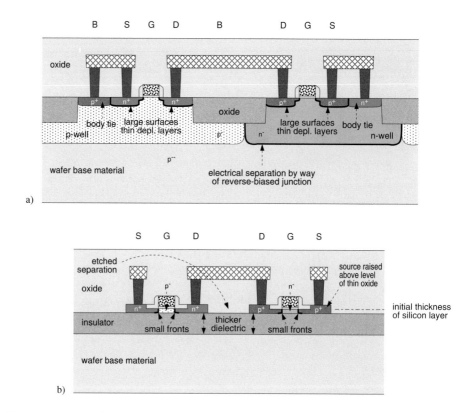

Fig. 14.32 Bulk devices (a) and ultra-thin-body SOI devices (b) compared (not drawn to scale).

In the past, SOI circuits had found only niche applications such as radiation-hard devices and high-temperature electronics, but this is about to change. SOI CMOS technology has many beneficial properties likely to offset the higher wafer costs in demanding applications.

+ The inherent electrical insulation of MOSFETs does away with the need for wells.
+ No parasitic BJTs and, hence, no exposure to latch-up.
+ No need for body ties and, hence, superior layout density.
+ Reduced sensitivity to radiation effects and higher operating temperatures.
− The poor thermal conductivity of the insulating layer impedes heat removal.

The omission of wells and body ties also entails a difficulty, however, because charges trapped in a MOSFET's body underneath the inversion channel cannot drain into vss or vdd. Any such charge thus acts like a back gate and influences the transistor's transfer characteristic in a history-dependent way, an undesirable phenomenon known as the **floating body effect**. The absence of body ties further makes it more difficult to firmly shut off the parasitic BJT that comes along with the MOSFET and that is implicitly connected in parallel.

Making the silicon layer extremely thin — on the order of 5 to 25 nm — minimizes those problems. This variation of SOI technology is known as **ultra-thin-body** (UTB) SOI or as depleted substrate transistors (DSTs). Another benefit is that leakage through reverse-biased drain and source junctions is greatly reduced [408]. Compare the junctions of a bulk device with an SOI device in fig.14.32. Note that ultra-thin-body technology does away with the entire bottom surface and so eliminates much of the subsurface current that would otherwise flow from or to the drain through the depths of the bulk material underneath. At room temperature, the subthreshold slope S is down to a mere 65 mV/decade from the 90 mV/decade more typical for bulk CMOS, which supports lower threshold voltages for the same amount of leakage.

From an electrical perspective, the ultra-shallow sheets that make up the drain and source greatly inflate the parasitic series resistance of each MOSFET. To partially compensate for this, the source and drain islands can be been made to protrude above their initial thickness by way of epitaxial growth as shown in fig.14.32b. This technique, promoted as **raised source and drain**, is reported to provide some 30% of extra current drive over a comparable flat arrangement [412]. On top of the benefits of SOI, ultra-thin-body technology thus provides:

+ Reduced floating body effect
+ Lower parasitic source and drain capacitances due to smaller junction surfaces and thicker buried oxide dielectric instead of the thin depletion layers.
+ No junction surface at the bottom, hence reduced leakage.
+ Steeper subthreshold slope (smaller S) and, hence, much lower subthreshold conduction.
+ Faster operation and/or better energy efficiency.
− Larger source and drain series resistance as raised sources and drains only partially offset the higher sheet resistance.

Example

The Cell processor co-developed by IBM, Sony, and Toshiba for gaming, multimedia, and server applications combines strained silicon on SOI wafers, low-permittivity ILD, and one layer of local interconnect plus eight additional levels of copper in a 90 nm CMOS technology. The circuit presented in 2005 is scheduled to enter mass production with Sony's PlayStation 3.
□

Chapter 15

Outlook

The driving force behind the rapid expansion of the microelectronics industry is its aspiration to offer ever more powerful circuits at lower unit prices. From different perspectives, sections 15.1 through 15.4 attempt to find out how and for how long this trend may be expected to continue into the future. The impact of the galloping progress of semiconductor fabrication technology on VLSI design practices is discussed in section 15.5.

15.1 | Evolution paths for CMOS technology

During the past decades, microelectronics has continuously and rapidly evolved according to the motto "smaller, faster, cheaper".[1] The reason why this has been possible is the **scaling property** of CMOS technology first stated by Robert Dennard and his colleagues in 1972 [414]. They observed that MOSFETs would continue to behave largely in the same way provided their geometric dimensions and voltage levels could be made to shrink in a linear fashion so as to maintain constant electric fields, see fig.15.1. Better still, they predicted that key figures of merit like gate delay and energy efficiency would greatly benefit from downscaling. The question is

"For how long can CMOS scaling continue and where does this trend lead to?"

15.1.1 Classic device scaling

The driving force behind moving from one process generation to the next is to lower fabrication costs per device and per circuit by shrinking geometries so as to obtain more paying circuits from a wafer of some given diameter. Other key objectives include:

---

[1] The aircraft industry offers an intriguing analogy. For many decades, new planes had been developed to go "higher, farther, faster" than their predecessors. Since the late 1960s, however, the top speeds and ceilings of both commercial and military aircraft have essentially remained the same. Although going faster would be possible technically, this makes little sense economically, and the priorities have changed to operating costs, payload, reliability, pollution control, etc.

- Smaller gate delays $t_{pd} \propto \frac{C_k \, U_{dd}}{I_{ds \, on}}$.
- Higher current drive $I_{ds \, on} \propto \frac{\mu \, \epsilon_{ox}}{t_{ox}} \frac{W}{L} (U_{dd} - U_{th})^\alpha$.[2]
- Lower dissipated energy per switching event $E_{ch \, k} \propto C_k U_{dd}^2$.
- Low leakage currents (subthreshold conduction, reverse-biased junctions, gate leakage).

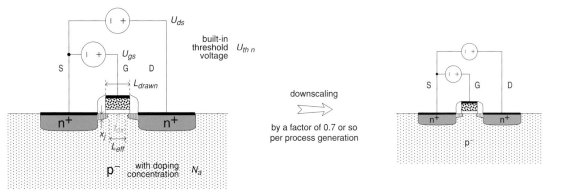

Fig. 15.1 Cross section of an n-channel MOSFET with quantities subject to classic scaling.

Downsizing layout geometry and supply voltage does not suffice, though, and other device parameters must be adjusted to preserve decent transistor characteristics and reliability figures. Table 15.1 shows the tradeoffs.

- Junction depths x_j must be reduced to mitigate short-channel effects (SCEs).
- Substrate doping concentrations N_a and N_d must be increased in the channel region to scale the depth of the depletion layer that forms underneath.
- Gate dielectrics must be made even thinner to maximize control of channel conductivity.
- Due to subthreshold leakage, threshold voltages $U_{th \, n}$ and $U_{th \, p}$ cannot be lowered proportionally with supply voltage U_{dd} as they ideally would.
- Doping profiles must become still more abrupt.

Numerical example

To get a better appreciation of the road ahead, let us do some back-of-the-envelope calculations for the prospective end-of-scaling transistor from table 15.1. For that purpose, we assume $L_{drawn} = 22$ nm > 10 nm $= L_g$ and $W = 3L_{drawn}$.

1) Number of molecular "layers" in an assumed gate dielectric of SiO$_2$:

$$n_l = \frac{t_{ox}}{t_{SiO_2}} \approx \frac{0.5 \text{ nm}}{0.25 \text{ nm}} = 2 \tag{15.1}$$

---

[2] The limiting effect of ohmic source and drain resistance is neglected here.

if we assume a thickness of 0.25 nm per layer as device physicists typically do.[3] It has been found that silicon dioxide remains a reliable dielectric down to 1.2 nm, which corresponds to 5 layers. Beyond that, a different gate dielectric material must be sought.

2) Number of extra electron charges on a gate electrode when fully turned "on":

$$n_e = \frac{Q_g}{q_e} = C_g \frac{U_{dd}}{q_e} \geq \epsilon_r \epsilon_0 \frac{W L_g}{t_{ox}} \frac{U_{dd}}{q_e} \approx 3.9 \cdot 8.85 \, \text{pF/m} \frac{3 \cdot 22 \, \text{nm} \cdot 10 \, \text{nm}}{0.5 \, \text{nm}} \frac{0.6 \, \text{V}}{0.16 \, \text{aC}} \approx 170 \quad (15.2)$$

which number suggests that quantum effects will impact electrical characteristics. 3) Number of doping atoms in the silicon volume underneath the MOSFET's gate dielectric:

$$n_d = V N_a \approx W L_g \, x_j N_a = 3 \cdot 22 \, \text{nm} \cdot 10 \, \text{nm} \cdot 5 \, \text{nm} \cdot 10^{19} / \text{cm}^3 \approx 33 \quad (15.3)$$

which indicates that it will no longer be possible to adjust threshold voltages by locally doping the body material underneath the gate as statistical fluctuations would otherwise grow to unacceptable heights. Instead, threshold voltages will have to be defined by selecting a gate material with an appropriate work function.

□

Current expectations are that field effect devices should continue to behave essentially like present-day MOSFETs down to a channel length of 10 nm or so. Yet, the above calculations underline that the traditional gate stack of horizontal layers of salicided polysilicon, SiO_2, and silicon is running out of steam. Figure 15.2 nicely summarizes the situation while the significance of new interconnect, dielectric, and gate materials has been discussed in section 14.3.

Observation 15.1. *Classic device scaling, that was limited to proportional resizing of a few geometric and electric parameters, is no longer sufficient to provide the progress that the public expects from semiconductors. Only the adoption of new materials within the basic CMOS fabrication flow has made it possible to continue the historical trend.*

The upcoming sections attempt to explore more profound changes that lie ahead.

---

[3] One can estimate the thickness of one molecular layer of silicon dioxide in either of two ways.
a) Assuming a perfect tetrahedral crystalline structure (quartz):

$$t_{SiO_2} = 0.185 \, \text{nm} \quad (15.4)$$

is obtained from the three-dimensional geometry of the crystal lattice [415].
b) Assuming a totally amorphous material (silica):

$$t_{SiO_2} \approx \sqrt[3]{V_{SiO_2}} = 0.356 \, \text{nm} \quad (15.5)$$

because

$$V_{SiO_2} = \frac{m_{Si} + 2m_O}{\varrho_{SiO_2} N_{Avo}} = \frac{28 \frac{g}{mol} + 2 \cdot 16 \frac{g}{mol}}{2.21 \frac{g}{cm^3} \cdot 6.022 \cdot 10^{23} \frac{1}{mol}} \approx 4.51 \cdot 10^{-29} \, \text{m}^3 \quad (15.6)$$

where V, m, and N_{Avo} stand for molecular volume, atomic mass, and Avogadro's number respectively. The density ϱ_{SiO_2} is 2.21 $\frac{g}{cm^3}$ for amorphous silica as compared with 2.65 $\frac{g}{cm^3}$ for quartz. Though the gate oxide gets thermally grown and annealed to become largely crystalline, rough interfaces obviate a perfect lattice, thereby casting doubt on the concept of molecular layers; see fig.14.30 for an illustration. Still, our simplistic calculations support the assumption 0.185 nm $< t_{SiO_2} \approx 0.25$ nm < 0.356 nm.

Table 15.1 Summary of general MOSFET scaling trends.

	technology parameter subject to scaling					
	gate length L_g [nm]	equiv. gate oxide thickness (EOT) t_{ox} [nm]	supply voltage U_{dd} [V]	threshold voltage U_{th} [V]	junction depth x_j [nm]	substrate/well doping concentr. N_a, N_d [cm^{-3}]
desirable property	advocated direction of scaling					
small gate delay[a]	↓	↓	↑	↓		
low switching energy		↑	↓			
low leakage	↑	↑		↑		
overall scaling trend	↓	↓	↓	↓	↓	↑
prime motivations	improve layout density, drive and speed	maintain current drive and gain factor	avoid gate dielectric breakdown, improve energy efficiency	maintain current drive	suppress short-channel effects (SCEs)	scale depth of depletion layer
limiting factors	lithography, SCE, power dissip. density	gate dielectric tunneling, reliability	threshold voltage, ramp times, prop. delays	off-state leakage	series resistance	loss of source and drain isolation
year	approximate numerical indications					
≈ 1974	6000	100	5	0.8	800	10^{16}
≈ 1997	250	5	2.5	0.55	190	$4 \cdot 10^{17}$
≈ 2004	100	3	1.5	0.4	40	10^{18}
≈ 2007	70	1.5	0.9	0.25	30	10^{18}
far future	10	0.5	0.6	0.2	5	10^{19}

[a] In full-swing CMOS, gate delay t_{pd} is essentially minimized by maximizing current drive $I_{ds\,on}$.

15.1.2 The search for new device topologies

Mitigating the short-channel effects in MOSFETs necessitates improving the influence of the gate voltage on the channel current. Various concepts are being studied to do so.

In a planar **double-gate** device (DG-MOSFET), a horizontal inversion channel is sandwiched between a pair of electrically connected gate electrodes in such a way as to steer the channel from both sides at the same time, see fig.15.3b. The main difficulty with this approach is to manufacture a bottom gate underneath the channel and to align it to the top gate [416].

In a **fin-FET**, a tiny vertical fin of silicon carries a gate electrode much as a horse carries a yoke [408] [417], see fig.15.3c. The electrical field emanating from the gate impinges on the channel from

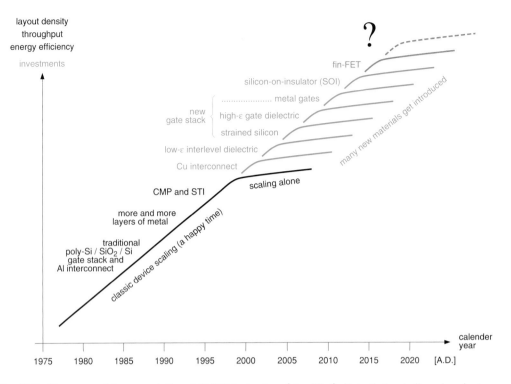

Fig. 15.2 The past and future evolution of CMOS technology (simplified). Note that not all semiconductor manufacturers necessarily introduce innovations at the same time and in the same order.

three sides, which justifies the synonym tri-gate device used by Intel, for instance. The fin's narrow width, which can be as small as 10 nm, results in an ultra-thin MOSFET body thickness while the fin's height of order 50 nm becomes the channel width; the channel length is defined by the yoke's thickness on the order of 20 nm.

The presumed **gate-all-around**, aka surround gate transistor and pillar FET, of fig.15.3d, finally, features a vertical channel fully enclosed by an annular gate electrode [418].

Currently, the most promising of these topologies seems to be the fin, several of which can be connected in parallel to obtain the equivalent of a wider FET with a better current drive, see fig.15.4. At the end of 2006, Infineon reported having successfully manufactured a circuit that includes more than 3000 multi-gate fin-FETs in 65 nm technology [419]. Quiescent currents were measured to be one order of magnitude below those of comparable planar single-gate devices. [420] observes that the ultra-thin body proper to the fin-FET provides gate control superior to that in a classical device. He further suggests avoiding the problem of random doping variations by leaving the body undoped and by using a single near-midgap metal as gate material for both n- and p-channel devices. From technology CAD (TCAD) and other analyses, he concludes that such devices should scale down to below 10 nm.

An argument that can hardly be overestimated is that all three topologies rely on much the same materials, fabrication procedures, and facilities as conventional CMOS, which makes it possible to leverage the present experience and equipment.

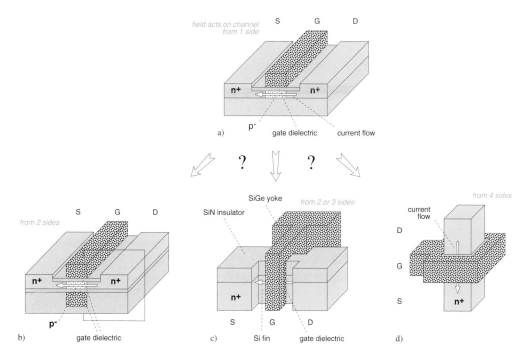

Fig. 15.3 Topological options for future devices. State-of-the-art ultra-thin-body SOI MOSFET (a), planar double-gate MOSFET (b), fin-FET (c), and gate-all-around transistor (d) (simplified).

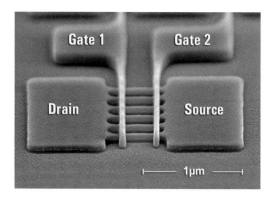

Fig. 15.4 Two multi-gate fin-FETs connected in series (source Infineon, reprinted with permission).

15.1.3 Vertical integration

Taking advantage of the third dimension holds the promise of increasing density much beyond what device scaling alone would permit. Yet, be cautioned that heat evacuation severely limits the amount of power that can possibly be dissipated in a truly three-dimensional volume.

Stacking one layer of MOSFETs on top of a layer underneath is essentially achieved by making the wafer surface perfectly flat by way of chemical mechanical polishing (CMP) before depositing and patterning another set of silicon, poly, and interlevel dielectric layers to form the next upper layer of devices. Making sure that the thin films of silicon deposited are monocrystalline and of good quality is a critical element of the process [421].

3D integration is more likely to be adopted for memories than for random logic because their tiled layouts does not ask for too much routing resources and because layout density remains an extremely strong competitive advantage.[4] Also, only parts sold in very high quantities can compensate for the larger mask count associated with vertical integration. At the time of writing (2007), antifuse-based PROMs are expected to hit the market soon. Prototypes of SRAMs where MOSFETs are stacked three levels high and interconnected using contact plugs and short metal straps have also been fabricated [422]. The 6-transistor bit cell is reported to occupy a mere $25F^2$. Others suggest using germanium-on-insulator (GOI) FETs with metal gates as a second layer of devices [409].

Alternative techniques that tap the third dimension essentially by stacking ordinary microchips are discussed in section 11.4.7 in the context of packaging.

15.1.4 The search for better semiconductor materials

HIGH-MOBILITY SEMICONDUCTORS

A MOSFET's gain factor and, hence, a CMOS circuit's operating speed directly depend on carrier mobility μ. Although strained silicon offers a substantial benefit over plain silicon, the search for semiconductor materials with still higher carrier mobilities continues.[5]

Germanium, from which early BJTs had been manufactured in the 1950s, features electron and hole mobilities more than twice as high as in silicon. What had led to the demise of Ge in favor of Si as base material for transistors and ICs were the lack of a stable oxide and significant leakage currents as a consequence of a narrower bandgap. Now that the traditional SiO_2 oxides are bound to be replaced anyway, Ge may, or might not, reappear along with high-permittivity gate dielectrics and metal gates, all the more so, as its lower processing temperatures tend to make it compatible with a wider range of materials. The usage of **germanium-on-insulator** (GOI) wafers is currently being investigated for the 45 nm and later generations with the idea of getting a grip on subthreshold leakage by slashing junction areas much as in ultra-thin-body SOI technology, see fig.14.32. A limiting factor is the comparatively low electrical field that Ge is able to sustain.

Even higher electron mobilities are observed in compound semiconductors such as **gallium arsenide** (GaAs, III–V) and **indium phosphide** (InP, III–V). Devices such as hetero FETs (HFET) and heterobipolar junction transistors (HBT) fabricated from these materials have been demonstrated to exhibit much higher transit frequencies than their silicon counterparts do. However, the fact that Si logic reaches a point where wiring rather than gate delay dominates a circuit's operating

---

[4] Three-dimensional structures have in fact been well established in DRAMs ever since buried trench capacitors made their appearance in the 1 Mibit generation. Manufacturing the access transistor vertically along the wall of the capacitor or on top of it appears a natural extension to further squeeze the bit cell.

[5] Key physical characteristics of semiconductor materials are listed in appendix D.3 while strained silicon and SiGe are discussed in section 14.3.4.

speed makes a transition fruitless and unlikely in the context of digital VLSI. Notable exceptions are physically small subcircuits such as prescalers, phase locked loops (PLL), voltage-controlled oscillators (VCOs), low-noise amplifiers (LNA), radio-frequency (RF) mixers, and the like.

The main impediment for all exotic base materials is that they sacrifice much of the investment and know-how in current silicon preparation and processing. Silicon germanium (SiGe) is a less disruptive alternative that combines improved mobilities with the manufacturability and economy of Si. As opposed to III–V and II–VI compounds, SiGe (IV–IV) capitalizes on the abundance of large wafers, mature production facilities, and low defect densities of silicon [423].

METALLIC MATERIALS

A material that exhibits an extremely small overlap between valence and conductance band is **graphene**, essentially a planar monocrystalline graphitic sheet. The authors of [424] were the first to build FETs from this material that had long been believed to be unstable in favor of soot, fullerenes, and carbon nanotabes (CNTs). At room temperature, experiments revealed ambipolar behavior[6] with carrier mobilities up to $10\,000$ cm^2/V s, more than six times as much as for silicon. Due to a thick SiO$_2$ gate dielectric of 300 nm, gate voltages were on the order of an impractical 100 V. More recent measurements on graphene FETs with a top gate electrode and a SiO$_2$ dielectric of 20 nm have brought down gate voltages to below 7 V while confirming better carrier mobilities than in ultra-thin-body (UTB) SOI MOSFETs [425].

WIDE-BANDGAP SEMICONDUCTORS

As an outgrowth of their large bandgaps in excess of 3 eV, both **gallium nitride** (GaN) and **silicon carbide** (SiC) can withstand much stronger electrical fields than Si and Ge. Another prime advantage of silicon carbide is that it can endure temperatures up to 600 °C. The first SiC devices to enter mass production were blue light emitting diodes (LEDs), to be displaced later by InGaN LEDs solely because of their superior efficiency. The most important application of SiC today are Schottky diodes capable of sustaining reverse voltages up to 1200 V; switching transistors and other power devices are expected to follow. SiC further is a promising material for automotive sensors. Current expectations are that SiC will remain confined to niche applications in power circuits and harsh environments, though. This is because SiC cannot be grown by conventional crystal pulling techniques as the material sublimes instead of melting at reasonable pressures. Also, its extraordinary hardness renders sawing and polishing difficult.

GaN has superior RF characteristics and is expected to compete with GaAs for applications such as microwave power amplifiers in radars and base stations for cellular telephony [426]. In either case, crystal growth and wafer processing technology are not mature yet for cost-effective mass fabrication of ICs. Also, GaN and SiC field effect transistors are built as **metal–semiconductor field effect transistors** (MESFETs), which implies that they come as depletion types. As opposed to silicon CMOS, complementary n- and p-channel FETs are not available so far. Though a few

---

[6] This is to say that electrons <u>and</u> holes act as mobile charge carriers depending on the electrical field applied. The conductivity of an **ambipolar device** is minimal for intermediate gate voltages and augments towards both larger and smaller values.

NMOS-style SSI circuits have been demonstrated, fabrication technology lags many years behind and neither material is likely to displace silicon anytime soon.

POLYMER SEMICONDUCTORS

A couple of organic polymers, such as pentacene, have been found to have an electronic structure that makes it easy for charge carriers to move through them. The idea behind **integrated plastic circuits** (IPC) is to create layered circuits similar to those of traditional silicon ICs by using such organic materials in conjunction with printing techniques instead of costly photolithography and wafer processing under clean room conditions, high temperatures, and vacuum. With minimum feature sizes of 10 μm or more and carrier mobilities on the order of 0.5 cm^2/V s,[7] plastic thin-film transistors will hardly displace silicon microelectronics but will have to target new fields.

Opportunities for polymer semiconductors exist in macroelectronics where a large active area is essential such as in flat panel displays, in organic light emitting diode (OLED) panels for lighting purposes, in photoelectric panels, and in certain sensor applications. This is because manufacturing costs per area are much lower than for silicon, while the opposite is true when costs are related to transistor count. Prospective applications also extend to cheap mass products, notably radio-frequency identification (RFID) tags for everyday goods such as clothing and luggage, parcels, documents, letters, tickets, price and inventory tags, and the like. The adoption of molecular self-assembled monolayers (SAMs) as gate dielectric has recently lowered the operating voltage to 1.5 to 3 V [427].

Challenges include the disparity of carrier mobilities in n- and p-type materials, gate delays on the order of 2 ms, and the rapid degradation when exposed to oxygen, humidity, electrical stress, or intense sunlight. It will be difficult to manufacture large polymer circuits that are light-weight, flexible, and do not break as long as they need to be protected with sheets of glass. Also, toxicity may stand in the way of adopting disposable electronics on a larger scale.

A novel material that might come to the rescue are thin films made of randomly arranged carbon nanotubes (CNTs) similar in structure to felt and other nonwoven textiles. Such films have been found to combine good conductivity, high carrier mobility, flexibility, mechanical and chemical robustness, and even optical transparency. Reports on the current state of polymer electronics and of carbon nanotube films are available in [428] and [489] respectively.

15.2 | Is there life after CMOS?

A variety of radically new technological concepts for information storage and processing have been proposed; some of them are briefly touched on below. Note, in particular, that a fast, non-volatile, energy-efficient, and cost-effective memory technology would allow computer architects to greatly simplify the present memory hierarchy and to do away with a lot of current drain.

---

[7] While mobilities in excess of 2 cm^2/V s have been reported in the research literature, these refer to vacuum deposition, not to low-cost solution-based printing techniques.

15.2.1 Non-CMOS data storage

Phase-change RAM (PRAM)

Certain alloys such as Sb_2Te_3 (antimony telluride), $Ge_2Sb_2Te_5$ (another chalcogenide), and GeSb (germanium antimony) exhibit rapid and reversible transitions between two stable phases with widely different optical and electrical properties. Electrical conductivity and optical reflectivity are good in the polycrystalline state while the opposite holds for the amorphous condition. Today's rewritable CDs and DVDs use lasers to store and to recover information in films of chalcogenide material by optical means. The phase-change memories currently under development take advantage of changes in conductivity. Phase transitions are induced by electrical heating with the orientation of the transition being controlled by applying distinct temperature vs. time profiles.

A PRAM is built from zillions of tiny volumes that can be addressed randomly with the aid of one access transistor per such volume. Two forms of implementation are being researched. In the so-called line-cell memories the chalcogenide material forms a narrow bridge that spans the short gap between two metal contacts (much like a PROM fuse that gets recrystallized but never blown) whereas a vertical arrangement is adopted in ovonic unified memories (OUM). Intel and Samsung have announced 128 Mibit and 512 Mibit PRAMs respectively as replacements for flash memories for the 2007/2008 timeframe.

Ferroelectric RAM (FeRAM)

A ferroelectric material behaves analogously to the magnetic behavior of iron. That is, one observes spatial domains inside which all electric dipole moments get aligned in the same direction. An FeRAM maps a logic 0 or 1 onto the opposing directions of polarization of one such domain. The construction of an FeRAM requires a dielectric material with strong remanent electrostatic polarization that is compatible with CMOS silicon processing. Candidates include lead zirconium titanate $PbZr_xTi_{1-x}O_3$ (PZT) and compounds of strontium, bismuth, and tantalum (SBT).[8] The smallest storage cells are obtained from combining one access transistor with a capacitor built around a dielectric of ferroelectric material. Data readout is destructive, but data storage is non-volatile with expected retention times of ten years and more [429]. Ramtron has announced a 4 Mibit FeRAM would become available in 2007.

Magnetic RAM (MRAM)

Several approaches to MRAM design are competing. What they have in common is that a bit of information is stored into some storage device by applying a magnetic field and retrieved by measuring its resistance.[9] MRAM designs differ in the exact nature of the physical effect they take advantage of for storage and in how the storage devices are accessed. A first effect is called giant **magnetoresistance** (GMR) [430]; $\frac{\Delta R}{R_0}$ ratios on the order of 0.3 have been reported. Tunneling magnetoresistance (TMR) forms the basis for other MRAM designs where each cell is built around a magnetic tunnel junction (MTJ) device [431]. The individual storage devices are arranged as a matrix and addressed with the aid of one or two access transistors [432] [433] much like conventional RAM cells. The inherently non-volatile cells hold the promise of fast operation, good layout density,

---

[8] PZT, as well as strontium titanate and barium titanate, are **perovskite** ceramics, a class of crystalline oxide materials that share a common lattice structure. Depending on the specific atomic elements incorporated, perovskites exhibit unique properties such as high permittivity and even ferroelectricity.

[9] As opposed to this, readout of the bygone magnetic core memories was destructive because it worked by having the cell remagnetized by another magnetic field and by measuring the transient current induced.

and unlimited cycling [434]. Industrial production has commenced in 2006 with a 4 Mibit MRAM by Freescale that works with MTJs.

Table 15.2 | Competing memory technologies.

	CMOS			extensions		
Desiderata	DRAM	SRAM	Flash	PRAM	FeRAM	MRAM
small bit cell, high density	yes	no	yes	yes	yes	yes
fast read/write operations	almost	yes	no	unclear	unclear	yes
non-volatile	no	no	yes	yes	yes	yes
static storage with no refresh	no	yes	yes	yes	yes	yes
low standby power	no	partly	yes	yes	yes	yes
non-destructive readout	no	yes	yes	yes	no	yes
unlimited endurance	yes	yes	no	almost	almost	yes
truly random access	yes	yes	no	yes	yes	yes
number of extra masks[a]	6–8	0–2	6–8	3–4	2	3–4
ready for mass production	yes	yes	yes	no	yes	yes
cost-efficient	yes	no	yes	unclear	not yet	not yet

[a] Numbers are from [259] and refer to the masks required on top of a CMOS high-performance logic process.

15.2.2 Non-CMOS data processing

Carbon nanotubes

Untreated carbon nanotubes (CNTs) electrically resemble p-type material with bandgap varying from zero (metallic) to narrow (semiconducting) depending on tube diameter and exact structure. Much like in a MOSFET channel, the voltage applied to an insulated electrode placed nearby can be made to modulate the current through a semiconducting CNT [435].

Logic gates combining field-controlled nanotubes with off-chip pull-down resistors on the order of 100 MΩ were manufactured for the first time in 2001 [436]. Each p-channel-like device consisted of a single nanotube deposited on top of an aluminum strip that would later act as gate electrode. The native Al_2O_3 layer formed a dielectric a few nanometers thick. Measured device characteristics were $I_{ds\,max} \approx -100$ nA and $R_{ds\,on} \approx 26$ MΩ with an on/off ratio of at least 10^5. Logic gates, bistable feedback loops, and even a ring oscillator have been built.

In the same year, IBM researchers selectively doped a nanotube with potassium (K) to obtain the equivalent of an n-channel transistor, thereby paving the way for complementary logic. They successfully fabricated a CMOS-like inverter in which the customary MOSFETs were replaced by a carbon nanotube, one section of which was doped with potassium while the second remained untreated. The inverter operated from a 4 V supply and, with a measured value of −1.6, was found to indeed exhibit a voltage amplification greater than unity.

In 2005, it was discovered that a Y-shaped multi-walled carbon nanotube (MWCNT) exhibits abrupt current switching characteristics similar to those of an AND gate [437]. This remarkable finding might open the path to nanoelectronics where entire logic operations are built from nanotubes with multiple taps rather than by substituting field-controlled nanotubes for Si-based MOSFETs in otherwise conventional MOS subcircuits. Future research will be needed to obtain a better understanding of the physical mechanisms involved, to obtain sufficient amplification in nanotube logic networks, and to intentionally modify the electronic properties of nanotubes.

Nanojunctions

Devices made of crossed nanowires are reminiscent of bipolar design. Nanowires of p-type Si and n-type GaN materials with diameters on the order of 10 to 30 nm can be made to form pn-junctions at their intersections. The successful manufacturing of diodes, of field-effect transistors, and of a few logic gates was also reported in 2001 [438].

Molecular electronics

[439] have built small combinational functions by placing carbon monoxide molecules in atomically precise configurations with the aid of a scanning tunneling microscope. The "circuits" that operate at a temperature of 5 K are orders of magnitude smaller than their silicon equivalents, but are also exceedingly slow.

While all these results are very exciting indeed, many hurdles remain before nanodevices can possibly match up with CMOS VLSI. These include the ability to deposit or grow nanotubes, custom molecules, or other nanodevices at specific locations, shrinking interconnects to the same scale as the devices, and increasing both fabrication yields and operating frequencies by orders of magnitude. Computing with molecule-size devices implies operating close to the thermal limit where computation becomes probabilistic, and will be subject to unprecedented levels of variability and unreliability. Replacing current architectures with new ones that permit one to build dependable systems from highly unreliable devices is an open issue for nano-scale computing.

Crossbar logic

Manufacturing layout details to individual specifications is a major obstacle to shrinking memory and logic circuits down to the scale of a few nanometers. Electrically programming prefabricated circuits of highly regular, periodic layout instead would clearly help by easing the requirements put on lithography. In addition, generous logic redundancy can be made to cope with a substantial amount of fabrication defects and variability. It is because of these two properties that crossbar-type arrangements — reminiscent of PLAs — are currently believed to be among the most promising candidates for going beyond transistor-based microelectronics.

Crossbar structures comprise two orthogonal layers of nanowires separated by a monolayer of molecules. Molecular connections on the basis of modified rotaxanes have been developed that can be switched electrically between a low- and a high-resistance state and that retain their state for three years or longer [440]. Certain material stacks exhibit uncommon current–voltage characteristics with an 8-shaped hysteresis. In 2004, [441] of Hewlett-Packard showed that it is possible to construct a bistable device from two such stacks, provided their switching thresholds can be adjusted to differ from each other in a controlled way. With its data storage, logic inversion, and level-restoration capabilities, such a device could be combined with molecular-scale resistor- or junction-type connections to form arbitrary logic arranged into crossbar-type architectures. This has earned it the name "crossbar latch", although the bistable is strictly unclocked. While the circuits operate at room temperature, they are, at present, of μm rather than nm size, operate at audio frequencies, and have limited lifetime.

Magnetic flux quantum device

Magnetic flux quantum devices rely on the quantum nature of magnetic flux in a superconducting circuit to represent a binary datum. Three-layer **Josephson junction** structures that operate at a temperature of approximately 4 K are being used for both data storage and data processing. The **rapid single flux quantum** (RSFQ) logic being investigated today is built around the voltage pulses that occur each time a magnetic flux quantum enters or leaves the superconducting loop. With a mere 6 eV or so dissipated to cycle one bit, energy efficiency is three to four orders of magnitude better than that of CMOS in spite of higher clock frequencies. Circuits of substantial size have already been fabricated [442], such as a high-resolution A/D converter that comprises 3000 Josephson junctions over an active area of 2 mm by 7 mm and that runs at 12.8 GHz. Another achievement is a 16 kibit **cryogenic RAM** with 400 ps access and 100 ps cycle time [443]. Superconducting RSFQ logic is likely to enter the hardware scene as one of the enabling technologies for future supercomputers [444].

Quantum cellular arrays

A QCA cell is constructed by arranging four quantum dots in a square. Two extra mobile electrons are allowed to tunnel between neighboring sites. As a consequence of Coulomb forces, the two electrons must occupy one out of two diagonal patterns at any time. In an isolated cell, the choice between these two states is arbitrary. For a larger array of cells, however, the state of each cell is determined by its interaction with neighboring cells through Coulomb forces. Although experimental verification is, at present, limited to a simple signal propagation path (wire), plans do exist for forks (fanout), inverters, and majority gates [445].

Quantum devices

Other quantum devices currently being investigated include

- single-electron transistors (SETs) and single-electron memories,
- resonant tunneling devices (RTDs), and
- quantum dots and quantum wires.

All quantum approaches suffer from shortcomings such as limited on/off ratio, lack of current drive capability, undesirable interaction with the environment, and critical noise margins. Extensive research is under way. So far, quantum devices have been commercially successful only where silicon could not do the job such as in emitting light (LEDs, laser diodes).

In any case, the key question is whether prospective alternative forms of logic can be smaller, cheaper, and faster, and dissipate less energy than future CMOS circuits. The authors of [446] conclude that there are several viable emerging memory technologies, but no alternative superior to CMOS in terms of density, speed, and energy. Revolutionary concepts are thus quite unlikely to take over before the progress of CMOS and its extensions slows down or comes to a standstill.

Fundamental bounds imposed by quantum physics, materials, switching devices, etc. are discussed in [447], while [487] [448] [449] focus on CMOS technology and practical limitations. Starting from conventional CMOS devices, [409] and [450] propose and evaluate various extensions for logic and memory applications whereas [451] expects silicon technology to come to a standstill in the year 2013.

15.3 | Technology push

15.3.1 The so-called industry "laws" and the forces behind them

What pushes the industry forward is the search for competitive advantages. The net result for semiconductor memories has been captured in Moore's legendary (first) "law".

Moore's law: The capacity of DRAMs quadruples approximately every three years.

Back in 1965, Gordon Moore — who was later to co-found Intel — had observed that memory capacity would double every year. This initial observation had remained valid until the late 1970s when the doubling period slowed down to 18 months, but the exponential growth remained. Keeping this stunning progress alive for four decades would not have been possible without ever heavier investments in fabrication equipment and in R&D. In fact, there is a second "law" by Moore which is as important, but not as popular as his first.

Moore's second law: The capital requirements for a DRAM fab (construction and equipment) grow by a factor of 1.8 over the three years that separate one memory generation from the next.

While the value of 1.8 is debatable, the continuous growth of the sums to be invested is not.

It is extremely important to note that none of the above is a law of science or nature. Rather, these are empirical estimations on the dynamics of the semiconductor industry that rest on observations from the past. Extrapolations into a distant future are likely to be misleading as such "predictions" do not account for changes in technology, consumer markets, and corporate finance that we must expect to see when the semiconductor industry approaches a point where progress becomes increasingly difficult and expensive.

An observation made by John Sturtevant and illustrated by the graph in fig.15.5 vividly documents the difficulties of predicting the future in high-tech industries.

Sturtevant's law: According to experts in the field, optical lithography has always been anticipated to come to an end six or seven years in the future.

Many times over, unforeseen innovations, including the various resolution enhancement techniques (RETs) discussed in section 14.2.5, have been able to extend the lifetime of this technology.

Back to Moore's (first) law, we are thus likely to witness one of three possible outcomes within our lifetime.

- o Moore's law will slow down and eventually come to a standstill by the time atomic and quantum scales are approached.
- o Moore's law will remain in place, but only by shifting over to some radically different kind of device technology (please refer to section 15.2 for potential contenders).

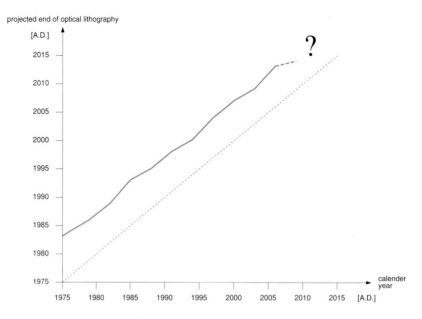

Fig. 15.5 The anticipated end of optical lithography over time (after John Sturtevant).

○ Moore's law will remain in place, but only by interpreting it at the system level, rather than at the device scale, with high-density packaging, 3D-integration technologies, and new assembly procedures and materials gradually gaining in importance.

That the future evolution of microelectronics technology must be consistent with the laws of physics goes without saying. Yet, any kind of industrial activity is governed by the laws of economics, and hence also depends on markets and capital. More specifically, there is a need to recover current investments before committing oneself to new ones. And if Moore's law is to live on, the challenge will be to maintain the decreasing **cost per function** that has always been the virtue of CMOS scaling.

Price competition forces semiconductor manufacturers to lower their unit costs. The fiercest competition is for **commodity chips**, that is for VLSI circuits of highly standardized functionality that get produced in very large quantities (primarily RAMs, but also microprocessors and many ASSPs). Lower unit price together with other benefits obtained from downscaling are the driving forces behind the relentless race towards new materials, next-generation lithography, more metal layers, larger wafers, and better yield. All this gave rise to Moore's first law and continues to support the general trend but says nothing on whether the actual doubling period is 12, 18 or 24 months.

The downside of this technological progress is an expanding capital need of the semiconductor industry as captured by Moore's second law. By 2005, construction and equipment of a state-of-the-art fabrication plant together ask for an investment between 2 and 5 GUSD. The sales volume required to make such an investment profitable is on the order of 5 GUSD/a, which corresponds to at least 5 million wafer starts per year [452].[10]

---

[10] For comparison, the capital requirements of a few large industrial and national undertakings are stated below.

Also note that expenses for fab construction and manufacturing equipment must be amortized over a period of four years or less in the semiconductor industry.

Examples

IBM operates a state-of-the-art 300 mm wafer processing facility in East Fishkill (NY) for circuits in silicon-on-insulator (SOI) CMOS technology with dual damascene additive copper metallization. For the 130 nm technology generation, investments in the manufacturing plant were on the order of 3 GUSD. 3.2 GUSD is the budget for the new fab AMD plans to erect in Malta (NY). Intel's current projections for a 450 mm fab are of 5 GUSD or more.

☐

Sturtevant's law notwithstanding, optical lithography (with refractive optics) will come to an end one day. Important investments are currently being made to develop new light sources, new photoresists, new masking techniques, new optics, new processes, and a new metrology. Post-optical lithography is, therefore, likely to require considerably longer pay-back periods. Cutting feature size in two, which currently takes place within six years, may then ask for ten years or more. Already today, industry is struggling to reuse equipment and skills at the present rate of introduction of new process flows, materials, and tools.

Observation 15.2. *Ever heavier investments tend to impose standardization of fabrication processes, global alliances, contract manufacturing, joint ventures, and accelerated concentration in the semiconductor industry.*

15.3.2 Industrial roadmaps

The most authoritative document on the plans of the semiconductor industry is the **International Technology Roadmap for Semiconductors** (ITRS), the result of a world-wide consensus-building process that attempts to look 15 years into the future and that gets completely revised every two years. An excerpt of the edition finalized late in 2005 [1] is given in table 15.3. Do not misinterpret the ITRS as predictions of the future. Rather, the roadmap identifies those technical capabilities that have to be developed so that industry can essentially stay on Moore's law and can continue to lower cost per function by 29% a year as it has done in the past.

- The budget for the Large Hadron Collider, a particle accelerator built from superconducting magnets to be installed in CERN's existing 27 km circular tunnel near Geneva, amounts to 4.7 GCHF = 3.8 GUSD.
- A terrestrial link across the Øresund connecting Malmö and Copenhagen via a suspension bridge, a tunnel, an artificial island, and a series of dams was opened in the year 2000. Closing the 16 km gap in the European traffic system is estimated to have cost some 2.6 GUSD.
- The Human Genome Project (HGP) carried out by a coalition of a dozen or so research institutions got funded by the US government with a budget of 3 GUSD.
- In 2001, a new trunk line for high-speed trains (TGV) was inaugurated in the Rhône valley from Lyon to Marseille and Nîmes. Total project costs for the 250 km stretch including bridges, tunnels, and new railway stations were 24.2 GFFR = 3.15 GUSD.
- 4.5 GUSD were spent on constructing the USS "Harry S. Truman", the world's largest aircraft carrier that was put into service in 1998.
- Developing the Airbus A380 high-capacity passenger aircraft has called for investments on the order of 12 GEUR.

Table 15.3 Targets for the overall technology characteristics, excerpt from the 2005 International Technology Roadmap for Semiconductors. In spite of minor inconsistencies, pre-2007 figures were kept for comparison. Storage capacities, overall transistor counts, and die sizes refer to volume production as opposed to the introduction of engineering samples. Also, where the Roadmap distinguishes between high performance, low operating power, and low standby power circuits, the figures for the first category are given unless stated otherwise.

						semiconductor technology in general						
year	min. half pitch [nm]	metal level	gate oxide thick.[a] [nm]	threshold voltage [mV]	supply voltage [V]	power dissipation[b] [W]	FET curr. drive[c] [A/m]	FET leakage curr.[d] [A/m]	inverter delay[e] [ps]	lithograph. field [mm²]	wafer diameter [cm]
2002	130	7	1.5		1.1–1.2	2.4–130	900	0.01	7.6	800	30
2004	90	10	1.2	200–500	0.9–1.2	2.2–158	1110	0.05	2.85	704	30
2007	65	11	1.1	165–524	0.8–1.2	3.0–189	1200	0.20	1.92	858	30
2010	45	12	0.7	151–502	0.7–1.1	3.0–198	2050	0.28	1.20	858	30
2013	32	13	0.6	167–483	0.6–1.0	3.0–198	2220	0.29	0.75	858	45
2016	22	13	0.5	195–487	0.5–1.0	3.0–198	2713	0.11	0.45	858	45
2019	16	14	0.5	205–488	0.5–1.0	3.0–198	2744	0.11	0.30	858	45

[a] Physical t_{ox} of SiO_2 or equivalent oxide thickness (EOT) in the case of a different material.
[b] Lower bound applies to battery-operated CPU, upper bound to high-performance CPU with forced cooling.
[c] Nominal n-channel "on"-state current $I_{ds\,on\,n}$ at 25 °C junction temperature. $I_{ds\,on\,p} \approx -(0.4 \text{ to } 0.5)\,I_{ds\,on\,n}$.
[d] Maximum "off"-state drain-to-source current $|I_{ds\,off}|$ at 25 °C junction temperature and nominal U_{dd}.
[e] Inverter delay $t_{pd\,inv} \approx 3\,C_n U_{dd}/I_{ds\,on\,n}$, that is three times the intrinsic NMOS delay given in the roadmap.

	DRAMs				microprocessors				ASICs		
year	bit cell area [μm²]	rel. cell area[b]	die size [mm²]	storage capacity [bit]	die size [mm²]	transistor count [M]	local clock freq.[a] [MHz]	total pins or balls[c]	transistor density [M/mm²]	rel. trans. area[d]	total pins or balls[e]
2002	0.130	8	127	512Mi	310	276	1684		0.89	66	
2004	0.065	8	110	1Gi	310	553	4171		1.78	69	
2007	0.0324	8	110	2Gi	310	1106	9285	1088	3.57	68	3371
2010	0.0122	6	93	4Gi	310	2212	15 079	1450	7.14	69	4015
2013	0.0061	6	93	8Gi	310	4424	22 980	1930	14.3	68	4736
2016	0.0038	6	93	16Gi	310	8848	39 683	2568	28.5	72	5483
2019	0.0015	6	93	32Gi	310	17 696	62 443	3418	57.1	68	6347

[a] Local on-chip clock, peripheral I/O clock is slower.
[b] Area of one bit cell in multiples of F^2, where F^2 stands for the area of one lithographic square.
[c] 1/3 for I/O signals, 2/3 for power and ground.
[d] Average area per transistor in multiples of F^2, where F^2 stands for the area of one lithographic square.
[e] 1/2 for I/O signals, 1/2 for power and ground.

15.4 | Market pull

The formidable projected growth of integration density strains the imagination of all involved in the microelectronics industry and business, but the most essential questions are

"How much computing and communication does the world really need?" and consequently "How much is it prepared to pay for that?"

Throughout the past fifty years, the thirst for computing power, memory capacity, and data rate has escalated in pace with the progress of microelectronics. Will this continue for future ULSI technologies even after today's market drivers (such as personal computing, personal and global communication, Internet services, multimedia, smartcards, and automotive components) are firmly in place? What kind of applications will ask and pay for 32 Gibit DRAMs and for 62 GHz CPUs in large quantities by the year 2019 so as to justify the necessary investments?

High tech per se is of little value for society. Demand in microelectronics is about the uses and the usefulness of information technology. All perceivable needs of the general public seem to fall into one of the categories below:

- Telecommunication (e.g. telephone, messaging, Internet, videoconferencing, and the like).
- Entertainment (e.g. radio, TV, audio, video, multimedia, Internet, games).
- Personal computing (e.g. document processing, spreadsheets, presentations, archiving).
- Business and administration (e.g. databases, billing, accounting, inventory management).
- Electronic payment systems (e.g. debit and credit cards, transaction processing).
- Automation and energy management (e.g. process control, building control).
- Transportation equipment (e.g. vehicle electronics, traffic control, logistics).
- Health care and medicine (e.g. diagnosis and therapy, pacemakers, hearing aids).
- Security and privacy (e.g. alarm systems, ciphering, personal identification).
- Rights management (e.g. software licensing, watermarking).
- Learning, information processing, and knowledge management in general.

At some point in the future, it may well be that many of these needs will be largely satisfied by the technology then available [453]. Customers for top-notch microelectronics might more and more become confined to a few companies and agencies that notoriously find themselves in bad need of supercomputing capabilities, e.g. for simulation of fluid dynamics and aerodynamics, weather forecasting and climate modelling, nuclear physics, pattern recognition, missile guidance, cryptanalysis, vehicle crash simulation, movie animation, ray tracing and wave propagation analysis, financial and economic forecasts, DNA and genome analysis, drug and materials research, and — ironically — semiconductor device and process simulation at the nanotechnology scale.

Observation 15.3. *For humans to develop any broader interest in a new technology, a product or service must ease their daily lives, offer them improved safety and comfort, provide them with economic advantages, improve their social status, or bring fun and entertainment.*

Examples and counterexamples

The reason why the CD has displaced the vinyl record was not just the better sound quality but also the smaller format, absence of wearout, and easier handling (troublefree loading, relative robustness, random access, program information). The fact that industry had agreed on one common standard reassured customers. Much the same story was later repeated with the DVD versus the VHS cassette.

Cellular phones (e.g. GSM) have been an immediate success in spite of their mediocre audio quality and higher charges when compared with the conventional telephone network. What the general public valued most were mobility, convenience, and a spontaneous lifestyle made possible thereby.

The wireless application protocol (WAP), in contrast, was a complete failure because the idea of handling complex transactions from a hand-held device with its tiny display, a down-sized keyboard, and slow data transmission made no sense to customers.

High-definition television (HDTV) still is slow to catch on (in Europe). Both consumers and broadcasters have little incentive to invest heavily in new equipment when they feel traditional analog-transmission TV is good enough to meet their expectations.
□

On more technical grounds, the non-recurring costs of IC manufacturing grow with each process generation. Ever more complex wafer manufacturing processes, below-wavelength lithography, and a generous number of metal layers all take their toll. Post-optical lithography is bound to further inflate up-front costs. This trend is becoming prohibitive for certain IC categories, such as USICs, and the gap is being filled by field-programmable logic (FPL) and software-controlled microprocessors or signal processors wherever possible.

15.5 | Evolution paths for design methodology

15.5.1   The productivity problem

Semiconductor fabrication technology outstrips our capabilities to design, verify, and test VLSI circuits. For microprocessors, the 2005 ITRS roadmap postulates that the maximum transistor count grows by a factor of 32 over fifteen years (i.e. by a factor of two every three years). To match this rate, design productivity must improve by 26% per year or close to 2% a month.

Example

Early microprocessors were designed by a handful of people in a couple of months. In spite of generous EDA resources, the design of a top-performance microprocessor today takes more than five years and involves up to a thousand people in its final phase. Nevertheless, processors reach market with dozens of bugs as do other VLSI chips.
□

Much of the microelectronics industry is subject to **consumerization**, which is to say that the traditional separation between high-tech computer and telecommunication equipment on the one hand and low-cost consumer products on the other has largely gone. Most of today's electronic goods are both high-tech and low-cost, just consider the PC, DVD, HDTV, ADSL, GSM, UMTS, and other multimedia platforms.

In this situation, a key question for the entire ASIC industry is that of

"How to design and test ever more complex circuits with less effort".

Excessive design efforts not only compromise the timely market introduction of new products, but also inflate up-front costs, shifting the break-even point of microelectronic circuits to still higher sales volumes.

IMPROVING CURRENT DESIGN PRACTICES

Focussing human attention on higher levels of abstraction and having design automation software take care of all lower-level details is a strategy with a long and highly successful tradition in digital VLSI design.[11] Forthcoming extensions of this strategy are expected to work from largely behavioral system models like MATLAB/SIMULINK and other visual formalisms such as statecharts and flow graphs that are more concise and intuitive than RTL HDL code.

A totally contrarian tendency is to hold VLSI designers responsible for all sorts of nasty details that are not really their core business.

- Standard cell libraries are on the market with no built-in body ties. Burdening digital designers with adding separate body tie cells as part of cell placement unnecessarily inflates the number of physical design iterations and holds a risk of mistaken choices.

- There is a trend towards asking circuit designers to include phase shift mask (PSM) and optical proximity correction (OPC) features in their layout drawings. Such details should definitely be confined to specialists such as library developers or, better still, to EDA tools capable of adjusting and fine-tuning layout data in an automatic post-processing step.

- Much the same also holds for protection against process-induced damage associated with reactive-ion etching, time-dependent dielectric breakdown (TDDB), and antenna rules.

- Most design flows require that DRC and LVS be carried out over the full set of layers. In cell-based design where designers do the metal routing but neither draw nor can control the layouts of the silicon layers underneath, this is a waste of effort as designers are unnecessarily forced to procure and process layout data at the full-detail level. By respecting a number of restrictions on cell layouts — which sub-wavelength lithography will impose anyway — it should be possible to isolate them from details and to draw a line of separation between their responsibilities and those of library developers.

Tool integration can still be improved. An obstacle that needs to be removed are the distinct specification languages and design representations currently being used by software and hardware engineers. Merging these two "ladders" into one will help to better support architectural decisions

---

[11] Just recall major milestones from the past:
- Mask generation from computer data,
- Automatic design rule check (DRC),
- Schematic entry (SPICE-type simulation),
- Automatic physical design (cell-based design, automatic place and route (P&R), macrocell generation)
- Logic synthesis (including automatic test pattern generation),
- HDLs and synthesis from RTL models, and the adoption of
- Virtual components (VC).

[454]. Architecture design might eventually be carried out with little human intervention by future electronic system-level (ESL) synthesis tools.

A successful concept is design reuse. Purchasing/licensing of entire subsystems is highly popular, and both the technical and business implications of VCs have been discussed in section 13.4. Another approach to design reuse is **incremental design**, aka model year development, whereby only part of a ULSI circuit is re-engineered from one product generation to the next. Incremental design is standard practice for PCs, GSM chipsets, and automotive electronics, to name just a few.

Among others things, VCs are hampered by the lack of standard interfaces. Ideally, connecting digital subsystems should be as convenient as connecting analog audio equipment. Absolute minimum requirements for a standard interface include

a) agreed-on data and message formats,
b) agreed-on mechanisms for exception handling,
c) agreed-on data transfer protocols, and
d) flawless timing.

Note that globally asynchronous locally synchronous (GALS) system operation addresses c) and d). Standardization efforts are undertaken and coordinated by the Virtual Socket Interface Alliance (VSIA). On-chip bus and interface standards such as the AMBA (Advanced Microprocessor Bus Architecture) family clearly help. The absence of industry-wide naming conventions for signals is a similar, yet much more mundane, impediment.

More than ever, we are designing circuits beyond our capabilities of verification. In spite of occupying an unreasonable proportion of the overall development effort, design verification cannot guarantee the absence of functional or electrical flaws. Simulation alone is clearly insufficient as too many ASICs and virtual components (VC) fail when put into service in a real environment. Formal methods are only slowly coming to help. Problems in need of major improvements include verification at the system level, executable specifications, and variation modelling.

PUSHING INTEGRATION DENSITY TO THE MAX DOES NOT ALWAYS MAKE SENSE

The idea behind the buzzword **system-on-a-chip** (SoC) is to integrate a complex system on a single die. While appealing in theory, there are a number of practical problems associated with developing, manufacturing, and marketing highly complex and highly specific ASICs.

- Design and verification take a lot of time and effort.
- Yields are likely to suffer as die sizes grow large.
- High power densities call for expensive cooling.
- Products cannot be scaled up and down to meet a variety of needs.
- Highly selective and narrow markets imply smaller sales volumes.
- All this boils down to high up-front costs and a high risk.

The problem becomes even more serious in true systems that ask for largely different circuit technologies, such as flash memory, optoelectronics, and bipolar RF circuits, to be integrated on the same die as the digital subsystem. As explained in section 11.4.7, a high-density package that combines a couple of dies with the necessary discrete components is more appropriate whenever a technologically heterogeneous product is to be manufactured in not-so-large quantities.

15.5.2 Fresh approaches to architecture design

WHAT TO DO WITH SO MANY DEVICES?

The bottom line of table 15.3 refers to the year 2019. It predicts semiconductor technology will be capable of fabricating memories with more than $32 \cdot 10^9$ transistors and top-end microprocessors with $17 \cdot 10^9$ or so transistors on a single chip.

"What is the best use industry can make of this rich proliferation of devices?"

Field-programmable logic (FPL) provides us with a vivid example of how an abundant and cheap resource, namely transistors, is turned into qualities that are valued much higher in the marketplace, namely agility and short turnaround times. Also, a fair amount of programmability and configurability is about the only way to successfully market highly complex parts. The alternative of putting more and more transistors into service to implement ever more specialized functions tends to narrow the application range and to reduce sales volume.

Instruction set processors go one step further in that they are not only highly flexible and universal but also provide an abstract machine, a simple mental model that serves as the starting point for software design and compilation. Application developers are thus freed from having to bother about hardware details and become free to focus on functional issues. Many transistors and much circuit acitivity are "wasted" in making this possible, though. Still, for the sake of productivity, standard processor cores plus on-chip firmware are likely to replace dedicated architectures in many ASICs where throughput and heat removal are not of foremost concern.

Another quality highly sought-after in the marketplace is energy efficiency. The question is how more transistors can possibly be put to service to lower static and dynamic power significantly and at the same time. In the 1980s, CMOS logic had displaced TTL and ECL in spite of its inferior switching speed because only CMOS circuits proved to be amenable to truly large-scale integration. This would have been impossible, had the much better energy efficiency of CMOS not allowed for more than compensating the loss in throughput from the slower devices with more sophisticated and more complex architectures. Equally important was the fact that VLSI slashed node capacitances, interconnect delays, and – above all — manufacturing costs. CMOS scaling further provided a perspective for future development. [455] thinks it should be possible to repeat this exploit by combining ultra-low-voltage operation with 3D integration.

As observed in [456], circuits with many billions of devices just give rise to other concerns:

"It is unlikely that all of these devices will work as anticipated by the designer."
"Nobody will be able to functionally test such a circuit to a reasonable degree."

Reliability and fault tolerance may be achieved by pursuing error correction, built-in self-test (BIST), self-diagnosis, redundant hardware, and possibly even self-repair. However, while these approaches can protect against fabrication defects and failures that occur during circuit operation, they fail to address faults or omissions made as part of the design process. Only the future will tell whether more utopian ideas such as self-programming and self-replication are technically and economically viable approaches for being embodied in semiconductor circuits, or whether they will remain confined to biological creatures [457] [458].

In view of engineering efficiency, future gigachips must be based on regular arrays of regularly connected circuits, or it is unlikely that their design and test will ever be completed. In addition, the circuits and connections will have to be (re)configurable to solve any problems from locally malfunctioning devices and interconnect [459]. Memory chips have long been designed along these principles but it is unclear how to apply them to information processing circuits.

CLOCK FREQUENCIES, CORE SIZES, AND THERMAL POWER CANNOT GROW INDEFINITELY

The domination of interconnect delay impacts architecture design because rapid interaction over chip-size distances has become impractical. Thermal and energy efficiency considerations further limit node activity budgets. As a result, CPU architecture design has moved towards multiprocessors since 2005 after a frenzied race towards multi-GHz clock rates and ever more complex uniprocessors. Fresh approaches are sought; others known for years may see a revival:

- Moving clock distribution from the chip to the package level where RC delays are much smaller as explained in section 11.4.7.[12]
- Combining fast local clocks (determined by a few gate delays) with a slower global clock (bounded by the longest global interconnect).
- Extensive clustering whereby an architecture is broken down into subsystems or clusters that operate concurrently with as little inter-cluster communication as possible [28]; [460] foresees a maximum cluster size of 50–100 kGE.[13] The approach can be complemented with programmable interconnect between clusters.
- Processing-in-memory (PiM) architectures attempt to do away with the memory bottleneck of traditional CPUs and cache hierarchies by combining many data processing units and memory sections on a single chip.[14]
- Globally asynchronous locally synchronous (GALS) and similar concepts where stallable subsystems exchange data via latency-insensitive communication protocols [461].
- Data flow architectures where execution is driven by the availability of operands.
- Networks on chip (NoC) whereby major subsystems exchange data via packet switching.
- Logic gates as repeaters (LGR) is a concept whereby cells from a design's regular netlist are extensively inserted into long wires in lieu of the extra inverters or buffers normally used as repeaters. Put differently, the functional logic gets distributed into the interconnect [462]. The goal is to minimize the longest path delay without the waste of area and energy incurred with pipelined interconnect.
- Systolic arrays and cellular automata with signals propagating as wavefronts.
- Neural-network-style architectures, aka biologically inspired computing or amorphous computing, where a multitude of primitive and initially identical cells self-organize into a more powerful network of a specific functionality.

---

[12] Remember that flip-chip techniques can connect to anywhere on a die, not just to the periphery.

[13] Not only the Cell microprocessor jointly developed by Sony, Toshiba, and IBM, but also Sun's Niagara CPU can be viewed as steps in this direction.

[14] The strict separation into a general-purpose CPU and a large memory system is a characteristic trait of the von Neumann and Harvard computer architectures and not normally found in the dedicated hardware architectures presented in chapter 2.

Observation 15.4. *Deep submicron architecture design and floorplanning essentially follow the motto "Plan signal distribution first, only then fill in the local processing circuitry!"*

The term **wire planning** describes an approach that begins by determining an optimal plan for global wiring that distributes the acceptable delays over functional blocks and their interconnects. Logic synthesis, and place and route, are then commissioned to work out the details taking advantage of timing slacks [463]. Wave steering is a related effort that integrates logic synthesis for pass transistor logic (PTL) with layout design [464].

CIRCUIT STYLE

As stated at the beginning of this chapter, it is the search for improvements in

+ layout density,
+ operating frequency, and
+ energy efficiency

that is driving the rush to ever smaller geometries. Yet, various electrical characteristics are bound to deteriorate as a consequence of shrinking device dimensions. These include

− "off"-state leakage current (drain to source),
− gate dielectric tunneling (gate to channel),
− drain junction tunneling (drain to bulk),
− "on"-to-"off"-state conductance ratio,
− parameter variations and device matching,
− transfer characteristic and voltage amplification of logic gates,
− cross-coupling effects,
− noise margins, and the
− susceptibility to all kinds of disturbances.[15]

While DRAMs are highly sensitive to leakage, static CMOS logic is less so. Fully complementary CMOS subcircuits are ratioless and level-restoring, two properties that render static CMOS logic fairly tolerant with respect to both systematic deterioration and random variability of device parameters. However, as the search for power efficiency mandates modest voltage swings and as supply voltages are expected to drop well below 1 V for technology reasons, differential signaling is bound to become more pervasive in order to maintain adequate noise margins.

15.6 | Summary

• What has fueled the spectacular evolution of CMOS into a high-density, high-performance, low-cost technology essentially was its scaling property. This trend will continue into the future

---

[15] Such as ground bounce, crosstalk, radiation, ESD, PTV, and OCV variations. On the positive side, latch-up will no longer be a problem with core <u>and</u> I/O voltages below 1.5 V or with the transition to SOI technology.

at the price of admitting new materials into the fabrication process (gate stacks, interlevel dielectrics, magnetoresistive or chalcogenide layers, etc.).

- The cost structure of VLSI has always favored high fabrication volumes and — at the same time — penalized products that sell in small quantities. The move to more sophisticated fabrication processes is going to further accentuate this trait because better lithographic resolution, new materials, more interconnect layers, more lithographic steps, larger wafers, better but more expensive process equipment, more complex circuits, more sophisticated engineering software, the purchase of VCs, and more onerous test equipment all contribute to inflating NRE costs. As a consequence, ASIC vendors are becoming more selective in accepting low-volume business; FPL and program-controlled processors fill the gap.

15.7 | Six grand challenges

As a final note, let us summarize what we consider the most challenging problems that the semiconductor industry as a whole currently faces. Note that addressing those problems involves rethinking across many levels: devices, circuits, architectures, operating system, application software, design methodology, EDA, testing, manufacturing, and business models.

1. How to make VLSI systems more energy-efficient in terms of both dynamic and static losses.
2. How to have design productivity keep pace with manufacturing capabilities.
3. How to verify (test) highly complex and/or heterogenous designs (circuits).
4. How to cope with increasing device and interconnect variabilities.
5. How to survive the upcoming transitions to post-optical lithography, 450 mm wafers, new device topologies, new materials, and nanotechnologies.
6. How to accommodate products that do not sell in huge quantities with more reasonable cost structures.

15.8 | Appendix: Non-semiconductor storage technologies for comparison

STORAGE DENSITIES

storage technology	approximate density	comment
Magnetic recording		
longitudinal recording	100 Gbit/inch^2 ≈ 0.0065 μm^2/bit	2005 production
perpendicular recording	130 Gbit/inch^2 ≈ 0.0050 μm^2/bit	2006 production
	230 Gbit/inch^2 ≈ 0.0028 μm^2/bit	2005 laboratory

STORAGE CAPACITIES

storage medium	approx. capacity	typical information content
Optical disks (single-sided single-layered, laser wavelength)		
compact disk (CD), 780 nm (extreme red)	6.3 Gbit	74 min uncompressed audio or 93 500 pages of text
digital versatile disk (DVD), 650 nm (red)	38 Gbit	133 min MPEG-2 compressed video or 564 000 pages of text
"blu-ray" disk (DVR), 405 nm (blue–violet)	216 Gbit	12.5 h MPEG-2 compressed video or 2 h HDTV video
Magnetic disks		
high-end personal computer 3.5 inch hard disk (2006)	6 Tbit	operating system, application software and user data
DNA sequences[a]		
Mycobacterium tuberculosis H37Rv	8.8 Mbit	4 411 529 base pairs of raw genome data
Homo sapiens	6 Gbit	approx. 3 G base pairs organized into 35 000 or so genes and distributed over 23 pairs of chromosomes
Human brain		
permanent memory	1.8 Gbit	knowledge and recollection
life-long speed reading	260 Gbit	printed text read at 5 letters/word, 1 kword/min, 8 h/d, 7 d/w for 60 a

[a] Keep in mind that the entire genome is repeated in every cell throughout an organism.

Appendix A

Elementary Digital Electronics

A.1 | Introduction

Working with electronic design automation (EDA) tools requires a good understanding of a multitude of terms and concepts from elementary digital electronics. The material in this chapter aims at explaining them, but makes no attempt to truly cover switching algebra or logic optimzation as gate-level synthesis is fully automated today. Readers in search of a more formal or more comprehensive treatise are referred to specialized textbooks and tutorials such as [465] [466] [25] [467] and the seminal but now somewhat dated [468].[1] Textbooks that address digital design more from a practical perspective include [146] [469] [470] [471].

Combinational functions are discussed in sections A.2 and A.3 with a focus on fundamental properties and on circuit organization respectively before section A.4 gives an overview on common and not so common bistable memory devices. Section A.5 is concerned with transient behavior, which then gets distilled into a few timing quantities in section A.6. At a much higher level of abstraction, section A.7 finally sums up the basic microprocessor data transfer protocols.

A.1.1 Common number representation schemes

Our familiar decimal number system is called a **positional number system** because each digit in a number contributes to the overall value with a weight that depends on its position (this was not so with the ancient Roman numbers, for instance). In a positional number system, there is a natural number $B \geq 2$ that serves as a base, e.g. $B = 10$ for decimal and $B = 2$ for binary numbers. Each digit position i is assigned a weight B^i so that when a non-negative number gets expressed

---

[1] Those with a special interest in mathematics may want to refer to appendix 2.11 where switching algebra is put into perspective with fields and other algebraic structures.

with a total of w digits, the value follows as a weighted sum

$$(a_l, a_{l-1}, ..., a_{r+1}, a_r)_B = \sum_{i=r}^{l} a_i B^i \tag{A.1}$$

where $l \geq r$ and $w = l - r + 1$. A decimal point is used to separate the integer part made up of all digits with index $i \geq 0$ from the fractional part that consists of those with index $i \leq -1$. When writing down an integer, we normally assume $r = 0$. As an example, 173_{10} stands for $1 \cdot 10^2 + 7 \cdot 10^1 + 3 \cdot 10^0$ ($w=3$, $l=2$, $r=0$). Similarly, the binary number 101.01_2 stands for $1 \cdot 2^2 + 0 \cdot 10^1 + 1 \cdot 10^0 + 0 \cdot 10^{-1} + 1 \cdot 10^{-2} = 5.25_{10}$ ($l=2$, $r=-2$, $w=5$). The leftmost digit position has the largest weight B^l while the rightmost digit has the smallest weight B^r. In the context of binary numbers, these two positions are referred to as the most (MSB) and as the least significant bit (LSB) respectively.

Table A.1 | Representations of signed and unsigned integers with four bits.

bit pattern MSB LSB $a_3\, a_2\, a_1\, a_0.$	interpreted as unsigned	interpreted as signed offset-binary O-B	2's complem. 2'C	1's complem. 1'C	sign & magn. S&M
1111.	15	7	-1	$(-)\,0$	-7
1110.	14	6	-2	-1	-6
1101.	13	5	-3	-2	-5
1100.	12	4	-4	-3	-4
1011.	11	3	-5	-4	-3
1010.	10	2	-6	-5	-2
1001.	9	1	-7	-6	-1
1000.	8	0	-8^a	-7	$(-)\,0$
0111.	7	-1	7	7	7
0110.	6	-2	6	6	6
0101.	5	-3	5	5	5
0100.	4	-4	4	4	4
0011.	3	-5	3	3	3
0010.	2	-6	2	2	2
0001.	1	-7	1	1	1
0000.	0	-8^a	0	0	0
bit weights	$2^3\, 2^2\, 2^1\, 2^0.$	same $-\,2^3$	$-2^3\, 2^2\, 2^1\, 2^0.$	$-(2^3\!-\!1)\, 2^2\, 2^1\, 2^0.$	$\pm(2^2\, 2^1\, 2^0).$
sign inversion	n.a.	$(\overline{a_3\, a_2\, a_1\, a_0}) + 1$	$\overline{a_3\, a_2\, a_1\, a_0}$	$\overline{a_3}\, a_2\, a_1\, a_0$	
VHDL type	unsigned	n.a.	signed	n.a.	n.a.

^{a} Has no positive counterpart, sign-inversion rule does not apply.

As for signed numbers, several schemes have been developed to handle them in digital circuits and computers. Table A.1 illustrates how the more common ones map between bit patterns and numbers. For the sake of conciseness, integers of only four bits are considered in the examples.

The leftmost bit always indicates whether a number is positive or negative. Except for that one bit, offset-binary and 2's complement encodings are the same. What they further have in common is that the most negative number has no positive counterpart (with the same number of bits). Conversely, two patterns for zero exist in 1's complement and in sign-and-magnitude representation, which complicates the design of arithmetic units such as adders, subtractors, and comparators. What makes the 2's complement format so popular is the fact that any adder circuit can be used for subtraction if arguments and result are coded in this way.

Observation A.1. *Digital hardware deals with bits exclusively. What gives a bit pattern a meaning as a character, as signed or unsigned, as an integer or fractional number, as a fixed-point or floating-point number, etc., essentially is the interpretation by humans, or by human-made software.*

A bit pattern remains absolutely devoid of meaning unless the pertaining number representation scheme is known.[2] Hardware description languages (HDL) provide digital designers with various data types and with index ranges to assist them in keeping track of number formats.

A.1.2   Notational conventions for two-valued logic

The restriction to two-valued or bivalent logic[3] seems to suggest that the two symbols 0 and 1 from switching algebra should suffice as a basis for mathematical analysis. This is not so, however, and two more logic values are needed so that we end up with a total of four symbols.[4]

- 0 stands for a logic zero.
- 1 stands for a logic one.
- X denotes a situation where a signal's logic state as 0 or 1 remains **unknown** after analysis.
- – implies that the logic state is left open in the specifications because it does not matter for the correct functioning of a circuit. One is thus free to substitute either a 0 or a 1 during circuit design, which explains why this condition is known as **don't care**.

The mathematical convention for identifying the **logic inverse**, aka Boolean complement, of a term is by overlining it, and we will adhere to that convention throughout this chapter. That is, if a is a variable, then its complement shall be denoted $\bar{a}$.[5] Most obviously, one has $\bar{0} = 1$, $\bar{1} = 0$, and $\bar{\bar{a}} = a$.

---

[2]   As an analogy, a pocket calculator handles only numbers and does not know about any physical unit involved, e.g. [m], [kg], [s], [μA], [kΩ] and [EUR]. It is up to the user to enter arguments in correct units and to know how to read the results. Incidentally, note that we do not want to go into floating-point numbers here as floating-point arithmetics is not very common in ASICs. A 32 bit and a 64 bit format are defined in the IEEE 754 standard, handy converters are available on the Internet.

[3]   Note that **binary** is almost always used instead of **bivalent**. This is sometimes misleading as the same term also serves to indicate that a function takes two arguments. The German language, in contrast, makes a distinction between "zweiwertig" (bivalent) and "zweistellig" (takes two arguments).

[4]   Actually, this is still insufficient for practical purposes of circuit design. A more adequate set of nine logic values has been defined in the IEEE 1164 standard and is discussed in full detail in section 4.2.3; what we present here is just a subset.

[5]   Unfortunately, this practice is not viable in the context of EDA software because there is no way to overline identifiers with ASCII characters. A more practical and more comprehensive naming scheme is proposed in section 5.7. Taking the complement is expressed by appending the suffix xB to the original name so that the Boolean complement of A is denoted as AxB (for "A bar").

A.2 | Theoretical background of combinational logic

A digital circuit is qualified as **combinational**, if its present output gets determined by its present input exclusively when in steady-state condition. This contrasts with **sequential** logic, the output of which depends not only on present but also on past input values. Sequential circuits must, therefore, necessarily keep their state in some kind of storage elements whereas combinational ones have no state. This is why the former are also referred to as state-holding or as **memorizing**, and the latter as state-less or as **memoryless**.

In this section, we confine our discussion to combinational functions and begin by asking

"How can we state a combinational function and how do the various formalisms differ?"

A.2.1 Truth table

Probably the most popular way to capture a combinational function is to come up with a truth table, that is with a list that indicates the desired output for each input.

Table A.2 | A truth table of three variables that includes don't care entries.

x	y	z	g
0	0	0	1
0	0	1	1
0	1	0	–
0	1	1	–
1	0	0	1
1	0	1	0
1	1	0	–
1	1	1	0

Let us calculate the number of possible logic functions of n variables. Observe that a truth table comprises 2^n fields, each of which must be filled either with a 0 or a 1 (don't care conditions do not contribute any extra functions). So there are 2^{2^n} different ways to complete a truth table and, hence, 2^{2^n} distinct logic functions.

n	functions	
1	4	
2	16	as listed in table A.4
3	256	
4	65 536	

A.2.2 The n-cube

A geometric representation is obtained by mapping a logic function of n variables onto the n-dimensional unit cube. This requires a total of 2^n nodes, one for each input value. Edges connect all node pairs that differ in a single variable. A drawing of the n-cube for truth table A.2 appears in fig.A.1. Note that the concept of n-cubes can be extended to arbitrary numbers of dimensions, although representing them graphically becomes increasingly difficult.

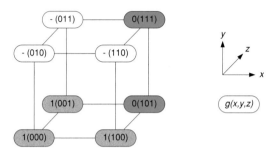

Fig. A.1 3-cube equivalent to table A.2.

A.2.3 Karnaugh map

The Karnaugh map, an example of which is shown in table A.3, is another tabular format where each field stands for one of the 2^n input values. The fields are arranged so as to preserve adjacency relations from the n-cube when the map is thought to be inscribed on a torus. Although extensions for five and six variables have been proposed, the merit of easy visualization which makes Karnaugh maps so popular tends to get lost beyond four variables.

Table A.3 | Karnaugh map equivalent to table A.2.

	g	00	01	11	10
				yz	
x	0	1	1	–	–
	1	1	0	0	–

A.2.4 Program code and other formal languages

Logic operations can further be described by way of a formal language, a medium that has become a focus of attention with the advent of automatic simulation and synthesis tools. A specification on the basis of VHDL is depicted in prog.A.1. Note that, while the function described continues to be combinational, its description is procedural in the sense that the processing of the associated program code must occur step by step.

Program A.1 | A piece of behavioral VHDL code that is equivalent to table A.2

```
entity gfunction is
   port (
       X : in Std_Logic;
       Y : in Std_Logic;
       Z : in Std_Logic;
       G : out Std_Logic );
end gfunction;

architecture procedural of gfunction is
begin
   process (X,Y,Z)
       variable temp: Std_Logic;
   begin
       temp := '-';
       if Y='0' then temp := '1'; end if;
       if X='1' and Z='1' then temp := '0'; end if;
       G <= temp;
   end process;
end procedural;
```

A.2.5 Logic equations

What truth tables, Karnaugh maps, n-cubes, and the VHDL code example shown in prog.A.1 have in common, is that they specify — essentially by enumeration — input-to-output mappings. Put differently, they all define a logic function in purely **behavioral** terms.

Logic equations, in contrast, also imply the operations to use and in what order to apply them. Each such equation suggests a distinct gate-level circuit and, therefore, also conveys information of **structural** nature. Even in the absence of don't care conditions, a great variety of logically equivalent equations exist that implement a given truth table. Since, in addition, it is always possible to expand a logic equation into a more complex one, we note

Observation A.2. *For any given logic function, there exist infinitely many logic equations and gate-level circuits that implement it.*

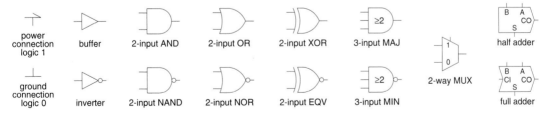

Fig. A.2 Schematic icons of common combinational functions.

Figure A.2 illustrates the symbols used in schematic diagrams to denote the subcircuits that carry out simple combinational operations. Albeit fully exchangeable from a purely functional point of view, two equations and their associated gate-level networks may significantly differ in terms of circuit size, operating speed, energy dissipation, and manufacturing expenses. Such differences often matter from the perspectives of engineering and economy.

Example

The Karnaugh map below defines a combinational function of three variables. Equations (A.2) through (A.11) all implement that very function. Each such equation stands for one specific gate-level circuit and three of them are depicted next. They belong to equations (A.5), (A.10), and (A.11) respectively. More circuit alternatives are shown in fig.A.4.

f		yz			
		00	01	11	10
x	0	1	1	0	0
	1	0	1	0	1

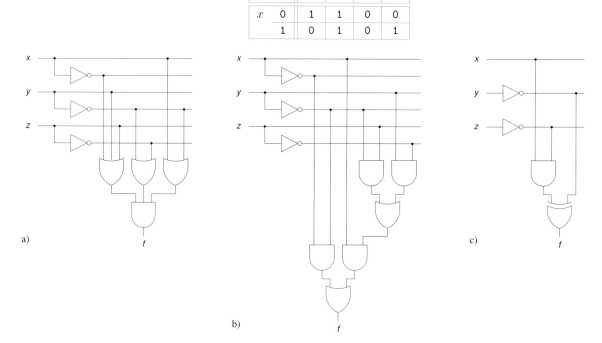

a)   b)   c)

Fig. A.3 A selection of three circuit alternatives for the same logic function.

□

A.2.6 Two-level logic

Sum-of-products

Any switching function can be described as a sum-of-products (SoP), where sum and product refer to logic OR and AND operations respectively, see equations (A.2) and (A.3), for instance.[6]

---

[6] We will denote the sum and product operators from switching algebra as $\vee$ and $\wedge$ respectively to minimize the risk of confusion with the conventional arithmetic operators $+$ and $\cdot$. However, for the sake of brevity, we will frequently drop the $\wedge$ symbol from product terms and write xyz when we mean $x \wedge y \wedge z$. In doing so, we imply that $\wedge$ takes precedence over $\vee$.

Disjunctive form is synonymous for sum-of-products. A product term that includes the full set of variables is called a **minterm** or a fundamental product. The name **canonical sum** stands for a sum-of-products expression that consists of minterms exclusively. The right-hand side of (A.2) is a canonical sum whereas that of (A.3) is not.

$$f = \overline{x}\,\overline{y}\,\overline{z} \vee \overline{x}\,\overline{y}\,z \vee x\,\overline{y}\,z \vee x\,y\,\overline{z} \tag{A.2}$$

$$f = \overline{x}\,\overline{y} \vee \overline{y}\,z \vee x\,y\,\overline{z} \tag{A.3}$$

PRODUCT-OF-SUMS

As the name suggests, product-of-sums (PoS) formulations are dual to SoP formulations. Not surprisingly, the concepts of conjunctive form, **maxterm**, fundamental sum, and **canonical product** are defined analogously to their SoP counterparts. Two PoS examples are given in (A.4) and (A.5); you may want to add the canonical product form yourself.

$$f = (\overline{x} \vee y \vee z)\,(\overline{x} \vee \overline{y} \vee \overline{z})\,(x \vee \overline{y})\,(x \vee \overline{y})\,(x \vee \overline{y} \vee z) \tag{A.4}$$

$$f = (\overline{x} \vee y \vee z)\,(\overline{y} \vee \overline{z})\,(x \vee \overline{y}) \tag{A.5}$$

OTHER TWO-LEVEL LOGIC FORMS

SoP and PoS forms are subsumed as **two-level logic**, aka two-stage logic, because they both make use of two consecutive levels of OR and AND operations. Any inverters required to provide signals in their complemented form are ignored as double-rail logic is assumed.[7] As a consequence, not only (A.2) through (A.5), but also (A.6) and (A.7) qualify as two-level logic.

$$f = \overline{\overline{(\overline{x}\,\overline{y})}\,\overline{(\overline{y}\,z)}\,\overline{(x\,y\,\overline{z})}} \tag{A.6}$$

$$f = \overline{x}\,\overline{y}\,\overline{z} \vee y\,z \vee \overline{x}\,y \tag{A.7}$$

Incidentally, observe that (A.6) describes a circuit that consists of NAND gates and inverters exclusively. As illustrated in fig.A.4, this formulation is easily obtained from (A.3) by applying the De Morgan theorem[8] followed by **bubble pushing**, that is by relocating the negation operators from all inputs of the second-level gates to the outputs of the first-level gates.

Observation A.3. *It is always possible to implement an arbitrary logic function with no more than two consecutive levels of logic operations.*

This is why two-level logic is said to be universal. The availability of manual minimization methods, such as the Karnaugh or the Quine–McCluskey method [468], the multitude of circuit alternatives to be presented in sections A.3.1 through A.3.3, and the — now largely obsolete — belief that propagation delay directly relates to the number of stages have further contributed to the popularity of two-level logic since the early days of digital electronics.

---

[7] The term **double-rail** logic refers to logic families where each variable is being represented by a pair of signals a and $\overline{a}$ that are of opposite value at any time (e.g. in CVSL). Every logic gate has two complementary outputs and pairwise differential inputs. Taking the complement of a variable is tantamount to swapping the two signal wires and requires no extra hardware.

    This situation contrasts with **single-rail** logic, where every variable is being transmitted over a single wire (e.g. in standard CMOS and TTL). A complement must be obtained explicitly by means of an extra inverter.

[8] The **De Morgan theorem** of switching algebra states $\overline{x} \vee \overline{y} = \overline{x\,y}\ (= \overline{x \wedge y})$ and $(\overline{x} \wedge \overline{y} =) \overline{x}\,\overline{y} = \overline{x \vee y}$.

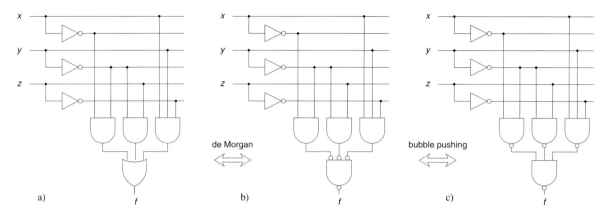

Fig. A.4 Translating an SoP logic (a) into a NAND-NAND circuit (c) or back.

A.2.7 Multilevel logic

Multilevel logic, aka multi-stage logic, differs from two-level logic in that logic equations extend beyond two consecutive levels of OR and AND operations. Examples include (A.8) and (A.9) where three stages of ORs and ANDs alternate; (A.10) with the same operations nested four levels deep also belongs to this class.

$$f = (\overline{x} \vee z)\,\overline{y} \vee x\,(y\,\overline{z}) \tag{A.8}$$

$$f = \overline{(x\,\overline{z} \vee y)\,(\overline{x} \vee \overline{y} \vee z)} \tag{A.9}$$

$$f = \overline{x}\,\overline{y} \vee x\,(\overline{y}z \vee y\overline{z}) \tag{A.10}$$

Equation (A.11) below appears to have logic operations nested no more than two levels deep as well, yet the inclusion of an exclusive-or function[9] makes it multilevel logic. This is because the XOR function is more onerous to implement than an OR or an AND and because substituting those for the XOR results in a total of three consecutive levels of logic operations.

$$f = x\,\overline{z} \oplus \overline{y} \tag{A.11}$$

The circuits that correspond to (A.10) and (A.11) are depicted in fig.A.3b and c respectively. Drawing the remaining two schematics is left to the reader as an exercise.

Originally somewhat left aside due to the lack of systematic and affordable procedures for its minimization, multilevel logic has become popular with the advent of adequate computer tools. VLSI also destroyed the traditional preconception that fewer logic levels would automatically bring about shorter propagation delays.

---

[9] The exclusive-or XOR is also known as the antivalence operation, and its negated counterpart as the equivalence operation EQV or XNOR. Please further note that OR and AND operations take precedence over XOR and EQV.

A.2.8 Symmetric and monotone functions

A logic function is said to be **totally symmetric** iff it remains unchanged for any permutation of its variables; **partial symmetry** exists when just a subset of the variables can be permuted without altering the function. A logic function is characterized as being **monotone** or unate iff it is possible to rewrite it as a sum-of-products expression where each variable appears either in true or in complemented form exclusively. If all variables are present in true form in the SoP, then the function is called **monotone increasing**, and conversely **monotone decreasing** if all variables appear in their complemented form.

Examples

$$c = xy \vee xz \vee yz \tag{A.12}$$

$$s = xyz \vee \overline{x}\,\overline{y}\,z \vee \overline{x}\,y\,\overline{z} \vee x\,\overline{y}\,\overline{z} \tag{A.13}$$

$$h = \overline{\overline{x} \vee \overline{y}\,\overline{z}} = xy \vee xz \tag{A.14}$$

$$m = xz \vee y\overline{z} \tag{A.15}$$

$$n = xy \vee x\overline{z} \tag{A.16}$$

$$o = \overline{wx \vee yz} = \overline{w}\,\overline{y} \vee \overline{w}\,\overline{z} \vee \overline{x}\,\overline{y} \vee \overline{x}\,\overline{z} \tag{A.17}$$

function	name	symmetric	monotone
c (A.12)	3-input majority (MAJ)	totally	increasing
s (A.13)	3-input exclusive or (XOR)	totally	no
h (A.14)	anonymous	partially	increasing
m (A.15)	2-way multiplexer (MUX)	no	no
n (A.16)	anonymous	no	yes
o (A.17)	anonymous	partially	decreasing

☐

A.2.9 Threshold functions

Many combinational functions can be thought to work by counting the number of variables that are at logic 1 and by producing either a 0 or a 1 at the output depending on whether that figure exceeds some fixed number or not.[10] Perforce, all such threshold functions are totally symmetric and monotone. Examples include OR and AND functions along with their inverses.

Probably more interesting are the **majority function** (MAJ) and its inverse the **minority function** (MIN) that find applications in adders and as part of the Muller-C element. MAJ and MIN gates always have an odd number of inputs of three or more. This is because majority and minority are mathematically undefined for even numbers of arguments and are of no practical interest for a single variable. In the case of a 3-input MAJ gate (A.12), the condition for a logic 1 at the output is $\#1s \geq 2$ as reflected by its icon in fig.A.6c.

---

[10] Incidentally, observe the relation to artificial neural networks that make use of similar threshold functions.

A.2.10 Complete gate sets

A set of logic operators is termed a (functionally) **complete gate set** if it is possible to implement arbitrary combinational logic functions from an unlimited supply of its elements.

Examples and counterexamples

Complete gate sets include but are not limited to the following sets of operations: {AND, OR, NOT}, {AND, NOT}, {OR, NOT}, {NAND}, {NOR}, {XOR, AND}, {MAJ, NOT}, {MIN}, {MUX}, and {INH}. As opposed to these, none of the sets {AND, OR}, {XOR, EQV}, and {MAJ} is functionally complete.[9]
□

Though any complete gate set would suffice from a theoretical point of view, actual component and cell libraries include a great variety of logic gates that implement up to one hundred or so distinct logic operations to better support the quest for density, speed, and energy efficiency.

Observe that several complete gate sets have cardinality one, which means that a single operator suffices to construct arbitrary combinational functions. One such gate that deserves special attention is the 4-way MUX. It is in fact possible to build any combinational operation with two arguments from a single such MUX without rewiring. Consider the circuit of fig.A.5, where the two operands are connected to the multiplexer's select inputs. For each 4-bit value that gets applied to data lines p_3 through p_0, the multiplexer then implements one out of the 16 possible switching functions listed in table A.4. The 4-way MUX thus effectively acts as a 2-input gate the functionality of which is freely programmable from externally.

Table A.4 The 16 truth tables and switching functions implemented by the circuit of fig.A.5.

	assignment				function implemented			assignment				function implemented	
	p_3	p_2	p_1	p_0				p_3	p_2	p_1	p_0		
x	1	1	0	0			x	1	1	0	0		
y	1	0	1	0			y	1	0	1	0		
p	setting				$q =$	name	p	setting				$q =$	name
0	0	0	0	0	0	null, never	15	1	1	1	1	1	unity, always
1	0	0	0	1	$\overline{x \vee y}$	NOR, Pierce	14	1	1	1	0	$x \vee y$	OR, sum
2	0	0	1	0	$\overline{x}\, y$	INH x, inhibit	13	1	1	0	1	$x \vee \overline{y}$	y implies x
4	0	1	0	0	$x\, \overline{y}$	INH y, inhibit	11	1	0	1	1	$\overline{x} \vee y$	x implies y
3	0	0	1	1	$\overline{x}$	NOT x	12	1	1	0	0	x	pass x
5	0	1	0	1	$\overline{y}$	NOT y	10	1	0	1	0	y	pass y
6	0	1	1	0	$x \oplus y$	XOR, antival.	9	1	0	0	1	$\overline{x \oplus y}$	EQV, equival.
7	0	1	1	1	$\overline{x\, y}$	NAND, Sheffer	8	1	0	0	0	$x\, y$	AND, product

A.2.11 Multi-output functions

All examples presented so far were single-output functions. We speak of a multi-output function when a **vector** of several bits is produced rather than just a **scalar** signal of cardinality one.

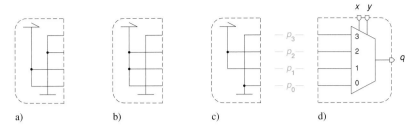

Fig. A.5 A programmable logic gate. 4-way multiplexer (d) with the necessary settings for making it work as an inverter (a), a 2-input NAND gate (b), and as an XOR gate (c).

Example

The **full adder** is a simple multi-output function of fundamental importance. It adds two binary digits and a carry input to obtain a sum bit along with a carry output. With x, y, and z denoting the three input bits, $(A.12)$ and $(A.13)$ together describe the logic functions for the carry-out bit c and for the sum bit s respectively.

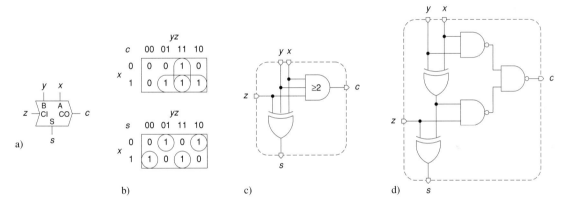

Fig. A.6 Full adder. Icon (a), Karnaugh maps (b), and two circuit examples (c,d). More sophisticated circuit examples are discussed in section 8.1.7

☐

A.2.12 Logic minimization

Given the infinitely many solutions, we must decide

"How to select an appropriate set of logic equations for some given combinational function"

METRICS FOR LOGIC COMPLEXITY AND IMPLEMENTATION COSTS

The goal of logic minimization is to find the most economic circuit for a given logic function under some speed and energy constraints. The criterion for economy depends on the technology targetted. Minimum package count used to be a prime objective at a time when electronics engineers were assembling digital systems from SSI/MSI components. Today, it is the silicon area occupied by gates and wiring together that counts for full-custom ICs. The number of gate equivalents (GEs) is more popular in the context of field-programmable logic (FPL) and semi-custom ICs.

From a mathematical point of view, the number of literals is typically considered as the criterion for logic minimization. By **literal** we refer to an appearance of a logic variable or of its complement. As an example, the right-hand side of (A.3) consists of seven literals that make up three composite terms although just three variables are involved.

An expression is said to contain a **redundant** literal if the literal can be eliminated from the expression without altering the truth table. Equation (A.18), the Karnaugh map of which is shown in fig.A.7a, contains several redundant literals. In contrast, none of the eleven literals can be eliminated from the right-hand side of (A.19), as illustrated by the Karnaugh map of fig.A.7b. The concept of redundancy applies not only to literals but also to composite terms.

Redundant terms and literals result in redundant gates and gate inputs in the logic network. They are undesirable due to their impact on circuit size, load capacitances, performance, and energy dissipation. What's more, **redundant logic** causes severe problems with testability, essentially because there is no way to tell whether a redundant gate or gate input is working or not by observing a circuit's behavior from its connectors to the outside world.

MINIMAL VERSUS UNREDUNDANT EXPRESSIONS

Unredundant and minimal are not the same. This is illustrated by (A.20), an equivalent but more economical replacement for (A.19) which gets along with just eight literals. Its Karnaugh map is shown in fig.A.7c.

Observation A.4. *While a minimal expression is unredundant by definition, the converse is not necessarily true.*

$$e = \overline{x}\,\overline{y}\,\overline{z}\,\overline{t} \vee \overline{x}\,\overline{z}\,t \vee \overline{x}\,y\,z\,t \vee xz \vee x\,\overline{y}\,z \tag{A.18}$$

$$e = \overline{x}\,\overline{y}\,\overline{z} \vee \overline{x}\,\overline{z}\,t \vee yzt \vee xz \tag{A.19}$$

$$e = \overline{x}\,\overline{y}\,\overline{z} \vee \overline{x}\,y\,t \vee xz \tag{A.20}$$

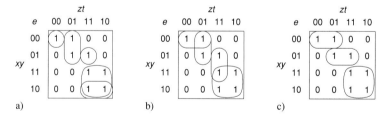

Fig. A.7 Three Karnaugh maps for the same logic function. Redundant form as stated in logic equation (A.18) (a), unredundant form as in (A.19) (b), and minimal form as in (A.20) (c).

Note that for obtaining the minimal expression (A.20) from the unredundant one (A.19), a detour via the canonical expression is required, during which product terms are first expanded and then regrouped and simplified in a different manner. Thus, there is more to logic minimization than eliminating redundancy.

Next consider function d tabulated in fig.A.8. There are two possible sum-of-products expressions shown in equations (A.21) and (A.22), both of which are minimal and use six literals. The minimal product-of-sums form of (A.23) also includes six literals. We conclude

Observation A.5. *A minimal expression is not always unique.*

$$d = \overline{x}\,\overline{y} \vee xz \vee y\,\overline{z} \tag{A.21}$$

$$d = xy \vee \overline{x}\,\overline{z} \vee \overline{y}\,z \tag{A.22}$$

$$d = (\overline{x} \vee y \vee z)(x \vee \overline{y} \vee \overline{z}) \tag{A.23}$$

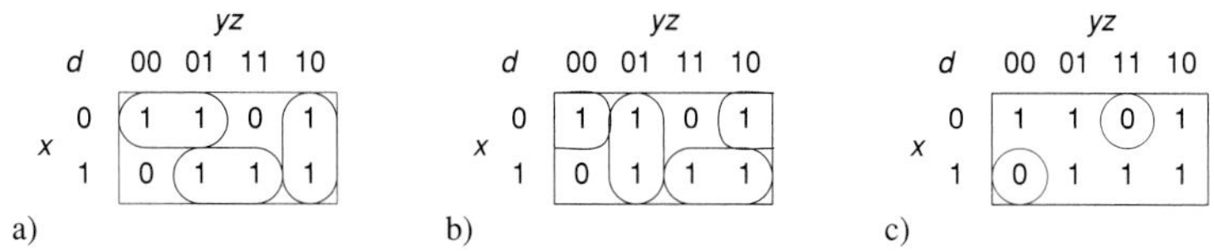

Fig. A.8 Karnaugh maps for equations (A.21) through (A.23), all of which implement the same function with the same number of literals.

MULTILEVEL VERSUS TWO-LEVEL LOGIC

Observation A.6. *While it is possible to rewrite any logic equation as a sum of products and as a product of sums, the number of literals required to do so grows exponentially with the number of input variables for certain logic functions.*

The 3-input XOR function (A.13), for instance, includes 12 literals. Adding one more argument t asks for 8 minterms, each of which takes 4 literals to specify, thereby resulting in a total of 32 literals. In general, an n-input parity function takes $2^{(n-1)} \cdot n$ literals when written in two-level logic form. Asymptotic complexity is not the only concern, however. Multilevel circuits are often faster and more energy-efficient than their two-level counterparts.

The process of converting a two-level into an equivalent multilevel logic equation is referred to as **factoring**, aka structuring, and the converse as **flattening**.

$$\overline{x}\,\overline{y}\,\overline{z} \vee \overline{x}\,\overline{y}\,z \vee x\,\overline{y}\,z \vee x\,y\,\overline{z} \underset{\text{flattening}}{\overset{\text{factoring}}{\rightleftharpoons}} \overline{x}\,\overline{y}\,(\overline{z} \vee z) \vee x\,(\overline{y}z \vee y\overline{z}) \tag{A.24}$$

MULTI-OUTPUT VERSUS SINGLE-OUTPUT MINIMIZATION

Probably the most important finding on multi-output functions is

Observation A.7. *Minimizing a vectored function for each of its output variables separately does not, in general, lead to the most economical solution for the overall network.*

This is nicely illustrated by the example of fig.A.9. Solution (a), which is obtained from applying the Karnaugh method one output bit at a time, requires a total of 15 literals (and 7 composite terms).

By reusing conjunctive terms for two or more bits, solution (b) makes do with only 9 literals (and 7 composite terms). In terms of gate equivalents, overall circuit complexity amounts to 12.5 and 9.5 GEs if all ORs and ANDs get remapped to NAND gates by way of bubble pushing.

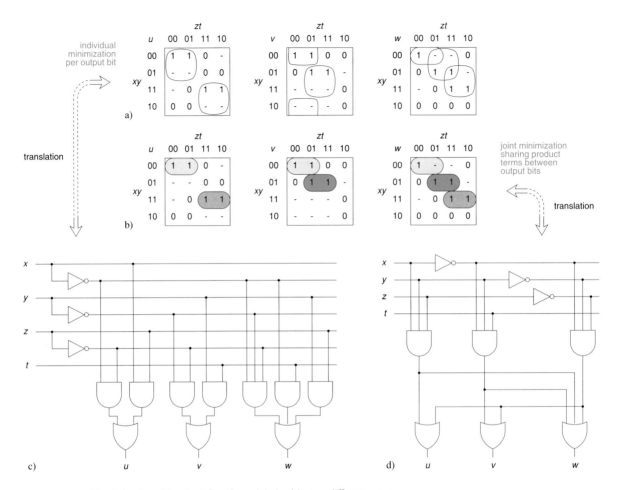

Fig. A.9 A multi-output function minimized in two different ways.

MANUAL VERSUS AUTOMATED LOGIC OPTIMIZATION

Observation A.8. *Manual logic optimization is not practical in VLSI design.*

For real-world multi-output multilevel networks, the solution space of this multi-objective optimization problem (area, delay, energy) is way too large to be explored by hand. Also, solutions are highly dependent on nasty details (external loads, wiring parasitics, cell characteristics, etc.) that are difficult to anticipate during logic design. Logic minimization on the basis of AND and OR gates with unit delays is a totally unacceptable oversimplification.

A.3 | Circuit alternatives for implementing combinational logic

A.3.1 Random logic

The term is misleading in that there is nothing undeterministic to it. Rather, **random logic** refers to networks built from logic gates the arrangement and wiring of which may appear arbitrary at first sight. Examples have been given earlier, see fig.A.9 for instance. Standard cells and gate arrays are typical vehicles for implementing random logic in VLSI.

As opposed to random logic, **tiled logic** exhibits a regularity immediately visible from the layout because subcircuits get assembled from a small number of abutting layout tiles. As tiling combines logic, circuit, and layout design in an elegant and efficient way, we will present the most popular tiled building blocks.

A.3.2 Programmable logic array (PLA)

A PLA starts from two-level logic and packs all operations into two adjacent rectangular areas referred to as AND- and OR-**plane** respectively. Each input variable traverses the entire AND-plane both in its true and in its complemented form, see fig.A.10. A product term is formed by placing or by omitting transistors that act on a common perpendicular line. Each input variable participates in a product in one of three ways:

true	1
complemented	0
not at all	–

Parallel product lines bring the intermediate terms to the OR-plane where they cross the output lines. The sums are then obtained very much like the products, the only difference being that products are not available in complemented form, so that any product enters a sum in either of two ways, namely

true	1
not at all	0

The general arrangement as two pairs of interacting meshes yields a very compact layout. What's more, a PLA is readily assembled from a small set of predefined layout tiles for any logic function. The criteria for logic minimization are not the same as for random logic. The PLA's width being fixed by the number of input and output variables, PLA minimization software[11] must act on the number of conjunctive terms to minimize layout height.

Example

Figure A.10 shows a PLA-style circuit that implements the logic function of fig.A.9c. For the sake of simplicity, switches and resistors have been substituted for the MOSFETs of an actual

---

[11] Such as the seminal ESPRESSO [472], which also introduced the above notation for capturing PLA codings.

circuit.[12] When comparing this with the random logic of fig.A.9d, keep in mind that a PLA gets penalized on such a small function by the overhead associated with complementing all inputs, distributing signals over both planes, and restoring voltages to proper levels.

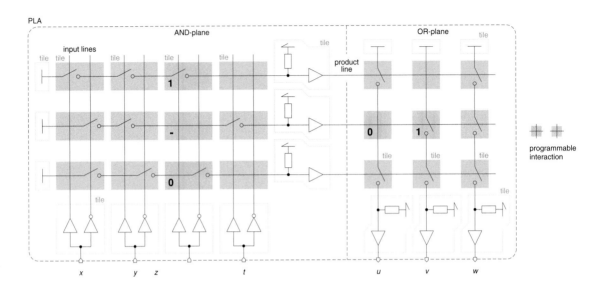

Fig. A.10 The circuit organization of a NAND-NAND-type PLA drawn with switches and resistors instead of transistors. The function implemented is the same as in fig.A.9.

Using the notation introduced before, the PLA's configuration and programming are expressed in a concise manner as follows.

AND-plane				OR-plane		
x	y	z	t	u	v	w
1	1	1	–	1	0	1
0	1	–	1	0	1	1
0	0	0	–	1	1	1

☐

Owing to their superior layout densities, PLAs used to be popular building blocks for combinational functions with many inputs and outputs. They lost momentum when automatic synthesis of random logic and multiple metal layers became available, but the underlying concepts continue to play an important role in field-programmable logic (FPL).

---

[12] In fig.A.10, the switches in both planes work against static pull-up loads, which circuit style is a departure from the truly complementary CMOS style presented in section 8.1. Transistor networks trimmed down in this or a similar fashion are typical for PLAs.

A.3.3 Read-only memory (ROM)

While both AND- and OR-planes are configurable in a PLA, programming of a ROM is confined to the OR-plane.[13] The role of the fixed AND-plane is assumed by a built-in address decoder that computes all possible minterms from the input variables, see fig.A.11. A ROM is an attractive option when full decoding of inputs is indeed required, but otherwise remains of limited appeal because its size doubles for each extra input or address bit.

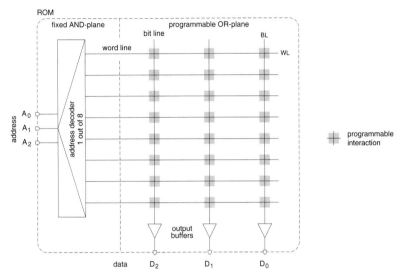

Fig. A.11   General ROM arrangement (8 words by 3 bits, grossly simplified).

A.3.4 Array multiplier

Two-level logic is extremely uneconomic for addition, multiplication, and other functions where the SoP includes many minterms that cannot be merged. Early IC designers have thus extended the tiling approach to multilevel logic and specifically to various forms of multipliers.

Circuit organization is patterned after the classic procedure tought at elementary school whereby multiplier and multiplicand are placed on two orthogonal sets of parallel lines, with a 1-digit by 1-digit multiplication being carried out at every intersection. Addition is distributed over the grid by including an adder at every intersection and by having each multiply–add operation propagate its sum and carry to two out of the four adjacent tiles for further processing: the sum towards the bottom and the carry towards the left. The entire multiplier thus consists of largely identical tiles arranged as a two-dimensional array. Communication within the array remains strictly local, which minimizes parasitic capacitances and interconnect delays.

---

[13] Make sure you understand that, in spite of its name, a ROM is a purely combinational function or — which is the same thing — a memoryless subcircuit unable to hold a state. Further note that the term programmable array logic (PAL) denotes a third breed of tiled two-level logic where the AND-plane is programmable and the OR-plane is predefined.

This is where the commonalities between the various types of array multiplier end. Popular examples include the **Braun multiplier** for unsigned and the **Bough–Wooley multiplier** for signed numbers. **Booth-recoded multipliers** have also been constructed along these lines. While tiled multipliers have fallen behind much as PLAs have, their basic circuit organization lives on in random logic implementations.

A.3.5 Digest

Figure A.12 summarizes the circuit options for implementing a combinational function. Tiled logic does not waste nearly as much resources for wiring as cell-based random logic does because signals are brought from one layout tile to the next directly. Another benefit of ROMs and PLAs is that any reprogramming is limited to minor modifications to one metal mask or two (as long as the array's overall capacity is not exceeded), whereas any modification to random logic necessitates redoing several, if not all, mask levels and is bound to affect the subcircuit's footprint.

Conversely, the area overhead associated with auxiliary circuits (such as input and output circuitry, decoders, and pull-ups) makes ROMs and PLAs uneconomical for small subfunctions. Developing a layout generator also represents an important investment, part of which must be renewed with each process generation. Only ROMs continue to be routinely supported as combinational macrocells today (together with RAMs for memory functions).

One reason is that the availability of multiple metal layers has narrowed down the difference between cell-based and tiled logic in terms of layout density. Probably the most important argument in favor of random logic is the fact that automatic HDL synthesis and multilevel logic optimization during the 1990s matured into powerful and comprehensive software tools that cover not only combinational but also sequential circuits.

Assembling look-up tables, arithmetic logic units, and the alike from abutting layout tiles is no longer an option unless the word widths involved are very large and unless maximum density, performance, and/or energy efficiency are sought in spite of limited wiring resources.

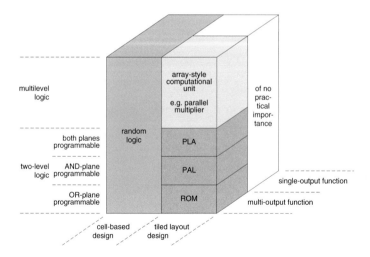

Fig. A.12 Design space for implementing combinational functions.

A.4 | Bistables and other memory circuits

Bistable subcircuits are essential building blocks of sequential circuits. What they all have in common is the ability to store one bit of information by assuming either of two stable states. An almost infinite variety of circuit implementations has been developed over the years using various design styles and fabrication technologies. As a consequence, designations for bistables proliferate, which continues to generate confusion today. Yet, this is totally unnecessary.

Observation A.9. *From the perspectives of architecture and logic design, it suffices to consider a bistable's behavior and to ignore everything about its internal structure and operation.*

We are going to present a simple taxonomy that rests on behavioral criteria exclusively before deriving a coherent and unambiguous naming convention from that.

Clocked storage elements clearly distinguish between input terminals that determine <u>what</u> state transitions shall take place and others which determine <u>when</u> such transitions must occur. Any terminal that belongs to the latter category is referred to as a **clock input**.[14]

Unclocked storage elements, in contrast, do not evidence such a separation.

Table A.5 | Taxonomy of bistables as a function of their behavior.

	Bistable		
	clocked		unclocked
Behavior	edge-triggered	level-sensitive	
Data inputs get evaluated	at any active clock edge	while clock is at active level	at any time
Name	**flip-flop**	**latch**	no single name
Examples	D-flip-flop E-flip-flop T-flip-flop JK-flip-flop	D-latch	SR-seesaw Muller-C MUTEX snapper[a]
Clock terminal	identified by ∧	identified by ⊓	none

[a] Discussed in section 8.4.1.

Clocked bistables must be subdivided further into edge-triggered and level-sensitive ones. In any **edge-triggered** bistable, it is a transition of the clock signal that causes the data present at the input terminal to be admitted into the circuit and to be stored there.

---

[14] Most clocked bistables are driven from a single clock. Although not really popular with circuit designers, some bistables require a double-rail clock of two signals CLK and $\overline{\text{CLK}}$ driven by complementary waveforms. We are not concerned with this subordinate detail here.

The behavior of **level-sensitive** memory circuits is slightly more complex in that such circuits may be in pass or in hold mode depending on the logic value of the clock. In **pass mode**, data are simply propagated from the input to the output, which is why this mode is also termed transparent mode. In **hold mode**, output data are kept frozen and input data are being ignored, which explains why the device is sometimes said to be opaque.

Throughout this text, we will consistently refer to a bistable that is both clocked and edge-triggered as a **flip-flop**. Conversely, we reserve the word **latch** for any bistable that is clocked and level-sensitive. The phrase **clocked bistable** is used as a generic term for both.[15]

Table A.5 puts the names and behaviors of all popular bistables into perspective.

A.4.1 Flip-flops or edge-triggered bistables

THE DATA OR D-TYPE FLIP-FLOP

The D-type flip-flop exhibits the simplest behavior an edge-triggered bistable can have. Most engineers find it easiest to think in terms of D-flip-flops and to convert their designs into other forms, if necessary.

The basic D-flip-flop has a clock terminal CLK and a data input D. The output datum is available in true form Q, in complemented form $\overline{Q}$, or both. Similarly, the clock can induce a state transition either on its rising or on its falling edge, referred to as **active edge**.[16] Please see fig.A.13 for the truth table of a basic rising-edge-triggered D-type flip-flop. Icon and signal waveforms are also shown along with the causality relation.

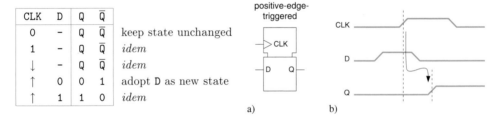

CLK	D	Q	$\overline{Q}$	
0	–	Q	$\overline{Q}$	keep state unchanged
1	–	Q	$\overline{Q}$	*idem*
↓	–	Q	$\overline{Q}$	*idem*
↑	0	0	1	adopt D as new state
↑	1	1	0	*idem*

Fig. A.13 Rising-edge-triggered D-flip-flop. Truth table (left), icon (a), and waveforms (b).

A vast collection of more elaborate flip-flop variations is obtained from the basic D-type by extending its functionality in numerous directions and by combining these new features.

Initialization facilities

An extra input found on most flip-flops makes it possible to put the circuit into some predefined start state. Such initialization mechanisms come in two flavors.

---

[15] Be warned that many sources use either word indiscriminatly for any kind of bistable. The term latch, in particular, is often meant to include some forms of unclocked bistables, see footnote 23. Also, the fact that latches have a pass mode is sometimes emphasized by calling them **transparent latches**, although this property is shared by all latches. For the sake of clarity and simplicity, we refrain from such practices.

[16] As shown in section 8.2.5, flip-flops that trigger on either edge exist, but these are not widely used.

Both synchronous **clear CLR** and synchronous **load LOD** inputs affect flip-flop operation solely on the active clock edge. As the names say, the former imposes logic state 0 while the latter brings the flip-flop into state 1. Either one can be considered as just another data input that masks the regular data input D, see table A.6. Any basic D-type flip-flop is easily upgraded to include a synchronous clear or load by adding one or two gates in front of it.

The asynchronous **reset RST** and the asynchronous **set SET**, in contrast, have an immediate effect on the flip-flop's operation because they directly act on the state-preserving memory loop with no intervention from the clock, see table A.6.[17] This also explains why the asynchronous (re)set mechanism must be incorporated into the elementary flip-flop circuit itself, there is no way to add it later. We therefore conclude

Observation A.10. *A D-type flip-flop with an asynchronous reset input forms a fundamental building block from which any more sophisticated flip-flop, counter slice, or other edge-triggered bistable can be assembled with the aid of a few extra logic gates.*[18]

Table A.6 Truth tables of rising-edge-triggered D-type flip-flops with synchronous clear (left) and with active-low asynchronous reset (right).

CLK	CLR	D	Q	$\overline{Q}$	
0	–	–	Q	$\overline{Q}$	keep state unchanged
1	–	–	Q	$\overline{Q}$	*idem*
↓	–	–	Q	$\overline{Q}$	*idem*
↑	1	–	0	1	enter state 0
↑	0	0	0	1	adopt D as new state
↑	0	1	1	0	*idem*

$\overline{RST}$	CLK	D	Q	$\overline{Q}$	
0	–	–	0	1	enter state 0
1	0	–	Q	$\overline{Q}$	keep state unchanged
1	1	–	Q	$\overline{Q}$	*idem*
1	↓	–	Q	$\overline{Q}$	*idem*
1	↑	0	0	1	adopt D as new state
1	↑	1	1	0	*idem*

If a bistable is equipped with two conflicting initialization inputs, i.e. with **CLR** and **LOD** or with **RST** and **SET**, it must be specified which of the two takes precedence over the other.[19] Although flip-flops with both asynchronous reset and asynchronous set inputs exist, there is no meaningful application for them in synchronous designs.

Scan facility

An effective and popular way to ensure the testability of sequential logic is to replace all ordinary D-type flip-flops with special scan flip-flops and to connect them in such a way as to make them cooperate like a shift-register while in scan test mode. A **scan flip-flop** essentially includes a select function at the input. Depending on the logic value present at the scan mode control terminal **SCM**,

---

[17] Incidentally, note that asynchronous (re)set signals often are of **active-low** polarity for better protection against noise and other fugitive events. See section 10.4.4 for more details.

[18] Examples of how this can be done are given in section 8.2.4 of the main text.

[19] Simultaneous activation of asynchronous set and reset inputs of equal precedence levels is disallowed as this could lead to irregular behavior — similarly to that observed when a seesaw is forced into the forbidden state — and/or to anomalous static power dissipation.

the data admitted into the flip-flop are taken either from the data input D (during normal operation) or from the scan input SCI (during scan test), see table A.7.[20]

Table A.7 | Truth table of a rising-edge-triggered scan flip-flop.

CLK	SCM	SCI	D	Q	Q̄	
0	–	–	–	Q	Q̄	keep state unchanged
1	–	–	–	Q	Q̄	*idem*
↓	–	–	–	Q	Q̄	*idem*
↑	0	–	0	0	1	adopt D as new state, "normal operation mode"
↑	0	–	1	1	0	*idem*
↑	1	0	–	0	1	adopt SCI as new state, "scan mode"
↑	1	1	–	1	0	*idem*

Enable/disable facility

Not all flip-flops in a circuit need to be updated at every active clock edge; many of them must conserve their state for many consecutive clock cycles. This requires that flip-flops be equipped with a mechanism to enable or disable state transitions via a special control input ENA, see table A.8. A data flip-flop is readily extended to become a so-called enable or **E-type flip-flop**; it suffices to add a multiplexer in front of its data input that feeds the uncomplemented output back as long as the enable signal remains inactive, see fig.6.26 for an illustration.

Table A.8 | Truth table of a rising-edge-triggered E-type flip-flop.

CLK	ENA	D	Q	Q̄	
0	–	–	Q	Q̄	keep state unchanged
1	–	–	Q	Q̄	*idem*
↓	–	–	Q	Q̄	*idem*
↑	0	–	Q	Q̄	*idem*
↑	1	0	0	1	adopt D as new state
↑	1	1	1	0	*idem*

THE TOGGLE OR T-TYPE FLIP-FLOP

A toggle flip-flop is a bistable that changes state at every active clock edge, see table A.9. It is obtained from a D-flip-flop by providing an inverting feedback from the true output back to the data input. Similarly to the D-type, the T-flip-flop is easily upgraded to include an enable input, in which case toggling takes place only if the enable is active at the active clock edge.

THE NOSTALGIA OR JK-TYPE FLIP-FLOP

In lieu of a single data input D, the JK-flip-flop has two inputs labeled J and K that, together, determine the state the bistable is going to enter at the next active clock edge, see table A.10. The

---

[20] Please refer to fig.6.6 in the main text for a schematic and the very basics of scan path testing.

Table A.9 Truth table of a rising-edge-triggered T-type flip-flop.

CLK	Q	$\overline{Q}$	
0	Q	$\overline{Q}$	keep state unchanged
1	Q	$\overline{Q}$	*idem*
↓	Q	$\overline{Q}$	*idem*
↑	$\overline{Q}$	Q	change state, "toggle"

JK-flip-flop is essentially a leftover from the days of SSI that is kept in today's cell libraries mainly for reasons of compatibility. What made it popular is its versatility. Permanently tying J and K to logic 1 results in a T-type flip-flop. A toggle flip-flop with enable is obtained when J and K are connected to form a common input. D-type behavior asks for K to be the inverse of J, which then serves as data input D.

Table A.10 Truth table of a rising-edge-triggered JK-type flip-flop.

CLK	J	K	Q	$\overline{Q}$	
0	–	–	Q	$\overline{Q}$	keep state unchanged
1	–	–	Q	$\overline{Q}$	*idem*
↓	–	–	Q	$\overline{Q}$	*idem*
↑	0	0	Q	$\overline{Q}$	*idem*
↑	0	1	0	1	adopt $J = \overline{K}$ as new state
↑	1	0	1	0	*idem*
↑	1	1	$\overline{Q}$	Q	change state, "toggle"

A.4.2   Latches or level-sensitive bistables

THE DATA OR D-TYPE LATCH

Very much like a basic flip-flop, a basic latch features just a data input D and a clock input CLK. Please note the latter terminal is often referred to as "enable" E or G in datasheets and icons. However, as this input carries the only signal that defines <u>when</u> the latch is to leave its present state and <u>when</u> it is to enter a new one, it clearly must be understood and handled as a clock.

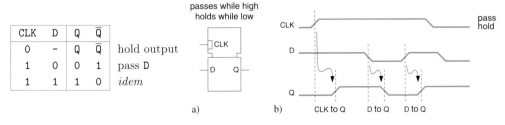

CLK	D	Q	$\overline{Q}$	
0	–	Q	$\overline{Q}$	hold output
1	0	0	1	pass D
1	1	1	0	*idem*

Fig. A.14 D-latch transparent on logic 1. Truth table (left), Icon (a), and waveforms (b).

D-type latches find applications not only as subcircuits of flip-flops but also as bistable memory devices in their own right when used in conjunction with a level-sensitive clocking scheme.[21]

A.4.3 Unclocked bistables

Unclocked bistables differ from their clocked counterparts in that there is no clock terminal that might trigger a state transition without, at the same time, also contributing towards defining the next state. Put in simple words, there is no distinction between <u>when</u> and <u>what</u> inputs.

Observation A.11. *Any information stored in an unclocked bistable is vulnerable to spurious events such as glitches, runt pulses, and other noise on the inputs.*

This is because the absence of a dedicated clock input implies that a transition at some data or control terminal may spark off a state change at any time. The situation sharply contrasts with clocked bistables, where there exists no input other than the clock and an optional asynchronous (re)set that can possibly cause the state to flip.

THE SR-SEESAW

Any two inverting gates interconnected in such a way as to form a loop exhibit bistable behavior because there is positive feedback and zero latency. Let us begin by analyzing two cross-coupled NOR gates. The truth table of the circuit shown in fig.A.15 exposes two alarming peculiarities.

Firstly, the input vector $S = R = 1$ causes the two output terminals to assume identical values, thereby violating the rule that outputs Q and $\bar{Q}$ must always assume complementary values. The situation is often referred to as the **forbidden state**.

Secondly, the outcome is unpredictable when the circuit is switched from the forbidden state to the data storage condition. In the occurrence, when inputs S and R simultaneously change from 1 back to 0, nodes Q and $\bar{Q}$ will eventually reassume complementary values, but there is no way to tell whether they will settle to 01 or to 10 after abandoning the forbidden 00 state.[22]

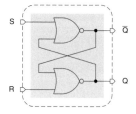

S	R	Q	$\bar{Q}$	
0	0	Q	$\bar{Q}$	maintain output, data storage condition
0	1	0	1	enter state 0, "reset"
1	0	1	0	enter state 1, "set"
1	1	0	0	noncomplementary output, "forbidden state"

Fig. A.15 Seesaw built from NOR gates. Truth table (left), circuit example (right).

---

[21] See sections 8.2.1, 6.2.4, and 6.2.5 which corroborate the notion of terminal CLK as a clock rather than as an enable input.

[22] This is because the two stable states are separated by a thin line of metastable equilibrium. Any bistable that is brought close to that line must revert to one stable state or the other before normal operation resumes. However, as explained in section 7.4, the course is undeterministic and the time it takes is unbounded.

An alternative circuit is obtained from NAND gates, fig.A.16, yet analysis yields analogous results. As both circuits exhibit bistable behavior but operate with no intervention from a clock, we refer to them as level-sensitive SR-type seesaws or simply as seesaws.[23]

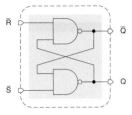

$\bar{R}$	$\bar{S}$	Q	$\bar{Q}$	
0	0	1	1	noncomplementary output, "forbidden state"
0	1	0	1	enter state 0, "reset"
1	0	1	0	enter state 1, "set"
1	1	Q	$\bar{Q}$	maintain output, data storage condition

Fig. A.16 Seesaw built from NAND gates. Truth table (left), circuit example (right).

Although seesaws are at the heart of many latch and flip-flop circuits, naked zero-latency feedback loops are unsuitable for data storage in synchronous designs because of numerous shortcomings they suffer from.[24] Safe and useful applications of seesaws are very limited and include the generation of non-overlapping clock signals and the debouncing of mechanical contacts.

THE EDGE-TRIGGERED SR-SEESAW

As opposed to the seesaws of figs.A.15 and A.16, the set and reset inputs are edge-triggered rather than level-sensitive. As a consequence, there is no such thing as a forbidden state with noncomplementary output values. Still, the outcome remains unpredictable when both inputs transit from 0 to 1 in overly rapid succession. Also, with two flip-flops, the circuit of fig.A.17b is more costly in terms of transistor count than any other bistable discussed in this section.

THE MULLER-C ELEMENT

The Muller-C element is a bistable with one output and two interchangeable inputs. The output immediately assumes whatever value the two inputs agree on, but preserves its past value when the two input values differ. The behavior can be likened to hysteresis or to a majority seesaw. The idea

---

[23] SR is just an acronym for set–reset. We have coined the name **seesaw** to avoid the confusion that arises when the more popular terms (asynchronous) SR-flip-flop, SR-latch, $\bar{S}\bar{R}$-latch, and NOR|NAND-latch are being used. As shown in table A.5, we make it a habit to distinguish among latches, flip-flops, and unclocked bistables.

[24] As will be explained in section 5.4, synchronous circuits boast a clear and consistent separation into signals that trigger state changes and others that determine the sequence of states. This separation is not supported by unclocked bistables, which makes them vulnerable to hazards and may render their behavior unpredictable.

[25] **Signal transition graphs** (STGs) are extremely helpful for describing the behavior of asynchronous circuits and controllers. Each node stands for an event such as a signal transition. A rising edge is identified by an appended + and a falling edge by a -. The updating of data and the withdrawal of data — returning to a high-impedance condition, for instance — are also considered to be events. Each solid edge captures a cause/effect relationship implemented within the subcircuit being modelled while a dashed edge indicates the waiting for a condition to be satisfied by the surrounding circuitry. The position of all marks reflects the present state of the circuit. STGs belong to a subclass of Petri nets known as "marked graphs" and obey the same rules: For an event to take place, each incoming edge must carry a mark. When the transition actually fires, those marks get absorbed and a new mark is placed on every outgoing edge. The number of marks does not, therefore, necessarily remain the same.

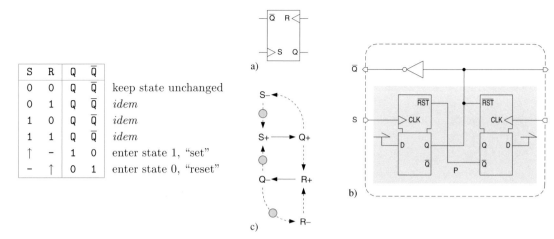

S	R	Q	$\bar{Q}$	
0	0	Q	$\bar{Q}$	keep state unchanged
0	1	Q	$\bar{Q}$	*idem*
1	0	Q	$\bar{Q}$	*idem*
1	1	Q	$\bar{Q}$	*idem*
↑	–	1	0	enter state 1, "set"
–	↑	0	1	enter state 0, "reset"

Fig. A.17 Edge-triggered seesaw. Truth table (left), icon (a), circuit example (b), and signal transition graph (STG) (c).[25]

is easily generalized to more than two inputs. One may also observe that the Muller-C behaves like an AND-gate while the output is low, and like an OR-gate while the output is high.

Table A.11 | Truth table of Muller-C element.

standard form				with B inverted			
A	B	C		A	B	C	
0	0	0	enter state 0, "reset"	0	0	C	maintain output
0	1	C	maintain output	0	1	0	enter state 0, "reset"
1	0	C	maintain output	1	0	1	enter state 1, "set"
1	1	1	enter state 1, "set"	1	1	C	maintain output

Some circuit implementations combine logic gates into a zero-latency feedback loop, the most elegant solution being with a 3-input majority gate as shown in fig.A.18c. Other circuits use a memory element of four transistors reminiscent of a snapper, see fig.A.18d.[26]

Similarly to seesaws, the Muller-C must not be used as part of synchronous designs because of the absence of a clock input. It finds useful applications for the processing of handshake signals in self-timed systems, however [474] [134].

THE MUTUAL EXCLUSION ELEMENT

Also known as interlock element, the mutual exclusion element — or MUTEX for short — features two symmetric inputs R1 and R2 that are associated with outputs G1 and G2 respectively. The two outputs are never active at the same time, see table A.12. At rest, both inputs are inactive, R1 = R2 = 0, and so are the outputs. When a positive impulse arrives on either input, it immediately

---

[26] [473] compares four alternative circuits and concludes that the majority gate implementation is superior to the weak feedback approach in terms of delay and, above all, energy efficiency.

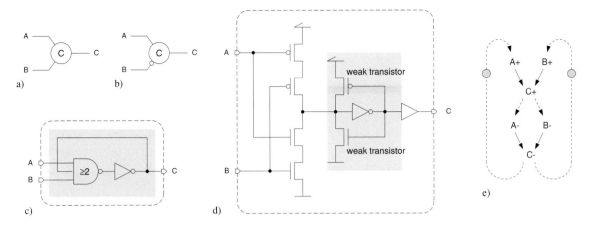

Fig. A.18 Muller-C element. Icon (a), alternative circuit examples (c,d), and signal transition graph (STG) (e). Icon for a Muller-C with one inverting input (b).

gets propagated to the pertaining output. Should a second impulse arrive on the other input later, it will not get passed on until the first impulse has come to an end. Impulses are thus propagated on a "first come, first served" basis unless two of them arrive simultaneously, in which case the circuit arbitrarily selects one to pass through and withholds the other.[27]

Table A.12 Truth table of the mutual exclusion element.

R1	R2	G1	G2	
0	0	0	0	wait
0	1	0	1	let R2 pass
1	0	1	0	let R1 pass
1	1	G1	G2	let the earlier impulse pass, resolve conflict in case of simultaneous arrival

A look back tells us that the seesaw already had the capability of discriminating between two events on the basis of their order of arrival. A comparison of the MUTEX truth table with those of figs.A.16 and A.15 reveals that they are in fact the same (except for swapped and inverted output or input terminals respectively). Not surprisingly, one finds a level-sensitive seesaw in the mutual exclusion circuit, see fig.A.19b for an example. What, then, are the four extra transistors good for?

Note, to begin with, that the seesaw waits in the forbidden state while R1 = R2 = 0. It then enters either the set or the reset state depending on which input switches from 0 to 1 first. If both inputs go high at the same time, or nearly so, the seesaw is subject to marginal triggering. The circuit then lingers in a state of metastable equilibrium before eventually returning to a stable state and letting one input through.[28] From a practical point of view, it is important to make sure that the output signals G1 and G2 remain logically unambiguous, free of glitches, and consistent with the truth table in spite of the seesaw hovering in an irregular condition.

---

[27] Note the analogy with two persons simultaneously arriving at the door of a closet or phone booth.
[28] See footnote 22.

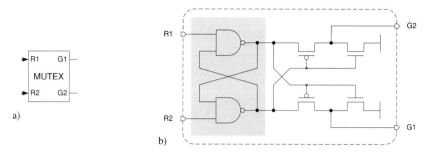

Fig. A.19 Mutual exclusion element. Icon (a) and circuit example (b).

The four transistors must, therefore, be understood to form some kind of filter that defers the circuit's response until the seesaw has recovered. An observation made in [141], namely that the closer in time a pair of rising transitions arrives on the inputs, the longer the MUTEX takes to decide which of them to propagate, comes as no surprise from this perspective.

The mutual exclusion element plays a key role in a subcircuit known as an arbiter and is instrumental in self-timed circuits.

A.4.4 Random access memories (RAMs)

As everyone knows, random access memories serve as short-term repositories for large quantities of data. A RAM essentially consists of a large array of elementary binary storage cells that share a common input/output or data port D, see fig.A.20. Any access has to occur one data word at a time and the address A serves to identify the data word currently being accessed. As an example, a 1 Mi × 4 bit RAM accepts and returns data as 4-bit quantities and requires a 20 bit address to select one out of the $2^{20} = 1\,048\,576$ memory locations available.[29]

The overall organization bears many common traits with ROMs in that both the bit cell array and the address decoder are assembled from a few layout tiles. What sets a RAM apart from a ROM are bistable storage cells the state of which can be changed from the data port in very little time. A write enable input WR/$\overline{\text{RD}}$ that controls the operation of the bidirectional input/output buffers for write and read operation is another important departure.

Two techniques for implementing two-valued memory cells prevail today. In a **static RAM** (SRAM), each bit of data is stored with the aid of two cross-coupled inverters that form a positive feedback loop. The two stable points of equilibrium so obtained are identified with logic 0 and 1 respectively.

As opposed to this, it is the presence or absence of an electrical charge on a small capacitor that reflects the binary information in a **dynamic RAM** (DRAM). The elementary bit cell is utterly simple and small, thereby maximizing the memory capacity available from some given piece of

---

[29] Kibi- (ki), mebi- (Mi), gibi- (Gi), and tebi- (Ti) are binary prefixes recommended by various standard bodies for 2^{10}, 2^{20}, 2^{30}, and 2^{40} respectively because the more common decimal SI prefixes kilo- (k), mega- (M), giga- (G), and tera- (T) give rise to ambiguity as $2^{10} \neq 10^3$. As an example, 1 MiByte = 8 Mibit = $8 \cdot 2^{20}$ bit.

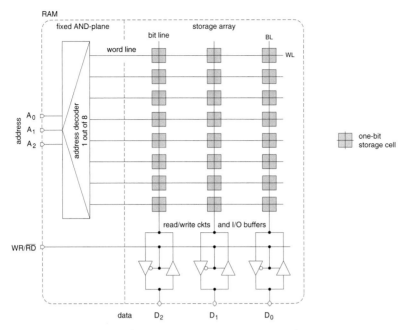

Fig. A.20 General RAM arrangement (8 words by 3 bits, grossly simplified).

silicon.[30] The ensuing low costs per storage bit are the main reason why DRAMs dominate the mass market for computer main memory in spite of their longer access times.

In either case, data storage is **volatile** exactly as for latches and flip-flops, which is to say that the information is lost upon disruption of the supply voltage.

To save on the overall pin count, the address is typically time-multiplexed over a single address port in commodity DRAM components, e.g. 20 bit as a pair of two 10 bit chunks. Some memories handle write and read transfers over separate input and output ports. Also available are dual-port RAMs that feature two independent I/O ports and that allow for two concurrent read/write transfers.

A.5 | Transient behavior of logic circuits

Our discussion of logic circuits has, so far, been concerned with steady-state conditions and with the end points of switching processes exclusively. As a consequence from delays and inertial effects, circuits assembled from transistors, wires, and other real-world components exhibit various transient phenomena, though. A key question is

"How do the outputs of digital (sub)circuits evolve from one value to the next?"

---

[30] Schematic diagrams are given in section 8.3 along with further details.

Let us first be concerned with transient waveforms as witnessed on an oscilloscope hooked to the output of a combinational circuit before turning our attention to the underlying mechanisms that cause them.

A.5.1 Glitches, a phenomenological perspective

One might naively expect that binary signals progress from one value to the next in monotonic ramps. Experiments with real circuits reveal that this is not always the case, though. Any kind of fugitive or nonmonotonic event on a binary signal is termed a **glitch** or a **hazard**.[31] Transients are sometimes catalogued from a purely phenomenological point of view, i.e. as a function of the signal waveform observed.

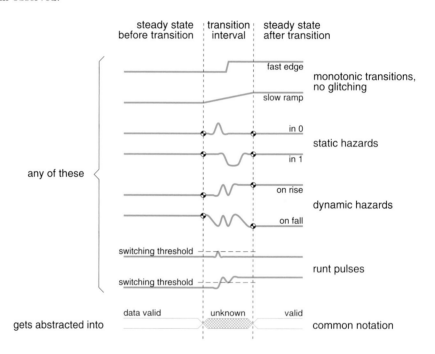

Fig. A.21 Transients classified according to their waveforms.

A **static hazard** manifests itself as a temporary deviation from its steady-state logic value that occurs in response to some other signal change. If a signal moves back and forth before eventually settling to a logic value opposite to the initial one, we speak of a **dynamic hazard**. As illustrated in fig.A.21, both static and dynamic hazards may be classified in further detail.

The voltage excursion during a glitch does not necessarily make a full swing, stunted pulses being observed as well. They occur when a signal has begun to change (order) just before an antagonistic effect (counterorder) sets in, thereby preventing the first transition from completing. Such spurious events, termed **runt pulses**, may render circuit operation unpredictable and irreproducible if they reverse direction in the vicinity of the logic family's switching threshold or if they cut across it for a very short lapse of time before turning back.

A.5.2 Function hazards, a circuit-independent mechanism

This first mechanism can be understood from the logic function alone and does not depend on any specific circuit implementation. Consider the Karnaugh map in table A.13, for instance. When the input vector xyz changes from 001 to 011, then the output h switches from 1 to 0. Conversely, the output stays 1 when the input goes from 001 to 000.

Table A.13 | A Karnaugh map that may give rise to function hazards.

h		yz			
		00	01	11	10
x	0	1	1	0	0
	1	0	1	1	0

What happens when two or more inputs change at a time? As before, the final output value is found in the appropriate field. The intermediate steps leading to that result will, however, depend on the actual sequence of events at the input. Assume input vector xyz is to change from 001 to 111. Depending on whether variable x or y is switching first, the intermediate input is 101 or 011. While in the first case output h remains at 1 throughout, it temporarily assumes value 0 in the latter case and so gives rise to a static hazard on 1.

Similarly, three inputs that change shortly one after the other may give rise to a dynamic hazard, e.g. when input xyz goes from 001 to 011 followed by 111 before settling on 110.

This type of transient output in response to two or more input transitions is termed **function hazard** since it depends on the logic function exclusively. Whether the resulting hazards are of static or of dynamic nature is immaterial in this context. Even the most humble binary functions, such as AND and OR, exhibit function hazards for an appropriate sequence of input transitions. n-cubes are most convenient for tracing function hazards, see fig.A.22.

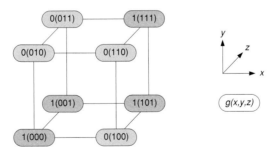

Fig. A.22 3-cube equivalent to table A.13.

One might argue that no intermediate steps and, therefore, also no function hazards would arise if the switching of the inputs were to occur simultaneously and instantaneously. However, due to dissimilar propagation delays along interconnect lines and within the logic circuitry itself, the result is very much the same as if inputs had switched with a timewise offset.

In practice, the waveforms that result from function hazards range from barely noticeable runt pulses to multiple full-blown glitches, depending on how much the input signals are skewed and depending on the logic and interconnect delays involved. Please note that the existence of a hazard can also pass totally unnoticed from the output waveform, a situation which is sometimes referred to as **near hazard**.[31]

A.5.3 Logic hazards, a circuit-dependent mechanism

The second mechanism differs from the first in that the switching of a single input variable suffices to generate unwanted transients. Also, the emergence of transients cannot be explained from the function alone but is related to a particular gate-level circuit structure. As an example, consider fig.A.23, a circuit implementing the Karnaugh map of table A.13.

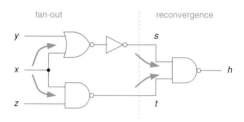

Fig. A.23 A simple combinational network exposed to logic hazards.

Let's assume that input vector xyz changes from **101** to **001**. The fact that the combinational function maps either input to logic **1** may lead us to believe that the output would steadily remain at that value. Yet, unequal propagation delays will cause the switching of inner nodes to be slightly skewed. In the occurrence, the additional inverter in the upper path might be responsible for delaying the switching of s with respect to t. Vector st then changes from **10** via **11** to **01** and output h switches from **1** via **0** back to **1**, which amounts to a static hazard. This behavior, which cannot be explained as a function hazard because initial and final node are adjacent in the n-cube, is referred to as a **logic hazard**.

Note, by the way, that the inverse input change, i.e. xyz from **001** to **101**, produces no glitch since vector st goes from **01** via **00** to **10**, three values all of which result in a **1** at the output. This is not a general characteristic of logic hazards, however.

From a more detailed point of view, logic hazards can be traced back to a combination of **reconvergent fanout** and a function hazard. Let us view the logic circuit of fig.A.23 as being composed of

---

[31] Strictly speaking, the term **glitch** refers to visible waveforms while **hazard** should be used when relating to the mechanisms that may, or might not, cause some combinational circuit to develop nonmonotonic transients. This distinction is not always maintained, though, as the term hazard is often meant to include the observable phenomena as well. Incidentally, note that a glitch can also have causes other than hazards such as ground bounce, crosstalk, or electrostatic discharge (ESD), for instance.

There is an analogy with medicine in that the clinical picture with its observable symptoms (phenotype) is what matters from a therapeutic and personal point of view. A more profound analysis, in contrast, is concerned with why some living creatures have a higher predisposition to being struck by certain diseases than others do (genotype). Whether the illness actually manifests itself in a given individual or not is of little interest from this epidemiologic perspective.

two subcircuits. For certain inputs yz, the subcircuit to the left will broadcast a change of variable x to both of its output nodes s and t. The subcircuit to the right where these two signals recombine is then confronted with multiple changing inputs, and output h may, or might not, develop a glitch, depending on the circuit's detailed timing characteristics.

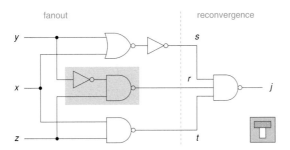

Fig. A.24 Same circuit as in fig.A.23 with logic hazards suppressed.

Reconvergent fanout is a necessary but not a sufficient condition for logic hazards. This is documented by the circuit of fig.A.24, a modification of fig.A.23 obtained from adding two gates. When xyz changes from 101 to 001 the new node r stays 0, thereby preventing output j from moving away from its steady-state value. Both the original and the modified circuits contain reconvergent fanout, but only the original one is glitching in response to a single input change. Exact conditions for the existence of logic hazards are being given in [468]. As a rule, multilevel networks tend to glitch more intensively than two-level networks because multiple paths of different lengths are more likely to coexist. Similarly to what was found for function hazards, the waveforms caused by logic hazards depend on timing and other circuit details.

The suppression of hazards in fig.A.24 has been bought at the expense of introducing redundant hardware. In fact, the added gates — represented by two extra literals in the logic equation — do not affect the logic function of the network and are, therefore, redundant. As redundant logic is almost impossible to test and entails superfluous switching activity, this approach is not recommended in the context of VLSI design.

A.5.4 Digest

Our findings on transient phenomena in combinational logic are best summed up as follows.

Observation A.12. *Hazards may, or might not, bring forth glitches, extra signal events unwanted and unaccounted for on the logic and higher levels of abstraction.*

Whether a hazard actually materializes as a rail-to-rail pulse, as a runt pulse, or not at all depends on circuit structure, gate and interconnect delays, load conditions, wiring parasitics, layout arrangement, operating conditions (PTV), on-chip variations (OCVs), and other implementation details of relatively minor importance.

Observation A.13. *A combinational circuit is susceptible to developing hazards*
○ *if two or more input variables change at a time,*[32] *or*
○ *if the circuit includes reconvergent fanout and if one or more inputs change at a time.*

In conclusion, all logic networks able to carry out some form of computation may give rise to hazards and glitches. Needless to say, this includes almost all digital circuits of technical interest.[33]

A.6 | Timing quantities

We have now learned about transient events that are associated with the working of combinational logic. Next we want to abstract all relevant inertial effects of digital components and subcircuits to a small set of quantities. Essentially, we want to unambiguously state when a signal is valid, when it is liable to evolve, and when it is safe to update it without worrying about inner details of the circuits concerned.

A.6.1   Delay quantities apply to combinational and sequential circuits

Note from fig.A.25 that it takes a <u>pair</u> of delay parameters to adequately describe how long transient phenomena persist at the output of a digital circuit in response to a change at one of the circuit's inputs. One of these two timing quantities is very popular, while the other is not.

t_{pd} **Propagation delay.**[34] The time required to process new input from applying a stable logic value at a (data or clock) input terminal until the output has settled on its final value, i.e. until all transients in response to that input change have died out.

t_{cd} **Contamination delay**, aka (output) retain delay and retain time.[35] The inertial time from altering the logic value at a (data or clock) input until a first change of value occurs at the output results, i.e. until transients start in response to that input change. By definition, $0 \leq t_{cd} < t_{pd}$ must hold for any physical component or (sub)circuit.

Contamination delay figures are rarely published in datasheets, though.[36] As a stopgap, safety-minded engineers often substitute the lower bound for the missing parameters by admitting $t_{cd} \equiv 0$. In doing so, causal behavior is taken for granted but any inertial effects are ignored.

---

[32] "At a time" here means before all transients in response to the earlier input change have died out.

[33] How to design safe circuits in spite of hazards is a major topic of section 5.3.

[34] Also referred to as "settling time" or "maximum delay" by some authors.

[35] The terms "internal delay", "output hold time", and "minimum (propagation) delay" are synonyms found in the literature. We do not support their usage as they lend themselves to confusions, however.

[36] There are various reasons for this. For one thing, the concept of contamination delay is virtually unknown to most practitioners and textbooks in digital electronics. For another thing, manufacturers are reluctant to commit themselves to <u>minimum</u> values for any delay parameter because they reserve the right to upgrade their fabrication processes at any time in search of better performance or lower manufacturing costs. Also, there is a common industry practice known as **down-binning**, which implies shipping a faster device against an order for a slower part. Because of its superior speed, a faster device is bound to exhibit a shorter contamination delay than the original part.

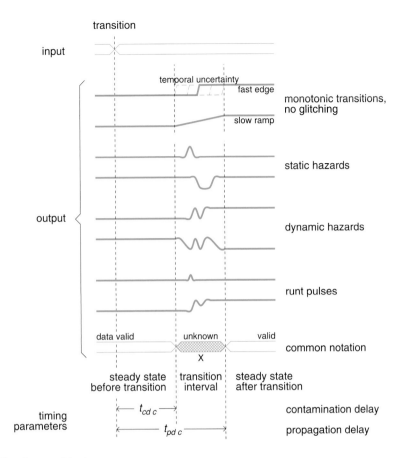

Fig. A.25 Transients revisited.

While this is a viable workaround in many cases, contamination delay is an essential quality of any flip-flop.[37]

Observation A.14. *Signal propagation along some path through a digital network is captured by a pair of timing parameters termed contamination and propagation delay respectively. Their numeric figures are chosen so as to enclose all transient phenomena (such as ramps, hazards, glitches and runt pulses) that might occur at the output in response to some input change.*

What does this mean for a digital circuit that features multiple input and/or output terminals, a very commonplace situation? If the propagation delay of such a circuit is to be characterized by a single quantity, it is obviously necessary to consider all of the terminals. Propagation delay of vectored inputs and outputs is, therefore, determined by whichever signal takes the uppermost

---

[37] Equating all contamination delays with zero is an inadmissible oversimplification that makes it impossible to gain a more profound understanding of how clocked circuits operate. The functioning of a simple shift register, for instance, cannot be explained under that assumption. Refer to section 6.2.2 for an in-depth analysis.

amount of time to traverse a circuit (longest path). Conversely, it is the quickest travel time through a logic network that determines the contamination delay of a circuit (shortest path).

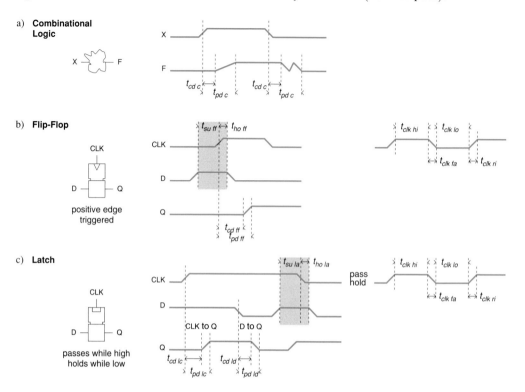

Fig. A.26 Timing characteristics of combinational subcircuits and of clocked bistables.

In the occurrence of a [level-sensitive] latch, the point of time when output data become valid depends on the data input in some situations and on the clock input in others. Two propagation delays are thus required to characterize the timing of a latch. $t_{pd\,ld}$ indicates the **data-to-output delay** (from input D to output Q) while $t_{pd\,lc}$ states the **clock-to-output delay** (from CLK to Q). The same argument obviously also applies to the contamination delay.

A.6.2 Timing conditions apply to sequential circuits only

The orderly functioning of sequential (sub)circuits such as flip-flops, latches, and memories requires that they be driven by a clock of clear-cut waveform. Ambiguous voltages, glitches, sluggish ramps, and stunted waveforms are unacceptable. Clocked circuits further impose dead times during which the clock must remain stable before it may be allowed to toggle again. Two pairs of timing conditions have been defined to capture the characteristics of an acceptable clock. Note that all four parameters relate to no other signal than to the clock itself.

$t_{pu\,clk\,min}$ **Clock minimum pulse width**. The time span during which the clock signal must firmly be kept either low or high before it is permitted to swing back to the opposite state. Shorter clock pulses must be avoided as a bistable is otherwise likely to behave in an unpredictable way. In practice, it is often necessary to distinguish between

$t_{lo\,clk\,min}$ clock minimum pulse width low (MPWL), which refers to logic 0 (pause), and
$t_{hi\,clk\,min}$ clock minimum pulse width high (MPWH), which refers to logic 1 (mark).

They combine into $t_{pu\,clk\,min} = \max(t_{lo\,clk\,min}, t_{hi\,clk\,min})$.

$t_{ra\,clk\,max}$ **Clock maximum ramp time**. The timewise allowance for the clock to ramp from one logic state to the opposite one. Driving a bistable with overly slow waveforms may cause inner nodes to float or to be placed under control of conflicting drivers, which may lead to an irrecoverable loss of data. To prevent this from happening, a pair of maximum transition or ramp times is imposed on the clock.[38]

$t_{ri\,clk\,max}$ clock maximum rise time refers to the transition from 0 to 1 while
$t_{fa\,clk\,max}$ clock maximum fall time is concerned with the inverse transition.

They combine into $t_{ra\,clk\,max} = \min(t_{ri\,clk\,max}, t_{fa\,clk\,max})$.

Any clocked (sub)circuit further imposes requirements on the timewise relationship between any of its data inputs and the clock that is driving that (sub)circuit.

t_{su} **Setup time**. The lapse of time immediately before the active clock edge during which an input is required to assume a fixed logic value of either 0 or 1 at the input of a clocked (sub)circuit. The setup condition is here to make sure all inner nodes have settled to values determined by new input data before the (sub)circuit locks into the corresponding state in response to the subsequent active clock edge. Violating the setup requiremetn must be avoided under any circumstance because bistables are otherwise likely to behave in an unpredictable way.

t_{ho} **Hold time**. The lapse of time immediately after the active clock edge during which data are required to remain logically unchanged at the input of a clocked (sub)circuit. The hold condition assures that all inner nodes have properly settled so that the new state is maintained even when the stimuli that caused the transitions in the first place have been removed. Violating the hold requirement must be avoided for the reasons explained before. Although either the setup or the hold time may assume a negative value for certain components or (sub)circuits,[39] $t_{su} + t_{ho} > 0$ always holds.

Observation A.15. *Setup time and hold time together demarcate a brief lapse of time in the immediate vicinity of the active clock edge. Their numerical values are chosen so as to guarantee that data get stored and/or processed as intended under all circumstances.*

---

[38] There is another reason for doing so. Timing data of clocked subcircuits vary with the waveform of the driving clock. Yet, for reasons of economy, it is common industrial practice to characterize flip-flops, latches, RAMs, and the like for one typical clock ramp time, e.g. for $t_{clk\,ra} = t_{clk\,ri} = t_{clk\,fa} = 50$ ps. Driving them with a clock waveform that exhibits much slower ramps will cause the actual parameter values of t_{pd}, t_{cd}, t_{su}, and t_{ho} to significantly deviate from the figures published in datasheets and simulation models.

[39] A negative hold time means that a data input is not required to preserve its value until <u>after</u> the active clock edge for being properly stored in the (sub)circuit, but is free to resume new transient activities <u>before</u> that time. Similarly, a negative setup time means that the input is allowed to switch until after the active clock edge.

Any ambiguous or changing input during the aperture so defined is likely to expose any bistable or other clocked (sub)circuit to marginal triggering and to cause it to fail in an unexpected and unpredictable way.[40] To prevent this from happening, the logic value must be kept constant and logically well-defined throughout. In the case of a circuit with multiple input bits, such as a register, every bit must remain stable either at 0 or at 1 throughout.

Bistables with extra control inputs and/or with multiple clocks impose additional timing conditions. More specifically, asynchronous set and reset signals must not be released in the immediate vicinity of an active clock edge otherwise exposing the bistable to marginal triggering.

$t_{su\,rst}$ **Recovery time**, aka release time. Indicates by how much time the deactivation of an asynchronous (re)set input must precede the active clock edge so as to allow a bistable to unlock properly before taking up normal operation.

$t_{ho\,rst}$ **Home time** is more suggestive than the popular synonym removal time. Indicates for how much time an asynchronous (re)set input must remain activated following the active clock edge in order to safely bring a bistable home into its reset state. If the asynchronous (re)set is released too early, then chances are that the bistable will fall into some undetermined state or a metastable condition.

Some datasheets explicitly specify recovery and home times while others refer to these quantities as particular cases of setup and hold times.[41]

A.6.3 Secondary timing quantities

The timing quantities introduced so far suffice for modelling digital circuits. Yet, it is sometimes convenient to define parameters that are derived from those primary quantities.

t_{cw} **Data-call window**, aka aperture time, setup-and-hold window, and sampling time. The overall time span during which data must maintain a constant and well-defined value at the input of a memorizing (sub)circuit $t_{cw} = t_{su} + t_{ho} > 0$.

t_{id} **Insertion delay**. As the name suggests, this term denotes the extra delay that is inflicted on a signal when a given (sub)circuit is being inserted into a signal's propagation path. In the case of a combinational circuit, insertion delay is the same as the propagation delay on the longest path, that is $t_{id\,c} = \max(t_{pd\,c}) = t_c$.

For flip-flops, one has $t_{id\,ff} = t_{su\,ff} + t_{pd\,ff} = t_{ff}$ (where $t_{pd\,ff}$ refers to the non-inverting output unless indicated otherwise) because this is the minimum lapse of time an edge-triggered bistable takes to store and propagate a data item provided the active clock edge is optimally timed.

---

[40] You may want to consult section 7.4.1 for more information on what exactly happens with a bistable circuit in this case. How to determine the setup and hold time figures of a given bistable is also explained there.

[41] The difference is that an asynchronous control signal must subsequently retain its passive value indefinitely unless the circuit is to be reinitialized to its start state, whereas any ordinary (synchronous) control signal is essentially free to switch after the hold time has expired. Using separate terms thus seems justified, but we will not insist on this.

$f_{to\,ff\,max}$ **Maximum toggling rate**. No flip-flop can be made to operate faster than its insertion delay and its minimum clock pulse widths permit. The utmost clock frequency thus is $f_{to\,ff\,max} = \frac{1}{T_{to\,ff\,min}}$, where $T_{to\,ff\,min} = \max(t_{id\,ff}, t_{hi\,clk\,min\,ff} + t_{lo\,clk\,min\,ff})$. Although this quantity is given much publicity in advertisements, it remains of little practical interest as it leaves no room for any data processing activity.

r_{sl} **Slew rate**. The average velocity of voltage change during a logic transition, that is $r_{sl} = \frac{\Delta u}{\Delta t}$. The quantity is positive for rising edges, $r_{sl\,ri} = \frac{U_h - U_l}{t_{ri}}$, and negative for falling ones, $r_{sl\,fa} = \frac{U_l - U_h}{t_{fa}}$.

δ **Duty cycle**. The average proportion of time during which a signal is active or a load is energized. For an active-high signal of period T, one has $\delta = \frac{t_{hi}}{T} = \frac{t_{hi}}{t_{hi} + t_{lo}}$ if the switching is so fast that ramp times can be ignored. $0 \leq \delta \leq 1$ holds by definition.

A.6.4 Timing constraints address synthesis needs

Timing constraints differ from delays and timing conditions in that they do not describe the timewise behavior of an existing component or design for the purposes of analysis and simulation, but serve to capture target characteristics of a circuit-to-be for the purposes of design and synthesis. As illustrated in fig.4.17, most timing constraints in practice specify an upper bound for some propagation delay, yet, asking for a minimum contamination delay sometimes also makes sense.

A.7 | Microprocessor input/output transfer protocols

While microcomputer architectures are beyond the scope of this text, we briefly review the three fundamentally different I/O transfer protocols because many ASICs are to interface with a microprocessor bus system. You may find this information useful as a preparation for reading section 8.5.4.

A peripheral device that wishes to deliver data to a microcomputer has to ask for an input transfer operation. Conversely, an output transfer is solicited when the peripheral needs to obtain data. In either case, the peripheral sets a **service request** flag. Three conceptually different ways exist for notifying the microcomputer about such an event and for handling the subsequent data transfer, see fig.A.27.

1. Polling. In this scheme, the CPU actively waits for service requests to arrive. The peripheral is wired to some port on a peripheral interface adapter (PIA) so that the CPU can examine its status by way of read operations on the pertaining PIA address. Whenever the operating system expects a peripheral device to ask for I/O operations in the near future, it enters a **program loop** which makes it periodically read that port and check the appropriate bit position there to find out whether a service request is pending or not.

If so, the program branches to a service routine which tells the CPU how many data words to transfer, where to get them from, how to process them, and where to deliver them to. It

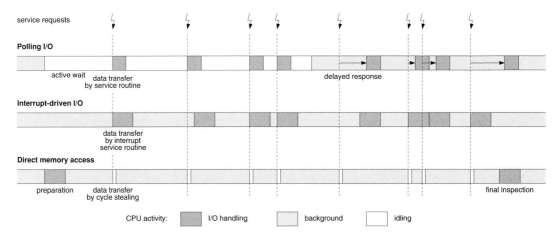

Fig. A.27 The three basic data transfer protocols for microcomputer input/output.

must also be made sure that the service request flag gets set back to its inactive state once the peripheral has been serviced.

If not so, the CPU proceeds with executing the loop's code, which makes it repeat read-and-check operations until the expected service request finally arrives.

Designing the polling loop always is a compromise, see fig.A.27. A tight loop ensures a fast response but leaves little room for doing anything useful while waiting, whereas an ampler loop provides room to carry out computations in the background, but results in prolonged response times as the request bit perforce gets examined less frequently.

2. **Interrupt-driven.** As opposed to polling, the CPU is not locked in a loop but is free to execute code in a background process. In order to make this possible, the request flag is brought out to a line that connects to a special input of the CPU termed **interrupt request**. Activation of this line diverts program execution to an interrupt service routine once execution of the current instruction has been completed. The routine first causes the CPU to properly suspend the current process by saving the contents of critical registers in memory or on a special stack. The remainder of the I/O operation occurs in much the same way as for polling. After having completed the interrupt service routine, the CPU resumes the suspended process.

Processor instruction sets typically include a pair of special instructions that allow programmers to temporarily suspend the interrupt mechanism in order to bring critical code sequences to an end without having their execution delayed or broken up. Withholding interrupt requests in this way is often referred to as interrupt masking.

3. **Direct memory access** (DMA). The CPU is freed from most of the burden associated with I/O transfers by delegating them to a special hardware unit termed a **DMA controller** that is hooked to the service request line in lieu of the CPU. Before a series of I/O operations can begin, the CPU in a preparatory step instructs the DMA controller how many data words to transfer and indicates their destination and/or source addresses. The CPU itself is not involved in the subsequent transfer operations, which contrasts sharply with polling

and interrupt-driven I/O. Instead, the DMA controller handles the actual data moves by stealing memory cycles from the CPU for its own memory accesses whenever notified by the peripheral that a data item is waiting to be accepted or delivered.

At the end of the commissioned series of transfers, the DMA controller typically informs the CPU by way of an interrupt. As part of the pertaining service routine, the CPU does then inspect some status register in the DMA controller to find out whether the transfer has been successfully completed or prematurely aborted.

Table A.14 | The basic I/O transfer protocols compared.

| | Input/Output Data Transfer Scheme | | |
	Polling	Interrupt-driven	DMA
Hardware overhead	close to none, PIA port bit	small, interrupt mechanism	moderate, DMA controller
CPU burden	high as CPU has to idle in a loop	moderate, once per data item	minimal, once per data series
Response time	unpredictable, depends on loop	a few instruction cycles unless masked	almost immediate, a few clock cycles
Transfer rate	moderate	fair	high

As an improvement to polling, it is possible to combine most of its simplicity with the efficiency of interrupt-driven I/O by having a timer periodically trigger the interrupt line. In the interrupt service routine, the CPU then polls one peripheral device after the other and branches to a service routine for those which have a request pending.

A.8 | Summary

- Digital hardware deals with bit patterns rather than with numbers. These patterns assume a meaning as numbers only when interpreted according to the specific number representation scheme that the designers had in mind.

- As zero-latency loops may give rise to unpredictable behavior, it is best to avoid them.

- Be prepared to observe unexpected transient pulses at the output of any combinational logic unless you have proof to the contrary.

- While all datasheets and textbooks mention setup time, hold time, and propagation delay, it is not possible to understand how a simple shift register works without introducing the concept of contamination delay.

- Any bistable is either an
  - Edge-triggered flip-flop, a
  - Level-sensitive latch, or an
  - Unclocked bistable.

It is most important to keep those apart technically and linguistically.

- Before starting up any kind of electronic design automation tool, it is important to

    - Know what its optimization criteria and limitations are,
    - Find out whether it does indeed apply to the problem at hand, and
    - Develop an understanding of the available options and controls so as to
    - Set all options and control knobs to suitable values.

- Three basic protocols are available for organizing the transfer of data between a peripheral device and a microcomputer:

    - Periodically polling the peripheral for service requests in a program loop.
    - Having a special signal from the peripheral interrupt regular program execution.
    - By bypassing the CPU with the aid of an extra direct memory access controller.

Appendix B

Finite State Machines

This chapter is divided into two major sections. Section B.1 reviews the classes of finite state machines used in electronics design and their equivalence relationships. Although this material is strongly related to automata theory — or actually part of it — no attempt is made to cover the theory since there are excellent and comprehensive textbooks on the subject. Rather, the emphasis is on a number of mathematical facts relevant to hardware design that are not normally found in such references. Section B.2 then looks at finite state machines more from an implementation point of view, yet without committing one to any specific technology.

B.1 | Abstract automata

Automata theory is a mathematical discipline concerned with fundamental issues of discrete computation such as formal languages and grammars, parsing, decidability, and computability. The underlying formal models are crude abstractions that essentially simplify computing equipment to transducers that, while changing from state to state, convert a given input string into some output string. Most issues relevant to digital design such as hardware architecture, computer arithmetics, parasitic states, state encoding, transient effects, delays, synchronization, etc. are neglected, which raises the question

"Why study the abstract subject of automata theory in the context of electronics design?"

The motivation is threefold:

Functional specification. Describing what a digital system has to do is not always easy. Automata theory often helps to specify the relationship between a circuit's inputs and outputs in a more formal way, especially for control- and protocol-oriented tasks.

Modelling and verification. When viewed from outside, the behavior of an entire system can be modelled as a single finite state machine. While usually not a very efficient approach to constructing a circuit, this abstraction proves useful for verifying a system's behavior by simulation and testing.

Synthesis. Almost any practical system is composed of a number of cooperating subsystems, each of which can in turn be modelled as a finite state machine. At a certain level of detail, any synchronous circuit is patterned after a specific type of automaton.

"What do we mean by finite state machine then?"

Definition B.1. *A deterministic automaton is a system, which at discrete moments of time* $t = 0, T, 2T, 3T, \ldots, kT, (k+1)T, \ldots$ *satisfies the following conditions:*

1. *At any of these moments of time, the input to the system can be chosen from a set of possible stimuli I.*
2. *At each of these moments of time, the system subjected to an input i can be in just one of a set of possible states S and outputs one out of a set of possible responses O.*
3. *At any of these moments of time, the state of the system and the input to it uniquely define the state which the system is going to assume in the next such moment of time.*

Throughout this text, we will stay within the framework of **discrete** and **deterministic** automata so defined. We will further limit our discussion to **finite state machines** (FSMs) where each of the three sets I, S, and O is restricted to a finite number of elements. Please note that, in general, I, S, and O do not have the same cardinality, although they may happen to do so.

B.1.1 Mealy machine

There is more than one way to model and to implement the behavior of systems that meet the criteria of finite state machines stated in the previous section. A first approach consists in describing such a system by way of two equations

$$o(k) = g(i(k), s(k)) \tag{B.1}$$
$$s(k+1) = f(i(k), s(k)) \tag{B.2}$$

where $i \in I$, $s \in S$, and $o \in O$. g is termed **output function** and f **transition function** or next state function. In addition, at $k = 0$ the automaton is assumed to be in a special state $s_0 \in S$ which is called **start state**.

Equations (B.1) and (B.2) together form a Mealy model. More precisely, we speak of a Mealy automaton if the present output depends on the present input. Many real-world examples are of more simple nature, which gets reflected by the absence of one or more terms from (B.1) and (B.2). See section B.1.5 for a detailed classification scheme.

Instead of stating transition function f and output function g as equations, any finite state machine can be completely described by listing next state and output values for any combination of present state and input. Such a list is termed a **state table**, see fig.B.1 for an example.

The well-known **state graph**, aka state transition diagram, is nothing else than a pictorial representation of a state table. Each state is represented by a vertex and each transition by a directed edge. A short arrow identifies the start state. Where an input symbol causes a state to persist, i.e. to fall back on itself, a loop is drawn.

So far, we have been concerned with FSM <u>behavior</u> exclusively, but we should also ask for a hardware <u>structure</u> that behaves accordingly. The result from expressing output function and transition function with a data dependency graph (DDG) is shown in fig.B.2 for a Mealy machine. As will be shown in section B.2.5, this straightforward solution is not necessarily also the most efficient one, however.

Example

$I = \{a,b\}$, $S = \{p,q,r,t\}$, and $O = \{0,1\}$.

$i(k)$	a	b	a	b
$s(k)$	$o(k)$		$s(k+1)$	
p	1	0	r	t
q	0	1	p	q
r	1	0	q	r
t	0	1	r	p

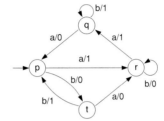

Fig. B.1 Mealy machine with four states. State table (left) and state graph (right).

☐

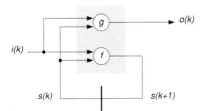

Fig. B.2 DDG of a Mealy automaton.

B.1.2 Moore machine

The Moore model differs from the Mealy model in that the present output depends on the present state exclusively; there is no input literal $i(k)$ in output function g. As a consequence, the output is allowed to change only as a result of a state transition.[1]

$$o(k) = g(s(k)) \tag{B.3}$$
$$s(k+1) = f(i(k), s(k)) \tag{B.4}$$

---

[1] Because state transitions are restricted to discrete moments of time kT, the Moore model is sometimes said to have a synchronous output. This interpretation commonly found in texts on automata theory abstracts from propagation delay. In practice, a Moore output will settle to its final value a couple of gate delays after the active clock edge and remain constant until the next active clock edge.

As shown below, the more restrictive formulation for the output function (B.3) gets reflected by omissions in the network structure, state table, and state graph of Moore automata.

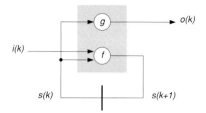

Fig. B.3 DDG of a Moore automaton.

Example

$I = \{a,b\}$, $S = \{u,v,w,x,y,z\}$, and $O = \{0,1\}$.

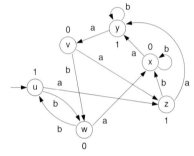

$i(k)$		a	b
$s(k)$	$o(k)$	$s(k+1)$	
u	1	z	w
v	0	z	w
w	0	x	u
x	0	y	x
y	1	v	y
z	1	y	x

Fig. B.4 Moore machine with six states. State table (left) and state graph (right).

☐

Both Mealy and Moore automata are pervasive in electronics circuits, primarily in controllers that govern sequential data processing and data exchange operations.

B.1.3 Medvedev machine

Some particular situations in electronic hardware design obviate logic operations in the output function.[2] Medvedev automata are a subclass of Moore machines where the output function g has essentially degenerated to the identity function.

$$o(k) = s(k) \tag{B.5}$$
$$s(k+1) = f(i(k), s(k))$$

Medvedev automata are also known as finite acceptors or as automata without output because studying state transitions alone is sufficient for many problems from automata and formal language

---

[2] E.g. because of the delays and transients associated with combinational networks, see section B.2.4. Also, the direct observability and controllability of Medvedev outputs via scan-path-type test structures is welcome in many controller applications.

theory. Most counters are practical examples of Medvedev automata, see fig.B.5 for a hardware structure.

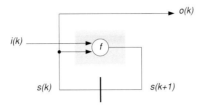

Fig. B.5 DDG of a Medvedev automaton.

B.1.4 Relationships between finite state machine models

Two questions of both theoretical and practical importance are

"Under what conditions is it possible to replace a finite state machine by another one?" and "How do the various classes of automata relate to each another?"

Since finite state machines can be viewed as transducers that convert some input string into some output string, the most natural way of defining **functional equivalence** is as follows.

Definition B.2. *Two automata are considered equivalent if they always yield identical strings of output symbols for any identical strings of input symbols.*

It is important to understand that the state graphs of equivalent automata need not be isomorphic. This is because the above definition refers to input and output quantities only. Put in other words, automata are abstracted to black boxes in the context of equivalence. Figures B.8 and B.9, for instance, show a pair of Mealy machines that are equivalent but not isomorphic.

EQUIVALENCE OF MEALY AND MOORE MACHINES IN THE CONTEXT OF AUTOMATA THEORY

While the above definition of equivalence works fine when comparing Mealy automata with Mealy automata or Moore automata with Moore automata, there is a technical problem when comparing machines across the two classes. A Mealy machine can respond to an input change at any time whereas the response of a Moore machine is necessarily deferred to after the next active clock edge. Put differently, Moore outputs have latency 1 and Mealy outputs latency 0. No matter what the first input $i(0)$ looks like, the first symbol in the output string from a Moore machine is $o(0) = g(s_0)$; only later can the input affect the output. This implies that Moore automata take $n + 1$ computation periods to process a total of n consecutive input symbols whereas Mealy automata require only n periods. The numbers of output symbols released from the two models differ accordingly.

As a workaround, the requirement for equivalence is relaxed as follows in the context of automata theory because that theory is primarily concerned with the mapping between symbol strings.

Definition B.3. *A Mealy and a Moore automaton are considered equivalent if they always yield identical strings of output symbols for any identical string of input symbols when the first output*

symbol — which is associated with the start state — is deleted from the output string of the Moore automaton.

This somewhat academic understanding gives rise to a well-known result

Theorem B.1. *For any Mealy automaton there exists an equivalent Moore automaton in the broad sense of definition B.3, and vice versa.*

At first sight this may seem surprising because the output function of the Mealy model is more general than that of the Moore model. We will sketch a constructive proof, i.e. two algorithms for converting a Moore machine into an equivalent Mealy machine, and vice versa. More details can be obtained from [475], an excellent textbook on abstract automata and formal languages.

Converting a Moore machine into an equivalent Mealy model is very easy. For every vertex of the state graph delete the output symbol associated with the vertex and attach it to all edges that enter that vertex. Clearly, the procedure will leave the number of states unchanged.

Coming up with a Moore model for a Mealy machine cannot simply follow the inverse procedure because any attempt to assign to a vertex the output symbol associated with its incoming edges must lead to a conflict unless all incoming edges agree in their outputs. Thus, whenever a conflict arises, the vertex is split into as many copies as there are distinct output symbols attached to its incoming edges. All output symbols can so be transferred from the incoming edges to their respective copy of the vertex. Each copy keeps the full set of outgoing edges attached to the original vertex. The process is repeated for the successor nodes until all vertices have been visited. Please note that the number of states may — and often will — increase dramatically when going from a Mealy to a Moore machine.

Incidentally, we conclude from the two conversion procedures that (a) for any Moore automaton there exists an equivalent Mealy automaton with a smaller or equal number of states, and (b) the above Moore-to-Mealy conversion algorithm does not, in general, lead to a solution with the minimum possible number of states.

Example

In the broad sense of automata theory, the two state machines described in figs.B.1 and B.4 are actually equivalent. Their respective output strings are opposed in table B.1 for two input strings chosen at random.

EQUIVALENCE OF MEALY AND MOORE MACHINES IN THE CONTEXT
OF HARDWARE DESIGN

From an engineering point of view, the extra output symbol $o(0)$ and the timewise offset of all subsequent symbols cannot be abstracted from. After all, few applications will tolerate replacing a finite state machine by another one the output of which lags or leads by a full clock cycle. Unless latency is indeed uncritical for the application at hand, equivalence must, therefore, be understood in the stricter sense of its original definition, which implies that

Table B.1 | Output strings of equivalent Mealy and Moore automata compared.

k	0	1	2	3	4	7	8
automaton	output						
input	a	b	b				
Mealy	1	0	0				
Moore	[1]	1	0	0			
input	b	b	a	a	b	a	
Mealy	0	1	1	1	1	0	
Moore	[1]	0	1	1	1	1	0

☐

Theorem B.2. *A Mealy automaton and a Moore automaton can never be equivalent in the more narrow sense of definition B.2.*

EQUIVALENCE OF MOORE AND MEDVEDEV MACHINES

Theorem B.3. *For any Moore automaton there exists an equivalent Medvedev automaton in the more narrow sense of definition B.2, and vice versa.*

The proof is by showing how to build an equivalent Medvedev automaton from a Moore automaton.[3] The problem with designing a Medvedev automaton is that $|O| = |S|$ while, in general, this is not the case for Moore automata, where $|O| \leq |S|$. To get around this difficulty, we allow the state symbol to be composed of two parts, namely a left part, which we will also use as output symbol, and a right part, that will remain hidden from the outside world. In any case their concatenation must yield a unique state symbol. Note that this trick does not introduce actual logic operations into the output function and is, therefore, consistent with our definition of Medvedev automata. However, it breaks the constraint that forced $O = S$.

The conversion algorithm works as follows. Consider the state graph of the Moore automaton. Assign every state its output symbol as left part of its state symbol. If all states are uniquely labeled, the right parts remain empty and conversion is completed since the automaton had already followed the Medvedev model in the first place. Otherwise, assign each state a right part so as to make it unique. The easiest way to do so is simply to copy the state symbols from the initial Moore automaton. When fed with the same stimuli, the resulting automaton will always output the same responses as the Moore automaton.

Perhaps a more intuitive conversion procedure is given in fig.B.6.

The number of states of a Medvedev automaton so obtained is always equal to that of the original Moore model. It takes more bits to encode the wider Medvedev state symbols, though.

---

[3] The inverse transform is trivial since any Medvedev automaton is a Moore automaton by definition.

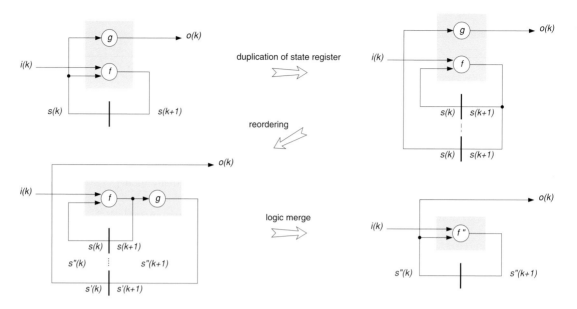

Fig. B.6 Turning a Moore machine into an equivalent Medvedev machine.

Example

The Medvedev automaton shown in fig.B.7 has been obtained from the Moore automaton of fig.B.4 by way of the constructive algorithm stated above.

$I = \{a, b\}$, $S = \{0c, 0d, 0e, 1c, 1d, 1e\}$, and $O = \{0, 1\}$.

$i(k)$		a	b
$s(k)$	$o(k)$	$s(k+1)$	
1c	1	1e	0d
0c	0	1e	0d
0d	0	0e	1c
0e	0	1d	0e
1d	1	0c	1d
1e	1	1d	0e

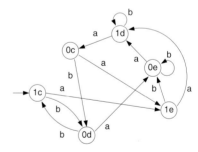

Fig. B.7 Medvedev machine with six states. State table (left) and state graph (right).

☐

B.1.5   Taxonomy of finite state machines

The table below, which is patterned after a Karnaugh map, classifies deterministic automata according to what actually determines their next state and their output.

Table B.2 Taxonomy of finite state machines.

				output function g depends on			
				—	state	input and state	input
transition			—	†	§	§	C
function		state	† ‡	A		‡
f	input and	state	† ‡	O	Y	‡
depends on	input		† ‡	D		‡
				Moore model		Mealy model	

Several subclasses, the fields of which are marked with special characters in table B.2, make no sense from a technical point of view. Here are the reasons why:

† Unobservable automata are useless in engineering applications.
‡ There is no point in controlling a state that exerts no influence on the output.
§ Having the output depend on a fixed and thus effectively inexistent state makes no sense.

Other subclasses have been given a name because they find widespread applications as digital function blocks. They are identified by capital letters as listed below.

	Subclass	Example
Y	Full Mealy automaton[4]	Controller
O	Full Moore automaton[4]	Controller, cell of cellular automaton
A	Autonomous automaton	Clock generator, pseudo-random-number generator
D	Delay automaton	Pipeline stage (combinational logic plus register)
C	Combinational logic	Full adder, unpipelined multiplier

B.1.6 State reduction

Consider the state table of some finite state machine and assume that two states have exactly the same entries in their present output fields and also in their next state fields. As an example, this applies to states "7" and "10" in fig.B.8. From a graph point of view, this means their outgoing edges are labeled in exactly the same way and point to exactly the same vertices. It is intuitively clear that any two such states must appear to be the same when nothing but the machine's inputs and outputs are observed.

Definition B.4. *Two states of a finite state machine are considered indistinguishable if the machine can be placed in either of the two and responds with identical strings of output symbols to any string of input symbols.*

Indistinguishable states are also known as equivalent states and as **redundant states**. Merging them has no effect on a machine's behavior. The new automaton so obtained will necessarily be equivalent to the original one, but simpler to implement. State reduction is the process of collapsing

---

[4] The word "full" is meant to imply that no term has been dropped from the general equations (B.1) (B.2) and (B.3) (B.4) respectively.

redundant states until no equivalent state machine with a smaller number of states exists. Collapsing all states that have identical state table entries does not suffice, however, as two states can have distinct next state fields and still be perfectly indistinguishable.

Theorem B.4. *Two states of a finite state machine are indistinguishable iff they have (a) identical outputs and (b) go to indistinguishable successor states for any possible input symbol.*

The difficulty with applying this theorem to state reduction directly lies in its recursiveness. A more practical approach is the **implication chart algorithm** due to Paull and Unger and nicely described in [476], for instance. Luckily, there is no need for designers do that manually as automatic state reduction is part of HDL synthesis. We thus refrain from presenting algorithmic details and are content to show an FSM before and after state reduction.

Example

The state graph depicted in fig.B.8 has been chosen for demonstration purposes with no particular application in mind.

$i(k)$	00	01	10	11	00	01	10	11
$s(k)$		$o(k)$				$s(k+1)$		
1	1	1	1	1	1	1	1	2
2	0	0	0	0	2	8	8	3
3	0	0	0	0	3	5	5	4
4	0	0	0	0	4	5	5	2
5	0	0	0	0	6	5	5	9
6	0	0	0	0	7	11	11	3
7	1	0	0	1	1	1	1	1
8	0	0	0	0	9	8	8	6
9	0	0	0	0	10	11	11	4
10	1	0	0	1	1	1	1	1
11	0	0	0	0	1	1	1	1

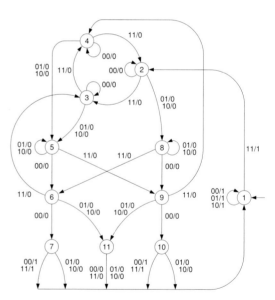

Fig. B.8 Original state machine. State table (left) and state graph (right).

Systematic state reduction not only confirms $(7, 10)$ as indistinguishable states, but also reveals equivalences for $(2, 3, 4)$, $(5, 8)$, and $(6, 9)$. While the new state machine shown in fig.B.9 preserves the input-to-output relationship, the number of states has dropped from 11 to 6.

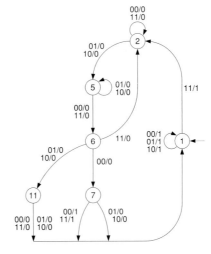

$i(k)$	00	01	10	11	00	01	10	11
$s(k)$		o	(k)			$s(k$	$+1)$	
1	1	1	1	1	1	1	1	2
2	0	0	0	0	2	5	5	2
5	0	0	0	0	6	5	5	6
6	0	0	0	0	7	11	11	2
7	1	0	0	1	1	1	1	1
11	0	0	0	0	1	1	1	1

Fig. B.9 Reduced state machine. State table (left) and state graph (right).

☐

B.2 | Practical aspects and implementation issues

How to turn finite state machines into electronic hardware is discussed in the main text. Yet, several practical problems can be understood from a mathematical background alone.

B.2.1 Parasitic states and symbols

Input symbols, states, and output symbols must ultimately be encoded as binary vectors. As it takes $w_x \geq \lceil \log_2 |X| \rceil$ bits to uniquely encode the $|X|$ elements of a set X, the code vector may assume 2^{w_x} distinct values. This implies that $2^{w_x} - |X| \geq 0$ code values exist that do not correspond to any element $x \in X$. Such unused values that result as a side effect from binary coding are termed **parasitic** or residual.

As a consequence, any finite state machine implemented with two-valued electronics will thus exhibit parasitic input symbols unless $2^{w_i} = |I|$, parasitic states unless $2^{w_s} = |S|$, and parasitic output symbols unless $2^{w_o} = |O|$. Being careful engineers, we ask ourselves

"What happens if, by accident, a finite state machine falls into some parasitic state or when it is presented with some parasitic input symbol?"[5]

From a mathematical perspective, neither transition function f nor output function g is defined. In practice, the circuit logic will generate some outputs in a deterministic but unspecified way. Thus, while the designer's intention is to build a state graph of $|S|$ vertices each of which has an out-degree

---

[5] Such unforeseen situations may occur as a consequence from interference, transmission errors, switching noise, poor synchronization, ionizing radiation, hot plug-in, or temporary sagging of power.

$|I|$, the actual result is a supergraph of 2^{w_s} vertices with out-degree 2^{w_i} that includes the original graph as a subgraph. Figure B.10 illustrates this by way of a Medvedev automaton that implements a controllable up/down counting function modulo 5.

Depending on the exact characteristics of the supergraph, a physical automaton may react in various ways in response to a parasitic input symbol: The automaton may show no discernible reaction, may produce just one mistaken output symbol, may move to some regular state via a transition that was unplanned for, or may fall into a parasitic state. In the last case, two different outcomes are possible: The state machine may either return to the regular subgraph after a number of clocks, or get trapped in a dead state or in a circular path forever, a dramatic situation known as **lock-up** condition. Figure B.10a shows all shades of how a physical circuit can fail in response to a parasitic input.

Example

$I = \{h, u, d\}$, $S = \{0, 1, 2, 3, 4\}$, and $O = \{0, 1, 2, 3, 4\}$.

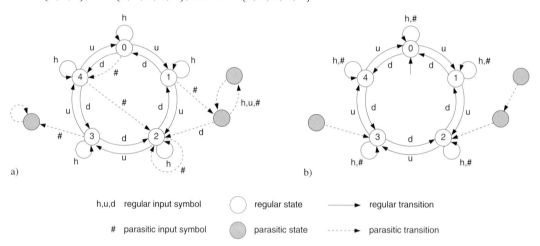

Fig. B.10 State graph of a modulo 5 up/down counter with parasitic states and input symbols left loose (a) compared with a safer version (b). Note that (b) leaves room for further improvement.

☐

Observation B.1. *In the presence of irregular conditions, parasitic states and symbols left undealt with hold the dangers of serious circuit malfunctioning and of permanent lock-up.*

"What can the digital designer do about parasitic finite state machine behavior?"

Broadly speaking, **fault tolerance** and **graceful degradation** are the goals of any engineering activity. They imply that a system confronted with irregular input data or some other form of disturbance shall

- absorb it with as little impact on its internal functioning as possible,
- continue to produce the most meaningful and/or least offensive output, and
- confine the consequences of any temporary failure to the shortest possible time span.

In the context of finite state machines, five measures must contribute towards these goals.

1. Collapse all parasitic input symbols to carefully selected regular ones.
2. Make sure that all parasitic states reconverge to the original subgraph. The standard practice is to explicitly indicate a regular successor state for each parasitic state before logic synthesis is undertaken.
3. Assign inoffensive output symbols to all parasitic states.
4. Provide some means for forcing the automaton into start state s_0 from any other state by adding an extra reset mechanism.

Finite state machines that adhere to these guidelines are sometimes qualified as **fail safe**. Note that, from a mathematical point of view, measures 1 and 2 extend the domain of f and g to include all parasitic values of i and s, whereas measure 4 depends on an ancillary mechanism that is independent of f and g. In the occurrence of the above example, a safer version of the modulo 5 up/down counter is shown in fig.B.10b. Albeit at a somewhat different level of abstraction, the measure below is as important as the ones mentioned before.

5. Notify the next higher level in the system hierarchy, e.g. by way of an error signal or message, whenever a parasitic state or input symbol has been detected. This avoids any innocent interpretation of corrupted FSM output and makes it possible for the superordinate system levels to decide on corrective action.

B.2.2   Mealy-, Moore-, Medvedev-type, and combinational output bits

The fact that output symbols get encoded as binary vectors gives rise to another subtlety. Consider an automaton where a subset of the output bits depends on the present state exclusively, that is, where some of the bits do not get affected by the present input. In analogy to the classification scheme for automata, such outputs are termed **Moore-type outputs**, with "decoded Moore-type outputs" and "unconditional outputs" being synonyms.

Clearly, the state machine as a whole remains a Mealy automaton as long as there exist other output bits — called **Mealy-type outputs**, aka "conditional outputs" — that actually are functions of the present input as well.

Similarly, a Mealy or a Moore machine may but need not include **Medvedev-type outputs**, aka "undecoded Moore-type outputs" and "direct outputs". As the name suggests, such output lines are nothing else than bits tapped from the state register either in direct or in complemented form. Their switching occurs essentially aligned to the clock.

Last but not least, an FSM may but need not feature **combinational outputs**, i.e. bits that depend on the present input exclusively. Combinational and Mealy-type outputs are sometimes subsumed as **through paths**; only Mealy machines can sport them.

Being knowledgeable about output types has been found to be useful not only during circuit design, but also during logic simulation, timing analysis, and prototype testing. An example is to follow soon.

B.2.3   Through paths and logic instability

The presence of a through path in a Mealy machine holds a serious danger. Any external circuitry that uses the FSM's present output to determine the FSM's present input may give rise to logic

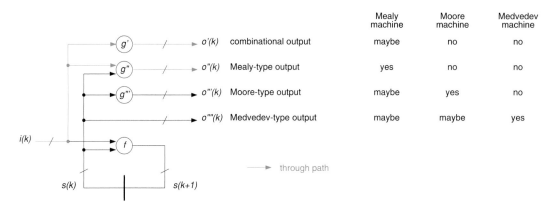

	Mealy machine	Moore machine	Medvedev machine
$o'(k)$ combinational output	maybe	no	no
$o''(k)$ Mealy-type output	yes	no	no
$o'''(k)$ Moore-type output	maybe	yes	no
$o''''(k)$ Medvedev-type output	maybe	maybe	yes

Fig. B.11 Finite state machine with output bits broken into four subsets. Each subset depends on $i(k)$ and $s(k)$ in a different way and is labeled accordingly.

contradictions. Uncontrolled oscillations may then develop because Mealy and combinational outputs instantly respond to new input. What makes the problem particularly treacherous is that contradictions and oscillations may actually occur for a limited subset of states and input values exclusively, while the design behaves in a totally inconspicuous way in all other situations.

Example

Consider the control loop below, where act, ini, and dcr are integer variables and ... stands for some unspecified data manipulations.

```
...
act := ini
repeat
   ...
   act := act - dcr
until act < 0
...
```

A possible hardware structure is depicted in fig.B.12a. A finite state machine interprets the carry/borrow bit from the ALU and controls ALU operation and data transport paths. Figure B.12b shows the tiny portion of a Mealy-type state graph relevant to the above loop computations. The intention is to subtract dcr from act until the ALU produces a borrow. Everything works fine as long as act≥dcr. When this relation ceases to hold, however, the circuit enters an oscillatory regime caused by the mutual and contradictory dependency of op2 and borrow in a feedback loop that encompasses the ALU, the controller, and the leftmost multiplexer. This unstable condition is made at all possible by the attempt to carry out subtraction and decision making in a single clock cycle, which concept differs from how a microprocessor would evaluate the above piece of code.

Observation B.2. *Instability may but need not develop in Mealy machines if the surrounding logic provides immediate feedback from the output to the input of the same machine. The existence of a*

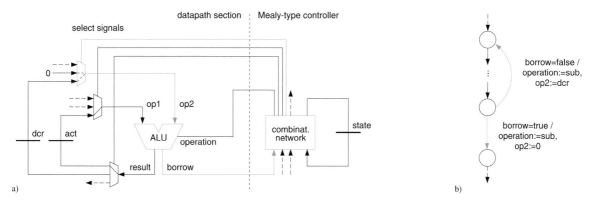

Fig. B.12 Instability in a Mealy-type controller subject to combinational feedback. Block diagram (a), portion of controller state graph (b).

□

zero-latency feedback path from one or more output bits of a through path to one or more input bits of the same through path is a necessary precondition for this to happen.

Basically there are four options for staying clear of instability in automata:

- Select an automaton with non-zero latency, see table B.5.
- Include latency by making feedback paths start from Moore or Medvedev bits exclusively.
- Add a latency register to the surrounding circuitry.
- Formally prove that no logic instability exists in spite of the zero-latency loop.

B.2.4 Switching hazards

From our discussion of transient effects in digital circuitry, we know that almost any combinational network has the potential of developing brief unwanted pulses known as hazards. As the output function g of an FSM is no exception, both Mealy and Moore automata must be suspected to generate hazards unless one has proof to the contrary. Medvedev machines, in contrast, cannot give rise to hazards because they lack a combinational network between state register and output.

Signals from any automaton can be made hazard-free by adding extra flip-flops at the output to align their switching to the clock if need be. The term **registered outputs** is often used to discern those bits that pass through such a **resynchronization register** from the normal ones that are taken from the output logic directly.

Practically speaking, when hazard-free outputs are to be combined with minimum latency, one can either add a resynchronization register to a Mealy automaton or build a Medvedev machine by encoding its state such that the output bits can be tapped from the state register directly; the conversion procedure of fig.B.6 should help. Either option also eliminates the risk of instability, see table B.5 for an overview.

B.2.5 Hardware costs

The costs of an FSM are given by the w_s bistables that maintain its state, the logic gates for computing transition function f and output function g, plus the necessary wiring. Note that f and g share part of the logic gates in typical hardware implementations. Although the costs of these resources in a full-custom IC are not the same as those in field-programmable logic (FPL), a number of observations can be made.

Concurrency, hierarchy, and modularity are key to efficiency

Classical state graphs tend to explode in size even when systems of very moderate complexity are being described.[6] The reasons are as follows.

- State graphs are flat with no levels of abstraction, they lack any notion of hierarchy.
- State graphs lack modularity, they do not distinguish between mechanisms.
- State graphs cannot model concurrent activities other than by a single global state.

More importantly, combinatorial explosion is also a problem for electronic hardware if a circuit is organized as one finite state machine. Beyond a certain complexity, it is much more efficient to partition the desired functionality into a bunch of smaller cooperating automata, a pattern which is sometimes also referred to as **linked state machines** (LSMs). This not only permits one to compose state and complex behavior from many simple models, but also makes it possible to use specialized and highly efficient subcircuits for implementing subfunctions such as counters, look-up tables (LUTs), en/decoders, etc.

Observation B.3. *While it is always possible to model a clocked sequential system as one Mealy-type automaton, a cluster of cooperating smaller automata (including counters, shift registers, etc.) is a much more efficient model for designing and verifying digital circuits.*

Example

A synchronous circuit is to generate a binary pseudo-random sequence of length 15 at a rate of 1/4 of its clock. Conceptually, what we need is an autonomous automaton with 60 states. There are two basic options. The entire functionality is either packed into a single FSM, or decomposed into a divide-by-4 counter plus a 4 bit linear feedback shift register (LFSR) working under control of that counter. Table B.3 shows the hardware costs. The structured approach saves 4/5 of the area when compared with a flat machine the states of which are randomly encoded.

---

[6] A more succinct visual formalism are the **statecharts** proposed by David Harel [477] [478]. In a nutshell

statecharts = state graphs + hierarchy + concurrency + interprocess communication

Statecharts help a lot to expose orthogonalities in behavioral models. CAE tools for editing and simulating state-charts are commercially available. Most of them are capable of generating program code for microprocessors, some of them also generate HDL code for further processing by synthesis tools. Yet, competition seems to disallow companies giving credit to Harel for proposing the statechart formalism that all their tools have in common.

Table B.3 | Hardware costs of a pseudo-random sequence generator organized in various ways.

structure	state assignment	flip-flops	std cells	nets	size [GE]	area [$M\lambda^2$]
counter & LFSR	native	6	9	13	60	0.15
flat Moore FSM	adjacent	6	26	33	90	0.23
„ „	random (typ.)	6	102	107	261	0.76
„ „	one hot	60	69	130	488	1.16

☐

STATE REDUCTION

State reduction, see section B.1.6, always has the benefit of eliminating unnecessary clutter from state tables and state graphs. In most cases, state reduction also pays off in terms of circuit complexity and performance because it introduces new don't care conditions, gives more room for finding a better state encoding, and, thereby, leads to a more economical solution.

Example

Table B.4 juxtaposes two standard cell implementations of the FSM specified earlier in figs.B.8 (unreduced) and B.9 (reduced). The relevant cost factor is the total area occupied after routing. Although the example is an artificial one, the figures indicate that the benefit from state reduction is mainly due to the simplification of the combinational network and not so much a matter of doing away with a flip-flop or two.

Table B.4 | Impact of state reduction on hardware costs for the finite state machine of fig.B.8.

number of states	state assignment	flip-flops	std cells	nets	size [GE]	area [$M\lambda^2$]
11	adjacent	4	35	38	94	0.26
6	adjacent	3	14	20	44	0.12

☐

Observation B.4. *It normally pays to eliminate redundant states from a state graph or state table prior to translating it into hardware.*

This need not always be true, however, since using fewer bits for storing a machine's state can sometimes increase the number of terms and literals in transition and/or output functions, inflate combinational circuitry, and offset the savings obtained from using fewer bistables.

STATE ENCODING

State encoding, aka state assignment, is the process of deciding on how the various states are going to be mapped onto vectors of binary digits. An obvious requirement is that each state is assigned a unique bit vector. The number of bits required for uniquely encoding $|S|$ states is

$$w_s \geq \lceil \log_2 |S| \rceil \tag{B.6}$$

where $\lceil x \rceil$ denotes the least integer not smaller than x. We speak of **minimum bit encoding** when (B.6) is made to hold with equality because the number of bistables is then minimal.

From a purely <u>functional</u> point of view, state encoding is immaterial because any unique state assignment necessarily leads to a correct circuit that is equivalent to those resulting from all other mappings.[7]

From an <u>efficiency</u> point of view, some state encodings will yield smaller and faster circuits than others. Energy dissipation and testability are also likely to differ. As the implications from the subsequent steps in the design process — logic synthesis, placement, and routing — are difficult to anticipate, one might be tempted to complete the design and to evaluate different schemes on the grounds of the final result. The number n_s of truly distinct state assignments, i.e. those that cannot be derived from others simply by permuting and/or complementing state bits, has been known since the late 1950s [479] and is given in (B.7). Numeric evaluation quickly tells us that, even for small automata, it is not computationally feasible to find the best state assignment using an enumerative approach.

$$n_s = \frac{(2^{w_s} - 1)!}{(2^{w_s} - |S|)! w_s!} \tag{B.7}$$

From an <u>engineering</u> point of view, we are not so much interested in the absolute optimum but in finding a good state assignment that results in a near-minimal hardware solution with reasonable effort. A variety of heuristic approaches have been devised, all of which attempt to carry out state encoding so as to minimize the number of literals in the FSM's logic equations in some way or another. The techniques differ in whether they target two-level or multilevel logic equations and in how they resolve conflicts between contradicting requirements.

Adjacent state assignment is the name for a class of heuristics that assumes two-level logic in sum-of-products form. The idea behind all such heuristics essentially is to lower the number of product terms and the number of literals by assigning states that have similar entries in the state table codes that differ in a single bit. Put differently, they make similar states adjacent in the Karnaugh map. Somewhat surprisingly, this approach has been found to have beneficial effects on multilevel logic implementations too and is, therefore, quite popular.

As the name suggests, **one-hot state encoding** uses a binary vector of length $|S|$ and assigns each state a code where a single digit is logic 1 and all others are 0.[8] When compared with minimum-bit state encoding, one-hot encoding typically results in many more bistables but, at the same time, also in a less combinational logic. Whether this pays off or not depends on the application; it certainly did not in the example of table B.3. In FPL devices with limited routing resources it sometimes is the only way to implement substantial finite state machines. On the negative side, one-hot encoding

---

[7] Designers typically leave state assignment to a synthesis tool and often ignore the actual binary codes selected. However, testing and debugging will involve gaining access to state registers, e.g. by way of a scan path or by physical probing. Interpreting those binary codes then requires knowledge of the state encoding chosen.

[8] Please make sure you understand that one-hot state encoding is not in contradiction to state reduction.

brings about a huge number of parasitic states, namely $2^{|S|} - |S|$. Making all of them reconverge to the main subgraph as described in section B.2.1 often proves unwieldy.

B.3 | Summary

- The key characteristics of the most common automata are collected below. Observe that responding to a new input within the on-going computation period (latency 0) and unconditional stability (no through path) are mutually exclusive.

Table B.5 Six types of automata and their key characteristics.

resynchronization register	class of automaton		
	Mealy	Moore	Medvedev
no	latency 0 possibly unstable hazards likely	latency 1 stable hazards likely	latency 1 stable hazard-free
yes	latency 1 stable hazard-free	latency 2 stable hazard-free	latency 2 stable hazard-free

- Designing with finite state machines involves the following steps:
  1. Partitioning the desired functionality into a cluster of cooperating automata whenever this is advantageous from an economy or modularity point of view.
  2. Selecting the appropriate type of automaton and the types for its outputs.
  3. Detailed specification and verification of statechart, state graph, or state table.
  4. State reduction.
  5. Deciding on how to safely handle parasitic inputs and states, if any.
  6. State assignment.
  7. Minimization of combinational functions.
  8. Designing the circuit logic from the components or library cells available.

Today's electronic design automation packages routinely cover steps 4 and 6 through 8. Also available are software tools for the editing of state graphs and for visualizing inputs, state transitions, and outputs at a behavioral level.

Appendix C

VLSI Designer's Checklist

C.1 | Design data sanity

☐ Are design data fully consistent? Have all design modifications always been propagated? Has the design data base never been tampered with (e.g. using a text or stream editor)?

☐ Are you sure that no source data such as HDL specifications, schematics, netlists, macrocell generator instructions, and the like have been modified after physical design was begun?

☐ Have all verification steps (simulation, ERC, timing verification, DRC, LVS, etc.) been carried out on the most recent version of the design?

☐ Do the cell libraries and/or transistor models being used indeed apply to the fabrication process and the operating conditions targetted?

☐ Have all library elements been fully characterized? Beware of "0" or other default entries sometimes entered by library developers for properties (such as area, propagation delay, power dissipation, etc.) the numerical values of which have not yet been established.

C.2 | Pre-synthesis design verification

☐ Is a bit-true and cycle-true behavioral model available (in HDL, C, or Matlab)? Has this circuit model been thoroughly tested in system-level simulations? Have the system designers checked and accepted the results so obtained?

☐ Do the logic gauges used in simulating a behavioral model systematically cover all modes and conditions under which the circuit is going to operate?

☐ Do the logic gauges also address uncommon situations, such as exceptional control flows, corrupt input data, numeric exceptions (e.g. divide by zero), overflows and underflows, truncation and rounding, data values outside of their habitual range, non-rational frequency ratios, and the like?

C.3 | Clocking

☐ Is a consistent clocking discipline being used? Which one?

☐ Are all bistables either edge-triggered (flip-flops) or level-sensitive (latches) throughout? Do not forget to check synthesizer-generated blocks (e.g. datapaths, finite state machines, boundary-scan logic), macrocells (e.g. register files, pipelined datapaths), and megacells (e.g. microprocessor cores, communication interfaces).

☐ Are there absolutely no violations of the dissociation principle of synchronous design? Trace all clock distribution paths and check all cells and terminals that depend on or interact with a clock signal.

☐ Do(es) the clock tree(s) exclusively end at clock inputs of
  ◇ flip-flops, or
  ◇ latches, and possibly also
  ◇ safe clock gating subcircuits?

☐ Is the circuit free of botches from asynchronous design such as
  ◇ cross-coupled gates (e.g. NAND- or NOR-type SR-seesaw),
  ◇ other combinational networks featuring a zero-latency feedback loop,
  ◇ one-shots and monoflops,
  ◇ clock choppers and other ad hoc frequency-multiplication subcircuits,
    (other than a clean PLL design operating on the main clock),
  ◇ crude clock gates (rather than safe clock gating subcircuits),
  ◇ delay lines,
  ◇ ring oscillators,
  ◇ hazard suppression networks (redundant logic, low pass filters, etc.)?

☐ Are all clock signals free of hazards under all circumstances?

☐ In single-edge-triggered one-phase designs, do all bistables operate on the same edge?

☐ If there are any edge-triggered macrocells, such as RAMs or megacells, do they operate on the same edge as well?

☐ In level-sensitive two-phase designs, do all logic paths begin at a latch driven by a first clock signal and end at a latch driven by the second clock?

☐ Do all driving clocks feature sufficiently short rise and fall times?

☐ Has the entire clock distribution network been balanced with respect to delay?

☐ Have rise time, fall time, and clock skew been (re)checked after layout?

☐ If an on-chip clock oscillator is being provided, is there an external clock input that bypasses it for testing?

☐ Is the circuit a true static CMOS network throughout? Do not forget to check synthesizer-generated blocks, macrocells, and megacells too.

☐ If dynamic logic is being used, what is the lowest admissible clock frequency at which the circuit will operate? How was that operating limit determined? Have charge retention, charge sharing, and charge decay been modelled properly?

C.4 | Gate-level considerations

☐ Is there a global reset? Does it operate synchronously or asynchronously? Do not forget to check macrocells, megacells, and virtual components too.

☐ Are all bistables connected to a reset pin? If not, how many clock cycles does it take to bring them all into a known state (homing sequence)? Counters, finite state machines, and bistables that are part of feedback loops require special attention. Multiple clock domains further complicate the issue.

☐ If there is a global asynchronous reset facility, is the user free to apply and release the reset signal at any time without risking timing violations or causing the circuit to come up in an inconsistent state? If there are constraints, are they acceptable?

☐ Are asynchronous (re)set inputs nowhere being misused for anything else than overall initialization? Is there no combinational logic acting on them? Trace all reset distribution paths and check all cells and terminals that depend on or interact with a reset signal.

☐ Compare the leaf cells of the reset tree with those of the clock distribution tree. They should essentially be the same. If exceptions exist, are they understood?

☐ If some asynchronous (re)set input depends on a signal other than the global reset, can that local reset be guaranteed to be free of hazard under any circumstance?

☐ Is the circuit free of zero-latency loops (combinational feedback)? If not, can race conditions be excluded under any circumstances? Note that Mealy machines connected to zero-latency logic networks need particular attention.

☐ Is the longest path known? Is its delay acceptable for the slowest PTV condition and the highest clock frequency with which the circuit will have to operate?

☐ Is the shortest path known? Is its delay acceptable for all PTV conditions under which the circuit will have to operate?

☐ Setup time, hold time, contamination delay, and propagation delay define the I/O timing.
    ◇ Are all four parameters known for all primary inputs and outputs?
    ◇ Do they provide reasonable data-valid windows for the external circuitry?

⋄ Does the resulting data-call window impose reasonable timing conditions?

⋄ How are data-valid and data-call intervals affected by PTV variations?

☐ Were there any warnings or error messages from the HDL analyzer, synthesizer, logic optimizer, or technology mapper? If so, are they fully understood?

☐ Do the nature and the number of bistables obtained from synthesis match your expectations? The presence of a latch in what is supposed to be an edge-triggered design points to a problem. The same applies to flip-flops with asynchronous set <u>and</u> reset inputs in synchronous designs.

☐ If snappers are being included, have you made sure they are never used for the purpose of data storage?

C.5 | Design for test

☐ Is there a built-in self-test of sufficient coverage?

☐ If not, are controllability and observability guaranteed by test structures such as scan path(s), block isolation, and boundary scan? Are there no large subcircuits being shadowed by on-chip RAMs and other poorly accessible circuit items?

☐ Are there any wide counters or accumulators that might ask for excessively many test vectors to reach some critical state (e.g. overflow)? If so, are they presettable from the primary inputs or via the scan path?

☐ Are all bistables part of a scan path (full scan)?

☐ If not, have test vectors been obtained before sending the design to fabrication? Has fault coverage been determined and is it satisfactory? Do test vectors allow one to locate faults with satisfactory resolution? Is the number of test vectors required acceptable?

☐ If partial scan is being used, do all unscanned storage elements (flip-flops, counters, RAMs, etc.) get disabled in scan mode?

☐ Does no bistable in a scan path act on the asynchronous (re)set of some other sequential subcircuit?

☐ Do all flip-flops that make part of one scan path get triggered by the same clock edge?

☐ Is all clock gating neutralized while in scan mode?

☐ Have skew margins also been verified in test mode? Remember that scan paths are particularly vulnerable to clock skew.

☐ Is one-at-a-time access to multi-driver nodes not only guaranteed during regular circuit operation, but in test mode as well? Recall that scanning in and scanning out may lead to states that are never reached during normal circuit operation.

☐ Has the design been resimulated after scan insertion and physical design? Do such simulations include scan-in and scan-out sequences?

☐ Are there any inputs of gates or cells that are permanently tied to 0 or 1? If so, have they been made controllable in test mode?

C.6 | Electrical considerations

☐ Is there no CMOS gate input or MOSFET gate electrode left open?

☐ Are all drivers sized to handle their loads? Are there no nodes with excessive rise and fall times? What about the reset distribution network?

☐ Are there pad drivers on all primary outputs? Have they been sized to provide adequate, but not excessive, driving capabilities for the off-chip loads under the given timing requirements? Can the core logic handle the load imposed by the pad drivers?

☐ If extra test pads are provided for probing, has each such pad been electrically decoupled from the core logic by way of an adequate drivers?

☐ Is there a level shifter at every point where a logic signal passes from one voltage domain to a second one? Have the level shifters on the primary inputs been chosen to correctly translate from off-chip to on-chip logic levels?

☐ Is the number of ground and power pads commensurate with switching currents? Remember clocking and simultaneously switching primary outputs contribute heavily to surge currents and to ground bounce.

☐ Have ground and power nets of padframe and core logic been decoupled from each other as much as possible?

☐ Does the chip include any analog subcircuits that need particular protection from noise coupling?

☐ Has a noise analysis (ground bounce and crosstalk) been carried out? Do noise margins and setup/hold margins provide sufficient headroom? Have layout and package parasitics been taken into account in these analyses?

☐ Does the circuit include multi-driver nodes, i.e. nodes that are driven by three-state outputs? If so, does the control logic exclude conflicts under all circumstances? In case multi-driver nodes are left undriven for a prolonged period of time (say more than a few hundred nanoseconds, for instance), are there pull-ups/downs or snappers preventing them from drifting away?

☐ Does the circuit make use of transmission gates? If so, are all of them electrically embedded so as to exclude unpleasant surprises such as conflicting drivers, floating nodes, poor signal levels, overly slow rise and fall times, backward signal propagation, charge sharing, and simulation results that are not consistent with reality?

☐ Are all pads equipped with protection networks against damage from electrostatic discharge (ESD)? Have the pads been qualified with respect to their ESD protection and latch-up avoidance characteristics? Is ESD resistance adequate for the anticipated storage, transport, handling, and operation conditions?

C.7 | Pre-layout design verification

☐ Have static verification techniques (code inspection, netlist screening, electrical rule check) been used to make sure that all netlists conform to standard integrity rules? Were there no error messages? Are the causes of all warnings reported fully understood, if there were any?

☐ Do the logic gauges used in simulating a netlist systematically cover all modes and conditions under which the circuit is going to operate?

☐ In addition to the coverage issues considered during pre-synthesis verification, detecting various lower-level weaknesses may require extra simulation vectors. Do simulations also address marginal timing conditions, imperfect signal waveforms (as a consequence from noise, jitter, poor edges, glitches, reflections, etc.), and the like?

☐ Is simulation output systematically verified by having the simulator compare actual against expected responses, or is this done by visual inspection only?

☐ Is the simulation period organized according to the standard four-phase scheme $\triangle \downarrow \top \uparrow$ (stimulus application, passive clock edge, response acquisition, active clock edge)? If so, have you made sure there are no unwitting inconsistencies? If not, are the vectors obtained from simulation independent of speed and truly portable?

☐ If the design includes three-state nodes or dynamic logic, has charge decay been modelled during simulations? Have realistic decay times been used?

☐ Were all simulations carried out with models that consistently check and report timing violations? Remember that SPICE-type models have no such capabilities. Have there been any reports of setup violations, hold violations, or other timing problems?

☐ Were all simulations carried out with the reporting of unsettled nodes being enabled and properly set up? Did any nodes get reported as unsettled immediately before the active clock edge or before output data being sampled?

☐ Have ramp times, shortest and longest path lengths, skew margins, and I/O timings been anticipated with the aid of static timing verification?

☐ If not, have simulation runs at least been carried out with distinct clock frequencies?

☐ For a library-based design, did all timing-related analyses refer to realistic PTV conditions? Beware of libraries characterized by their vendors with a derating factor of 1 (nominal) for a junction temperature of 25 °C or so, and/or for a typical process outcome.

☐ For a transistor-level design, do the timing-related analyses consider distinct process outcomes for n- and p-channel transistors (e.g. typical/typical, slow/slow, fast/fast, slow/fast, and fast/slow)? Did they also cover the relevant range of temperature and supply voltage conditions?

☐ In the final netlist, have you rechecked the presence of test structures such as scan paths, extra logic for the generation of auxiliary test signals, and other non-functional structures after logic optimization?

C.8 | Physical considerations

☐ Have you traced all ground, power, clock, and global reset nets? Are there no shorts? Are there no accidental discontinuities? Do not forget to check the padframe too.

☐ Are all supply lines routed on thick high-level metals as much as possible? Are they free of unnecessary contacts, vias, and geometric constrictions?

☐ Has the overall power distribution network been found to be adequate in view of
◇ current densities (electromigration),
◇ parasitic resistance (ohmic losses), and
◇ parasitic inductance (ground bounce).

☐ Are clock lines and long or critical signal lines routed in metal throughout?

☐ Have all heavily loaded lines been sized accordingly? Remember a central clock driver imposes a heavier burden on the clock distribution network than a clock tree does.

☐ If vias and/or contacts must be used on critical and, therefore, wide wires, are they adequately sized and stippled?

☐ Have all regular pads been instantiated from the appropriate library? Recall that only the uppermost metal is accessible for bonding, which makes the correct choice of pads depend on the metal options of the target process being used.

☐ If extra test pads are to be provided for probing, are all such pads made of the uppermost metal? Are there openings in the overglass layer? Is there a sufficient overlap between overglass and top-level metal to produce a hermetic seal?

☐ Are all wells properly tied to VSS (p-well) or VDD (n-well)? Do all library cells used include body ties (well and substrate contacts)? If not, is the density of body ties adequate? Do they provide the low-resistance paths required to prevent latch-up?

☐ Are n- and p-channel MOSFETs consistently separated in the layout so as to spoil the lateral BJTs?

☐ Does the layout include any structures that violate layout rules or that might overtax photographic equipment for mask making (e.g. curved lines and arbitrary angles)? What about texts and company logos?

☐ Is the bonding pattern consistent with manufacturing rules?

C.9 | Post-layout design verification

☐ Have all of the following verification steps been carried out?
◇ Layout rule check (DRC).
◇ Layout extraction (extract).

◇ Comparison of layout netlist versus schematic netlist (LVS).
◇ Post-layout timing verification (ramp times, clock skew, critical paths, I/O timing).
◇ Post-layout simulation.

☐ Have layout parasitics, process, temperature and voltage (PTV) variations, on-chip variations (OCVs), and noise-induced jitter been taken into account?

☐ Are you sure that all delay and energy parameters have been assigned their best possible estimates during back-annotation? Mismatches between the names of formal and actual parameters have been reported to cause back-annotation to go astray.

☐ If any warnings or errors have been reported, are the underlying causes fully understood and can they safely be considered to have no impact on the correct functioning of the circuit and on its fabrication yield?

☐ Have you carried out sufficient random inspections and verifying calculations to protect yourself against things that go wrong without producing any message?

☐ Is there any area that has been excluded from layout rule checking or from layout extraction by way of some pseudolayer? If so, have you made sure the layout exempted from verification is indeed correct?

☐ If hand-crafted layout is being used, is there a schematic for each cell against which to compare the netlist obtained from layout extraction?

C.10 | Preparation for testing of fabricated prototypes

☐ Have the test vectors intended for prototype verification been obtained from functional simulation runs and have the following points been observed in those simulations?
◇ Standard four-phase scheme, possibly with a longer clock period,
  e.g. △ @ 10 ns ↓ @ 50 ns ⊤ @ 90 ns ↑ @ 100 ns.
◇ Is the homing sequence limited to a few clock cycles?
◇ Are there no internal nodes that have been forced, charged, or otherwise initialized
  to known states by way of simulator commands?
◇ If multi-driver nodes are included, have simulations been carried out
  with charge decay time zero?
◇ If snappers are being used, have they been disconnected for simulation purposes?
◇ Do all primary inputs show relevant activity during simulation?
◇ Does every primary output toggle at least once?
◇ Do all primary inputs, all primary outputs, all three-state control signals, and
  all direction control signals appear in the trace file?
  Note that driver controls must be observed independently from whether they are
  internal or external to the chip.
◇ Were all signals in the trace file acquired at the correct time e.g. at $(90 \text{ ns} + k \cdot 100 \text{ ns})$?
◇ Does the trace file format adhere to the ATE[1] file format?

---

[1] ATE is an acronym for automatic test equipment.

☐ Are there waveform plots from simulation that cover the initialization phase for comparison with measured signals?

☐ Have you documented for each primary output in what way it depends on the primary inputs and/or on the circuit's logic state, i.e. whether it behaves as a Mealy, a Moore, or a Medvedev output?

C.11 | Thermal considerations

☐ Has overall power dissipation been estimated? Have the contributions from off-chip loads been included in this analysis?

☐ What temperature range is anticipated for a die under all expected operating conditions with the bonding method, package type, and cooling system used?

☐ Does die temperature stay within bounds that are acceptable for the semiconductor technology, the cell libraries, the package, and the mounting techniques being used?

☐ What impact do these temperature variations have on chip performance? If there are analog subcircuits on the same chip, how are they going to be affected?

C.12 | Board-level operation and testing

☐ Is the pinout scheme (pin1) clearly identified on the IC package and on the PCB board (datasheets and other extra documents will inevitably get lost)?

☐ Is there a decoupling capacitor for each IC? Does it have a sufficiently high resonance frequency? Is it wired in such a way as to provide a low inductance path?

☐ Are all ground and power pins connected to VSS and VDD respectively?

☐ Are noise voltages on ground and power nets acceptable?

☐ Are there any board-level lines that need termination?

☐ Have you checked signal integrity using an oscilloscope before moving to a logic analyzer for troubleshooting? Logic analyzers and other binary-valued instruments are inadequate for detecting and locating electrical problems such as overload, slow edges, noise, ringing, reflections, jitter, driver conflicts, mestastability, floating nodes, clock skew, and the like.

C.13 | Documentation

☐ Is there a datasheet which provides all information needed by the user to put the chip into operation? An adequate datasheet includes

⋄ A functional description (behavior, input-to-output relationship).

⋄ A block diagram (high-level structure).

⋄ A description of internal registers, data formats, data flow, state diagrams, and the like.

⋄ A declaration of all unsupported situations at control and data inputs (illegal combinations of modes, parasitic values, limitations of numeric range, etc.).

⋄ A declaration of all supported situations at control and data outputs.

⋄ A pad/pin list with signal name, signal type (in/out/bidir/three-state), signal polarity (active high/low level or rising/falling/either edge), and signal function.

⋄ Input and output voltage levels, drive capabilities, and pull-up/down resistors.

⋄ Preliminary signal timing information including waveform plots that do state on which basis they have been obtained (from worst/typical/best-case simulation or from actual measurements; if so, from how many parts and under what PTV conditions).

⋄ Absolute maximum ratings (voltages, temperature, resistance to ESD, etc.).

☐ Is there a record of all data possibly required by peer engineers for review, testing, qualifying, bug fixing, modification, and reuse of the design? An adequate documentation includes

⋄ Information on the fabrication process and the cell library used, if any.

⋄ A reference to the transistor and/or cell models used.

⋄ A cell hierarchy diagram that comprises information on
- whether a cell is memorizing or memoryless,
- the type of the cell (standard cell, macrocell, megacell, compound),
- how it has been obtained (synthesis, schematic entry, hand layout),
- the clocking scheme used, and
- the test structures incorporated.

⋄ All schematic diagrams and HDL source files.

⋄ The logic gauges, testbenches, etc. that have been used and are available for functional simulation, fault simulation, and test.

Appendix D

Symbols and constants

D.1 | Mathematical symbols used

Quantity	Unit	Explanation
α	1	MOSFET velocity saturation index
α	1	defect clustering factor
α_k	1	node activity
β	1	BJT current gain
β	A/V^2	MOSFET gain factor
$\beta_\square$	A/V^2	process gain factor
Γ	1	cycles per data item
δ	1	duty cycle
ϵ_0	A s/V m	permittivity of vacuum, also known as electric constant
ϵ_r	1	relative permittivity, also known as dielectric constant
$\theta_{a,c,j}$	K or °C	temperature (of ambient air, case, and junction respectively)
Θ	s^{-1}	data throughput
λ	1	MOSFET channel length modulation factor
Λ	m	pitch of virtual layout grid
μ	m^2/Vs	carrier mobility
μ_0	V s/A m	permeability of vacuum, also known as magnetic constant
μ_r	1	relative permeability
ϱ	kg/m^3	density
ρ	Ω m	resistivity
σ_k	1	crossover energy factor
Ψ	W/Hz = J	dissipated power per switching rate
A	m^2	area or, as a generalization, circuit size in gate equivalents [GEs]
AT	m^2 s	size–time product, alternatively in [GEs]
B	1	base in a positional number system
c	USD	cost, occasionally in [EUR] or [CHF]

c	m/s	speed of light in a medium
c_0	m/s	speed of light in vacuum
c_{ox}	F/m^2	gate capacitance per area
C	F	capacitance
d	1	iterative decomposition factor
d	m	diameter
D	m^{-2}	defect density
E	m	enclosure of one layout structure around another
E	J	energy, occasionally in [eV]
E_{ch}	J	energy dissipated for charging and discharging
E_{cr}	J	energy dissipated because of crossover currents
E_{gap}	J or eV	bandgap energy
E_{lk}	J	energy dissipated because of leakage
E_{rr}	J	energy dissipated because of resistive loads
E	V/m	electric field strength
f_{clk}	Hz	clock frequency
f_{cp}	Hz	computation rate
f_d	Hz	edge rate
$f_{to\,ff}$	Hz	flip-flop toggling rate
F	m	minimum pitch of a line with staggered contacts
F	V/m	electric field strength
G	1	total gate count
G, g	S	conductance
h	m	geometric height, thickness
H	m	half pitch
i	1	input symbol for a finite state machine
I	1	set of input symbols for a finite state machine
I, i	A	current
I_d	A	MOSFET drain current
I_f	A	forward current
I_{oup}	A	output current
I_r	A	reverse current
I_s	A	MOSFET source current
J	A/m^2	current density
k	J/K or eV/K	Boltzmann constant
$K_{P,\theta,V}$	1	derating factor (for process, temperature, and voltage respectively)
K_1	s	metastability parameter of a bistable
K_2	Hz	metastability parameter of a bistable
l	m	geometric length
L	1	latency in computation periods
L	H	inductance
L	m	length of MOSFET gate
L_{eff}	m	effective MOSFET gate length
L_{drawn}	m	drawn MOSFET gate length
m	1	MOSFET body effect coefficient
$m_{\text{chem. El.}}$	g/mole	atomic mass

M	m	minimum feature size
n	1	number of chips
n	1	index of refraction
$N_{A,D}$	m^{-3}	doping concentration (of acceptors and donors respectively)
N_{Avo}	mole^{-1}	Avogadro's number
o	1	output symbol from a finite state machine
O	1	set of output symbols from a finite state machine
p	1	pipelining factor, loop unfolding factor
P	1	number of lithographic patterning steps
P	W	power
q	1	replication factor
q_e	C	elementary charge
Q	C	charge
r_{cap}	s/F $= \Omega$	load factor
r_{sl}	V/s	slew rate
R	Ω	(electrical) resistance
$R_{\square}$	Ω	sheet resistance
R_θ	K/W	thermal resistance
s	1	time-sharing factor
s	1	state of a finite state machine
s_0	1	start state of a finite state machine
s_{ra}	1	slope sensitivity factor
S	1	set of states of a finite state machine
S	m	spacing between two layout structures
S	V	MOSFET subthreshold slope
t_{al}	s	time allowed to recover from metastability
t_{cd}	s	contamination delay
t_{di}	s	clock distribution delay
t_{fa}	s	fall time
t_{hi}	s	high time
t_{ho}	s	hold time
t_{id}	s	insertion delay
t_{it}	s	intrinsic delay
t_{jt}	s	clock jitter
t_{lo}	s	low time
t_{lp}	s	longest path delay
t_{mr}	s	metastability resolution time
t_{MTBE}	s	mean time between errors
t_{ox}	m	gate dielectric thickness
t_{pd}	s	propagation delay
t_{pu}	s	pulse width
t_{ra}	s	ramp width
t_{ri}	s	rise time
t_{sk}	s	clock skew
t_{sp}	s	shortest path delay
t_{su}	s	setup time

T	s	time per data item, period
T_{clk}	s	clock period
T_{cp}	s	computation period
U,u	V	voltage[1]
U_θ	V	thermal voltage
U_{bi}	V	junction built-in voltage
U_{bs}	V	MOSFET body-to-source voltage
U_{dd}	V	supply voltage
U_{ds}	V	MOSFET drain-to-source voltage
U_f	V	forward voltage
U_{gs}	V	MOSFET gate-to-source voltage
U_{ih}	V	input high voltage
U_{il}	V	input low voltage
U_{inp}	V	input voltage
U_{inv}	V	inverter threshold voltage
U_{nm}	V	noise margin
U_{oh}	V	output high voltage
U_{ol}	V	output low voltage
U_{oup}	V	output voltage
U_{pn}	V	junction anode-to-cathode voltage
U_r	V	reverse voltage
U_{th}	V	MOSFET threshold voltage
U_{trip}	V	trip voltage
v	1	voltage amplification
V	m^3	volume
$w_{i,o,s}$	1	word width (for input, output, and state respectively)
w	m	geometric width
W	m	width of MOSFET gate or other layout structure
X	m	extension of one layout structure beyond another
y	1	fabrication yield
$\#_{items}$	1	number of items

D.2 | Abbreviations

aka	also known as
ckt	circuit
iff	if and only if
wrt	with respect to

For technical and scientific acronyms please check the Index.

---

[1] American writers usually write V for the quantity voltage and V for the unit volts. In accordance with recommendations by the International Electrotechnical Commission (IEC) and the Système International d'Unités (SI), we use U as quantity symbol for voltage and V as unit symbol for volt to clearly distinguish the two.

D.3 | Physical and material constants

Table D.1 | Selected physical constants.

Avogadro's number N_{Avo}	$6.022 \cdot 10^{23}$/mole
Boltzmann constant k	$13.81 \cdot 10^{-24}$ J/K $= 86.17 \cdot 10^{-6}$ eV/K
Planck constant $\hbar = \frac{h}{2\pi}$	$0.1055 \cdot 10^{-33}$ J s $= 0.6582 \cdot 10^{-15}$ eV s
absolute zero	0 K $= -273.15$ °C
elementary or electron charge q_e	$0.1602 \cdot 10^{-18}$ C
permittivity of vacuum ϵ_0	$8.854 \cdot 10^{-12}$ A s/V m $(= F/m)$
permeability of vacuum μ_0	$4\pi \cdot 10^{-7}$ V s/A m $= 1.257 \cdot 10^{-6}$ V s/A m $(= H/m)$
speed of light in vacuum c_0	$299.8 \cdot 10^6$ m/s
thermal voltage $U_\theta = \frac{k\theta_j}{q_e}$	25.9 mV @ 300 K junction temperature

Note: The properties of thin films may considerably differ from those of bulk materials.

Table D.2 | Key properties of selected materials (mostly after [385] and [426]).

	semimetal	semiconductors			
material	C^2	Ge	Si	InP	GaAs
crystallographic variety[a]	G	D	D	Z	Z
bandgap energy E_{gap} at 300 K [eV]	≈ 0	0.66	1.12	1.35	1.42
relative permittivity ϵ_r		16.0	11.9	12.4	13.1
approx. breakdown field $[\frac{kV}{mm}]$		10	30	50	40
electron mobility μ_e at 300 K $[\frac{cm^2}{Vs}]$	$\leq 10\,000$	3900	1500	4600	8500
hole mobility μ_h at 300 K $[\frac{cm^2}{Vs}]$		1900	450		400
saturated electron velocity $[10^6 \frac{cm}{s}]$			10	10	13

---

[2] What makes carbon, C, so special is that it comes in many allotropic variations. **Diamond** forms a tetrahedral crystal lattice where carbon atoms sit in the corners and are held together by covalent bonds exactly as in monocrystalline silicon. This spatial arrangement with four strong bonds oriented around each nucleus renders diamond extremely hard and durable. The large bandgap of about 5.5 eV makes it an electrical insulator and optically transparent.

Graphene is the name given to a single layer of carbon atoms where each nucleus sits in the corner of a hexagon and is covalently bonded to three neighbors in such a way as to form a planar lattice reminiscent of chicken wire [480]. **Graphite** is a three-dimensional structure where many such sheets are held together by much weaker Van der Waals forces, which explains why graphite appears soft and slick. The bandgap is almost zero, which means that valence and conduction bands barely touch. As a consequence, graphite has a mediocre electrical conductivity and is sometimes termed a semimetal or metalloid.

Carbon nanotubes (CNTs) can be thought of as graphene sheets that have rolled up. They exist as multi-walled and as single-walled hollow cylinders, straight or twisted. Diameter is a little over 1 nm for a single-walled nanotube, and up to 50 nm for multi-walled nanotubes. CNTs of different sizes can have bandgaps as low as zero (as metal), as high as that of silicon, and almost anywhere in between. **Fullerenes** are spherical macromolecules that resemble soccer balls. Carbon nanotubes and fullerenes share an extraordinary strength and stability.

The soot and lampblack deposits that form when organic fuels are burned with a lack of sufficient oxygen largely consist of **amorphous carbon** with no long-range pattern of atomic positions.

Table D.2 (*cont.*)

atomic mass m $[\frac{g}{mole}]$	72.64	28.09	145.79	144.63
density ϱ $[\frac{g}{cm^3}]$	5.327	2.328	4.787	5.320
lattice spacing [nm]	0.564 61	0.543 09	0.586 87	0.565 33
melting point [°C]	937	1415	1057	1238
thermal conductivity $[\frac{W}{cm\,K}]$	0.6	1.5	0.7	0.46

	wide-gap semiconductors				insulator
material	SiC	SiC	GaN	C[2]	SiO$_2$
crystallographic variety	6H	4H		D	amorph.
bandgap energy E_{gap} at 300 K [eV]	3.03	3.26	3.49	5.47	8–9
relative permittivity ϵ_r	9.66	9.7	9.0	5.68	≈3.9
approx. breakdown field $[\frac{kV}{mm}]$	300	300	300	1000	500–1000
electron mobility μ_e at 300 K $[\frac{cm^2}{Vs}]$	b	700	<2000	4500	
saturated electron velocity $[10^6 \frac{cm}{s}]$	20	20	13	27	
atomic mass m $[\frac{g}{mole}]$	40.10	40.10	83.73	12.01	60.08
density ϱ $[\frac{g}{cm^3}]$	3.2	3.2	6.1	3.520	2.27
lattice spacing [nm]	0.308	0.308		0.356 68	
melting point [°C]	2830	2830	600	3800	≈1700
thermal conductivity $[\frac{W}{cm\,K}]$	5	5	>1.5	20	0.014

[a] D = Diamond, Z = Zincblende, G = graphene (planar monocrystalline sheet).
[b] Highly anisotropic.

Table D.3 Selected conductor materials.

material	resistivity ρ $[10^{-9}\ \Omega\ m]$	melting temp. [°C]	
W	56.5	3410	used in contact/via plugs
Al	26.5	660	pure metal
Al 0.5% Cu	≈30	n.a.	used for interconnect lines
Cu	16.7	1083	"
Ti	420	1668	given for comparison
Ta	125	2996	"
Mo	52	2610	"
Au	23.5	1063	"
Ag	15.9	961	"

Table D.4 Selected dielectric materials.

material	relative permittivity ϵ_r	
ceramic Al_2O_3	7–8	used for packages
epoxy resins	≈ 4.2	” and circuit boards
silicate glass SiO_2	1.8 (@ 75% porosity) to 3.9	inorganic ILD
fluorinated silicate glass (FSG) SiOF	3.0–3.7	”
hydrogen silsesquioxane (HSQ)	3.0–2.7	”
carbon-doped oxide (CDO) SiOC	≈ 2.4 (nanoporous) to 3.3	
organosilicate glass (SiCOH)	≈ 1.8 (nanoporous) to 2.9	
polyimides	3.0–3.6	organic ILD
parylene	2.6	”
benzocyclobutane	2.6	”
Teflon^{TM} family	≤ 2.0	”

References

1. International Sematech. International Technology Roadmap for Semiconductors, 2005. http://www.itrs.net/Common/2005ITRS/Home2005.htm.

2. Bill McClean. 2001 IC Industry at the Crossroads. *Semiconductor International*, 24(1), January 2001.

3. Fabless Semiconductor Association. http://www.fsa.org.

4. Charles E. Stroud. *A Designer's Guide to Built-in Self-Test*. Springer, 2002.

5. Michael L. Bushnell and Vishwani Agrawal. *Essentials of Electronic Testing*. Kluwer Academic Publishers, 2000.

6. Alberto Sangiovanni-Vincentelli. The Tides of EDA. *IEEE Design & Test of Computers*, 20(6):59–75, November–December 2003.

7. David Andrews, Douglas Niehaus, and Peter Ashenden. Programming Models for Hybrid CPU/FPGA Chips. *IEEE Computer*, 37(1):118–120, 2004.

8. OptiMagic Inc. The FPGA Site. http://www.fpga-site.com.

9. Markus Wannemacher. Die aufzu Halbleiterhersteller-Ecke. http://www.aufzu.de/semi/halbleit.html.

10. Jonathan Rose, Abbas El Gamal, and Alberto Sangiovanni-Vincentelli. Architecture of Field-Programmable Gate Arrays. *Proceedings of the IEEE*, 81(7):1013–1029, July 1993.

11. Bob Zeidman. *Designing with FPGAs and CPLDs*. CMP Books, Lawrence KS, 2002.

12. Clive Maxfield. *The Design Warrior's Guide to FPGAs*. Newnes, Burlington MA, 2004.

13. A. Curiger, H. Bonnenberg, R. Zimmermann, N. Felber, H. Kaeslin, and W. Fichtner. VINCI: VLSI Implementation of the new Secret-Key Block Cipher IDEA. In *Proceedings of the IEEE 1993 Custom Integrated Circuits Conference*, pages 15.5.1–4, San Diego CA, 1993. IEEE.

14. Nirmal R. Saxena *et al.* Dependable Computing and Online Testing in Adaptive and Configurable Systems. *IEEE Design & Test of Computers*, 17(1):29–41, January–March 2000.

15. Ingrid Verbauwhede, Patrick Schaumont, and Henry Kuo. Design and Performance Testing of a 2.29 Gb/s Rijndael Processor. *IEEE Journal on Solid State Circuits*, 38(3):569–571, March 2003. (Encryption only, max. throughput is 2.29 Gbit/s with 256 bit blocks.)

16. Giovanni De Micheli and Rajesh K. Gupta. Hardware/Software Co-Design. *Proceedings of the IEEE*, 85(3), March 1997.

17. Ingrid Verbauwhede and Alireza Hodjat. High-Throughput Programmable Cryptoprocessor. *IEEE Micro*, 24(3):34–45, May/June 2004.

18. Tilman Glökler, Andreas Hoffmann, and Heinrich Meyr. Methodical Low-Power ASIP Design Space Exploration. *Journal of VLSI Signal Processing Systems*, 33(3):229–246, March 2003.

19. Sven Woop, Joerg Schmittler, and Philipp Slusallek. RPU: A Programmable Ray Processing Unit for Realtime Ray Tracing. In *Proceedings of the ACM SIGGRAPH conference*, pages 434–444, Los Angeles, July/August 2005. ACM.

20. Russell Tessier and Wayne Burleson. Reconfigurable Computing for Digital Signal Processing: A Survey. *Journal of VLSI Signal Processing Systems*, 28(1/2):7–27, May/June 2001.

21. Neil Jacobson. *The in-system configuration handbook: a designer's guide to ISC*. Kluwer Academic Publishers, Hingham MA, 2004.

22. John Villasenor and Brad Hutchings. The Flexibility of Configurable Computing. *IEEE Signal Processing*, 15(5):67–84, September 1998.

23. P.H.W. Leong *et al.* Pilchard — A Reconfigurable Computing Platform with Memory Slot Interface. In *Proceedings of the IEEE Symposium on Field-Programmable Custom Computing Machines (FCCM)*, Rohnert Park, CA, 2001. IEEE.

24. John L. Hennessy and David A. Patterson. *Computer Architecture, a Quantitative Approach*. Morgan Kaufmann Publishers, San Mateo CA, fourth edition, 2007.

25. Mohamed Rafiquzzaman. *Fundamentals of Digital Logic and Microcomputer Design*. John Wiley & Sons, Hoboken NJ, 2005.

26. Nick Tredennick. Microprocessor-Based Computers. *IEEE Computer*, pages 27–37, October 1996.

27. Yale Patt. Requirements, Bottlenecks and Good Fortune: Agents for Microprocessor Evolution. *Proceedings of the IEEE*, 89(11):1553–1559, November 2001.

28. Artur Klauser. Trends in High-Performance Microprocessor Design. *Telematik Journal*, 7(1):12–21, April 2001.

29. Doug Burger and James R. Goodman. Billion-Transistor Architectures: There and Back Again. *IEEE Computer*, 37(3):22–28, March 2004.

30. Doug Burger *et al.* Scaling to the End of Silicon with EDGE Architectures. *IEEE Computer*, 37(7):44–54, July 2004.

31. Randy Goldberg and Lance Riek. *A Practical Handbook of Speech Coders*. CRC Press, Boca Raton FL, 2000.

32. Yu Hen Hu. CORDIC-Based VLSI Architectures for Digital Signal Processing. *IEEE Signal Processing Magazine*, 9(3):16–35, July 1992.

33. John Stephen Walther. The Story of Unified CORDIC. *Journal of VLSI Signal Processing*, 25(2):107–112, June 2000. Part of special issue on CORDIC.

34. A. Burg, M. Borgmann, M. Wenk, M. Zellweger, W. Fichtner, and H. Bölcskei. VLSI Implementation of MIMO Detection using the Sphere Decoder Algorithm. *IEEE Journal of Solid-State Circuits*, 40(7):1566–1577, 2005.

35. Jay R. Southard. MacPitts: An Approach to Silicon Compilation. *IEEE Computer*, 16(12):74–82, December 1983.

36. Keshab K. Parhi. *VLSI Digital Signal Processing Systems*. John Wiley & Sons, New York, 1999.

37. Boaz Porat. From Academe to Industry (or from Writing Papers to Making Chips): Experiences and Conclusions. *IEEE Signal Processing Magazine*, 20(4):8–11, July 2003.

38. X. Lai, J. L. Massey, and S. Murphy. Markov Ciphers and Differential Cryptanalysis. In *Advances in Cryptology — EUROCRYPT '91*, pages 8–13. Springer Verlag, Berlin, 1991.

39. M. S. Hrishikesh *et al.* The Optimal Logic Depth per Pipeline Stage is 6 to 8 FO4 Inverter Delays. In *Proceedings of 29th International Symposium on Computer Architecture*, pages 14–24, 2002.

40. H. Bonnenberg, A. Curiger, N. Felber, H. Kaeslin, and X. Lai. VLSI Implementation of a New Block Cipher. In *Proceedings of the International Conference on Computer Design*, Cambridge MA, October 1991. IEEE.

41. Robert G. Swartz. Ultra-High Speed Multiplexer/Demultiplexer Architectures. *International Journal of High Speed Electronics*, 1(1):73–99, 1990.

42. Marc Biver and Hubert Kaeslin and Carlo Tommasini. In-Place Updating of Path Metrics in Viterbi Decoders. *IEEE Journal of Solid-State Circuits*, 24(4):1158–1160, August 1989.

43. Jarmo Takala and Konsta Punkka. Scalable FFT Processors and Pipelined Butterfly Units. *Journal of VLSI Signal Processing Systems*, 43(2/3):113–123, June 2006.

44. Keshab K. Parhi. Approaches to Low-Power Implementations of DSP Systems. *IEEE Transactions on Circuits and Systems I: Fundamental Theory and Applications*, 48(10):1214–1224, October 2001.

45. Doris Keitel-Schulz and Norbert Wehn. Embedded DRAM Development: Technology, Physical Design, and Application Issues. *IEEE Design & Test of Computers*, 18(3):7–15, May/June 2001.

46. Kiyoo Itoh. *VLSI Memory Chip Design*. Springer Verlag, New York, 2001.

47. Sung-Mo Kang and Yusuf Leblebici. *CMOS Digital Integrated Circuits*. McGraw Hill, Boston MA, 2003.

48. John E. Ayers. *Digital Integrated Circuits, Analysis and Design*. CRC Press, Boca Raton FL, 2004.

49. Kiat-Seng Yeo and Kaushik Roy. *Low-Voltage, Low-Power VLSI Subsystems*. McGraw-Hill, New York, 2005.

50. Roberto Bez, Emilio Camerlenghi, Alberto Modelli, and Angelo Visconti. Introduction to Flash Memory. *Proceedings of the IEEE*, 91(4):489–502, April 2003.

51. Takashi Kobayashi, Hideaki Kurata, and Katustaka Kimura. Trends in High-Density Flash Memory Technology. *IEICE Transactions*, E87-C(10):1656–1663, October 2004.

52. Narendra Shenoy. Retiming: Theory and Practice. *Integration, the VLSI journal*, 22(1–2):1–21, August 1997.

53. Sachin Sapatnekar. *Timing*. Kluwer Academic Publishers, Boston MA, 2004.

54. Charles E. Leiserson and James B. Saxe. Retiming Synchronous Circuitry. *Algorithmica*, 6(1):5–35, 1991.

55. Hervé J. Touati and Robert K. Brayton. Computing the Initial State of Retimed Circuits. *IEEE Transactions on Computer-Aided Design*, 12(1):157–162, January 1993.

56. Nikolay Petkov. *Systolic Parallel Processing*. North-Holland, Amsterdam, 1993.

57. Charles E. Leiserson and James B. Saxe. Optimizing Synchronous Systems. *Journal of VLSI and Computer Systems*, 1(1):41–67, 1983.

58. Katsuhiko Hayashi, Kaushal K. Dhar, Kazunori Sugahara, and Kotaro Hirano. Design of High-Speed Digital Filters Suitable for Multi-DSP Implementation. *IEEE Transactions on Circuits and Systems*, 33(2):202–217, February 1986.

59. Peter M. Kogge. *The Architecture of Pipelined Computers*. McGraw-Hill Book Company, New York, 1981.

60. Keshab K. Parhi. Finite word effects in pipelined recursive filters. *IEEE Transactions on Acoustics, Speech and Signal Processing*, 39(6):1451–1454, June 1991.

61. Mehdi Hatamian and Keshab K. Parhi. A 85-MHz Fourth-Order Programmable IIR Digital Filter Chip. *IEEE Journal of Solid-State Circuits*, 27(2):175–183, February 1992.

62. Horng-Dar Lin and David G. Messerschmitt. Finite State Machine has Unlimited Concurrency. *IEEE Transactions on Circuits and Systems*, 38(5):465–475, May 1991.

63. Jun Ma, Keshab K. Parhi, and Ed F. Deprettere. A Unified Algebraic Transformation Approach for Parallel Recursive and Adaptive Filtering and SVD Algorithms. *IEEE Transactions on Signal Processing*, 49(2):424–437, February 2001.

64. R. Zimmermann, A. Curiger, H. Bonnenberg, H. Kaeslin, N. Felber, and W. Fichtner. A 177 Mb/s VLSI Implementation of the International Data Encryption Standard. *IEEE Journal of Solid-State Circuits*, 29(3):303–307, March 1994.

65. Gilles Privat and Alain D. Wittmann. Pipelined recursive filter architectures for subband image coding. *Integration, the VLSI Journal*, 14(3):361–379, February 1993.

66. Kai Hwang. *Computer Arithmetics*. John Wiley & Sons, New York, 1979.

67. Reto Zimmermann. *Binary Adder Architectures for Cell-Based VLSI and their Synthesis*. PhD thesis, Swiss Federal Institute of Technology, Zurich, 1998.

68. OpAr, Notes and Exercices of Arithmetics.
http://tima-cmp.imag.fr/~guyot/Cours/Oparithm/english/Op_Ar2.htm.

69. Jean-Pierre Deschamps, Géry J. A. Bioul, and Gustavo D. Sutter. *Synthesis of Arithmetic Circuits*. John Wiley & Sons, Hoboken NJ, 2006.

70. Peter B. Denyer and David Renshaw. *VLSI Signal Processing: A Bit-Serial Approach*. Addison-Wesley Publishing Company, Wokingham, 1985.

71. Stewart G. Smith and Peter B. Denyer. *Serial-Data Computation*. Kluwer Academic Publishers, Boston MA, 1988.

72. Stanley A. White. Applications of Distributed Arithmetic to Digital Signal Processing: A Tutorial Review. *IEEE Acoustics, Speech, and Signal Processing Magazine*, 6(3):4–19, July 1989.

73. Distributed Arithmetic Laplacian Filter. XCELL No. 20, 1996. Xilinx Inc.

74. Les Mintzer. FIR Filters with Field-Programmable Gate Arrays. *Journal of VLSI Signal Processing*, 6(2):119–127, August 1993.

75. Kyung-Saeng Kim and Kwyro Lee. Low-Power and Area-Efficient FIR Filter Implementation Suitable for Multiple Taps. *IEEE Transactions on Very Large Scale Integration (VLSI) Systems*, 11(1):150–153, February 2003.

76. G. Fettweis, L. Thiele, and H. Meyr. Algorithm transformations for unlimited parallelism. In *Proc. of the International Symposium on Circuits and Systems*. IEEE, New Orleans, 1990.

77. Bernard Carré. *Graphs and Networks*. Clarendon Press, Oxford, 1979.

78. Bernard Sklar. How I Learned to Love the Trellis. *IEEE Signal Processing Magazine*, 20(3):87–102, May 2003.

79. Alan Allan *et al.* 2001 Technology Roadmap for Semiconductors. *Computer*, 35(1):42–53, January 2002.

80. Mark L. Chang and Scott Hauck. Précis: A Usercentric Word-Length Optimization Tool. *IEEE Design & Test of Computers*, 22(4):349–361, July/August 2005.

81. Frank K. Gürkaynak. *GALS System Design: Side-Channel-Attack-Secure Cryptographic Accelerators*. PhD thesis, ETH Zürich, 2006.

82. Michael C. McFarland. Formal Verification of Sequential Hardware: A Tutorial. *IEEE Transactions on Computer-Aided Design*, 12(5):633–654, May 1993.

83. J. L. Lions *et al.* Ariane 5 Flight 501 Failure. Report by the Inquiry Board, European Space Agency, 1996.

84. Dick Price. Pentium FDIV flaw — lessons learned. *IEEE Micro*, 15(2):86–88, April 1995.

85. David Goldberg. Computer Arithmetic. In David A. Patterson and John L. Hennessy, editors, *Computer Architecture, a Quantitative Approach*. Morgan Kaufmann Publishers, San Mateo CA, second edition, 1996.

86. H. P. Sharangpani and M. L. Barton. Statistical Analysis of Floating Point Flaw in the Pentium Processor. Technical bulletin, Intel Corporation, 1994.

87. Niklaus Wirth. *Compilerbau*. B. G. Teubner, Stuttgart, 1977.

88. Janick Bergeron. *Writing Testbenches, Functional Verification of HDL Models*. Kluwer Academic Publishers, Boston MA, 2000.

89. Pradip Bose, Thomas M. Conte, and Todd M. Austin. Challenges in Processor Modelling and Validation. *IEEE Micro*, 19(3):9–14, May/June 1999.

90. Gregg D. Lahti. Test Benches: The Dark Side of IP Reuse. In Synopsys User Group, editor, *SNUG 2000*. Synopsys, San Jose, 2000. http://www.snug-universal.org/papers.htm.

91. Samir Palnitkar. *Design Verification with "e"*. Prentice-Hall, Upper Saddle River NJ, 2004.

92. Maher N. Mneimneh and Karem A. Sakallah. Principles of Sequential Equivalence Checking. *IEEE Design & Test of Computers*, 22(3):248–257, May/June 2005.

93. E. Clarke, O. Grumberg, and D. Peled. *Model Checking*. MIT Press, Cambridge MA, 2000.

94. Christoph Kern and Mark R. Greenstreet. Formal verification in hardware design: A survey. *ACM Transaction on Design Automation of Electronic Systems*, 4(2):123–193, April 1999.

95. Farn Wang. Formal verification of timed systems: A survey and perspective. *Proceedings of the IEEE*, 92(8):1283–1305, August 2004.

96. Zainalabedin Navabi. *Verilog Digital System Design*. McGraw-Hill, New York, 2006.

97. Jayaram Bhasker. *Verilog HDL Synthesis*. Star Galaxy Publishing, Allentown PA, 1998.

98. Stephen Bailey. Comparison of VHDL, Verilog and SystemVerilog, 2006. http://www.model.com/technical_portal/default.asp.

99. Sandi Habinc and Peter Sinander. Using VHDL for Board Level Simulation. *IEEE Design & Test of Computers*, 13(3):66–78, 1996.

100. Eugen Röhm. Latest Benchmark Results of VHDL Simulation Systems. In *Proceedings of the '95 European Design Automation Conference*, pages 406–411. IEEE, Brighton, 1995.

101. Eugen Röhm. "Ihr seid durchschaut!", Vergleich verschiedener VHDL-Simulatoren anhand von Benchmarks. *Elektronik*, (9), 1996.

102. Dirkjan Jongeneel and Ralph H. J. W. Otten. Technology Mapping for Area and Speed. *Integration, the VLSI Journal*, 29(1):45–66, March 2000.

103. IEEE, New York. *IEEE Standard VHDL Language Reference Manual*, 2002. IEEE Standard 1076-2002.

104. IEEE, New York. *IEEE Standard VHDL Language Reference Manual*, 1994. IEEE Standard 1076-1993.

105. Reto Zimmermann. VHDL AMS Syntax (IEEE Standard 1076.1-1999). http://dz.ee.ethz.ch/support/ic/vhdl/vhdlams_syntax.html.

106. Reto Zimmermann. VHDL Syntax (IEEE Standard 1076-1993). http://dz.ee.ethz.ch/support/ic/vhdl/vhdl93_syntax.html.

107. Jayaram Bhasker. *A Guide to VHDL Syntax*. Prentice-Hall, Englewood Cliffs NJ, 1995.

108. J. M. Bergé *et al*. *VHDL '92*. Kluwer, Boston MA, 1993.

109. Volnei A. Pedroni. *Circuit Design with VHDL*. MIT Press, Cambridge MA, 2004.

110. Paul Molitor and Jörg Ritter. *VHDL, eine Einführung*. Pearson Studium, Munich, 2004.

111. Peter J. Ashenden, Gregory D. Peterson, and Darrell A. Teegarden. *The System Designer's Guide to VHDL-AMS*. Morgan Kaufmann, San Francisco CA, 2003.

112. Peter J. Ashenden. *The Designer's Guide to VHDL*. Morgan Kaufmann Publishers, San Francisco, 2nd edition CA, 2002.

113. Sudhakar Yalamanchili. *Introductory VHDL: From Simulation to Synthesis*. Prentice-Hall, Upper Saddle River NJ, 2001.

114. James R. Armstrong and F. Gail Gray. *VHDL Design Representation and Synthesis*. Prentice-Hall PTR, Upper Saddle River NJ, 2000.

115. Ulrich Heinkel *et al*. *The VHDL Reference*. John Wiley & Sons, Chichester, 2000.

116. Mark Zwolinski. *Digital System Design with VHDL*. Prentice-Hall, Harlow, 2000.

117. K. C. Chang. *Digital Systems Design with VHDL and Synthesis, An Integrated Approach*. IEEE Computer Society Press, Los Alamitos CA, 1999.

118. K.C. Chang. *Digital Design and Modeling with VHDL and Synthesis*. IEEE Computer Society Press, Los Alamitos CA, 1997.

119. Ben Cohen. *VHDL Coding Styles and Methodologies*. Kluwer Academic Publishers, Boston MA, 2nd edition, 1999.

120. Peter J. Ashenden. *The Student's Guide to VHDL*. Morgan Kaufmann Publishers, San Francisco CA, 1998.

121. Zainalabedin Navabi. *VHDL Analysis and Modeling of Digital Systems*. McGraw-Hill, New York, 2nd edition, 1998.

122. Stefan Sjoholm and Lennart Lindh. *VHDL for Designers*. Prentice-Hall, London, 1997.

123. Ben Cohen. *VHDL Answers to Frequently Asked Questions*. Kluwer Academic Publishers, Boston MA, 1997.

124. Jayaram Bhasker. *A VHDL Synthesis Primer*. Star Galaxy Publishing, Allentown PA, 1996.

125. Yu-Chin Hsu, Kevin F. Tsai, Jessie T. Liu, and Eric S. Lin. *VHDL Modelling for Digital Design Synthesis*. Kluwer Academic Publishers, Boston MA, 1995.

126. Douglas E. Ott and Thomas J. Wilderotter. *A Designer's Guide to VHDL Synthesis*. Kluwer Academic Publishers, Boston MA, 1994.

127. R. Airiau, J.M. Bergé, and V. Olive. *Circuit Synthesis with VHDL*. Kluwer Academic Publishers, Boston MA, 1994.

128. Jayaram Bhasker. *A VHDL Primer*. Prentice-Hall, Englewood Cliffs NJ, 1994.

129. Himanshu Bhatnagar. *Advanced ASIC Chip Synthesis Using Synopsys Design Compiler and PrimeTime*. Kluwer Academic Publishers, Boston MA, 1999.

130. Klaus Lagemann. The Hamburg VHDL Archive. http://tams-www.informatik.uni-hamburg.de/vhdl.

131. Peter J. Ashenden and Philip A. Wilsey. Protected Shared Variables in VHDL: IEEE Standard 1076a. *IEEE Design & Test of Computers*, 16(4):74–83, October, November, December 1999.

132. Alain Vachoux. Analog and Mixed-Signal Extensions to VHDL. *Analog Integrated Circuits and Signal Processing*, 16(2):97–112, June 1998.

133. François Pêcheux, Christophe Lallement, and Alain Vachoux. VHDL-AMS and Verilog-AMS as Alternative Hardware Description Languages for Efficient Modeling of Multidiscipline Systems. *IEEE Transactions on Computer-Aided Design*, 24(2):204–225, February 2005.

134. Scott Hauck. Asynchronous Design Methodologies: An Overview. *Proceedings of the IEEE*, 83(1):69–93, January 1995.

135. Jens Sparsø and Steve Furber. *Principles of Asynchronous Circuit Design, A Systems Perspective*. Kluwer Academic Publishers, Boston MA, 2001.

136. Alain J. Martin and Mika Nyström. Asynchronous Techniques for System-on-Chip Design. *Proceedings of the IEEE*, 94(6):1089–1120, June 2006.

137. Peter Alfke. Just Say NO to Asynchronous Design. In *User Guide and Tutorials*. Xilinx Inc., San Jose, CA, 1991.

138. Steve Knapp. KISS those asynchronous-logic problems good-bye. *Personal Engineering and Instrumentation News*, pages 53–55, November 1997. http://www.fpga-site.com/kiss.html.

139. Kees van Berkel *et al.* Asynchronous Does Not Imply Low Power, But ... In Anantha Chandrakasan and Robert Brodersen, editors, *Low-Power CMOS Design*, pages 227–232. IEEE Press, Piscataway NJ, 1998.

140. Alex Kondratyev and Kelvin Lwin. Design of Asychronous Circuits Using Synchronous CAD Tools. *IEEE Design & Test of Computers*, 19(4):107–117, July/August 2002 (incl. a note by Alain J. Martin on practical asynchronous circuits).

141. Kenneth Y. Yun and Ryan P. Donohue. Pausible Clocking: A First Step Toward Heterogeneous Systems. In *Proceedings of the ICCD-96*, pages 118–123, 1996.

142. Frank Gürkaynak, Stephan Oetiker, Hubert Kaeslin, Norbert Felber, and Wolfgang Fichtner. GALS at ETH Zürich: Success or Failure? In *Proceedings of the 12th IEEE International Symposium on Asynchronous Circuits and Systems (ASYNC 2006)*, Grenoble, March 2006.

143. Fenghao Mu and Christer Svensson. Self-tested self-synchronization circuit for mesochronous clocking. *IEEE Transactions on Circuits and Systems II: Analog and Digital Signal Processing*, 48(2):129–140, February 2001.

144. Ingemar Söderquist. Globally Updated Mesochronous Design Style. *IEEE Journal of Solid-State Circuits*, 38(7):1242–1249, July 1993.

145. Andreas Burg, Frank K. Gürkaynak, Hubert Kaeslin, and Wolfgang Fichtner. Variable Delay Ripple Carry Adder with Carry Chain Interrupt Detection. In *Proc. of the IEEE International Symposium on Circuits and Systems*, Bangkok, May 2003. IEEE.

146. James E. Buchanan. *BiCMOS/CMOS Systems Design*. McGraw-Hill, New York, 1991.

147. Reto Zimmermann and Rod Whitby. Emacs VHDL Mode Home Page. http://opensource.ethz.ch/emacs/vhdl-mode.html.

148. Jose Luis Neves and Eby G. Friedman. Buffered Clock Tree Synthesis with Non-Zero Clock Skew Scheduling for Increased Tolerance to Process Parameter Variations. *Journal of VLSI Signal Processing*, 16(2/3):149–161, June/July 1997.

149. Xun Liu, Marios C. Papaefthymiou, and Eby G. Friedman. Retiming and Clock Scheduling for Digital Circuit Optimization. *IEEE Transactions on Computer-Aided Design*, 21(2):184–203, February 2002.

150. Jeng-Liang Tsai *et al.* Yield-Driven False-Path-Aware Clock Skew Scheduling. *IEEE Design & Test of Computers*, 22(3):214–222, May/June 2005.

151. Tim Horel and Gary Lauterbach. UltraSparc-III. *IEEE Micro*, 19(3):73–85, May/June 1999.

152. Antonio G. M. Strollo, Ettore Napoli, and Carlo Cimino. Analysis of Power Dissipation in Double Edge-Triggered Flip-Flops. *IEEE Transactions on Very Large Scale Integration (VLSI) Systems*, 8(5):624–629, October 2000.

153. Shi-Zheng Eric Lin, Chieh Changfan, Yu-Chin Hsu, and Fur-Shing Tsai. Optimal Time Borrowing Analysis and Timing Budgeting Optimization for Latch-Based Design. *ACM Transactions on Design Automation of Electronic Systems*, 7(1):217–230, January 2002.

154. James J. Engel *et al.* Design methodology for IBM ASIC products. *IBM Journal of Research and Development*, 40(4):387–406, July 1996.

155. J. D. Warnock. The Circuit and Physical Design of the POWER4 Microprocessor. *IBM Journal of Research and Development*, 46(1), 2002. http://www.research.ibm.com/journal/rd/461/warnock.html.

156. Derek C. Wong, Giovanni De Micheli, Michael J. Flynn, and Robert E. Huston. A Bipolar Population Counter Using Wave Pipelining to Achieve 2.5× Normal Clock Frequency. *IEEE Journal of Solid-State Circuits*, 27(5):745–753, May 1992.

157. Wayne P. Burleson, Maciej J. Ciesielski, Fabian Klass, and Wentai Liu. Wave Pipelining: A Tutorial and Research Survey. *IEEE Transactions on Very Large Scale Integration Systems*, 6(3):464–474, September 1998.

158. Kenneth D. Wagner. Clock System Design. *IEEE Design & Test of Computers*, pages 9–27, October 1988.

159. William J. Dally and John W. Poulton. *Digital Systems Engineering*. Cambridge University Press, Cambridge, 1998.

160. Eby G. Friedman. Introduction to Clock Distribution Networks in VLSI Circuits and Systems. In Eby G. Friedman, editor, *Clock Distribution Networks in VLSI Circuits and Systems*. IEEE Press, New York, 1995.

161. Daniel Dobberpuhl. A 200 MHz 64 Bit Dual Issue CMOS Microprocessor. In *Proceedings of the International Solid-State Circuits Conference*, San Francisco, February 1992. IEEE. DEC 21064 alpha chip.

162. Phillip J. Restle *et al.* A Clock Distribution Network for Microprocessors. *IEEE Journal of Solid-State Circuits*, 36(5):792–799, May 2001.

163. Eric S. Fetzer. Using Adaptive Circuits to Mitigate Process Variations in Microprocessor Design. *IEEE Design & Test of Computers*, 23(6):438–451, November/December 2006. Itanium 2 montecito.

164. Ana Sonia Leon *et al.* A Power-Efficient High-Throughput 32-Thread SPARC Processor. *IEEE Journal of Solid-State Circuits*, 42(1):7–16, January 2007. Niagara, ultrasparc t1, energy losses due to leakage 25

165. D. Pham *et al.* The Design and Implementation of a First-Generation CELL Processor. In *International Solid-State Circuits Conference Digest of Technical Papers*, 2005.

166. Jose Alvarez, Hector Sanchez, Gianfranco Gerosa, and Roger Countryman. A Wide-Bandwidth Low-Voltage PLL for PowerPC Microprocessors. *IEEE Journal of Solid-State Circuits*, 30(4):383–391, April 1995.

167. Bruno W. Garlepp *et al.* A Portable Digital DLL for High-Speed CMOS Circuits. *IEEE Journal of Solid-State Circuits*, 34(5):632–644, May 1999.

168. Simon Tam *et al.* Clock Generation and Distribution for the First IA-64 Microprocessor. *IEEE Journal of Solid-State Circuits*, 35(11):1545–1552, November 1990.

169. Masafumi Nogawa and Yusuke Ohtomo. A Data-Transition Look-Ahead DFF Circuit for Statistical Reduction in Power Consumption. *IEEE Journal of Solid-State Circuits*, 33(5):702–706, May 1998.

170. Adrianus M. G. Peeters. *Single-Rail Handshake Circuits*. PhD thesis, Eindhoven University of Technology, Eindhoven, 1996.

171. Andrea Pfister. *Metastability in Digital Circuits with Emphasis on CMOS Technology*. PhD thesis, ETH Zürich, Zurich, 1989.

172. Jens U. Horstmann, Hans W. Eichel, and Robert L. Coates. Metastability Behavior of CMOS ASIC Flip-Flops in Theory and Test. *IEEE Journal on Solid State Circuits*, 24(1):146–157, February 1989.

173. Peter Alfke. *Metastability Delay and Mean Time Between Failure in Virtex-II Pro FFs*. Xilinx, San Jose CA, 2002.

174. Takahiko Kozaki *et al.* A 156-Mb/s Interface CMOS LSI for ATM Switching Systems. *IEICE Transactions on Communications*, E76-B(6):684–693, June 1993.

175. Ran Ginosar. Fourteen Ways to Fool Your Synchronizer. In *Proceedings Ninth IEEE International Symposium on Asynchronous Circuits and Systems (ASYNC 2003)*, pages 89–96, Vancouver, May 2003.

176. J. Juan-Chico, M. J. Bellido, A. J. Acosta, M. Valencia, and J. L. Huertas. Analysis of Metastable Operation in a CMOS Dynamic D-Latch. *Analog Integrated Circuits and Signal Processing*, 14:143–157, 1997.

177. Suk-Jin Kim, Jeong-Gun Lee, and Kiseon Kim. A Parallel Flop Synchronizer and the Handshake Interface for Bridging Asynchronous Domains. *IEICE Transactions on Fundamental of Electronics, Communications, and Computer Science*, E87-A(12):3166–3173, December 2004.

178. Lindsay Kleeman and Antonio Cantoni. Metastable Behavior in Digital Systems. *IEEE Design & Test of Computers*, 4(6):4–19, December 1987.

179. David J. Kinniment, Alexandre Bystrov, and Alex V. Yakovlev. Synchronization Circuit Performance. *IEEE Journal on Solid State Circuits*, 37(2):202–209, February 2002.

180. Thaddeus J. Gabara, Gregory J. Cyr, and Charles E. Stroud. Metastability of CMOS Master/Slave Flip-Flops. *IEEE Transactions of Circuits and Systems II: Analog and Digital Signal Processing*, 39(10):734–740, October 1992.

181. Neil Weste and Kamran Eshragian. *Principles of CMOS VLSI Design*. Addison-Wesley, Reading MA, second edition, 1993.

182. Michael John Sebastian Smith. *Application-Specific Integrated Circuits*. Addison-Wesley, Reading MA, 1997.

183. Sanjay Dabral and Timothy J. Maloney. *Basic ESD and I/O Design*. John Wiley & Sons, New York, 1998.

184. Kerry Bernstein *et al. High Speed CMOS Design Styles*. Kluwer Academic Publishers, Boston MA, 1999.

185. Paul Gronowski. Issues in Dynamic Logic Design. In Anantha Chandrakasan, William J. Bowhill, and Frank Fox, editors, *Design of High-Performance Microprocessor Circuits*. IEEE Press, Piscataway NJ, 2001.

186. Jan M. Rabaey. *Digital Integrated Circuits*. Prentice-Hall, Upper Saddle River, NJ, 1996.

187. Yuan Taur and Tak H. Ning. *Fundamentals of Modern VLSI Devices*. Cambridge University Press, Cambridge, 1998.

188. Adnan Kabbani, Dhamin Al-Khalili, and Asim J. Al-Khalili. Technology-Portable Analytical Model for DSM CMOS Inverter Transition-Time Estimation. *IEEE Transaction on Computer-Aided Design*, 22(9):1177–1187, September 2003.

189. Rochit Rajsuman. Iddq Testing for CMOS VLSI. *Proceedings of the IEEE*, 88(4):544–566, April 2000.

190. Sagar S. Sabade and Duncan M. Walker. I_{DDX}-based Test Methods: A Survey. *ACM Transactions on Design Automation of Electronic Systems*, 9(2):159–198, April 2004.

191. Narsingh Deo. *Graph Theory with Applications to Engineering and Computer Science*. Prentice-Hall Inc., Englewood Cliffs NJ, 1974.

192. Takao Uehara and William M. van Cleemput. Optimal Layout of CMOS Functional Arrays. *IEEE Transactions on Computers*, C-30(5):305–312, May 1981.

193. Christian Piguet *et al.* Low power–low voltage standard cell libraries. In *Low Voltage–Low Power Workshop at the 21nd European Solid-State Circuits Conference*, 1995.

194. Reto Zimmermann and Wolfgang Fichtner. Low-power Logic Styles: CMOS versus Pass Transistor Logic. *IEEE Journal on Solid-State Circuits*, 32(7):1079–1090, July 1997.

195. Sumeer Goel, Ashok Kumar, and Magdy A. Bayoumi. Design of Robust, Energy-Efficient Full Adders for Deep-Submicron Design Using Hybrid-CMOS Logic Style. *IEEE Transactions on Very Large Scale Integration (VLSI) Systems*, 14(12):1309–1321, December 2006.

196. Massimo Alioto and Gaetano Palumbo. Analysis and Comparison on Full Adder Block in Submicron Technology. *IEEE Transactions on Very Large Scale Integration (VLSI) Systems*, 10(6):806–823, December 2002.

197. Flavio Carbognani *et al.* Transmission Gates Combined with Level-Restoring CMOS Gates Reduce Glitches in Low-Power Low-Frequency Multipliers. *IEEE Transactions on Very Large Scale Integration (VLSI) Systems*, 2008 To be published.

198. Hung Tien Bui, Yuke Wang, and Yiangtao Jiang. Design and Analysis of Low-Power 10-Transistor Full Adders Using Novel XOR-XNOR Gates. *IEEE Transactions on Circuits and Systems II: Analog and Digital Signal Processing*, 49(2):25–30, February 2002.

199. Behrooz Parhami. *Computer Arithmetic: Algorithms and Hardware Designs*. Oxford University Press, Oxford, 2004.

200. Miloš D. Ercegovac and Tomás Lang. *Digital Arithmetic*. Morgan Kaufmann Publishers, San Francisco CA, 2000.

201. Computer Arithmetic: Principles, Architectures, and VLSI Design. http://www.iis.ee.ethz.ch/zimmi/arith_lib.html.

202. David Goldberg. What Every Computer Scientist Should Know About Floating-Point Arithmetic. *ACM Computing Surveys*, 23(1):5–48, March 1991.

203. Paul E. Gronowski *et al.* High-Performance Microprocessor Design. *IEEE Journal on Solid-State Circuits*, 33(5):676–686, May 1998. DEC 21064, 21164, 21264 Alpha chips.

204. Bai-Sun Kong, Sam-Soo Kim, and Young-Hyun Jun. Conditional Capture D-Flip-Flop for Statistical Power Reduction. *IEEE Journal of Solid-State Circuits*, 36(8):1263–1271, August 2001.

205. Vladimir Stojanovic, Vojin G. Oklobdzija, and Raminder Bajwa. A Unified Approach in the Analysis of Latches and Flip-Flops for Low-Power Systems. In *Proceedings 1998 International Symposium on Low Power Electronics and Design*, Monterey, CA, August 1998.

206. Uming Ko and Poras T. Balsara. High-Performance Energy-Efficient D-Flip-Flop Circuits. *IEEE Transactions on Very Large Scale Integration (VLSI) Systems*, 8(1):94–98, February 2000.

207. Stephen H. Unger. Double Edge Triggered Flip-Flops. *IEEE Transactions on Computers*, C-30(6):447–451, June 1981.

208. Razak Hossain, Leszek D. Wronski, and Alexander Albicki. Low Power Design Using Double Edge Triggered Flip-Flops. *IEEE Transactions on Very Large Scale Integration (VLSI) Systems*, 2(2):261–265, June 1994.

209. Ying-Haw Shu, Shing Tenqchen, Ming-Chang Sun, and Wu-Shiung Feng. XNOR-based Double-Edge-Triggered Flip-Flops for Two-Phase Pipelines. *IEEE Transactions on Circuit and Systems II: Express Briefs*, 53(2):138–142, February 2006.

210. Vojin G. Oklobdzija, Vladimir M. Stojanovic, Dejan M. Markovic, and Nikola M. Nedovic. *Digital System Clocking, High-Performance and Low-Power Aspects*. John Wiley & Sons, Hoboken NJ, 2003.

211. Nikola M. Nedovic and Vojin G. Oklobdzija. Dual-Edge-Triggered Storage Elements and Clocking Strategy for Low-Power Systems. *IEEE Transactions on Very Large Scale Integration (VLSI) Systems*, 13(5):577–590, May 2005.

212. Paul Naish and Peter Bishop. *Designing ASICs*. Ellis Horwood Ltd., Chicester, 1988.

213. Ashok K. Sharma. *Semiconductor Memories, Technology, Testing, and Reliability.* IEEE Press, New York, 1997.

214. Betty Prince. *High Performance Memories.* John Wiley & Sons, Chichester, 1999.

215. Brent Keeth and R. Jacob Baker. *DRAM Circuit Design, a Tutorial.* IEEE Press, New York, 2000.

216. Bharadwaj S. Amrutur and Mark A. Horowitz. Fast Low-Power Decoders for RAMs. *IEEE Journal of Solid-State Circuits*, 36(10):1506–1515, October 2001.

217. Tadaaki Yamauchi and Michihiro Yamada. Embedded DRAM. In Anantha Chandrakasan, William J. Bowhill, and Frank Fox, editors, *Design of High-Performance Microprocessor Circuits.* IEEE Press, Piscataway NJ, 2001.

218. Sanghoon Hong *et al.* Low-Voltage DRAM Sensing Scheme with Offset-Cancellation Sense Amplifier. *IEEE Journal of Solid-State Circuits*, 37(10):1356–1360, October 2002.

219. Harry J. M. Veendrick. *Deep-Submicron CMOS ICs, from Basics to ASICs.* Kluwer Academic Publishers, Deventer, 2000.

220. David A. Hodges, Horace G. Jackson, and Resve A. Saleh. *Analysis and Design of Digital Integrated Circuits in Deep Submicron Technology.* McGraw-Hill, Boston MA, third edition, 2003.

221. Takayuki Tawahara. Low-Voltage Embedded RAMs in Nanometer Era. *IEICE Transactions on Electronics*, E90-C(4):735–742, April 2007.

222. Jatuchai Pangjun and Sachin S. Sapatnekar. Low-Power Clock Distribution Using Multiple Voltages and Reduced Swings. *IEEE Transactions on Very Large Scale Integration (VLSI)*, 10(3):309–318, June 2002.

223. Ming-Dou Ker, Shih-Lun Chen, and Chia-Sheng Tsai. Overview and Design of Mixed-Voltage I/O Buffers with Low-Voltage Thin-Oxide CMOS Transistors. *IEEE Transactions on Circuits and Systems I: Regular Papers*, 53(9):1934–1945, September 2006.

224. Ming-Dou Ker and Shih-Lun Chen. Design of Mixed-Voltage I/O Buffer by Using NMOS-Blocking Techniques. *IEEE Journal on Solid State Circuits*, 41(10):2324–2333, October 2006.

225. Mohammad Maymandi-Nejad and Manoj Sachdev. A Monotonic Digitally Controlled Delay Element. *IEEE Journal of Solid-State Circuits*, 40(11):2212–2219, November 2005.

226. Hiromasa Noda *et al.* An On-Chip Clock-Adjusting Circuit with Sub-100-ps Resolution for a High-Speed DRAM Interface. *IEEE Transactions on Circuits and Systems II: Analog and Digital Signal Processing*, 47(8):771–775, August 2000.

227. Guang-Kaai Dehng, June-Ming Hsu, Ching-Yuan Yang, and Shen-Iuan Liu. Clock-Deskew Buffer Using a SAR-Controlled Delay-Locked Loop. *IEEE Journal of Solid-State Circuits*, 35(8):1128–1136, August 2000.

228. James D. Lyle. *SBus, Information, Applications, and Experience.* Springer-Verlag, New York, 1992.

229. Yukiya Miura and Hiroshi Yamazaki. An Analysis of the Relationship between $I_{dd\,q}$ Testability and D-Type Flip-Flop Structure. *IEICE Transactions on Information and Systems*, E81-D(10):1072–1078, October 1998.

230. C. T. Sah. Characteristics of the Metal–Oxide–Semiconductor Transistors. *IEEE Transactions on Electron Devices*, ED-11:324–345, July 1964.

231. Narain Arora. *MOSFET Models for VLSI Circuit Simulation.* Springer-Verlag, Vienna, 1993.

232. Harold Shichman and David A. Hodges. Modeling and Simulation of Insulated-Gate Field-Effect Transistor Switching Circuits. *IEEE Journal on Solid-State Circuits*, SC-3(3):285–289, September 1968.

233. Takayasu Sakurai and A. Richard Newton. Delay Analysis of Series-Connected MOSFET Circuits. *IEEE Journal on Solid-State Circuits*, 26(2):122–131, February 1991.

234. Yannis Tsividis. *Operation and Modelling of the MOS Transistor.* McGraw-Hill Book Company, New York, 1987.

235. H. Craig Casey. *Devices for Integrated Circuits: Silicon and III–V Compound Semiconductors.* John Wiley & Sons, New York, 1999.

236. David J. Roulston. *An Introduction to the Physics of Semiconductor Devices.* Oxford University Press, New York, 1999.

237. Josef Watts *et al.* Advanced Compact Models for MOSFETs. *Technical Proceedings of the 2005 Workshop on Compact Modeling*, pages 3–12, 2005. www.nsti.org.

238. Felix Bürgin, Flavio Carbognani, and Martin Hediger. Low-Power Architectural Trade-Offs in a VLSI Implementation of an Adaptive Hearing Aid Algorithm. In *Proceedings of the 43rd Design Automation Conference*, San Francisco, CA, July 2006.

239. Jürgen Wassner. *Data Statistics and Low-Power Digital VLSI.* PhD thesis, ETH Zürich, 2001.

240. José Luis Rosselló and Jaume Segura. Charge-based Analytical Model for the Evaluation of Power Consumption in Submicron CMOS Buffers. *IEEE Transactions on Computer-Aided Design*, 21(4):433–448, April 2002. Revised since 2nd review in Jan/Feb 2001.

241. Gary K. Yeap. *Practical Low Power Digital Design.* Kluwer Academic Publishers, Boston MA, 1998.

242. Siva Narendra *et al.* Full-Chip Subthreshold Leakage Power Prediction and Reduction Techniques for Sub-0.18 μm CMOS. *IEEE Journal of Solid-State Circuits*, 501–510, March 2004.

243. Alice Wang and Anantha Chandrakasan. Energy-Efficient DSPs for Wireless Sensor Networks. *IEEE Signal Processing Magazine*, 19(4):68–78, July 2002.

244. Eric A. Vittoz. Weak Inversion for Ultimate Low-Power Logic. In Christian Piguet, editor, *Low-Power CMOS Circuits.* CRC Press, Boca Raton FL, 2006.

245. Azeez Bhavnagarwala, Blanca Austin, and James Meindl. Projections for High Performance, Minimum Power CMOS ASIC Technologies: 1998–2010. In *Proceedings of the Tenth Annual IEEE International ASIC Conference*, pages 185–188. IEEE, Portland OR, September 7–10 1997.

246. Dan Ernst *et al.* Razor: Circuit-Level Correction of Timing Errors for Low-Power Operation. *IEEE Micro*, 24(6):10–20, November/December 2004.

247. André Meyer and Thomas Peter. Channel Estimation for MIMO-OFDM; Impact of Fixed-Point Arithmetics Wordwidths on Area Requirements and Power Consumption. Semester thesis plus private communication, Integrated Systems Laboratory, ETH Zürich, 2006.

248. Claude Arm, Jean-Marc Masgonty, and Christian Piguet. Double-Latch Clocking Scheme for Low-Power I.P. Cores. In *Proceedings of Patmos 2000 (International Workshop on Power and Timing Modeling, Optimization and Simulation)*, Göttingen, September 2000.

249. Qi Wang, Sarma Vrudhula, Gary Yeap, and Shantanu Ganguly. Power Reduction and Power-Delay Trade-Offs Using Logic Transformations. *ACM Transactions on Design Automation of Electronic Systems*, 4(1):97–121, January 1999.

250. Vishwani Agrawal. Low-Power Design by Hazard Filtering. In *10th International Conference on VLSI Design*, 1997.

251. Myungchul Yoon and Byeong hee Roh. A Novel Low-Power Bus Design for Bus-Invert Coding. *IEICE Transactions on Electronics*, E90-C(4):731–734, April 2007.

252. Youngsoo Shin, Kiyoung Choi, and Young-Hoon Chang. Narrow Bus Encoding for Low-Power DSP Systems. *IEEE Transactions on Very Large Scale Integration (VLSI) Systems*, 9(5):656–660, October 2001.

253. Jun Yang, Rajiv Gupta, and Chuanjun Zhang. Frequent Value Encoding for Low Power Data Buses. *ACM Transactions on Design Automation of Electronic Systems*, 9(3):354–384, July 2004.

254. Neil Weste and David Harris. *CMOS VLSI Design.* Pearson Education, Boston MA, third edition, 2005.

255. Shunzo Yamashita *et al.* Pass-Transistor/CMOS Collaborated Logic: The Best of Both Worlds. In *1997 Symposium on VLSI Circuits, Digest of Technical Papers*, Kyoto, June 1997.

256. Flavio Carbognani *et al.* A Low-Power Audio Transmission-Gate-based 16-bit Multiplier for Digital Hearing Aids. *Analog Integrated Circuits and Signal Processing*, 2008. To be published.

257. Tadahiro Kuroda and Takayasu Sakurai. Low-Voltage Technologies and Circuits. In Anantha Chandrakasan and Robert Brodersen, editors, *Low-Power CMOS Design*, pages 61–65. IEEE Press, Piscataway NJ, 1996.

258. Tadahiro Kuroda *et al.* A 0.9 V, 150 MHz, 10 mW, 4 mm^2 2-D Discrete Cosine Transform Core Processor with Variable Threshold Voltage (VT) Scheme. *IEEE Journal of Solid-State Circuits*, 1770–1779, November 1996.

259. William Krenik, Dennis D. Buss, and Peter Rickert. Cellular Handset Integration — SiP versus SoC? *IEEE Journal on Solid State Circuits*, 40(9):1839–1846, 2005.

260. Tadahiro Kuroda, Tetsuya Fujita, Fumitoshi Hatori, and Takayasu Sakurai. Variable Threshold-Voltage CMOS Technology. *IEICE Transactions on Electronics*, 1705–1715, November 2000.

261. Masanao Yamaoka, Ryuta Tsuchiya, and Takayuki Kawahara. SRAM Circuit with Expanded Operating Margin and Reduced Stand-By Leakage Current Using Thin-BOX FD-SOI Transistors. *IEEE Journal on Solid-State Circuits*, 41(11):2366–2372, November 2006.

262. James T. Kao and Anantha P. Chandrakasan. Dual-Threshold Voltage Techniques for Low-Power Digital Circuits. *IEEE Journal of Solid-State Circuits*, 1009–1018, July 2000.

263. Nikhil Jayakumar and Sunil P. Khatri. A Predictably Low-Leakage ASIC Design Style. *IEEE Transactions on Very Large Scale Integration (VLSI) Systems*, 15(3):276–285, March 2007.

264. Hiroshi Kawaguchi, Koichi Nose, and Takayasu Sakurai. A Super Cut-Off CMOS (SCCMOS) Scheme for 0.5 V Supply Voltage with Picoampere Stand-By Current. *IEEE Journal of Solid-State Circuits*, 35(10):1498–1501, October 2000.

265. Mircea R. Stan. CMOS Circuits with Subvolt Supply Voltages. *IEEE Design & Test of Computers*, 34–43, March–April 2002.

266. Lawrence T. Clark, Franco Ricci, and Manish Biyani. Low Standby Power State Storage for Sub-130 nm Technologies. *IEEE Journal of Solid-State Circuits*, 40(2):498–506, February 2005.

267. Pietro Babighian, Luca Benini, Alberto Macii, and Enrico Macii. Low-Overhead State-Retaining Elements for Low-Leakage MTCMOS Design. In *Proceedings of the ACM Great Lakes Symposium on VLSI*, pages 367–370, Chicago IL, 2005.

268. P. R. van der Meer and A. van Staveren. Standby-Current Reduction for Deep-Submicron VLSI CMOS Circuits: Smart Series Switch. In *Proceedings of the ProRISC/IEEE Workshop*, pages 401–404, 2000.

269. James T. Kao and Anantha P. Chandrakasan. MTCMOS Sequential Circuits. In *Proceedings of the IEEE European Solid-State Circuits Conference 2001*, pages 332–335, Neuchâtel, September 2001.

270. Burcin Baytekin and Boris Murmann. A Comparative Study of Standby-mode Multi Threshold CMOS Data Paths. http://www-inst.eecs.berkeley.edu/~bmurmann/ee241.

271. Kouichi Kumagai *et al.* A Novel Powering-down Scheme for Low Vt CMOS Circuits. In *1998 Symposium on VLSI Circuits Digest of Technical Papers*, pages 44–45, 1998.

272. Bipul C. Paul, Amit Agarwal, and Kaushik Roy. Low-power design techniques for scaled technologies. *Integration, the VLSI Journal*, 39(2):64–88, March 2006.

273. Narender Hanchate and Nagarajan Ranganathan. LECTOR: A Technique for Leakage Reduction in CMOS Circuits. *IEEE Transactions on Very Large Scale Integration (VLSI) Systems*, 12(2):196–205, February 2004.

274. Trevor Mudge. Power: A First-Class Architectural Design Constraint. *Computer*, 34(4):52–58, April 2001.

275. Luca Benini, Giovanni De Micheli, and Enrico Macii. Designing Low-Power Circuits: Practical Recipes. *IEEE Circuits and Systems Magazines*, 1(1):6–25, first quarter 2001.

276. Tadahiro Kuroda. Low Power CMOS Design Challenges. *IEICE Transactions on Electronics*, 1021–1028, August 2001.

277. Anantha P. Chandrakasan *et al.* Power Aware Wireless Microsensor Systems. In *Proceedings of the 28th European Solid-State Circuits Conference 2002*, 47–54, Florence, September 2002.

278. Anantha P. Chandrakasan, Miodrag Potkonjak, Renu Mehra, Jan Rabaey, and Robert W. Brodersen. Optimizing Power Using Transformations. *IEEE Transactions on Computer-Aided Design*, 14(1):12–31, January 1995.

279. Hirotsugu Kojima, Satoshi Tanaka, and Katsuro Sasaki. Half-Swing Clocking Scheme for 75% Power Saving in Clocking Circuitry. *IEEE Journal of Solid-State Circuits*, 30(4):432–435, April 1995.

280. Siva Narendra and Anantha Chandrakasan. *Leakage in Nanometer CMOS Technologies*. Springer, 2006.

281. Hisato Oyamatsu, Masaaki Kinugawa, and Masakazu Kakumu. Design Methodology of Deep Submicron CMOS Devices for 1V Operation. *IEICE Transactions on Electronics*, E79-C(12):1720–1725, December 1996.

282. Pankay Pant, Vivek K. De, and Abhijit Chatterjee. Simultaneous Power Supply, Threshold Voltage, and Transistor Size Optimization for Low-Power Operation of CMOS Circuits. *IEEE Transactions on Very Large Scale Integration (VLSI) Systems*, 6(4):538–545, December 1998.

283. Dennis Sylvester and Himanshu Kaul. Power-Driven Challenges in Nanometer Design. *IEEE Design & Test of Computers*, 18(6):12–22, November/December 2001.

284. Benton H. Calhoun and Anantha P. Chandrakasan. Standby Power Reduction Using Dynamic Voltage Scaling and Canary Flip-Flop Structures. *IEEE Journal of Solid-State Circuits*, 1504–1511, September 2004.

285. Atila Alvandpour, Per Larsson-Edefors, and Christer Svensson. Impact of Miller Capacitances on Power Consumption. In *Proceedings of Patmos '98 (International Workshop on Power and Timing Modeling, Optimization and Simulation)*, Lyngby, October 1998.

286. Alice Wang and Anantha Chandrakasan. A 180 mV Subthreshold FFT Processor Using a Minimum Energy Design Methology. *IEEE Journal of Solid-State Circuits*, 40(1):310–319, January 2005.

287. Kouichi Kanda, Hattori Sadaaki, and Takayasu Sakurai. 90% Write Power-Saving SRAM Using Sense-Amplifying Memory Cell. *IEEE Journal of Solid-State Circuits*, 39(6):927–933, June 2004.

288. Massimo Alioto and Gaetano Palumbo. Performance Evaluation of Adiabatic Gates. *IEEE Transactions on Circuits and Systems I: Fundamental Theory and Applications*, 47(9):1297–1308, September 2000.

289. Massimo Alioto and Gaetano Palumbo. NAND/NOR Adiabatic Gates: Power Consumption Evaluation and Comparison Versus the Fan-In. *IEEE Transactions on Circuits and Systems I: Fundamental Theory and Applications*, 49(9):1253–1262, September 2002.

290. William C. Athas, Nestoras Tzartzanis, Lars J. Svensson, and Lena Peterson. A Low-Power Microprocessor Based on Resonant Energy. *IEEE Journal of Solid-State Circuits*, 32(11):1693–1701, November 1997.

291. Dusan Suvakovica and C. Andre T. Salama. Energy Efficient Adiabatic Multiplier–Accumulator Design. *Journal of VLSI Signal Processing*, 33(1/2):83–103, January/February 2003.

292. Yibin Ye and Kaushik Roy. QSERL: Quasi-Static Energy Recovery Logic. *IEEE Journal of Solid-State Circuits*, 36(2):239–248, February 2001.

293. Suhwan Kim, Conrad H. Ziesler, and Marios C. Papaefthymiou. Charge-Recovery Computing on Silicon. *IEEE Transactions on Computers*, 54(6):651–659, June 2005.

294. Jouko Marjonen and Markku Åberg. A Single Clocked Adiabatic Static Logic — A Proposal for Digital Low Power Applications. *Journal of VLSI Signal Processing*, 27(3):253–268, March 2001.

295. Bart Desoete and Alexis De Vos. A Reversible Carry–Look-Ahead Adder Using Control Gates. *Integration, the VLSI Journal*, 33(1–2):89–104, December 2002.

296. Flavio Carbognani, Felix Bürgin, Norbert Felber, Hubert Kaeslin, and Wolfgang Fichtner. Two-Phase Resonant Clocking for Ultra-Low-Power Hearing Aid Applications. In *Proc. of the Design, Automation and Test in Europe (DATE) Conference*, Munich, March 2006.

297. Xiaonan Zhang. Coupling Effects on Wire Delay, Challenges in Deep Submicron VLSI Design. *IEEE Circuits & Devices*, 12(6):12–18, November 1996.

298. Shannon V. Morton. Techniques for Driving Interconnect. In Anantha Chandrakasan, William J. Bowhill, and Frank Fox, editors, *Design of High-Performance Microprocessor Circuits*. IEEE Press, Piscataway NJ, 2001.

299. Mohamed A. Elgamel and Magdy A. Bayoumi. Interconnect Noise Analysis and Optimization in Deep Submicron Technology. *IEEE Circuits and Systems Magazine*, 3(4):6–17, fourth quarter 2003. Interconnect, noise, coupling, buffer insertion.

300. Junmopu Zhang and Eby G. Friedman. Crosstalk Modeling for Coupled RLC Interconnects with Applications to Shield Insertion. *IEEE Transactions on Very Large Scale Integration (VLSI) Systems*, 14(6):641–646, June 2006. Interconnect model, inductance, crosstalk.

301. C. A. Steidel. Assembly Techniques and Packaging. In Simon M. Sze, editor, *VLSI Technology*. McGraw-Hill, New York, 1983.

302. Masakazu Shoji. *CMOS Digital Circuit Technology*. Prentice-Hall, New York, 1988.

303. Adnan Kabbani and Asim J. Al-Khalili. Estimation of Ground Bounce Effects on CMOS Circuits. *IEEE Transactions on Components and Packaging Technologies*, 22(2):316–325, June 1999.

304. Andrey Mezhiba and Eby G. Friedman. Inductive properties of High-Performance Power Distribution Grids. *IEEE Transactions on Very Large Scale Integration (VLSI)*, 10(6):487–493, December 2002.

305. National Semiconductor, Santa Clara CA. *Understanding and Minimizing Ground Bounce*. Application note 640.

306. Richard Ulrich and Leonard Schaper. Putting Passives in Their Place. *IEEE Spectrum*, 40(7):26–30, July 2003. Resistor, capacitor, printed circuit board.

307. Atsushi Muramatsu, Masanori Hashimoto, and Hidetoshi Onodera. Effects of On-Chip Inductance on Power Distribution Grid. *IEICE Transactions on Fundamental of Electronics, Communications, and Computer Science*, E88(12):3564–3576, December 2005.

308. Anil Jain *et al.* A 1.2 GHz Alpha Microprocessor with 44.8 GB/sec of Chip Pin Bandwidth. In *International Solid-State Circuits Conference Digest of Technical Papers*, San Francisco, February 2001. IEEE.

309. Qing K. Zhu. *Power Distribution Network Design for VLSI*. John Wiley & Sons, Hoboken NJ, 2004.

310. Stefan Rusu and Gadi Singer. The First IA-64 Microprocessor. *IEEE Journal of Solid-State Circuits*, 35(11):1539–1544, November 1990. Itanium.

311. David Jones. PCB Layout Design Tutorial, 2006. http://www.pcb123.com.

312. Howard W. Johnson and Martin Graham. *High-Speed Digital Design, a Handbook of Black Magic*. Prentice-Hall PTR, Englewood Cliffs NJ, 1993.

313. Henry W. Ott. *Noise Reduction Techniques in Electronic Systems*. John Wiley & Sons, New York, 1988.

314. Wilhelm G. Spruth. *The Design of a Microprocessor*. Springer-Verlag, Berlin, 1989.

315. R. Senthinathan and J. L. Prince. Application Specific CMOS Output Driver Circuit Design Techniques to Reduce Simultaneous Switching Noise. *IEEE Journal of Solid-State Circuits*, 28(12):1383–1388, December 1993.

316. T. Gabara and D. Thompson. Ground bounce control in CMOS integrated circuits. In *Proceedings of the International Solid-State Circuits Conference*. IEEE, February 1988.

317. Changsik Yoo. A CMOS Buffer Without Short-Circuit Power Consumption. *IEEE Transactions on Circuits and Systems II: Analog and Digital Signal Processing*, 47(9):935–937, September 2000.

318. L. Benini, P. Vuillod, A. Bogliolo, and G. De Micheli. Clock Skew Optimization for Peak Current Reduction. *Journal of VLSI Signal Processing*, 16(2/3):117–130, June/July 1997.

319. Erico Guizzo and Harry Goldstein. Expressway to your Skull. *IEEE Spectrum*, 43(8):32–37, August 2006. Differential signaling, memory interface, playstation 3.

320. Stefanos Sidiropoulos, Chih-Kong Ken Yang, and Mark Horowitz. High-Speed Inter-Chip Signaling. In Anantha Chandrakasan, William J. Bowhill, and Frank Fox, editors, *Design of High-Performance Microprocessor Circuits*. IEEE Press, Piscataway NJ, 2001.

321. Mingdeng Chen *et al.* Low-Voltage Low-Power LVDS Drivers. *IEEE Journal on Solid-Sate Circuits*, 40(2):472–479, February 2005. LVDS.

322. Francesco Piazza and Qiuting Huang. A Low Power CMOS Dual Modulus Prescaler for Frequency Synthesizers. *IEICE Transactions on Electronics*, E80-C(2):314–319, February 1997.

323. Massimo Alioto and Gaetano Palumbo. Power-Aware Design Techniques for Nanometer MOS Current-Mode Logic Gates: A Design Framework. *IEEE Circuits & Systems Magazine*, 6(4):40–59, fourth quarter 2006.

324. J. Briaire and K. S. Krisch. Principles of Substrate Crosstalk Generation in CMOS Circuits. *IEEE Transactions on Computer-Aided Design*, 19(6):645–653, 2000.

325. Willy M. C. Sansen. *Analog Design Essentials*. Springer, Dordrecht, 2006.

326. Ali Afzali *et al.* Substrate Noise Coupling in SoC Design: Modeling, Avoidance, and Validation. *Proceedings of the IEEE*, 94(12):2109–2138, December 2006.

327. John Baliga. Packaging Provides Viable Alternatives to SOC. *Semiconductor International*, 23(8):168–181, July 2000.

328. Etienne Hirt and Rolf Schmid. High Density Packaging, der schnelle Weg zur Miniaturisierung von Elektronik. *Bulletin des Schweizerischen Elektrotechnischen Vereins*, September 15 2000.

329. Rolf Schmid, Michael Scheffler, and Ralf Haller. High Density Packaging: Hohe Integration, niedrige Entwicklungskosten. *Elektronik* (12), 2004.

330. George Harman. *Wire Bonding in Microelectronics*. McGraw-Hill, New York, 1997.

331. Rolf Schmid, Michael Scheffler, and Ralf Haller. High Density Packaging oder ASIC? *Bulletin SEV*, 96(1):37–39, 2005.

332. Rao R. Tummala *et al.* SOP: What Is It and Why? A New-Microsystem-Integration Technology. *IEEE Transactions on Advanced Packaging*, 27(2):241–249, May 2004.

333. Peter Rickert and William Krenik. Cell Phone Integration: SiP, Soc, and PoP. *IEEE Design & Test of Computers*, 23(3):188–195, 2006.

334. Tezzaron Semiconductor. FaStack Stacking Technology, February 2004. http://www.tezzaron.com/technology/FaStack.htm.

335. Christian Val *et al.* Very-High-Speed 3D SiP. *HDI, the Magazine of High Density Interconnect*, 4(5):22–29, May 2001.

336. W. Rhett Davis *et al.* Demystifying 3D ICs: The Pros and Cons of Going Vertical. *IEEE Design & Test of Computers*, 22(6):498–509, November–December 2005.

337. Harry Goldstein. Packages. *IEEE Spectrum*, 38(8):46–51, August 2001.

338. Yangdong Deng and Wojciech P. Maly. 2.5-Dimensional VLSI System Integration. *IEEE Transactions on Very Large Scale Integration (VLSI) Systems*, 13(6):668–677, June 2005.

339. Carver Mead and Lynn Conway. *Introduction to VLSI Systems*. Addison-Wesley, Reading MA, 1980.

340. Peng Bai *et al.* A 65 nm Logic Technology Featuring 35 nm Gate Length, Enhanced Channel Strain, 8 Cu Interconnect Layers, Low-k ILD and 0.57 μm^2 SRAM Cell. In *2004 IEEE International Electron Devices Meeting Technical Digest*, pages 657–660, December 2004. http://www.intel.com/technology/silicon/65nm_technology.htm.

341. Dan Clein. *CMOS IC Layout Concepts, Methodologies, and Tools*. Butterworth-Heinemann, Boston MA, 1999.

342. J. T. Yue. Reliability. In C. Y. Chang and Simon M. Sze, editors, *ULSI Technology*. McGraw-Hill, New York, 1996.

343. Ulf Wedemeyer. Evaluierung unterschiedlicher Wafer-Level Elektromigrations-Testverfahren und unterschiedlicher Teststrukturen in Bezug auf eine Korrelation zum etablierten Elektromigrations-Test. Master's thesis, Technische Fakultät der Universität Kiel, Kiel, 1998.

344. J. Joseph Clement. Electromigration reliability. In Anantha Chandrakasan, William J. Bowhill, and Frank Fox, editors, *Design of High-Performance Microprocessor Circuits*. IEEE Press, Piscataway NJ, 2001.

345. Jens Lienig. Interconnect and Current Density Stress – An Introduction to Electromigration-Aware Design. In *Proc. of SLIP '05*, pages 81–88, 2005.

346. Andreas D. Stricker. *Technology Computer Aided Design of ESD Protection Devices*. PhD thesis, ETH Zürich, 2001.

347. Joachim C. Reiner. *Latent Gate Oxide Damage Induced by Ultra-Fast Electrostatic Discharge*. PhD thesis, ETH Zürich, 1995.

348. ESD Association Standards Committee. *ESD Association Advisory Handbook*. ESD Association, Rome NY, 1994.

349. James E. Vinson and Juin J. Liou. Electrostatic Discharge in Semiconductor Devices: Protection Techniques. *Proceedings of the IEEE*, 88(12):1878–1900, December 2000.

350. Markus P. J. Mergens. *On-Chip ESD Protection in Integrated Circuits: Device Phyiscs, Modelling, Circuit Simulation*. PhD thesis, ETH Zürich, 2001.

351. Ajith Amerasekera and Charvaka Duvvury. *ESD in Silicon Integrated Circuits*. John Wiley & Sons, Chichester, 1995.

352. James E. Vinson and Juin J. Liou. Electrostatic Discharge in Semiconductor Devices: An Overview. *Proceedings of the IEEE*, 86(2):399–418, February 1997.

353. Steven H. Voldman. *ESD Physics and Devices*. John Wiley & Sons, New York, 2004.

354. Ronald R. Troutman. *Latchup in CMOS Technology, The Problem and Its Cure*. Kluwer Academic Press, Boston MA, 1986.

355. M. G. Pecht and R. Radojcic. *Guidebook for Managing Silicon Chip Reliability*. CRC Press, Boca Raton FL, 1998.

356. Giovanni De Micheli. *Synthesis and Optimization of Digital Circuits*. McGraw-Hill Inc., New York, 1994.

357. Dennis Sylvester and Kurt Keutzer. Rethinking Deep-Submicron Circuit Design. *IEEE Computer*, 32(11):25–33, November 1999.

358. Alfred K. Wong. Some Thoughts on the IC Design–Manufacture Interface. *IEEE Design & Test of Computers*, 22(3):206–213, May/June 2005.

359. Christopher Spence. Full-chip lithography simulation and design analysis: how OPC is changing IC design. In *Proceedings of SPIE*, volume 5751, pages 1–14, San Jose CA, 2005. The International Society for Optical Engineering. Plenary Paper.

360. Mehdi Hatamian. Understanding Clock Skew in Synchronous Systems. In Stuart K. Tewksbury, editor, *Concurrent Computations, Algorithms, Architecture, and Technology*, pages 87–96. Plenum Press, New York, 1988.

361. Yehia Massoud and Yehea Ismail. Grasping the Impact of On-Chip Inductance. *IEEE Circuits & Devices*, 17(4):14–21, July 2001. Interconnect model, inductance, crosstalk, ramp time, power dissipation.

362. Yehea Ismail. On-Chip Inductance Cons and Pros. *IEEE Transactions on Very Large Scale Integration (VLSI) Systems*, 10(6):685–694, December 2002.

363. Song-Ra Pan, Tai-Chen Chen, and Yao-Wen Chang. Timing Modeling and Optimization under the Transmission Line Model. *IEEE Transactions on Very Large Scale Integration (VLSI) Systems*, 12(1):28–41, January 2004.

364. Michael Riordan and Lillian Hoddeson. *Crystal Fire, the Birth of the Information Age*. W. W. Norton & Company, New York, 1997.

365. Ross Knox Bassett. *To the Digital Age: Research Labs, Start-up Companies, and the Rise of MOS Technology*. Johns Hopkins University Press, Baltimore MA, 2002.

366. Bob Johnstone. *We were burning: Japanese entrepreneurs and the forging of the electronic age*. Basic Books, New York, 1999.

367. Michael Keating and Pierre Bricaud. *Reuse Methodology Manual for Systems-on-a-Chip Designs*. Kluwer Academic Publishers, Boston MA, 1998.

368. Wilfred Corrigan. ASIC Challenges: Emerging from a Primordial Soup. *IEEE Design & Test*, 15(3):4–7, July–September 1998.

369. Charles Shelor. Designing with Intellectual Property. *VHDL Times*, 6(3):1–15, 1997.

370. Ian Kuon and Jonathan Rose. Measuring the Gap Between FPGAs and ASICs. *IEEE Transactions on Computer-Aided Design*, 26(2):203–215, February 2007.

371. William G. Howard Jr. and Bruce R. Guile. *Profiting from Innovation*. The Free Press, MacMillan Inc., New York, 1992.

372. Howard Anderson. Why Big Companies Can't Invent. *IEEE Engineering Management Review*, 32(3):77–79, 2004.

373. James E. Gover. Strengthening the Competitiveness of U.S. Microelectronics. *IEEE Transactions on Engineering Management*, 40(1), February 1993.

374. Shane Greenstein. The paradox of commodities. *IEEE Micro*, 24(2):73–75, March/April 2004.

375. Gadi Kaplan and Tekla S. Perry. Engineers as Entrepreneurs. *IEEE Spectrum*, 35(8):14–23, 1998.

376. In unserer Marktnische konnten wir als Kleinunternehmen gross werden. Produktinnovation und Wettbewerbsvorsprung durch Mikroelektronik Nr.4, Microswiss, 1995.

377. Odd Rønning. ASIC in door lock. *ChipShop Newsletter*, (NL7):8, 1995.

378. Terry Thomas. Technology for IP Reuse and Portability. *IEEE Design & Test of Computers*, 16(4):7–13, October, November, December 1999.

379. Will Strauss. Digital Signal Processing, The New Semiconductor Technology Driver. *IEEE Signal Processing Magazine*, 17(2):52–56, March 2000.

380. Herman Beke. System-on-a-chip IP revolution. *European Semiconductor*, 20(8):33–34, August 1998.

381. James D. Plummer, Michael D. Deal, and Peter B. Griffin. *Silicon VLSI Technology*. Prentice-Hall, Upper Saddle River NJ, 2000.

382. Keith Leaver. *Microelectronic Devices*. Imperial College Press, London, 1996.

383. Donald A. Neamen. *Semiconductor Physics and Devices*. McGraw-Hill, Boston MA, 2003.

384. Bart van Zeghbroeck. Principles of Semiconductor Devices, 2004. http://ece-www.colorado.edu/bart/book/book.

385. Educational Applets in Solid State Materials. http://jas.eng.buffalo.edu/index.html.

386. Stanley Wolf. *Silicon Processing for the VLSI Era, Deep Submicron Process Technology*, volume 4. Lattice Press, Sunset Beach CA, 2002.

387. Dietrich Widmann, Hermann Mader, and Hans Friedrich. *Technology of Integrated Circuits*. Springer, Berlin, 2000.

388. Peter van Zant. *Microchip Fabrication, a practical guide to semiconductor processing*. McGraw-Hill, New York, fourth edition, 2000.

389. C. Y. Chang and Simon M. Sze. *ULSI Technology*. McGraw-Hill, New York, 1996.

390. Palash Das and Richard L. Sandstrom. Advances in Excimer Laser Technology for Sub-0.25 μm Lithography. *Proceedings of the IEEE*, 90(10):1637–1652, 2002.

391. Marc D. Levenson. Using Destructive Optical Interference in Semiconductor Lithography. *OPN Optics & Photonics News*, 17(4):30–35, April 2006.

392. Kevin Lucas *et al.* Logic Design for Printability Using OPC Methods. *IEEE Design & Test of Computers*, 23(1):30–37, January–February 2006.

393. Christopher Spence. Mask Data Preparation Issues for the 90 nm Node: OPC Becomes a Critical Manufacturing Technology, February 2004. http://www.future-fab.com/documents.asp?d_ID=2318.

394. Keith Diefendorff. Extreme Lithography. *Microprocessor Report Online*, June 2000. www.MPRonline.com.

395. Lutz Aschke, Stefan Mengel, and Jenspeter Rau. Lithographie am Limit. *c't Magazin für Computertechnik*, June 2003.

396. Lloyd R. Harriott. Limits of Lithography. *Proceedings of the IEEE*, 89(3):366–374, 2001.

397. Michael Fritze *et al.* Enhanced Resolution for Future Fabrication. *IEEE Circuits & Devices*, 19(1):43–47, January 2003.

398. Alfred K. Wong. Microlithography: Trends, Challenges, Solutions, and their Impact on Design. *IEEE Micro*, 23(2):12–21, March–April 2003.

399. Linda Geppert. Chip Making's Wet New World. *IEEE Spectrum*, 41(5):21–25, May 2004.

400. Ban P. Wong, Anurag Mittal, Yu Cao, and Greg Starr. *Nano-CMOS Circuit and Physical Design*. John Wiley & Sons, Hoboken NJ, 2004.

401. P.K. Vasudev and P.M. Zeitzoff. Si-ULSI with a Scaled-Down Future. *IEEE Circuits & Devices*, 14(2):19–29, March 1998. Special Issue on NTRS '97.

402. David A. Muller. A sound barrier for silicon? *Nature Materials*, 4(9):645–647, September 2005.

403. Yee-Chia Yeo, Tsu-Jae King, and Chenming Hu. MOSFET Gate Leakage Modeling and Selection Guide for Alternative Gate Dielectrics Based on Leakage Considerations. *IEEE Transactions on Electron Devices*, 50(4):1027–1035, April 2003. Oxynitride, gate oxide leakage.

404. J. Zhu, Z. G. Liu, and Y. R. Li. HfAlON films fabricated by pulsed laser ablation for high-k gate dielectric applications. *Materials Letters*, 59(7):821–825, March 2005. HfAlON, gate oxide.

405. Chee Wee Liu, S. Maikap, and C.-Y. Yu. Mobility-Enhancement Technologies. *IEEE Circuits & Devices*, 21(3):21–36, May/June 2005.

406. James D. Plummer and Peter B. Griffin. Material and Process Limits in Silicon VLSI Technology. *Proceedings of the IEEE*, 89(3):240–258, 2001.

407. Suman Datta *et al.* High Mobility Si/SiGe Strained Channel MOS Transistors with HfO_2/TiN Gate Stack. *IEEE Electron Device Letters*, 408–410, 2004.

408. Leland Chang *et al.* Moore's Law Lives On. *IEEE Circuits & Devices*, 19(1):35–42, January 2003.

409. Albert Chin and Séan P. McAlister. The Power of Functional Scaling. *IEEE Circuits & Devices*, 21(1):27–35, January/February 2005.

410. Kerry Bernstein and Norman J. Rohrer. *SOI Circuit Design Concepts*. Kluwer Academic Publishers, Boston MA, 2000.

411. Silke H. Christiansen, Rajendra Singh, and Ulrich Gösele. Wafer Direct Bonding: From Advanced Substrate Engineering to Future Applications in Micro/Nanoelectronics. *Proceedings of the IEEE*, 94(12):2060–2106, December 2006.

412. Gerald Marcyk and Robert Chau. New Transistors for 2005 and Beyond, 2001. http://www.intel/com/research/silicon/micron.htm.

413. Ching-Te Chuang *et al.* Scaling Planar Silicon Devices. *IEEE Circuits & Devices*, 2(1):6–19, January/February 2004.

414. Mark Bohr. A 30 Year Retrospective on Dennard's MOSFET Scaling Paper. *IEEE Solid-State Circuits Society Newsletter*, 12(1):11–23, 2007.

415. Simon Sze. *Semiconductor Devices, Physics and Technology*. John Wiley & Sons, New York, 1985.

416. P. M. Solomon *et al.* Two Gates are Better than One. *IEEE Circuits & Devices*, 19(1):48–62, January 2003.

417. Edward J. Novak *et al.* Turning Silicon on Its Edge, Overcoming silicon scaling barriers with double-gate and FinFET technology. *IEEE Circuits & Devices*, 2(1):20–31, January/February 2004.

418. D. J. Eaglesham. 0.18 μm CMOS and beyond. In *Proceedings 36th Design Automation Conference*, pages 703–708, New Orleans, 1999. ACM.

419. Infineon Technologies AG. Infineon leverages Multi-Gate Technology to achieve breakthrough results , December 2006. http://www.infineon.com/cgi-bin/ifx/portal/ep/redirectPage.do.

420. Jerry G. Fossum. Physical insights on nanoscale multi-gate CMOS design. *Solid-State Electronics*, 51(2):188–194, February 2007.

421. Thomas H. Lee. A Vertical Leap for Microchips. *Scientific American*, 286(1):50–57, January 2002.

422. Soon-Moon Jung *et al.* Highly Area Efficient and Cost Effective Double Stacked S^3 (Stacked Single-crystal Si) Peripheral CMOS SSTFT and SRAM Cell Technology for 512 Mbit Density SRAM. In *International Electron Devices Meeting Technical Digest*, pages 11.2.1–11.2.4, 2004.

423. David C. Ahlgren and Basanth Jagannathan. SiGe for mainstream semiconductor manufacturing. *Solid State Technology*, 43(1):535–58, January 2000.

424. K. S. Novoselov *et al.* Electric Field Effect in Atomically Thin Carbon Films. *Science*, 306(5696):666–671, October 22, 2004.

425. Max C. Lemme *et al.* A Graphene Field-Effect Device. *IEEE Electron Device Letters*, 28(4):282–287, 2007.

426. Lester F. Eastman and Umesh K. Mishra. The Toughest Transistor Yet. *IEEE Spectrum*, 39(5):28–33, May 2002.

427. Hagen Klauk *et al.* Ultralow-power organic complementary circuits. *Nature*, 445(7129):745–748, February 15, 2007.

428. Arokia Nathan and Babu R. Chalamala. Special Issues on Flexible Electronics Technology. *Proceedings of the IEEE*, 93(7 & 8), July & August 2005.

429. Gary F. Derbenwick and Alan F. Isaacson. Ferroelectric Memory: On the Brink of Breaking Through. *IEEE Circuits & Devices*, 17(1):20–30, January 2000.

430. Romney R. Katti. Giant Magnetoresistive Random-Access Memories Based on Current-in-Plane Devices. *Proceedings of the IEEE*, 91(5):687–702, May 2003.

431. Saied Tehrani *et al.* Magnetoresistive Random-Access Memory Using Magnetic Tunnel Junctions. *Proceedings of the IEEE*, 91(5):703–714, May 2003.

432. Hans Werner Schumacher. Schnelle MRAM dank ballistischer Bitansteuerung. *Bulletin SEV/VSE*, 97(11):17–20, May 2006.

433. Noboru Sakimura *et al.* MRAM Cell Technologies for Over 500-MHz SoC. *IEEE Journal of Solid-State Circuits*, 42(4):830–838, April 2007.

434. William H. Butler and Arunava Gupta. Magnetic Memory: A signal boost is in order. *Nature Materials*, 3(12):846–847, December 2004.

435. Philip G. Collins and Phaedon Avouris. Nanotubes for Electronics. *Scientific American*, 283(48):38–45, December 2000.

436. Adrian Bachtold, Peter Hadley, Takeshi Nakanishi, and Cees Dekker. Logic Circuits with Carbon Nanotube Transistors. *Science*, 294(5545):1317–1320, November 9, 2001.

437. P. R. Banduro, C. Daraio, S. Jin, and A. M. Rao. Novel electrical switching behaviour and logic in carbon nanotube Y-junctions. *Nature Materials*, 4(9):663–666, September 2005.

438. Yu Huang *et al.* Logic Gates and Computation from Assembled Nanowire Building Blocks. *Science*, 294(5545):1313–1317, November 9, 2001.

439. A. J. Heinrich, C. P. Lutz, J. A. Gupta, and D. M. Eigler. Molecule Cascades. *Science*, 298(5597):1381–1387, November 2002. Carbon monoxide, two-input sorter, molecular electronics.

440. Philip J. Kuekes, Gregory S. Snider, and R. Stanley Williams. Crossbar Nanocomputers. *Scientific American*, 293(5):50–55, November 2005.

441. Philip J. Kuekes, Duncan R. Stewart, and Stanley Williams. The crossbar latch: Logic value storage, restoration and inversion in crossbar circuits. *Journal of Applied Physics*, 97:034301/1–5, December 2004. Crossbar, molecular electronics.

442. Darren K. Brock, Elie K. Track, and John M. Rowell. Superconductor ICs: the 100-GHz second generation. *IEEE Spectrum*, 37(12):40–46, September 2000.

443. Darren K. Brock. RSFQ Technology: Circuits and Systems. *International Journal of High Speed Electronics and Systems*, 11(1):307–362, March 2001.

444. Thomas Sterling. How to Build a Hypercomputer. *Scientific American*, 285(1):28–35, July 2001. See http://htmt.jpl.nasa.gov/intro.html for a more technical report.

445. John Baliga. QCA Devices may take over when CMOS is done. *Semiconductor International*, 22(12):48, October 1999.

446. Ralph Cavin and Victor Zhirnov. Generic device abstractions for information processing technologies. *Solid-State Electronics*, 50(4):520–526, 2006.

447. James D. Meindl. Gigascale Integration: Is the Sky the Limit? *IEEE Circuits & Devices*, 12(6):19–32, November 1996.

448. David Frank, Robert Dennard, Edward Nowak, Paul Solomon, Yuan Taur, and Hon-Sum Wong. Device Scaling Limits of Si MOSFETs and their Application Dependencies. *Proceedings of the IEEE*, 89(3):259–288, 2001.

449. M. Alam, B. Weir, and P. Silverman. A Future of Function or Failure. *IEEE Circuits & Devices*, 18(2):42–48, March 2002.

450. H.-S. Wong, D. Frank, P. Solomon, C. Wann, and J. Welser. Nanoscale CMOS. *Proceedings of the IEEE*, 87(4):537–570, 1999.

451. Ted Lewis. The Next 10000_2 Years. *IEEE Engineering Management Review*, 25(2):48–64, 1997.

452. Michael J. Flyn and Patrick Hung. Microprocessor Design Issues: Thoughts on the Road Ahead. *IEEE Micro*, 25(3):16–31, May/June 2005.

453. James Surowiecki. Technology and Happiness, Why More Gadgets Don't Necessarily Increase Our Well-Being. *IEEE Engineering Management Review*, 33(2):68–75, 2005.

454. Frank Vahid. The Softening of Hardware. *IEEE Computer*, 36(4):27–34, April 2003.

455. Tze-Chiang Chen. Where CMOS is Going: Trendy Hype vs. Real Technology. *IEEE Solid-State Circuits Society Newsletter*, 20(3):5–9, September 2006.

456. D. K. Ferry, R. O. Grondin, and L. A. Akers. Two-Dimensional Automata in VLSI. In Roderick K. Watts, editor, *Submicron Integrated Circuits*. John Wiley & Sons, New York, 1989.

457. Lukas Sekanina. *Evolvable Components, from Theory to Hardware Implementations*. Springer, London, 2004.

458. Xin Yao. Following the Path of Evolvable Hardware. *Communications of the ACM*, 42(4):47–49, April 1999.

459. R. Iris Bahar *et al.* Architectures for Silicon Nanoelectronics and Beyond. *IEEE Computer*, 40(1):25–32, January 2007.

460. Dennis Sylvester and Kurt Keutzer. Impact of Small Process Geometries on Microarchitectures in Systems on a Chip. *Proceedings of the IEEE*, 89(4):467–489, April 2001.

461. Luca. P. Carloni, Kenneth L. McMillan, and Alberto L. Sangiovanni-Vincentelli. Theory of Latency-Insensitive Design. *IEEE Transactions on Computer-Aided Design*, 20(9):1059–1076, September 2001.

462. Michael Moreinis *et al.* Logic Gates as Repeaters (LGR): An Area-Efficient Timing Optimization. *IEEE Transactions on Very Large Scale Integration (VLSI) Systems*, 14(11):1276–1286, November 2006.

463. Ralph H. J. M. Otten and Robert K. Brayton. Performance Planning. *Integration, the VLSI Journal*, 29(1):1–24, March 2000.

464. Arindam Mukherjee and Małgorzata Marek-Sadowska. Wave Steering to Integrate Logic and Physical Syntheses. *IEEE Transactions on Very Large Scale Integration (VLSI) Systems*, 11(1):105–120, February 2003.

465. Sajjan G. Shiva. *Introduction to Logic Design*. Marcel Dekker, Inc., New York, 1998.

466. Tsutomu Sasao. *Switching Theory for Logic Synthesis*. Kluwer Academic Publishers, Boston MA, 1999.

467. Jaakko T. Astola. *Fundamentals of Switching Theory and Logic Design.* Springer, Dordrecht, 2006.

468. Edward J. McCluskey. *Logic Design Principles.* Prentice-Hall, Englewood Cliffs NJ, 1986.

469. James E. Buchanan. *Signal and Power Integrity in Digital Systems: TTL, CMOS, & BiCMOS.* McGraw-Hill, New York, 1996.

470. Thomas L. Floyd. *Digital Fundamentals.* Prentice-Hall International Inc., London, 2000.

471. John F. Wakerly. *Digital Design, Principles and Practice.* Pearson Prentice Hall, Upper Saddle River NJ, 2006.

472. Robert K. Brayton, Gary D. Hachtel, Curtis T. McMullen, and Alberto L. Sangiovanni-Vincentelli. *Logic Minimization Algorithms for VLSI Synthesis.* Kluwer Academic Publishers, Boston MA, 1984.

473. Maitham Shams, Jo C. Ebergren, and Mohamed I. Elmasry. Modeling and Comparing CMOS Implementations of the C-Element. *IEEE Transactions on Very Large Scale Integration (VLSI) Systems,* 6(4):563–567, December 1998.

474. Teresa H. Meng. *Synchronization Design for Digital Systems.* Kluwer Academic Publishers, Boston MA, 1991.

475. Daniel I. A. Cohen. *Introduction to Computer Theory.* John Wiley & Sons, New York, 1986.

476. Douglas Lewin and David Protheroe. *Design of Logic Systems.* Chapman & Hall, London, 1992.

477. David Harel. Statecharts: A Visual Formalism for Complex Systems. *Science of Computer Programming,* 8(3):231–274, June 1987.

478. David Harel. On Visual Formalisms. *Communications of the ACM,* 31(5):514–530, May 1988.

479. Edward J. McCluskey and Stephen H. Unger. A note on the number of internal variable assignments for sequential switching circuits. *IRE Transactions on Electronic Computers,* EC-8:439–440, December 1959.

480. A. K. Geim and K. S. Novoselov. The rise of graphene. *Nature Materials,* 6(3):183–191, March 2007.

481. John Rogers, Calvin Plett, and Foster Dai. *Integrated Circuit Design for High-Speed Frequency Synthesis.* Artech House, Norwood MA, 2006.

482. R. Scott Mackay, Henry Kamberian, and Yuan Zhang. Methods to reduce lithography costs with reticle engineering. *Microelectronic Engineering.* 83(4–9):914–918, April–September 2006.

483. Charles A. Harper. *Electronic Packaging and Interconnection Handbook.* McGraw-Hill, New York, 2004.

484. Weifeng Liu and Michael Pecht. *IC Component Sockets.* John Wiley & Sons, Hoboken NJ, 2004.

485. Christopher Saint and Judy Saint. *IC Mask Design, Essential Layout Techniques.* McGraw-Hill, New York, 2002.

486. Christopher Saint and Judy Saint. *IC Layout Basics: a Practical Guide.* McGraw-Hill, New York, 2002. (The title notwithstanding, this is more an introduction to microelectronic concepts for layout persons than a guide to layout design techniques for circuit engineers.)

487. Robert H. Dennard, Jin Cai, and Arvind Kumar. A Perspective on Today's Scaling Challenges and Possible Future Directions. *Solid-State Electronics.* 51(4):518–525, April 2007.

488. Brian Santo. Plans for Next-Gen Chips Imperiled. *IEEE Spectrum.* 44(8):8–10, August 2007.

489. George Gruner. Carbon nanotube films for transparent and plastic electronics. *Journal of Materials Chemistry.* 16(35):3533–3539, 2006.

490. Kanak Agarwal *et al.* Parametric Yield Analysis and Optimization in Leakage Dominated Technologies. *IEEE Transactions on Very Large Scale Integration (VLSI) Systems,* 15(6):613–623, June 2007.

Index